LABORATORY MANUAL FOR
Anatomy & Physiology

THIRD EDITION

CAT VERSION

MICHAEL G. WOOD

with

William C. Ober, M.D.
Art Coordinator and Illustrator

Claire W. Garrison, R.N.
Illustrator

Ralph T. Hutchings
Biomedical Photographer

Shawn Miller
Organ and Animal Dissector

Mark Nielsen
Organ and Animal Dissection Photographer

PEARSON
Benjamin Cummings

San Francisco Boston New York
Cape Town Hong Kong London Madrid Mexico City
Montreal Munich Paris Singapore Sydney Tokyo Toronto

Publisher: *Daryl Fox*
Executive Editor: *Leslie Berriman*
Assistant Editor: *Michael Roney*
Editorial Assistant: *Blythe Robbins*
Managing Editor: *Deborah Cogan*
Production Supervisor: *Caroline Ayres*
Production and Composition: *Carlisle Publishers Services*
Production Editor: *Amy Hackett*
Design Manager: *Mark Ong*
Interior Designer: *tani hasegawa*
Cover Designer: *tani hasegawa*
Senior Manufacturing Buyer: *Stacey Weinberger*
Executive Marketing Manager: *Lauren Harp*
Cover Photo Credit: *Scott Markewitz/Getty Images*

Safety Notification
The Author and Publisher believe that the lab experiments described in this
publication, when conducted in conformity with the safety precautions described
herein and according to the school's laboratory safety procedures, are reasonably safe
for the student to whom this manual is directed. Nonetheless, many of the described
experiments are accompanied by some degree of risk, including human error, the
failure or misuses of laboratory or electrical equipment, mismeasurement, chemical
spills, and exposure to sharp objects, heat, bodily fluids, blood, or other biologics. The
Author and Publisher disclaim any liability arising from such risks in connection with
any of the experiments contained in this manual. If students have any questions or
problems with materials, procedures, or instructions on any experiment, they should
always ask their instructor for help before proceeding.

PEARSON

Benjamin
Cummings

ISBN 0-8053-7369-1
10 9 8 7 6 5 4 3 2 CRK 09 08
www.aw-bc.com

Contents

See inside front cover for Brief Contents.

Preface

This laboratory manual is designed to serve the laboratory course that accompanies the two-semester anatomy and physiology lecture course. It provides students with comprehensive coverage of anatomy and physiology, beautiful four-color art and photographs, and an intuitive pedagogical framework. The primary goals of this manual are to provide students with hands-on experiences that reinforce the information they learn in the lecture course and to help them understand three-dimensional relationships, visualize complex structures, and comprehend intricate physiological processes.

The manual is written to correspond to all current two-semester anatomy and physiology textbooks, although those students and instructors using Frederic H. Martini's *Fundamentals of Anatomy & Physiology*, Seventh Edition will recognize here some of the superb art from that text by William Ober and Claire Garrison, Martini's renowned biomedical illustrators.

With this third edition, the manual is now available in three separate versions. The Main Version covers the full two-semester A&P curriculum, including dissections of the cow eye and of the sheep heart, brain, and kidney. The Cat Version includes all of the same material plus an additional section of nine cat dissection exercises encompassing the major body systems. The Pig Version, similarly, includes all of the material from the Main Version with a separate section of nine fetal pig dissection exercises. The Cat and Pig Versions will make the manual more useful to instructors who have their students do animal dissections in the laboratory.

Organization

The laboratory manual contains 46 exercises, plus the 9 additional dissection exercises in each of the Cat and Pig Versions. Large systems, such as the skeletal, muscular, and nervous systems, appear across several exercises, the first serving as an overview exercise that introduces the major anatomical organization of the system. Programs with limited laboratory time might use the overview exercises for a hands-on summary of these organ systems that can be completed during a short laboratory period.

Each exercise is divided into a series of activities covering a range of topics. Every exercise begins with a list of the Laboratory Activities it contains and a set of Objectives for student learning. A general introduction to the exercise gives students a preview of what they are about to learn; then, individual activities focus on more specific study. The activities are self-contained, and instructors may easily assign only certain activities within an exercise.

Each Laboratory Activity section begins with an introduction to the activity and a review of the concepts necessary for understanding it. These are followed by two or three QuickCheck Questions that students can use to gauge their comprehension of the material before proceeding. The activity itself consists of a clearly marked list of Materials for the activity and the Procedures for carrying it out. These are placed side by side for easy use.

Finally, after the set of activities within an exercise, the exercise concludes with a Laboratory Report that includes questions to assess and reinforce student learning.

Content Updates and Pedagogical Changes

The third edition of this laboratory manual has been substantially revised from its previous edition to make it precisely suited for the A&P laboratory course that parallels the two-semester A&P lecture course.

Expanded Anatomy and Physiology Coverage

In this third edition, many exercises have been enhanced with additional activities. On the anatomy side, the histology exercises include more tissues and new micrographs of each tissue. The manual now features more comprehensive coverage of the skeletal and muscular systems, including additional bony features and muscles, updated art, and an enhanced labeling system.

Physiology coverage has also been expanded. For example, new sensory activities direct students to map the density of touch and temperature receptors, examine referred pain, and use oculoscopes and otoscopes for observations of the eyes and ears. Building upon the success of the three lessons using the Biopac Student Lab System in the second edition, the third edition includes five additional BIOPAC lessons, enhancing physiological explorations of the muscular, nervous, cardiovascular, and respiratory systems. For more information on BIOPAC activities, see the special section later in the Preface.

Almost every exercise in the manual has been upgraded in both content and pedagogy. The following outline details the major content updates in the third edition:

Exercise 4, "Use of the Microscope," has a new use of the microscope section and expanded depth of field activity, both supported with new art.

Exercise 5, "Anatomy of the Cell and Cell Division," has new coverage of perioxisomes.

Exercise 6, "Movement Across Cell Membranes," has a new activity, observing phagocytosis by *Amoeba proteus*.

Exercise 7, "Epithelial Tissue," now has broader coverage with the addition of stratified cuboidal epithelium and stratified columnar epithelium.

Exercise 8, "Connective Tissues," has been expanded to include embryonic connective tissues, dense irregular connective tissue, and elastic tissue. A new activity on fluid connective tissues guides students through observations of red and white blood cells and platelets.

Exercise 12, "Body Membranes," has a new activity on observing mucous membranes.

Exercise 13, "Skeletal System Overview," now includes a guided activity on bone markings to prepare students for studies of skeletal anatomy in the subsequent exercises.

Exercise 14, "The Axial Skeleton," has been made more comprehensive with the addition of the following features of the skull: supraorbital notch, lacrimal fossa, inferior temporal line, zygomatic arch, middle clinoid process, hypophyseal fossa, inferior orbital fissure, pterygoid plates, olfactory foramina, lateral mass, and ethmoidal air cells. A new activity on the fetal skull describes each fontanel. Most activities have new introductions.

Exercise 15, "The Appendicular Skeleton," now includes the radial groove on the humerus and several features on the coxal bone: the lunate surface, acetabular notch, pelvic outlet, and pelvic angle. A new activity for the patella includes learning how to differentiate the right and left patellae. The exercise features a new figure detailing the ulna.

Exercise 16, "Articulations," is updated with a new section on the anatomy of the knee, including the seven ligaments, the patellar retinaculae, and the menisci.

Exercise 17, "Organization of Skeletal Muscles," uses revised terminology and includes the proteins nebulin and titin.

Exercise 18, "Muscles of the Head and Neck," now describes the lateral and medial pterygoid muscles.

Exercise 19, "Muscles of the Chest, Abdomen, Spine, and Pelvis," includes more than a dozen additional muscles and structures, including the tendinous inscriptions; the intertransversarii, rotators, and interspinalis muscles of the vertebral column, and the scalene, spinalis, and semispinalis muscle groups.

Exercise 21, "Muscles of the Pelvis, Leg, and Foot," is enhanced with inclusion of the quadratus femoris, plantaris, popliteus, and flexor digitorus longus muscles.

Exercise 22, "Muscle Physiology," builds on the electromyography (EMG) lesson of the last edition with a new BIOPAC activity on muscle recruitment and fatigue, providing students with the tools necessary to explore muscle contractions.

Exercise 23, "Organization of the Nervous System," investigates reaction time and learning with a new BIOPAC activity.

Exercise 24, "The Spinal Cord, Spinal Nerves, and Reflexes," has new activities on biceps brachii and triceps brachii reflexes.

Exercise 25, "Anatomy of the Brain," contains a new activity that allows students to test the function of each cranial nerve.

Exercise 26, "Autonomic Nervous System," reinforces ANS pathways with a drawing activity in the Laboratory Report.

Exercise 27, "General Senses," is a new exercise with several new activities, including mapping the density of tactile and thermoreceptors, inducing referred pain with ice water, and studying proprioception.

Exercise 28, "Special Senses: Olfaction and Gustation," is a new exercise with activities that include microscopic observations, a study of olfactory adaptation, and an investigation of the interrelationship between olfaction and gustation.

Exercise 29, "Anatomy of the Eye," now includes observation of the retina with an ophthalmoscope.

Exercise 30, "Physiology of the Eye," newly incorporates the BIOPAC system to record electrooculograms (EOG) during a variety of eye movements. This study of saccades, coupled with the study of nystagmus, helps students understand the role of eye movements for visual tracking and equilibrium.

Exercise 31, "Anatomy of the Ear," has been updated with an activity on observing the tympanic membrane using an otoscope.

Exercise 34, "Blood," has new procedures to assist students in distinguishing the various types of cells in blood.

Exercise 35, "Anatomy of the Heart," has new procedures for identifying the anatomy of the heart and includes new discussions of marginal arteries, the moderator band, and the posterior cardiac vein.

Exercise 36, "Anatomy of the Systemic Circulation," includes a new activity on fetal circulation that covers the coronary and vascular structures of the fetus.

Exercise 37, "Cardiovascular Physiology," has revised BIOPAC activities with new screen-capture images to assist students in their investigations of the electrocardiogram (ECG) and pulse.

Exercise 40, "Physiology of the Respiratory System," explores respiratory volumes, capacities, rate, and depth with two integrated BIOPAC activities. These computerized activities provide students with tools to accurately measure respiratory functions.

Exercise 41, "Anatomy of the Digestive System," has new activities for microscopic examinations of the esophagus, duodenum, and ileum.

Exercise 42, "Digestive Physiology," has new procedures for the carbohydrate digestion experiment and new art to support all digestive physiology investigations.

Exercise 44, "Physiology of the Urinary System," includes new art that enhances the microscopic examination of urine sediments.

Exercise 45, "Anatomy of the Reproductive System," now includes an activity on spermatogenesis and oogenesis.

Resequencing of Material

In some cases, exercises have been rearranged to respond to reviewer suggestions and more closely match the topical order of the major A&P textbooks. In addition, a few exercises have been combined or reshuffled for a more natural progression through the topics.

The most major changes include the following: First, histology is now covered in four exercises instead of five. Second, the two brain anatomy exercises of the previous edition have now been combined into a single exercise. Third, the general senses and olfaction/gustation are now covered in separate exercises. Fourth, all BIOPAC lessons are now integrated into the appropriate exercises as laboratory activities rather than as alternate exercises as in the previous edition. Finally, as discussed earlier, cat dissections now appear only in the Cat Version of the manual and have been moved to a separate Cat Dissection section, with each organ system dissection as a stand-alone exercise.

The following detailed outline summarizes the major organizational changes:

Exercise 4, "Use of the Microscope," is reorganized to facilitate better integration of the study of microscope parts and microscope usage.

Exercise 5, "Anatomy of the Cell and Cell Division," is reorganized, with organelles grouped into nonmembranous and membranous categories.

Exercise 7, "Epithelial Tissue," combines several new tissues into the "Introduction to Tissues" exercise of the previous edition.

Exercise 13, "Skeletal System Overview," resequences activities on the gross anatomy of bone, bone tissue, an overview of the skeletal system, and description of bone markings.

Exercise 17, "Organization of Skeletal Muscles," has a tighter organization.

Exercise 19, "Muscles of the Chest, Abdomen, Spine, and Pelvis," is reorganized into more natural muscle groupings.

Exercise 22, "Muscle Physiology," now incorporates the BIOPAC activity that was an alternate lesson in the previous edition.

Exercise 25, "Anatomy of the Brain," is completely reorganized, combining the last edition's Exercises 26, "External Anatomy of the Brain," and 27, "Internal Anatomy of the Brain," into a single, integrated exercise with a regional approach to the anatomy.

Exercise 27, "General Senses," and Exercise 28, "Special Senses: Olfaction and Gustation," have been split out from the last edition's Exercise 31, "General Senses, Gustation, Olfaction."

Exercise 36, "Anatomy of the Systemic Circulation," is completely redesigned with new sequences of vessel descriptions for ease of use and learning.

Exercise 37, "Cardiovascular Physiology," now incorporates the two BIOPAC activities that were alternate exercises in the previous edition.

Exercise 44, "Physiology of the Urinary System," has been reorganized for ease of use and clarity.

Labeling and Drawing Activities

All labeling and drawing activities were reexamined to ensure that they work effectively with beginning A&P students for teaching key anatomical structures. In response to reviewer suggestions, the amount and level of labeling in every figure in the manual were assessed and, where necessary, reduced. In addition, word banks have been added to all unlabeled figures as a reference for student labeling. Finally, some labeling figures and drawing activities have been moved to Laboratory Reports to prevent overloading of art and labeling in the exercises themselves.

Laboratory Reports

The Laboratory Reports that follow each exercise have also been extensively updated. The Matching sections were evaluated for appropriate rigor and rewritten to be more difficult. New Analysis and Application questions are now included in the reports; these questions require students to analyze or think critically about what they learned in the laboratory and apply it to new situations. Finally, all write-in spaces for recording observations, data, and drawings have been moved from the laboratory activities to the Laboratory Reports that students hand in.

New and Revised Features

Significant changes have been made to create a more prominent and consistent features program. For those familiar with the second edition of the manual, a brief description of the changes follows; more complete descriptions of the features appear later in the Preface.

- A new feature, Safety Alert, ⚠, has been created to call out important safety information for students.
- The Study Tip, *Study Tip*, and Clinical Application, *Clinical Application*, features have been expanded.
- Two to three new QuickCheck Questions, *QuickCheck Questions*, appear at the end of most activity introductions, to help students evaluate their understanding of the material before undertaking the laboratory activity. These replace the last edition's QuickCheck review sheets.
- Definitions from the second edition's Word Power sections at the beginnings of exercises have been incorporated into the text so that students can learn each relevant term in context.
- The design has been enhanced to make all features stand out more clearly from the text, and they have been treated to give each feature a unique identity.

Revised Narrative

For this third edition, special attention was paid to looking over every word with an eye toward clarity, consistency, and conciseness. In particular, instructions for doing the laboratory activities incorporate a number of suggestions from reviewers and are now significantly clearer and more straightforward.

Anatomical Terminology

In this new edition, anatomical terminology has been updated to match that used in *Fundamentals of Anatomy & Physiology*, Seventh Edition, by Frederic H. Martini, which draws from the standard terminology adopted by the International Federation of Associations of Anatomists in *Terminologia Anatomica*.

Art Program

A major feature of this laboratory manual is the outstanding full-color art program with illustrations by William Ober and Claire Garrison and photographs by Ralph Hutchings. Here, as with their renowned work on the various anatomy and physiology textbooks by Frederic H. Martini, they have designed the art to guide students from familiar to detailed structures. The text and the art of the laboratory manual are carefully woven together to bring together reading and visual learning modalities. Figures include detailed legends, reference icons, pinpoint labeling, and multi-dimensional views of the human body. Numerous figures incorporate written labeling assignments as part of the laboratory activities.

A number of significant changes have been made to the art for this new edition of the manual. This third edition incorporates a substantial number of pieces of the new art from the simultaneously published seventh edition of *Fundamentals of Anatomy & Physiology* by Frederic H. Martini, the A&P textbook known for its stellar art program by Ober and Garrison. The Main Version of this laboratory manual also includes 12 original pieces and 7 heavily revised pieces from Ober and Garrison.

The Cat and Pig Versions of the manual contain all-new dissection photos, taken specifically for this laboratory manual by the expert dissector/photographer team of Shawn Miller and Mark Nielsen at the University of Utah, placed alongside all-new Ober and Garrison cat and pig art. For a more complete description of the new dissection sections, including their art, see below.

In addition, all histology images have been replaced with new ones that show students what they will actually see in the laboratory, rather than unreasonably perfect stock slides.

Finally, all of the art for the three preexisting and five new BIOPAC activities—more than 25 pieces in all—has been redrawn for clarity.

Design

One of the main goals of this revision was to create a manual that is easy for instructors and students to use. To that end, the completely new design of this third edition is not only attractive but also straightforward, intuitive, and functional. Foremost among the features of the new design is the use of color and headings as signposts that make navigating within exercises and activities logical and easy.

Within each exercise, the beginning of a new activity is indicated by a blue "Laboratory Activity" heading, LAB ACTIVITY , accompanied by a green number for that activity, 2 . After the introductory content material for that activity, the green activity number is then repeated next to the paired Materials and Procedures headings that mark the beginning of actual laboratory work. Finally, the end of the activity is marked by a closing green icon, ■. This sequence of colored headings and icons within each activity is useful for the instructor in assigning specific activities within an exercise and helps students turn immediately to the appropriate section, whether they wish to read an introduction or carry out a lab activity.

Icons are also used throughout the manual to call out important items. Activities using the BIOPAC Student Lab System are highlighted with an icon ◔ placed in the margin next to the introductory heading for that activity. In the Cat and Pig Versions of the manual, colors and icons are used together to indicate the Dissection Exercises. The exercise opener has a uniquely colored screen to differentiate it from the exercises in the rest of the manual, and a circular icon showing a cat or pig in profile, ◑ ◑, provides a further visual indication of the exercise's nature.

Features

This third edition contains a host of features to give students the edge in learning anatomy and physiology.

Clinical Applications

To tie students' learning more tightly to potential future careers in the allied health field, many exercises include relevant discussions of particular diseases and injuries in Clinical Application boxes, *Clinical Application*. Thought-provoking critical-thinking questions are featured in many of these boxed sections.

Study Tips

Where appropriate, Study Tip, *Study Tip*, boxes give students handy mnemonic devices, tricks for learning certain relationships, or general suggestions for retaining the complex concepts that underlie the study of A&P. Content-specific Study Tip boxes appear throughout the manual, while general laboratory and Study Tip boxes appear primarily in the first third of the book, where students need them most.

Safety Alerts

New to this edition, Safety Alerts, , highlight critical safety information where appropriate. These descriptions of safe procedures include correct usage of instrumentation, proper precautions for undertaking particular activities, and safe handling and disposal of chemical and biological wastes. In addition, Exercise 1 covers laboratory safety and should be completed by all students at the beginning of a semester.

QuickCheck Questions

The review and introduction sections to activities are followed, as appropriate, by two or three QuickCheck Questions, *QuickCheck Questions*, that students can use to review key concepts and test their understanding before proceeding to the laboratory work.

Word Power and Pronunciation Guides

To promote student awareness in the language of science, many of the important terms are followed by a brief explanation of relevant prefixes, suffixes, and root words. In addition, difficult words are followed by a pronunciation guide. Each word is divided into syllables by hyphens, and the main accented syllable of multisyllabic words is in capital letters. An overbar above letters indicates long vowel sounds:

"a," as in "tray," (trā)
"e," as in "ear," (ēr)
"i," as in "fly," (flī)
"o," as in "bone," (bōn)
"u," as in "you," (ū)

Laboratory Reports

Comprehensive Laboratory Reports are located at the conclusion of each exercise. The reports include a variety of tasks, such as matching exercises, short-answer questions, fill-in-the-blanks, labeling activities, and drawing exercises, each designed to reinforce the objectives for learning that are detailed at the beginning of the exercise. The new Analysis and Application questions described earlier push students to think beyond the confines of a particular exercise and connect their learning with the rest of their A&P work so far. Instructors may utilize the Laboratory Reports for after-lab follow-up, quizzes, and out-of-class assignments. The manual is designed so that students can easily remove and hand in Laboratory Reports.

BIOPAC Activities

Since the second edition, this manual has featured exercises using the Biopac Student Lab System, an integrated suite of hardware and software that provides students with powerful tools for studies in physiology. In this edition, the three BIOPAC experiments that appeared as alternate exercises in the second edition have been integrated as Laboratory Activities within the appropriate body-system exercise. In addition, five new BIOPAC investigations have been incorporated into the manual. All eight of these BIOPAC activities feature new full-color art and screen shots to walk students through the procedures. BIOPAC activities can be easily identified by the BIOPAC logo to the left of the activity title.

The BIOPAC activities include step-by-step instructions to guide students in the setup and use of the hardware and software. All components of the system are designed for a quick setup that can easily be configured for various physiological investigations. The Biopac Student Laboratory System works in both Macintosh™ and Windows™ operating environments.

Cat and Pig Dissection Exercises

Dissection gives students perspectives on the texture, scale, and relationships of anatomy. For those instructors who choose to teach dissection in their laboratories, this manual is available in two new versions, the Cat Version and the Pig Version, featuring sections at the back of the manual detailing the dissection of the cat or fetal pig. Each of these versions includes nine exercises that progress through the major body systems, with the goal of relating these exercises to students' study of the human body. These exercises appear together in a section following the main body of the manual. Safety guidelines and disposal methods are incorporated into each dissection exercise.

One of the anatomy field's premier dissector/photographer duos, Shawn Miller and Mark Nielsen of the University of Utah, was hired to do the cat and pig dissection photos for these new versions of the manual. Carefully following detailed specifications, they provided a selection of exquisite photos that were reviewed by a large team of content and photography experts.

Bill Ober and Claire Garrison provided new illustrations to be placed alongside the new dissection photos so that students can compare an idealized artistic view with a more realistic photograph.

Instructor's Manual

For this edition, the *Instructor's Manual* has been greatly improved to make it more useful. Like the second edition, it includes answers to the labeling exercises and Laboratory Reports. With this revision, however, it also adds teaching tips, laboratory preparation requirements, a master materials list for each exercise, answers to QuickCheck Questions, and media references to InterActive Physiology® (IP) and PhysioEx™.

Instructor's Resource CD-ROMs

New to the third edition, the Instructor's Resource CD-ROMs include the illustrations and photographs from the manual in an easy-to-use media package. The images can be used for presentation, for comparison/contrast with what students are seeing on their lab tables, or for guiding students in locating particular structures or seeing particular processes during a lab exercise.

Acknowledgments

I thank all the talented and creative individuals at Benjamin Cummings. I especially credit my editor, Michael Roney, for his expertise in organizing and coordinating the enormous number of details in the project and managing a challenging production schedule. I was very fortunate to have Irene Nunes as the development editor on the text. Her experienced eye for consistency and accuracy immensely improved the cohesiveness of the third edition. Leslie Berriman, Executive Editor, provided valuable guidance and participated in pulling together a resourceful team of editors, dissectors, photographers, and illustrators whose outstanding work shines in the third edition. I appreciate the participation of Claire Alexander and Mary Ann Murray in the development of the content and photographs for the cat and pig Dissection Exercises. Caroline Ayres, Production Supervisor, and Deborah Cogan, Managing Editor, masterfully coordinated all aspects of the production process. I thank Amy Hackett and her fine team at Carlisle Publishers Services for their creative layout and attention to detail.

I also thank tani hasegawa for her outstanding interior design, which gives this complex assemblage of text, art, photographs, and procedures a user-friendly feel and has greatly improved access to the manual's content. tani also designed the lovely covers of all three versions of this manual. Mark Ong oversaw the design process and provided crucial insight into our design complexities.

Lauren Harp, Executive Marketing Manager, and the entire Addison Wesley/Benjamin Cummings sales team deserve thanks for their fine efforts in presenting this manual to A & P instructors.

Foremost, I am grateful to a number of people for this edition's excellent art and photography. Frederic H. Martini, author of *Fundamentals of Anatomy & Physiology*, Seventh Edition, deserves credit for his insight and creativity in visualizing anatomical and physiological concepts with the talented biomedical illustrators William Ober and Claire Garrison, whose beautiful art I have utilized throughout. For this edition, William Ober and Claire Garrison also worked closely with me to create many new pieces for the manual. Shawn Miller and Mark Nielsen of the University of Utah are a gifted dissector/photographer team whose meticulous work is coupled with the Ober and Garrison art in the spectacular new cat, pig, and organ dissection plates. The award-winning human photographs in the manual are by biomedical photographer Ralph Hutchings.

Many individuals made significant contributions to the first and second editions that are still incorporated in this third edition. Thank you again to Edward J. Greding, Jr., Anil Rao, Joe Wheeler, Michael Timmons, Michael McKinley, Esmeralda Salazar, and Gerardo Cobarruvias for their generosity. Special thanks are extended to my colleagues Albert Drumright, III, Lillian Bass, Billy Bob Long, Joel McKinney, and Joyce Germany for their suggestions, encouragement, and support of the manual.

My gratitude is also extended to the many students who have provided suggestions and comments to me over the years.

I thank BIOPAC Systems, Inc., and especially Frazer Findlay and Jocelyn Kremer, for their continued support and partnership with Pearson Education and assistance in incorporating their state-of-the-art instrumentation into the third edition.

Reviewers and users of the first two editions provided insightful and experienced-based suggestions, many of which helped guide the revision into this third edition. I thank the reviewers for their time and devotion to the project.

Second Edition Reviewers:

Mary Bracken, *Trinity Valley Community College*
Theresia Elrod, *Manatee Community College*
Tejandra Gill, *University of Houston*
David Harris, *University of Southern Maine Lewiston-Auburn College*
Arlene Luckock, *De Anza College*
Laura Mastrangelo, *Hudson Valley Community College*
Elizabeth McMahon, *Warren County Community College*
Kerri Vierling, *South Dakota School of Mines and Technology*
Francis Wray, *University of West Florida*

First Edition Reviewers:

Terri DiFiori, *Pasadena City College*
Michael A. Dorset, *Cleveland State Community College*
Linda Griffin Gingerich, *St. Petersburg Junior College*
Theresa Hornstein, *Lake Superior College*
Stephen Lebsack, *Linn-Benton Community College*
Wendy Rounds, *Harford Community College*
Kelly J. Sexton, *Northlake College*

Most importantly, I am deeply grateful to my wife, Laurie, and children, Abi and Beth, for enduring months of my late-night writing and the pressures of never-ending deadlines.

Any errors or omissions in this third-edition publication are my responsibility and are not a reflection on the editorial and review team. Comments from faculty and students are welcomed and may be directed to me at the addresses below. I will consider each suggestion in the preparation of the fourth edition.

Michael G. Wood, *Del Mar College*, 101 Baldwin Blvd., Corpus Christi, TX 78404, mwood@delmar.edu

About the Author

Michael G. Wood received his Master's of Science in 1986 at Pan American University, now the University of Texas at Pan American. His graduate research examined the ecological impact of an introduced species used widely in aquaculture, the blue tilapia, on the native fish population in the Rio Grande drainage of South Texas. Today he is a Professor of Biology at Del Mar College in Corpus Christi, Texas, where he has taught anatomy and physiology and biology for more than 20 years. He has received the "Educator of the Year" and "Teacher of the Year" awards from the Del Mar College student body and from the local business community. He is a member of the Human Anatomy and Physiology Society. Professor Wood also serves on planning committees with the local school district and has a passion for reading and playing the guitar. The author and his family enjoy camping, dancing, and a yard full of cats, dogs, and plants.

Dedication

To my wife, Laurie: Her devotion, encouragement, and sacrifices made this extended project possible.

Laboratory Safety

OBJECTIVES

On completion of this exercise, you should be able to:

- Locate all safety equipment in the laboratory.
- Show how to handle glassware safely, including insertion and removal of glass rods used with stoppers.
- Demonstrate how to clean up and dispose of broken glass safely.
- Demonstrate how to plug in and unplug electrical devices safely.
- Explain how to protect yourself from and dispose of body fluids.
- Demonstrate how to mix solutions and measure chemicals safely.
- Describe the potential dangers of each laboratory instrument.
- Discuss disposal techniques for glass, chemicals, body fluids, and other hazardous materials.

Experiments and exercises in the anatomy and physiology laboratory are, by design, safe. Some of the hazards are identical to those found in your home, such as broken glass and the risk of electrical shock. The major hazards can be grouped into six categories: glassware, electrical, body fluids, chemical, laboratory instruments, and preservatives. The following is a discussion of the hazards each category poses and a listing of safety guidelines you should follow to prevent injury to yourself and others while in the laboratory. Proper disposal of biological and chemical wastes ensures that these contaminants will not be released into your local environment.

Laboratory Safety Rules

The following guidelines are necessary to ensure that the laboratory is a safe environment for students and faculty alike.

1. No unauthorized persons are allowed in the laboratory. Only students enrolled in the course and faculty are to enter the laboratory.
2. Never perform an unauthorized experiment; unless you have your instructor's permission, never make changes to any experiment that appears either in this manual or in a class handout.
3. Do not smoke, eat, chew gum, or drink in the laboratory.

4. Always wash your hands before and after each laboratory exercise involving chemicals, preserved materials, or body fluids and immediately after cleaning up spills.

5. Wear shoes at all times while in the laboratory.

6. Be alert to unsafe conditions and to unsafe actions by other individuals in the laboratory. Call attention to those conditions or activities. Someone else's accident can be as dangerous to you as one that you cause.

7. Glass tubes called *pipettes* are commonly used to measure and transfer solutions. Never pipette a solution by mouth. Always use a pipette bulb. Your instructor will demonstrate how to use the particular type of bulb available in your laboratory.

8. Immediately report all spills and injuries to the laboratory faculty.

9. Inform the laboratory faculty of any medical condition that may limit your activities in the laboratory.

Location of Safety Equipment

Write here the location of each piece of safety equipment as your instructor explains how and when to use it:

nearest telephone _____
first aid kit _____
fire exits_____
fire extinguisher _____
eye wash station _____
chemical spill kit_____
fan switches _____
biohazard container_____

Glassware

Glassware is perhaps the most dangerous item in the laboratory. Broken glass must be cleaned up and disposed of safely. Other glassware-related accidents can occur when a glass rod or tube breaks while you are attempting to insert it into a cork or rubber stopper.

Broken Glass

• Sweep up broken glass immediately. Never use your hands to pick up broken glass. Instead, use a whisk broom and dustpan to sweep the area clear of all glass shards.

• Place broken glass in a box, tape the box shut, and write "BROKEN GLASS INSIDE" in large letters across it. This will alert custodians and other waste collectors to the hazard inside. Your laboratory instructor will arrange for disposal of the sealed box.

Inserting Glass into a Stopper

• Never force a dry glass rod or tube into the hole cut in a cork or rubber stopper. Use a lubricant such as glycerin or soapy water to ease the glass through the stopper.

• When inserting a glass rod or tube into a stopper, always push on the rod/tube near the stopper. Doing so reduces the length of glass between the stopper and your hand and greatly decreases the chance of your breaking the rod and jamming glass into your hand.

Electrical Equipment

Electrical hazards in the laboratory are similar to those in your home. A few commonsense guidelines will almost eliminate the risk of electrical shock.

- Uncoil an electrical cord completely before plugging it into an electrical outlet. Electrical cords are often wrapped tightly around the base of a microscope, and users often unwrap just enough cord to plug in the microscope. Moving the focusing mechanism of the microscope may pinch the part of the cord still wrapped around the base and shock the user. Inspect the cord for fraying and the plug for secure connections.

- Do not force an electrical plug into an outlet. If the plug does not easily fit into the outlet, inform your laboratory instructor.

- Unplug all electrical cords by pulling on the plug, not the cord. Pulling on the cord may loosen wires inside the cord, which can cause an electrical short and possibly an electrical shock to anyone touching the cord.

- Never plug in or unplug an electrical device in a wet area.

Body Fluids

The three body fluids most frequently encountered in the laboratory are saliva, urine, and blood. Because body fluids can harbor infectious organisms, safe handling and disposal procedures must be followed to prevent infecting yourself and others.

- Work only with your own body fluids. It is beyond the scope of this manual to explain proper protocol for collecting and experimenting on body fluids from another individual.

- Never allow a body fluid to touch your unprotected skin. Always wear gloves and safety glasses when working with body fluids—even though you are using your own fluids.

- Always assume that a body fluid can infect you with a disease. Putting this safeguard into practice will prepare you for working in a clinical setting where you may be responsible for handling body fluids from the general population.

- Clean up all body-fluid spills with either a 10% bleach solution or a commercially prepared disinfectant labeled for this purpose. Always wear gloves during the cleanup, and dispose of contaminated wipes in a biohazard container.

Chemicals

Most chemicals used in laboratories are safe. Following a few simple guidelines will protect you from chemical hazards.

- Read the label describing the chemical you are going to work with. Be aware of chemicals that may irritate skin or stain clothing. Chemical containers are usually labeled to show contents and potential hazards. Most laboratories and chemical stockrooms keep copies of technical chemical specifications, called *Material Safety Data Sheets (M.S.D.S.)*. These publications from chemical manufacturers detail the proper use of the chemicals and the known adverse effects they may cause. All individuals have a federal right to inspect these documents. Ask your laboratory instructor for more information on M.S.D.S.

- Never touch a chemical with unprotected hands. Wear gloves and safety glasses when weighing and measuring chemicals and during all experimental procedures involving chemicals.

- Always use a spoon or spatula to take a dry chemical from a large storage container. Do not shake a dry chemical out of its jar; doing so may result in your dumping the entire container of chemical onto yourself and your workstation.

- When pouring out some volume of a solution kept in a large container, always pour the approximate amount required into a smaller beaker first and then pour from this beaker to fill your glassware with the solution. Attempting to pour from a large storage container directly into any glassware other than a beaker may result in spilled solution coming into contact with your skin and clothing.

- To keep from contaminating a storage container, do not return the unused portion of a chemical to its original container. Dispose of the excess chemical as directed by your instructor. Do not pour any chemicals—unused or used—down the sink unless directed to do so by your instructor.

- When mixing solutions, always add a chemical to water; never add water to the chemical. By adding the chemical to the water, you reduce the chance of a strong chemical reaction occurring.

Laboratory Instruments

You will use a variety of scientific instruments in the anatomy laboratory. Safety guidelines for specific instruments are included in the appropriate exercises. This discussion concerns the instruments most frequently used in laboratory exercises.

- **Microscope** The microscope is the main instrument you will use in the study of anatomy, and Exercise 4 of this manual is devoted to the use and care of this instrument. A few simple safety rules will prevent injury to yourself and damage to the microscope.
  1. Always carry a microscope with two hands, and do not swing the instrument as you carry it.
  2. Use only the special lens paper and cleaning solution provided by your laboratory instructor to clean the microscope lenses. Other papers and cloths may scratch the optical coatings on the lenses. An unapproved cleaning agent may dissolve the adhesives used in the lenses.
  3. Always unwrap the electrical cord completely before plugging in the microscope.
  4. Unplug the microscope by pulling on the plug, not by tugging on the electrical cord.

- **Dissection Tools** Working with sharp blades and points always presents the possibility of injury. Always cut away from yourself, and never force a blade through a tissue. Use small knife strokes for increased blade control rather than large cutting motions. Always use a sharp blade, and dispose of used blades in a specially designated "sharps" container. Carefully wash and dry all instruments upon completion of each dissection.

  Special care is necessary while changing disposable scalpel blades. Your instructor may demonstrate the proper technique for blade replacement. Always wash the used blade before removing it from the handle. Examine the handle and blade, and determine how the blade fits onto the handle. Do not force the blade off the handle. If you have difficulty changing blades, ask your instructor for assistance.

- **Water Bath** A water bath is used to incubate laboratory samples at a specific temperature. Potential hazards involving water baths include electrical shock due to contact with water and burn-related injuries caused by touching hot surfaces or spilling hot solutions. Electrical hazards are minimized by following the safety rules concerning plugging and unplugging electrical devices. Burns are avoided by using tongs to immerse or remove samples from a water bath. Point the open end of all glassware containing a sample away from yourself and others. If the sample boils, it could splatter out and burn your skin. Use a water-bath rack to support all glassware, and place hot samples removed from a water bath in a cooling rack. Monitor the temperature and water level of all water baths. Excessively high temperatures increase the chance of burns and usually ruin an experiment. When using boiling water baths, add water frequently, and do not allow all the water to evaporate.

- **Microcentrifuge** A microcentrifuge is used for blood and urine analyses. The instrument spins at thousands of revolutions per minute. Although the moving parts are housed in a protective casing, it is important to keep all loose hair, clothing, and jewelry away from the instrument. Never open the safety lid while the centrifuge is on or spinning. Do not attempt to stop a spinning centrifuge with your hand. The instrument has an internal braking mechanism that stops it safely.

Preservatives

Most animal and tissue specimens used in the laboratory have been treated with chemicals to prevent decay. These preservatives are irritants and should not contact your skin or your mucous membranes (linings of the eyes, nose, and mouth, and urinary, digestive, and reproductive openings). The following guidelines will protect you from these hazards.

- If you are pregnant, limit your exposure to all preservatives. Discuss the laboratory exercise with your instructor. Perhaps you can observe rather than perform the dissection.
- Always wear gloves and safety glasses when working with preserved material.
- Your laboratory may be equipped with exhaust fans to ventilate preservative fumes during dissections. Do not hesitate to ask your instructor to turn on the fans if the preservative odor becomes bothersome.
- Many preservatives are either toxic or carcinogenic, and all require special handling. Drain as much preservative as possible from a specimen before beginning a dissection. Pour the drained preservative either into the specimen storage container or into a dedicated container provided by your instructor. Never pour preservative down the drain.
- Promptly wipe up all spills and clean your work area when you have completed a dissection. Keep your gloves on during the cleanup, and dispose of gloves and paper towels in the proper biohazard container.

Disposal of Chemical and Biological Wastes

To safeguard the environment and individuals employed in waste collection, it is important to dispose of all potentially hazardous wastes in specially designed containers. State and federal guidelines detail the storage and handling procedures for chemical and biological wastes. Your laboratory instructor will manage the wastes produced in this course.

- **Body Fluids** Objects contaminated with body fluids are considered a high-risk biohazard and must be disposed of properly. Special biohazard containers will be available during exercises that involve body fluids. A special biohazard sharps container may be provided for glass, needles, and lancets.

- **Chemical Wastes** Most chemicals used in undergraduate laboratories are relatively harmless and may be diluted in water and poured down the drain. Your instructor will indicate during each laboratory session which chemicals can be discarded in this manner. Other chemicals should be disposed of in a dedicated waste container.
- **Preservatives and Preserved Specimens** Dispose of preservatives in a central storage container maintained for that purpose. As noted earlier, never pour preservative solutions down the drain. Dispose of all preserved specimens by wrapping them in a plastic bag filled with an absorbent material such as cat litter and placing the bag in a designated area for pickup by a hazardous-waste company.

Laboratory Safety

Name _____

Date _____

Section _____

A. Short-Answer Questions

1. Discuss how to protect yourself from body fluids, such as saliva and blood.

2. Why should you consider a body fluid capable of infecting you with a disease?

3. Describe how to dispose of materials contaminated with body fluids.

4. Explain how to safely plug and unplug an electrical device.

5. Discuss how to protect yourself from preservatives used on biological specimens.

6. Why are special biohazard containers used for biological wastes?

7. Explain how to clean up broken glass.

8. List the location of the following safety items in the laboratory:
   first aid kit _____
   nearest telephone _____
   eye wash station _____
   fire exits _____
   fire extinguisher _____
   chemical spill kit _____

1

fan switches _____

biohazard container _____

9. Your instructor informs you that a chemical is not dangerous. How should you dispose of the chemical?

10. What precautions should you take while using a centrifuge?

11. How are preservatives correctly discarded?

12. Discuss how to safely measure and mix chemicals.

Introduction to the Human Body

OBJECTIVES

On completion of this exercise, you should be able to:

- Define *anatomy* and *physiology* and discuss the specializations of each.

- Describe each level of organization in the body.

- Describe anatomical position and its importance in anatomical studies.

- Use directional terminology to describe the relationships of the surface anatomy of the body.

- Describe and identify the major planes and sections of the body.

- Locate all abdominopelvic quadrants and regions on laboratory models.

- Locate the organs of each organ system.

- Describe the main function of each organ system.

Knowledge of what lies beneath the skin and how the body works has been slowly amassed over a span of nearly 3,000 years. It may be obvious to us now that any logical practice of medicine depends on an accurate knowledge of human anatomy, yet people have not always realized this. Through most of human history, corpses were viewed with superstitious awe and dread. Observations of anatomy by dissection were illegal, and medicine therefore remained an elusive practice that often harmed rather than helped the unfortunate patient. Despite these superstitions and prohibitions, however, there have always been scientists who wanted to know the human body as it really is rather than how it was imagined to be.

The founder of anatomy was the Flemish anatomist and physician Andreas Vesalius (1514–1564). Vesalius set about to describe human structure accurately. In 1543 he published his monumental work *De Humani Corporis Faberica (On the Structure of the Human Body),* the first meaningful text on human anatomy. In this work he corrected more than 200 errors of earlier anatomists and produced drawings that are still useful today. Vesalius' work laid the foundation for all future knowledge of the human body. Merely imagining the body's internal structure at last became unacceptable in medical literature.

Many brilliant anatomists and physiologists since Vesalius' time have contributed significantly to the understanding of human form and function. If you are interested in this history, your campus and community libraries have numerous books on the history of medicine and science.

LAB ACTIVITY **1** Organization of the Body

Anatomy is the study of body structures. Early anatomists described the body's **gross anatomy**, the large parts such as muscles and bones. As knowledge of the body advanced and scientific tools permitted more detailed observations, the field of anatomy began to diversify into such areas as **microanatomy**, the study of microscopic structures; **cytology**, the study of cells; and **histology**, the study of **tissues**, which are groups of cells coordinating their effort toward a common function.

Physiology is the study of organ function. As with anatomy, physiology may be investigated from the molecular level to the organism level of organization. Physiology may be considered the work that cells must do to keep the body stable and operating efficiently. **Homeostasis** (hō-mē-ō-STĀ-sis; *homeo-*, unchanging + *stasis*, standing) is the maintenance of a relatively steady internal environment through physiological work. Stress, inadequate diet, and disease disrupt the normal physiological processes and may, as a result, lead to either serious health problems or death.

The various **levels of organization** at which anatomists and physiologists study the body are reflected in the fields of specialization in anatomy and physiology. Each higher level increases in structural and functional complexity, progressing from chemicals to cells, tissues, organs, and finally the organ systems that function to maintain the organism.

Figure 2.1 uses the cardiovascular system to illustrate these levels of organization. The simplest is the **molecular level**, sometimes also called the *chemical level* and shown at the bottom of the figure. Atoms, such as carbon and hydrogen, bond together and form molecules. The heart, for instance, contains protein molecules that are involved in contraction of the cardiac muscle. Molecules are organized into cellular structures called *organelles*, which have distinct shapes and functions. The organelles collectively constitute the next level of organization, the **cellular level**. Cells are the fundamental level of biological organization because it is cells, not molecules, that are alive. Different types of cells working together constitute the **tissue level**. Although tissues lack a distinct shape, they are distinguishable by cell type, such as the cardiac muscle cells of the heart. Tissues function together at the **organ level**; at this level, each organ has a distinct three-dimensional shape and a range of functions that is broader than the range of functions for individual cells or tissues. The **organ system level** includes all the organs interacting to accomplish a common goal. The heart and blood vessels, for example, constitute the cardiovascular organ system and physiologically work to move blood through the body. All organ systems make up the individual, which is referred to as the **organism level**.

QuickCheck Questions

1.1 What is the lowest living level of organization in the body?

1.2 What is homeostasis?

---

1 *Materials*

☐ Variety of objects and object sets, each representing a level of organization

Procedures

1. Classify each object or object set as to the level of organization it represents. Write your answers in the spaces provided.

   • Molecular level

   • Cellular level

   • Tissue level

   • Organ level

   • Organ system level

   • Organism level ■

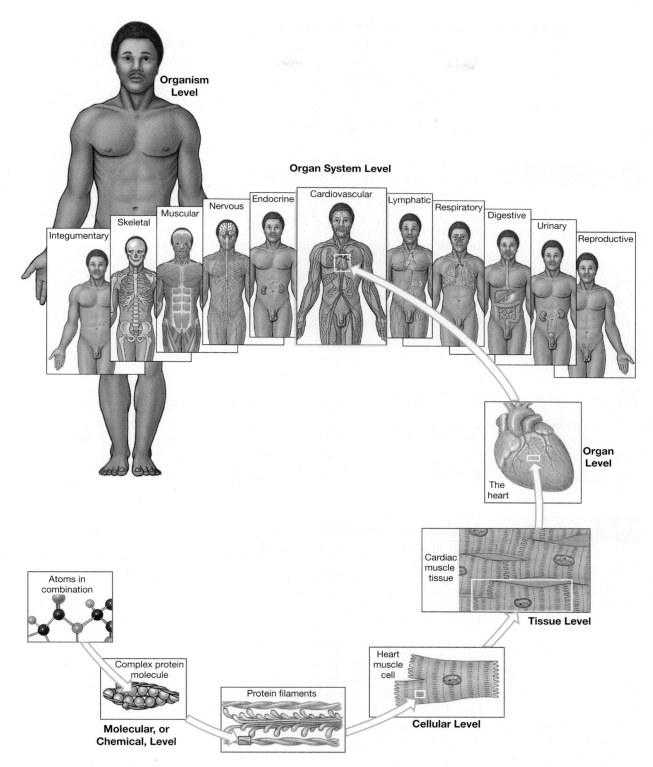

Figure 2.1 Levels of Organization

Interacting atoms form molecules that combine in the protein fibers of heart muscle cells. These cells interlock, creating heart muscle tissue that constitutes most of the walls of a three-dimensional organ, the heart. The heart is one component of the cardiovascular system, which also includes the blood and blood vessels. The combined organ systems form an organism, a living human being.

Study Tip

Getting Organized for Success

You will benefit more from your laboratory studies if you prepare for each laboratory meeting. Before class, read the appropriate exercise(s) in this manual and complete the labeling of as many figures as possible. If possible, relate the laboratory exercises to the theory concepts in the lecture textbook. Approaching the laboratory in this manner will maximize your hands-on time with laboratory materials and improve your understanding. ●

LAB ACTIVITY 2 Introduction to Organ Systems

The human body comprises 11 **organ systems,** each responsible for a specific function. The organs of a system coordinate their activities to maintain homeostasis. Most anatomy and physiology courses are designed to progress through the lower levels of organization first and then examine each organ system. Because organ systems work together to maintain the organism, it is important that you have a basic understanding of the function of each one. To this end, examine Figure 2.2 to learn the major organs and the basic function of each organ system.

QuickCheck Questions

2.1 What is a function of the endocrine system?

2.2 Name the major organs of the integumentary system.

2 Materials

- ☐ Torso models with internal organs
- ☐ Anatomical models
- ☐ Charts

Procedures

1. Locate the principal organs of each organ system on the models. If available, use a model that permits you to remove and examine the various organs. Practice returning each organ to its anatomical location.

2. On your own body, identify the general location of as many organs as possible. ■

LAB ACTIVITY 3 Anatomical Position

The human body can bend and stretch in a variety of directions. Although this flexibility allows us to move and manipulate objects in our environment, it can cause difficulty when describing and comparing structures. For example, what is the correct relationship between the wrist and the elbow? If your arm is raised above your head, you might reply that the wrist is above the elbow. With your arms at your sides, you would respond that the wrist is below the elbow. Each response appears correct, but which is the proper anatomical relationship?

To avoid confusion, the body is always referred to as being in a universal position called **anatomical position**. In anatomical position, the individual is standing erect with the feet pointed forward, the eyes straight ahead, and the palms of the hands facing forward with the arms at the sides (Figure 2.3). Stand up and assume the anatomical position. The entire front of your body should be oriented forward. Notice that anatomical position is not the natural posture of the body, because the arms and hands have been rotated forward to bring the forearms and palms to the front. An individual in anatomical position is said to be **supine** (soo-PĪN) when lying on the back and **prone** when lying face down.

When you are observing a person in anatomical position, that person's right is on your left and his or her left is on your right. For example, when shaking hands, your right hand crosses over to meet the other person's right hand. The positioning of right and left is important to remember when viewing structures in which the right side differs from the left side.

THE INTEGUMENTARY SYSTEM

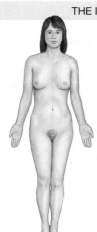

Gross Anatomy:
- Skin
- Hair
- Sweat glands
- Nails

Functions:
- Protects against environmental hazards
- Helps regulate body temperature
- Provides sensory information

THE NERVOUS SYSTEM

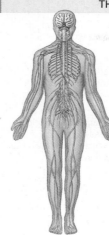

Gross Anatomy:
- Brain
- Spinal cord
- Peripheral nerves
- Sense organs

Functions:
- Directs immediate responses to stimuli
- Coordinates or moderates activities of other organ systems
- Provides and interprets sensory information about external conditions

THE SKELETAL SYSTEM

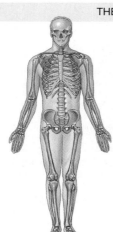

Gross Anatomy:
- Bones
- Cartilage
- Associated ligaments
- Bone marrow

Functions:
- Provides support and protection for other tissues
- Stores calcium and other minerals
- Forms blood cells

THE ENDOCRINE SYSTEM

Gross Anatomy:
- Pituitary gland
- Thyroid gland
- Pancreas
- Adrenal glands
- Gonads (testes and ovaries)
- Endocrine tissues in other systems

Functions:
- Directs long term changes in the activities of other organ systems
- Adjusts metabolic activity and energy use by the body
- Controls many structural and functional changes during development

THE MUSCULAR SYSTEM

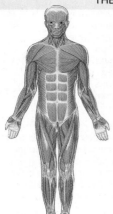

Gross Anatomy:
- Skeletal muscles and associated ligaments

Functions:
- Enables movement
- Provides protection and support for other tissues
- Generates heat that maintains body temperature

THE CARDIOVASCULAR SYSTEM

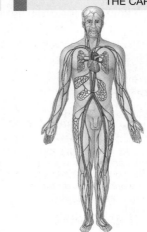

Gross Anatomy:
- Heart
- Blood
- Blood vessels

Functions:
- Distributes blood cells, water, and dissolved materials, including nutrients, waste products, oxygen, and carbon dioxide
- Distributes heat and assists in control of body temperature

Figure 2.2 An Introduction to Organ Systems

13

THE LYMPHATIC SYSTEM

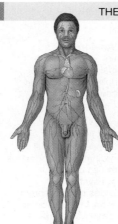

Gross Anatomy:
- Spleen
- Thymus
- Lymphatic vessels
- Lymph nodes
- Tonsils

Functions:
- Defends against infection and disease
- Returns tissue fluids to the bloodstream

THE URINARY SYSTEM

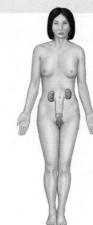

Gross Anatomy:
- Kidneys
- Ureters
- Urinary bladder
- Urethra

Functions:
- Excretes waste products from the blood
- Controls water balance by regulating volume of urine produced
- Stores urine prior to voluntary elimination
- Regulates blood ion concentrations and pH

THE RESPIRATORY SYSTEM

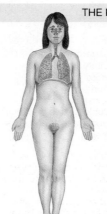

Gross Anatomy:
- Nasal cavities
- Sinuses
- Larynx
- Trachea
- Bronchi
- Lungs
- Alveoli

Functions:
- Delivers air to alveoli (sites in lungs where gas exchange occurs)
- Provides oxygen to bloodstream
- Removes carbon dioxide from bloodstream
- Produces sounds for communication

THE MALE REPRODUCTIVE SYSTEM

Gross Anatomy:
- Testes
- Epididymis
- Ductus deferens
- Seminal vesicles
- Prostate gland
- Penis
- Scrotum

Functions:
- Produces male sex cells (sperm) and hormones

THE DIGESTIVE SYSTEM

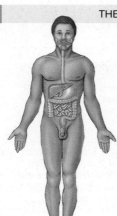

Gross Anatomy:
- Teeth
- Tongue
- Pharynx
- Esophagus
- Stomach
- Small Intestine
- Large intestine
- Liver
- Gallbladder
- Pancreas

Functions:
- Processes and digests food
- Absorbs and conserves water
- Absorbs nutrients (ions, water, and the breakdown products of dietary sugars, proteins, and fats)
- Stores energy reserves

THE FEMALE REPRODUCTIVE SYSTEM

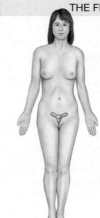

Gross Anatomy:
- Ovaries
- Uterine tubes
- Uterus
- Vagina
- Labia
- Clitoris
- Mammary glands

Functions:
- Produces female sex cells (oocytes) and hormones
- Supports developing embryo from conception to delivery
- Provides milk to nourish newborn infant

Figure 2.2 An Introduction to Organ Systems *(Continued)*

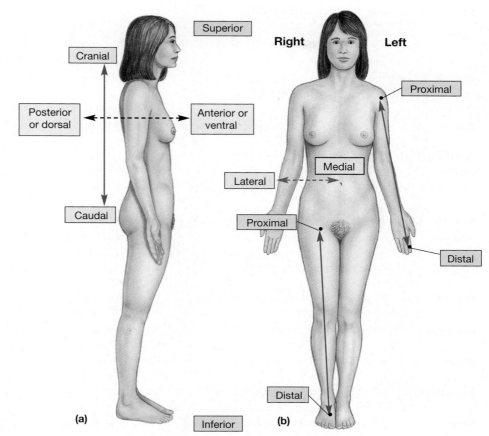

Figure 2.3 Directional References
Important directional terms used in this text are indicated by arrows; definitions and descriptions are included in Table 2.1. **(a)** A lateral view. **(b)** An anterior view.

QuickCheck Questions

3.1 Why is anatomical position important in anatomical studies?

3.2 Is the natural stance while standing considered anatomical position?

3 *Materials*

☐ Yourself or a laboratory partner

Procedures

1. Stand and assume anatomical position. Consider how this orientation differs from your normal stance. ■

LAB ACTIVITY 4 Directional Terminology

Imagine attempting to give someone directions if you could not use terms like "north" and "south" or "left" and "right." These words have a unique meaning and guide the traveler toward his or her destination. Describing the body also requires specific terminology. Expressions such as "near," "close to," or "on top of" are too vague for anatomical descriptions. To prevent misunderstandings, precise terms are used to describe the locations and associations of anatomy. These terms have their roots in the Greek and Latin languages. Figure 2.3 and Table 2.1 display the most frequently used directional terms. Notice that most of them can be grouped into opposing pairs, or antonyms.

• **Superior** and **inferior** describe vertical positions. *Superior* means above, and *inferior* means below. For example, the head is superior to the shoulders, and the knee is inferior to the hip. (Remember that the body is always assumed to be in anatomical position when we use these directional terms.) **Cranial** refers to the head or toward the head region. The cranial cavity, for instance, is the internal space of the skull.

Table 2.1 Directional Terms (see Figure 2.3)

Term	Region or Reference	Example
Anterior	The front; before	The navel is on the *anterior* surface of the trunk.
Ventral	The belly side (equivalent to anterior when referring to human body)	In humans, the navel is on the *ventral* surface.
Posterior	The back; behind	The shoulder blade is located *posterior* to the rib cage.
Dorsal	The back (equivalent to posterior when referring to human body)	The *dorsal* body cavity encloses the brain and spinal cord.
Cranial or cephalic	The head	The *cranial,* or *cephalic,* border of the pelvis is on the side toward the head rather than toward the thigh.
Superior	Above; at a higher level (in human body, toward the head)	In humans, the cranial border of the pelvis is *superior* to the thigh.
Caudal	The tail (coccyx in humans)	The hips are *caudal* to the waist.
Inferior	Below; at a lower level	The knees are *inferior* to the hips.
Medial	Toward the body's longitudinal axis; toward the midsagittal plane	The *medial* surfaces of the thighs may be in contact; moving medially from the arm across the chest surface brings you to the sternum.
Lateral	Away from the body's longitudinal axis; away from the midsagittal plane	The thigh articulates with the *lateral* surface of the pelvis; moving laterally from the nose brings you to the eyes.
Proximal	Toward an attached base	The thigh is *proximal* to the foot; moving proximally from the wrist brings you to the elbow.
Distal	Away from an attached base	The fingers are *distal* to the wrist; moving distally from the elbow brings you to the wrist.
Superficial	At, near, or relatively close to the body surface	The skin is *superficial* to underlying structures.
Deep	Farther from the body surface	The bone of the thigh is *deep* to the surrounding skeletal muscles.

- **Anterior** and **posterior** refer to front and back. *Anterior* means in front of or forward. The anterior surface of the body includes all front surfaces, including the palms of the hand. Structures on the back surface of the body are said to be posterior. The heart, for example, is posterior to the breastbone and anterior to the spine.

- In four-legged animals, anatomical position is with all four limbs on the ground, and therefore the meanings of some directional terms change. *Superior* now refers to the back, or **dorsal**, surface, and *inferior* refers to the belly, or **ventral,** surface. **Cephalic** means anterior in four-legged animals, and **caudal** refers to posterior structures.

- **Medial** and **lateral** describe positions relative to the body's midline, the vertical middle of the body or of a structure. *Medial* has two meanings. It describes one structure as being closer to the body's midline than some other structure; for instance, the ring finger is medial to the middle finger when the hand is held in anatomical position. *Medial* also describes a structure that is permanently between others, as the nose is medial to the eyes. *Lateral* means either farther from the body's midline or permanently to the side of some other structure; the eyes are lateral to the nose, and, in anatomical position, the middle finger is lateral to the ring finger.

- **Proximal** refers to parts near another structure. **Distal** describes structures that are distant from other structures. These terms are frequently used to describe the proximity of a structure to its point of attachment on the body. For example, the thigh bone (femur) has a proximal region where it attaches to the hip and a distal region toward the knee.

• **Superficial** and **deep** describe layered structures. *Superficial* refers to parts on or close to the surface. Underneath an upper layer are deep, or bottom, structures. The skin is superficial to the muscular system, and bones are usually deep to the muscles.

Some directional terms seem to be interchangeable, but there is usually a precise term for each description. For example, *superior* and *proximal* both describe the upper region of arm and leg bones. When discussing the point of attachment of a bone, *proximal* is the more descriptive term. When describing the location of a bone relative to an inferior bone, the term *superior* is used.

QuickCheck Questions

4.1 What is the relationship of the upper arm at the shoulder joint to the forearm?

4.2 What is the relationship of muscles to the skin?

4   *Materials*

☐ Torso models
☐ Charts
☐ Anatomical models

Procedures

1. Review each directional term presented in Figure 2.3 and Table 2.1.

2. Use the laboratory models and your own body to practice using directional terms while comparing anatomy. The Laboratory Report at the end of this exercise may be used as a guide for comparisons. ■

LAB ACTIVITY   5   Regional Terminology

Approaching the body from a regional perspective simplifies the learning of anatomy. Body surface features are used as anatomical landmarks to assist in locating internal structures, and as a result many internal structures are named after an overlying surface structure. For example, the back of the knee is called the popliteal (pop-LIT-ē-al) region, and the major artery in the knee is the popliteal artery. Figure 2.4 and Table 2.2 present the major regions of the body.

The head is referred to as the **cephalon** and consists of the **cranium,** or skull, and the **face**. The neck is the **cervical** region. The main part of the body is the **trunk,** which attaches the neck, arms, and legs. The upper trunk is the chest, or **pectoral,** region. Below the chest is the **abdominal** region, which narrows at the **pelvis**. The back surface of the trunk, the **dorsum,** includes the **loin,** or lower back, and the **gluteal** region of the buttock. The side of the trunk below the ribs is the **flank**.

The shoulder, or **scapular,** region attaches the **upper limb**, or arm, to the trunk and forms the **axilla,** the armpit. The upper part of the arm is the **brachium**, the **antebrachium** is the forearm. Between the brachium and antebrachium is the **antecubitis** region, the elbow. The wrist is called the **carpus,** and the inside surface of the hand is the **palm**.

The pelvis attaches the **lower limb**, or leg, to the trunk at the **inguinal** area, or **groin**. The upper leg is the **thigh,** the back of the knee is the **popliteal** region, and the lower leg is the calf, or **sura. Tarsus** refers to the ankle, and the sole of the foot is the **plantar** surface.

Reference to the position of internal abdominal organs is simplified by partitioning the trunk into four equal **quadrants**. Observe in Figure 2.5a the vertical and horizontal planes used to delineate the quadrants. Quadrants are used to describe the positions of organs. The stomach, for example, is mostly located in the left upper quadrant.

For more detailed descriptions, the abdominal surface is divided into nine **abdominopelvic regions**, shown in Figure 2.5b. Four planes are used to define the regions: two vertical and two transverse planes arranged in the familiar tic-tac-toe

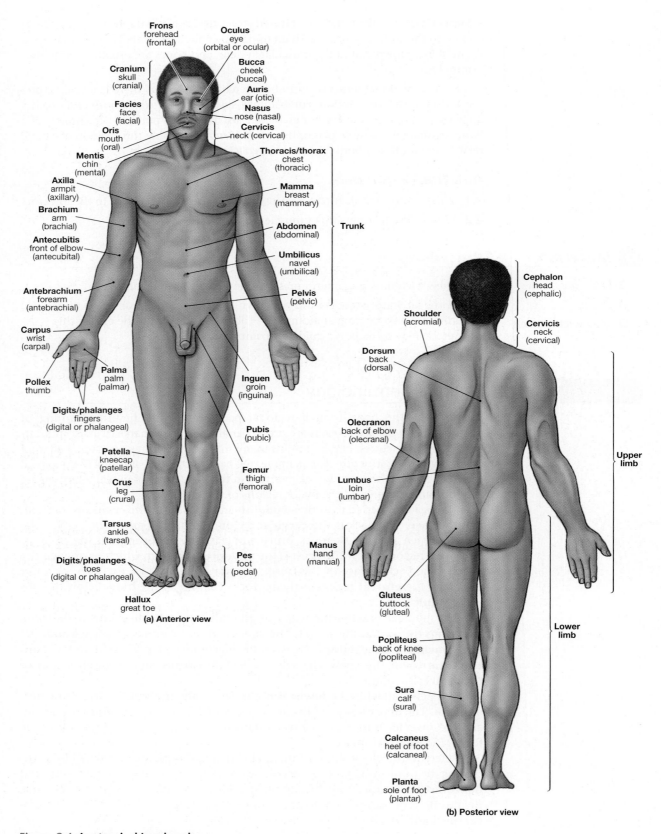

Frons
forehead
(frontal)

Oculus
eye
(orbital or ocular)

Cranium
skull
(cranial)

Bucca
cheek
(buccal)

Auris
ear (otic)

Nasus
nose (nasal)

Facies
face
(facial)

Cervicis
neck (cervical)

Oris
mouth
(oral)

Mentis
chin
(mental)

Thoracis/thorax
chest
(thoracic)

Axilla
armpit
(axillary)

Mamma
breast
(mammary)

Brachium
arm
(brachial)

Abdomen
(abdominal)

Antecubitis
front of elbow
(antecubital)

Umbilicus
navel
(umbilical)

Trunk

Antebrachium
forearm
(antebrachial)

Pelvis
(pelvic)

Carpus
wrist
(carpal)

Palma
palm
(palmar)

Pollex
thumb

Inguen
groin
(inguinal)

Digits/phalanges
fingers
(digital or phalangeal)

Pubis
(pubic)

Patella
kneecap
(patellar)

Femur
thigh
(femoral)

Crus
leg
(crural)

Tarsus
ankle
(tarsal)

Digits/phalanges
toes
(digital or phalangeal)

Pes
foot
(pedal)

Hallux
great toe

(a) Anterior view

Cephalon
head
(cephalic)

Shoulder
(acromial)

Cervicis
neck
(cervical)

Dorsum
back
(dorsal)

Olecranon
back of elbow
(olecranal)

Upper
limb

Lumbus
loin
(lumbar)

Manus
hand
(manual)

Gluteus
buttock
(gluteal)

Lower
limb

Popliteus
back of knee
(popliteal)

Sura
calf
(sural)

Calcaneus
heel of foot
(calcaneal)

Planta
sole of foot
(plantar)

(b) Posterior view

Figure 2.4 Anatomical Landmarks

Anatomical terms are shown in boldface type, common names in plain type, and anatomical adjectives in parentheses.

18

Table 2.2 **Regions of the Human Body (see Figure 2.4)**

Structure	Region
Cephalon (head)	Cephalic region
Cervicis (neck)	Cervical region
Thoracis (thorax, or chest)	Thoracic region
Brachium (arm)	Brachial region
Antebrachium (forearm)	Antebrachial region
Carpus (wrist)	Carpal region
Manus (hand)	Manual region
Abdomen	Abdominal region
Lumbus (loin)	Lumbar region
Gluteus (buttock)	Gluteal region
Pelvis	Pelvic region
Pubis (anterior pelvis)	Pubic region
Inguen (groin)	Inguinal region
Femur (thigh)	Femoral region
Crus (anterior leg)	Crural region
Popliteus (back of knee)	Popliteal region
Sura (calf)	Sural region
Tarsus (ankle)	Tarsal region
Pes (foot)	Pedal region
Planta (sole)	Plantar region

pattern. The vertical planes, called the right and left **lateral planes**, are positioned slightly medial to the nipples, each plane on the side of the nipple that is closer to the body center. The lateral planes divide the trunk into three nearly equal vertical regions. A pair of transverse planes crosses the vertical planes to isolate the nine regions. The **transpyloric plane** is superior to the umbilicus (navel) at the level of the pylorus, the lower region of the stomach. The **transtubercular plane** is inferior to the umbilicus and crosses the abdomen at the level of the superior hips.

The nine abdominopelvic regions are as follows. The **umbilical region** surrounds the umbilicus. Lateral to this region are the right and left **lumbar regions**. Above the umbilicus is the **epigastric region** containing the stomach and much of the liver. The right and left **hypochondriac** (hī-pō-KON-drē-ak; *hypo*, under + *chondro*, cartilage) **regions** are lateral to the epigastric region. Inferior to the umbilical region is the **hypogastric,** or **pubic, region**. The right and left **iliac,** or **inguinal, regions** border the hypogastric region laterally.

QuickCheck Questions

5.1 What are the major regions of the upper limb?

5.2 How is the abdominal surface divided into different regions?

5 Materials

- ☐ Torso models
- ☐ Charts

Procedures

1. Review the regional terminology in Figure 2.4 and Table 2.2.
2. Identify on a laboratory model or yourself the regional anatomy as presented in Figure 2.5.
3. Label each abdominopelvic quadrant and region in Figure 2.5b and on laboratory models. ∎

epigastric region
hypogastric (pubic) region
left hypochondriac region
left inguinal (iliac) region
left lumbar region
right hypochondriac region
right inguinal (iliac) region
right lumbar region
umbilical region

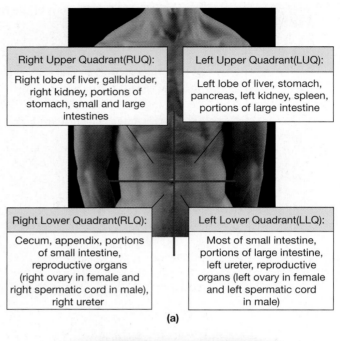

Right Upper Quadrant(RUQ):

Right lobe of liver, gallbladder, right kidney, portions of stomach, small and large intestines

Left Upper Quadrant(LUQ):

Left lobe of liver, stomach, pancreas, left kidney, spleen, portions of large intestine

Right Lower Quadrant(RLQ):

Cecum, appendix, portions of small intestine, reproductive organs (right ovary in female and right spermatic cord in male), right ureter

Left Lower Quadrant(LLQ):

Most of small intestine, portions of large intestine, left ureter, reproductive organs (left ovary in female and left spermatic cord in male)

(a)

1. _____
2. _____
3. _____
4. _____
5. _____
6. _____
7. _____
8. _____
9. _____

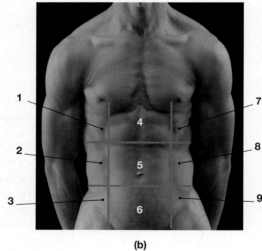

(b)

Figure 2.5 Abdominopelvic Quadrants and Regions

(a) Abdominopelvic quadrants divide the area into four sections. These terms, or their abbreviations, are most often used in clinical discussions. (b) Abdominopelvic regions divide the same area into nine sections, providing more precise regional descriptions. (c) Overlapping quadrants and regions and the relationship between superficial anatomical landmarks and underlying organs.

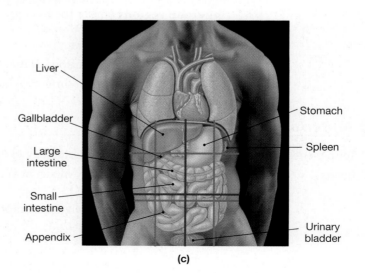

Liver

Gallbladder

Large intestine

Small intestine

Appendix

Stomach

Spleen

Urinary bladder

(c)

LAB ACTIVITY 6 Planes and Sections

The body must be cut in order for us to study its internal organization. The process of cutting the body is called **sectioning**. Most structures, such as the trunk, knee, arm, and eyeball, can be sectioned. The imaginary line of a section is called a **plane**. The orientation of the plane of section determines the shape and appearance of the exposed internal region. Imagine cutting one soda straw crosswise and another straw lengthwise. The former produces a circle, and the latter produces a concave rectangular surface.

Three major types of sections are used in the study of anatomy: two vertical sections and one transverse section (Figure 2.6 and Table 2.3). **Transverse** sections are perpendicular to the vertical orientation of the body. (The crosswise cut you made on the imaginary straw yielded a transverse section.) Transverse sections are often called cross-sections because they go across the body axis. A transverse section divides superior and inferior structures. **Vertical** sections are parallel to the vertical axis of the body and include sagittal and frontal sections. A **sagittal** section divides a body or organ into right and left portions. A **midsagittal** section equally divides structures, and a **parasagittal** section produces nearly equal divisions. A **frontal**, or **coronal**, section separates anterior and posterior structures.

QuickCheck Questions

6.1 Which type of section would separate the kneecap from the leg?

6.2 Amputation of the forearm is performed by which type of section?

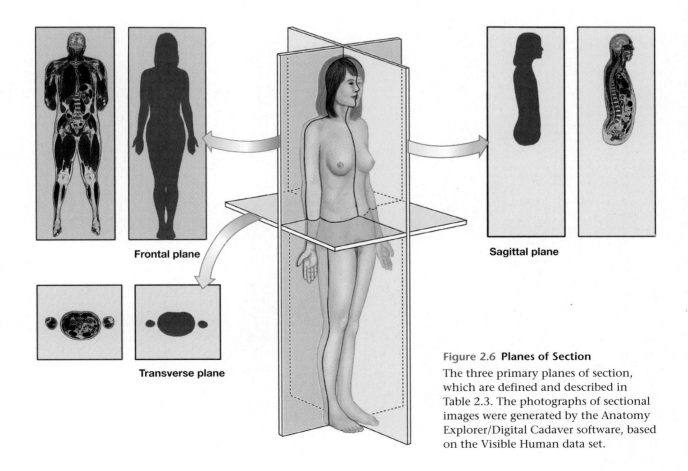

Frontal plane

Sagittal plane

Transverse plane

Figure 2.6 Planes of Section

The three primary planes of section, which are defined and described in Table 2.3. The photographs of sectional images were generated by the Anatomy Explorer/Digital Cadaver software, based on the Visible Human data set.

Table 2.3 Terms That Indicate Planes of Section (see Figure 2.6)

Orientation of Plane	Plane	Directional Reference	Description
Perpendicular to long axis	Transverse or horizontal	Transversely or horizontally	A *transverse*, or *horizontal*, *section* separates superior and inferior portions of the body.
Parallel to long axis	Sagittal	Sagittally	A *sagittal section* separates right and left portions. You examine a sagittal section, but you section sagittally.
	Midsagittal		In a *midsagittal section* the plane passes through the midline, dividing the body in half and separating right and left sides.
	Parasagittal		A *parasagittal section* misses the midline, separating right and left portions of unequal size.
	Frontal or coronal	Frontally or coronally	A *frontal*, or *coronal*, *section* separates anterior and posterior portions of the body; *coronal* usually refers to sections passing through the skull.

6 *Materials*

☐ Anatomical models with various sections

Procedures

1. Review each plane and section shown in Figure 2.6 and Table 2.3.
2. Identify the sections on models and other materials presented by your instructor.
3. Cut several common objects, such as an apple and a hot dog, along their sagittal and transverse planes. Compare the exposed arrangement of the interior. ∎

Introduction to the Human Body

Name _____

Date _____

Section _____

A. Matching

Match each directional term on the left with the correct description on the right.

_____	**1.** anterior	**A.** to the side
_____	**2.** lateral	**B.** away from a point of attachment
_____	**3.** proximal	**C.** close to the body surface
_____	**4.** inferior	**D.** front
_____	**5.** posterior	**E.** away from the body surface
_____	**6.** medial	**F.** above, on top of
_____	**7.** distal	**G.** toward a point of attachment
_____	**8.** superficial	**H.** below, a lower level
_____	**9.** superior	**I.** back
_____	**10.** deep	**J.** toward the middle

B. Labeling

Label the regions of the body in Figure 2.7.

C. Short-Answer Questions

1. Describe the levels of organization in the body.

2. Why is anatomical position important when describing structures?

3. List the nine abdominopelvic regions and the location of each.

4. Compare the study of anatomy with that of physiology.

2

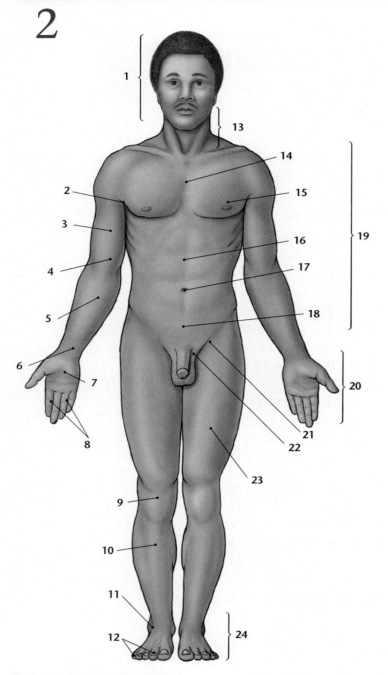

Figure 2.7 **Regional Terminology**

1. _____
2. _____
3. _____
4. _____
5. _____
6. _____
7. _____
8. _____
9. _____
10. _____
11. _____
12. _____
13. _____
14. _____
15. _____
16. _____
17. _____
18. _____
19. _____
20. _____
21. _____
22. _____
23. _____
24. _____

D. Directional and Regional Terminology

Use the correct directional term to show the relationship between the following structures:

1. The chin is _____ to the nose.

2. In either vertical plane, the brachium is _____ to the antecubitis.

3. The index finger is _____ to the ring finger.

4. The skin is _____ to the muscles.

5. The trunk is _____ to the pubis.

6. The middle toe is _____ to the little toe.

7. Where it attaches to the elbow, the upper humerus is _____ to the elbow.

8. The ears are _____ to the eyes.

9. The buttock is _____ to the pubis.

10. Relative to where the leg attaches to the trunk, the knee is _____ to the hip.

E. Drawing

1. Draw two pictures of a donut (torus) sectioned by a plane. In one drawing, make the sectioning plane parallel to the circular surface of the donut; in the other drawing, make the sectioning plane perpendicular to that surface.

2. Sketch the body trunk and the planes that designate the abdominopelvic regions.

2

F. Analysis and Application

1. Describe how the cardiovascular system and the urinary system work together to maintain homeostasis.

2. Which organ systems protect the body from infection?

3. Long-term coordination of body functions is regulated by which organ system?

4. Which organ system stores minerals for the body?

Body Cavities and Membranes

OBJECTIVES

On completion of this exercise, you should be able to:

- Describe and identify the location of the dorsal cavity and its divisions.

- Describe and identify the location of the ventral cavity and its divisions.

- Describe and identify the serous membranes of the body.

Body cavities are internal spaces that house internal organs, such as the brain in the cranium and the digestive organs in the abdomen. The walls of a body cavity support and protect the soft organs contained in the cavity. In the trunk, large cavities are subdivided into smaller cavities that contain individual organs. The smaller cavities are enclosed by thin sacs, such as those around the heart, lungs, and intestines. Most of the space in a cavity is occupied by the enclosed organ and by a thin film of liquid.

There are two major body cavities, dorsal and ventral. The dorsal cavity contains the central nervous system, and the ventral cavity houses the thoracic and abdominopelvic organs. In the ventral body cavity, the heart, lungs, stomach, and intestines are covered with a double-layered **serous** (SĒR-us; *seri-*, watery) **membrane**. Each serous membrane isolates one organ and reduces friction and abrasion on the organ surface. In the dorsal body cavity, the brain and spinal cord are contained within the meninges, a protective three-layered membrane that we shall study later with the nervous system.

LAB ACTIVITY 1   Dorsal Body Cavity

The **dorsal body cavity** has two regions: the **cranial cavity** within the oval part of the skull and the **spinal cavity** inside the vertebral column (Figure 3.1). The brain is located in the cranial cavity, and this part of the skull, called the *cranium,* encases and protects the delicate brain. The spinal cavity is a long canal that passes through the spine. The vertebrae of the spine have holes called *vertebral foramina* that collectively form the walls of the spinal cavity and protect the spinal cord. The cranial and spinal cavities are continuous with each other and join at the base of the skull, where the spinal cord meets the brain. The membranes of the brain pass into the spinal cavity and cover the spinal cord.

QuickCheck Question

1.1 What structures form the walls of the dorsal body cavity?

1 Materials

☐ Torso models
☐ Articulated skeleton
☐ Charts

Procedures

1. Label and review each component of the dorsal body cavity in Figure 3.1.

2. Locate the dorsal body cavity and each subdivision on your own body.

3. Identify each part of the dorsal body cavity on the laboratory models and articulated skeleton.

4. Identify the organ(s) of each part of the dorsal body cavity on the laboratory models. ■

LAB ACTIVITY 2   Ventral Body Cavity

The **ventral body cavity**, also called the **coelom** (SĒ-lÙm; *koila,* cavity), is the entire space of the body trunk anterior to the vertebral column and posterior to the breastbone and the abdominal muscle wall. Using Figure 3.1 as a guide, trace the outline of the ventral body cavity on the anterior surface of your own body. This large cavity is divided into two major cavities, the **thoracic** (*thorax,* chest) **cavity** and the **abdominopelvic cavity**. These cavities, in turn, are further divided into the specific cavities that surround individual organs.

The walls of the thoracic cavity, or chest cavity, are muscle and bone. This cavity contains the heart, lungs, trachea, larynx, esophagus, thymus gland, and many large blood vessels. Main subdivisions of the thoracic cavity are the **pericardial** (*peri-,* around + *kardia,* heart) **cavity**, which contains the heart, and two **pleural cavities**, one for each lung. The part of the thoracic cavity that contains the heart—plus its large blood vessels, the thymus gland, the trachea, and the esophagus—is called the **mediastinum** (mē-dē-as-TĪ-num or mē-dē-AS-tĭ-num; *media-,* middle). The lungs are located outside the mediastinum.

The abdominopelvic cavity is separated from the thoracic cavity by a dome-shaped muscle, the diaphragm. Locate the approximate position of the diaphragm on your anterior surface and in the various views of Figure 3.1. The abdominopelvic cavity is the space between the diaphragm and the floor of the pelvis. This cavity is subdivided into the abdominal cavity and the pelvic cavity. The **abdominal cavity** contains most of the digestive organs, such as the liver, gallbladder, stomach, pancreas, kidneys, and small and large intestines. The **pelvic cavity** is the small cavity enclosed by the pelvic girdle of the hips. This cavity contains the internal reproductive organs, parts of the large intestine, the rectum, and the urinary bladder.

QuickCheck Questions

2.1 Name the various subdivisions of the ventral body cavity.

2.2 What organs are in each smaller cavity contained in the thoracic cavity?

2 Materials

☐ Torso models
☐ Articulated skeleton
☐ Charts

Procedures

1. Review the location of the ventral cavity and its subdivisions by labeling Figure 3.1.

2. With your finger, trace the location of the various divisions of the ventral body cavity on the anterior of your trunk.

3. Locate the ventral cavity and each of its subdivisions on laboratory models, mannequins, charts, and skeleton.

4. Identify the organ(s) within the various cavities of the ventral body cavity on the laboratory models. ■

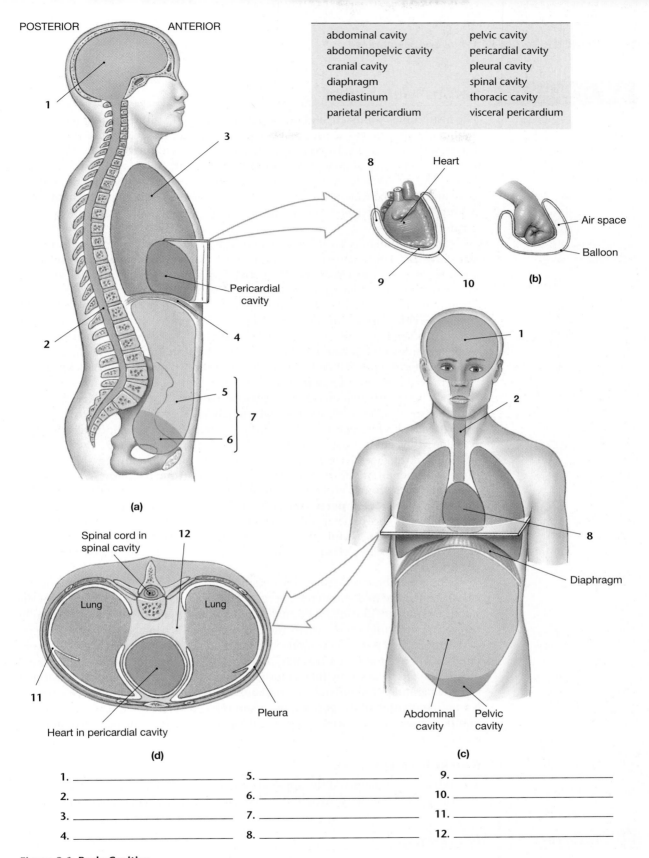

POSTERIOR   ANTERIOR

abdominal cavity   pelvic cavity
abdominopelvic cavity   pericardial cavity
cranial cavity   pleural cavity
diaphragm   spinal cavity
mediastinum   thoracic cavity
parietal pericardium   visceral pericardium

Heart

8

Pericardial
cavity

9   10

Air space

Balloon

(b)

(a)

Spinal cord in
spinal cavity   12

Lung   Lung

11

Pleura

Heart in pericardial cavity

(d)

Diaphragm

Abdominal   Pelvic
cavity   cavity

(c)

1. _____   5. _____   9. _____
2. _____   6. _____   10. _____
3. _____   7. _____   11. _____
4. _____   8. _____   12. _____

Figure 3.1 **Body Cavities**

(a) The dorsal body cavity is bounded by the bones of the skull and vertebral column. The muscular diaphragm divides the ventral body cavity into a superior thoracic (chest) cavity and an inferior abdominopelvic cavity. The pericardial cavity is inside the chest cavity. (b) The heart is suspended within the pericardial cavity like a fist pushed into a balloon. The attachment site, corresponding to the wrist of the hand, lies at the connection between the heart and major blood vessels. (c) An anterior view of and (d) a transverse section through the ventral body cavity, showing the central location of the pericardial cavity within the chest cavity. The sectional plane shows how the mediastinum divides the thoracic cavity into two pleural cavities.

LAB ACTIVITY 3 — Serous Membranes

The heart, lungs, stomach, and intestines are encased in serous membranes, which are double-layered, as mentioned earlier, and have a minuscule fluid-filled cavity between the two layers. Directly attached to the exposed surface of an internal organ is the **visceral** (VIS-er-al; *viscera*, internal organ) **layer** of the serous membrane. The **parietal** (pah-RĪ-e-tal; *pariet-*, wall) **layer** is superficial to the visceral layer and lines the wall of the body cavity. The **serous fluid** between these layers is a lubricant that reduces friction and abrasion between the layers as the enclosed organ moves.

Figure 3.1b highlights the anatomy of the serous membrane of the heart, the **pericardium**. This membrane is composed of an outer **parietal pericardium** and an inner **visceral pericardium**. The parietal pericardium is a fibrous sac attached to the diaphragm and supportive tissues of the thoracic cavity. The visceral pericardium is attached to the surface of the heart. Between these two serous layers is the **pericardial cavity**. Imagine pushing your fist into a water-filled balloon, as in Figure 3.1b; your fist is the heart and the balloon is the pericardium. The balloon immediately surrounding your hand is the visceral pericardium, and the outer layer of the balloon is the parietal pericardium. The water-filled space is the pericardial cavity, with water representing serous fluid.

Each lung is isolated in a separate **pleural cavity**. The **parietal pleura** (PLOO-rah) lines the thoracic wall, and the **visceral pleura** is attached to the surface of the lung. Because each lung is contained inside a separate cavity, a puncture wound on one side of the chest usually collapses only the corresponding lung.

Most of the digestive organs are encased in the **peritoneum** (per-i-tō-NĒ-um), the serous membrane of the abdomen. The **parietal peritoneum** has numerous folds that wrap around and attach the abdominal organs to the posterior abdominal wall. The **visceral peritoneum** lines the organ surfaces. The peritoneal cavity is between the parietal and visceral peritoneal layers. The peritoneum has many blood vessels, lymphatic vessels, and nerves that support the digestive organs. The kidneys are **retroperitoneal** (*retro-*, behind) and are located outside the peritoneum.

Serous membranes may become inflamed and infected as a result of bacterial invasion or damage to the underlying organ. Fluids often build up in the cavity of the serous membrane, causing additional complications. **Peritonitis** is an infection of the peritoneum that occurs when the digestive tract is damaged—often by ulceration, rupture, or a puncture wound—in a way that permits intestinal bacteria to contaminate the peritoneum. **Pleuritis**, or **pleurisy**, is an inflammation of the pleural membrane often caused by tuberculosis, pneumonia, or thoracic abscess. Breathing is made painful as the inflamed membranes move during respiration. **Pericarditis** is an inflammation of the pericardium resulting from infection, injury, heart attack, or other causes. In advanced stages, a buildup of fluid causes the heart to compress, a condition that results in decreased cardiac function.

QuickCheck Question

3.1 What would happen if the appendix burst and intestinal contents leaked into the peritoneal cavity? Would there be an immediate effect on the kidneys?

3 *Materials*

- ☐ Torso models
- ☐ Charts

Procedures

1. Review and label each serous membrane in Figure 3.1.
2. Identify the pericardium, pleura, and peritoneum on the models and charts.
3. Identify the organ(s) of each body cavity on the laboratory models. ∎

Body Cavities and Membranes

Name _____

Date _____

Section _____

A. Matching

Match each term on the left with the correct description on the right.

_____	**1.** serous membrane	**A.**	in reference to an internal organ
_____	**2.** parietal	**B.**	serous membrane of abdomen
_____	**3.** pleura	**C.**	in reference to a body wall
_____	**4.** visceral	**D.**	double-layered protective membrane
_____	**5.** pericardium	**E.**	serous membrane of lungs
_____	**6.** peritoneum	**F.**	serous membrane of heart

B. Fill in the Blanks

1. The heart is located in a small cavity called the _____, which is inside a larger cavity, the _____.

2. The _____ surrounds the digestive organs in the abdominal cavity.

3. The kidneys are _____ because they are located outside the _____.

4. A plane at the top of the hips separates the abdominal cavity from the _____.

5. The inner membrane layer surrounding a lung is the _____.

6. The brain and spinal cord are contained in the body cavity called the _____.

7. A lubricating substance in body cavities is called _____.

8. The large medial area of the chest is called the _____.

9. The muscle that divides the ventral body cavity horizontally is the _____.

10. The outer layer of a serous membrane is the _____ layer.

C. Labeling

Label the nine numbered cavities in Figure 3.2.

D. Analysis and Application

1. Nicole has a respiratory infection that has caused her right pleura to dry out. Describe the symptoms that could be related to this condition.

2. Describe the cavities that protect the brain and spinal cord.

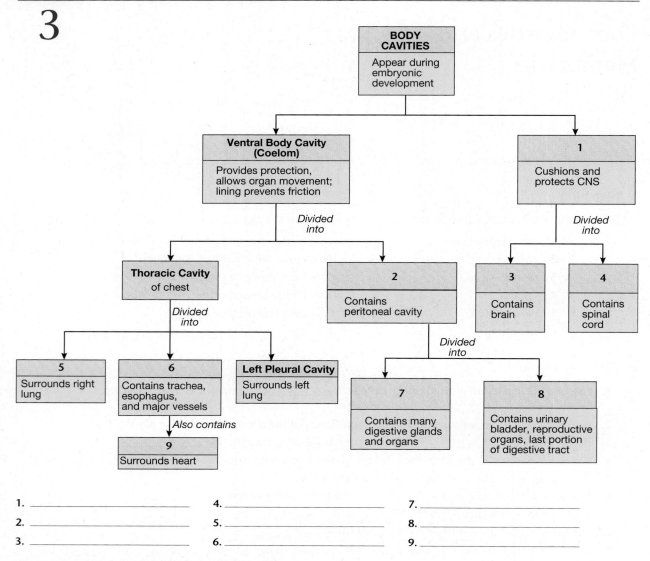

1. _____   4. _____   7. _____
2. _____   5. _____   8. _____
3. _____   6. _____   9. _____

Figure 3.2 **Relationships of the Various Body Cavities**

3. Why are organs in the ventral body cavity surrounded by double-layered membranes instead of a single-layered membrane?

Use of the Microscope

OBJECTIVES

On completion of this exercise, you should be able to:

- Describe how to properly carry, clean, and store a microscope.

- Identify the parts of a microscope.

- Focus a microscope on a specimen and adjust the magnification.

- Adjust the light source of a microscope.

- Calculate the total magnification for each objective lens.

- Measure the field diameter at each magnification.

- Make a wet-mount slide.

As a student of anatomy and physiology, you will explore the organization and structure of cells, tissues, and organs. The basic research tool for your observations is the microscope. The instrument is easy to use once you learn its parts and how to adjust them to produce a clear image of a specimen. To this end, it is important that you complete each activity in this exercise and that you are able to use a microscope effectively by the end of the laboratory period.

The compound microscope uses several lenses to direct a narrow beam of light through a thin specimen mounted on a glass slide. Focusing knobs move either the lenses or the slide to bring the specimen into focus within the round viewing area of the lenses, an area called the **field of view**. Lenses magnify objects so that the objects appear larger than they actually are. As magnification increases, the viewer can more easily see details that are close together. It is this increase in **resolution**—the ability to distinguish between two objects located close to each other—that makes the microscope a powerful observational tool.

Microscope Care and Handling

Because the microscope is a precision scientific instrument with delicate optical components, you should always observe the following guidelines when using one.

1. Carry the microscope with two hands, one hand on the arm and the other hand supporting the base. (See Figure 4.1 for the parts of the microscope.) Do not swing the microscope as you carry it to your laboratory bench, because

such movement could cause a lens to fall out. Avoid bumping the microscope as you set it on the laboratory bench.

2. If the microscope has a built-in light source, completely unwind the electric cord before plugging it in.

3. To clean the lenses, use only the lens-cleaning fluid and lens paper provided by your instructor. Facial tissue is unsuitable for cleaning because it is made of small wood fibers that will damage the special optical coating on the lenses.

4. Store the microscope with the cord wrapped neatly around the base, the low-power objective lens in position, and the stage in the uppermost position. Either return the microscope to the storage cabinet or cover it with a dust cover.

LAB ACTIVITY   1   Parts of the Compound Microscope

Figure 4.1 shows the parts of a typical compound microscope. Your laboratory may be equipped with a different type; if so, your instructor will discuss the type of microscope you will use.

• The **body tube** is the cylindrical tube that supports the ocular lens and extends down to the nosepiece. A microscope has one body tube if it has one ocular lens and two body tubes if it has two ocular lenses.

• The **arm** is the supportive frame of the microscope. It joins the body tube to the base. As noted earlier, the microscope is correctly carried with one hand on the arm and the other on the base.

• The **base** is the broad, flat lower support of the microscope.

• The **ocular lens** is the eyepiece where you place your eye(s) to observe the specimen. The magnification of most ocular lenses is 10X. This results in an image 10 times larger than the actual size of the specimen. **Monocular** microscopes have a single ocular lens; **binocular** microscopes have two ocular lenses, one for each eye. It is necessary to adjust binocular microscopes so that the two ocular lenses form a single image. This is easily accomplished by adjusting the distance between the two lenses so that you see a single image through them. Move the body tubes apart and look into the microscope. If two images are visible, slowly move the body tubes closer together until you see a single circular field of view with both eyes.

• The **nosepiece** is a rotating disk at the base of the body tube where several objective lenses of different lengths are attached. Turning the nosepiece moves an objective lens into place over the specimen being viewed.

• The **objective lenses** are mounted on the nosepiece. Magnification is determined by the choice of objective lens. The longer the objective lens, the greater its magnifying power. Most student microscopes have 4X, 10X, and 40X objective lenses. Your microscope may also have an **oil-immersion objective lens**, which is usually 100X. With this lens, a small drop of oil is used on the slide. The optical density of the air surrounding the lens and slide is different from the optical density of the glass of which the lens and slide are made, and this difference in optical density reduces resolution. The oil eliminates the air between the lens and the slide and thereby improves the resolution of the microscope.

• The **stage** is a flat, horizontal shelf under the objective lenses that supports the microscope slide. The center of the stage has an **aperture**, or hole, through which light passes to illuminate the specimen on the slide. A pair of **stage clips** (or, in some models, a mechanical slide mechanism) holds the slide steady while you move the slide by hand. Some microscopes have a **mechanical stage** that moves the slide with more precision than is possible manually. The mechanical

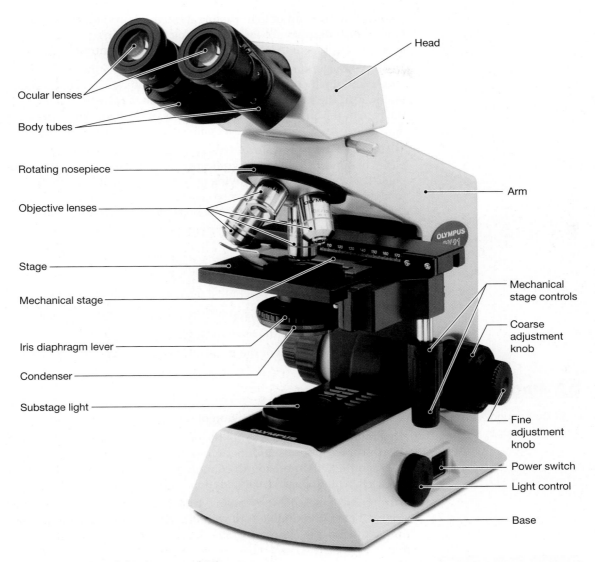

Figure 4.1 Parts of the Compound Microscope
Source: Courtesy of Olympus America, Inc.

stage has two **controls** on the side that move the slide around on the stage. One knob moves the slide horizontally, and the other moves it closer to and farther from the objective lens.

- Focusing knobs move the lenses into position to produce a sharp image. The **coarse adjustment knob** is the large dial on the side of the microscope that moves the objective lenses. Use this knob only at low magnification to find the initial focus on a specimen. The small dial on the side of the microscope is the **fine adjustment knob**. This knob moves the objective lens for precision focusing after coarse focus has been achieved. The fine adjustment knob is used at all magnifications and is the only focusing knob used at magnifications greater than low.

- The **condenser** is the small lens under the stage that narrows the beam of light and directs it through the specimen on the slide. A condenser adjustment knob moves the condenser vertically. For most microscope techniques, the condenser should be in the uppermost position, close to the stage aperture.

- The **iris diaphragm** is a series of flat metal plates found at the base of the condenser. The plates slide together and create an aperture in the condenser that regulates the amount of light passing through the condenser. Most microscopes have a small **diaphragm lever** extending from the iris, which is used to open or close the diaphragm to adjust the light for optimal contrast and minimal glare.

- A **lamp** provides illumination that passes through the specimen, through the lenses, and finally into your eyes. Most microscopes have a built-in light source underneath the stage. The **light control knob**, a rheostat dial located on either the base or the arm, controls the brightness of the light. Microscopes without a light source use a mirror to reflect ambient light into the condenser.

- Microscopes use a **compound lens system**, with each lens consisting of many pieces of optical glass. The magnification is stamped on the barrel of each objective lens, as is the magnifying power of the ocular lens. To calculate the total magnification of the microscope at a particular lens setting, you multiply the ocular lens magnification by the objective lens magnification. For example, a 10X ocular lens used with a 10X objective lens produces a total magnification of 100X.

QuickCheck Questions

1.1 What is the proper way to hold a microscope while carrying it?

1.2 Why is a facial tissue not appropriate for cleaning microscope lenses?

1.3 How do you change the magnification of a microscope ?

1.4 What is the function of the iris diaphragm on a microscope?

1 Materials

☐ Compound microscope

Procedures

1. Identify and describe the function of each part of the microscope.

2. Determine the magnification of the ocular lens and each objective lens on the microscope. Enter this information in the second and third columns of Table 4.1, and then fill in the fourth column by calculating the total magnification for each ocular/objective combination.

3. Use a ruler to measure the working distance between objective lens and slide for each magnification. Record your data in column 5 of Table 4.1. ∎

LAB ACTIVITY 2   Using the Microscope

The cross-sectional view of Figure 4.2 illustrates the optical components of a compound microscope. When light passes through the microscope lenses, the image becomes reversed. When you move the slide away and to your right, the image in the microscope moves left and toward you. (Microscopes that have a mechanical stage allow you to move the slide intuitively without having to consider image reversal.) After light passes through the specimen, the image is magnified by the lenses. To

Table 4.1 **Microscope Data**

	Ocular Lens	Objective Lens	Total Magnification	Working Distance	Field Diameter
Low Power	_____	_____	_____	_____	_____
Medium Power	_____	_____	_____	_____	_____ *
High Power	_____	_____	_____	_____	_____ *
Oil Immersion	_____	_____	_____	_____	_____ *

*Calculated field diameter

Figure 4.2 The Optical Components of a Compound Microscope

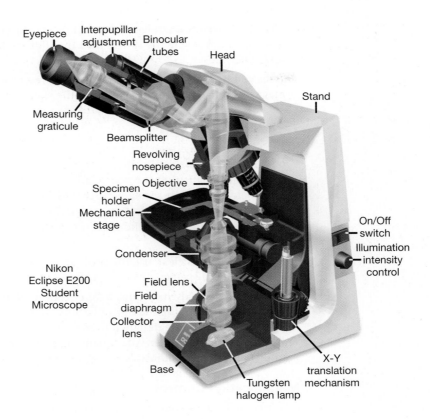

Eyepiece
Interpupillar adjustment
Binocular tubes
Head
Stand
Measuring graticule
Beamsplitter
Revolving nosepiece
Specimen holder
Objective
Mechanical stage
On/Off switch
Illumination intensity control
Condenser
Nikon Eclipse E200 Student Microscope
Field lens
Field diaphragm
Collector lens
Base
Tungsten halogen lamp
X-Y translation mechanism

change magnification, turn the nosepiece to rotate a different objective lens into place over the specimen.

The focusing knobs are used to move an objective lens closer to or farther from the slide to achieve a clear image. The coarse adjustment knob is used first to move the low-power objective lens until a focused image of the specimen is visible in the field of view. Minimal focusing is needed at higher magnifications because the space between objective lens and specimen is very small. The fine adjustment knob is used at higher magnifications to observe individual layers of a thick specimen. Although a specimen may appear very thin, it usually has many layers of cells. Using coarse focus at higher magnifications would move the objective lens too much and possibly force it into the slide.

Light intensity, or brightness, requires adjustment at each magnification in order to produce high-contrast, low-glare images. Correct illumination also reduces eyestrain. The condenser focuses the light into a narrow beam that passes through an iris before it illuminates the specimen. Adjusting the iris changes the amount of light entering the lens.

Four basic steps are involved in successfully viewing a specimen under the microscope: (1) setup, (2) focusing, (3) magnification control, and (4) light intensity control.

Setup

- If the microscope has an internal light source, plug in the electrical cord and turn the lamp on. If the microscope does not have a built-in light source, adjust the mirror to reflect light into the condenser.

- Check the position of the condenser; it should be in the uppermost position, near the stage aperture.

- Rotate the nosepiece to swing the low-magnification objective lens into position over the aperture. Having this lens in place provides maximum clearance between objective lens and stage for the placement of a slide.

- Place the slide on the stage and use the stage clips or mechanical slide mechanism to secure the slide. Move the slide so that the specimen is over the stage aperture.
- After you have finished your observations, reset the microscope to low magnification, remove the slide from the stage, and store the microscope.

Focusing

- To focus on a specimen, first move the low-power objective lens to its lowest position. Next, look into the ocular lens and raise the low-power objective lens by slowly turning the coarse adjustment knob. The image should come into focus. Note that some microscopes focus by moving the stage rather than the objective lens. To focus a specimen on this type of microscope, move the stage closer to the objective lens by turning the coarse adjustment knob.
- Once the image is clear, use the fine adjustment knob to examine the detailed structure of the specimen.
- When you are ready to change magnification, do not move the adjustment knobs before changing the objective lens. Most microscopes are **parfocal**, which means they are designed to stay in focus when you change from one objective lens to another. After changing magnification, use only the fine adjustment knob to adjust the objective lens.

Magnification Control

- Notice that the higher-power objective lenses are longer than the low-power objective lens. Thus, as magnification is increased, the distance between specimen and lens decreases. Therefore, always rotate the low-power objective lens into place before inserting or removing a slide from the stage. This provides ample working distance between lens and stage to adjust the slide.
- Always use low magnification during your initial observation of a slide. You will see more of the specimen and can quickly select areas on the slide for detailed studies at higher magnification.
- To examine part of the specimen at higher magnification, move that part of the specimen to the center over the aperture before changing to a higher-magnification objective lens. This repositioning keeps the specimen in the field of view at the higher magnification. Because a higher-magnification lens is closer to the slide, less of the slide is visible in the field of view. The image of the specimen enlarges to fill the field of view.

Light Intensity Control

- Use the light control knob to regulate the intensity of light from the bulb. Adjust the brightness so that the image has good contrast and no glare.
- Adjust the iris diaphragm by moving the diaphragm lever side to side. Notice how the field illumination is changed by different settings of the iris.
- At higher magnifications, increase illumination and open the iris diaphragm.

QuickCheck Questions

2.1 When is the coarse adjustment knob used on a microscope?

2.2 What is the typical view position for the condenser?

Figure 4.3 Preparing a Wet Mount
(a) Place the object to be mounted in a drop of water or stain on the slide. **(b)** Using tweezers or your fingers, touch the water or stain with the edge of the coverslip. **(c)** Finally, carefully lower the coverslip until it rests flat on the slide. Use a paper towel to absorb excess fluid that has leaked out from under the coverslip.

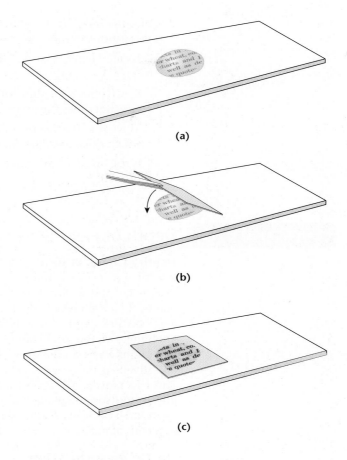

(a)

(b)

(c)

2 *Materials*

- ☐ Compound microscope, slide, and coverslip
- ☐ Newspaper cut into small pieces
- ☐ Dropper bottle containing water

Procedures

1. Make a wet-mount slide of a small piece of newspaper as follows:

   a. Obtain a slide, a coverslip, and a small piece of newspaper that has printing on it.

   b. Place the paper on the slide and add a small drop of water to it.

   c. Put the coverslip over the paper as shown in Figure 4.3. The coverslip will keep the lenses dry.

2. Move the low-magnification objective lens into position (if it is not already there), and place the slide on the stage.

3. Use the coarse adjustment knob to move the objective lens as close to the specimen as possible without touching the slide.

4. Move the slide until the printing is directly over the stage aperture. Look into the ocular lens and slowly turn the coarse adjustment knob until you see the fibers of the newspaper. Once they are in focus, adjust the light source for optimal contrast and resolution.

5. Use the fine adjustment knob to bring the image into crisp focus. Remember, the microscope you are using is a precise instrument and produces a clear image when adjusted correctly. Be patient and keep at it until you get a perfectly clear image.

6. Once the image is correctly focused, do the following and record your observations in the spaces provided.

   a. Locate a letter "a" or "e." Describe the ink and the paper fibers. _____.

   b. Slowly move the slide forward with the mechanical stage knob. In which direction does the image move? _____.

   c. Move the slide horizontally to the left using the mechanical stage knob. In which direction does the image move? _____.

   d. Is the image of the letter oriented in the same direction as the real letter on the slide? _____. ■

LAB ACTIVITY 3 Depth of Field Observation

Depth of field, or **focal depth**, is a measure of how much depth (thickness) of a specimen is in focus, and the in-focus thickness is called a **focal plane**. Depth of field is greatest at low power and decreases as magnification increases. In other words, the focal plane is thicker at low power and thinner at higher powers (Figure 4.4a). Because depth of field is reduced at higher power, you use the fine adjustment knob to move the focal plane up and down through the thickness of the specimen and in this way scan the specimen layers. As you turn the fine adjustment knob, the objective lens moves either closer to or farther from the slide surface (Figure 4.4b). This lens movement causes the focal plane to move through the layers of the specimen. Most specimens are many cell layers thick. By slowly rotating the fine adjustment knob back and forth, you will see different layers of the specimen come into or go out of focus.

QuickCheck Question

3.1 What is depth of field in a microscope?

3 Materials

- ☐ Compound microscope
- ☐ Slide of colored threads (or slide of hairs from different students if thread slides are unavailable)

Procedures

To see how depth of field works, you will examine a slide of overlapping colored threads (or hairs). In examining your slide, notice how the threads are layered and how much of each thread is in focus at each magnification.

1. Move the low-power objective lens into position, and place the slide on the stage with the threads over the aperture.

2. Use the coarse adjustment knob to bring the threads into focus. Find the area where the threads overlap.

3. Rotate the nosepiece to select the medium-power objective lens.

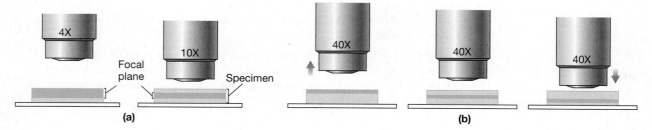

Figure 4.4 Focal Plane and Magnification
(a) Focal-plane thickness decreases as magnification increases. (b) Using the fine adjustment knob changes the distance between the objective and the stage, thereby moving the focal plane up and down through the specimen.

4. Use the fine adjustment knob to focus through the overlapping threads. After determining which thread is on top, which is in the middle, and which is on the bottom, write your observations in the space provided:

   Top thread _____.

   Middle thread _____.

   Bottom thread _____.

LAB ACTIVITY 4 Relationship Between Magnification and Field Diameter

At low magnification, the diameter of the field of view is large and most of the slide specimen is visible. As magnification increases, the field diameter decreases. This is because at higher power, the objective lens is closer to the slide and magnifies a smaller area. Figure 4.5 reviews the relationship between magnification and field diameter.

Study Tip | *Field Diameter*

You can demonstrate the relationship between magnification and field diameter by curling your fingers until the thumb of each hand overlaps the index and middle fingers of the same hand. The space enclosed by the curled fingers of each hand forms the barrel of a "lens." Place these two "lenses" to your eyes, and, while sitting up straight in your chair, look at this page. Notice that you can see the entire page at this "low magnification." Now slowly bend forward until the "lenses" are just a few inches away from the page. In this "high-magnification" view, the field of view is much smaller, and you can see only part of the page. ●

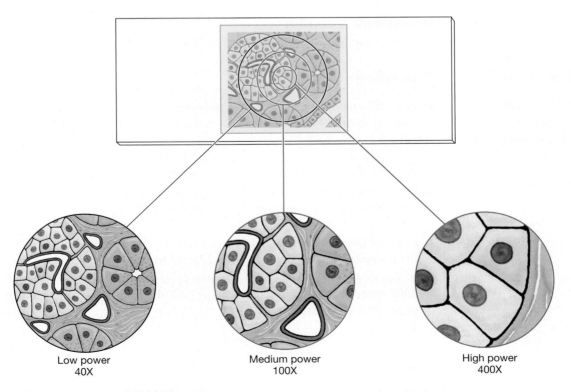

Low power
40X

Medium power
100X

High power
400X

Figure 4.5 Magnification and Field Diameter
Each circle on the slide illustrates the field diameter for a particular magnification; the corresponding circle outside the slide represents that magnification.

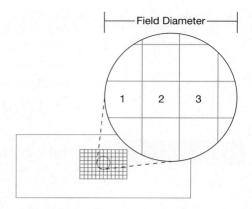

Figure 4.6 Calculation of Field Diameter Using Millimeter Graph Paper
In this sample, the field is approximately 3.5 mm in diameter.

Field diameter at low and medium magnifications can be measured using millimeter graph paper glued to a microscope slide. By aligning a vertical marking on the paper with the edge of the field and then counting the number of millimeter (mm) squares across the field, you can determine the diameter (Figure 4.6). Knowing the diameter of the field of view enables you to estimate the actual size of an object. For example, if the field diameter is 4 mm and an object occupies one-half of the field, the object is approximately 2 mm wide.

Once you know the field diameter for one magnification—we call this lens A in the following formula—you can calculate the field diameter for other magnifications (lens B) using the formula

$$\text{Field diameter of lens B} = \frac{\text{Field diameter of lens A} \times \text{total magnification of lens A}}{\text{Total magnification of lens B}}$$

QuickCheck Question

4.1 What happens to the field diameter as magnification increases?

4 Materials

☐ Compound microscope
☐ Graph-paper slides

Procedures

1. Place the graph-paper slide on the microscope stage and focus at low magnification. Position the slide so that a vertical line on the paper lines up with the edge of the field.

2. Count the number of millimeters across the field to measure the field diameter. Record this value in column 6 of Table 4.1.

3. Use your low-power measured field diameter to calculate the field diameter for the microscope set at medium power, and record this value in column 6 of Table 4.1.

4. Use your low-power measured field diameter to calculate the field diameter for the microscope set at high power, and record this value in column 6 of Table 4.1.

5. If your microscope has an oil-immersion objective lens, use any of the three field-diameter values you have listed in Table 4.1 to calculate the field diameter of the oil-immersion lens. ■

Use of the Microscope

Name _____

Date _____

Section _____

A. Matching

Match the part of the microscope on the left with the correct description on the right.

_____	**1.** ocular lens	**A.** used for precise focusing
_____	**2.** aperture	**B.** lower support of microscope
_____	**3.** body tube	**C.** narrows beam of light
_____	**4.** mechanical stage	**D.** hole in stage
_____	**5.** fine adjustment knob	**E.** used only at low power
_____	**6.** base	**F.** has knobs to move slide
_____	**7.** objective lens	**G.** special paper for cleaning
_____	**8.** coarse adjustment knob	**H.** eyepiece
_____	**9.** condenser	**I.** holds ocular lens
_____	**10.** lens paper	**J.** lens attached to nosepiece

B. Labeling

Label the parts of the microscope in Figure 4.7.

C. Short-Answer Questions

1. Which parts of a microscope are used to regulate the intensity and contrast of light? What is the function of each of these parts?

2. How is magnification controlled in a microscope?

3. Why should you always view a slide at low power first?

4. Briefly explain how to care for a microscope.

5. Describe when to use the coarse adjustment knob and when to use the fine adjustment knob.

4

1. _____
2. _____
3. _____
4. _____
5. _____
6. _____
7. _____
8. _____
9. _____
10. _____

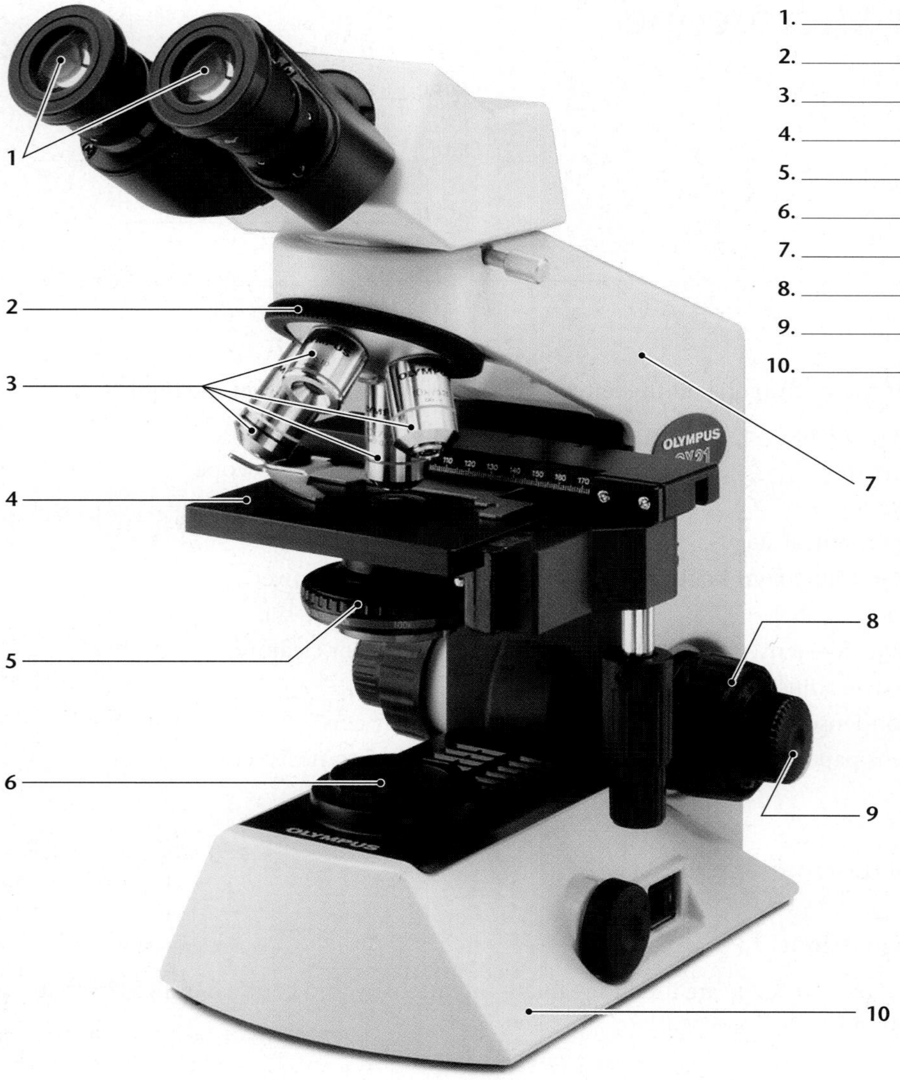

Figure 4.7 Parts of the Compound Microscope

Source: Courtesy of Olympus America, Inc.

D. Application and Analysis

1. You are looking at a slide in the laboratory and observe a cell that occupies one-quarter of the field of view at high magnification. Use your field diameter calculation from Laboratory Activity 4 to estimate the size of this cell.

2. Describe how the field diameter changes when magnification is increased.

Anatomy of the Cell and Cell Division

OBJECTIVES

On completion of this exercise, you should be able to:

- Identify cell organelles on charts, models, and other laboratory material.

- State a function of each organelle.

- Discuss a cell's life cycle, including the stages of interphase and mitosis.

- Identify the stages of mitosis using the whitefish blastula slide.

Cells were first described in 1665 by a British scientist named Robert Hooke. Hooke examined a thin slice of tree cork with a microscope and observed that it contained many small open spaces, which he called *cells*. Over the next two centuries, scientists examined cells from plants and animals and formulated the **cell theory**, which states that (1) all plants and animals are composed of cells, (2) all cells come from preexisting cells, (3) cells are the smallest living units that perform physiological functions, (4) each cell works to maintain itself at the cellular level, and (5) homeostasis is the result of the coordinated activities of all the cells in an organism.

Your cells are descendants of your parents' sperm and egg cells that combined to create your first cell, the zygote. You are now composed of trillions of cells, more than you could count in your lifetime. These cells must coordinate their activities to maintain homeostasis for your entire body. If a population of cells becomes dysfunctional, disease may result. Some organisms, like amebas, are composed of a single cell that performs all functions necessary to keep the organism alive. In humans and other multicellular organisms, cells are diversified, which means that different cells have different specific functions. This specialization leads to dependency among cells. For example, muscle cells are responsible for movement of the body. Because movement requires a large amount of energy, muscle cells rely on the cells of the cardiovascular system to distribute blood rich with oxygen and nutrients to them.

Although the body has a variety of cell types, a generalized composite cell is used to describe cell structure. All cells have an outer boundary, the cell membrane. Cells also have a nucleus and other internal structures called *organelles*. In this exercise you will examine the structure of the cell and how cells reproduce to create new cells that can be used for growth and repair of the body.

LAB ACTIVITY **1**   The Cell Membrane

The **cell membrane**, also called the *plasma membrane,* is the physical boundary of the cell, separating the **extracellular fluid** surrounding the cell from the **cytosol**, the fluid inside the cell. The cell membrane regulates the movement of ions, molecules, and other substances into and out of the cell. In Exercise 6 you will study transport of materials across cell membranes.

Cells interact with one another and with the surrounding environment via the cell membrane. Hormones, enzymes, and other regulatory molecules bind to specific receptor proteins in the cell membrane. Muscle and nerve cell membranes are excitable and produce electrical currents called *action potentials.* Some cells have **microvilli**, or folds in the cell membrane, which increase the surface area for absorption of materials into the cell.

The cell membrane is composed of a **phospholipid bilayer**, which is a double layer of phospholipid molecules, plus several other structural components, such as cholesterol molecules and glycolipid molecules (Figure 5.1). Each phospholipid molecule consists of a **hydrophilic** (*hydro-,* water + *philic,* loving) **head** and two **hydrophobic** (*phobic,* fearing) **tails**. In a cell membrane, the phospholipids are arranged in a double sheet of molecules with the hydrophilic heads facing the watery internal and external environments of the cell. The hydrophobic tails are sandwiched between the phospholipid heads.

Proteins are another major component of the cell membrane. Floating like icebergs in the phospholipid bilayer are a variety of **integral proteins**. These proteins have **channels** that regulate the passage of specific ions through the membrane. Some of these channels are **gated**, which means they can open and close to regulate ion flow across the cell membrane. Loosely attached to the external and internal surfaces of the membrane are **peripheral proteins**. Both integral proteins and

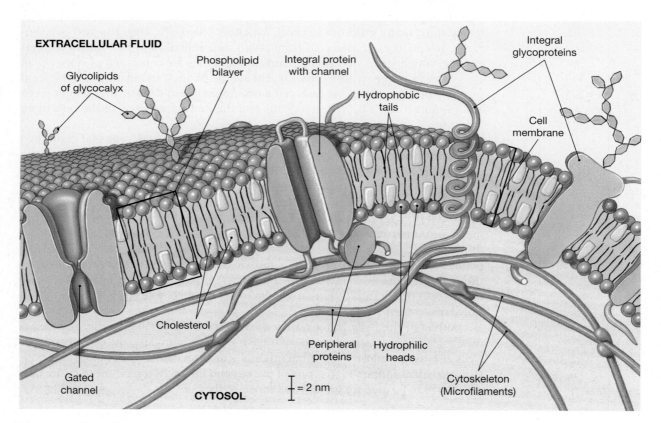

Figure 5.1 The Cell Membrane

peripheral proteins occur in a wide variety of shapes and have diverse functions. One class of integral proteins is the receptor proteins, which are sensitive to specific chemicals in the extracellular fluid. Other proteins and glycoproteins that are part of each cell membrane function in cell-to-cell recognition and as enzymes for intracellular and extracellular reactions.

QuickCheck Question

1.1 What are the major chemical components of the cell membrane?

1 **Materials**	**Procedures**
☐ Cell model	**1.** Review the structure of the cell membrane in Figure 5.1.
	2. Locate the cell membrane on a cell model. ■

LAB ACTIVITY 2   Organelles

Within the cell are **organelles** (or-gan-ELZ), the functional pieces of biological machinery. As represented in Figure 5.2 and Table 5.1, each organelle has a distinct anatomical organization and is specialized for a specific function. Organelles are suspended in the cytosol. The organelles and cytosol together compose the **cytoplasm** of the cell.

Organelles are grouped into two broad classes: nonmembranous and membranous. **Nonmembranous organelles** lack an outer membrane and are directly exposed to the cytosol. Ribosomes, microvilli, centrioles, the cytoskeleton, cilia, and flagella are nonmembranous organelles. **Membranous organelles** are enclosed in a phospholipid membrane that isolates the organelle from the cytosol. The nucleus, endoplasmic reticulum, Golgi apparatus, vesicles, peroxisomes, and mitochondria are membranous organelles.

Keep in mind while studying cell models that most organelles are not visible with a light microscope. The nucleus typically is visible as a dark-stained oval. It encases and protects the **chromosomes**, which store genetic instructions for protein production by the cell.

Study Tip

Information Linking

Practice connecting information together rather than memorizing facts and terms. Approach organelles with the goal of integrating structure with function. As you identify organelles on cell models, consider the function of each organelle. Once you have an understanding of each organelle, begin to associate them with each other as functional teams. For example, vesicles transport molecules from the endoplasmic reticulum to the Golgi apparatus. This assimilation of information improves your ability to apply knowledge in a working context. ●

Nonmembranous Organelles

- **Microvilli** are small folds in the cell membrane that increase the surface area of the cell. With more membrane surface, the cell can absorb extracellular materials, such as nutrients, at a greater rate.

- **Microtubules** are small hollow tubes made of the protein tubulin. These protein pipes are found in all cells, and their functions are to anchor organelles and give rigidity to the cell membrane. Centrioles, the cytoskeleton, cilia, and flagella are made of microtubules.

- **Centrioles** are paired organelles composed of microtubules; the **centrosome** is the area surrounding the centrioles. When a cell divides, the two centrioles in a pair migrate to opposite poles of the nucleus, and a series of spindle fibers

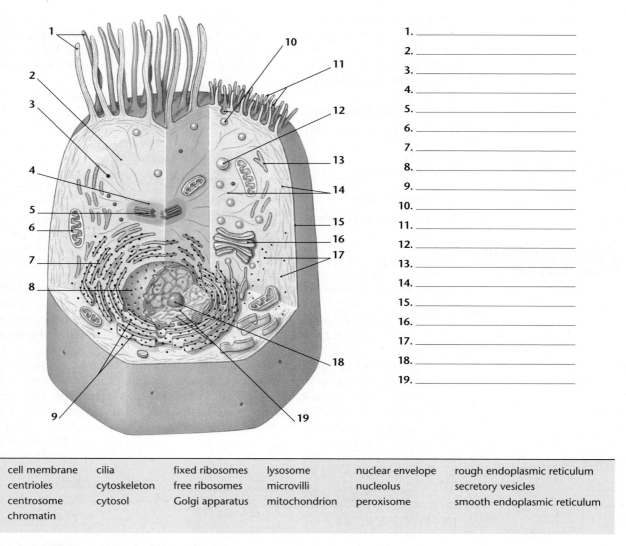

1. _____
2. _____
3. _____
4. _____
5. _____
6. _____
7. _____
8. _____
9. _____
10. _____
11. _____
12. _____
13. _____
14. _____
15. _____
16. _____
17. _____
18. _____
19. _____

cell membrane	cilia	fixed ribosomes	lysosome	nuclear envelope	rough endoplasmic reticulum
centrioles	cytoskeleton	free ribosomes	microvilli	nucleolus	secretory vesicles
centrosome	cytosol	Golgi apparatus	mitochondrion	peroxisome	smooth endoplasmic reticulum
chromatin					

Figure 5.2 The Anatomy of a Composite Cell
See Table 5.1 for a summary of the functions associated with the various cell structures.

radiate from them. The **spindle fibers** pull the chromosomes of the nucleus apart to give each forming cell a full complement of genetic instructions.

• All cells have a **cytoskeleton** for structural support and anchorage of organelles. Many cells of the respiratory and reproductive system have **cilia**, which are short, hairlike projections extending from the cell membrane. One type of human cell, the spermatozoon, has a single, long **flagellum** (fla-JEL-um) for locomotion.

• **Ribosomes** direct protein synthesis. Instructions for making a protein are stored in deoxyribonucleic acid (DNA) molecules in the cell nucleus. The "recipe" for a protein is copied from a segment of DNA onto a molecule of messenger RNA that then carries the instructions out of the nucleus and to the ribosome. Each ribosome consists of one large subunit and one small subunit, with both subunits clamping around the messenger RNA molecule to coordinate protein synthesis. Ribosomes occur either as **free ribosomes** in the cytoplasm or as **fixed ribosomes** attached to the endoplasmic reticulum (ER).

Table 5.1 Summary of Cell Organelles and Their Function

Appearance	Structure	Composition	Function(s)
	CELL MEMBRANE	Lipid bilayer, containing phospholipids, steroids, and proteins	Isolation; protection; sensitivity; support; control of entrance/exit of materials
	CYTOSOL	Fluid component of cytoplasm	Distributes materials by diffusion
	NONMEMBRANOUS ORGANELLES		
	Cytoskeleton: Microtubule Microfilament	Proteins organized in fine filaments or slender tubes	Strength and support; movement of cellular structures and materials
	Microvilli	Membrane extensions containing microfilaments	Increase surface area to facilitate absorption of extracellular materials
	Centrosome Centriole	Cytoplasm containing two centrioles, at right angles: Each centriole is composed of 9 microtubule triplets	Essential for movement of chromosomes during cell division; organization of microtubules in cytoskeleton
	Cilia	Membrane extensions containing microtubule doublets in a 9 + 2 array	Movement of materials over cell surface
	Ribosomes	RNA + proteins; fixed ribosomes bound to rough endoplasmic reticulum, free ribosomes scattered in cytoplasm	Protein synthesis
	MEMBRANOUS ORGANELLES		
	Endoplasmic reticulum (ER)	Network of membranous channels extending throughout the cytoplasm	Synthesis of secretory products; intracellular storage and transport
	Rough ER (RER)	Has ribosomes bound to membranes	Modification and packaging of newly synthesized proteins
	Smooth ER (SER)	Lacks attached ribosomes	Lipid and carbohydrate synthesis
	Golgi apparatus	Stacks of flattened membranes (cisternae) containing chambers	Storage, alteration, and packaging of secretory products and lysosomal enzymes
	Lysosomes	Vesicles containing digestive enzymes	Intracellular removal of damaged organelles and pathogens
	Peroxisomes	Vesicles containing degradative enzymes	Catabolism of fats and other organic compounds; neutralization of toxic compounds generated in the process
	Mitochondria	Double membrane, with inner membrane folds (cristae) enclosing important metabolic enzymes	Produce 95% of the ATP required by the cell
Nuclear pore	**Nucleus**	Nucleoplasm containing nucleotides, enzymes, nucleoproteins, and chromatin; surrounded by double membrane (nuclear envelope)	Control of metabolism; storage and processing of genetic information; control of protein synthesis
Nuclear envelope	**Nucleolus**	Dense region in nucleoplasm containing DNA and RNA	Site of rRNA synthesis and assembly of ribosomal subunits

Membranous Organelles

- The **nucleus** controls the activities of the cell, such as protein synthesis, gene action, cell division, and metabolic rate. The material responsible for the dark appearance of the nucleus in a stained specimen is **chromatin**, uncoiled chromosomes consisting of DNA and protein molecules. A **nuclear envelope** surrounds the nuclear material and contains pores through which instruction molecules from the nucleus pass into the cytosol. A darker-stained region inside the nucleus, the **nucleolus**, produces ribosomal RNA molecules for the creation of ribosomes.

- Surrounding the nucleus is the **endoplasmic reticulum** (en-dō-PLAZ-mik re-TIK-ū-lum). Two types of ER occur: **rough ER**, which has ribosomes attached to its surface; and **smooth ER**, which lacks ribosomes. Generally, the ER functions in the synthesis of organic molecules, transport of materials within the cell, and storage of molecules. Materials in the ER may pass into the Golgi apparatus for eventual transport out of the cell. Proteins produced by ribosomes on the rough ER surface enter the ER and assume the complex folded shape characteristic of the ER. Smooth ER is involved in the synthesis of many organic molecules, such as cholesterol and phospholipids. In reproductive cells, smooth ER produces sex hormones. In liver cells, it synthesizes and stores glycogen, while in muscle and nerve cells it stores calcium ions. Intracellular calcium ions are stored in the smooth ER in muscle, nerve, and other types of cells.

- The **Golgi** (GŌL-jē) **apparatus** is a series of flattened saccules adjoining the ER. The ER can pass protein molecules in transport vesicles to the Golgi apparatus for modification and secretion. Cell products such as mucus are synthesized, packaged, and secreted by the Golgi apparatus. In a process called **exocytosis**, small **secretory vesicles** pinch off the saccules, fuse with the cell membrane, and then rupture to release their contents into the extracellular fluid. The phospholipid membranes of the empty vesicles contribute to the renewal of the cell membrane.

- **Lysosomes** (LĪ-sō-sōm; *lyso-*, dissolution + *soma,* body) are vesicles produced by the Golgi apparatus. They are filled with powerful enzymes that digest worn-out cell components and destroy microbes. As certain organelles become worn out, lysosomes dissolve them, and some of the materials are used to rebuild the organelles. White blood cells trap bacteria with cell membrane extensions and pinch the membrane inward to release a vesicle inside the cell. Lysosomes fuse with the vesicle and release enzymes to digest the bacteria. Injury to a cell may result in the rupture of lysosomes, followed by destruction or autolysis of the cell. Autolysis is implicated in the aging of cells owing to the accumulation of lysosomal enzymes in the cytosol.

- **Peroxisomes** are vesicles filled with enzymes that break down fatty acids and other organic molecules. Metabolism of organic molecules can produce free-radical molecules, such as hydrogen peroxide (H_2O_2), that damage the cell. Peroxisomes protect cell structure by metabolizing hydrogen peroxide to oxygen and water.

- **Mitochondria** (mī-tō-KON-drē-uh) produce useful energy for the cell. Each mitochondrion is wrapped in a double-layered phospholipid membrane. The inner membrane is folded into fingerlike projections called **cristae** (the singular is *crista*). The region of the inner membrane between cristae is the **matrix**. To provide the cell with energy, molecules from nutrients are passed along a series of **metabolic enzymes** in the cristae to produce a molecule called *adenosine triphosphate (ATP),* the energy currency of the cell. The abundance of mitochondria varies greatly among cell types. Muscle and nerve cells have large numbers of mitochondria that supply energy for contraction and generate nerve impulses, respectively. Mature red blood cells lack mitochondria and subsequently have a low metabolic rate.

Table 5.2 **Cell Organelles and Their Functions**

Organelle	Organelle Structure	Organelle Function
1. Mitochondria	_____	_____
2. _____	_____	Contains digestive enzymes
3. _____	Phospholipid bilayer with embedded proteins	_____
4. Endoplasmic reticulum	_____	_____
5. _____	_____	Exocytosis of cellular products
6. _____	Composed of two subunit molecules	_____
7. _____	_____	Regulates cell activities

Clinical Application

Maternal Mitochondria

Mitochondria are unique in that they contain mitochondrial DNA and evidently do not rely on the DNA in the cell nucleus for replication instructions. Your mitochondria were inherited from your mother's egg; the sperm from your father contributed no mitochondria and few other organelles. Suppose a woman inherited a mitochondrial disorder from her mother. Could she pass this trait on to a son? In general, how would the health of a cell be affected if the mitochondria were not fully functional? ▶

QuickCheck Questions

2.1 What are the two major categories of organelles?

2.2 Which organelles are involved in the production of protein molecules?

2 Materials

☐ Cell models and charts

Procedures

1. Label the organelles in Figure 5.2 and review the organelles in Table 5.1.
2. Identify the organelles on a laboratory cell model.
3. Complete Table 5.2 by filling in the blanks. ■

LAB ACTIVITY 3   Cell Division

Cells must reproduce if an organism is to grow and repair damaged tissue. During cell reproduction, a cell divides its genes equally and then splits into two identical cells. The division involves two major "events": mitosis and cytokinesis. During **mitosis** (mī-TŌ-sis), the genetic material of the nucleus condenses into chromosomes and is equally divided between the two forming cells. Toward the end of mitosis, **cytokinesis** (sī-tō-ki-NĒ-sis; *cyto-*, cell + *kinesis*, motion) separates the cytoplasm to produce the two daughter cells. The daughter cells have the same number of chromosomes as the parent cell.

Interphase

Examine the cell life cycle in Figure 5.3. Most of the time a cell is not dividing and is in **interphase**. This is not a resting period for the cell because during this phase the cell carries out various functions and prepares for the next cell division. Distinct phases occur during interphase, each related to cell activity. At this time the nucleus is visible, as is the darker nucleolus. During the G_0 **phase** the cell performs its specialized functions. The G_1 **phase** is a time for protein synthesis, growth, and replication

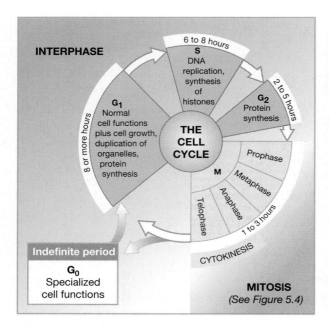

Figure 5.3
The Cell Life Cycle

of organelles. Replication of DNA occurs during the **S phase**. After DNA replication, each chromosome is double-stranded and consists of two **chromatids**, the original strand and an identical copy. The chromatids are held together by a **centromere**. The **G₂ phase** is another time for protein synthesis.

Mitosis

The **M phase** of the cell cycle is the time of mitosis, during which the nuclear material divides. After chromosomes are duplicated in the S phase of interphase, the double-stranded chromosomes migrate to the middle of the cell, and thin microtubules called *spindle fibers* attach to each chromatid. Chromosomes are divided when the spindle fibers drag sister chromatids to opposite ends of the cell. The division is complete when the cell undergoes cytokinesis and pinches inward to distribute the cytoplasm and chromosomes into two new daughter cells. The four parts of mitosis are prophase, metaphase, anaphase, and telophase.

• **Prophase**: Mitosis starts with prophase (PRŌ-fāz; *pro-*, before), when chromosomes become visible in the nucleus (Figure 5.4). In early prophase the chromosomes are long and disorganized, but as prophase continues the nuclear membrane breaks down and the chromosomes shorten and move toward the middle of the cell. In the cytoplasm, the centrioles begin to move to opposite

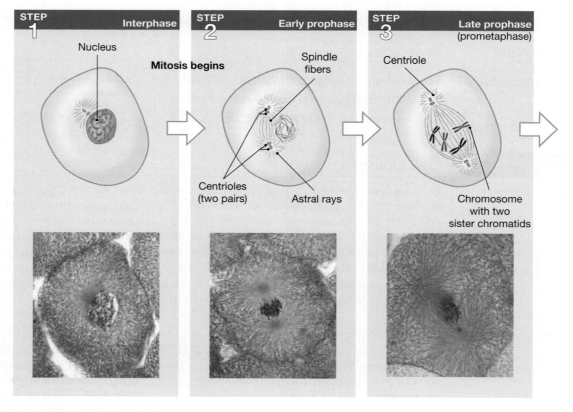

Figure 5.4 Interphase, Mitosis, and Cytokinesis
Diagrammatic and microscopic views of representative cells undergoing cell division.

sides of the cell. Between the centrioles, microtubules fan out as spindle fibers and extend across the cell.

- **Metaphase**: Metaphase (MET-a-fāz; *meta-*, change) occurs when the chromosomes line up in the middle of the cell at the **metaphase plate**. Spindle fibers extend across the cell from one pole to the other and attach to the centromeres of the chromosomes. The cell is now prepared to partition the genetic material and give rise to two new cells.

- **Anaphase**: Separation of the chromosomes is called *anaphase* (AN-a-fāz; *ana-*, apart). Spindle fibers pull apart the chromatids of a chromosome and drag them toward opposite poles of the cell. Once apart, individual chromatids are considered chromosomes. Toward the end of anaphase, cytokinesis begins and a **cleavage furrow** develops along the metaphase plate. Cytokinesis begins, and the cell membrane pinches. Cytokinesis continues into the next stage of mitosis, telophase.

- **Telophase**: Mitosis nears completion during telophase (TĒL-ō-fāz; *telo-*, end) as each batch of chromosomes unwinds inside a newly formed nuclear membrane. Each daughter cell has a set of organelles and a nucleus containing a complete set of genes. Telophase ends as the cleavage furrow deepens along the metaphase plate and separates the cell into two identical daughter cells. These daughter cells are in interphase and, depending on their cell type, may divide again.

Clinical Application | ### *Cell Division and Cancer*

A tumor is a mass of cells produced by uncontrolled cell division. The mass replaces normal cells, and cellular and tissue functions are compromised. If metastasis (me-TAS-ta-sis), or spreading of the abnormal cells, occurs, secondary tumors may develop. Cells that metastasize are often malignant and can cause cancer. ▶

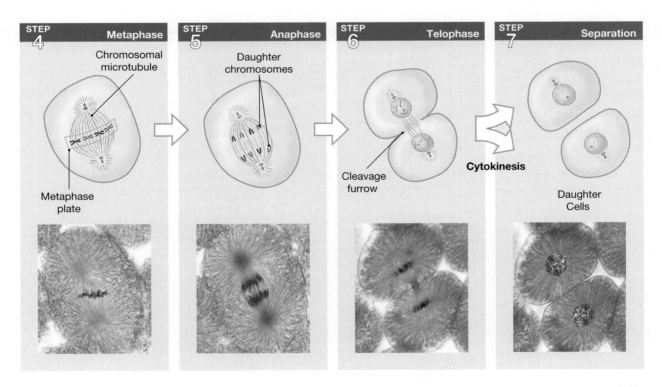

Figure 5.4 *(Continued)*

QuickCheck Questions

3.1 What are the major phases in the cell life cycle?

3.2 What occurs during the S phase of interphase?

3 *Materials*

- ☐ Compound microscope
- ☐ Whitefish blastula slide

Procedures

1. Obtain a slide of the whitefish blastula. A **blastula** is a stage in early development when the embryo is a rapidly dividing mass of cells that is growing in size and, eventually, in complexity. For microscopic observation of the cells, the whitefish embryo is sectioned and stained. A typical slide preparation usually has several sections, each with cells in various stages of mitosis.

2. Scan the slide at low power, and observe the numerous cells of the blastula.

3. Slowly scan a group of cells at medium power to locate a nucleus, centrioles, and spindle fibers. The chromosomes appear as dark, thick structures in the cell.

4. Using Figure 5.4 as a reference, locate cells in the following phases:

   - Interphase with a distinct nucleus
   - Prophase with disorganized chromosomes
   - Metaphase with equatorial chromosomes attached to spindle fibers
   - Anaphase with chromosomes separating toward opposite poles
   - Telophase with nuclear membranes forming around each set of genetic material
   - Cytokinesis in late anaphase and telophase

5. Draw and label cells in each stage of mitosis in the space provided.

Interphase	Prophase	Metaphase

Anaphase	Telophase	Daughter Cells

Anatomy of the Cell and Cell Division

5

Name _____

Date _____

Section _____

A. Matching

Match each cellular structure on the left with the correct description on the right.

_____	**1.** cell membrane	**A.**	copy of a chromosome
_____	**2.** centrioles	**B.**	component of cell membranes
_____	**3.** ribosome	**C.**	short, hairlike cellular extensions
_____	**4.** smooth ER	**D.**	part of phospholipid molecule
_____	**5.** chromatid	**E.**	intracellular fluid
_____	**6.** lysosomes	**F.**	involved in mitosis
_____	**7.** integral protein	**G.**	folds of the inner mitochondrial membrane
_____	**8.** cytoplasm	**H.**	composed of a phospholipid bilayer
_____	**9.** cristae	**I.**	stores calcium ions in muscle cells
_____	**10.** hydrophilic head	**J.**	site for protein synthesis
_____	**11.** cytosol	**K.**	vesicles with powerful digestive enzymes
_____	**12.** cilia	**L.**	intracellular fluid and the organelles

B. Labeling

Label the structures of the cell membrane in Figure 5.5.

C. Fill in the Blanks

1. Replication of genetic material results in chromosomes consisting of two

_____.

2. A cell in metaphase has chromosomes located in the _____ of the cell.

3. Division of the cytoplasm to produce two daughter cells is called _____.

4. Double-stranded chromosomes separate during the _____ stage of mitosis.

5. During interphase, DNA replication occurs in the _____ phase.

6. Microtubules called _____ attach to chromatids and pull them apart.

7. Chromosomes become visible during the _____ stage of mitosis.

8. The last stage of mitosis is _____.

9. Division of the nuclear material is called _____.

10. Matching chromatids are held together by a _____.

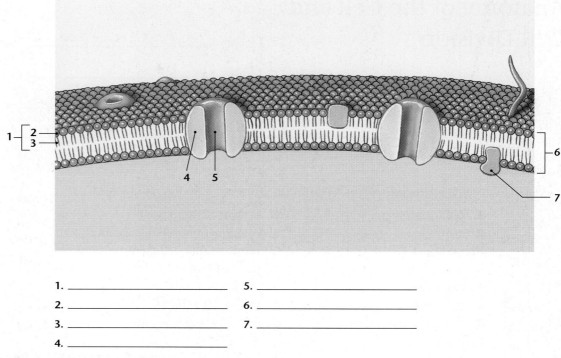

1. _____ 5. _____

2. _____ 6. _____

3. _____ 7. _____

4. _____

Figure 5.5 The Cell Membrane

D. Short-Answer Questions

1. What is the purpose of cell division?

2. Describe a phospholipid molecule and its interaction with water.

3. What is the function of the spindle fibers during mitosis?

4. What structures in the cell membrane regulate ion passage?

E. Drawing

1. Draw and label the cell membrane. Include details of the phospholipid bilayer and integral proteins.

2. Draw and label the internal organization of a mitochondrion.

3. Draw and label a cell with six chromosomes during metaphase. How will the chromosomes appear during anaphase and telophase?

LABORATORY REPORT

F. Analysis and Application

1. Describe how the nucleus, ribosomes, rough ER, Golgi apparatus, and cell membrane interact to produce and release a protein molecule from the cell.

2. Lysosomes are sometimes referred to as "suicide bags." Describe what would happen to a cell if its lysosomes ruptured.

3. What happens in a cell during the S portion of interphase?

4. Describe how chromosomes are evenly divided during mitosis.

5. Identify where in a cell the production of protein, carbohydrate, and lipid molecules occurs.

Movement Across Cell Membranes

OBJECTIVES

On completion of this exercise, you should be able to:

• Describe the two main processes by which substances move into and out of cells.

• Explain what Brownian movement is and how it can be shown.

• Discuss osmosis, diffusion, concentration gradients, and equilibrium in a solution.

• Describe the effect on cells of isotonic, hypertonic, and hypotonic solutions.

• Discuss the effects of solute concentration on the rate of diffusion and osmosis.

Cells are the functional living units of the body. In order for them to survive, materials must be transported across the cell membrane. Cells import nutrients, oxygen, hormones, and other regulatory molecules from the extracellular fluid and export wastes and cellular products to the extracellular fluid. Cells rely on the selectively permeable cell membrane to regulate the passage of these materials. Small molecules, such as water and many ions, cross the membrane without assistance from the cell. This movement is called **passive transport** and requires no energy expenditure by the cell. Diffusion and osmosis are the primary passive processes in the body and will be studied in this laboratory exercise. Other materials, like proteins and other macromolecules, are too big to pass through channels in the cell membrane. Movement of these larger molecules requires the use of carrier molecules in a process called **active transport**, a cell function that consumes a cell's energy.

LAB ACTIVITY 1 Brownian Movement

Molecules in gases and liquids are in a constant state of **Brownian movement**, motion that causes them to bump into adjacent molecules. (Although the molecules in a solid have this motion, they are held in place by chemical bonds; as a result, the bulk of the molecules remain in place and the solid retains its shape.) The more closely the molecules in a gas or liquid are packed together, the more frequently they collide with one another. Because of Brownian collisions, molecules initially packed together spread out and move toward an equal distribution throughout the container holding them. Brownian movement supplies the **kinetic energy** for passive transport mechanisms.

QuickCheck Questions

1.1 How does the kinetic energy of Brownian movement cause molecules to spread out in a container?

1.2 Why do solids retain their shape?

1 Materials

- ☐ Compound microscope
- ☐ Microscope slide and coverslip
- ☐ Small dropper bottle with eyedropper
- ☐ Tap water
- ☐ Waterproof ink
- ☐ Powdered kitchen cleanser

Procedures

1. Fill the dropper bottle three-fourths full of tap water and add a small amount of cleanser powder and waterproof ink. Add 10 to 15 ml of additional tap water.

2. Shake the bottle gently to mix the contents. Place a drop of the mixture on a microscope slide and place a coverslip over the drop.

3. Focus on the slide and locate the small granules of cleanser. Observe how the particles move and occasionally collide with one another.

4. Describe the movement observed under the microscope in C.1 of the Laboratory Report. ■

LAB ACTIVITY 2 Diffusion of a Liquid

Diffusion is the net movement of substances from a region of greater concentration to a region of lesser concentration. Simply put, diffusion is the spreading out of substances owing to collisions between moving molecules. Diffusion occurs throughout the body—in extracellular fluid, across cell membranes, and in the cytosol of cells. Examples of diffusion include oxygen moving from the lungs into pulmonary capillaries, odor molecules moving through the nasal lining to reach olfactory cells, and ions moving in and out of nerve cells to produce electrical impulses. Molecules diffuse through cells by two basic mechanisms: lipid-soluble molecules diffuse through the phospholipid bilayer of the cell membrane, and small, water-soluble ions and molecules pass through the channels of integral proteins. Molecules like proteins are too large to enter the membrane channels and therefore do not diffuse across the membrane.

Cells cannot directly control diffusion; it is a passive transport process much like a ball rolling downhill. If a substance is unequally distributed, a **concentration gradient** exists, and one region will have a greater concentration of the substance than other regions. The substance will diffuse until an equal distribution occurs, at a point called **equilibrium**. Figure 6.1 illustrates diffusion with a cube of colored sugar. Before the cube is placed in the water, the sugar molecules are concentrated

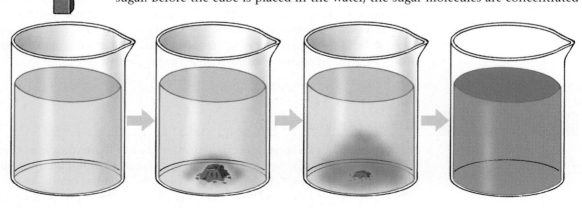

Figure 6.1 Diffusion

Placing a colored sugar cube in a glass of water establishes a steep concentration gradient. As the cube dissolves, many sugar and dye molecules are in one location, and none are elsewhere. As diffusion occurs, the molecules spread through the solution. Eventually, diffusion eliminates the concentration gradient. The sugar cube has dissolved completely, and the molecules are distributed evenly. Molecular motion continues, but there is no net directional movement.

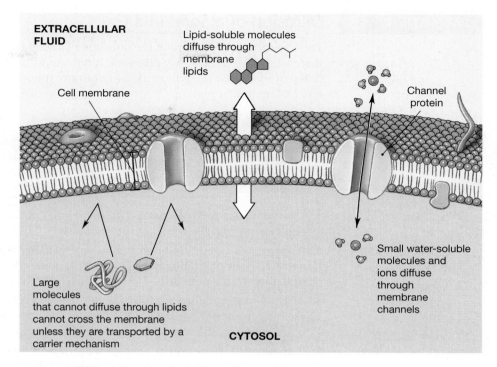

EXTRACELLULAR FLUID

Lipid-soluble molecules diffuse through membrane lipids

Cell membrane

Channel protein

Large molecules that cannot diffuse through lipids cannot cross the membrane unless they are transported by a carrier mechanism

Small water-soluble molecules and ions diffuse through membrane channels

CYTOSOL

Figure 6.2 **Diffusion Across the Cell Membrane**

in it. Once submerged, the sugar dissolves, and the molecules disperse as they bump into other sugar molecules and water molecules. Eventually, the colored molecules become evenly distributed, and the solution is in equilibrium. Even at equilibrium, the molecules are in motion. When one molecule bumps another out of position, the shift forces some other molecule into the vacated space. One movement cancels the other, and no net movement occurs.

Temperature, pressure, and concentration gradient are some factors that influence the rate of diffusion. Brownian movement slows as temperature decreases. Therefore, diffusion is also slower at lower temperatures, because the more slowly a molecule is moving, the longer it takes that molecule to travel far enough to collide with another molecule.

Figure 6.2 summarizes diffusion across the cell membrane. (For additional background information, review the structure of the cell membrane in Exercise 5.)

QuickCheck Questions

2.1 Is diffusion a passive process or an active process?

2.2 Describe a solution that is at equilibrium.

2 Materials

- ☐ 2 pyrex beakers, 250 ml or larger
- ☐ Tap water
- ☐ Ice
- ☐ Hotplate or microwave oven
- ☐ Food coloring dye

Procedures

1. Fill one beaker three-fourths full with tap water and small ice chips.

2. Fill the other beaker three-fourths full with tap water and warm the water in a microwave oven or on a hotplate. Do not boil the water.

3. Leave both beakers undisturbed for several minutes to let the water settle. Remove any remaining ice chips from the chilled water.

4. Carefully add two drops of food coloring dye to each beaker.

5. Observe for several minutes as the dye diffuses. Continue observing the beakers every three to four minutes until equilibrium is reached. ■

Diffusion of a Solid in a Gel

This experiment demonstrates the diffusion of a solid chemical in a thick gelatinous material. As the solid slowly dissolves, it diffuses into the gel. Two chemicals with different molecular masses are used to illustrate the relationship between diffusion rate and molecular mass.

QuickCheck Question

3.1 Make a prediction: Which would you expect to diffuse at a faster rate, a molecule that has a high molecular mass or one that has a low molecular mass?

3 Materials

- ☐ Petri dish with plain agar
- ☐ Cork bore or soda straw
- ☐ Potassium permanganate crystals
- ☐ Iodine crystals
- ☐ Ruler

Procedures

1. Use the bore or straw to punch two small holes in the agar, approximately equidistant from the center of the petri dish.

2. Place a small amount of potassium permanganate crystals in one hole and an equal amount of iodine crystals in the other hole. Do not spill crystals on any other part of the petri dish.

3. After 30 to 45 minutes, measure the distance each chemical has diffused. Measure from the edge of the hole farthest from the center of the dish to the outer boundary of the diffusion area.

4. Dispose of the petri dish as instructed by your laboratory instructor. ∎

Osmosis

Osmosis (oz-MŌ-sis; *osmos*, thrust) is the net movement of water through a selectively permeable membrane, from a region of greater water concentration to a region of lesser water concentration. We can define osmosis as the *diffusion* of water through a selectively permeable membrane. It occurs when two solutions of different solute concentrations are separated by a selectively permeable membrane. A **solution** is the result of dissolving a **solute** in a **solvent**. In a 1% aqueous solution of some salt, for example, the salt is the solute and occupies 1% of the solution volume; the solvent—water—makes up the remaining 99% of the solution volume. As solute concentration increases, the space available for water molecules decreases, and we can think of this as the water concentration decreasing. For osmosis to occur in a cell, there must be a difference in water concentrations on the two sides of the cell membrane. This difference in concentration establishes the concentration gradient for osmosis.

Study Tip

Water, Ions, and Membranes

Only water moves across the membrane during osmosis. If the membrane were permeable to solute molecules, those molecules would move across the membrane until solute equilibrium was reached. Once the solute molecules were in equilibrium, the water molecules would also be in equilibrium. The water concentration gradient would be eliminated, and with no concentration gradient, there can be no osmosis. ●

Figure 6.3 shows a U-shaped pipe with a selectively permeable membrane located at the bottom where the two arms of the U meet. There are identical molecules on either side of the membrane, but in different concentrations. The small blue dots represent water molecules, and the large pink dots are solute molecules that cannot cross the membrane. (The pale blue background also represents water molecules, but you should concentrate just on the ones represented by the dots.) The numbers of blue dots in step 1 tell you that, before our experiment begins, arm A has more water molecules and fewer solute molecules than arm B. Note that in step 1, the water

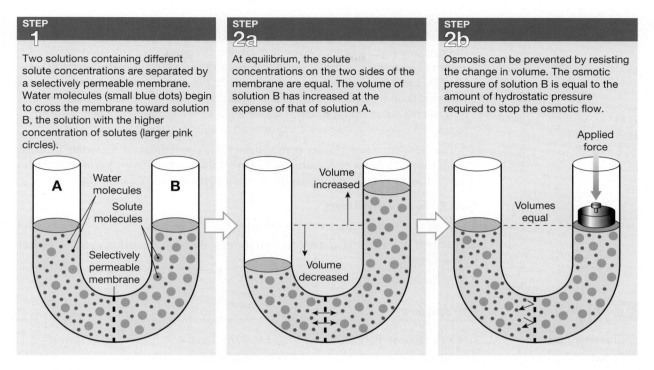

Figure 6.3 Osmosis

Step 1: Two solutions containing different solute concentrations are separated by a selectively permeable membrane. Water molecules (small blue dots) begin to cross the membrane toward arm B, which contains the solution with the higher concentration of solutes (larger pink circles). *Step 2a:* At equilibrium, the solute concentrations on the two sides of the membrane are equal. The volume of the solution in arm B has increased at the expense of that in arm A. *Step 2b:* Osmosis can be prevented by resisting the volume change. The osmotic pressure of solution B is equal to the amount of hydrostatic pressure required to stop the osmotic flow.

level is the same in the two arms. As water moves from arm A to arm B through the selectively permeable membrane, the volume in arm B increases until equilibrium is reached in step 2a. Water and solute concentrations are now equal on the two sides of the membrane.

Solutions have an **osmotic pressure** because of the presence of solute. The greater the solute concentration, the greater the osmotic pressure of the solution. During osmosis, the solution with the greater osmotic pressure causes water to move toward it. In effect, "water follows solute," and osmotic pressure is a "pulling" pressure that draws water toward the higher solute concentration. Notice in step 2b of Figure 6.3 that a force applied to arm B will stop osmosis if the pressure resulting from that force is equal to the osmotic pressure causing the osmosis.

Drinking water is often purified by **reverse osmosis**, a process in which the pressure applied to arm B is greater than the osmotic pressure. This increased external pressure forces water molecules across the membrane from right to left in Figure 6.3. As more and more water is forced into arm A, the concentration of solute molecules in that arm gets lower and lower until at some point the solute concentration is so low that we consider the water to be pure.

Clinical Application | *Dialysis*

Dialysis is a passive process similar to osmosis except that, besides water, small solute particles can pass through a selectively permeable membrane. Large particles are unable to cross the membrane, and thus particles can be separated by size during dialysis. Dialysis does not occur in the body, but it is used in the medical procedure called **kidney dialysis** to remove wastes from the blood of a patient whose kidneys are not functioning properly. Blood from an artery passes into thousands of minute selectively permeable tubules in a dialysis cartridge. A dialyzing solution having the same concentration of materials to remain in the blood (nutrients and certain electrolytes) is pumped into the cartridge to flow over the tubules. As blood flows through the tubules, wastes diffuse from the blood, through the selectively permeable tubules, and into the dialyzing solution. Once waste levels in the blood have been reduced to a safe level, the patient is disconnected from the dialysis apparatus. ▶

The following experiment demonstrates the movement of materials through dialysis tubing. Like a cell membrane, dialysis tubing is selectively permeable. Small pores in the tubing allow the passage of small particles but not large ones.

QuickCheck Questions

4.1 How is osmosis different from diffusion?

4.2 Which has a greater water concentration, a 1% solute solution or a 2% solute solution?

4.3 What is osmotic pressure?

4 Materials

- ☐ Dialysis tubing
- ☐ Two dialysis tubing clips or two pieces of thread
- ☐ Gram scale
- ☐ 500-ml beaker
- ☐ Distilled water
- ☐ 5% starch solution
- ☐ Lugol's iodine solution

Procedures

1. Cut a strip of dialysis tubing 15 cm (6.0 in.) long.

2. Add approximately 100 ml of distilled water to the beaker. Soak the dialysis tubing in the water for three to four minutes and then remove it from the beaker.

3. Fold one end of the tubing over and seal it securely with a tubing clip or a piece of thread, forming what is called a dialysis bag. Rub the unclipped end of the bag between your fingers to open the tubing.

4. Fill the bag approximately three-quarters full with starch solution, then fold the end of the tubing over. Clip or tie this end closed without trapping too much air inside.

5. Rinse the bag to remove traces of starch solution from its outside surface. Dry the outside of the filled bag and weigh it to determine its mass. Record your mass measurement in the Initial Observations column of Table 6.1.

6. Submerge the bag completely in the beaker of water, as shown in Figure 6.4. Add enough Lugol's solution to discolor the water in the beaker, and then complete the Initial Observations column in Table 6.1.

7. After 60 minutes:

   a. Without disturbing the setup, examine the beaker and bag and decide if starch, iodine, or water moved either way across the tubing membrane. Record your observations in Table 6.2 and in the second, third, and fourth rows of the Final Observations column of Table 6.1.

   b. Remove the bag from the beaker, dry the outer surface, and determine the mass of the bag plus contents. Record your measurement in the Final Observations column of Table 6.1. ■

Table 6.1 Dialysis Experiment Observations

Dialysis Bag	Initial Observations	Final Observations
Mass of bag plus starch solution		
Shape of filled bag		
Color of starch solution		
Color of beaker water		

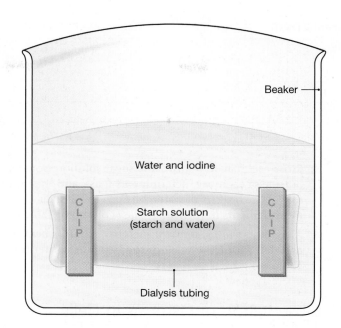

Figure 6.4 Osmosis Setup Using Dialysis Membrane
Note: The tubing may be tied with thread if clips are unavailable.

Table 6.2 Dialysis Experimental Results

Substance	Movement (in, out, none)	Process (diffusion, osmosis)
Water	_____	_____
Starch	_____	_____
Iodine	_____	_____

LAB ACTIVITY 5   Concentration Gradients and Osmotic Rate

This experiment demonstrates the relationship between concentration gradient and rate of osmosis. Molecules at greater concentrations are packed closer together and have a higher incidence of collisions with neighboring molecules. By comparing changes in mass in a series of dialysis bags, you will measure the osmotic rate at different solute concentrations.

5 Materials

- ☐ Three 15-cm (6.0-in.) strips of dialysis tubing
- ☐ Six dialysis tubing clips or six pieces of thread
- ☐ Three 500-ml beakers
- ☐ Distilled water
- ☐ 1%, 5%, and 10% sugar solutions

Procedures

1. Add approximately 100 ml of distilled water to each beaker. Place one tubing strip in each beaker, soak for three to four minutes to loosen the tubing, and then remove the strips from the beakers.

2. Fold one end of one piece of tubing over and seal it securely with a tubing clip or a piece of thread, forming a dialysis bag. Rub the unclipped end of the bag between your fingers to open the tubing.

3. Fill the bag approximately three-quarters full with the 1% sugar solution, then fold the end of the tubing over. Clip or tie this end closed without trapping too much air inside.

4. Prepare two other bags with the remaining two pieces of tubing. Fill one with the 5% sugar solution and the other with the 10% solution.

Table 6.3 Osmosis Experimental Data

Dialysis Bag	Initial Mass	Final Mass
1% sugar		
5% sugar		
10% sugar		

5. Rinse each bag to remove any sugar solution from the outside surface. Dry the outside of each bag, determine its mass, and then submerge it completely in one of the beakers of water, one bag to a beaker. Record your mass measurements in the Initial Mass column of Table 6.3.

6. After 60 minutes, remove each bag from its beaker, dry the outer surface, and determine the mass of the bag plus contents. Record the final masses in Table 6.3. ■

LAB ACTIVITY 6 Observation of Osmosis in Cells

A solution that has the same solute concentrations as a cell is an **isotonic solution**. If the solute concentrations are the same, the solvent concentrations are also the same. A solution containing more solute (and therefore less solvent) than a cell is a **hypertonic solution**, and a solution containing less solute than a cell is a **hypotonic solution**. The cell is the reference point; solute concentrations in solutions are compared with solute concentrations in the cell. Sitting in a hypertonic solution, a cell will lose water as a result of osmotic movement and will shrink, or **crenate**. Sitting in a hypotonic solution, a cell will gain water and perhaps burst, or **lyse**. **Hemolysis** is the process of a blood cell rupturing in hypotonic solution.

Your laboratory instructor may choose to use plant cells rather than blood cells to study **tonicity**, the effect of solutions on cells. Plant cells have a thick outer cell wall that provides structural support for the plant. Pushed against the inner surface

Figure 6.5 Osmotic Flow across Cell Membranes

(a) Because these red blood cells are immersed in an isotonic saline solution, no osmotic flow occurs and the cells have their normal appearance. (b) Immersion in a hypotonic saline solution results in the osmotic flow of water into the cells. The swelling may continue until the cell membrane ruptures. (c) Exposure to a hypertonic solution results in the movement of water out of the cells. The red blood cells shrivel and become crenated.

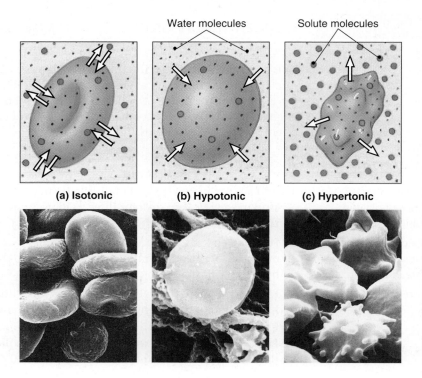

Water molecules   Solute molecules

(a) Isotonic   (b) Hypotonic   (c) Hypertonic

of the cell wall is the cell membrane. To study osmosis in plant cells, observe the distribution of the cell's organelles and attempt to locate the cell membrane. In a hypertonic solution, the plant cell loses water and the cell membrane shrinks away from the cell wall.

6 *Materials*

- ☐ Blood (supplied by instructor) or live aquatic plant (Elodea)
- ☐ Microscope slides and coverslips, microscope, eyedroppers, wax pencil
- ☐ 0.90% saline solution (isotonic)
- ☐ 2.0% saline solution (hypertonic)
- ☐ Distilled water (hypotonic)
- ☐ Gloves and safety glasses

Procedures—Blood Cells

1. With the wax pencil, write along one of the shorter slide edges. Label one slide "Iso," one "Hypo," and one "Hyper."
2. Put on safety glasses and disposable gloves before handling any blood.
3. Add a small drop of blood to each slide. Do not touch the blood. Place a coverslip over each slide.
4. Add a drop of isotonic solution to the outer edge of the coverslip of the "Iso" slide. Repeat with the other slides and solutions.
5. Use the microscope to observe changes in cell shape as osmosis occurs. Compare your results with the cells in Figure 6.5.
6. Dispose of materials contaminated with blood in a biohazard waste container. ■

Procedures—Elodea Leaf

1. With the wax pencil, write along one of the shorter slide edges. Label one slide "Iso," one "Hypo," and one "Hyper."
2. Place one Elodea leaf flat on each slide. Place a coverslip over each leaf.
3. Add a drop of isotonic solution to the outer edge of the coverslip of the "Iso" slide. Repeat with the other slides and solutions.
4. Use the microscope to observe changes in cell shape as osmosis occurs. Compare your results with the cells in Figure 6.5.
5. Rinse and clean the slides or dispose of them in a sharps box for glass. ■

LAB ACTIVITY 7   Active Transport Processes

Cells use carrier molecules to move nondiffusible materials through the cell membrane. Unlike passive processes, this carrier-assisted movement may occur against a concentration gradient. Movement of this type is called *active transport* and requires the cell to use energy. Whereas the passive processes of diffusion and osmosis may occur in both living and dead cells, only living cells can supply the energy necessary for active transport.

Endocytosis is the active transport of materials into a cell. Figure 6.6 illustrates **phagocytosis**, the movement of large particles into the cell. The figure shows a cell ingesting a bacterium. The cell forms extensions of its cell membrane, called *pseudopodia,* to capture the bacterium. When the pseudopodia touch one another, they fuse and trap the bacterium in a membrane vesicle. Inside the cell, lysosomes surround and empty their powerful enzymes into the vesicle and destroy the bacterium. During the process called **pinocytosis**, the cell invaginates a small area of the cell membrane and traps not the large particles of phagocytosis but rather small particles and fluid. The forming vesicle continues to pinch inward.

Exocytosis is the active transport of materials out of the cell. An intracellular vesicle fills up with materials, fuses with the cell membrane, and releases its contents into the extracellular fluid. The Golgi apparatus secretes cell products by pinching off small secretory vesicles that fuse with the cell membrane for exocytosis. Cells also eliminate debris and excess fluids by exocytosis. Table 6.4 summarizes transport mechanisms across the cell membrane.

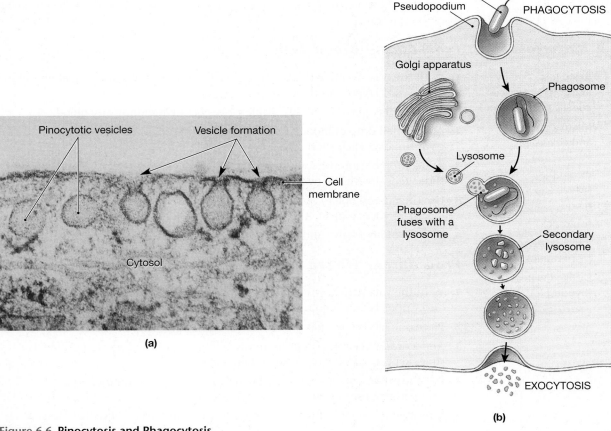

Figure 6.6 Pinocytosis and Phagocytosis
(a) An electron micrograph showing pinocytosis at the bases of microvilli in a cell lining the intestinal tract.
(b) Material brought into the cell by phagocytosis is enclosed in a phagosome and subsequently exposed to lysosomal enzymes. After nutrients are absorbed from the vesicle, the residue is discharged by exocytosis.

QuickCheck Questions

7.1 How do active and passive transports differ?

7.2 Give an example of endocytosis.

7 *Materials*

- ☐ *Amoeba proteus* culture, starved for 48 hours
- ☐ *Tetrahymena* culture
- ☐ Eyedropper
- ☐ Microscope depression slide and coverslip
- ☐ Compound microscope

Procedures

1. Add a drop of the *Amoeba proteus* culture in the well of the depression slide. Place the well with a coverslip.

2. Examine the slide at low magnification and verify that the culture has living Amoeba. Reduce the intensity of the microscope's light to keep the slide cool.

3. Lift the coverslip and introduce a drop of the *Tetrahymena* culture. *Tetrahymena* are small ciliated protists that the Amoeba will ingest.

4. Observe the activity of the Amoeba feeding on the *Tetrahymena*. Select an Amoeba ingesting a *Tetrahymena* and observe the process of pseudopodia formation and movement at low and medium powers.

5. Upon completion of your observations, rinse and dry the depression slide and coverslip. ■

Table 6.4 Summary of Mechanisms Involved in Movement across Cell Membranes

Mechanism	Process	Factors Affecting Rate	Substances Involved (location)
Diffusion	Molecular movement of solutes; direction determined by relative concentrations	Size of gradient; size of molecules; charge; lipid solubility, temperature	Small inorganic ions, lipid-soluble materials (all cells)
Osmosis	Movement of water molecules toward solution containing relatively higher solute concentration; requires selectively permeable membrane	Concentration gradient; opposing osmotic or hydrostatic pressure	Water only (all cells)
Filtration	Movement of water, usually with solute, by hydrostatic pressure; requires filtration membrane	Amount of pressure; size of pores in filter	Water and small ions (blood vessels)
Carrier-Mediated Transport			
Facilitated diffusion	Carrier proteins passively transport solutes across a membrane down a concentration gradient	Size of gradient, temperature, and availability of carrier protein	Glucose and amino acids (all cells, but several different regulatory mechanisms exist)
Active transport	Carrier proteins actively transport solutes across a membrane regardless of any concentration gradients	Availability of carrier, substrate, and ATP	Na^+, K^+, Ca^{2+}, Mg^{2+} (all cells); other solutes by specialized cells
Secondary active transport	Carrier proteins passively transport two solutes, with one (normally Na^+) moving down its concentration gradient; the cell must later expend ATP to eject the Na^+	Availability of carrier, substrates, and ATP	Glucose and amino acids (specialized cells)
Vesicular Transport			
Endocytosis	Creation of membranous vesicles containing fluid or solid material	Stimulus and mechanics incompletely understood; requires ATP	Fluids, nutrients (all cells); debris, pathogens (specialized cells)
Exocytosis	Fusion of vesicles containing fluids and/or solids within the cell membrane	Stimulus and mechanics incompletely understood; requires ATP	Fluids, debris (all cells)

Movement Across Cell Membranes

Name _____

Date _____

Section _____

A. Matching

Match each term in the left column with its correct description from the right column.

_____ **1.** osmosis	**A.** movement resulting from molecular collisions
_____ **2.** diffusion	**B.** diffusion of water through a selectively permeable membrane
_____ **3.** concentration gradient	**C.** substance dissolved into a solution
_____ **4.** solute	**D.** shrinking of cells due to water loss
_____ **5.** solvent	**E.** difference in solute concentration between two solutions
_____ **6.** active transport	**F.** uniform distribution of a substance
_____ **7.** crenation	**G.** bursting of a red blood cell
_____ **8.** equilibrium	**H.** movement from region of high concentration to region of low concentration
_____ **9.** Brownian movement	**I.** substance that dissolves other substances in a solution
_____ **10.** hemolysis	**J.** movement requiring use of cellular energy

B. Definitions

Define the following terms.

1. osmotic pressure

2. dialysis

3. hypertonic solution

4. hemolysis

5. exocytosis

6. Brownian movement

C. Results

Laboratory Activity 1

1. Describe the movement you observed under the microscope.

Laboratory Activity 2

1. Why does the dye diffuse in the water?

2. Was there a difference in diffusion rates between the chilled and warmed water?

3. How did temperature influence the diffusion rate?

Laboratory Activity 3

1. Which crystal diffused farther?

2. Which crystal has the larger molecular mass? How does molecular mass affect diffusion rate?

Laboratory Activity 4

1. Which had the greater osmotic pressure, the starch solution in the dialysis bag or the iodine solution in the beaker?

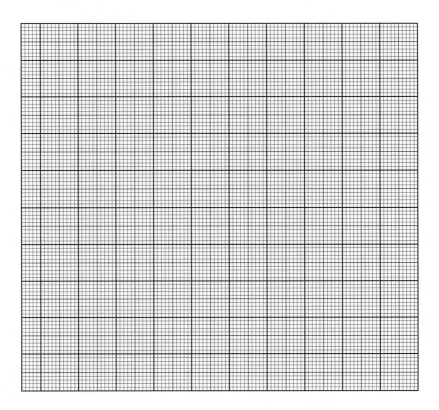

2. Use osmotic pressure to explain what happened in this experiment.

3. How did you detect whether it was the starch solution that moved or the iodine solution?

Laboratory Activity 5

1. Which solution had the greatest osmotic pressure? Because of this highest pressure, what happened more with this solution than with the other two solutions?

2. Describe the relationship between solution concentration and osmosis rate.

3. On graph paper or below, plot the increase in mass for each bag on the vertical axis and the solution concentration on the horizontal axis. This graph shows how solute concentration affects osmosis rate.

Laboratory Activity 6

1. Describe the appearance of the blood (or plant) cells in the hypotonic solution. What has happened to these cells?

2. Describe the appearance of the blood (or plant) cells in the hypertonic solution. What has happened to these cells?

3. Why is blood plasma isotonic to the cytosol of a cell?

D. Short-Answer Questions

1. Describe the components of a 2% sugar solution.

2. After a long soak in the tub, you notice that your skin has become wrinkled and that your fingers and toes feel bloated. Describe why this occurs.

E. Analysis and Application

1. Describe how molecules of a gaseous substance can diffuse through the air.

2. A blood cell is placed in a 1.5% salt solution. Will osmosis, diffusion, or both occur across the cell membrane? Why?

3. Why is a concentration gradient necessary for passive transport?

Epithelial Tissue

OBJECTIVES

On completion of this exercise, you should be able to:

- List the characteristics used to classify epithelia.
- Describe how epithelia are attached to the body.
- Describe the microscopic appearance of each type of epithelia.
- List the location and function of each type of epithelia.
- Identify each type of epithelia under the microscope.

Histology is the study of tissues. A **tissue** is a group of similar cells working together to accomplish a specific function. It may be difficult for us to appreciate how individual cells contribute to the life of the entire organism, but we readily see the effect of tissues in our bodies. Consider how much effort is focused on reducing fat tissue and exercising muscle tissue. An understanding of histology is vital for the study of organ function. The stomach, for example, plays major digestive roles, as it secretes digestive juices and is involved in the mixing and movement of food. Each of these functions is performed by specialized tissue.

Figure 7.1 is an overview of tissues of the body. Molecules and atoms combine to form cells, which secrete materials into the surrounding extracellular fluid. The cells and their secretions compose the various tissues of the body. There are four major categories of tissues in the body: **epithelial**, **connective**, **muscle**, and **neural**. Each category includes specialized tissues that have specific locations and functions. Many tissues form organs, such as the stomach, a muscle, or a bone. Organs working together to accomplish major processes (such as digestion, movement, or protection) constitute an organ system.

During your microscopic observations of tissues in the following laboratory activities, it is important to scan the entire slide to examine the tissue at low power. A slide may have several tissues, and you must survey the specimen to locate a particular tissue. Once you have located the tissue, increase the magnification and observe the individual cells of the tissue. Take your time when studying a tissue; a quick glance through the microscope is not sufficient to learn enough to be able to identify a tissue on a laboratory examination.

This exercise discusses epithelial tissue; Exercises 8, 9, and 10 cover connective tissue, muscle tissue, and neural tissue, respectively.

Figure 7.1 An Orientation to the Tissues of the Body

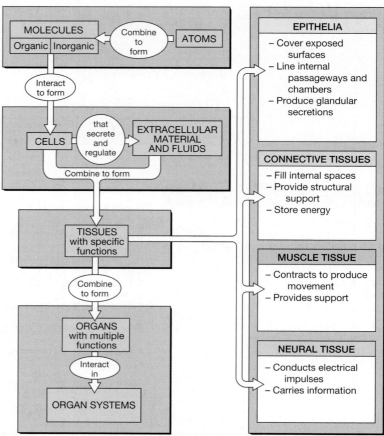

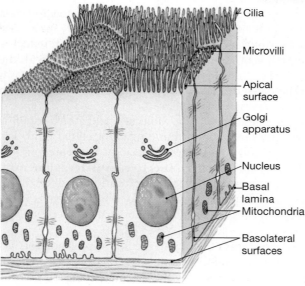

Cilia
Microvilli
Apical surface
Golgi apparatus
Nucleus
Basal lamina
Mitochondria
Basolateral surfaces

Figure 7.2
The Polarity of Epithelial Cells
Many epithelial cells have an uneven distribution of organelles between the free surface (here, the top) and the basal lamina. In many cases, the free surface bears microvilli; in some cases, this surface has cilia or (very rarely) stereocilia. *(All three would not normally be on the same group of cells but are depicted here for purposes of illustration.)* In some epithelia, such as the lining of the kidney tubules, mitochondria are concentrated near the base of the cell, probably to provide energy for the cell's transport activities.

Introduction to Epithelia

Epithelia (e-pi-THĒ-lē-a; singular **epithelium**), or epithelial tissues, are lining and covering tissues. They are the only tissues visible on the body. The respiratory, digestive, reproductive, and urinary systems all have openings to the external environment, and each is lined with an epithelium. The entire body surface is covered with an epithelium in the form of the top layer of the skin. Epithelia always have one free surface where the epithelial cells are exposed either to the external environment or to an internal passageway or cavity. Because epithelia are surface and lining tissues, they do not contain blood vessels; they are **avascular** tissues. Epithelial cells obtain nutrients and other necessary materials by diffusion from the underlying connective tissue. Glide your tongue over your inner cheeks and feel the epithelial lining.

Each epithelium is attached to the body by a **basal lamina** (LA-mi-nah; *lamina*, plate), or basement membrane, located between the epithelium and a connective tissue layer (Figure 7.2). The basal lamina is two layers of cellular glue that anchor the epithelium to the underlying connective tissue layer. The base of the epithelium is in contact with the basal-lamina layer called the *lamina lucida*

(LA-mi-nah LOO-si-dah; *lucida,* clear), which contains glycoproteins and other protein molecules that prevent substances in the adjoining connective tissue from seeping into the epithelia. The basal-lamina layer facing the connective tissue is the *lamina densa*. Thick protein fibers in this layer entwine with the proteins in the lamina lucida for secure attachment of the epithelium.

Study Tip | ### *Looking at Epithelia*

When observing epithelia microscopically, it might be difficult to locate the basal lamina. Simple epithelia often have a conspicuous basal lamina. Find the free surface of the tissue, and then look on the opposite edge of the cells. The basal lamina appears as a dark line between the epithelial cells and the connective tissue. You will not be able to distinguish between the lamina lucida and lamina densa with the microscopes available in your laboratory. ●

Epithelia have a wide range of functions, each dependent on the type of cells in the tissue. On exposed surfaces, thick layers of **stratified epithelium** protect against excessive friction, prevent dehydration, and keep microbes and chemicals from invading the body. Thin, one-layered **simple epithelium** provides a surface for exchange of materials, such as the exchange of gases between the lungs and blood. Absorption, secretion, and diffusion all occur across simple epithelia. The epithelial tissue that covers the body surface contains many of the body's sensory organs. In some epithelial tissue, such as that associated with glands, the cells are short-lived, and in these cases **stem cells** must constantly produce new cells to replenish the tissue.

Epithelium occurs in sheets of cells that are tightly joined together, much like pieces to a jigsaw puzzle. The cells are attached to one another by strong intercellular connections. Some cell membranes of adjoining epithelial cells are glued together by protein molecules known as cell-adhesion molecules. Cell junctions occur where adjoining cell membranes connect. Some cell junctions have pores through which materials can move from one cell to another. Four types of connections occur: gap junctions, tight junctions, intermediate junctions, and desmosomes. Explore your lecture textbook and the Internet for more information about cell junctions.

Classification of Epithelia

Epithelial tissues are classified according to cell shape and cell organization (Figure 7.3). These characteristics are used together to name a tissue.

Cell Shape

- **Squamous** (SKWĀ-mus; *squama,* scale) epithelial cells are irregularly shaped, flat, and scalelike. These cells, depending on how they are organized, function either in protection or in secretion and diffusion.

- **Cuboidal** epithelial cells are squarish and have a large central nucleus. They are found in the tubules of the kidneys and in many glands. The cube-shaped cells can secrete and absorb materials across the tubular wall.

- **Columnar** epithelial cells are taller than they are wide, like the columns of an ancient Greek temple. The surface of a columnar cell may be covered with microscopic plasma membrane extensions called microvilli (Figure 7.2), which increase the cellular surface area for absorption and secretion. Other extensions are hairlike cilia, which move mucus and remove debris. Many locations in the respiratory and digestive systems are lined with columnar cells.

Cell Organization

- As noted earlier, simple epithelium is a single layer of cells attached to an underlying connective tissue by the basal lamina. In simple epithelial tissues, all cells touch the basal lamina and are therefore exposed to the upper surface of

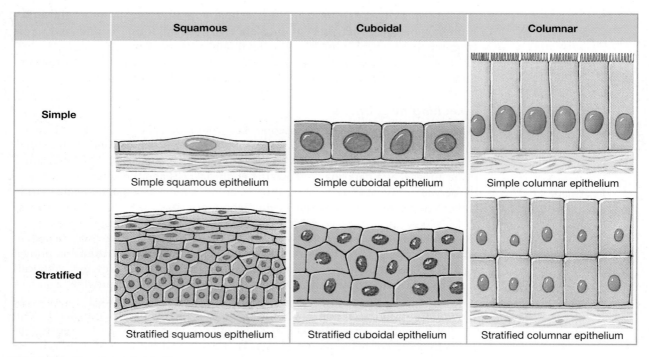

Figure 7.3 Classifying Epithelia
All epithelia are classified by the number of cell layers and by the shape of the cells.

the underlying connective tissue. Simple epithelium is found wherever a thin surface is required for diffusion, such as between the air sacs and the blood capillaries in the lungs.

- **Pseudostratified epithelium**, as its name implies, is "falsely stratified." As in simple epithelium, the cells all touch the basal lamina, but not all cells in the epithelial layer reach the upper surface of the underlying connective tissue. Cells in a pseudostratified epithelium grow to different heights, and taller cells grow over and cover shorter cells. Parts of the respiratory tract are lined with ciliated pseudostratified epithelium.

- Stratified epithelium has layers of cells, one stacked upon another. The cells in the bottom layer are attached to the basal lamina, and only cells in the upper layer are exposed to the free surface of the epithelium. This type of epithelial tissue is found in areas exposed to friction and abrasion. The top layer of the skin (called the *epidermis*), the lining of the mouth, parts of the pharynx, and the anus are covered with stratified epithelium. **Transitional epithelium** is a special kind of stratified epithelium with cells of many shapes that permit the tissue to stretch and recoil.

LAB ACTIVITY 1   Simple Epithelia

The cells in simple epithelia can be squamous, cuboidal, or columnar. General functions of simple epithelia are diffusion, absorption, and secretion. At the free surface, extensions of the plasma membrane called microvilli increase surface area for absorption. To protect the tissue at the free surface, cells called **goblet cells** secrete mucus that coats the cells. Other cells in a simple epithelial layer have hairlike **cilia** that sweep the mucus along the free surface to remove debris. Cilia are composed of microtubules that protrude from the cell.

QuickCheck Questions

1.1 How is epithelium organized and classified?

1.2 What are the functions of simple epithelia?

1 Materials

☐ Prepared microscope slides of simple squamous epithelium, simple cuboidal epithelium, and simple columnar epithelium

Procedures

1. Examine each simple epithelium under the microscope at low, medium, and high magnification.

2. Refer to the photomicrographs in Figures 7.4 and 7.5 and locate the featured structures.

3. Draw each tissue in the space provided in Figure 7.5.

   a. **Simple squamous epithelium** (Figure 7.4a) is found in serous membranes, in the lining of blood vessels and the heart, and in the air sacs of the lungs. A superficial preparation of this epithelium viewed under the microscope appears as a sheet of cells that look like ceramic tiles on a floor.

   b. **Simple cuboidal epithelium** preparations are often from the epithelia of either kidney tubules or thyroid follicles. In Figure 7.4b, the arc of cube-shaped cells at the top of each panel represents part of a tubule, and the two rows of cells above and below the white horizontal region are part of a tubule that has been sectioned longitudinally. (On many slides, the tubules have been transversely sectioned.)

   c. **Simple columnar epithelium** lines most of the digestive tract and the gallbladder. Figure 7.4c shows the lining of the small intestine. The wall of the intestine is folded to increase the surface area available for digestion and absorption of nutrients. Simple columnar epithelium covers the folded wall and is in direct contact with the contents of the intestine. Notice in the figure that the nuclei are uniformly located at the base of the cells. The large oval cells interspersed among the columnar cells are goblet cells. The uterine tubes and sinuses of the skull are lined with ciliated columnar epithelium. In the uterine tubes, the cilia transport released eggs to the uterus. ■

LAB ACTIVITY 2   Pseudostratified and Transitional Epithelia

Pseudostratified epithelium appears to be a stratified tissue because the nuclei are scattered in the cells, giving the appearance of a stratified tissue. All of the cells adhere to the basal lamina, but, as noted earlier, some cells grow taller and cover other cells at the free surface.

The specialized stratified epithelium known as transitional epithelium occurs in the urinary bladder, among other places in the body. The tissue allows the bladder to fill and empty. These tissues are detailed in Figure 7.6.

QuickCheck Questions

2.1 How is pseudostratified epithelium different from simple epithelium?

2.2 What is the function of transitional epithelium and where is it found in the body?

2 Materials

☐ Prepared microscope slides of pseudo-stratified epithelium and transitional epithelium

Procedures

1. Examine each epithelium under the microscope at low, medium, and high magnification.

2. Refer to the photomicrographs in Figures 7.6 and 7.7 and locate the featured structures.

SIMPLE SQUAMOUS EPITHELIUM

LOCATIONS: Mesothelia lining ventral body cavities; endothelia lining heart and blood vessels; portionsof kidney tubules (thin sections of loops of Henle); inner lining of cornea; alveoli of lungs

FUNCTIONS: Reduces friction; controls vessel permeability; performs absorption and secretion

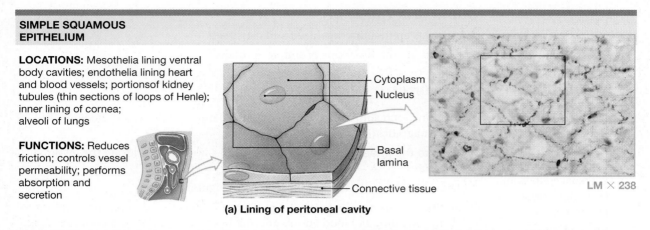

(a) Lining of peritoneal cavity

LM × 238

SIMPLE CUBOIDAL EPITHELIUM

LOCATIONS: Glands; ducts; portions of kidney tubules; thyroidgland

FUNCTIONS: Limited protection, secretion, absorption

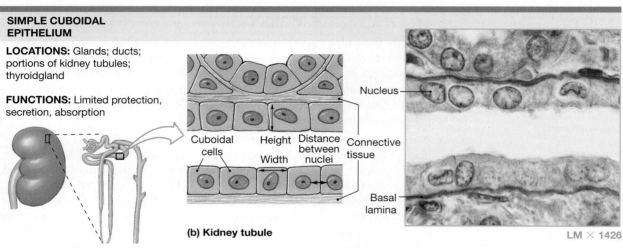

(b) Kidney tubule

LM × 1426

SIMPLE COLUMNAR EPITHELIUM

LOCATIONS: Lining of stomach, intestine, gallbladder, uterine tubes, and collecting ducts of kidneys

FUNCTIONS: Protection, secretion, absorption

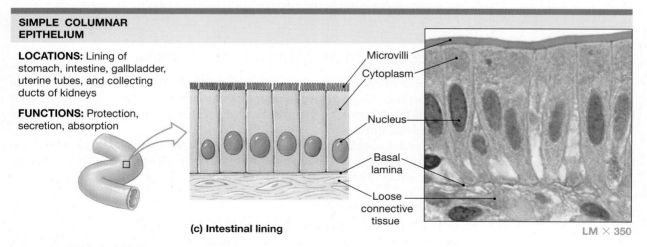

(c) Intestinal lining

LM × 350

Figure 7.4 Simple Epithelia

(a) A superficial view of the simple squamous epithelium (mesothelium) that lines the peritoneal cavity. The three-dimensional drawing shows the epithelium in superficial and sectional views. **(b)** A section through the simple cuboidal epithelial cells of a kidney tubule. **(c)** The simple columnar epithelium lining the small intestine.

Figure 7.5 Simple Epithelia
(a) Simple squamous epithelium (LM × 100) and sketch. **(b)** Simple cuboidal epithelium (LM × 100) and sketch. **(c)** Simple columnar epithelium (LM × 100) and sketch.

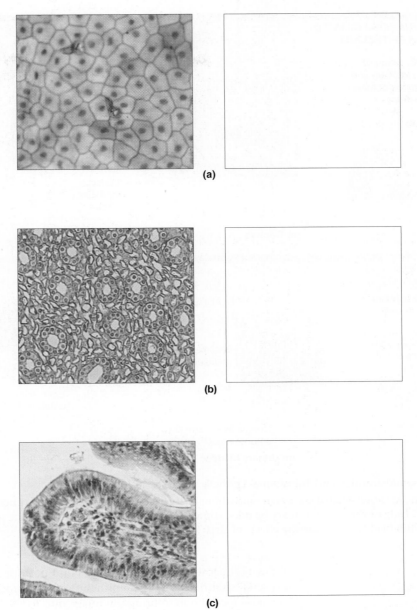

(a)

(b)

(c)

3. Draw each tissue in the space provided in Figure 7.7.

   a. Pseudostratified (ciliated) columnar epithelium lines the trachea and bronchi of the respiratory system and parts of the male reproductive tract. Although the tissue appears stratified, its organization is actually that of simple epithelium. All cells touch the basal lamina, but some do not reach the free surface (Figure 7.6a). Between the columnar cells are goblet cells that secrete mucus onto the epithelial free surface. The mucus traps dust and other particles in the inhaled air. Cilia sweep the mucus up to the throat, where it is swallowed and disposed of in the digestive tract.

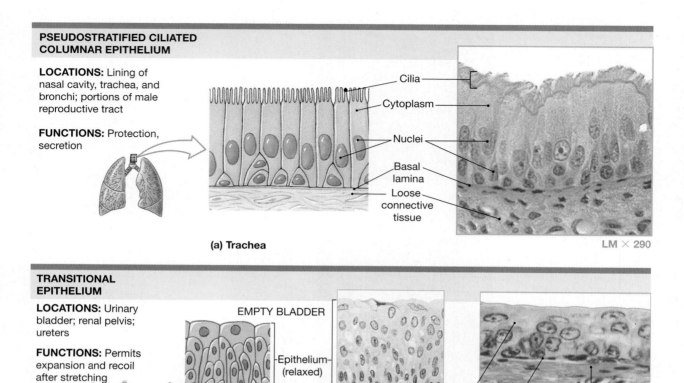

PSEUDOSTRATIFIED CILIATED COLUMNAR EPITHELIUM

LOCATIONS: Lining of nasal cavity, trachea, and bronchi; portions of male reproductive tract

FUNCTIONS: Protection, secretion

Cilia
Cytoplasm
Nuclei
Basal lamina
Loose connective tissue

(a) Trachea

LM × 290

TRANSITIONAL EPITHELIUM

LOCATIONS: Urinary bladder; renal pelvis; ureters

FUNCTIONS: Permits expansion and recoil after stretching

EMPTY BLADDER

Epithelium (relaxed)
Basal lamina
Connective tissue and smooth muscle layers

LM × 394

Epithelium (stretched)   Basal lamina   Connective tissue and smooth muscle layers

FULL BLADDER

(b) Urinary bladder

Figure 7.6 Pseudostratified and Transitional Epithelia
(a) The pseudostratified ciliated columnar epithelium of the respiratory tract. Note the uneven layering of the nuclei. **(b)** *Left:* The lining of the empty urinary bladder, showing a transitional epithelium in the relaxed state. *Right:* The lining of the full bladder, showing the effects of stretching on the appearance of cells in the epithelium.

- Swing the low-power objective lens into position, and place the slide of pseudostratified columnar epithelium on the microscope stage. Rotate the coarse focus knob to bring the image into focus.

- Notice how the nuclei are unevenly distributed, creating a stratified appearance. At high magnification, slowly rock the fine focus knob back and forth approximately one-quarter turn and examine the tissue surface for cilia.

- Examine how the epithelial cells are attached to the underlying connective tissue. Can you see the basal lamina?

- Identify the goblet cells interspersed between the columnar cells. Be sure to include a few goblet cells in your sketch.

b. Transitional epithelium lines organs, such as the urinary bladder, that must stretch and shrink (Figure 7.6b). The cells have a variety of shapes and sizes, and not all of them touch the basal lamina. Most transitional-tissue slides are prepared from relaxed transitional tissue, and the tissue appears thick, with many cells stacked one upon another. If the organ is stretched, the transitional epithelium is thin.

- Move the low-power objective lens into position, and place the transitional epithelium slide on the stage. Use the coarse focus knob to bring the image into focus.

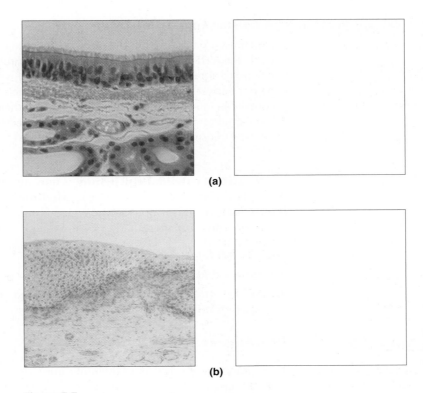

(a)

(b)

Figure 7.7
Pseudostratified and Transitional Epithelia
(a) Pseudostratified epithelium (LM × 100) and sketch. **(b)** Transitional epithelium (LM × 100) and sketch.

- Observe the thickness of the epithelium. Is your slide specimen made from an empty, relaxed bladder or from a full, stretched bladder?
- Describe the shape of the cells. ■

LAB ACTIVITY 3 Stratified Epithelia

Stratified epithelium is a multilayered tissue with only the bottom layer of cells in contact with the basal lamina and only the upper cells exposed to the free surface (Figure 7.8). The tissue occurs in areas exposed to abrasion and friction, such as the body surface and upper digestive tract. A variety of cell shapes occur in stratified epithelium, and the type of cells in the free surface layer is used to describe and classify the tissue.

Stratified squamous epithelium forms the epidermis. In the epidermis, new cells form at the basal lamina and are pushed toward the surface by the next group of new cells. The epithelial cells manufacture the protein **keratin** (KER-a-tin; *keros*, horn), which toughens the cells but also kills them. The cells then dehydrate and interlock into a broad sheet, forming a dry protective barrier against abrasion, friction, chemical exposure, and even infection. Stratified squamous epithelium of the skin is thus said to be **keratinized** and has a dry surface. Stratified squamous epithelium also lines the tongue, mouth, pharynx, esophagus, anus, and vagina. The lining epithelium in these regions is kept moist by lining cells on the tissue surface. This moist tissue is **nonkeratinized** stratified squamous epithelium.

Clinical Application

A Barrier Against Infection

Infectious organisms enter the body by penetrating lining epithelia and covering epithelia. AIDS, caused by the HIV virus, is transmitted when the virus enters a wound or abrasion in the epithelium. The purpose of "safe sex" is to maintain a barrier between your epithelia and other people's body fluids, which may contain an infectious biological agent. The epithelial linings of the mouth, vagina, urethra, and rectum should all be protected to prevent exposure to a sexually transmitted disease. ▶

Stratified cuboidal epithelium is uncommon. It is found in ducts of certain sweat glands. **Stratified columnar epithelium** is found only in parts of the mammary glands, in ducts from salivary gland, and in small regions of the pharynx, epiglottis, anus, and urethra.

QuickCheck Questions

3.1 How is stratified epithelium organized and classified?

3.2 What is the difference between keratinized and nonkeratinized epithelium?

3 Materials

☐ Prepared microscope slides of stratified squamous epithelium, stratified cuboidal epithelium, and stratified columnar epithelium

Procedures

1. Examine each stratified epithelium under the microscope at low, medium, and high magnification.

2. Refer to the photomicrographs in Figures 7.8 and 7.9 and locate the featured structures.

3. Draw each tissue in the space provided in Figure 7.9.

   a. Stratified squamous epithelium is usually stained red or purple on a microscope slide. Close observation reveals that not all cells are squamous and that some cuboidal and columnar cells are in the middle layers. Cells near the free surface of the tissue, however, are squamous. As usual, the bottom of the tissue is attached to the underlying connective tissue by the basal lamina.

      • Move the low-power objective lens into position, and place the slide on the stage. Use the coarse focus knob to bring the image into focus.

      • Locate the stratified squamous cells at the free surface. This is the light red or purple band of tissue.

      • Examine the tissue at medium and high powers. Are the cells all the same shape?

      • How many layers of cells are visible?

   b. Stratified cuboidal epithelium lines the ducts of sweat glands. The tissue is normally only two cell layers thick.

      • Move the low-power objective lens into position, and place the slide on the stage. Use the coarse focus knob to bring the image into focus.

      • Locate a small duct of a sweat gland. With its thick wall, it will look like a donut.

      • Is the basal lamina visible?

      • How many layers of cells are visible?

STRATIFIED SQUAMOUS EPITHELIUM

LOCATIONS: Surface of skin; lining of mouth, throat, esophagus, rectum, anus, and vagina

FUNCTIONS: Provides physical protection against abrasion, pathogens, and chemical attack

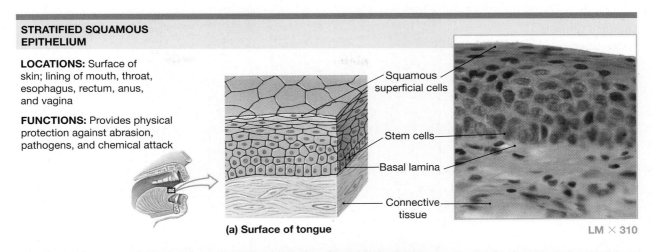

Squamous superficial cells

Stem cells

Basal lamina

Connective tissue

(a) Surface of tongue

LM × 310

STRATIFIED CUBOIDAL EPITHELIUM

LOCATIONS: Lining of some ducts (rare)

FUNCTIONS: Protection, secretion, absorption

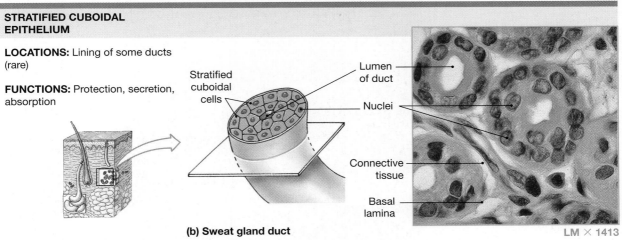

Stratified cuboidal cells

Lumen of duct

Nuclei

Connective tissue

Basal lamina

(b) Sweat gland duct

LM × 1413

STRATIFIED COLUMNAR EPITHELIUM

LOCATIONS: Small areas of the pharynx, epiglottis, anus, mammary gland, salivary gland ducts, and urethra

FUNCTION: Protection

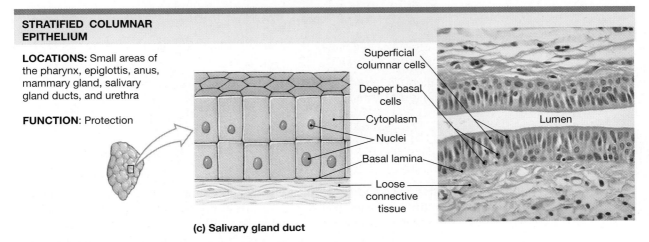

Superficial columnar cells

Deeper basal cells

Cytoplasm

Nuclei

Basal lamina

Loose connective tissue

Lumen

(c) Salivary gland duct

Figure 7.8 Stratified Epithelia

(a) Sectional and diagrammatic views of the stratified squamous epithelium that covers the tongue. **(b)** A sectional view of the stratified cuboidal epithelium that lines a sweat-gland duct in the skin. **(c)** A stratified columnar epithelium occurs along some large ducts, such as this salivary-gland duct.

Figure 7.9
Stratified Epithelia

(a) Stratified squamous epithelium (LM × 100).
(b) Stratified cuboidal epithelium (LM × 100). (c) Stratified columnar epithelium (LM × 100).

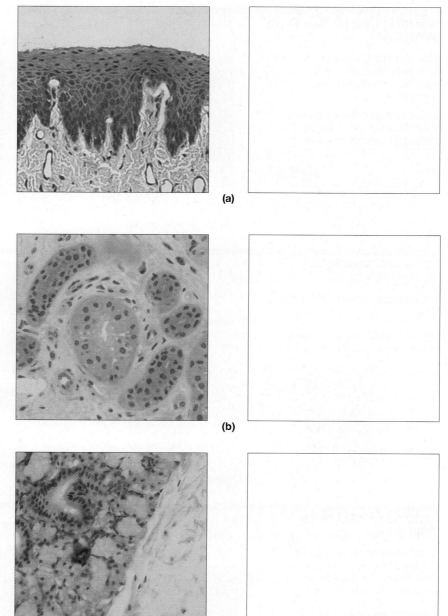

(a)

(b)

(c)

c. Stratified columnar epithelium is found lining the ducts of salivary glands, in the mammary glands, and in parts of the digestive tract. The tissue is normally only a few cell layers thick.

- Move the low-power objective lens into position, and place the slide on the stage. Use the coarse focus knob to bring the image into focus.
- Locate a small duct of a salivary gland. With its thick wall, it will look like a donut.
- Is the basal lamina visible?
- How many layers of cells are visible? ■

Epithelial Tissue

Name _____

Date _____

Section _____

A. Matching

Match each term in the left column with its correct description from the right column.

_____ 1. simple cuboidal epithelium
_____ 2. stratified squamous epithelium
_____ 3. transitional epithelium
_____ 4. simple columnar epithelium
_____ 5. simple squamous epithelium

A. lines urinary bladder
B. forms serous membranes
C. contains goblet cells
D. forms top layer of skin
E. forms tubules in kidneys

B. Fill in the Blanks

Complete the following statements:

1. Epithelium that occurs in a single layer is _____ epithelium.
2. The tissue under epithelium is _____.
3. Epithelium that stretches and relaxes is _____ epithelium.
4. The section of the basal lamina that contains glycoproteins and attaches epithelium to the underlying connective tissue is the _____.
5. Cells that secrete mucus are called _____.

C. Short-Answer Questions

1. Describe the different kinds of cells and layering of epithelia.

2. Compare the function of simple epithelium with that of stratified epithelium.

3. What type of epithelium occurs where organs must stretch and expand?

4. How is epithelium attached to underlying tissue?

LABORATORY REPORT

7

D. Completion

Complete Table 7.1 by filling in the empty boxes.

Table 7.1 **Summary of Epithelial Tissues**

		Description	*Location*	*Function*
Simple Epithelia	Simple squamous epithelium	Single layer of scalelike cells packed close together.	Alveoli of lungs, serous membranes	1.
	2.	Cube-shaped cells with large central nucleus, organized into tubules. Basal lamina may be visible.	3.	Secretion and absorption
	Simple columnar epithelium (nonciliated)	4.	5.	Absorption of nutrients and secretion of digestive juices
	6.	Single layer of cells all touching basal lamina but not all touching free surface. Nuclei distributed throughout, causing stratified appearance.	Trachea of respiratory system (ciliated), ducts of male reproductive system	7.
	Transitional epithelium	Single layer of cells all touching basal lamina but not all touching free surface. Nuclei distributed to cause stratified appearance.	8.	9.
Stratified Epithelia	Stratified squamous epithelium	10.	11.	Protection of underlying structures
	Stratified cuboidal epithelium	12.	Rare; may line sweat-gland ducts	Protection, secretion
	13.	Multiple layers of columnar cells.	14.	Protection

E. Analysis and Application

1. Suppose you are examining the inner lining of the stomach. Describe the tissue you are observing and relate its structure to its function in that location.

2. Compare and contrast the epithelium covering the skin with the epithelium lining the inner surface of the cheeks.

Connective Tissues

OBJECTIVES

On completion of this exercise, you should be able to:

- List the major types of connective tissue and the characteristics of each.
- Discuss the composition of the matrix of each type of connective tissue.
- Describe the structure and function of embryonic connective tissue.
- Describe the location and function of each type of connective tissue.
- Identify each type of connective tissue and its cell and matrix structure under the microscope.

Connective tissue provides the body with structural support and with a means of joining various structural components to one another. Unlike the cells in epithelia, cells in connective tissue are widely scattered throughout the tissue. These cells produce and secrete protein **fibers** and a **ground substance** that together form an extracellular **matrix**. The ground substance is composed mainly of glycoprotein and polysaccharide molecules that surround the cells as either a thick, syrupy liquid; a gelatinous layer; or a solid, crystalline material. Suspended in the ground substance are **collagen fibers**, which give tissues strength, and **elastic fibers**, which provide flexibility. As we age, cells secrete fewer protein fibers into the matrix, resulting in brittle bones and wrinkled skin. Leather is mostly collagen fibers from the dermis of animal skins that have been tanned and preserved. **Reticular fibers** are interwoven proteins found in reticular connective tissue; they provide a framework for support of internal soft organs, such as the liver and spleen.

The matrix of a connective tissue determines the physical nature of the tissue. Blood, for example, has a liquid matrix called *blood plasma* that allows the blood to flow freely through vessels. Fat tissue has a thick liquid matrix that is syrupy, like honey. Cartilage has a thick, gelatinous matrix that allows this connective tissue to slide easily over other structures. Bone has a solid matrix and provides the structural framework for the body.

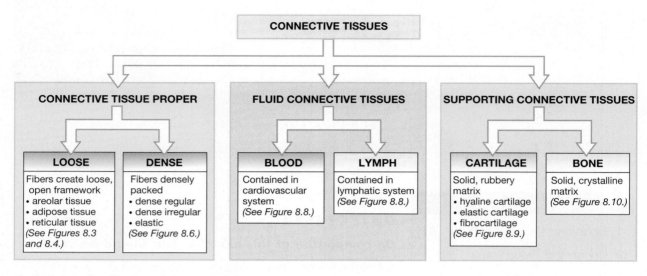

Figure 8.1 A Classification of Connective Tissues

Connective tissues are classified into three broad groups, distinguished primarily by cellular composition and the characteristics of the extracellular matrix (Figure 8.1). **Connective tissue proper** has a thick liquid matrix and a variety of cell types. **Fluid connective tissues** are liquid tissues that flow through blood vessels and lymphatic vessels. **Supportive connective tissues** have a strong gelatinous or solid matrix that acts as support for other tissues.

All connective tissue is produced in the embryo from an unspecialized tissue called **mesenchyme**. Cells in this embryonic connective tissue differentiate into cartilage, bone, and other types of connective tissue. Each tissue group is discussed in the following activities. Figure 8.1 and Table 8.1 present an overview of the connective tissue groups.

Table 8.1 Summary of Connective Tissues

Tissue	Description
Connective Tissue Proper	**Syrup matrix, various cell types**
Areolar connective tissue	Fibroblasts and mast cells, matrix with collagen and elastic fibers
Adipose tissue	Adipocytes with contents pushed to edge of cells
Reticular tissue	Network of reticular fibers supporting cells of liver, spleen, and lymph nodes
Dense connective tissue	Fibroblasts located between parallel bundles of yellow collagen fibers
Fluid Connective Tissues	**Fluid matrix, circulates in vessels**
Blood	Red blood cells transport respiratory gases, white cells provide immunity, platelets clot blood
Lymph	Fluid matrix with scattered lymphocytes
Supportive Connective Tissues	**Surrounded by outer cell-producing membrane, cells trapped in lacunae in matrix**
Hyaline cartilage	Perichondrium with chondroblasts, chondrocytes in lacunae
Elastic cartilage	Perichondrium with chondroblasts, chondrocytes in lacunae, elastic fibers visible in matrix
Fibrocartilage	Chondrocytes stacked up in columns within lacunae
Bone tissue	Concentric lamellae form osteons surrounding central canals, osteocytes trapped in lacunae, canaliculi interconnect osteocytes to central canal

LAB ACTIVITY 1   Embryonic Connective Tissue

Early in the third week of embryonic development, mesenchyme tissue (Figure 8.2) appears and produces the specialized cells to construct mature connective tissues. For example, mesenchyme differentiates into bone-producing cells in the area where the skeleton is forming. In the adult body, connective tissue contains only a few mesenchyme cells, which assist in tissue repair.

Mesenchyme tissue is a loose meshwork of many star-shaped cells. Unlike adult connective tissue, mesenchyme has no visible protein fibers in the ground substance that makes up the matrix. **Mucous connective tissue**, also called **Wharton's jelly**, is embryonic connective tissue of the umbilical cord and other parts of the embryo.

QuickCheck Questions

1.1 Why is mesenchyme more abundant in embryos and infants than in adults?

1.2 What type of mesenchyme is found in the umbilical cord?

1 *Materials*

☐ Compound microscope
☐ Prepared microscope slide of mesenchyme

Procedures

1. Place the slide of mesenchyme on the microscope stage and move the low-power objective lens into position. Rotate the coarse focus knob to bring the image into focus.

2. Scan the tissue at low power, and then increase the magnification first to medium and then to high.

   a. What is the shape of the mesenchymal cells?

   b. Are fibers visible in the matrix?

3. Compare the microscope image of the tissue with Figure 8.2. ■

LAB ACTIVITY 2   Connective Tissue Proper

Connective tissue proper includes two groups of tissues: loose and dense. **Loose connective tissue** has an open network of protein fibers in a thick, syrupy ground substance. Areolar, adipose, and reticular tissues are loose connective tissue. **Dense connective tissue** has thick bundles of collagen and elastic fibers with widely scattered cells. There are three types of dense connective tissue: dense regular, dense irregular, and elastic.

Figure 8.2 Connective Tissues in Embryos

(a) Mesenchyme, the first connective tissue to appear in an embryo.
(b) Mucous connective tissue from the umbilical cord of a fetus. Mucous connective tissue in this location is also known as *Wharton's jelly.*

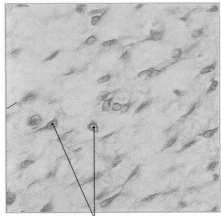

Mesenchymal cells   LM × 136

(a) Mesenchyme

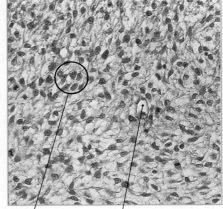

Mesenchymal cells   Blood vessel   LM × 136

(b) Mucous connective tissue

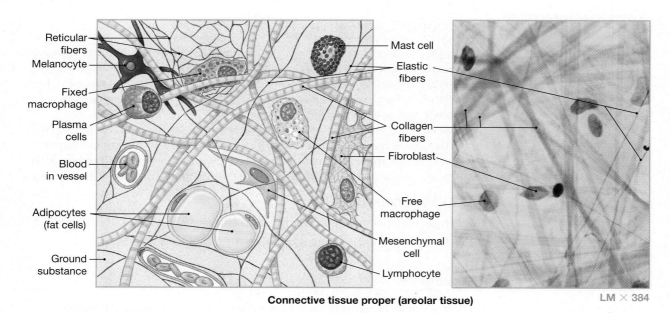

Connective tissue proper (areolar tissue)

LM × 384

Figure 8.3 The Cells and Fibers of Connective Tissue Proper
Diagrammatic and histological views of the cell types and fibers of connective tissue proper.

As shown in Figure 8.3, connective tissue proper contains a variety of cell types in addition to the collagen fibers and elastic fibers described above. **Fibroblasts** (FĪ-brō-blasts) are fixed (stationary) cells that secrete proteins that join other molecules in the matrix to form the collagen and elastic fibers. Phagocytic **macrophage** (MAK-rō-fā-jez; *phagein,* to eat) cells patrol these tissues, ingesting microbes and dead cells. Macrophages are mobilized during an infection or injury, migrate to the site of disturbance, and phagocytize damaged tissue cells and microbes. **Mast cells** release histamines that cause an inflammatory response in damaged tissues. **Adipocytes** (AD-i-pō-sīts) are fat cells and contain vacuoles for the storage of lipids.

Clinical Application

Liposuction

The surgical procedure called *liposuction* removes unwanted adipose tissue with a suction wand. The treatment is dangerous and may damage blood vessels or nerves near the site of fat removal. Overlying skin may appear pocketed and marbled after the procedure. Considering that connective tissues contain mesenchyme cells, what would most likely occur if a liposuction patient consumed excessive calories after the surgery? ▶

QuickCheck Questions

2.1 Which cell in connective tissue proper manufactures the protein fibers for the matrix?

2.2 What fiber types are common in the matrix of connective tissue proper?

2 *Materials*

☐ Compound microscope

☐ Prepared microscope slides of areolar connective tissue, adipose tissue, dense regular connective tissue, dense irregular connective tissue, elastic tissue, and reticular tissue

Procedures

1. **Areolar connective tissue** is distributed throughout the body. This tissue fills spaces between structures for support and protection, much as packing material surrounds an object in a box. For example, muscles are separated from the surrounding anatomy by areolar connective tissue. This tissue is very flexible and permits muscles to move freely without pulling on the skin. Figure 8.3 details the organization of areolar connective tissue. Most of the cells are oval-shaped fibroblasts that usually stain light. Mast cells are small and filled with dark-stained granules of histamine and heparin which cause inflammation. Collagen and elastic fibers are clearly visible in the matrix.

   a. Scan the slide of areolar connective tissue at low and medium magnifications.

   b. Which types of fibers do you see?

   c. Identify the fibroblasts, mast cells, and macrophages.

   d. How can you distinguish each cell type?

   e. In Figure 8.5, draw a section of areolar connective tissue at medium magnification. Label both types of extracellular fibers and at least two types of cells.

2. **Adipose tissue** (also called *fat tissue*) is distributed throughout the body and is abundant under the skin and in the buttocks, breasts, and abdomen. Adipocytes are packed more closely together than are the cells in other types of connective tissue proper (Figure 8.4a). The distinguishing feature of adipose tissue is displacement of the nucleus and cytoplasm due to the storage of lipids. When an adipocyte stores fat, its vacuole expands with lipid and fills most of the cell while pushing the organelles and cytosol to the periphery. With the cytoplasm at the edge of the cells, adipocytes look like graduation rings.

   a. Scan the slide of adipose tissue at low and medium magnifications.

   b. How are the cells arranged?

   c. Describe the distribution of cytoplasm in the cells.

   d. At high magnification, observe individual adipocytes.

   e. Draw several cells in the space provided in Figure 8.5.

3. **Reticular tissue**, shown in Figure 8.4b, forms the internal framework for soft organs, such as the spleen, liver, and lymphatic organs. The tissue is composed of an extensive network of reticular fibers with small, oval reticulocytes.

   a. Observe reticular tissue at low, medium, and high magnifications.

   b. Draw a section of the tissue in the space provided in Figure 8.5 and label the reticular fibers and reticulocytes.

4. **Dense regular connective tissue** (Figure 8.6a) contains small fibroblasts between parallel bundles of thick collagen fibers. This strong tissue forms tendons, which join muscle to bone, and ligaments, which join bone to bone. Because tendons and ligaments conduct pulling forces mainly from one direction, the collagen fibers in dense regular tissues are parallel. Flat layers of

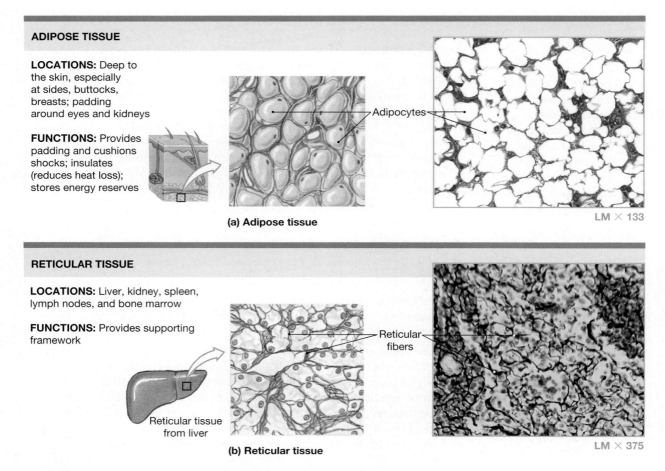

ADIPOSE TISSUE

LOCATIONS: Deep to the skin, especially at sides, buttocks, breasts; padding around eyes and kidneys

FUNCTIONS: Provides padding and cushions shocks; insulates (reduces heat loss); stores energy reserves

Adipocytes

(a) Adipose tissue

LM × 133

RETICULAR TISSUE

LOCATIONS: Liver, kidney, spleen, lymph nodes, and bone marrow

FUNCTIONS: Provides supporting framework

Reticular tissue from liver

Reticular fibers

(b) Reticular tissue

LM × 375

Figure 8.4 Adipose and Reticular Tissues
(a) Adipose tissue is a loose connective tissue dominated by adipocytes. In standard histological preparations, the tissue looks empty because the lipids in the fat cells dissolve in the alcohol used in tissue processing. (b) Reticular tissue has an open framework of reticular fibers, which are usually very difficult to see because of the large numbers of cells around them.

dense connective tissue called *fascia* protect and isolate muscles from surrounding structures and allow muscle movement. On slides with limited stain, the profusion of collagen fibers makes this tissue appear yellow under the microscope.

a. Examine dense regular connective tissue at low and medium magnifications.

b. Describe the composition of the matrix.

c. How are the fibroblasts arranged in this tissue?

d. Draw and label a section of dense connective tissue in Figure 8.7.

5. **Dense irregular connective tissue** (Figure 8.6b) is a mesh of collagen fibers with interspersed fibroblasts. Irregular tissue is located in the dermis of the skin and in the layers surrounding cartilage and bone. The kidneys, liver, and spleen are protected inside a capsule of dense irregular connective tissue. With its meshwork of collagen fibers, this connective tissue supports areas that receive stress from many directions.

a. Examine dense irregular connective tissue at low and medium magnifications.

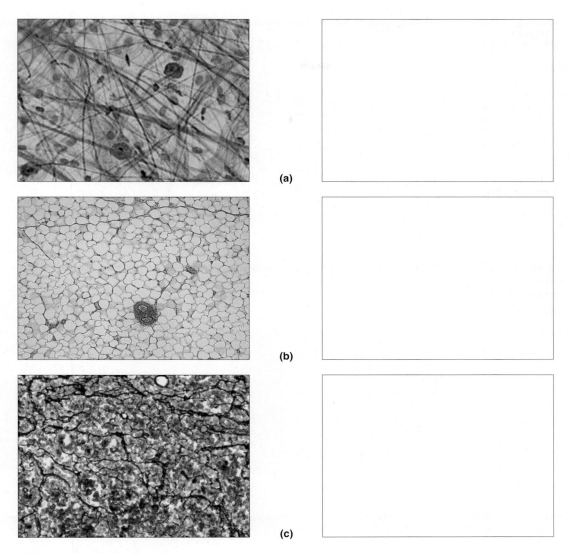

(a)

(b)

(c)

Figure 8.5 Loose Connective Tissues
(a) Areolar connective tissue (LM × 100) and sketch. **(b)** Adipose tissue (LM × 100) and sketch. **(c)** Reticular tissue (LM × 100) and sketch.

    b. Describe the organization of the collagen fibers.

    c. How are the fibroblasts arranged in the tissue?

    d. Draw and label a section of dense irregular connective tissue in Figure 8.7.

6. Elastic tissue (Figure 8.6c) is a dense regular connective tissue with elastic fibers in the matrix rather than collagen fibers. The elastic fibers are thicker than collagen fibers and are in large bundles. Elastic tissue supports the vertebrae of the spine as elastic ligaments and occurs in the blood chambers in the penis.

    a. Examine elastic tissue at low and medium magnifications.

    b. Describe the organization of the elastic fibers.

    c. How are the fibroblasts arranged in the tissue?

    d. Draw and label a section of elastic connective tissue in Figure 8.7. ■

DENSE REGULAR CONNECTIVE TISSUE

LOCATIONS: Between skeletal muscles and skeleton (tendons and aponeuroses); between bones or stabilizing positions of internal organs (ligaments); covering skeletal muscles; deep fasciae

FUNCTIONS: Provides firm attachment; conducts pull of muscles; reduces friction between muscles; stabilizes relative positions of bones

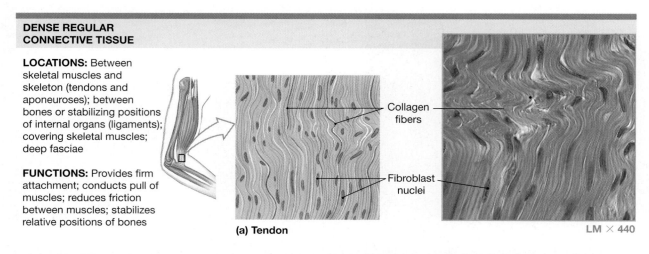

Collagen fibers

Fibroblast nuclei

(a) Tendon

LM × 440

DENSE IRREGULAR CONNECTIVE TISSUE

LOCATIONS: Capsules of visceral organs; periostea and perichondria; nerve and muscle sheaths; dermis

FUNCTIONS: Provides strength to resist forces applied from many directions; helps prevent overexpansion of organs such as the urinary bladder

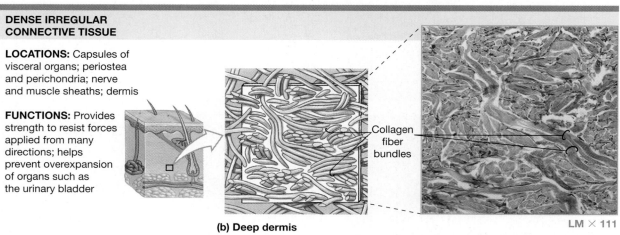

Collagen fiber bundles

(b) Deep dermis

LM × 111

ELASTIC TISSUE

LOCATIONS: Between vertebrae of the spinal column (ligamentum flavum and ligamentum nuchae); ligaments supporting penis; ligaments supporting transitional epithelia; in blood vessel walls

FUNCTIONS: Stabilizes positions of vertebrae and penis; cushions shocks; permits expansion and contraction of organs

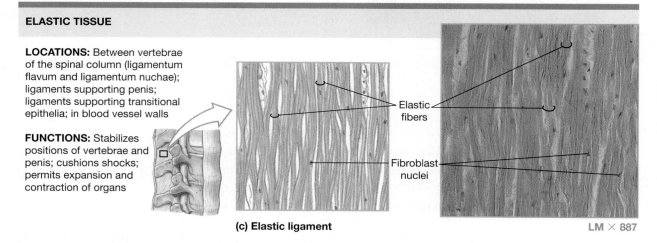

Elastic fibers

Fibroblast nuclei

(c) Elastic ligament

LM × 887

Figure 8.6 Dense Connective Tissues

(a) The dense regular connective tissue in a tendon. Notice the densely packed, parallel bundles of collagen fibers. The fibroblast nuclei can be seen flattened between the bundles. **(b)** The deep dermis of the skin contains a thick layer of dense irregular connective tissue. **(c)** An elastic ligament, an example of elastic tissue. Elastic ligaments extend between the vertebrae of the vertebral column. The bundles are fatter than those of a tendon or a ligament composed of collagen.

Figure 8.7 Dense Connective Tissues

(a) Elastic connective tissue (LM × 400).
(b) Dense irregular connective tissue (LM × 100). **(c)** Dense regular connective tissue (LM × 400).

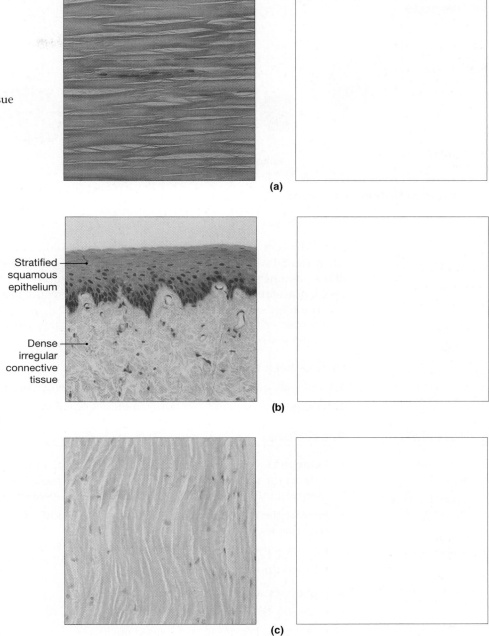

(a)

Stratified squamous epithelium

Dense irregular connective tissue

(b)

(c)

LAB ACTIVITY **3** Fluid Connective Tissue

Fluid connective tissue includes **blood** and **lymph** tissues. These tissues have a liquid matrix and circulate in blood vessels or lymphatic vessels. Blood is composed of cells collectively called the *formed elements* (Figure 8.8), which are supported in a liquid ground substance called blood **plasma**. Protein fibers are dissolved in the matrix of both blood and lymph tissues. During blood clotting, in a process called *coagulation*, fibers in blood produce a fibrin net to trap cells as they pass through the wound. Fibers in blood also regulate the viscosity, or thickness, of the blood. The formed elements are grouped into three general categories: red blood cells, white blood cells, and platelets. Red blood cells, called **erythrocytes**, transport blood

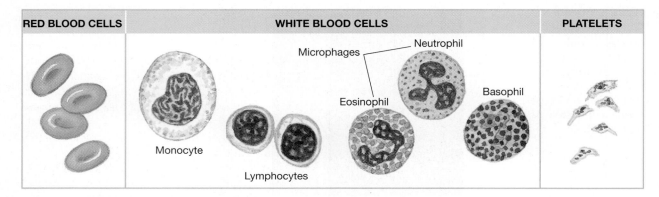

Figure 8.8 **Formed Elements of the Blood**

gases. The cells are biconcave discs, with a center so thin that it often looks hollow when viewed with a microscope. White blood cells, called **leukocytes**, are the cells of the immune system and protect the body from infection. Upon injury to a blood vessel, **platelets** become sticky and form a plug to reduce bleeding.

The most common cells of the lymphatic system are lymphocytes, which are white blood cells produced in lymphoid tissues. A detailed study of these tissues is presented in Exercises 34 and 38.

QuickCheck Questions

3.1 Why are blood and lymph classified as fluid connective tissues?

3.2 Describe the ground substance and fibers of blood.

3 Materials

- ☐ Compound microscope
- ☐ Prepared microscope slide of blood

Procedures

1. Swing the low-power objective lens into position, and then place the blood slide on the microscope stage. Rotate the coarse focus knob and bring the specimen into focus. Use the fine focus knob for precise focus.

2. Increase the magnification to medium power. Locate the numerous erythrocytes.
   - What is the shape of an erythrocyte?
   - Do erythrocytes have nuclei?

3. Leukocytes stain darker than erythrocytes.
   - Identify the leukocytes.
   - Do the leukocytes have a stained nucleus?
   - Are the leukocytes as numerous as the erythrocytes?

4. Look closely between the erythrocytes and leukocytes.
   - What are the small cell fragments between the larger blood cells?
   - What is the function of these cells? ■

LAB ACTIVITY 4   Supportive Connective Tissue

Cartilage and bone are supportive connective tissues and contain a strong matrix of fibers capable of supporting body weight and stress. **Cartilage** is a rubbery, avascular tissue with a gelatinous matrix and many fibers for structural support. Materials diffuse through the matrix to reach the cartilage cells. **Bone** has a solid matrix composed of calcium phosphate and calcium carbonate. These salts crystallize on collagen fibers and form a hard material called **hydroxyapatite**.

A membrane surrounds all supportive connective tissues to protect the tissue and supply new tissue-producing cells. The **perichondrium** (pe-rē-KON-drē-um) surrounds cartilage and produces **chondroblasts** (KON-drō-blasts; *chondros,* cartilage), which secrete the fibers and ground substance of the cartilage matrix. Eventually, chondroblasts become trapped in the matrix in small spaces called **lacunae** (la-KOO-nē, *lacus,* pool) and lose the ability to produce additional matrix. These cells are then called **chondrocytes** and function in maintenance of the mature tissue. Examine Figure 8.9 and locate the features of cartilage.

As shown in Figure 8.10, bone tissue is surrounded by the **periosteum** (pe-rē-OS-tē-um), which contains **osteoblasts** (OS-tē-ō-blasts) for bone growth and repair. Like chondroblasts, osteoblasts secrete the organic components of the matrix, become trapped in lacunae, and mature into **osteocytes**. Compact bone is characterized by the presence of columns of tissue called **osteons**. Each osteon surrounds a **central canal** containing blood vessels. Rings of matrix, called **concentric lamellae** (lah-MEL-lē; *lamella,* thin plate), surround the central canal. **Canaliculi** (kan-a-LIK-ū-lē; little canals) are small channels in the lamellae that provide passageways through the solid matrix for diffusion of nutrients and wastes. Other bone cells, called **osteoclasts**, secrete small quantities of carbonic acid to dissolve portions of the bone matrix and release calcium ions into the blood for various chemical processes.

QuickCheck Questions

4.1 Describe the matrix of cartilage.

4.2 How are cartilage and bone tissues similar to each other?

4 | Materials

- ☐ Compound microscope
- ☐ Prepared microscope slides of hyaline cartilage, elastic cartilage, fibrocartilage, and bone

Procedures

1. **Hyaline** (HĪ-uh-lin; *hyalus,* glass) **cartilage** (Figure 8.9a, b) is the most common cartilage in the body. It is located in most joints of the skeletal system, in the nasal septum, in the larynx, and in lower respiratory passageways. The gelatinous matrix provides flexible support and reduces friction between bones in a joint. Examine the micrograph of hyaline cartilage in Figure 8.9a. The tissue is distinguishable from other cartilages by the apparent lack of fibers in the matrix. Hyaline cartilage does contain elastic and collagen fibers, but they do not stain and therefore are not visible. Along the outer perimeter of the cartilage is a perichondrium containing chondroblasts. Chondrocytes occur deeper in the tissue and are surrounded by lacunae.

   a. Scan the slide of hyaline cartilage at low, medium, and high magnifications.

   b. Describe the location of the perichondrium and chondroblasts.

   c. Examine the deeper middle region of the cartilage.

   d. How do cartilage cells in the middle region differ from cells in the perichondrium?

   e. In the space provided in Figure 8.11, draw a section of hyaline cartilage and label perichondrium, chondroblasts, chondrocytes, and lacunae.

2. **Elastic cartilage** (Figure 8.9c) has many elastic fibers in the matrix and is therefore easily distinguished from hyaline cartilage. Chondrocytes are trapped in lacunae in the middle of the tissue, and the perichondrium and chondroblasts are found on the periphery. The elastic fibers permit considerable binding and twisting of the tissue. The pinna (flap) of the ear, the larynx, and the tip of the nose all contain elastic cartilage. Bend the pinna of

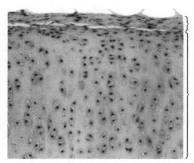

Perichondrium

Hyaline cartilage

(a)

Figure 8.9 Types of Cartilage

(a) A perichondrium separates cartilage from other tissues. **(b)** Hyaline cartilage. Note the translucent matrix and the absence of prominent fibers. **(c)** Elastic cartilage. The closely packed elastic fibers are visible between the chondrocytes. **(d)** Fibrocartilage. The collagen fibers are extremely dense, and the chondrocytes are relatively far apart.

HYALINE CARTILAGE

LOCATIONS: Between tips of ribs and bones of sternum; covering bone surfaces at synovial joints; supporting larynx (voice box), trachea, and bronchi; forming part of nasal septum

FUNCTIONS: Provides stiff but somewhat flexible support; reduces friction between bony surfaces

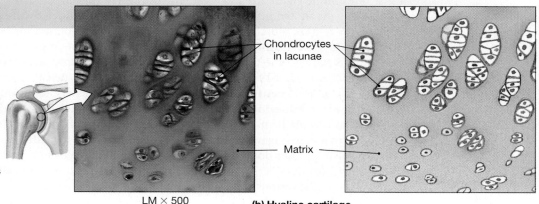

Chondrocytes in lacunae

Matrix

LM × 500

(b) Hyaline cartilage

ELASTIC CARTILAGE

LOCATIONS: Auricle of external ear; epiglottis; auditory canal; cuneiform cartilages of larynx

FUNCTIONS: Provides support, but tolerates distortion without damage and returns to original shape

Chondrocyte in lacuna

Elastic fibers in matrix

LM × 358

(c) Elastic cartilage

FIBROCARTILAGE

LOCATIONS: Pads within knee joint; between pubic bones of pelvis; intervertebral discs

FUNCTIONS: Resists compression; prevents bone-to-bone contact; limits relative movement

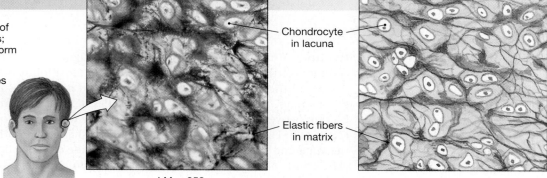

Collagen fibers in matrix

Chondrocyte in lacuna

LM × 750

(d) Fibrocartilage

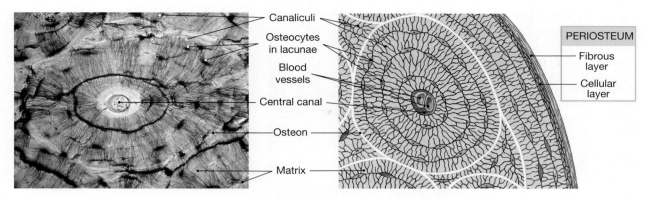

Canaliculi

Osteocytes
in lacunae

Blood
vessels

Central canal

Osteon

Matrix

PERIOSTEUM

Fibrous
layer

Cellular
layer

Figure 8.10 Bone
The osteocytes in bone are usually organized in groups around a central space that contains blood vessels. For the photomicrograph, a sample of bone was ground thin enough to become transparent. Bone dust filled the lacunae and the central canal, making them appear dark.

one of your ears and notice how the elastic cartilage returns it to the original shape after distortion.

  a. Observe the elastic cartilage slide at low, medium, and high magnifications.

  b. Identify the perichondrium with its many small chondroblasts and the chondrocytes trapped in lacunae deeper in the tissue.

  c. What is visible in the matrix of this cartilage?

  d. In the space provided in Figure 8.11, draw a section of elastic cartilage and label perichondrium, chondroblasts, chondrocytes, lacunae, and elastic fibers in the matrix.

3. **Fibrocartilage** contains irregular collagen fibers that are visible in the matrix and chondrocytes that are stacked on one another much like coins (Figure 8.9d). This cartilage is very strong and durable, and its function is to cushion joints and limit bone movement. The intervertebral disks of the spine, the pubic symphysis of the pelvis, and the pads in the knee joint are all exposed to stress forces of the skeleton. Fibrocartilage functions as a shock absorber in these areas.

  a. Observe the slide of fibrocartilage cartilage at low, medium, and high magnifications.

  b. Identify the groups of chondrocytes stacked up in lacunae.

  c. How does the arrangement of cells in fibrocartilage differ from the cell distribution in hyaline and elastic cartilages?

  d. Can you see collagen fibers in the matrix (not visible on all slide preparations)?

  e. In the space provided in Figure 8.11, draw a section of fibrocartilage and label the chondrocytes.

4. Bone tissue is uniquely organized into osteons, which are columns of bone tissue that are distinct when viewed microscopically (Figure 8.10). Osteocytes occur in the margins between concentric lamellae. The canaliculi branch through the solid matrix and connect with the central canal, where blood vessels are located. The main functions of bone are support of the body, attachment of skeletal muscles, and protection of internal organs. Bone tissue is studied in more detail in laboratory Exercise 13.

  a. Scan the bone slide at low and medium magnifications to observe the pattern of osteons.

Figure 8.11 Supportive Connective Tissue
(a) Hyaline cartilage (LM × 100) and sketch. (b) Elastic cartilage (LM × 100) and sketch. (c) Fibrocartilage (LM × 100) and sketch. (d) Bone tissue (LM × 100) and sketch.

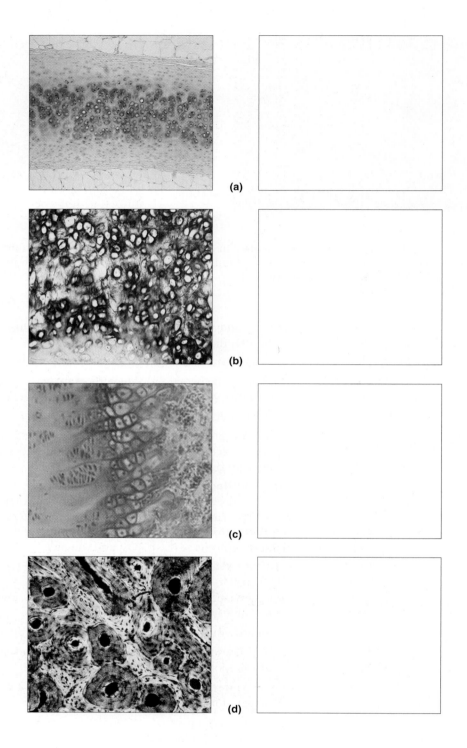

(a)

(b)

(c)

(d)

b. Increase the magnification to high, and observe the detailed structure of a single osteon.

c. Identify the rings of tissue of the osteon. What are these rings called?

d. Locate the center of an osteon. What is the hole called?

e. How is bone tissue similar and dissimilar to cartilage tissue?

f. In the space provided in Figure 8.11, draw several osteons and label central canal, canaliculi, lamellae, and osteocytes. ■

Connective Tissues

Name _____

Date _____

Section _____

A. Matching

Match each term in the left column with its correct description from the right column.

_____ 1. collagen fiber   **A.** extracellular material

_____ 2. perichondrium   **B.** column of bone tissue

_____ 3. osteon   **C.** produces matrix fibers

_____ 4. lacuna   **D.** nutrient channels in bone matrix

_____ 5. fibroblast   **E.** protein fiber for flexibility

_____ 6. matrix   **F.** small space surrounding cell

_____ 7. elastic fiber   **G.** outer membrane of cartilage

_____ 8. ground substance   **H.** causes inflammation

_____ 9. mast cell   **I.** syrupy fluid of matrix

_____ 10. canaliculi   **J.** protein fiber for strength

B. Identification

Identify the connective tissue described in each statement.

1. Cells in which nucleus and cytoplasm are pushed against cell membrane _____

2. Cell types in this tissue include mast cells, fibroblasts, and macrophages _____

3. Solid matrix, lamellae surrounding central canals _____

4. Found in intervertebral disks, chondrocytes stacked inside lacunae _____

5. Parallel bundles of collagen fibers with fibroblasts between fibers _____

6. Gelatinous matrix, chondrocytes in lacunae, elastic fibers in matrix _____

7. Many cells among network of reticular fibers _____

C. Completion

List the cell type(s) found in the following connective tissues.

1. hyaline cartilage _____

2. bone tissue _____

3. adipose tissue _____

4. areolar connective tissue _____

5. elastic cartilage _____

6. dense regular connective tissue _____

7. reticular tissue _____

D. Short-Answer Questions

1. Which tissue in the embryo is a precursor of adult connective tissue?

2. What type of fibers are embedded in loose connective tissue?

3. What is the matrix composed of in elastic cartilage?

4. List the three major groups of connective tissue, and give an example of each.

E. Application and Analysis

1. Why are tendons made of dense regular connective tissue but the dermis contains dense irregular connective tissue?

2. How does connective tissue differ from epithelial tissue?

3. Describe the ground substance and fibers found in each type of connective tissue.

Muscle Tissue

OBJECTIVES

On completion of this exercise, you should be able to:

- List the three types of muscle tissue and describe a function of each.

- Describe the histological appearance of each type of muscle tissue.

- Identify each type of muscle tissue in microscope preparations.

There are three types of muscle tissue, each named for its location in the body. **Skeletal muscle** is attached to bone and provides the means by which the body skeleton moves, as in walking or moving the head. **Cardiac muscle** forms the walls of the heart and pumps blood through the vascular system. **Smooth**, or **visceral**, **muscle** is found inside hollow organs, such as the stomach, intestines, blood vessels, and uterus; this muscle type controls such functions as the movement of material through the digestive system, the diameter of blood vessels, and uterine contraction during labor.

Muscle tissue specializes in contraction. Muscle cells shorten during contraction, and this shortening produces a force, or tension, that causes movement. During the contraction phase of a heartbeat, for example, blood is pumped into blood vessels. The pressure generated by cardiac muscle contraction forces blood to flow through the vascular system to supply cells with oxygen, nutrients, and other essential materials.

LAB ACTIVITY **1** Skeletal Muscle

Skeletal muscle tissue is attached to bones of the skeleton by **tendons** made of dense regular connective tissue proper. When skeletal muscle tissue contracts, it pulls on a tendon that, in turn, pulls and moves a bone. The functions of skeletal muscle tissue include movement for locomotion, facial expressions, and speech; maintenance of body posture and tone; and heat production during shivering.

Figure 9.1a highlights skeletal muscle tissue. The tissue is composed of long structures called **muscle fibers**. During development, a number of embryonic cells called **myoblasts** (*myo-,* muscle + *-blast,* precursor) fuse into one large cellular structure that is the muscle fiber; because each fiber forms from numerous embryonic cells, it has many nuclei and is said to be **multinucleated**. The nuclei are clustered under the **sarcolemma** (sar-cō-LEM-uh; *sarco,* flesh), which is the muscle fiber's cell membrane. Muscle fibers are **striated** with a distinct banded pattern resulting from the repeating organization of internal contractile proteins called **filaments**. Skeletal muscle tissue may be consciously stimulated to contract and is therefore under **voluntary control**.

Study Tip

Muscle Terminology

In reference to muscle control, the terms *voluntary* and *involuntary* are used more for convenience than for description. Skeletal muscle is said to be voluntary, and yet you cannot stop your muscles from shivering when you are cold. Additionally, once you voluntarily start a muscle contraction, the brain assumes control of the muscle activity. The heart muscle is involuntary, but some individuals can control their heart rate. Generally, the term *voluntary* is associated with skeletal muscle and *involuntary* refers to cardiac and smooth muscle tissues.

The terms muscle fiber and muscle cell might also seem confusing. These two terms mean essentially the same thing, but the general convention is to say "muscle fiber" when referring to skeletal muscles and "muscle cell" when referring to cardiac and smooth muscles. We follow that convention in this manual. ●

QuickCheck Questions

1.1 Where is skeletal muscle tissue located in the body?

1.2 What are the functions of skeletal muscle tissue?

1 Materials

Procedures

- ☐ Compound microscope
- ☐ Skeletal muscle slide (striated muscle or voluntary muscle)

1. Place the slide of skeletal muscle tissue on the microscope stage and move the low-power objective lens into position. Rotate the coarse adjustment knob to bring the image into focus.

2. Using Figures 9.1 and 9.2 for reference, examine the tissue at low magnification.
   - Identify an individual skeletal muscle fiber.
   - How many nuclei does it have?

3. Change to medium magnification and observe the skeletal muscle fibers again.
   - Can you see striations across the fibers?
   - How are the nuclei positioned in the fibers?

4. If both transverse and longitudinal sections are on your slide, compare the appearance of the skeletal muscle fibers in the two sections.
   - How do the muscle fibers appear in transverse section?
   - How are the nuclei positioned in the fibers?

5. Draw and label the microscopic structure of skeletal muscle tissue in the space provided in Figure 9.2. ■

SKELETAL MUSCLE TISSUE

Cells are long, cylindrical, striated, and multinucleate; generally referred to as muscle fibers

LOCATIONS: Combined with connective tissues and neural tissue in skeletal muscles

FUNCTIONS: Moves or stabilizes the position of the skeleton; guards entrances and exits to the digestive, respiratory, and urinary tracts; generates heat; protects internal organs

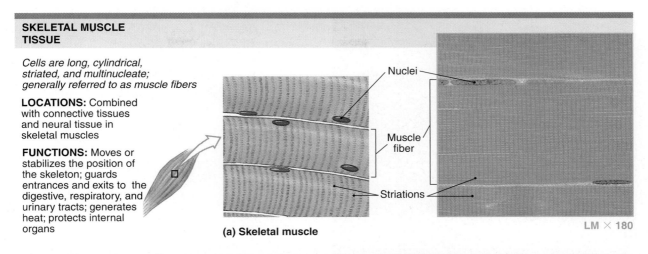

Nuclei

Muscle fiber

Striations

(a) Skeletal muscle

LM × 180

CARDIAC MUSCLE TISSUE

Cells are short, branched, and striated, usually with a single nucleus; cells are interconnected by intercalated discs

LOCATION: Heart

FUNCTIONS: Circulates blood; maintains blood (hydrostatic) pressure

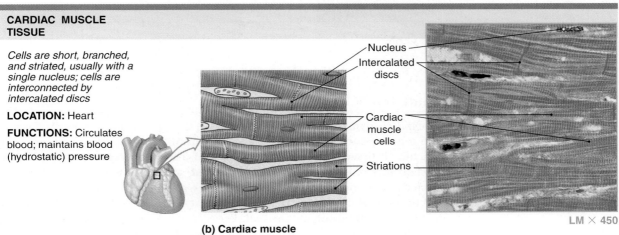

Nucleus

Intercalated discs

Cardiac muscle cells

Striations

(b) Cardiac muscle

LM × 450

SMOOTH MUSCLE TISSUE

Cells are short, spindle-shaped, and nonstriated, with a single, central nucleus

LOCATIONS: Found in the walls of blood vessels and in digestive, respiratory, urinary, and reproductive organs

FUNCTIONS: Moves food, urine, and reproductive tract secretions; controls diameter of respiratory passageways; regulates diameter of blood vessels

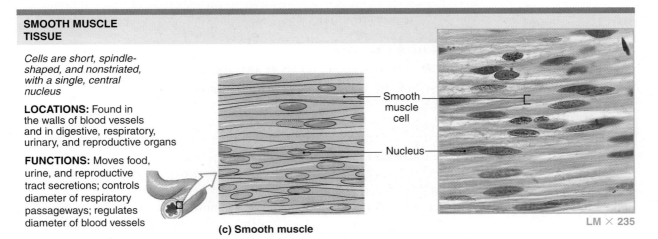

Smooth muscle cell

Nucleus

(c) Smooth muscle

LM × 235

Figure 9.1 Muscle Tissue

(a) Skeletal muscle fibers are large; have multiple, peripherally located nuclei; and exhibit prominent striations (banding) and an unbranched arrangement. **(b)** Cardiac muscle cells differ from skeletal muscle fibers in three major ways: size (cardiac muscle cells are smaller), organization (cardiac muscle cells branch), and number and location of nuclei (a typical cardiac muscle cell has one centrally placed nucleus). Both skeletal and cardiac muscle cells contain actin and myosin filaments in an organized array that produces striations. **(c)** Smooth muscle cells are small and spindle-shaped, with a central nucleus. They do not branch or have striations.

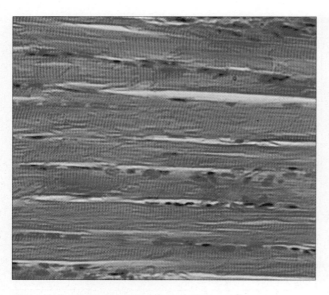

Figure 9.2 Skeletal Muscle Tissue (LM × 100)

LAB ACTIVITY 2 Cardiac Muscle

Cardiac muscle tissue occurs only in the walls of the heart. Compare the cardiac and skeletal muscle tissues in Figure 9.1a and 9.1b. Cardiac muscle tissue is striated like skeletal muscle tissue. Each **cardiac muscle cell**, also called a **cardiocyte**, has a single nucleus (the cell is said to be **uninucleated**) and is branched. Cardiocytes are connected to one another by **intercalated discs**, special gap junctions that conduct contraction stimuli from one cardiocyte to the next. Unlike skeletal muscles, cardiac muscle is under **involuntary control**. For example, when you exercise, nerves of the autonomic nervous system cause an increase in your heart rate in order to deliver more blood to the active tissues. When you relax or sleep, the autonomic nervous system lowers your heart rate.

QuickCheck Questions

2.1 What is the function of cardiac muscle tissue?

2.2 How are cardiocytes connected to one another?

---

2 Materials

☐ Compound microscope
☐ Cardiac muscle slide

Procedures

1. Place the slide of cardiac muscle tissue on the microscope stage and swing the low-power objective lens into position. Rotate the coarse adjustment knob to bring the image into focus.

2. Using Figures 9.1 and 9.3 for reference, examine the heart muscle at low and medium magnifications. If your cardiac muscle slide has different sections, observe the longitudinal section first.

   • How many nuclei are in each cardiac muscle cell?

   • How do cardiac muscle cells compare in size with skeletal muscle fibers?

3. Increase the magnification to high and observe several cardiocytes.

   • Do you see striations and branching?

   • What structure connects adjacent cells?

   • Draw and label the microscopic structure of cardiac muscle tissue in the space provided in Figure 9.3. ■

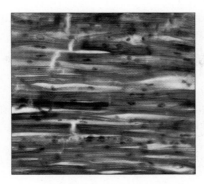

Figure 9.3 **Cardiac Muscle Tissue (LM × 100)**

LAB ACTIVITY 3 Smooth Muscle

Figures 9.1c and 9.4 show smooth muscle tissue. The muscle cells are **nonstriated** and lack the bands found in skeletal and cardiac muscle tissue. Each smooth muscle cell is **uninucleated** and spindle-shaped, thick in the middle and tapered at the ends like a toothpick. The tissue usually occurs in double sheets of muscle with one sheet positioned at a right angle to the other. This arrangement permits the tissue to shorten structures and decrease the diameter of vessels and passageways. Smooth muscle is under involuntary control.

QuickCheck Questions

3.1 Where is smooth muscle tissue located in the body?

3.2 How are smooth muscle cells different from skeletal muscle fibers and cardiac muscle cells?

3 *Materials*

☐ Compound microscope
☐ Smooth muscle slide (visceral muscle; tissues may be teased apart)

Procedures

1. Place the slide of smooth muscle tissue on the microscope stage and swing the low-power objective lens into position. Rotate the coarse adjustment knob to bring the image into focus.

2. Using Figures 9.1 and 9.4 for reference, examine the smooth muscle tissue at low magnification. Figure 9.4 shows smooth muscle that has been teased apart to separate the cells.

Figure 9.4 **Smooth Muscle Tissue teased (LM × 100)**

3. Locate the smooth muscle cells and center them in the microscope field. Increase the magnification to medium power.

- Where is the nucleus located in a typical cell?

- What is the shape of the smooth muscle cells?

- Do you see any striations?

- Draw and label the microscopic structure of smooth muscle tissue in the space provided in Figure 9.4. ■

Muscle Tissue

9

Name _____

Date _____

Section _____

A. Matching

Match each structure in the left column with its correct description from the right column. Each term in the right column may be used more than once.

_____ **1.** muscle fiber membrane	**A.** sarcolemma
_____ **2.** cellular connections between cardiocytes	**B.** intercalated disc
_____ **3.** striated, uninucleated cells	**C.** cardiac muscle
_____ **4.** muscle tissue in tip of tongue	**D.** skeletal muscle
_____ **5.** muscle tissue in artery	**E.** smooth muscle
_____ **6.** voluntary muscle	
_____ **7.** nonstriated cells	
_____ **8.** involuntary, striated cells	

B. Short-Answer Questions

1. Why are skeletal muscle fibers multinucleated?

2. Which types of muscle tissue are striated?

3. Where in the body does smooth muscle occur?

4. What is the function of intercalated discs in cardiac muscle?

5. Which muscle tissues are controlled involuntarily?

C. Analysis and Application

1. Give an example that illustrates the involuntary control of cardiac muscle tissue.

2. How are smooth muscle cells similar to skeletal muscle fibers? How are they different from skeletal muscle fibers?

3. How are skeletal and cardiac muscle tissues similar to each other? How do these two types of muscle tissue differ from each other?

Neural Tissue

OBJECTIVES

On completion of this exercise, you should be able to:

- List the basic functions of neural tissue.
- Describe the two basic types of cells found in neural tissue.
- Identify a neuron and its basic structure under the microscope.
- Describe how neurons communicate with other cells across a synapse.
- List several functions of glial cells.

To maintain homeostasis, the body must constantly evaluate internal and external conditions and respond quickly and appropriately to environmental changes. The nervous system processes information from sensory organs and responds with motor instructions to muscles and glands, which are collectively called the body's **effectors**. Cells responsible for receiving, interpreting, and sending the electrical signals of the nervous system are called **neurons**. Neurons are excitable, which means they can respond to environmental changes by processing stimuli into electrical impulses called **action potentials**. Sensory neurons detect changes in the environment and communicate these changes to the central nervous system (CNS), which consists of the brain and spinal cord. The CNS responds to the sensory input with motor commands to glands and muscle tissues. This constant monitoring and adjustment play a vital role in homeostasis.

The most numerous cells in the nervous system are **glial cells** (*glia,* glue), which make up a network of cells and fibers called the **neuroglia**. Large populations of one type of glial cells support and anchor neurons. Other types of glial cells wrap around neurons, creating a myelin sheath that greatly increases the communication speed of the neuron. Neurons and glial cells are collectively referred to as either *nerve tissue* or *neural tissue*. (The two terms are synonyms, and you will see both in textbooks and in scientific literature.)

LAB ACTIVITY 1   Neuron Structure

A typical neuron has distinct cellular regions. Examine Figure 10.1 and locate the **cell body** surrounding the **nucleus**. This area, also known as the **soma**, contains most of the neuron's organelles. Many fine extensions, called **dendrites** (DEN-drīts; *dendron,* a tree), receive information from other cells and send impulses toward the soma. The signal is then conducted into a single **axon** that carries information away from the soma, either to other neurons or to effector cells. At the end of the axon is an enlarged synaptic knob that contains membranous **synaptic vesicles**. These vesicles contain **neurotransmitter molecules**, which are chemical messengers used either to excite or to inhibit other cells. The axon releases neurotransmitters onto an adjacent neuron across a specialized junction called the **synapse** (SIN-aps; *synap,* union). Notice in the bottom drawing of Figure 10.1 the three axons on the left synapsing on the dendrites of the main neuron shown in blue. At the synapse, cells do not touch; they are separated by a small **synaptic cleft**. When an action potential reaches the end of a *presynaptic* axon, the neurotransmitter molecules released by the synaptic vesicles diffuse across the synaptic cleft and either excite or inhibit the *postsynaptic* cell.

Study Tip | *Identifying Axons*

On most neuron slides, it is difficult to distinguish axons from dendrites. Locate one neuron that is isolated from the others and examine the soma for a large extension. This is most likely the axon. ●

QuickCheck Questions

1.1 What part of a neuron sends information to another neuron?

1.2 In general, how do neurons communicate with other cells?

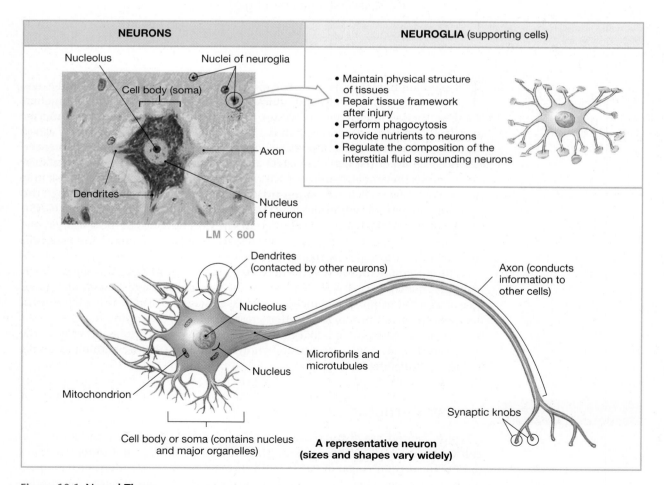

NEURONS	**NEUROGLIA** (supporting cells)

Nucleolus
Nuclei of neuroglia
Cell body (soma)
Axon
Dendrites
Nucleus of neuron
LM × 600

- Maintain physical structure of tissues
- Repair tissue framework after injury
- Perform phagocytosis
- Provide nutrients to neurons
- Regulate the composition of the interstitial fluid surrounding neurons

Dendrites (contacted by other neurons)
Axon (conducts information to other cells)
Nucleolus
Microfibrils and microtubules
Nucleus
Mitochondrion
Synaptic knobs
Cell body or soma (contains nucleus and major organelles)
A representative neuron (sizes and shapes vary widely)

Figure 10.1 Neural Tissue

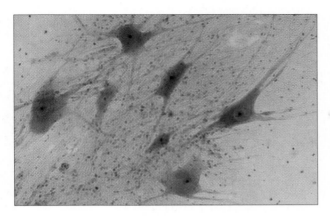 

Figure 10.2 **Neurons (LM × 100)**

1 *Materials*

- ☐ Compound microscope
- ☐ Prepared microscope slide of neurons

Procedures

1. Scan the slide at low magnification to locate the neurons. Select one neuron to examine more closely. Center this neuron in the field of view and increase the magnification. Adjust the light setting of the microscope if necessary.

2. On the neuron you have chosen, identify the soma, nucleus, dendrites (thin extensions), and axon (thicker extension).

3. Draw and label several neurons in the space provided in Figure 10.2. ∎

LAB ACTIVITY 2 Neuroglia

The glial cells of the neuroglia are the supportive cells of the nervous system. There are six types of glial cells, each with a specific function. The glial cells of the CNS are astrocytes, microglia, ependymal cells, and oligodendrocytes. **Astrocytes** attach blood vessels to neurons or anchor neurons in place. Phagocytic **microglia** are responsible for housekeeping chores in the nervous system. **Ependymal cells** line the spaces of the brain and spinal cord; they assist in the production and circulation of cerebrospinal fluid. **Oligodendrocytes** protect neurons by wrapping around them to isolate them from chemicals present in the interstitial fluid.

The part of the nervous system outside the CNS is called the *peripheral nervous system,* and the glial cells here are Schwann cells and satellite cells. **Schwann cells** wrap around peripheral neurons to increase the speed at which they transmit action potentials. Repair of peripheral neural tissue, which is any neural tissue outside the brain and spinal cord, is made possible by Schwann cells that build a "repair tube" to reconnect the severed axons. **Satellite cells** help regulate the environment around peripheral neural tissue.

In this exercise you will examine the most common glial cell in the nervous system, the astrocyte, shown in Figure 10.3. Astrocytes are major structural cells of the brain and spinal cord and serve a variety of functions. They provide a framework to support neurons. Cytoplasmic extensions of astrocytes, called *feet,* wrap around capillaries and form the blood-brain barrier that protects the brain and regulates the composition of the extracellular fluid.

QuickCheck Questions

2.1 What are the two major types of cells in neural tissue?

2.2 What are the major functions of astrocytes?

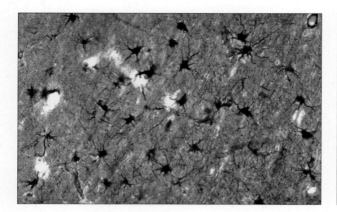

Figure 10.3 Astrocytes (LM × 100)

2 *Materials*

☐ Compound microscope

☐ Prepared microscope slide of astrocytes

Procedures

1. Move the low-power objective lens into position. Set the astrocyte slide on the stage and slowly turn the coarse adjustment knob until you can clearly see the specimen. Now use the fine focus adjustment as you examine the tissue.

2. Scan the slide and locate a star-shaped astrocyte. Center the cell in the field of view, and then increase the magnification to medium. Notice the numerous feet extending from the cell.

3. Draw and label several astrocytes in the space provided in Figure 10.3. ■

Neural Tissue

Name _____

Date _____

Section _____

A. Matching

Match each cell structure in the left column with its correct description from the right column.

_____ **1.** soma

_____ **2.** synaptic vesicles

_____ **3.** dendrite

_____ **4.** neurotransmitter

_____ **5.** axon

_____ **6.** synaptic cleft

A. membranous organelles containing neurotransmitters

B. chemical messenger released at synapse

C. cell body surrounding nucleus

D. space between two communicating neurons

E. propagates action potentials

F. receiving branch of neuron

B. Labeling

Label the anatomy of the neuron in Figure 10.4.

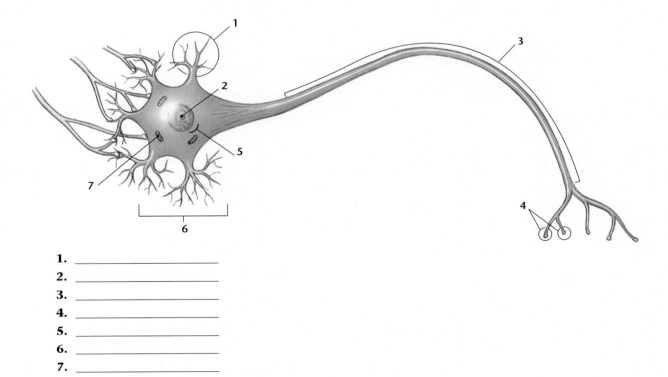

1. _____

2. _____

3. _____

4. _____

5. _____

6. _____

7. _____

Figure 10.4 **Neural Tissue**

10

C. Short-Answer Questions

1. What are the basic functions of neural tissue?

2. How do neurons communicate with other cells?

3. Which part of a neuron conducts an impulse toward the soma?

4. In which direction does an action potential travel in an axon?

D. Analysis and Application

1. List the cellular structures over which an impulse would pass, starting at the dendrites and traveling to where neurotransmitter molecules are released.

2. What are the major functions of glial cells?

The Integumentary System

OBJECTIVES

On completion of this exercise, you should be able to:

- Identify the two layers of the skin.

- Identify the layers of the epidermis.

- Distinguish between the papillary and reticular layers of the dermis.

- Identify the accessory structures of the skin.

- Identify a hair follicle, the parts of a hair, and an arrector pili muscle.

- Distinguish between sebaceous and sweat glands.

- Describe three sensory organs of the integument.

The **integumentary** (in-TEG-ū-MEN-ta-ree) **system** is the most visible organ system of the human body. The integument (in-TEG-ū-ment), or skin, is classified an organ system because it is composed of many different types of tissues and organs. Organs of the skin include oil-, wax-, and sweat-producing glands; sensory organs for touch; muscles attached to hair follicles; and blood and lymphatic vessels.

Skin seals the body in a protective barrier that is flexible yet resistant to abrasion and evaporative water loss. People interact with the external environment with the skin. Caressing a baby's head, feeling the texture of granite, and testing the temperature of bath water all involve sensory organs in the skin. Sweat glands in the skin cool the body to regulate body temperature. When exposed to sunlight, the skin manufactures vitamin D_3, a vitamin essential in calcium and phosphorus balance.

LAB ACTIVITY 1   Epidermis and Dermis

There are two principal tissue layers in the skin: a superficial layer of epithelium, called the **epidermis**, and a deeper layer of connective tissue, the **dermis** (Figure 11.1). These layers are also referred to collectively as the **cutaneous membrane**. Isolating the dermis from underlying structures is a fatty layer of connective tissue, the **hypodermis,** or **subcutaneous layer**.

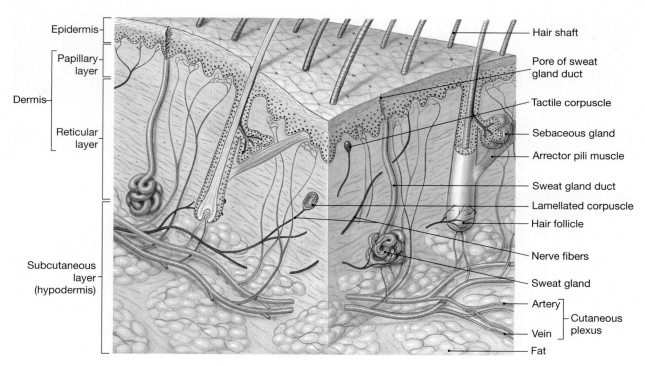

Epidermis

Dermis — Papillary layer, Reticular layer

Subcutaneous layer (hypodermis)

Hair shaft

Pore of sweat gland duct

Tactile corpuscle

Sebaceous gland

Arrector pili muscle

Sweat gland duct

Lamellated corpuscle

Hair follicle

Nerve fibers

Sweat gland

Artery
Vein — Cutaneous plexus

Fat

Figure 11.1 Components of the Integumentary System
Relationships among the major components of the integumentary system (with the exception of nails, which are shown in Figure 11.6).

The epidermis consists of stratified squamous epithelium organized into many distinct cell bands, or strata, as shown in Figure 11.2. Thick-skinned areas, such as the palms of the hands and soles of the feet, have five layers; thin-skinned areas have only four. Cells called **keratinocytes** are produced in the basal region of the epidermis and pushed upward toward the surface of the skin. During this migration, the keratinocytes synthesize and accumulate the protein keratin, which reduces water loss across the cell membrane. The uppermost layer of the epidermis consists of dead, dry, scalelike cells.

The five layers that occur in the epidermis of thick-skinned areas of the body are shown in Figure 11.2.

- The **stratum germinativum** (STRA-tum jer-mi-na-TĒ-vum), or **stratum basale**, is a single cell layer between the base of the epidermis and the upper surface of the dermis. The stem cells of this stratum are in a constant state of mitosis, replacing cells that have rubbed off the epidermal surface. As they are produced by stem cells, new keratinocytes push previously formed cells toward the surface. It takes from 15 to 30 days for a cell to migrate from the stratum germinativum to the top of the epidermis. Other cells in this layer, called **melanocytes,** produce the pigment **melanin** (ME-la-nin), which protects deeper cells from the harmful effects of ultraviolet (UV) radiation from the sun. Prolonged exposure to UV light causes an increase in melanin synthesis, resulting in a darkening, or tanning, of the skin.

- Superficial to the stratum germinativum is the **stratum spinosum**, which consists of five to seven rows of cells interconnected by thickened cell membranes called **desmosomes**. When a slide of epidermal tissue is being prepared, cells in this layer often shrink, but the desmosome bridges between cells remain intact. This results in cells with a spiny outline; hence the name "spinosum."

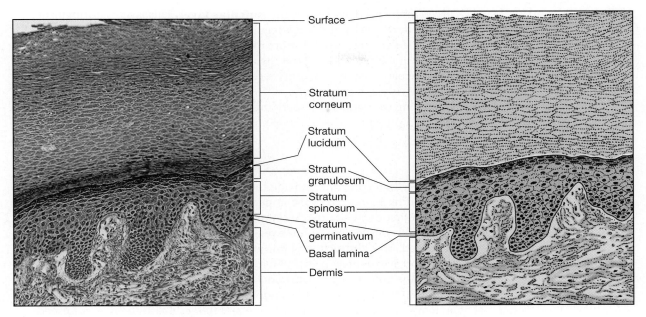

Figure 11.2 Layers of the Epidermis
A light micrograph through a portion of the epidermis, showing the major stratified layers of epidermal cells (LM × 90).

- Superior to the stratum spinosum is a layer of darker cells that make up the **stratum granulosum**. As cells from the stratum germinativum are pushed upward, they synthesize the protein **keratohyalin** (ker-a-tō-HĪ-a-lin), which increases durability and reduces water loss from the skin. Keratohyalin granules stain dark and give this layer its color.

- In thick skin, a thin, transparent layer of cells called the **stratum lucidum** lies superficial to the stratum granulosum. In the stratum lucidum, keratohyalin is converted to an intermediate protein called **eleidin** (a-lē-a-din), which eventually becomes keratin. Only the thick skin of the palms and the soles of the feet have the stratum lucidum; the rest of the skin is considered thin and lacks this layer.

- The **stratum corneum** (KOR-nē-um; *cornu,* horn) is the uppermost layer of the epidermis and contains many layers of flattened, dead cells. As cells from the stratum granulosum move upward, keratohyalin granules are converted to the fibrous protein keratin. As keratinocytes fill with keratin, the internal organization of the cell is disrupted and the cells die. These keratinized cells on the surface of the epidermis are resistant to dehydration and friction. Cells of the stratum corneum are constantly being shed or worn off and replaced by new cells produced in and pushed up from deeper layers. Cells in the stratum corneum accumulate the yellow-orange pigment **carotene**, which is common in light-skinned individuals.

Below the stratum germinativum is the dermis, a band of connective tissues that anchors the epidermis in place (Figure 11.1). The dermis is divided into two layers: papillary and reticular. Although there is no distinct boundary between these layers, the upper fifth of the dermis is designated the **papillary layer**. This layer consists of areolar connective tissue containing numerous collagen and elastic fibers. Folds in the connective tissue are called dermal **papillae** (pa-PIL-la; *papilla,* a small cone). These folds project into the epidermis, and we see them as the swirls of fingerprints. Fingerprint patterns are genetically based and do not change during an individual's life. Within the papillae are small sensory receptors for light touch, **Meissner's corpuscles**.

Deep to the papillary layer is the **reticular layer.** This layer is distinguished by widely scattered cells and dense irregular connective tissue proper interlaced with collagen fibers that are more distinct than those in the papillary layer. Large blood vessels, sweat glands, and adipose tissue in the reticular layer are less visible than those in the papillary layer. Sensory receptors in the reticular layer, called **Pacinian corpuscles**, detect deep pressure.

Attaching the dermis to underlying structures is the subcutaneous layer, which is composed primarily of adipose tissue and areolar connective tissue proper.

Clinical Application

Burns

Burns are classified by the damage they cause to the layers of the skin. First-degree burns injure cells of the upper epidermis. Sunburns and other topical burns are first-degree burns. Second-degree burns destroy the entire epidermis and portions of the dermis; however, hair follicles and glands of the dermis are not injured. This destruction of portions of the dermis causes blistering, and the wound is extremely painful. Third-degree burns penetrate completely through the skin, severely damaging the epidermis, dermis, and subcutaneous structures. This type of wound cannot heal because the restorative layers of the epidermis are lost. To prevent infection and to reestablish the barrier formed by the skin, a skin graft is used to cover the wound. Nerves are usually damaged by third-degree burns, with the result that these more serious burns may not be as painful as first- and second-degree burns. ▶

QuickCheck Questions

1.1 Describe the two layers of the skin.

1.2 Why does the epidermis constantly replace its cells?

1 Materials

- ☐ Compound microscope
- ☐ Prepared slide of the scalp (cross-section)

Procedures

1. Place the scalp slide on the microscope and focus on the specimen at low magnification.
2. Scan the slide vertically and identify the epidermis, dermis, and hypodermis.
3. Increase the magnification to medium and examine the epidermis. Locate the epidermal layers, beginning with the deepest layer, the stratum germinativum.
   - What is the shape of cells in the stratum spinosum?
   - What color is the stratum granulosum?
   - Does the scalp specimen have a stratum lucidum?
   - What is the top layer of cells called? Are these cells alive?
4. Study the dermis at low, medium, and high magnifications.
   - Distinguish between the papillary and reticular layers.
   - Are Meissner's corpuscles visible at the papillary folds?
   - What type of connective tissue is in the reticular layer? ■

LAB ACTIVITY 2   Accessory Structures of the Skin

During embryonic development, the epidermis produces accessory structures called **epidermal derivatives**, which include oil and sweat glands, hair, and nails. These structures are exposed on the surface of the skin and project deep into the dermis.

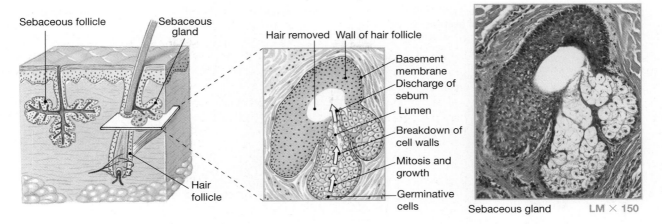

Figure 11.3 Sebaceous Glands and Follicles
The structure of sebaceous glands and sebaceous follicles in the skin.

- **Sebaceous** (se-BĀ-shus) **glands** are located in two types of follicles found in the epidermis: sebaceous follicles and hair follicles. These glands secret the oily substance **sebum**, which coats hair shafts and the epidermal surface. Observe in Figure 11.3 the infolding of epithelium to form the hair follicle, the sebaceous follicle, and the sebaceous gland. Notice how the gland cells empty onto the follicles. The sebaceous glands are **holocrine** exocrine glands, which means the gland cells rupture to release sebum into the duct of the gland. Cells at the base of the glands divide to replace the ruptured cells. New oil-filled cells in the glands are pushed toward the lumen of the duct. Once on the free surface of the lumen, the sebaceous cell membranes rupture, and sebum is discharged into the duct. The sebum formed in the hair follicles coats the hair shaft, to reduce brittleness, and prevents excessive drying of the scalp.

- **Sebaceous follicles** secrete sebum onto the surface of the skin. These follicles are not associated with hair and are distributed on the face, most of the trunk, and the male reproductive organs. Secretions from sebaceous follicles lubricate the skin and provide limited antibacterial action.

Clinical Application

Acne

Most teenagers have dealt with the skin blemishes called *acne*. When sebaceous ducts become blocked, sebum becomes trapped in the ducts. As the sebum accumulates, it causes the skin to rise, resulting in a pimple. A pimple with a white head shows that the duct is closed and that sebum has accumulated. A blackhead forms when an open sebaceous duct contains solid material infected with bacteria. ▶

- **Sweat glands**, or **sudoriferous** (sū-dor-IF-er-us) **glands**, are scattered throughout the dermis of most of the skin. They are exocrine glands that secrete their liquid either into sweat ducts leading directly to the skin surface or into ducts that empty into hair follicles, as shown in Figure 11.4.
    Two types of sweat glands occur in the dermis: apocrine and merocrine. **Apocrine sweat glands** are found in the groin, nipples, and axillae. These glands secrete a thick **sweat** into ducts associated with hair follicles. Bacteria on the hair metabolize the sweat and produce the characteristic body odor of, for example, axillary sweat. When body temperature increases, **merocrine** (MER-ō-krin) **sweat glands**, also called *eccrine* (EK-rin) *glands,* secrete onto the body surface a thinner sweat containing electrolytes, proteins, urea, and other

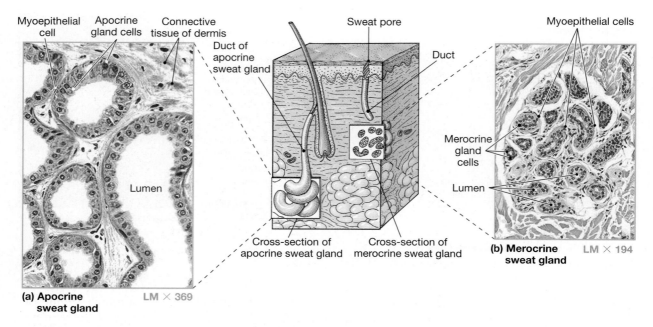

Myoepithelial cell
Apocrine gland cells
Connective tissue of dermis
Duct of apocrine sweat gland
Sweat pore
Duct
Myoepithelial cells
Lumen
Merocrine gland cells
Lumen
Cross-section of apocrine sweat gland
Cross-section of merocrine sweat gland

(a) Apocrine sweat gland   LM × 369

(b) Merocrine sweat gland   LM × 194

Figure 11.4 Sweat Glands

(a) Apocrine sweat glands, located in the axillae, groin, and nipples, produce a thick, odorous fluid by apocrine secretion (LM × 369). **(b)** Merocrine sweat glands produce a watery fluid by merocrine secretion (LM × 194).

compounds. The sweat absorbs body heat and evaporates from the skin, cooling the body. It also contributes to body odor because of the presence of urea and other wastes. Merocrine glands are not associated with hair follicles and are distributed throughout most of the skin.

- **Hair** covers most of the skin. Only the lips, nipples, portions of the external genitalia, soles, palms, fingers, and toes are without hair. Three major types of hair occur across the skin. **Terminal hairs** are the thick, heavy hairs on the scalp, eyebrows, and eyelashes. **Vellus** hairs are lightly pigmented and distributed over much of the skin as fine "peach fuzz." **Intermediate hairs** are the heavier hairs on the arms and legs. Hair generally serves a protective function. It cushions the scalp and prevents foreign objects from entering the eyes, ears, and nose. Hair also serves as a sensory receptor. Wrapped around the base of each hair is a **root hair plexus**, a sensory neuron sensitive to movement of the hair.

  Note in Figure 11.5 that each hair is embedded in a hair follicle. Deep in the follicle is the **hair root**. At the root tip is a **papilla** containing nerves and blood vessels and the living, proliferative part of the hair, the **matrix**. Cells in the matrix undergo mitotic divisions that cause elongation and growth of the hair. Above the matrix, keratinization of the hair cells causes them to harden and die. The resulting **hair shaft** contains an outer **cortex** and an inner **medulla**. A smooth muscle called the **arrector pili** (a-REK-tor PI-lē) **muscle** is attached to each hair follicle. When fur-covered animals are cold, these muscles contract to raise the hair and trap a layer of warm air next to the skin. In humans, the muscle has no known thermoregulatory use, because humans do not have enough hair to gain an insulation benefit. We do have arrector pili muscles, though, and their contracting when we are cold is what produces "gooseflesh." These muscles also respond to emotional stimuli, such as the sound of beautiful music or the fear you might feel walking down a dark alley at night. Arrector pili muscles in animals also are involved in emotional response: animals raise their fur to look bigger when they feel threatened.

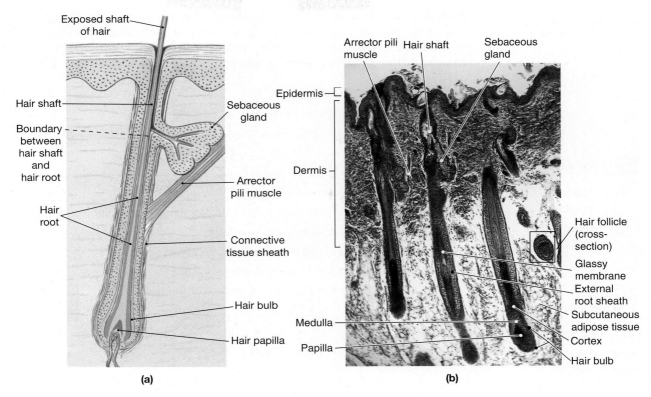

Figure 11.5 Structure of a Hair

(a) The anatomy of a single hair. (b) A light micrograph showing the sectional appearance of the skin of the scalp (LM × 73).

- **Nails**, which protect the dorsal surface and tips of the fingers and toes (Figure 11.6), consist of tightly packed keratinized cells. The visible elongated body of the nail, called the **nail body**, protects the underlying **nail bed** of the skin. Blood vessels underneath the nail body give the nail its pinkish color. The **free edge** of the nail body extends past the end of the digit. The nail root is at the base of the nail and is where new growth occurs. The **lunula** (LOO-nū-la; *luna*, moon) is a whitish portion of the proximal nail body where blood vessels do not

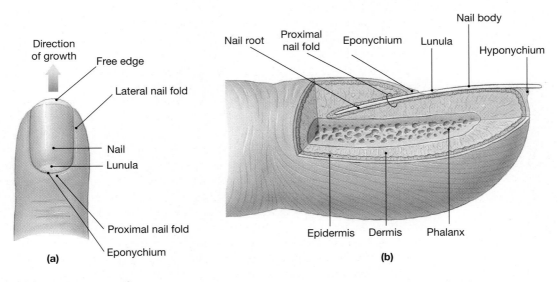

Figure 11.6 Structure of a Nail

These drawings illustrate the prominent features of a typical fingernail as viewed from the surface and in section.

show through. The **cuticle** around the nail is called the **eponychium** (ep-ō-NIK-ē-um; *epi,* over + *onyx,* nail). It is composed of a band of epidermis that seals the nail groove to the epidermis. Under the free edge of the nail is the **hyponychium** (hī-pō-NIK-ē-um), a thicker region of the epidermis.

QuickCheck Questions

2.1 List the accessory structures of the skin.

2.2 What are the two major types of glands in the skin?

2 Materials

☐ Compound microscope

☐ Prepared slide of the scalp (cross-section)

Procedures

1. On the scalp slide, locate a hair follicle.
   - What is the shape of the hair follicle? In which layer of the skin is it found?
   - Identify the hair shaft, cortex, and medulla.
   - Identify a sebaceous gland. Where does it empty its secretions?
2. Scan the dermis of the slide for a sudoriferous gland.
   - Trace the duct from the gland to the surface of the skin. ■

The Integumentary System

Name _____

Date _____

Section _____

A. Matching

Match each skin structure in the left column with its correct description from the right column.

_____	**1.** sebaceous gland	**A.** layer in thickened areas of epidermis
_____	**2.** apocrine sweat gland	**B.** deep to dermis, contains adipose tissue
_____	**3.** keratin	**C.** sweat gland associated with hair follicle
_____	**4.** arrector pili	**D.** produces new epidermal cells
_____	**5.** stratum corneum	**E.** deep layer of dermis
_____	**6.** papillary layer	**F.** protein that reduces water loss
_____	**7.** stratum germinativum	**G.** functions in thermoregulation
_____	**8.** reticular layer	**H.** surface layer of epidermis
_____	**9.** subcutaneous layer	**I.** muscle attached to hair follicle
_____	**10.** stratum lucidum	**J.** folded layer of dermis next to epidermis
_____	**11.** eccrine sweat gland	**K.** produces sebum

B. Labeling

Label the structures of the skin in Figure 11.7.

C. Short-Answer Questions

1. Describe the layers of epidermis in an area where the skin is thick.

2. How does the skin tan when exposed to sunlight?

3. List the types of sweat glands associated with the skin.

4. What is the function of arrector pili muscles in animals other than humans?

11

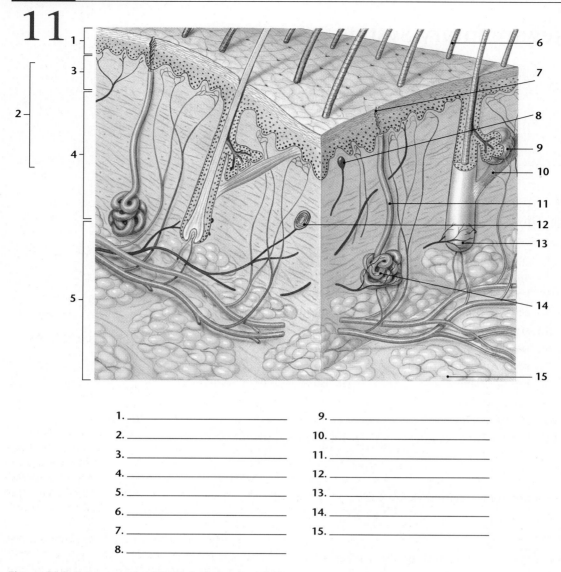

1. _____
2. _____
3. _____
4. _____
5. _____
6. _____
7. _____
8. _____

9. _____
10. _____
11. _____
12. _____
13. _____
14. _____
15. _____

Figure 11.7 **Components of the Integumentary System**

D. Analysis and Application

1. Why is the epidermis keratinized?

2. What is the main cause of acne, and in which part of the skin does it occur?

3. How are cells replaced in the epidermis?

Body Membranes

OBJECTIVES

On completion of this exercise, you should be able to:

- List and provide examples of the four types of body membranes.

- Discuss the components of each type of body membrane.

- Describe the histological organization of each type of body membrane.

The term *membrane* refers to a variety of anatomical structures, but in this exercise, **body membrane** refers to any sheet of tissue that wraps around an organ. Some body membranes cover structures and isolate them from the surrounding anatomy; other body membranes act as a barrier to prevent infections from spreading from one organ to another. Most body membranes produce a liquid that keeps the cells on the exposed, or free, surface of the membrane moist. Absorption may occur across these moist membranes.

Body membranes are composed of epithelium and connective tissue. Epithelium occurs on the exposed surface of the membrane and is supported by underlying connective tissue. There are four major types of membranes, as shown in Figure 12.1, classified by location. **Mucous membranes** occur where an opening is exposed to the external environment; **serous membranes** wrap around organs in the ventral body cavities; the **cutaneous membrane** is the skin; and **synovial** (sin-Ō-vē-ul) **membranes**, which do not contain true epithelium, line the cavities of movable joints.

LAB ACTIVITY 1 Mucous Membranes

The digestive, respiratory, urinary, and reproductive systems are all protected by mucous membranes, which are sometimes called *mucosae* (singular *mucosa*). The epithelium of a mucous membrane may be simple columnar, stratified squamous, pseudostratified, or transitional. It is always attached to the **lamina propria** (LA-mi-nuh PRO-prē-uh), a sheet of loose connective tissue that anchors the epithelium in place (Figure 12.2b). **Mucus**, the thick liquid that protects the epithelium of a mucous membrane, is secreted either by goblet cells in the

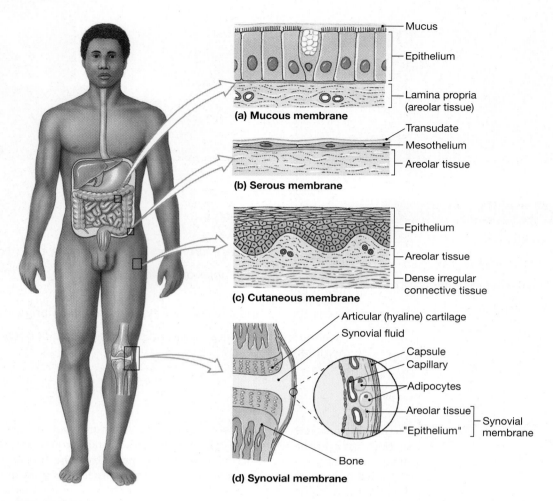

Mucus

Epithelium

Lamina propria
(areolar tissue)

(a) Mucous membrane

Transudate
Mesothelium
Areolar tissue

(b) Serous membrane

Epithelium

Areolar tissue

Dense irregular
connective tissue

(c) Cutaneous membrane

Articular (hyaline) cartilage
Synovial fluid
Capsule
Capillary
Adipocytes
Areolar tissue
"Epithelium" } Synovial membrane
Bone

(d) Synovial membrane

Figure 12.1 Membranes
(a) Mucous membranes are coated with the secretions of mucous glands. These membranes line the digestive, respiratory, urinary, and reproductive tracts. **(b)** Serous membranes line the ventral body cavities (the peritoneal, pleural, and pericardial cavities). **(c)** The cutaneous membrane, or skin, covers the outer surface of the body. **(d)** Synovial membranes line joint cavities and produce the fluid within the joint.

epithelium or by glands in the underlying **submucosa**. Mucus is viscous and contains a glycoprotein called **mucin**, salts, water, epithelial cells, and white blood cells. Mucin gives mucus its slippery and sticky attributes. Figure 12.1a shows the mucus coating the lining epithelium of a mucous membrane.

The mucous membrane of the digestive system has regional specializations. Where digestive contents are liquid, such as in the stomach and intestines, the epithelium of the mucous membrane is simple columnar with goblet cells. This mucous membrane functions in absorption and secretion. The mouth, pharynx, and rectum process either solid food or waste materials, both of which are abrasive to the epithelial lining. In these areas, the protective mucous membrane has a stratified squamous epithelium.

Most mucous membranes are constantly replacing epithelial cells. As materials move through a lumen, the exposed epithelial cells are scraped off the mucous membrane. Simple columnar cells in the small intestine, for instance, live for approximately 48 hours before they are replaced. The old cells are shed into the intestinal lumen and added to the feces.

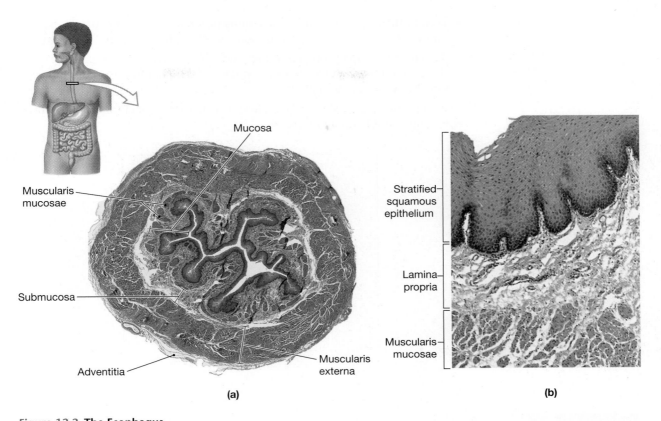

Figure 12.2 The Esophagus
(a) A transverse section through the esophagus. (b) The esophageal mucosa (LM × 77).

The nasal cavity, trachea (windpipe), and large bronchi of the lungs are lined with a mucous membrane in which the epithelium is populated with goblet cells that secrete mucus to trap particulate matter in inhaled air. This epithelium is pseudostratified ciliated columnar and transports the dust-trapping mucus upward for removal from the respiratory system.

Portions of the male and female reproductive systems are lined with a ciliated mucous membrane to help in the transport of sperm or eggs.

The mucous membrane of the urinary bladder has a transitional epithelium that enables the bladder wall to stretch and recoil as the bladder fills and empties. Urine keeps the membrane moist and prevents dehydration.

QuickCheck Questions

1.1 Where do mucus membranes occur in the body?

1.2 What type of epithelium occurs in the mucous membrane of the nasal cavity?

1 Materials

- ☐ Compound microscope
- ☐ Prepared esophagus slide (transverse section)

Procedures

1. Place the slide on the microscope stage and focus at low magnification.
2. Move the slide to the center of the tissue section and locate the lumen, the open space of the esophagus.
   - Notice the stratified squamous epithelium lining the folds along the wall. This is the epithelial part of the mucous membrane.

- Deep to the epithelium, note the lamina propria, which is mostly areolar connective tissue with numerous blood and lymphatic vessels.
- Deep to the lamina propria is a thick submucosal layer with esophageal glands that secrete mucus onto the surface of the pharynx. Scan the slide for a duct of an esophageal gland.

3. Draw a portion of the slide in the space provided here. Label the lumen, lining epithelium, and lamina propria. ∎

LAB ACTIVITY 2 Serous Membranes

Serous membranes are double-layered membranes that cover internal organs and line the ventral body cavities to reduce friction between organs. The **visceral layer** of the membrane is in direct contact with the particular organ, and the **parietal layer** lines the wall of the cavity. Between the two layers of the membrane is a minute space filled with **serous fluid**, a slippery lubricant. Different from the mucus secreted by mucous membranes, serous fluid is a thin, watery secretion similar to blood plasma. A thin layer of specialized simple epithelium called **mesothelium** covers the exposed surface of the pleura where it faces the pleural cavity. Interstitial fluid from the underlying connective tissue transudes, or passes through, the mesothelium to form the serous fluid.

The three serous membranes of the body are the **pericardium** of the heart, the **pleurae** of the lungs, and the **peritoneum** of the abdominal organs, all discussed in Exercise 3. Figure 12.3 details the pericardial and pleural serous membranes.

QuickCheck Questions

2.1 Where do serous membranes occur?

2.2 Describe the structure of a serous membrane.

2 Materials

- ☐ Compound microscope
- ☐ Prepared serous membrane slide (pleura)
- ☐ Laboratory models

Procedures

1. Place the slide on the microscope stage and focus at low power.
2. Locate the surface of the tissue.
   - On the surface is the epithelium called mesothelium. Describe the appearance of this tissue.
   - Locate the underlying connective tissue components of the pleura.

3. Draw a portion of the slide in the space provided here. Label the visceral layer, the serous cavity, and the mesothelium.

4. Identify the three serous membranes on laboratory models. ■

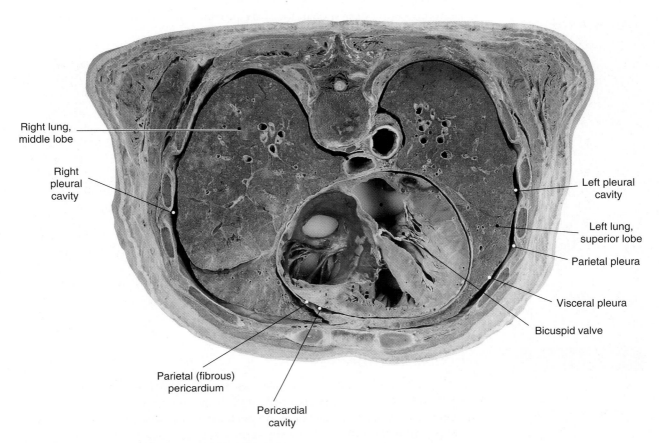

Right lung, middle lobe

Right pleural cavity

Left pleural cavity

Left lung, superior lobe

Parietal pleura

Visceral pleura

Bicuspid valve

Parietal (fibrous) pericardium

Pericardial cavity

Figure 12.3 Serous Membranes of Thoracic Cavity
The heart and lungs are encased in serous membranes.

LAB ACTIVITY 3   Cutaneous Membrane

The cutaneous membrane is the skin, which consists of the epidermis and dermis that cover the exterior surface of the body, as discussed in Exercise 11. The epidermis consists of stratified squamous epithelium, and the dermis is made up of a variety of connective tissues. Protection against abrasion and against the entrance of microbes is a major function of the cutaneous membrane. Unlike all mucous, serous, and synovial membranes, which are kept moist, the cutaneous membrane is dry. A process called **keratinization** waterproofs the skin's surface. The epithelial cells of the cutaneous membrane synthesize the hard protein keratin, which kills the cells, leaving a dry protective layer of tough, scaly dead cells on the skin surface. The cells are eventually shed and are constantly replaced.

QuickCheck Questions

3.1 What are the two layers that make up the cutaneous membrane?

3.2 What type of epithelium is found in the cutaneous membrane?

3 Materials

- ☐ Compound microscope
- ☐ Prepared slide of the skin (transverse section)
- ☐ Skin model

Procedures

1. Place the slide on the microscope stage and focus at low power.
2. Locate the epithelium of the epidermis.
   - What type of tissue is it composed of?
   - Identify the connective tissue of the dermis.
3. Draw a portion of the slide in the space provided here. Label the epidermis, stratified squamous epithelium, and dermis.

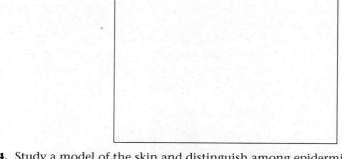

4. Study a model of the skin and distinguish among epidermis, dermis, and hypodermis. Which of these layers make up the cutaneous membrane? ■

LAB ACTIVITY 4   Synovial Membranes

Freely movable joints, such as the knee and elbow, are lined with a synovial membrane (Figure 12.4). The mobility of the joint is due primarily to the presence of a small joint cavity between the bones. An **articular cartilage** covers the surfaces of the bones in the joint cavity. The walls of the cavity are encapsulated with synovial membranes. This type of membrane is unique because the connective tissue is not covered with epithelium. In a synovial membrane, loose connective tissue produces a liquid that seeps from the tissue and fills the synovial cavity. This **synovial fluid**—

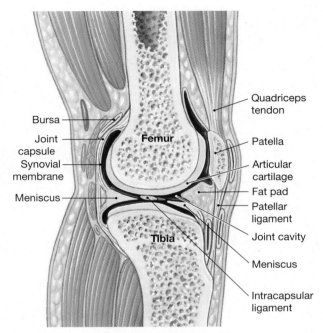

Figure 12.4 The Structure of a Synovial Joint
A sectional view of the knee joint. The synovial membrane
secretes fluid into the joint cavity.

a clear, lubricating solution containing mucin, salts, albumins, and nutrients—
passes between a patchwork of lining cells similar to epithelium. Unlike epithelia,
there is no basal lamina anchoring the lining cells to the underlying connective tissue. Fluid exchange between blood vessels and the loose connective tissue maintains
and replenishes the synovial fluid. Movement helps synovial fluid to circulate in the
joint; if a joint is immobilized for too long, both the cartilage and the synovial membrane may degenerate.

Joints such as the shoulder have additional synovial membranes called
bursae (singular **bursa**), which cushion structures such as tendons and ligaments. Bursae also occur over bones where the skin is thin, such as the elbow, to
reduce abrasion to the skin.

QuickCheck Questions

4.1 Where in the body do synovial membranes occur?

4.2 What is a bursa?

---

4 **Materials**

☐ Longitudinally
sectioned cow joint
or model of knee in
sagittal section

Procedures

1. Examine the gross anatomy of the cow joint or the knee model.
2. Locate the articular cartilage at the ends of the articulating bones, the synovial
   membrane, and bursae. ∎

Body Membranes

Name _____

Date _____

Section _____

A. Matching

Match each structure in the left column with its correct description from the right column.

_____ **1.** serous fluid   **A.** deep serous membrane layer

_____ **2.** lamina propria   **B.** produces mucus

_____ **3.** goblet cell   **C.** superficial serous layer

_____ **4.** visceral layer   **D.** propels mucus

_____ **5.** cilia   **E.** functions as a lubricant

_____ **6.** parietal layer   **F.** connective tissue of mucous membrane

B. Short-Answer Questions

1. Where in the body do serous membranes occur?

2. What type of tissue occurs on the surface of all body membranes?

3. What layers of the skin constitute the cutaneous membrane?

C. Analysis and Application

1. How are serous and synovial fluids produced?

2. What is the function of goblet cells, and in which type of membrane do they occur?

3. A layer called the lamina propria occurs in which type of membrane?

Skeletal System Overview

OBJECTIVES

On completion of this exercise, you should be able to:

- List the components of the axial skeleton and those of the appendicular skeleton.

- Describe the gross anatomy of a long bone.

- Describe the histological organization of compact bone and of spongy bone.

- List the five shapes of bones and give an example of each type.

- Describe the bone markings visible on the skeleton.

The skeletal system serves many functions. Bones support the body's soft tissues and protect vital internal organs. Calcium, lipids, and other materials are stored in the bones, and blood cells are manufactured in the bones' red marrow. Bones serve as levers that allow the muscular system to produce movement or maintain posture. In this exercise, you will study the gross structure of bone and the individual bones of the skeletal system.

Two types of bone tissue are found in the skeleton: compact and spongy. **Compact bone**, which is also called **dense bone**, seals the outer surface of bones and is found wherever stress arrives from one direction on the bone. **Spongy bone**, or **cancellous tissue**, is found inside the compact-bone envelope.

LAB ACTIVITY 1 Bone Structure

Bones are encapsulated in a tough, fibrous membrane called the **periosteum**. This membrane appears shiny and glossy and is sometimes visible on a chicken bone or on the bone in a steak. Histologically, the periosteum is composed of two layers: an outer fibrous layer where muscle tendons and bone ligaments attach and an inner cellular layer that produces cells called **osteoblasts** (OS-tē-ō-blasts) for bone growth and repair. **Osteocytes** are mature bone cells that maintain the mineral and protein components of bone matrix.

A long bone, such as the femur of the thigh, has a shaft, called the **diaphysis** (dī-AF-i-sis), with an **epiphysis** (ē-PIF-i-sis) on each end of the shaft (Figure 13.1).

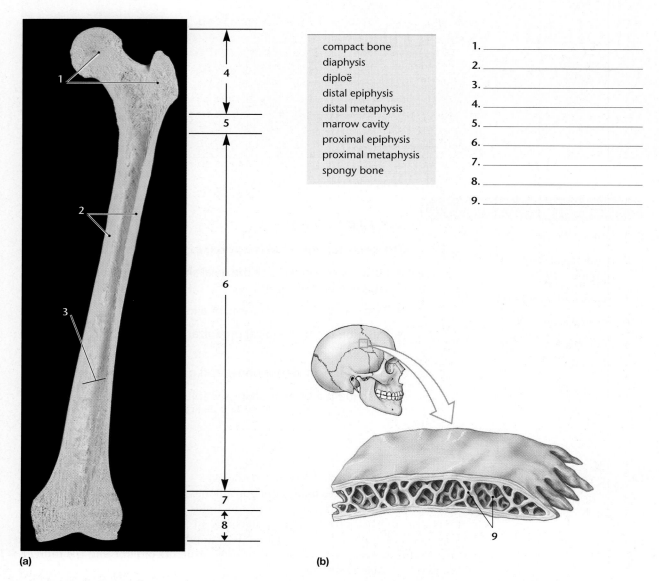

compact bone
diaphysis
diploë
distal epiphysis
distal metaphysis
marrow cavity
proximal epiphysis
proximal metaphysis
spongy bone

1. _____
2. _____
3. _____
4. _____
5. _____
6. _____
7. _____
8. _____
9. _____

(a)   (b)

Figure 13.1 Bone Structure
(a) The structure of a representative long bone in longitudinal section. (b) The structure of a flat bone.

The proximal epiphysis is on the superior end of the diaphysis, where the bone is attached to a muscle or tendon, and the distal epiphysis is on the inferior end. The diaphysis is compact bone, and the epiphysis is spongy bone. Wherever an epiphysis articulates with another bone, a layer of hyaline cartilage, the **articular cartilage**, covers the epiphysis. The interior of the diaphysis is hollow, forming a space called either the **medullary cavity** or the **marrow cavity.** This cavity contains spongy bone, and here lipid is stored as **yellow marrow**. An inner membrane called the **endosteum** (en-DOS-stē-um) lines the bone tissue facing the medullary cavity. **Osteoclast cells** in the endosteum secrete a weak acid that dissolves bone matrix to tear down bone either so that it can be remodeled or repaired or so that minerals stored in the bone can be released into the blood.

Between the diaphysis and either epiphysis is the **metaphysis** (me-TAF-i-sis). In a juvenile's bone, the metaphysis is called the **epiphyseal plate** and consists of a plate of hyaline cartilage that allows the bone to grow longer. By early adulthood, the rate of mitosis in the cartilage plate slows, and ossification fuses the epiphysis to the diaphysis. Bone growth stops when all the cartilage in the metaphysis has been replaced by bone. This bony remnant of the growth plate is now called the **epiphyseal line**.

Flat bones, such as the frontal and parietal bones of the skull, are thin bones with no marrow cavity. Flat bones have a layer of spongy bone sandwiched between layers of compact bone (Figure 13.1b). The compact bone layers are called the *external* and *internal tables* and are thick in order to provide strength for the bone. The spongy bone layer is filled with marrow and is called the **diploë** (DIP-lō-ē).

QuickCheck Questions

1.1 Describe the location of the two membranes that cover long bones.

1.2 Where is spongy bone found?

1 Materials

☐ Preserved long bone or fresh long bone from butcher shop

☐ Blunt probe

☐ Disposable examination gloves

☐ Safety glasses

Procedures

1. Before observing the long bone, complete the labeling of Figure 13.1.

2. Put on the safety glasses and examination gloves before you handle the bone.

3. Examine the long bone and locate the periosteum. Does it appear shiny? Are any tendons or ligaments attached to it? _____

4. If the bone has been sectioned, observe the internal bone tissue of the diaphysis. Is the bone tissue similar in all regions of the sectioned bone?

   _____

5. Locate an epiphysis and its articular cartilage. What is the function of the cartilage? _____

6. Locate the metaphysis. Most likely the bone has an epiphyseal line rather than a epiphyseal plate. Why? _____ ∎

LAB ACTIVITY 2 Histological Organization of Bone

Compact bone has supportive columns called either **osteons** or *Haversian systems* (Figure 13.2). Each osteon consists of many rings of calcified matrix called **concentric lamellae** (lah-MEL-lē; *lamella,* a thin plate). Between the lamellae, in small spaces in the matrix called **lacunae,** are mature bone cells called osteocytes.

Bone requires a substantial supply of nutrients and oxygen. Nerves, blood vessels, and lymphatic vessels all pierce the periosteum and enter the bone in a **perforating canal** oriented perpendicular to the osteons. This canal interconnects with **central canals** positioned in the center of osteons. Radiating outward from a central canal are small diffusion channels called **canaliculi** (kan-a-LIK-ū-lī) that facilitate nutrient, gas, and waste exchange with the blood.

To maintain its strength and weight-bearing ability, bone tissue is continuously being remodeled in a process that leaves distinct structural features in compact bone. Old osteons are partially removed, and the concentric rings of lamellae are fragmented, resulting in **interstitial lamellae** between intact osteons. Typically, the distal end of a bone is extensively remodeled throughout life, whereas areas of the diaphysis may never be remodeled. Other lamellae occur underneath the periosteum and wrap around the entire bone. These **circumferential lamellae** are added as a bone grows in diameter.

Unlike compact bone, spongy bone is not organized into osteons; instead, it forms a lattice, or meshwork, of bony struts called **trabeculae** (tre-BEK-ū-lē). Each trabecula is composed of layers of lamellae that are intersected with canaliculi. Filling the spaces between the trabeculae is red marrow, the tissue that produces most blood cells. Spongy bone is always sealed with a thin outer layer of compact bone.

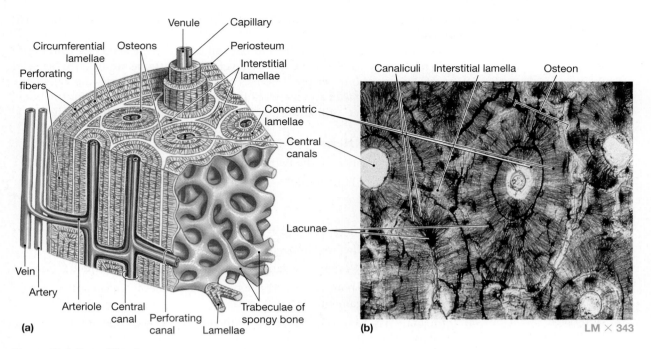

Figure 13.2 Bone Histology
(a) The organization of compact and spongy bone. **(b)** A thin section through compact bone. The intact matrix and central canals appear white, and the lacunae and canaliculi appear black.

QuickCheck Questions

2.1 Describe the three types of lamellae.

2.2 How is blood supplied to an osteon?

2 Materials

☐ Bone model

☐ Compound microscope

☐ Prepared slide of bone tissue (transverse section)

Procedures

1. Review the structures in Figure 13.2.

2. Examine a bone model and locate each structure shown in Figure 13.2.

3. Obtain a prepared microscope slide of bone tissue. Most slides are a transverse section through bone that is ground very thin. This preparation process removes the bone cells but leaves the bone matrix intact for detailed studies.

4. At low magnification, observe the overall organization of the bone tissue. How many osteons can you locate? _____

5. Select an osteon and observe it at a higher magnification. Identify the central canal, canaliculi, and lacunae. What is the function of the canaliculi? _____

6. Locate an area of interstitial lamellae. How do these lamellae differ from the concentric lamellae? _____ ■

LAB ACTIVITY 3 Classification of Bones

Bones may be grouped according to their shape, as illustrated in Figure 13.3. Already discussed in this exercise are **long bones,** which are greater in length than in width, and **flat bones,** which are thin and platelike. Additionally, bones of the wrist and ankle are **short bones,** almost as wide as they are long. The vertebrae of the spine are **irregular bones** that are not in any of the just-named categories. **Sesamoid**

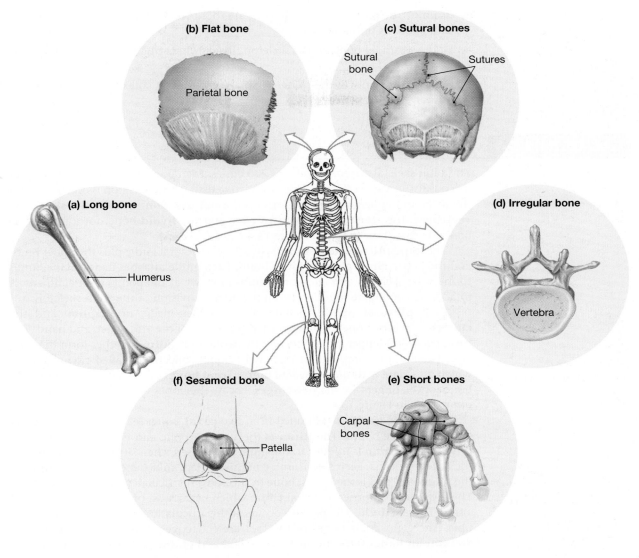

Figure 13.3 Shapes of Bones

bones form inside tendons. The patella (kneecap) develops inside the tendons anterior to the knee and is the largest sesamoid bone. **Sutural,** or **Wormian, bones** occur along suture lines of the skull wherever the suture has branched and surrounded a small piece of bone. The number of sutural bones varies from one person to another and is not included when counting the number of bones in the skeletal system.

QuickCheck Questions

3.1 Give examples of two types of short bones.

3.2 Give an example of an irregular bone.

| **3** **Materials** | **Procedures** |

☐ Articulated skeleton

1. Locate the long bones of the skeleton. How many do you find? _____

2. What types of bones are in the wrists and ankles of the skeleton?

   _____

3. Examine the bones of the skull. How are these bones classified? Do you see any sutural bones in the skull? _____

4. Find a sesamoid bone on the skeleton. Inside what structure does this bone develop? _____

5. Observe the bones of the spine. What are these bones collectively called, and how are they classified? _____ ■

LAB ACTIVITY 4   The Skeleton

The adult skeletal system consists of 206 bones. Each bone is an organ and includes osseous tissue, cartilage, and other connective tissues. The skeleton is organized into the axial and appendicular divisions. The **axial division** includes the **skull**, the **vertebrae**, the **sternum**, 12 pairs of **ribs**, and the **hyoid** bone, for a total of 80 bones. Locate these bones in Figure 13.4.

The **appendicular division** consists of the pectoral girdle plus the upper limbs and the pelvic girdle plus the lower limbs. Each girdle attaches its respective limbs to the axial skeleton and allows the limbs mobility at the points of attachment. A total of 126 bones are found in the appendicular division of the adult skeleton.

Each **pectoral girdle** includes the shoulder blade, or **scapula**; and the **clavicle**, or collar bone. Each **upper limb** consists of the arm, wrist, and hand. The **humerus** is the upper arm bone, and the **ulna** and **radius** together form the forearm. The hand is composed of eight wrist bones collectively called **carpal bones**, which articulate with the elongated **metacarpal bones** of the palm. The individual bones of the fingers are the **phalanges**. Locate the bones of the pectoral girdle and upper limb in Figure 13.4.

The **pelvic girdle** is fashioned from two **coxal bones**, each of which is an aggregate of three bones. The **ilium** is the superior bone and is the hip area of the torso. The **ischium** is inferior to the ilium and is used when sitting. The **pubis** is in the anterior pelvis. Each lower limb comprises a leg, ankle, and foot. The **femur** is the thighbone and is the largest bone in the body. It articulates with the coxal bone. The lower leg comprises a medial **tibia**, which bears most of the body weight, and a thin, lateral **fibula**. The **patella** occurs at the articulation between the femur and tibia. The eight ankle bones of the foot are collectively called the **tarsal bones**. **Metatarsal bones** form the arch of the foot, and **phalanges** form the toes. Locate the bones of the pelvic girdle and lower limb in Figure 13.4.

QuickCheck Questions

4.1 What are the two major divisions of the skeleton?

4.2 A rib belongs to which division of the skeletal system?

4 Materials

☐ Articulated skeleton

Procedures

1. Using Figure 13.4 as a guide, locate the bones of the axial division of the skeleton. List the major components of the axial division.

2. Using Figure 13.4 as a guide, locate the major components of the appendicular division of the skeleton.

3. What bones are found in the shoulder and upper limb?

4. What three bones fuse to form a coxal bone?

5. List the bones of the lower limb. ■

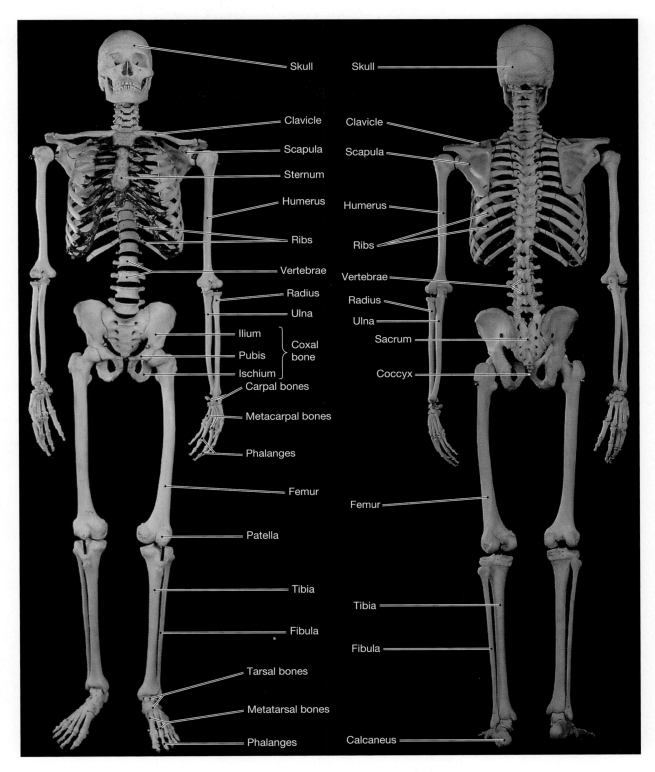

Figure 13.4 **The Skeleton**
(a) Anterior view. (b) Posterior view.

LAB ACTIVITY 5   Bone Markings

Each bone has certain anatomical features on its surface, called either **bone markings** or surface markings. A particular bone marking may be unique to a single bone or may occur throughout the skeleton. Table 13.1 illustrates examples of bone markings and organizes the markings into five groups. The first group includes general anatomical structures, and the second group lists bony structures for tendon and ligament attachment. The third group contains structures that occur at sites of articulation with other bones. The last two groups include depressions and openings.

5 *Materials*

☐ Articulated skeleton

Procedures

1. Using Table 13.1 for reference, locate on the skeleton
    *   A *foramen* in the skull. Describe this structure. _____
    *   A *fossa* on a humerus. How is the fossa different from a foramen? _____
    *   A *head* on a femur and humerus. Which other bones have a head? _____
    *   A *condyle* on two different bones. Describe this structure. _____
    *   A *tuberosity* on a humerus. What is the texture of this structure? _____

2. Locate one instance of each of the other marking types on the skeleton: process, ramus, trochanter, tubercle, crest, line, spine, neck, trochlea, facet, sulcus, canal, fissure, sinus. ■

Table 13.1 An Introduction to Skeletal Terminology

General Description	Anatomical Term	Definition
Elevations and projections (general)	Process	Any projection or bump
	Ramus	An extension of a bone making an angle with the rest of the structure
Processes formed where tendons or ligaments attach	Trochanter	A large, rough projection
	Tuberosity	A smaller, rough projection
	Tubercle	A small, rounded projection
	Crest	A prominent ridge
	Line	A low ridge
	Spine	A pointed process
Processes formed for articulation with adjacent bones	Head	The expanded articular end of an epiphysis, separated from the shaft by the neck
	Neck	A narrow connection between the epiphysis and the diaphysis
	Condyle	A smooth, rounded articular process
	Trochlea	A smooth, grooved articular process shaped like a pulley
	Facet	A small, flat articular surface
Depressions	Fossa	A shallow depression
	Sulcus	A narrow groove
Openings	Foramen	A rounded passageway for blood vessels or nerves
	Canal	A passageway through the substance of a bone
	Fissure	An elongated cleft
	Sinus or antrum	A chamber within a bone, normally filled with air

Femur

Skull

Humerus

Pelvis

Skeletal System Overview

Name _____

Date _____

Section _____

A. Matching

Match each structure in the left column with its correct description from the right column.

_____ 1. lacuna
_____ 2. trabecula
_____ 3. articular cartilage
_____ 4. diaphysis
_____ 5. epiphyseal plate
_____ 6. osteon
_____ 7. periosteum
_____ 8. interstitial lamellae
_____ 9. epiphysis
_____ 10. endosteum
_____ 11. medullary cavity
_____ 12. epiphyseal line
_____ 13. concentric lamellae
_____ 14. osteoblast

A. bone shaft
B. found in juvenile bones
C. bony column
D. contains yellow marrow
E. bone tip
F. lines marrow cavity
G. found in adult bones
H. forms osteon
I. bony projection
J. produces bone matrix
K. cellular space in bone matrix
L. remodeled osteons
M. outer membrane of bone
N. cartilage on epiphysis

B. Matching

Match each bone with the correct division and part of the skeleton. Each question may have more than one answer, and each choice can be used more than once.

_____ 1. scapula
_____ 2. coxal bone
_____ 3. patella
_____ 4. hyoid
_____ 5. radius
_____ 6. metacarpal
_____ 7. vertebra
_____ 8. clavicle
_____ 9. rib
_____ 10. femur
_____ 11. sternum
_____ 12. carpal

A. axial division
B. appendicular division
C. pectoral girdle
D. upper limb
E. pelvic girdle
F. lower limb

13

C. Labeling

Label Figure 13.5.

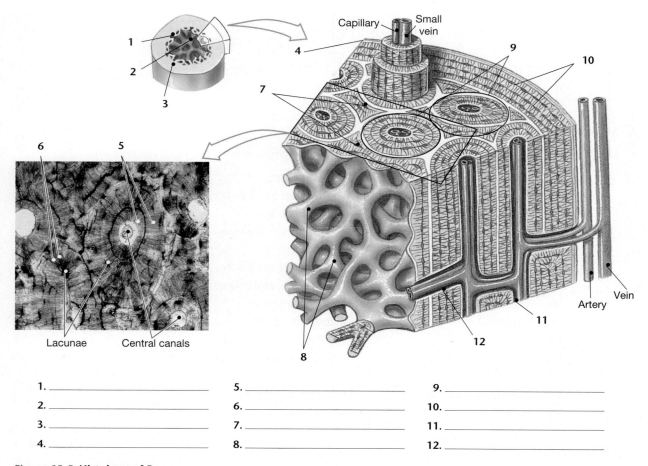

1. _____	5. _____	9. _____
2. _____	6. _____	10. _____
3. _____	7. _____	11. _____
4. _____	8. _____	12. _____

Figure 13.5 **Histology of Bone**

D. Short-Answer Questions

1. List the components of the axial skeleton.

2. List the components of the appendicular skeleton.

3. Describe the five types of surface markings on bones.

4. List the different shapes of bones.

E. Analysis and Application

1. Where does spongy bone occur in the skeleton?

2. How are the upper limbs attached to the axial skeleton?

3. Where does growth in length occur in a long bone?

The Axial Skeleton

OBJECTIVES

On completion of this exercise, you should be able to:

- Identify the components of the axial skeleton.
- Identify the cranial and facial bones of the skull.
- Identify the surface features of the cranial and facial bones.
- Describe the skull of a fetus.
- Describe the four regions of the vertebral column and distinguish among the vertebrae of each region.
- Identify the features of a typical vertebra.
- Discuss the articulation of the ribs with the thoracic vertebrae.
- Identify the components of the sternum.

The human skeleton consists of 206 bones, 80 of which are in the axial division. As noted in Exercise 13, the axial skeleton includes the skull, thoracic cage, and vertebral column (Figure 14.1). The axial skeleton provides a central framework for attachment of the appendicular skeleton (the shoulders, arms, hands, hips, legs, and feet). Internal organs such as the brain, heart, and lungs, are protected by the bones of the axial skeleton.

LAB ACTIVITY **1** Overview of the Skull

Of the 29 bones of the skull, 22 are organized into 14 **facial bones** and 8 **cranial bones** encasing the brain (Figure 14.2). The seven other bones of the skull, referred to as the *associated bones,* are the six bones of the middle ear (three auditory ossicles per ear) and the hyoid bone. The auditory ossicles are described in Exercise 31, "The Anatomy of the Ear."

The Cranium

The **frontal** bone of the cranium extends from the forehead posteriorly to the **coronal suture** and articulates with the two **parietal** bones (Figure 14.3). The parietal bones are joined superiorly by the **sagittal suture**. The **occipital** bone meets the parietals at the **lambdoid** (LAM-doyd) **suture**, completing the posterior wall of the cranium. The lambdoid suture is sometimes called the *occipitoparietal suture.*

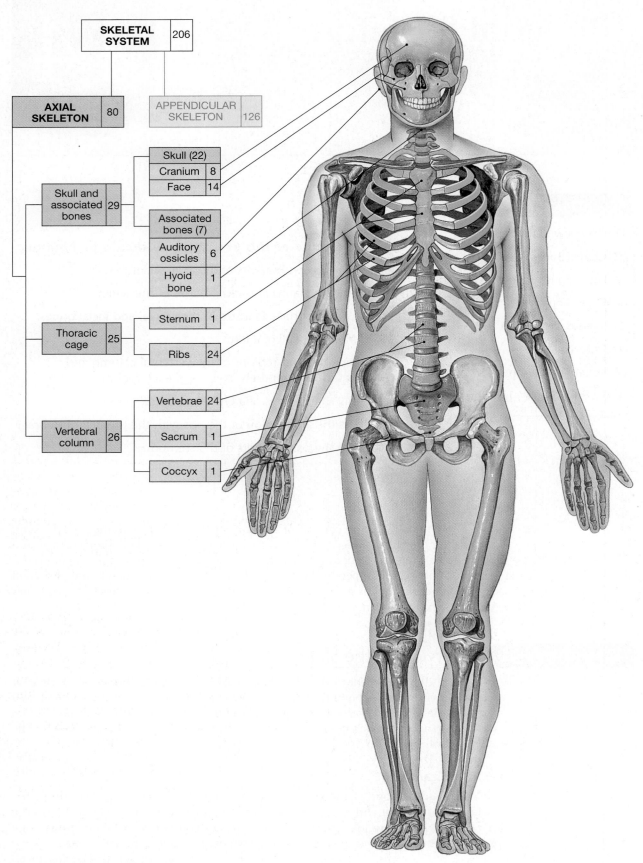

SKELETAL SYSTEM	206
AXIAL SKELETON	80
APPENDICULAR SKELETON	126

| Skull and associated bones | 29 |

Skull (22)	
Cranium	8
Face	14

Associated bones (7)	
Auditory ossicles	6
Hyoid bone	1

| Thoracic cage | 25 |

| Sternum | 1 |
| Ribs | 24 |

| Vertebral column | 26 |

Vertebrae	24
Sacrum	1
Coccyx	1

Figure 14.1 The Axial Skeleton

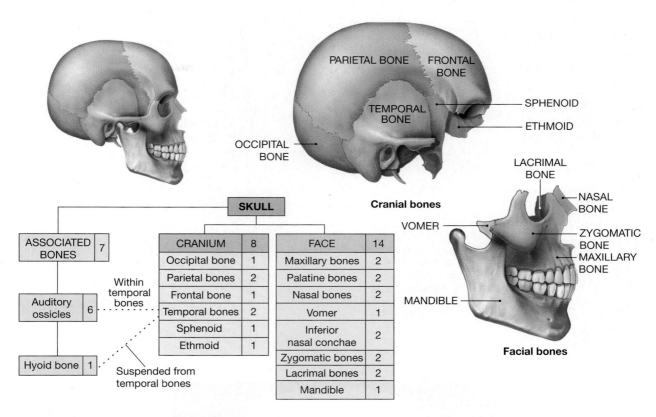

Figure 14.2 Cranial and Facial Subdivisions of the Skull
The seven associated bones are not illustrated.

The **temporal** bone articulates with the parietal bone at the **squamous suture**. The squamous and coronal sutures are linked by the **sphenoparietal suture**. Below this suture, the unpaired **sphenoid** is partly visible in Figure 14.3 as a vertical rectangle of bone. The sphenoid and parts of the frontal, temporal, and occipital bones form the floor of the cranium. Although it horizontally spans the floor of the cranium, the sphenoid is visible only on the lateral surface of the skull and in the back wall of the eye orbit. The single **ethmoid** is a small, rectangular bone deep in the eye orbit, behind the bridge of the nose (Figure 14.4). It forms the medial wall of both orbits.

The Face

The face is constructed of 14 bones: 2 nasal, 2 maxillary, 2 lacrimal, 2 zygomatic, 2 palatine, 2 inferior nasal conchae, the vomer, and the mandible. The small **nasal** bones form the bridge of the nose (Figure 14.4). Lateral to the nasals are the **maxillary** bones, or maxillae; these bones form the floor of the eye orbits and extend inferiorly to form the upper jaw. Below the eye orbits are the **zygomatic** bones, commonly called the cheekbones. At the bridge of the nose, lateral to each maxillary bone, are the small **lacrimal** bones of the medial eye orbitals. Through each lacrimal bone passes a small canal that allows tears to drain into the nasal cavity. The **inferior nasal conchae** (KONG-kē) are the lower shelves of bone in the nasal cavity. The other conchae in the nasal cavity are part of the ethmoid. The bone of the lower jaw is the **mandible**.

On the inferior surface of the skull, shown in Figure 14.5, the **palatine** bones form the posterior roof of the mouth next to the last molar tooth. A thin bone called the **vomer** divides the nasal cavity.

QuickCheck Questions

1.1 List the eight bones of the cranium.

1.2 List the 14 facial bones.

1.3 Describe the five main sutures of the skull.

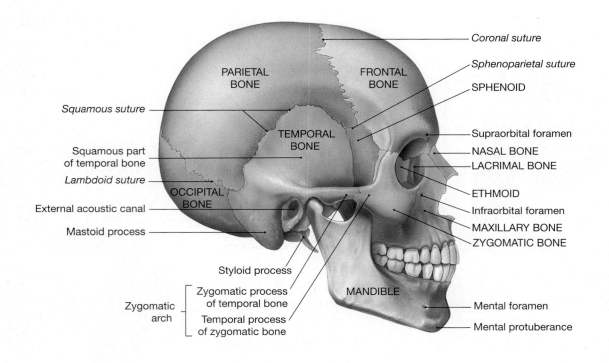

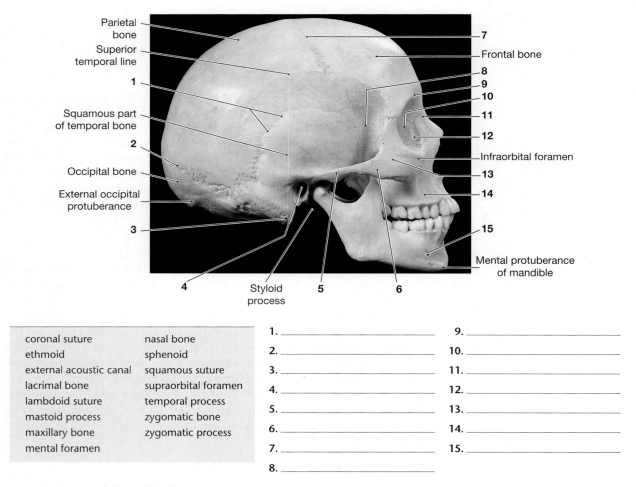

Figure 14.3 **Lateral View of Skull**

coronal suture	nasal bone
ethmoid	sphenoid
external acoustic canal	squamous suture
lacrimal bone	supraorbital foramen
lambdoid suture	temporal process
mastoid process	zygomatic bone
maxillary bone	zygomatic process
mental foramen	

1. _____
2. _____
3. _____
4. _____
5. _____
6. _____
7. _____
8. _____

9. _____
10. _____
11. _____
12. _____
13. _____
14. _____
15. _____

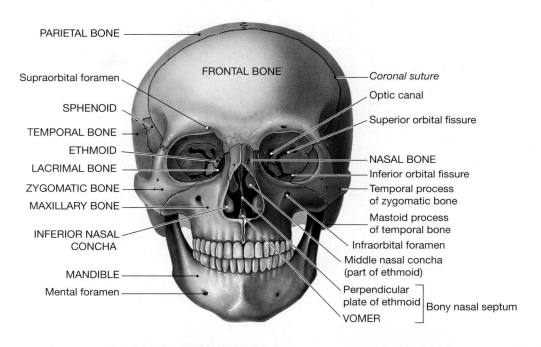

PARIETAL BONE

FRONTAL BONE

Supraorbital foramen

SPHENOID

TEMPORAL BONE

ETHMOID

LACRIMAL BONE

ZYGOMATIC BONE

MAXILLARY BONE

INFERIOR NASAL CONCHA

MANDIBLE

Mental foramen

Coronal suture

Optic canal

Superior orbital fissure

NASAL BONE

Inferior orbital fissure

Temporal process of zygomatic bone

Mastoid process of temporal bone

Infraorbital foramen

Middle nasal concha (part of ethmoid)

Perpendicular plate of ethmoid

VOMER

Bony nasal septum

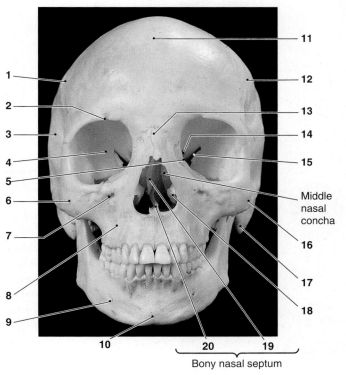

Middle nasal concha

Bony nasal septum

coronal suture
frontal bone
inferior nasal concha
infraorbital foramen
lacrimal bone
mastoid process
maxillary bone
mental foramen
mental protuberance
nasal bone
optic canal
parietal bone
perpendicular plate
sphenoid
superior orbital fissure
supraorbital foramen
temporal bone
temporal process
vomer
zygomatic bone

1. _____
2. _____
3. _____
4. _____
5. _____
6. _____
7. _____

8. _____
9. _____
10. _____
11. _____
12. _____
13. _____
14. _____

15. _____
16. _____
17. _____
18. _____
19. _____
20. _____

Figure 14.4 Anterior View of Skull

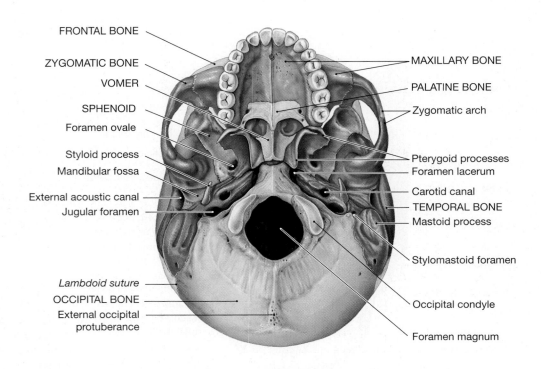

FRONTAL BONE
ZYGOMATIC BONE
VOMER
SPHENOID
Foramen ovale
Styloid process
Mandibular fossa
External acoustic canal
Jugular foramen
Lambdoid suture
OCCIPITAL BONE
External occipital protuberance

MAXILLARY BONE
PALATINE BONE
Zygomatic arch
Pterygoid processes
Foramen lacerum
Carotid canal
TEMPORAL BONE
Mastoid process
Stylomastoid foramen
Occipital condyle
Foramen magnum

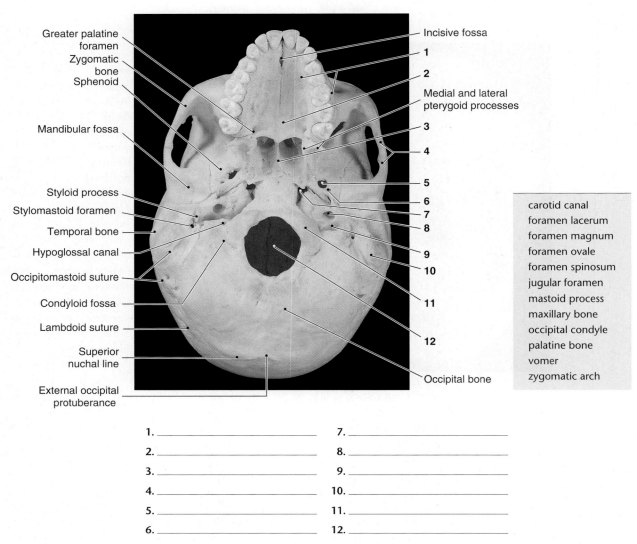

Greater palatine foramen
Zygomatic bone
Sphenoid
Mandibular fossa
Styloid process
Stylomastoid foramen
Temporal bone
Hypoglossal canal
Occipitomastoid suture
Condyloid fossa
Lambdoid suture
Superior nuchal line
External occipital protuberance

Incisive fossa
1
2
Medial and lateral pterygoid processes
3
4
5
6
7
8
9
10
11
12
Occipital bone

carotid canal
foramen lacerum
foramen magnum
foramen ovale
foramen spinosum
jugular foramen
mastoid process
maxillary bone
occipital condyle
palatine bone
vomer
zygomatic arch

1. _____   7. _____
2. _____   8. _____
3. _____   9. _____
4. _____   10. _____
5. _____   11. _____
6. _____   12. _____

Figure 14.5 Inferior View of Skull

158

1 *Materials*

□ Skull (unsectioned)

Procedures

1. Identify each cranial bone on a skull.
2. Identify each facial bone on a skull.
3. Determine the general location of each skull bone on your own head.
4. Insert your thumb halfway into one eye orbit of the study skull and your forefinger halfway into the other eye orbit. Gently pinch the bone between the orbits, which is the ethmoid. Remove your fingers and examine the medial wall of each orbit. ■

LAB ACTIVITY   2   Cranial Bones

The floor of the cranium has three depressions called *fossae* (Figure 14.6). The **anterior cranial fossa** is mainly the depression that forms the base of the frontal bone. Small portions of the ethmoid and sphenoid also contribute to the floor of this area. The **middle cranial fossa** is a depressed area extending over the sphenoid and the temporal and occipital bones. The **posterior cranial fossa** is found in the occipital bone.

Frontal Bone

The frontal bone forms the roof, walls, and floor of the anterior cranium (Figure 14.7). The **frontal squama** is the flattened expanse commonly called the forehead. In the midsagittal plane of the squama is the **metopic suture**, where the two frontal bones fused in early childhood (typically by the time the child was eight years old). As natural remodeling of bone occurs, this suture often disappears by age 30. The frontal bone forms the upper portion of the eye orbit. Superior to the orbit is the **supraorbital foramen,** which on some skulls occurs not as a complete hole but rather as a small notch, the **supraorbital notch.** In the anterior and medial regions of the orbit, the frontal bone forms the **lacrimal fossa**, an indentation for the lacrimal gland, which moistens and lubricates the eye.

Occipital Bone

The occipital bone forms the posterior floor and wall of the skull (Figure 14.8a). The most conspicuous structure of the occipital bone is the **foramen magnum**, the large hole where the spinal cord enters the skull and joins the brain. Along the lateral margins of the foramen magnum are flattened **occipital condyles** that articulate with the first vertebra of the spine. Passing under each occipital condyle is the **hypoglossal canal**, a passageway for the hypoglossal nerve, which controls muscles of the tongue and throat.

The occipital bone has many external surface marks that show where muscles and ligaments attach. The **external occipital crest** is a ridge that extends posteriorly from the foramen magnum to a small bump, the **external occipital protuberance**. Wrapping around the occipital bone lateral from the crest and protuberance are the **superior** and **inferior nuchal** (NOO-kul) **lines**, surface marks indicating where muscles of the neck attach to the skull.

Parietal Bones

The parietal bones form the posterior crest of the skull and are joined by the sagittal suture. The bones are smooth and have few surface features. The low ridges of the **superior and inferior temporal lines**, shown in Figure 14.8b, are above the squamosal suture, where a muscle for chewing attaches. No major foramina pass through the parietal bones.

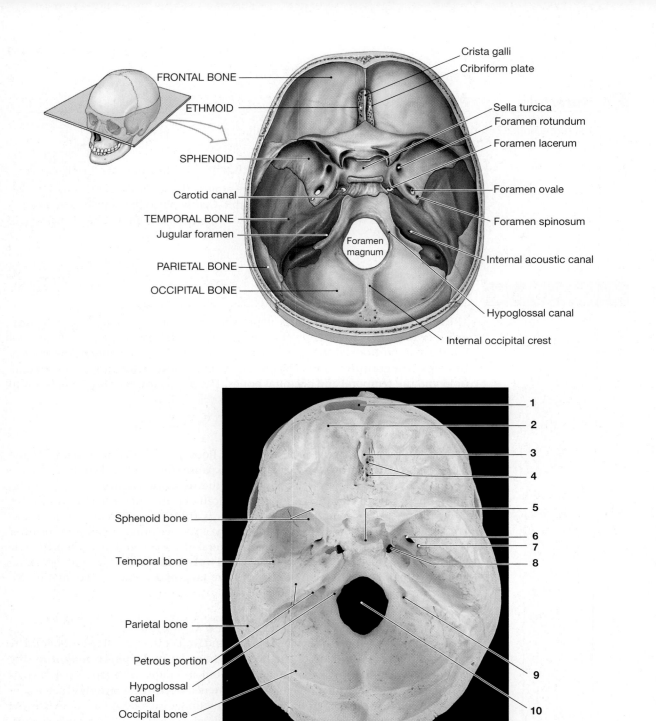

Crista galli
Cribriform plate
FRONTAL BONE
ETHMOID
Sella turcica
Foramen rotundum
Foramen lacerum
SPHENOID
Carotid canal
Foramen ovale
TEMPORAL BONE
Jugular foramen
Foramen spinosum
Foramen
magnum
PARIETAL BONE
Internal acoustic canal
OCCIPITAL BONE
Hypoglossal canal
Internal occipital crest

Sphenoid bone
Temporal bone
Parietal bone
Petrous portion
Hypoglossal canal
Occipital bone

1
2
3
4
5
6
7
8
9
10

cribriform plate	foramen spinosum
crista galli	frontal bone
foramen lacerum	frontal sinus
foramen magnum	jugular foramen
foramen ovale	sella turcica

1. _____
2. _____
3. _____
4. _____
5. _____

6. _____
7. _____
8. _____
9. _____
10. _____

Figure 14.6 Sectional Anatomy of the Skull
Horizontal section through the skull, showing the floor of the cranial cavity.

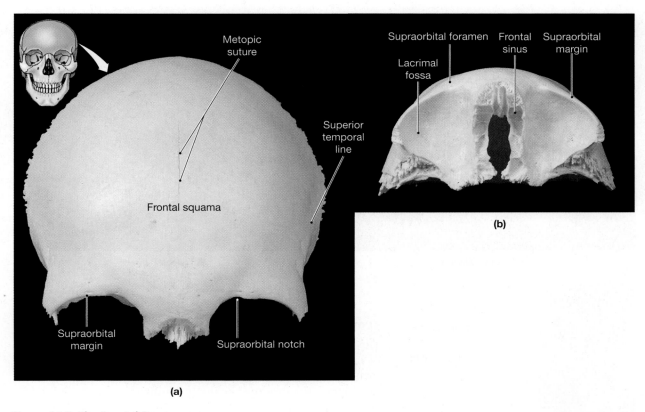

Figure 14.7 The Frontal Bone
(a) Anterior surface. (b) Inferior (orbital) surface.

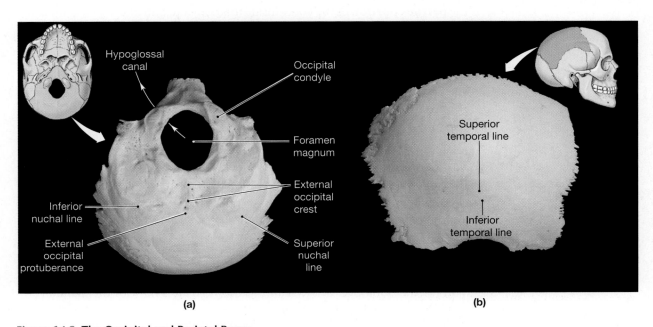

Figure 14.8 The Occipital and Parietal Bones
(a) Occipital bone, inferior view. (b) Right parietal bone, lateral view.

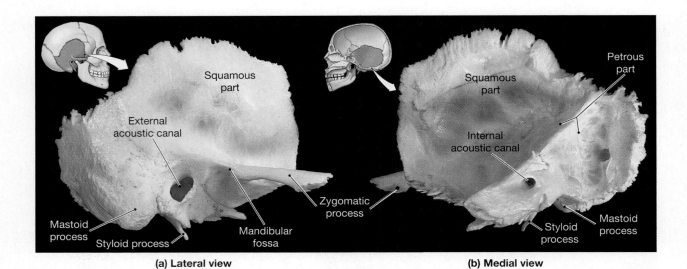

(a) Lateral view   **(b) Medial view**

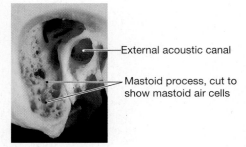

(c) Mastoid air cells

Figure 14.9

The Temporal Bones

(a, b) The right temporal bone. **(c)** A cutaway view of the mastoid air cells.

Temporal Bones

The temporal bones form the lower lateral walls and part of the floor of the middle cranial fossa. Each temporal bone is complex, with many surface features. Figure 14.9a details the lateral aspect of a temporal bone. The **external acoustic (auditory) canal** is a tube for conducting sound waves toward the eardrum. The large, flattened area above the canal is the **squamous portion**. Directly posterior to the external acoustic canal is the **mastoid process,** where a muscle that moves the head attaches. Anterior to the external acoustic canal is a thin extension of bone, the **zygomatic process**. The name of this structure refers to the bone it articulates with, not the bone it occurs on. This process joins a process of the zygomatic bone and forms the **zygomatic arch** located just below the temple.

Figure 14.9b shows a large crest arising from the floor of the temporal bone. This is the **petrous portion** of the bone. Inside this bony ridge are the organs for hearing and equilibrium and the tiny bones of the ear, the auditory ossicles. The **internal acoustic (auditory) canal** is on the posterior medial surface of the petrous portion. The internal acoustic canal is not continuous with the external acoustic canal. Directly posterior to the internal acoustic canal is the large **jugular foramen**. This foramen is an exit hole for cranial nerves and the jugular vein, a large vein that drains blood from the brain. On the anterior side of the petrous portion, at the medial border of the temporal bone, is the **carotid canal**, an entrance into the skull for the internal carotid artery that supplies oxygenated blood to the brain. The carotid canal tunnels medially through the temporal bone and exits the floor of the skull.

Study Tip | ***Petrous Portion***

On the floor of the temporal bone, the petrous portion separates the jugular foramen posteriorly from the anterior carotid canal. Imagine the petrous portion as a mountain ridge with a passageway on each side. Notice, however, how close the jugular foramen and carotid canal are on the inferior view of the skull shown in Figure 14.5. ●

In the inferior aspect of the skull, shown in Figure 14.5 with the mandible removed, a good anatomical landmark on the temporal bone is the long, needle-like **styloid** (STĪ-loyd; *stylos,* pillar) **process** located medial and anterior to the mastoid process. The delicate styloid process is a site for muscle attachment and is frequently damaged on study skulls. Between the styloid and the mastoid processes is a small foramen, the **stylomastoid foramen**, where the facial nerve exits the cranium.

(a)

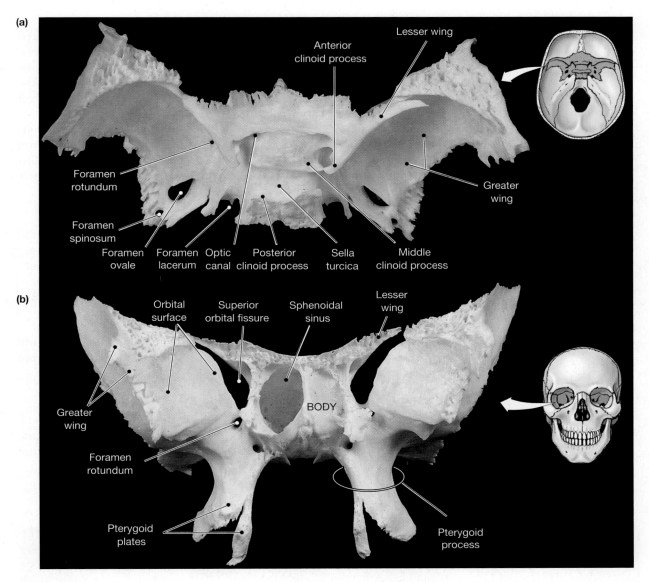

Figure 14.10 The Sphenoid
(a) Superior surface. (b) Anterior surface.

Anterio-inferior to the external acoustic canal is the **mandibular fossa,** a shallow depression where the mandible bone articulates with the temporal bone.

The Sphenoid

All cranial bones articulate with the sphenoid, which is the hub of the cranium. The sphenoid can be seen in all views of the skull (Figures 14.3–14.6). The easiest aspect of this bone to work with is the floor, shown in Figures 14.6 and 14.10. The anterior margin of the bone is the bat-shaped **lesser wing.** The **greater wing** is posterior to the lesser wing and contributes to the floor of the middle cranial fossa. The greater wing is also visible on the lateral surface of the skull, just inferior to the sphenoparietal suture. In the middle of the sphenoid is a raised platform, the **sella turcica** (TUR-si-kuh), commonly called the Turk's saddle. The depression in the sella turcica, where the pituitary gland of the brain sits, is the **hypophyseal** (hī-pō-FIZ-ē-ul) **fossa.** The **anterior, middle,** and **posterior clinoid** (KLĪ-noyd; *kline,* a bed) **processes** form the upper rims of the sella turcica.

Four pairs of foramina are aligned to each side of the sella turcica and serve as passageways for blood vessels and nerves. The **foramen ovale** (ō-VAH-lē; oval) is the oval hole. Posterior to the foramen ovale is a small **foramen spinosum**. These foramina are passageways for parts of the trigeminal nerve of the head. The **foramen rotundum** is the foramen with the tubular entrance anterior to the foramen ovale. Directly medial to the foramen ovale, where the sphenoid joins the temporal bone, is the **foramen lacerum** (LA-se-rum; *lacerare,* to tear), where the auditory (eustachian) tube enters the skull. Frequently, the carotid canal merges with the nearby foramen lacerum to form a single passageway.

Anterior to the foramen rotundum is a cleft in the sphenoid, the **superior orbital fissure**, where nerves to the ocular muscles pass. The **inferior orbital fissure** is the crevice at the lower margin of the sphenoid near the maxillary and other bones of the orbital floor. At the base of the anterior clinoid process is the **optic canal,** where the optic nerve enters the skull to carry visual signals to the brain.

Study Tip | ### Using Foramina as Landmarks

Notice the pattern of how the foramina line up along the sphenoid. Use the foramen ovale as a landmark, because it is easy to identify by its oval shape. Anterior to the foramen ovale is the foramen rotundum; posterior is the foramen spinosum. Medial to the foramen ovale is the foramen lacerum with the nearby carotid canal. ●

Extending vertically from the inferior surface of the sphenoid are the **pterygoid** (TER-i-goyd; *pterygion,* wing) **processes.** Each process divided into two **pterygoid plates,** where muscles of the mouth attach.

The Ethmoid

The ethmoid is a single rectangular bone immediately posterior to the lacrimal bone (Figure 14.4). It forms the medial walls of the orbits of the eyes, part of the septum of the nose, and part of the cranial floor. Observe in Figure 14.11 the vertical crest

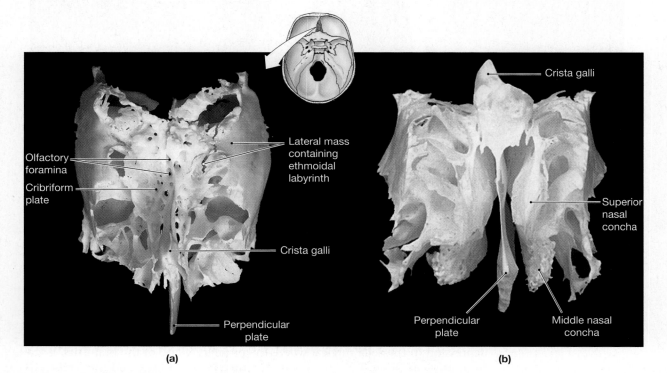

Olfactory foramina

Cribriform plate

Lateral mass containing ethmoidal labyrinth

Crista galli

Perpendicular plate

(a)

Crista galli

Superior nasal concha

Perpendicular plate

Middle nasal concha

(b)

Figure 14.11 The Ethmoid
(a) Superior surface. (b) Posterior surface.

of bone called the **crista galli** (*crista*, crest + *gallus*, chicken; cock's comb), where membranes that protect and support the brain are attached. At the base of the crista galli is a screen-like **cribriform** (*cribrum*, sieve) **plate** with many small **olfactory foramina** that serve as passageways for branches of the olfactory nerve. The inferior ethmoid has a thin sheet of vertical bone, the **perpendicular plate**, that contributes to the septum of the nasal cavity. On each side of the perpendicular plate are the **lateral masses** that enclose **ethmoidal air cells** (also called the *ethmoidal sinuses*), which connect to the nasal cavity. Extending inferiorly into the nasal cavity from the lateral masses are the **superior** and **middle nasal conchae**.

QuickCheck Questions

2.1 How do bones of the cranium articulate with the sphenoid?

2.2 Describe the location of the ethmoid in the orbit of the eye.

2.3 Explain the difference between a foramen, a canal, a fossa, a condyle, and a process.

2 Materials

- ☐ Skull sectioned horizontally
- ☐ Disarticulated ethmoid

Procedures

1. Review and label the cranial structures in Figures 14.3–14.6.
2. Locate the frontal bone on the skull.
   - Identify the frontal squama, supraorbital foramen, and lacrimal fossa.
   - Is the metopic suture visible on the skull?
3. Locate the parietal bones on the skull. Examine the lateral surface of a parietal bone and locate the superior and inferior temporal lines.
4. Identify the occipital bone on the skull.
   - Locate the foramen magnum, the occipital condyles, and the hypoglossal canal.
   - Locate the external occipital crest, external occipital protuberance, and superior and inferior nuchal lines.
5. Examine the temporal bones on the skull.
   - Locate the squamous and petrous portions, mastoid processes, and zygomatic processes. Can you feel the mastoid process on your own skull?
   - Find the mandibular fossa.
   - Identify the major passageways of the temporal bone: external and internal auditory canals, jugular foramen, and carotid canal.
   - Identify the styloid process and the stylomastoid foramen.
6. Examine the sphenoid and determine its borders with other bones.
   - Identify the lesser wings, greater wings, and sella turcica.
   - Observe the structure of the sella turcica, which includes the anterior, middle, and posterior clinoid processes.
   - Identify each foramen of the sphenoid: ovale, spinosum, rotundum, and lacerum.
   - Locate the optic canal and the superior and inferior orbital fissure.
   - On the inferior sphenoid, identify the pterygoid processes and the pterygoid plates.
7. Identify the ethmoid on the skull. Closely examine its location within the orbit.
   - Observe on the floor of the skull the crista galli, cribriform plate, and olfactory foramina.
   - Examine the perpendicular plate in the nasal cavity. Examine a disarticulated ethmoid and identify the lateral masses and the superior and middle nasal conchae. ■

LAB ACTIVITY   3   Facial Bones

Maxillary Bones

The paired maxillary bones are the foundation of the face (Figure 14.12). Inferior to the orbit is the **infraorbital foramen**. The **alveolar process** consists of the U-shaped processes where the upper teeth are embedded in the maxillary. From the inferior aspect, the **palatine process** of the maxillary is visible. This bony shelf forms the anterior hard palate of the mouth. At the anterior margin of the palatine process is the **incisive fossa**.

Palatine Bones

The palatine bones are small bones in the roof of the mouth posterior to the palatine processes. The palatine bones and maxillary bones together form the roof of the mouth and separate the oral cavity from the nasal cavity. This separation of cavities allows us to chew and breathe at the same time. Each palatine bone has a **greater palatine foramen** on the lateral margin, as detailed in Figure 14.5b. Only the inferior portion of the palatine bone is completely visible. The superior surface forms the floor of the nasal cavity and supports the vomer bone.

Zygomatic Bones

The cheekbones are the zygomatic bones (Figure 14.12). These bones also contribute to the floor and lateral walls of the orbit. Lateral and slightly inferior to the orbit is the small **zygomaticofacial foramen**. The posterior margin of the zygomatic bone narrows inferiorly to the **temporal process,** which joins the temporal bone zygomatic process to complete the zygomatic arch.

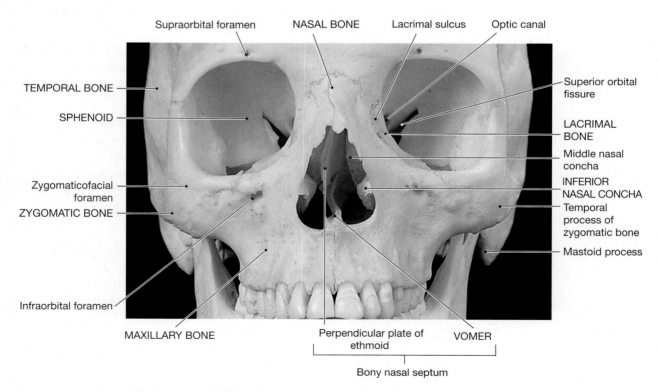

Figure 14.12 The Smaller Bones of the Face

Study Tip | ***Zygomatic Arch***

Each process of the zygomatic arch is named according to the bone with which it articulates. The *temporal* process is on the *zygomatic* bone and articulates with the *zygomatic* process of the *temporal* bone. ●

Lacrimal Bones

The lacrimal bones are the anterior portions of the medial orbital wall, illustrated in Figure 14.12. Each lacrimal bone is named after the lacrimal glands that produce tears. Tears flow medially across the eye and drain into the inferior **lacrimal fossa,** which transports them to the nasal cavity.

Nasal Bones

As Figure 14.12 shows, the nasal bones are the bridge of the nose. These bones articulate superiorly with the frontal bone, laterally with the maxillary bones, and posteriorly (internally) with the ethmoid.

The Vomer

The vomer separates the nasal chamber into right and left cavities. This thin sheet of bone is best viewed from the inferior aspect of the skull looking into the nasal cavities, as in Figure 14.12. The midsagittal section of the skull shown in Figure 14.13 illustrates the relationship between the bones of the nasal septum. The perpendicular plate of the ethmoid forms the superior portion of the septum, and the vomer is the inferior part of the septum.

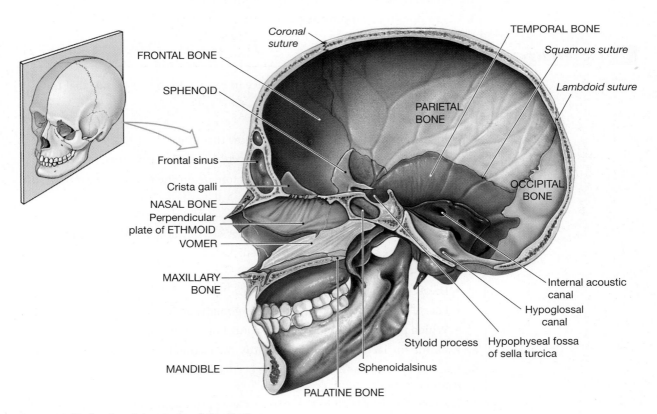

Figure 14.13 Sectional Anatomy of the Skull
Medial view of a sagittal section through the skull.

Inferior Nasal Conchae

The inferior nasal conchae are shelves that extend medially from the lower lateral portion of the nasal wall (Figure 14.12). They cause inspired air to swirl in the nasal cavity so that the moist mucous membrane lining can warm, cleanse, and moisten the air. Similar shelves of bone occur on the lateral walls of the ethmoid bone.

The Mandible

The mandible is the U-shaped bone of the lower jaw, detailed in Figures 14.14a and b. The horizontal **body** bends posteriorly at the **angle** to a raised projection, the **ramus**, which terminates at a U-shaped **mandibular notch**. Two processes extend upward from the notch, the anterior **coronoid** (kuh-RŌ-noyd) **process** and a posterior **condylar process**. The smooth condylar process, also called the **mandibular condyle**, articulates in the mandibular fossa on the temporal bone to form the **temporomandibular joint** (TMJ). Open and close your mouth to feel this articulation. The **alveolar process** is the crest of bone where the lower teeth articulate with the mandible bone. Lateral to the chin, or **mental protuberance** (*mental*, chin), is the mental foramen. The medial mandibular surface, shown in Figure 14.14b, features the **mandibular groove,** where the submandibular salivary gland rests against the bone. At the posterior end of the groove is the **mandibular foramen**, a passageway for a sensory nerve from the lower teeth and gums.

QuickCheck Questions

3.1 Which facial bones contribute to the orbit of the eye?

3.2 Which facial bones form the roof of the mouth?

3.3 How does the mandible bone articulate with the cranium?

3 *Materials*	*Procedures*

☐ Skull

1. Review the skeletal features of the face in Figures 14.3–14.14, and label the numbered elements in Figure 14.14.

2. Locate the maxillary bones on a skull.
   - Identify the infraorbital foramen below the orbit.
   - Locate the alveolar process, palatine process, and incisive fossa.
   - Feel your hard palate by placing your tongue on the roof of your mouth just behind your upper teeth.

3. Examine the palatine bones.
   - With which part of the maxillary bones do they articulate?
   - Identify the greater palatine foramen.

4. Identify the zygomatic bones.
   - Locate the zygomaticofacial foramen.
   - Locate the temporal process of the zygomatic arch.
   - Which part of the temporal bone contributes to the zygomatic arch?

5. Examine the lacrimal bone and identify the lacrimal fossa.

6. Locate the nasal bones. Which bone occurs between a nasal bone and a lacrimal bone?

7. Locate the vomer both in the inferior view of the skull and in the nasal cavity.

8. Identify the inferior nasal conchae in the nasal cavity.

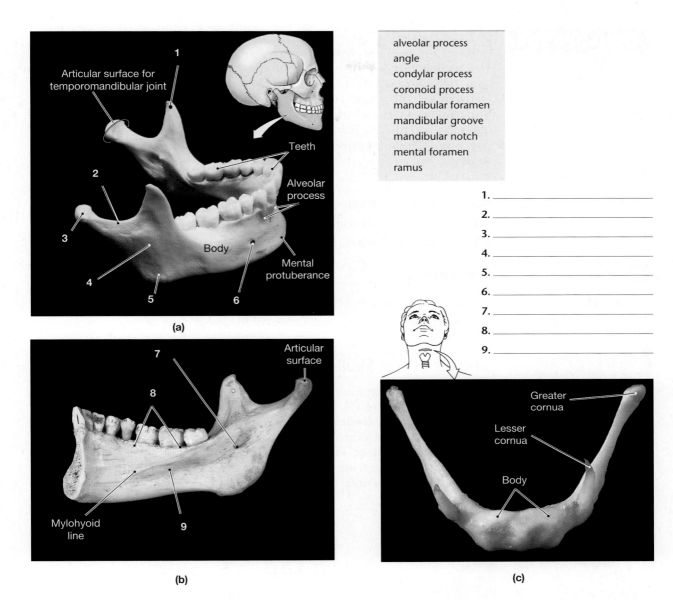

alveolar process
angle
condylar process
coronoid process
mandibular foramen
mandibular groove
mandibular notch
mental foramen
ramus

1. _____
2. _____
3. _____
4. _____
5. _____
6. _____
7. _____
8. _____
9. _____

Figure 14.14 The Mandible and Hyoid Bone
(a) A lateral view of the mandible. (b) A medial view of the right mandible. (c) An anterior view of the hyoid bone.

9. Examine the mandible. Disarticulate this bone from the skull if allowed to do so by your instructor.

- Identify the body, angle, and ramus.
- Identify the coronoid process, mandibular notch, and condylar process.
- Note how the mandible articulates with the temporal bone at the temporomandibular joint.
- Locate the alveolar process and the mental protuberance.
- On the medial surface of the mandible, locate the mandibular groove and the mandibular foramen. ■

LAB ACTIVITY 4   Hyoid Bone

The hyoid bone is a U-shaped bone inferior to the mandible (Figure 14.14c). This bone is unique because it does not articulate with any other bones. You cannot feel your hyoid bone because it is surrounded by ligaments and muscles of the throat and neck. Two horn-like processes for muscle attachment occur on the hyoid bone, an anterior pair of **lesser cornua** (KORN-ū-uh; *cornu-*, horn) and a larger pair of posterior **greater cornua**. These bony projections are also called the lesser and greater horns. Muscles that move the tongue and larynx attach to these pegs.

QuickCheck Questions

4.1 Where is the hyoid bone located?

4.2 Does the hyoid bone articulate with other bones?

4 Materials

☐ Articulated skeleton

Procedures

1. Examine the hyoid bone on an articulated skeleton.
2. Identify the greater and lesser cornua of the hyoid bone. ■

LAB ACTIVITY 5   Sinuses of the Skull

The skull contains four cavities called **paranasal sinuses** that connect with the nasal cavity (Figure 14.15). Like the nasal cavity, the sinuses are lined with a mucous membrane. The **frontal sinus** extends laterally over the orbit of the eyes. The **sphenoidal sinus** is located in the sphenoid directly inferior to the sella turcica. The ethmoid is full of **ethmoidal air cells** that collectively constitute the **ethmoidal sinus**. Each maxillary bone contains a large **maxillary sinus** situated lateral to the nasal cavity.

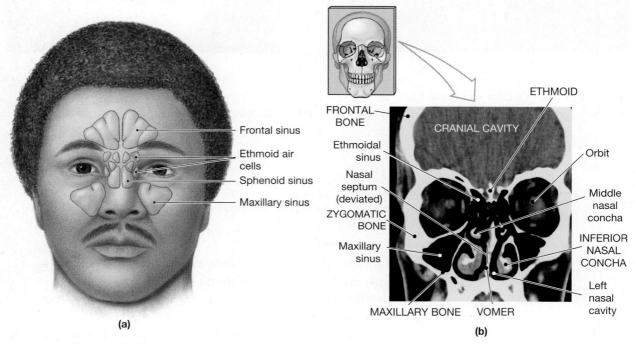

(a)

(b)

Figure 14.15 Paranasal Sinuses

(a) Location of the four paranasal sinuses. **(b)** An MRI scan showing a frontal section through the ethmoidal and maxillary sinuses.

Clinical Application | **Sinus Congestion**

In some individuals, allergies or changes in the weather can make the sinus membranes swell and secrete more mucus. The resulting congestion blocks connections with the nasal cavity, and the increased sinus pressure is felt as a headache. The sinuses also serve as resonating chambers for the voice, much like the body of a guitar amplifies its music, and when the sinuses and nasal cavity are congested, the voice sounds muffled. ▶

QuickCheck Questions

5.1 List the four types of paranasal sinuses.

5.2 What is the function of the sinuses?

5 Materials

☐ Skull (midsagittal section)

Procedures

1. Compare the frontal sinus on several sectioned skulls. Is the sinus the same size on each skull?

2. Locate the sphenoidal sinus on a sectioned skull. Under which sphenoidal structure is this sinus located?

3. Examine the maxillary sinus on a sectioned skull. How does the size of this sinus compare with the sizes of the other three sinuses?

4. Identify the ethmoidal air cells. What sinus do these cells collectively form? ■

LAB ACTIVITY 6   Fetal Skull

The fetal skull has many bones that are unfused and incompletely ossified. These bones develop as unfused patches in membranes of fibrous connective tissue. The soft spots, called **fontanels** (fon-tuh-NELZ), allow the skull to expand as the brain increases in size. At birth the skull is still not completely formed, and the fontanels enable it to flex and squeeze through the birth canal during delivery. By the age of four years, most of the bones have ossified across their fontanels.

Four major fontanels are present at birth (shown in Figure 14.16):

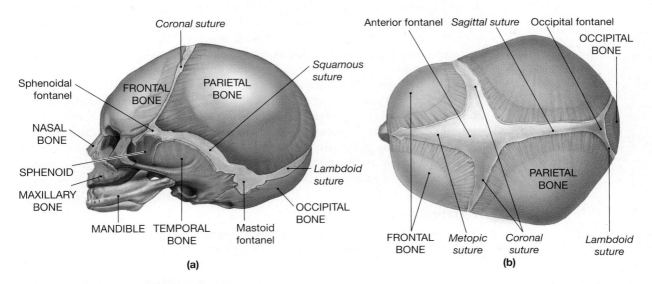

Figure 14.16 The Skull of an Infant

(a) A lateral view. The skull of an infant contains more individual bones than that of an adult. Many of the bones eventually fuse; thus, the adult skull has fewer bones. The flat bones of the skull are separated by areas of fibrous connective tissue, allowing for cranial expansion and the distortion of the skull during birth. The large fibrous areas are called fontanels. By about age four, these areas will disappear. (b) A superior view.

- The **anterior fontanel**, the largest, is located at the intersection of the frontal and parietal bones, where the frontal suture (between the two pieces of the immature frontal bone) meets the coronal and sagittal sutures.

- The **sphenoidal fontanel** is located at the intersection of the coronal and squamous sutures.

- The **mastoid fontanel** is located where the squamous and lambdoid sutures meet.

- The **occipital fontanel** is located at the junction of the sagittal and lambdoid sutures.

QuickCheck Questions

6.1 Why does the fetal skull have fontanels?

6.2 Why does the adult skull lack fontanels?

6 Materials

☐ Fetal skull

☐ Adult skull

Procedures

1. Identify each fontanel on a fetal skull, using Figure 14.16 as a guide.

2. Compare the fetal and adult skulls. Which has more bones? ■

LAB ACTIVITY 7 Vertebral Column

The vertebral column, or spine, is a flexible chain of 33 vertebrae. It articulates with the skull superiorly, the pelvic girdle inferiorly, and the ribs laterally. Vertebrae are grouped into four regions based on location and anatomical features (Figure 14.17). The first seven vertebrae are the **cervical** vertebrae of the neck. Twelve **thoracic** vertebrae articulate with the ribs. The lower back has five **lumbar** vertebrae, and a **sacrum** joining the hips is fashioned from five fused vertebrae. The **coccyx** (KOK-siks) is the tailbone and consists of three to five fused coccygeal vertebrae.

Notice in Figure 14.17 that the vertebral column is not straight but curved. Spinal curves are necessary to balance the body weight when a person is standing. At birth, the spinal column is generally rounded posteriorly like a bow, with **primary curvatures** in the thoracic and sacral regions. A few months after birth, **secondary curvatures** begin to develop in the cervical and lumbar regions. The cervical vertebrae start to curve anteriorly to support the head, which the baby can now hold up. The lumbar region begins to bend anteriorly to balance the body weight for standing. Once the child is approximately 10 years old, the spinal curves are established and the fully developed column has alternating secondary and primary curvatures.

From the cervical through the lumbar region are **intervertebral discs,** cushions of fibrocartilage between the articulating vertebrae. Each disc consists of an outer layer of strong fibrocartilage, the **annulus fibrosus,** surrounding an inner mass, the **nucleus pulposus**. Water and elastic fibers in the gelatinous mass of the nucleus pulposus absorb stresses that arise between vertebrae whenever a person is either standing or moving.

Vertebral Anatomy

Figure 14.18 illustrates the anatomical features of a typical vertebra, which consists of a large anterior vertebral body, a vertebral arch, and elongated processes that extend posteriorly.

Although there is regional specialization of the vertebrae, they share many surface features:

- The **body** is the thick, disc-shaped anterior portion; it is also called the **centrum**.

- The **spinous process** is a long, single extension of the posterior vertebral wall.

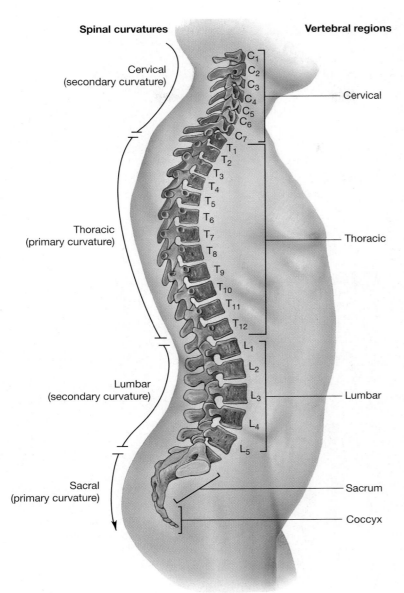

Spinal curvatures

Cervical (secondary curvature)

Thoracic (primary curvature)

Lumbar (secondary curvature)

Sacral (primary curvature)

Vertebral regions

C_1
C_2
C_3
C_4
C_5
C_6
C_7
T_1
T_2
T_3
T_4
T_5
T_6
T_7
T_8
T_9
T_{10}
T_{11}
T_{12}
L_1
L_2
L_3
L_4
L_5

Cervical

Thoracic

Lumbar

Sacrum

Coccyx

- The **transverse process** is an extension lateral to the spinous process.
- The **lamina** (LA-mi-na) is a flat plate of bone between the transverse and spinous processes.
- The **vertebral arch** is the span of bone formed by joined laminae.
- The **pedicle** (PE-di-kul) is a strut of bone extending posteriorly from the body to a transverse process.
- The **vertebral foramen** is a large hole posterior to the body; the anterior wall of the foramen is formed by the body, the lateral walls by the pedicles, and the roof by the lamina.
- The **inferior vertebral notch** is an inverted "U" on the inferior surface of the pedicle.
- The **superior articular process** is a projection on the superior surface of the pedicle.
- The **superior articular facet** is a smooth articular surface on the posterior tip of the superior articular process.
- The **inferior articular process** is a downward projection of the inferior lamina wall.
- The **inferior articular facet** is a smooth articular surface on the anterior tip of the inferior articular process.

Figure 14.17

The Vertebral Column

The major divisions of the vertebral column, showing the four spinal curvatures.

Vertebral Articulations

The vertebral column moves much like a gooseneck lamp: each joint moves slightly, but collectively they permit a wide range of motion. The greatest flexibility is in the cervical region for head movement. Closely examine the articulations between the vertebrae in Figure 14.18. Observe how the inferior articular process of the top vertebra articulates with the superior articular process of the vertebra immediately below. Also locate the **intervertebral foramen**, the hole created by the inverted U-shaped opening of the inferior vertebral notch on the upper vertebra and the roof formed by the pedicle of the lower vertebra.

Cervical Vertebrae

Seven cervical vertebrae are located in the neck. These vertebrae are recognizable by the presence of a **transverse foramen** on each transverse process (Figure 14.19). The vertebral artery travels up the neck through these foramina to enter the skull. The first two cervical vertebrae are modified for special articulations with the skull. The tip of the spinous process is **bifid** (branched) in vertebrae C_2–C_6.

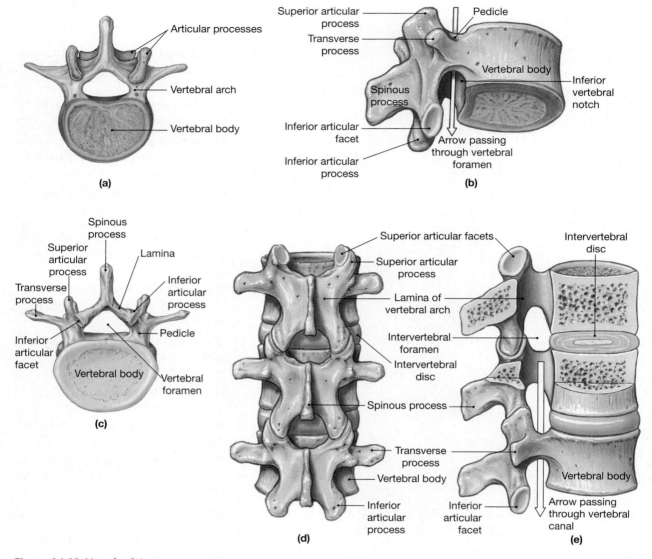

Figure 14.18 Vertebral Anatomy
The anatomy of a typical vertebra and the arrangement of articulations between vertebrae. **(a)** A superior view of a vertebra. **(b)** A lateral and slightly inferior view of a vertebra. **(c)** An inferior view of a vertebra. **(d)** A posterior view of three articulated vertebrae. **(e)** A lateral and sectional view of three articulated vertebrae.

The first cervical vertebra, C_1, is called the **atlas**, named after the Greek mythological character who carried the world on his shoulders. The atlas is the only vertebra that articulates with the skull. Notice in Figure 14.19b that the superior articular facets of the atlas are greatly enlarged. The occipital condyles of the occipital bone fit into the facet like two spoons nested together. When you nod your head, the atlas remains stationary while the occipital condyles glide in the facets.

The atlas is unusual in that it lacks a body and a spinous process. A small, rough structure—the posterior tubercle—occurs where the spinous process normally resides. A long spinous process would interfere with the occipitoatlas articulation. Without a body, the atlas has a very large vertebral foramen.

The **axis** is the second cervical vertebra, C_2. It is specialized to articulate with the atlas. A peglike **dens** (DENZ; *dens*, tooth), or **odontoid process**, arises superiorly from the body of the axis (Figure 14.19c). It fits against the anterior wall of the vertebral foramen and provides the atlas with a pivot point for when the

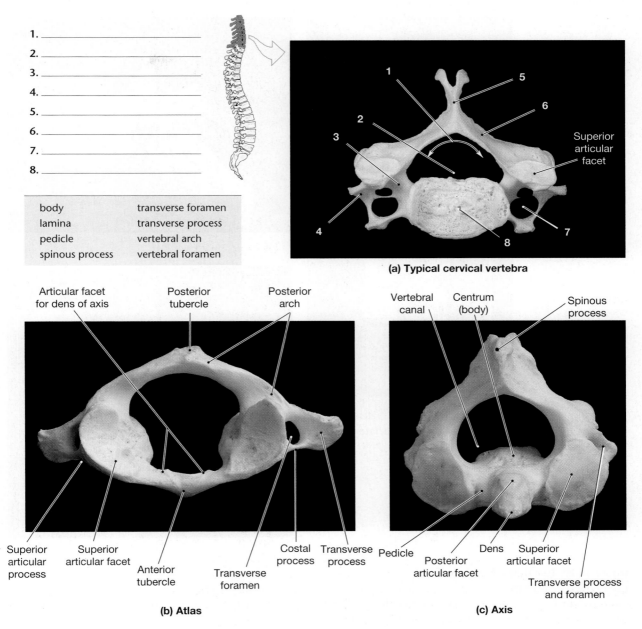

1. _____

2. _____

3. _____

4. _____

5. _____

6. _____

7. _____

8. _____

body	transverse foramen
lamina	transverse process
pedicle	vertebral arch
spinous process	vertebral foramen

Superior articular facet

(a) Typical cervical vertebra

Articular facet for dens of axis   Posterior tubercle   Posterior arch

Vertebral canal   Centrum (body)   Spinous process

Superior articular process   Superior articular facet   Anterior tubercle   Transverse foramen   Costal process   Transverse process

Pedicle   Posterior articular facet   Dens   Superior articular facet   Transverse process and foramen

(b) Atlas

(c) Axis

Figure 14.19 The Cervical Vertebrae
(a) Features of a typical cervical vertebra highlighting the transverse foramen, superior view. (b) The atlas (C_1) in a superior view. (c) Superior view of the axis (C_2).

head is turned laterally and medially. A **transverse ligament** secures the atlas around the dens.

Thoracic Vertebrae

The 12 thoracic vertebrae are larger than the cervical vertebrae and increase in size as they approach the lumbar region (Figure 14.20). The thoracic vertebrae articulate with the 12 pairs of ribs. Most ribs attach at two sites; on **transverse costal facets** (**articular facets**) at the tip of the transverse process and on a **demifacet** located on the posterior of the body (Figure 14.20c). Two demifacets usually are present on the same vertebral body, a superior and an inferior demifacet. Facets and demifacets

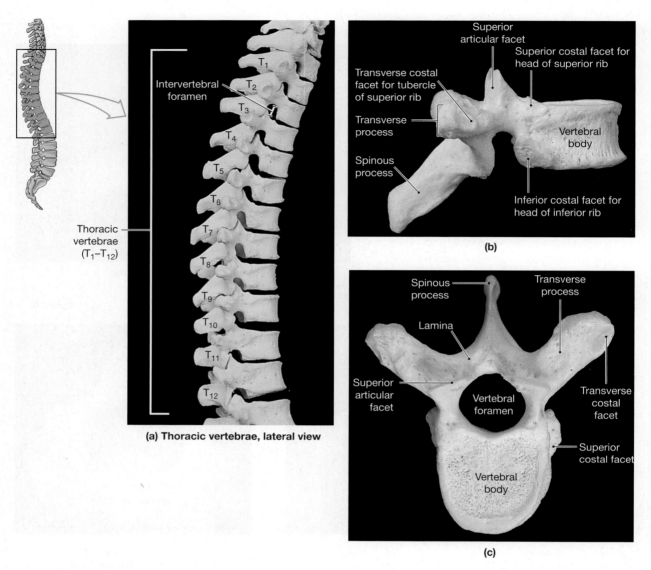

Figure 14.20 The Thoracic Vertebrae
(a) A lateral view of the thoracic region of the vertebral column. The vertebra prominens (C_7) resembles T_1 but lacks facets for rib articulation. Vertebra T_{12} resembles the first lumbar vertebra (L_1) but has a facet for rib articulation. **(b)** Thoracic vertebra, lateral view. **(c)** Thoracic vertebra, superior view.

are found only on thoracic vertebrae, and there is variation in where these features occur along the thoracic region.

Lumbar Vertebrae

The five lumbar vertebrae are large and heavy in order to support the weight of the head, neck, and trunk. Compared with thoracic vertebrae, lumbar vertebrae have a wider body, a blunt and horizontal spinous process, and shorter transverse processes (Figure 14.21). The vertebral foramen is smaller than that in thoracic vertebrae. To prevent the back from twisting when objects are being lifted or carried, the superior articular process is turned medially and the inferior articular processes are oriented laterally to interlock the lumbar vertebrae. No facets or transverse foramina occur on the lumbar vertebrae.

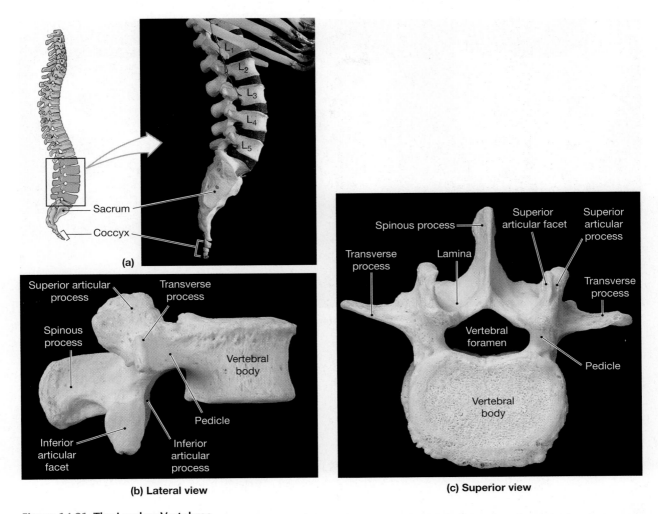

(b) Lateral view

(c) Superior view

Figure 14.21 The Lumbar Vertebrae
(a) A lateral view of the lumbar vertebrae and sacrum. **(b)** A lateral view of a typical lumbar vertebra. **(c)** A superior view of the same vertebra.

Sacral and Coccygeal Vertebrae

As noted earlier, the sacrum is a single bony element composed of five fused sacral vertebrae (Figure 14.22). It articulates with the ilium of the pelvic girdle to form the posterior wall of the pelvis. Fusion of the sacral bones before birth consolidates the vertebral canal into the **sacral canal**. On the fifth sacral vertebra, the sacral canal opens as the sacral hiatus (hī-Ā-tus). Along the lateral margin of the fused vertebral bodies are **sacral foramina**. The spinous processes fuse to form an elevation called the **median sacral crest**. A **lateral sacral crest** extends from the lateral margin of the sacrum. The sacrum articulates with each hip bone at the large **sacral tuberosity** located on the lateral border.

The coccyx (Figure 14.22) articulates with the fifth fused sacral vertebra at the **coccygeal cornua**. There may be anywhere from three to five coccygeal bones, but most people have four.

QuickCheck Questions

7.1 List the major regions of the vertebral column and the number of vertebrae in each region.

7.2 Describe three features found on all vertebrae.

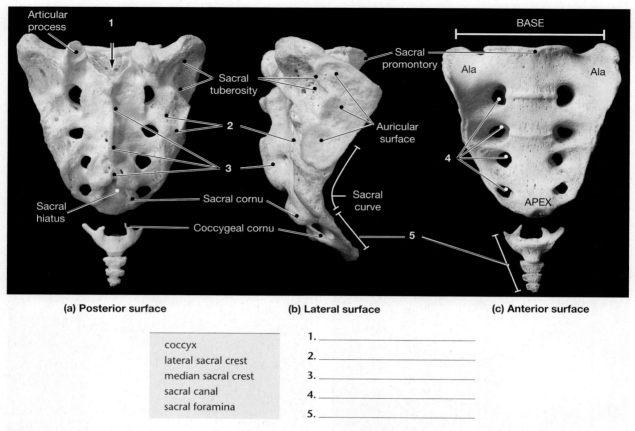

(a) Posterior surface   (b) Lateral surface   (c) Anterior surface

coccyx
lateral sacral crest
median sacral crest
sacral canal
sacral foramina

1. _____
2. _____
3. _____
4. _____
5. _____

Figure 14.22 The Sacrum and Coccyx
(a) A posterior view. (b) A lateral view from the right side. (c) An anterior view.

7   *Materials*

☐ Articulated skeleton
☐ Articulated vertebral
  column
☐ Disarticulated
  vertebral column

Procedures

1. Review the vertebral anatomy presented in Figures 14.17–14.22, and label the numbered elements in Figures 14.19 and 14.22.

2. Identify the four regions of the vertebral column on an articulated skeleton.

3. Describe the type of curvature found in each region.

4. Describe the anatomy of a typical vertebra. Locate each feature on a disarticulated vertebra.

   • Distinguish the anatomical differences among cervical, thoracic, and lumbar vertebrae.

   • Identify the unique features of the atlas and the axis. How do these two vertebrae articulate with the skull and with each other?

   • Discuss how a lumbar vertebra differs from a thoracic vertebra.

5. Describe the anatomy of the sacrum and the coccyx. ■

LAB ACTIVITY 8   Thoracic Cage

There are 12 pairs of ribs in both males and females. They articulate with the thoracic vertebrae posteriorly and the sternum anteriorly to enclose the thoracic organs in a protective rib cage. In breathing, muscles move the ribs to increase or decrease the size of the thoracic cavity and cause air to move into or out of the lungs.

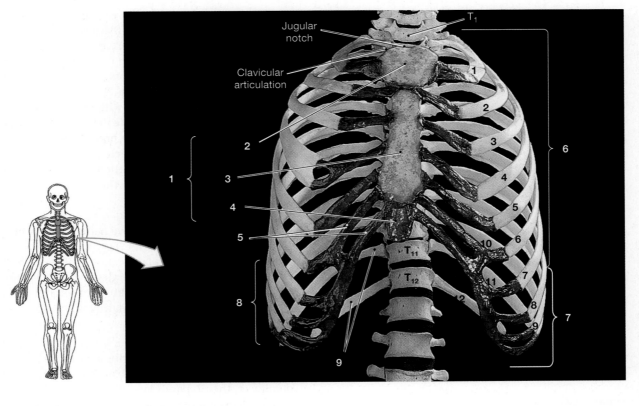

Figure 14.23   **The Thoracic Cage**
An anterior view of the thoracic cage and sternum.

body	1. _____
costal cartilage	2. _____
false ribs	3. _____
floating (vertebral) rib	4. _____
manubrium	5. _____
sternum	6. _____
true (vertebrosternal) ribs	7. _____
vertebrochondral ribs	8. _____
xiphoid process	9. _____

Sternum

The **sternum** is composed of three bony elements: a superior **manubrium** (ma-NOO-brē-um), a middle **body**, and an inferior **xiphoid** (ZĪ-foyd) **process** (Figure 14.23). The manubrium is triangular and articulates with the first pair of ribs and the clavicle. Muscles that move the head and neck attach to the manubrium. The body is elongated and receives the costal cartilage of ribs 2 through 7. The xiphoid process is shaped like an arrowhead and projects inferiorly off the **sternal body**. This process is cartilaginous until late adulthood, when it completely ossifies.

Ribs

Ribs, also called **costae**, are classified according to how they articulate with the sternum. The first seven pairs are called **vertebrosternal**, or *true*, ribs because their cartilage, the **costal cartilage**, attaches directly to the sternum. The remaining five pairs are called *false ribs* because their costal cartilage does not connect directly with the sternum. Ribs 8–10 are **vertebrochondral** ribs, and their cartilage fuses with the cartilage

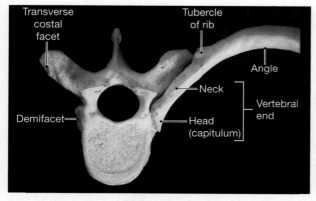

(a) Superior view

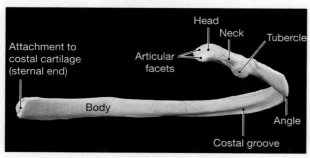

(b) Posterior view

Figure 14.24

The Ribs

(a) Details of rib structure and the articulations between the ribs and thoracic vertebrae. (b) A posterior view of the head of a representative rib from the right side (ribs 2–9).

of rib 7. Pairs 11 and 12 are called **vertebral** ribs and do not articulate with the sternum. Because the vertebral ribs do not join with the sternum, they are also called *floating ribs.*

Figure 14.24 presents the gross anatomy of a rib. A rib has a **head**, or **capitulum** (ka-PIT-ū-lum), with two **articular facets** for articulating with the demifacets of thoracic vertebrae, and a **tubercle** that articulates with the costal facet of the transverse process. Between the head and tubercle is a slender **neck**.

Differences in the way ribs articulate with the thoracic vertebrae are reflected in the variation of the demifacets.

- Vertebrae T_1 through T_8 all have paired demifacets, one superior and one inferior. The first rib articulates with an articular facet of T_1. The second rib articulates with the inferior demifacet of T_1 and the superior demifacet of T_2. Ribs 3 through 9 continue this pattern of articulating with two demifacets.

- Vertebrae T_9 through T_{12} have a single facet on the body, and the corresponding ribs articulate entirely on the one facet. After each rib articulates on a demifacet, the rib bends laterally and articulates with the costal facet on the transverse process. The last two pairs of ribs do not articulate on costal facets.

QuickCheck Questions

8.1 Which part of the sternum articulates with the clavicle?

8.2 Which ribs are true ribs, which are false ribs, and which are floating ribs?

8 Materials

- ☐ Articulated skeleton
- ☐ Articulated vertebral column with ribs
- ☐ Disarticulated vertebral column and ribs

Procedures

1. Label Figure 14.23.

2. Identify the manubrium, body, and xiphoid process of the sternum.

3. Discuss the anatomy of a typical rib.

   - How many pairs of ribs do human males have? How many pairs do human females have?

   - Describe the anatomical features involved in the articulation of a rib on a thoracic vertebra.

   - Identify the differences of articular facets along the thoracic region and relate this to how each rib articulates with the vertebrae. ∎

The Axial Skeleton

Name _____

Date _____

Section _____

A. Matching

Match each skull structure with the correct bone. Each choice may be used more than once.

_____	**1.** sella turcica	**A.** sphenoid
_____	**2.** crista galli	**B.** maxillary bone
_____	**3.** external acoustic canal	**C.** frontal bone
_____	**4.** foramen magnum	**D.** parietal bone
_____	**5.** zygomatic process	**E.** occipital bone
_____	**6.** condylar process	**F.** ethmoid
_____	**7.** petrous portion	**G.** nasal bone
_____	**8.** lesser wing	**H.** zygomatic bone
_____	**9.** mandibular fossa	**I.** mandible
_____	**10.** styloid process	**J.** temporal bone
_____	**11.** coronoid process	
_____	**12.** jugular foramen	
_____	**13.** superior nuchal line	
_____	**14.** superior temporal line	
_____	**15.** supraorbital foramen	

B. Matching

Match each structure of the vertebral column and rib cage with the correct description. Each choice may be used more than once.

_____	**1.** spinous process	**A.** all vertebrae
_____	**2.** transverse foramen	**B.** second cervical vertebra
_____	**3.** manubrium	**C.** thoracic vertebrae
_____	**4.** capitulum	**D.** head of rib
_____	**5.** vertebrosternal rib	**E.** true rib
_____	**6.** pedicle	**F.** body of vertebra
_____	**7.** tubercle	**G.** articulates with facet
_____	**8.** xiphoid process	**H.** false rib
_____	**9.** centrum	**I.** ribs 11 through 12
_____	**10.** vertebral foramen	**J.** sternum
_____	**11.** vertebrochondral rib	
_____	**12.** axis	
_____	**13.** dens	
_____	**14.** vertebral ribs	
_____	**15.** facet	

C. Labeling

Label Figure 14.25, a typical lumbar vertebra.

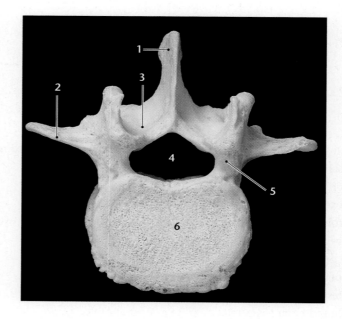

1. _____
2. _____
3. _____
4. _____
5. _____
6. _____

Figure 14.25 Typical Vertebra of the Lumbar Region
Vertebra is shown in superior view.

D. Short-Answer Questions

1. List the components of the axial skeleton.

2. How many bones are found in the cranium and the face?

3. Describe the three cranial fossae and the bones that form the floor of each.

4. List the sutures of the skull and the articulating bones.

5. Describe the four regions of the vertebral column.

E. Analysis and Application

1. Name two passageways in the floor of the skull for major blood vessels of the brain.

2. Describe the skeletal features where the vertebral column articulates with the skull.

3. Compare the articulation on the thoracic vertebrae of rib pairs 7 and 10.

4. Describe the bony orbit of the eye.

5. Describe the bony features of the nasal cavity.

The Appendicular Skeleton

OBJECTIVES

On completion of this exercise, you should be able to:

- Identify the bones and surface features of the pectoral girdle and upper limb.

- Articulate the clavicle with the scapula.

- Articulate the scapula, humerus, radius, and ulna.

- Identify the bones of the hand.

- Identify the bones and surface features of the pelvic girdle and lower limb.

- Articulate the coxal bones with the sacrum to form the pelvis.

- Articulate the coxa, femur, tibia, and fibula.

- Identify the bones of the foot.

The appendicular division of the skeletal system is attached to the vertebral column and sternum of the axial division (Figure 15.1). The appendicular division provides the bony structure of the limbs, permitting us to move around and to interact with our surroundings. This division consists of a pectoral girdle and the attached upper limbs and a pelvic girdle and the attached lower limbs. The shoulder, or pectoral girdle, portion of the appendicular division is loosely attached to the sternum to allow the shoulder and arm, or upper limb, a wide range of movement. The pelvic girdle portion of the appendicular division is securely attached to the sacrum of the spine and allows the leg, or lower limb, its wide range of motion.

LAB ACTIVITY 1   The Pectoral Girdle

On each side of the pectoral girdle is a collarbone, called the *clavicle* (KLAV-i-kul), and a shoulder blade, the *scapula* (SKAP-ū-la). The scapula rests against the posterior surface of the rib cage, and the clavicle connects the scapula to the sternum. The pectoral girdle attaches the arm and provides an anchor for arm muscles to pull against.

The Clavicle

The S-shaped clavicle is the only connection between the pectoral girdle and the axial skeleton. The round **sternal end** articulates with the sternum, and the **acromial**

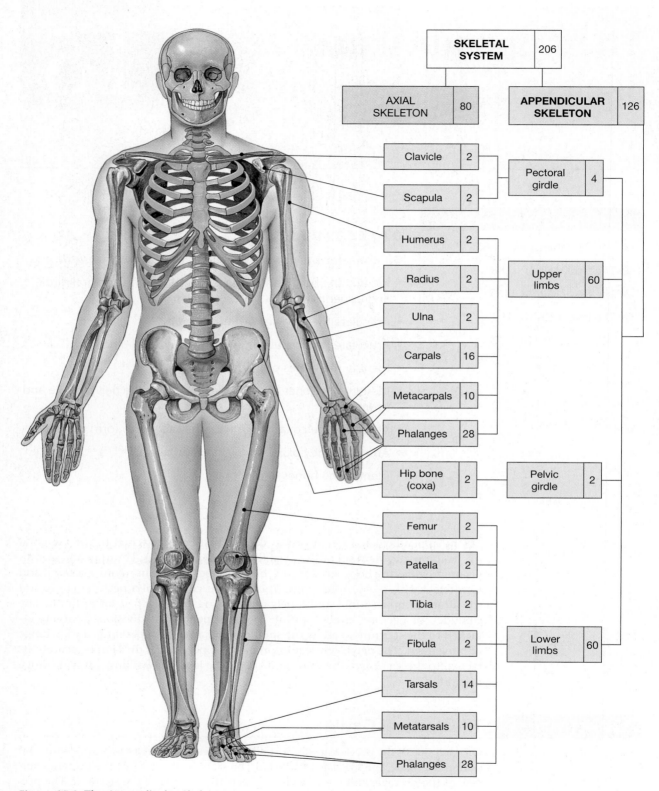

Figure 15.1 **The Appendicular Skeleton**

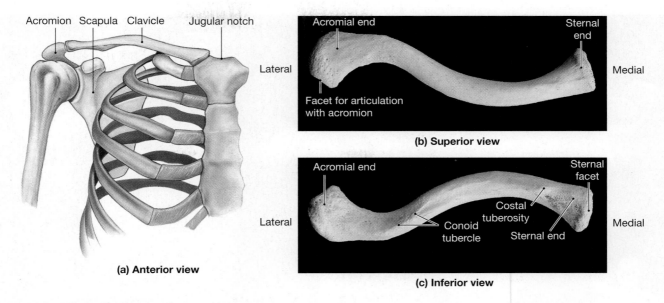

Figure 15.2 The Clavicle
(a) The position of the clavicle, anterior view. **(b)** Superior and **(c)** inferior views of the right clavicle.

(a-KRŌ-mē-al) **end** joins the scapula (Figure 15.2). Inferiorly, toward the acromial end, where the clavicle bends, is the **conoid tubercle**, an attachment site for shoulder muscles.

The Scapula

The scapula, shown in Figure 15.3, is composed of a triangular **body**. The long edges of the scapular body are the **superior**, **medial**, and **lateral borders**. An indentation in the superior border is the **scapular notch**. The corners where the borders meet are the **superior**, **lateral**, and **inferior angles**. The **subscapular fossa** is the smooth, triangular surface where the anterior of the scapula articulates with the ribs.

A prominent ridge, the **spine**, extends across the scapula body on the posterior surface (Figure 15.3c). At the lateral tip of the spine is the **acromion** (a-KRŌ-mē-on), which hangs over the **glenoid cavity** (glenoid fossa) where the humerus articulates. Superior to the glenoid cavity is the beak-shaped **coracoid** (KOR-uh-koyd) **process**. Observe the scapula in the lateral view of Figure 15.3b, and notice how the coracoid process extends over the anterior surface of the scapula. The **scapular neck** is the ring of bone around the base of the coracoid process and the glenoid cavity.

QuickCheck Questions

1.1 Which bones are part of a pectoral girdle?

1.2 Where does the arm articulate with the scapula?

---

1 *Materials*	*Procedures*

☐ Articulated skeleton
☐ Disarticulated skeleton

1. Locate a clavicle on the study skeleton and review the anatomy shown in Figure 15.2.
   • Identify the sternal and acromial ends of the clavicle. Can you feel these ends on your own clavicles?
   • Identify the conoid tubercle of the clavicle.

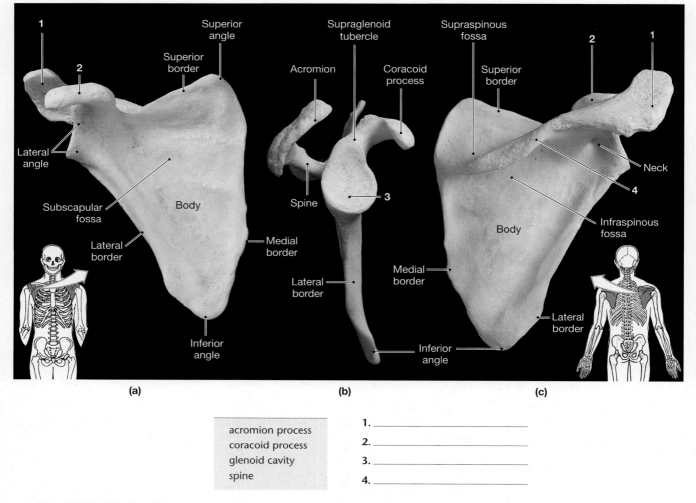

(a)   (b)   (c)

acronion process
coracoid process
glenoid cavity
spine

1. _____
2. _____
3. _____
4. _____

Figure 15.3 The Scapula
(a) Anterior, (b) lateral, and (c) posterior views of the right scapula.

  2. Locate a scapula on the study skeleton. Review the surface features of the scapula and label Figure 15.3.
  - Identify the borders, angles, and fossae of the scapula.
  - Identify the spine, the acromion, the coracoid process, and the glenoid cavity.
  - Can you feel the spine and acromion on your own scapula?

  3. Place a clavicle from the disarticulated skeleton on your shoulder and determine how it would articulate with your scapula. ∎

LAB ACTIVITY 2   The Upper Limb

An upper limb includes the bones of the arm, wrist, and hand—a total of 30 bones with all but 3 of them in the wrist and hand. The upper arm bone is the humerus, the forearm has the lateral radius and the medial ulna, which both articulate with the humerus at the elbow. The wrist comprises eight carpal bones. The hand contains 5 long, slender bones in the palm, called *metacarpals,* and the 14 phalanges of the fingers.

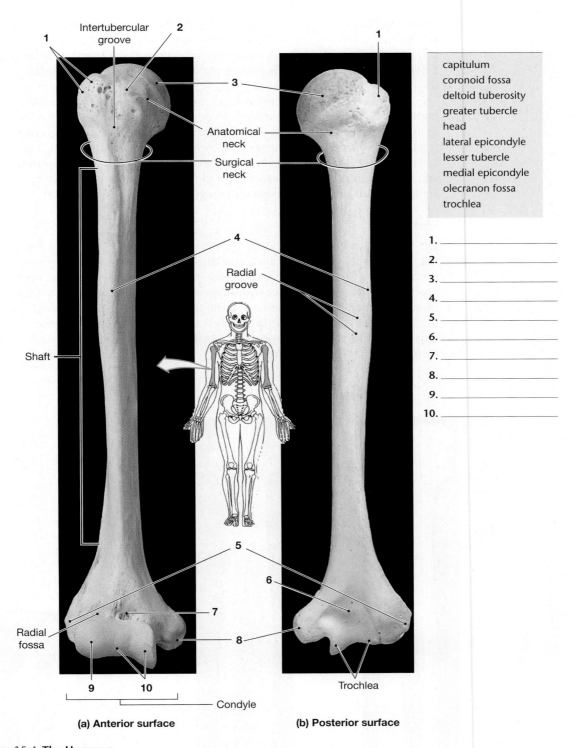

Intertubercular groove

1

2

3

Anatomical neck

Surgical neck

Shaft

Radial groove

4

Radial fossa

5

6

7

8

9   10

Condyle

Trochlea

(a) **Anterior surface**

(b) **Posterior surface**

1

1

capitulum
coronoid fossa
deltoid tuberosity
greater tubercle
head
lateral epicondyle
lesser tubercle
medial epicondyle
olecranon fossa
trochlea

1. _____
2. _____
3. _____
4. _____
5. _____
6. _____
7. _____
8. _____
9. _____
10. _____

Figure 15.4 The Humerus
(a) The anterior and **(b)** posterior surfaces of the right humerus.

The Humerus

The proximal **head** of the humerus articulates with the glenoid cavity of the scapula (Figure 15.4). Lateral to the head is the **greater tubercle**, and medial to the head is the **lesser tubercle**, both sites for muscle attachment. The **intertubercular groove** separates the tubercles. Between the head and the tubercles is the

anatomical neck, and inferior to the tubercles is the **surgical neck**. Inferior to the greater tubercle is the rough **deltoid tuberosity**, where the deltoid muscle of the shoulder attaches. At the inferior termination of the tuberosity is the **radial groove**, a depression that serves as the passageway through which runs the radial nerve to the posterior muscles of the arm.

The distal end of the humerus has a specialized **condyle** to accommodate two joints: the hingelike elbow joint and a pivot joint of the forearm, the latter used when doing such movements as turning a doorknob. The condyle has a round **capitulum** (*capit*, head) on the lateral side and the cylindrical **trochlea** (*trochlea*, a pulley) located medially. Above the trochlea are two depressions, the anterior **coronoid fossa** and the posterior triangular **olecranon** (ō-LEK-ruh-non) **fossa**. Above the condyle are the **medial** and larger **lateral epicondyles**.

The Ulna

The forearm has two parallel bones, the ulna and the radius. The ulna is easy to identify by the conspicuous U-shaped **trochlear notch**, as shown in Figure 15.5. The notch is like a C clamp with two processes that attach to the humerus, the superior **olecranon** and the inferior **coronoid process**. The processes fit into their corresponding fossae on the humerus. On the lateral surface of the coronoid process is the flat **radial notch**. Below the notch is the rough **ulnar tuberosity**. The distal ulna has a pointed **styloid process**. The **interosseous membrane** extends between the ulna and radius to support the bones.

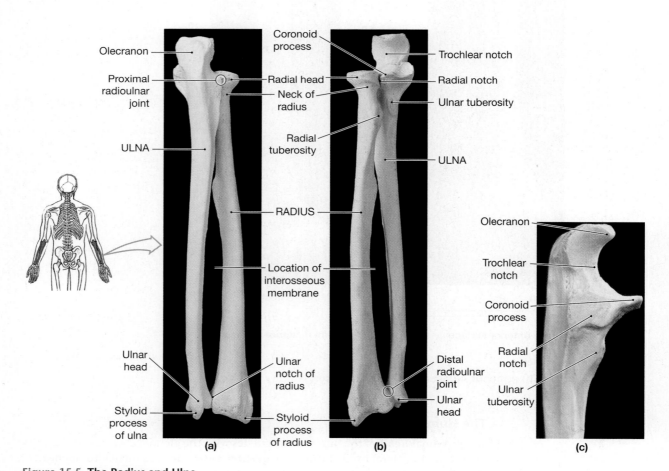

Figure 15.5 The Radius and Ulna
The radius and ulna are shown in (**a**) posterior and (**b**) anterior view. (**c**) A lateral view of the right ulna.

Study Tip | ***Elbow Terminology***

Notice that the terminology of the elbow is consistent in the humerus and ulna. The trochlear notch of the ulna fits into the trochlea of the humerus. The coronoid process and olecranon fit into their respective fossae on the humerus. ●

The Radius

The radius (Figure 15.5) has a disk-shaped **radial head** that pivots in the radial notch of the ulna. Below the head is the **neck**, and inferior to the neck is the **radial tuberosity**. On the distal portion, the **ulnar notch** on the medial surface articulates with the ulna. The **styloid process** of the radius is larger and not as pointed as the styloid process of the ulna.

The Wrist and Hand

Each wrist and hand contains 27 bones (Figure 15.6). The wrist is eight **carpals** (KAR-pulz) arranged in two rows of four, the proximal and the distal carpals. An easy method of identifying the carpals is to use the anterior wrist and start with the carpal next to the styloid process of the radius. From this reference point moving medially, proximal carpals are the **scaphoid bone**, **lunate bone**, **triquetrum**, and small **pisiform bone**. Returning on the lateral side, the four distal carpals are the **trapezium bone**, **trapezoid**, **capitate bone**, and **hamate bone**.

The five long bones of the palm are **metacarpals**. Each metacarpus is numbered with a roman numeral, with the thumb metacarpus being I. The 14 bones of the fingers are called **phalanges**. Each finger has three phalanges; the thumb, or **pollex**, has two.

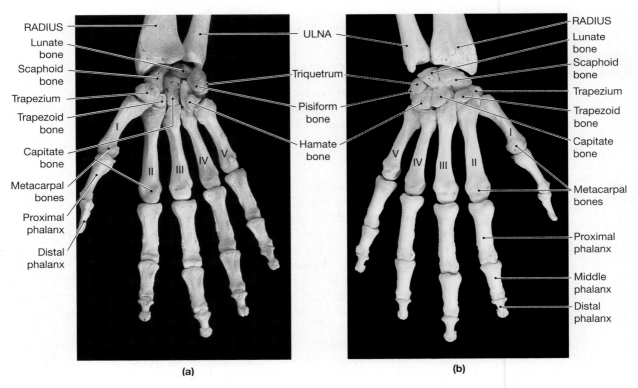

Figure 15.6 Bones of the Wrist and Hand
(a) Anterior and (b) posterior views of the right hand.

QuickCheck Questions

2.1 Which bones make up the arm?

2.2 What are the major bones of the wrist and hand?

2 Materials

☐ Articulated skeleton
☐ Disarticulated skeleton

Procedures

1. Locate a humerus of your study skeleton and review its surface features.
   - Label the anatomy of the humerus in Figure 15.4.
   - Identify the head, surgical and anatomical necks, tubercles, and intertubercular groove.
   - Identify the deltoid tuberosity and radial groove.
   - Identify the epicondyles and the condyle. Can you feel the epicondyles on your own humerus?
   - Identify the capitulum, trochlea, and fossae of the distal humerus.

2. Locate an ulna and radius of your study skeleton and review their surface features.
   - Study the processes that fit into the corresponding fossae of the humerus.
   - Identify the articulating anatomy between the ulna and radius.

3. Review Figure 15.6 and locate the bones of the wrist and hand on the study skeleton.
   - Identify the four proximal carpals and the four distal carpals.
   - Identify the metacarpals and the phalanges. Do all the fingers have the same number of phalanges?

4. Articulate the bones of the upper limb with those of the pectoral girdle. ■

LAB ACTIVITY 3   The Pelvic Girdle

The **pelvic girdle** forms the hips and is the anchoring point of the legs and vertebral column. Each hip, or **coxal bone**, is formed by the fusion of three bones: the **ilium** (IL-ē-um), **ischium** (IS-kē-um), and **pubis** (PŪ-bis). The **pelvis** is the bowl formed by the sacrum of the axial skeleton and the coxal bones.

The Coxal Bone

As Figure 15.7 shows, the ilium, ischium, and pubis that form the coxal bone (which is also called the os coxa) fuse along a horizontal axis passing through the middle of the **acetabulum** (a-se-TAB-ū-lum), the deep socket where the head of the femur articulates. The smooth inner wall of the acetabulum is the **lunate surface**. The anterior and inferior rims of the acetabulum are not continuous; instead, there is an open gap between them, the **acetabular notch**.

The superior ridge of the ilium is the **iliac crest**. It is shaped like a shovel blade, with the anterior and posterior **superior iliac spines** at each end. The large indentation below the posterior iliac spine is the **greater sciatic** (sī-A-tik) **notch**.

The ischium is the bone we sit on. The greater sciatic notch terminates at a bony point, the **ischial spine**. Inferior to this spine is the **lesser sciatic notch**. The **ischial tuberosity** is in the most inferior portion of the ischium and is a site for muscle attachment.

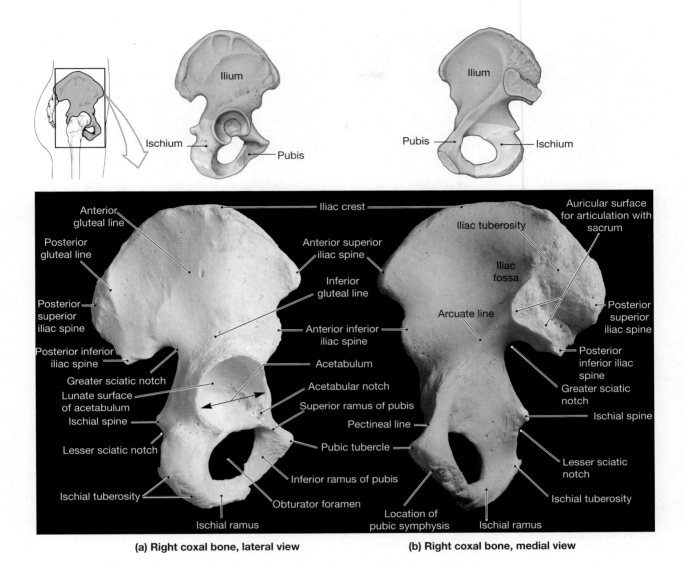

Ilium

Ischium — Pubis

Ilium

Pubis — Ischium

(a) Right coxal bone, lateral view   **(b) Right coxal bone, medial view**

Anterior gluteal line
Posterior gluteal line
Posterior superior iliac spine
Posterior inferior iliac spine
Greater sciatic notch
Lunate surface of acetabulum
Ischial spine
Lesser sciatic notch
Ischial tuberosity
Ischial ramus

Iliac crest
Anterior superior iliac spine
Inferior gluteal line
Anterior inferior iliac spine
Acetabulum
Acetabular notch
Superior ramus of pubis
Pectineal line
Pubic tubercle
Inferior ramus of pubis
Obturator foramen

Auricular surface for articulation with sacrum
Iliac tuberosity
Iliac fossa
Arcuate line
Posterior superior iliac spine
Posterior inferior iliac spine
Greater sciatic notch
Ischial spine
Lesser sciatic notch
Ischial tuberosity
Location of pubic symphysis
Ischial ramus

Figure 15.7 The Pelvic Girdle
(a) Lateral view. **(b)** Medial view.

The pubis bone forms the anterior portion of the coxal bone. It joins the ischium just anterior to the ischial tuberosity, creating the **obturator** (OB-tū-rā-tor) **foramen**. The most anterior region of the pubis is the pointed **pubic tubercle**.

A conspicuous feature on the posterior iliac crest is the rough **auricular surface** where the **sacroiliac joint** attaches the pelvic girdle to the sacrum of the axial skeleton. Anteriorly, the pubis bones join at the **pubic symphysis**, a strong joint containing fibrocartilage.

The pelvis of the male differs anatomically from that of the female (Figure 15.8). The female pelvis has a wider **pelvic outlet**, the space between the ischial spines. Additionally, the **pubic angle** at the pubis symphysis is wider in the female and more U-shaped. This angle is V-shaped in the male. The wider female pelvis provides a larger passageway for childbirth.

QuickCheck Questions

3.1 Which bones make up the pelvic girdle?

3.2 What structure does the femur articulate with at the hip?

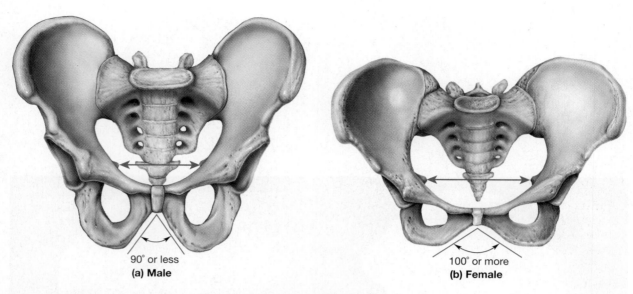

Figure 15.8 Anatomical Differences in the Male and Female Pelvis
Note the much sharper angle in the pelvis of a male **(a)** than in that of a female **(b)**.

3 Materials

- ☐ Articulated skeleton
- ☐ Disarticulated skeleton

Procedures

1. Locate a coxal bone on your study skeleton and review the anatomy in Figure 15.7.
2. Identify the ilium, ischium, and pubis bones.
   - Are sutures visible where these bones fused?
   - Locate the acetabulum and obturator foramen.
   - Identify other features of the ilium shown in Figure 15.7.
3. Trace along the iliac crest and down the posterior surface.
   - Identify the greater and lesser sciatic notches and the ischial spine.
   - What is the large rough area on the inferior ischium called?
   - Identify other features of the ischium and pubis shown in Figure 15.7.
4. Locate the auricular surface of the sacroiliac joint and the pubic symphysis.
5. Articulate the two coxal bones and the sacrum to form the pelvis.
6. Examine the pelvis on several articulated skeletons in the laboratory. How can you distinguish a male pelvis from a female pelvis? ■

LAB ACTIVITY 4   The Lower Limb

The lower limb consists of the **femur** in the thigh, a patella (kneecap), and the **tibia** (TI-bē-uh) and **fibula** (FIB-ū-la) of the lower leg. Each ankle contains 7 **tarsals**, and the foot has 5 **metatarsals** and 14 **phalanges**. Locate each bone of the lower limb in Figure 15.1.

The Femur

The femur is the largest bone of the skeleton (Figure 15.9). It supports the body's weight and bears the stress from the legs. The smooth, round **head** fits into the acetabulum of the coxal bone and permits the femur a wide range of movement. Below the head is a narrow **neck** that joins the head to the diaphysis. Lateral to the

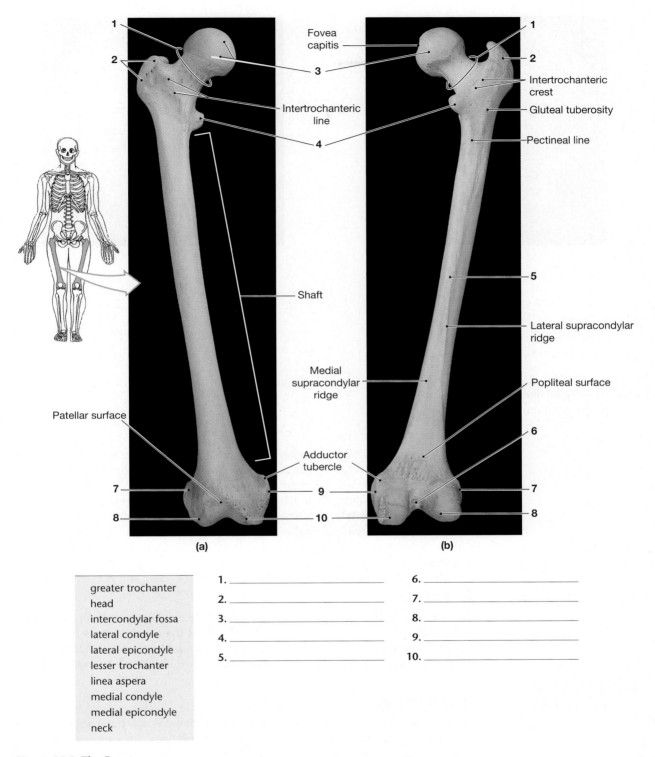

Fovea capitis

Intertrochanteric line

Intertrochanteric crest

Gluteal tuberosity

Pectineal line

Shaft

Lateral supracondylar ridge

Medial supracondylar ridge

Popliteal surface

Patellar surface

Adductor tubercle

(a)   (b)

greater trochanter	1. _____	6. _____
head	2. _____	7. _____
intercondylar fossa	3. _____	8. _____
lateral condyle	4. _____	9. _____
lateral epicondyle	5. _____	10. _____
lesser trochanter		
linea aspera		
medial condyle		
medial epicondyle		
neck		

Figure 15.9 **The Femur**

(a) Anterior surface. (b) Posterior surface.

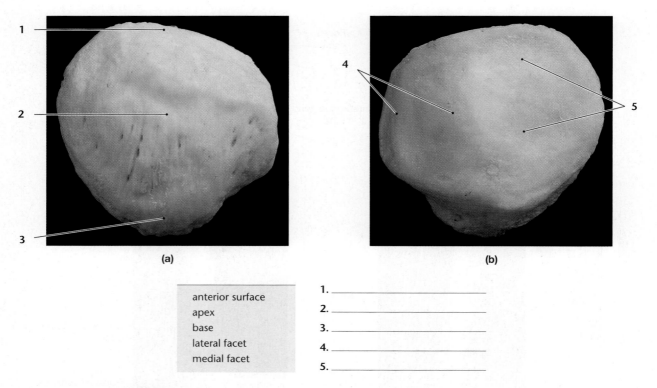

anterior surface
apex
base
lateral facet
medial facet

1. _____
2. _____
3. _____
4. _____
5. _____

Figure 15.10 The Right Patella
(a) Anterior surface. **(b)** Posterior surface.

head is a large stump, the **greater trochanter** (trō-KAN-ter); the **lesser trochanter** is on the medial surface. These large processes are attachment sites for powerful hip and thigh muscles. Along the middle of the posterior diaphysis is the **linea aspera**, a rough line for thigh muscle attachment. At the distal end of the femur are the **lateral** and **medial condyles** that articulate with the tibia. The condyles are separated posteriorly by the **intercondylar fossa**. A smooth patellar surface spans the condyles and serves as a gliding platform for the patella. Superior to the condyles are the **lateral** and **medial epicondyles**.

The Patella

The patella is encased within the tendons of the anterior thigh muscles. The superior border of the patella is the flat **base**; the **apex** is at the inferior tip (Figure 15.10). Tendons attach to the rough anterior surface, and the smooth posterior facets glide over the condyles of the femur. The **medial facet** is narrower than the **lateral facet**.

Study Tip | ***Patella Pointers***

It is easy to distinguish a right patella from a left one. Lay the bone on its facets, and point the apex away from you. Notice that the bone leans to one side. Because the lateral facet is larger, the bone will tilt and lean on that facet. Therefore, if the patella leans to the left, it is a left patella. ●

The Tibia

The lower leg bones are illustrated in Figure 15.11. The tibia is the large medial bone of the lower leg. The proximal portion of the tibia flares to develop the **lateral** and **medial condyles** that articulate with the corresponding femoral condyles. Separating the tibial condyles is a ridge of bone, the **intercondylar eminence**. This emi-

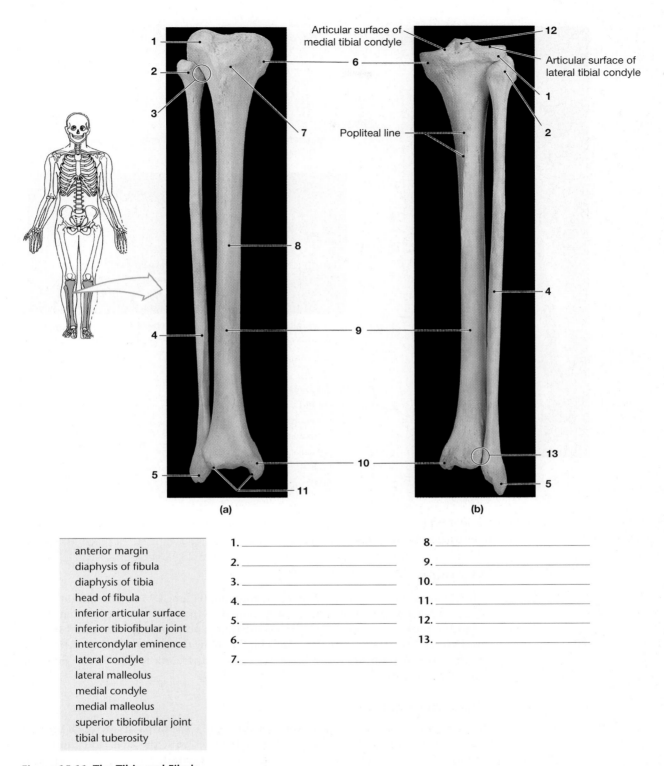

Figure 15.11 The Tibia and Fibula
(a) Anterior view. (b) Posterior view.

anterior margin
diaphysis of fibula
diaphysis of tibia
head of fibula
inferior articular surface
inferior tibiofibular joint
intercondylar eminence
lateral condyle
lateral malleolus
medial condyle
medial malleolus
superior tibiofibular joint
tibial tuberosity

1. _____
2. _____
3. _____
4. _____
5. _____
6. _____
7. _____

8. _____
9. _____
10. _____
11. _____
12. _____
13. _____

nence fits into the intercondylar fossa of the femur. Anteriorly below the condyles is the large **tibial tuberosity**, where thigh muscles attach. The anterior diaphysis forms a ridge, the **anterior margin**, along most of the shaft's length. The distal tibia is constructed to articulate with the foot. A large wedge, the **medial malleolus** (ma-LĒ-ō-lus), helps to stabilize the ankle joint. The inferior **articular surface** is smooth so that it can slide over the talus of the ankle.

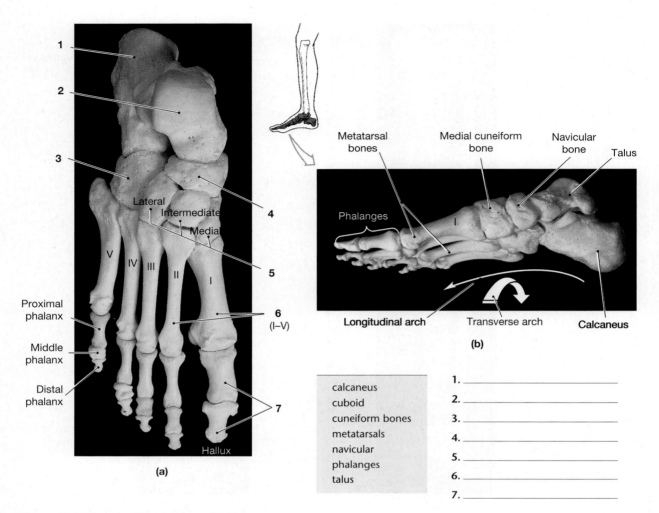

Lateral
Intermediate
Medial

V IV III II I

Proximal
phalanx

Middle
phalanx

Distal
phalanx

Hallux

(a)

Metatarsal
bones

Medial cuneiform
bone

Navicular
bone

Talus

Phalanges

I

Longitudinal arch

Transverse arch

Calcaneus

(b)

calcaneus
cuboid
cuneiform bones
metatarsals
navicular
phalanges
talus

1. _____
2. _____
3. _____
4. _____
5. _____
6. _____
7. _____

Figure 15.12 Bones of the Ankle and Foot.
(a) Superior view, right foot. **(b)** Medial view, right foot.

The Fibula

The fibula is the slender bone lateral to the tibia (Figure 15.11). The proximal and distal regions of the fibula appear very similar at first, but closer examination reveals the proximal head to be more rounded (less pointed) than the distal **lateral malleolus**. The head of the fibula articulates below the lateral condyle of the tibia at the **superior tibiofibular joint**. The distal articulation creates the **inferior tibiofibular joint**.

The Ankle and Foot

The ankle comprises seven tarsal bones (Figure 15.12). The **talus** sits on top of the heel bone, the **calcaneus** (kal-KĀ-nē-us), and articulates with the tibia and the lateral malleolus of the fibula. Anterior to the talus is the **navicular**, which articulates with three **cuneiform bones**: medial, intermediate, and lateral. Lateral to the navicular and the third cuneiform is the **cuboid**, which articulates posteriorly with the calcaneus. The arch of the foot is formed by five **metatarsals** spanning the arch. They are named with roman numerals I through V, from medial to lateral. Like the fingers of the hand, each toe has three **phalanges** except the **hallux**, or big toe, which has only two phalanges.

Study Tip | ### Hands and Feet

Because their names are so similar, it is easy to confuse the carpals and metacarpals of the hand with the tarsals and metatarsals of the foot. Just remember, when you listen to music you clap your carpals and tap your tarsals! ●

QuickCheck Questions

4.1 Which bones make up the lower limb?

4.2 What are the major bones of the foot?

4 Materials

☐ Articulated skeleton

☐ Disarticulated skeleton

Procedures

1. Locate the femur, tibia, and fibula on your study skeleton. Also locate these bones on your own body.

2. Identify the surface features of the femur and label Figure 15.9.
   • Locate the head, neck, and greater and lesser trochanters.
   • Trace your hand along the posterior of the diaphysis and feel the linea aspera. What attaches to this rough structure?
   • On the distal end of the femur, identify the epicondyles, condyles, and intercondylar fossa.

3. Identify the surface features of the patella and label Figure 15.10.
   • Examine the two facets of the patella. Are they the same size?
   • How can the facets be used to determine whether the patella is from a right leg or a left one?

4. Review the anatomy of the tibia and fibula and label 15.11.
   • What is the ridge on the tibial head called?
   • On the tibia, locate the condyles and the tibial tuberosity.
   • On the distal tibia, locate the medial malleolus.
   • On the fibula, distinguish between the proximal and distal ends.
   • Locate the lateral malleolus of the fibula. How does its shape differ from the fibular head?

5. Identify the bones of the ankle and foot. Label Figure 15.12.
   • Which bone directly receives the weight from the leg?
   • Which bones form the arch of the foot?
   • Do the toes all have the same number of phalanges?

6. Articulate the bones of the lower limb with the pelvis. ■

The Appendicular Skeleton

Name _____

Date _____

Section _____

A. Matching

Match each surface feature in the left column with its correct bone from the right column.
Each choice from the right column may be used more than once.

_____	**1.** acromion	**A.** clavicle
_____	**2.** intercondylar fossa	**B.** patella
_____	**3.** trochlea	**C.** fibula
_____	**4.** glenoid cavity	**D.** humerus
_____	**5.** ulnar notch	**E.** femur
_____	**6.** deltoid tuberosity	**F.** scapula
_____	**7.** greater trochanter	**G.** tibia
_____	**8.** sternal end	**H.** radius
_____	**9.** lateral malleolus	
_____	**10.** linea aspera	
_____	**11.** capitulum	
_____	**12.** medial malleolus	
_____	**13.** intercondylar eminence	
_____	**14.** base	

B. Fill in the Blanks

Complete each statement by filling in the blank with the correct directional term.

1. The humerus is _____ to the radius.

2. The fibula is _____ to the tibia.

3. The talus is _____ to the calcaneus.

4. The clavicle is _____ to the scapula.

5. The patella is _____ to the femur.

6. The ilium is _____ to the ischium.

7. The ulna is _____ to the radius.

8. The metacarpals are _____ to the carpals.

15

C. Labeling

Label the surface features of the coxal bone in Figure 15.13.

1. _____
2. _____
3. _____
4. _____
5. _____
6. _____
7. _____
8. _____

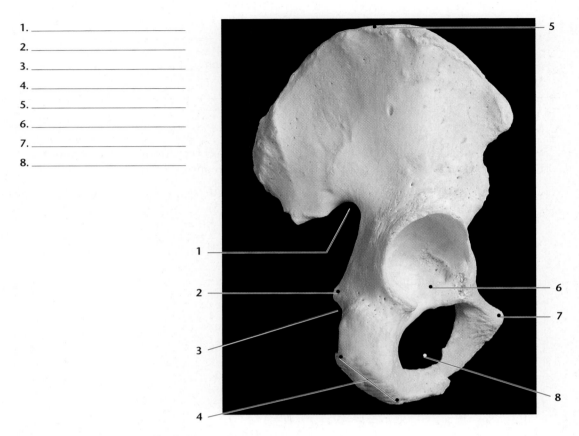

Figure 15.13 Lateral View of the Pelvic Girdle

D. Short-Answer Questions

1. List the bones of the pectoral girdle and upper limb.

2. Describe the coxal bone.

3. List the bones of the lower limb.

4. Compare the pelvis of males and females.

E. Analysis and Application

1. Describe the condyle of the humerus where the ulna and radius articulate.

2. Compare the bones of the wrist and hand with the bones of the ankle and foot.

3. Describe how the fibula articulates with the tibia and the foot.

4. Name a tuberosity for shoulder muscle attachment and a tuberosity for thigh muscle attachment.

Articulations

OBJECTIVES

On completion of this exercise, you should be able to:

• List the three types of functional joints and give an example of each.

• List the four types of structural joints and give an example of each.

• Describe the three types of diarthroses and the movement each produces.

• Describe the anatomy of a typical synovial joint.

• Describe and demonstrate the various movements of synovial joints.

Arthrology is the study of the structure and function of joints. An **articulation** occurs wherever two or more bones fit together and form a joint. If you were asked to identify joints in your body, you would most likely name those that allow a large range of movement, such as your knee or hip joint. These joints have a joint cavity between the bones that permits free movement. In some joints of the body, however, the bones are held closely together, a condition that allows no movement; an example of this type of nonmoving joint is found in the bones of the cranium.

Some individuals have more movement in a particular joint than most other people and are called "double jointed." Of course, they do not have two joints; the additional movement is a result of either the anatomy of the articulating bones or the position of tendons and ligaments around the joint.

LAB ACTIVITY 1   Joint Classification

Two classification schemes are commonly used for articulations. The functional scheme groups joints by the amount of movement permitted, and the structural scheme groups joints by the type of connective tissue between the articulating bones.

Functional Classification

The functional classification scheme divides joints into three groups: immovable joints, the *synarthroses;* semimovable joints, the *amphiarthroses;* and freely movable joints, the *diarthroses.* Table 16.1 summarizes these three groups.

1. **Synarthroses** (sin-ar-THRŌ-sēz; *syn-,* together + *arthros,* joint) have bones that are either closely fitted together or else surrounded by a strong ligament. Three types of synarthroses are found in the skeleton: fibrous joints, synchondroses, and synostoses.

   - **Fibrous** joints, which are strong synarthroses with fibrous connective tissue between the bones, are of two types. **Sutures** (*sutura,* a sewing together) occur in the skull wherever the bones interdigitate, or interlock together. One example is the **gomphosis** (gom-FŌ-sis; *gompho,* a peg or nail), a fibrous joint between the teeth and the alveolar bone of the jaw. This joint is lined with a strong periodontal ligament that holds the teeth in place and permits no movement.

   - **Synchondroses** (sin-kon-DRŌ-sēz; *syn-,* together + *condros,* cartilage) are synarthroses that have cartilage between the bones making up the joints. Two examples of this type of synarthrosis are the epiphyseal plate in a child's long bones and the cartilage between the ribs and sternum.

   - **Synostoses** (sin-os-TŌ-sēz; *-osteo,* bone) are synarthroses formed by the fusion of two bones, as in the frontal bone, coxal bones, and mandible bone. The joint between the diaphysis and either epiphysis of a mature long bone is also a synostosis.

Table 16.1 A Functional Classification of Articulations

Functional Category	Structural Category	Description	Example
Synarthrosis (no movement)	**Fibrous** *Suture*	Fibrous connections plus interlocking projections	Between the bones of the skull
	Gomphosis	Fibrous connections plus insertion in alveolar process	Between the teeth and jaws
	Cartilaginous *Synchondrosis*	Interposition of cartilage plate	Epiphyseal plates
	Bony fusion *Synostosis*	Conversion of other articular form to solid mass of bone	Portions of the skull, epiphyseal lines
Amphiarthrosis (little movement)	**Fibrous** *Syndesmosis*	Ligamentous connection	Between the tibia and fibula
	Cartilaginous *Symphysis*	Connection by a fibrocartilage pad	Between right and left halves of pelvis; between adjacent vertebral bodies along vertebral column
Diathrosis (free movement)	**Synovial**	Complex joint bounded by joint capsule and containing synovial fluid	Numerous; subdivided by range of movement
	Monaxial	Permits movement in one plane	Elbow, ankle
	Biaxial	Permits movement in two planes	Ribs, wrist
	Triaxial	Permits movement in all three planes	Shoulder, hip

2. **Amphiarthroses** (am-fē-ar-THRŌ-sēz) are joints held together by strong connective tissue; they are capable of only minimal movement. The two types are syndesmoses and symphyses.

   • **Syndesmoses** (sin-dez-MŌ-sēz; *syn-,* together + *desmo-,* band) occur between the parallel bones of the forearm and lower leg. A ligament of fibrous connective tissue forms a strong band that wraps around the bones. The syndesmosis thus formed prevents excessive movement in the joint.

   • **Symphyses** are amphiarthroses characterized by the presence of fibrocartilage between the articulating bones. The intervertebral disks, for instance, construct a symphysis between any two articulating vertebrae. Another symphysis in the body occurs where the two coxal bones unite at the pubis. This strong joint, called the pubic symphysis, limits flexion of the pelvis. During childbirth, a hormone softens the fibrocartilage to widen the pelvic bowl.

3. **Diarthroses** (dī-ar-THRŌ-sēz) are joints in which the bones are separated by a small joint cavity lined with a synovial (sin-NŌ-vē-ul) membrane. The cavity allows a wide range of motion. Movements are classified according to the number of planes the bones move through. **Monaxial** (mon-AX-ē-ul) joints, like the elbow, move in one plane. **Biaxial** (bī-AX-ē-ul) joints allow movement in two planes; move your wrist up and down and side to side to demonstrate biaxial movement. **Triaxial** (trī-AX-ē-ul) joints occur in the ball-and-socket joints of the shoulder and hip and permit movement in three planes. **Nonaxial** joints, also called **multiaxial**, are glide joints where the articulating bones can move slightly in a variety of directions. The anatomy of a diarthrotic joint is examined in more detail later in this exercise.

Structural Classification

The structural classification scheme for joints is important when discussing joint anatomy rather than movement. As Table 16.2 summarizes, four types of structural joints occur in the skeleton.

1. **Bony fusion** occurs where bones have fused together, and this type of joint permits no movement. A good example is the frontal bone. Humans are born with two frontal bones that, by the age of eight, fuse into a single frontal bone. The old articulation site is then occupied by bony tissue to form a bony fusion joint (a synostosis in the functional classification scheme).

2. **Fibrous joints** have strong fibrous connective tissue between the articulating bones, and as a result little to no movement occurs in these joints. There are three main types of fibrous joints: suture synarthroses, the gomphosis synarthrosis, and syndemosis amphiarthroses. (Skull fibrous tissue is removed whenever a skull is prepared for laboratory work and therefore is not present on the study skulls you will see.)

3. **Cartilaginous joints**, as their name implies, have cartilage between the bones. The type of cartilage—hyaline or fibrocartilage—determines the type of cartilaginous joint. Between the ribs and the sternum is a synchondrosis, a cartilaginous joint containing hyaline cartilage. The pubic symphysis and intervertebral discs are cartilaginous joints containing fibrocartilage between the bones.

4. **Synovial joints** have a joint cavity lined by a **synovial membrane**. All the free-moving joints—in other words, the diarthroses—are synovial joints. The three types are the monaxial, biaxial, and triaxial joints, described earlier.

Table 16.2 **A Structural Classification of Articulations**

Structure	Type	Functional Category	Example
Bony fusion	Synostosis (illustrated)	Synarthrosis	Metopic suture (fusion) — Frontal bone
Fibrous joint	Suture (illustrated) Gomphosis Syndesmosis	Synarthrosis Synarthrosis Amphiarthrosis	Lambdoid suture — Skull
Cartilaginous joint	Synchrondrosis Symphysis (illustrated)	Synarthrosis Amphiarthrosis	Pubic symphysis — Pelvis
Synovial joint	Monaxial Biaxial Triaxial (illustrated)	Diarthroses	Synovial joint

Clinical Application

Arthritis

Arthritis, a disease that destroys synovial joints by damaging the articular cartilage, comes in two forms. **Rheumatoid arthritis** is an autoimmune disease that occurs when the body's immune system attacks the cartilage and synovial membrane of the joint. As the disease progresses, the joint cavity is eliminated and the articulating bones fuse, resulting in painful disfiguration of the joint and loss of joint function. **Osteoarthritis** is a degenerative joint disease that often occurs due to age and wearing of the joint tissues. The articular cartilage is damaged, and bone spurs may project into the joint cavity. Osteoarthritis tends to occur in the knee and hip joints, whereas rheumatoid arthritis is more common in the smaller joints of the hand. ▶

QuickCheck Questions

1.1 What is the difference between the functional classification scheme for joints and the structural classification scheme?

1.2 List the three types of functional joints and how much movement each allows.

1.3 List the four types of structural joints and the type of connective tissue found in each.

1   Materials

☐ Articulated skeleton

Procedures

1. Locate on an articulated skeleton or on your body a joint from each functional group and one from each structural group.

2. Identify on your body and give an example of each of the following joints:
   - synarthrosis _____
   - amphiarthrosis _____
   - diarthrosis _____
   - syndesmosis _____
   - synchondrosis _____
   - synostosis _____
   - symphysis _____
   - suture _____

3. Identify two monaxial joints, two biaxial joints, and two triaxial joints on your body.
   - monaxial joints _____
   - biaxial joints _____
   - triaxial joints _____ ■

LAB ACTIVITY 2   ## Anatomical Structure of Synovial Joints

As mentioned previously, all diarthrotic joints are capable of free movement. This large range of motion is due to the anatomical organization of the joint: between the bones of every diarthrotic joint is a cavity lined with a synovial membrane.

Figure 16.1a highlights the important features of a typical synovial joint. The **joint cavity** is found between the bones and is bounded by the **articular cartilage** attached to the end of each bone. This cartilage is hyaline cartilage and provides a slippery, gelatinous surface should the two bones make contact across the cavity. The **synovial membrane** produces **synovial fluid**. Injury to a joint may cause inflammation of this membrane and lead to excessive fluid production. Notice in Figure 16.1a how the periosteum of each bone is continuous with the strong **articular capsule** that encases the joint.

Figure 16.1b illustrates the general features of the knee joint, which is a synovial joint. Notice the **bursa** (BUR-sa; *bursa,* a pouch) between the patella and the skin. Bursae are similar to synovial membranes, except that, instead of lining joint cavities, bursae provide padding between bones and other structures.

QuickCheck Questions

2.1 Where is the synovial membrane located in a joint?

2.2 Where is cartilage found in a synovial joint?

---

2   Materials

☐ Fresh beef joint

Procedures

1. Review the anatomy of the synovial joint in Figure 16.1.

2. On the fresh beef joint, locate and describe the joint cavity.

3. Identify the articular cartilage and articular capsule. ■

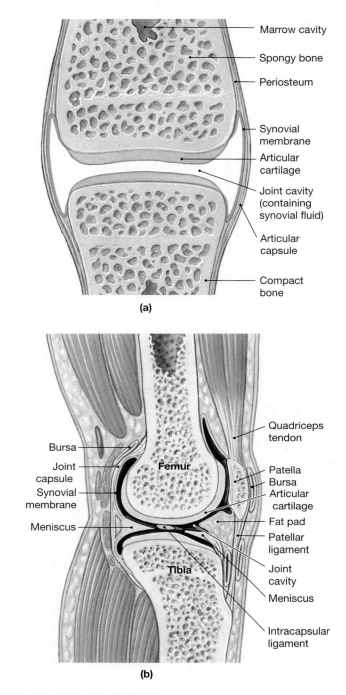

Figure 16.1 Structure of a Synovial Joint
(a) Basic structure of a typical synovial joint. (b) Simplified sectional view of the knee joint.

LAB ACTIVITY 3   Types of Diarthroses

Six types of diarthroses (synovial joints) occur in the skeleton. Each type permits a certain amount of movement owing to the joining surfaces of the articulating bones. Figure 16.2 details each type of joint and includes a mechanical representation of each joint to show planes of motion.

Types of Synovial Joints	Movement	Examples
Gliding joint 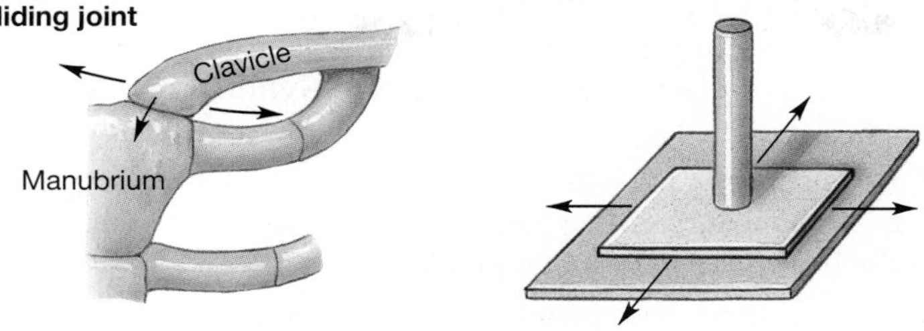	Slight nonaxial or multiaxial	• Acromioclavicular and claviculosternal joints • Intercarpal and intertarsal joints • Vertebrocostal joints • Sacroiliac joints
Hinge joint 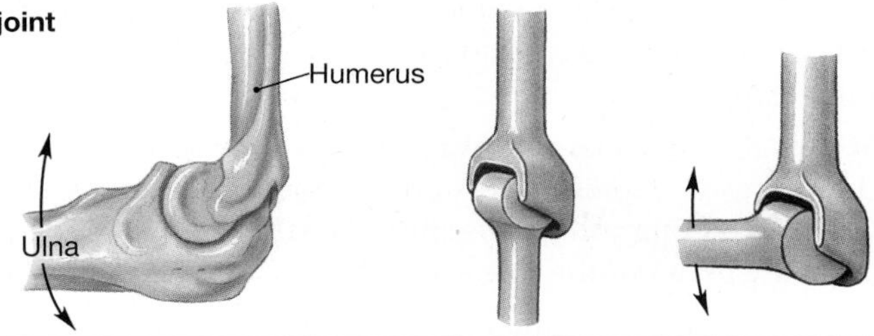	Monaxial	• Elbow joint • Knee joint • Ankle joint • Interphalangeal joint
Pivot joint 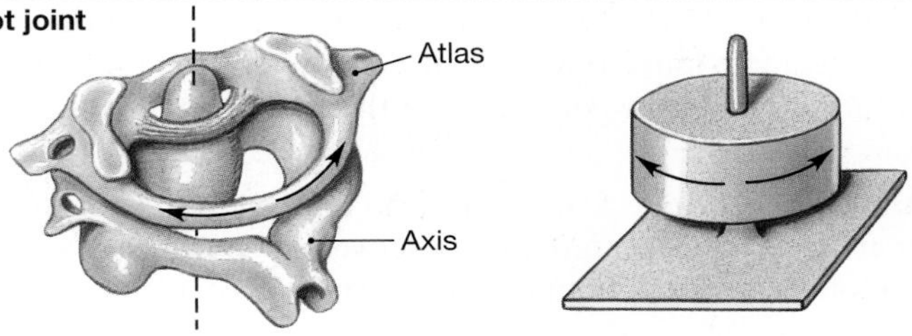	Monaxial (rotation)	• Atlas/axis • Proximal radioulnar joint
Ellipsoidal joint 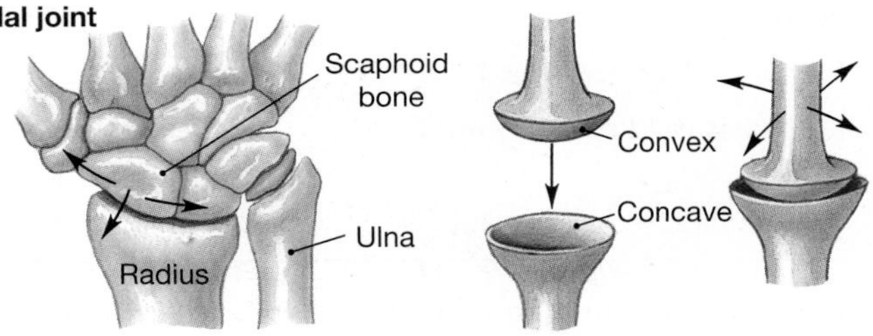	Biaxial	• Radiocarpal joint • Metacarpophalangeal joints 2–5 • Metatarsophalangeal joints
Saddle joint 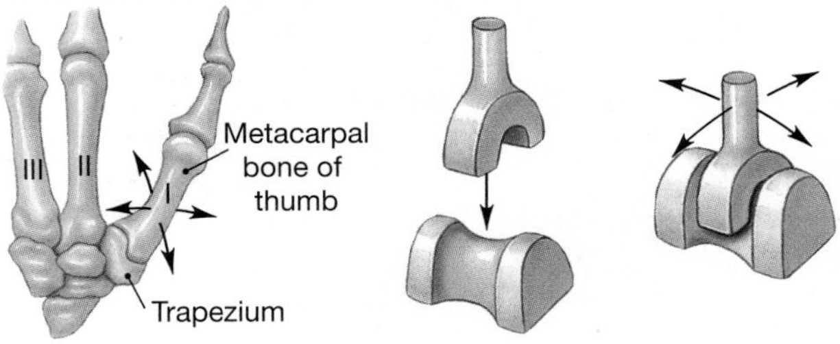	Biaxial	• First carpometacarpal joint
Ball-and-socket joint 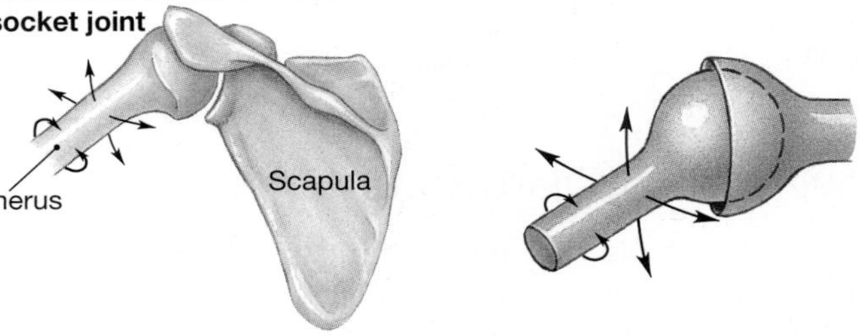	Triaxial	• Shoulder joint • Hip joint

Figure 16.2 A Functional Classification of Diarthroses

- **Gliding joints** are common where flat articular surfaces, such as in the wrist, slide by neighboring bones. The movement is typically nonaxial. In addition to the wrist, glide joints also occur between bones of the sternum and between the tarsals. When you place your open hand, palm facing down, on your desktop and press down hard, you can observe the gliding of your wrist bones.

- **Hinge joints** are monaxial, operating like a door hinge, and are located in the elbows, fingers, toes, and knees. Bending your legs at the knees and your arms at the elbows is possible because of your hinge joints.

- **Pivot joints** are monaxial joints that permit one bone to rotate around another. Shake your head "no" to operate the pivot joint between your first two cervical vertebrae. The first cervical vertebra (the atlas), pivots around the second cervical vertebra (the axis).

- **Ellipsoidal joints** are also called **condyloid joints**. A convex surface of one bone articulates in a concave depression of another bone. This concave-to-convex spooning of articulating surfaces permits biaxial movement. The articulation between the bones of the forearm and wrist is an ellipsoidal joint.

- The **saddle joint** is a biaxial joint found only at the junction between the thumb metacarpus and the trapezium bone of the wrist. Place a finger on your lateral wrist and feel the saddle joint move as you touch your little finger with your thumb. This joint permits you to oppose your thumb to grasp and manipulate objects in your hand.

- **Ball-and-socket joints** occur where a spherical head of one bone fits into a cup-shaped fossa of another bone, as in the joint between the humerus and the scapula. This triaxial joint permits dynamic movement in many planes.

QuickCheck Questions

3.1 What are the six types of diarthroses?

3.2 What type of diarthrosis is a knuckle joint?

| 3 | *Materials* | *Procedures* |

☐ Articulated skeleton

1. Locate each type of synovial joint on an articulated skeleton or on your body. On the skeleton, notice how the structure of the joining bones determines the amount of joint movement.

2. Give an example of each type of synovial joint:
   - gliding _____
   - hinge _____
   - pivot _____
   - ellipsoidal _____
   - saddle _____
   - ball-and-socket _____ ■

LAB ACTIVITY 4   Skeletal Movement at Diarthrotic Joints

The diversity of bone shapes and joint types permits the skeleton to move in a variety of ways. Figure 16.3 illustrates angular movements, which occur either front to back in the anterior/posterior plane or side to side in the lateral plane. For clarity, the figure includes a small dot at the joint where a demonstrated movement is described. Tables 16.3 and 16.4 (on pages 214–215) summarize articulations of the axial and appendicular division of the skeleton.

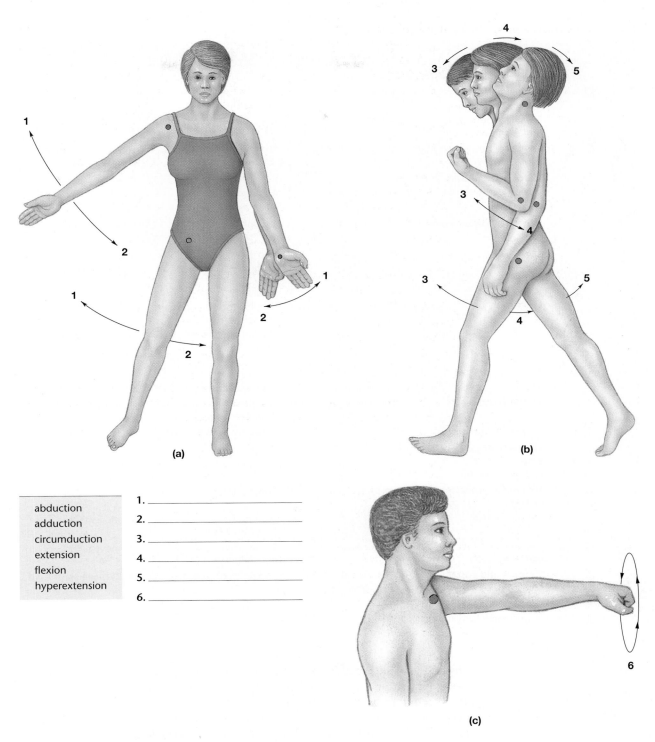

abduction
adduction
circumduction
extension
flexion
hyperextension

1. _____
2. _____
3. _____
4. _____
5. _____
6. _____

(a)

(b)

(c)

Figure 16.3 Angular Movements

- **Abduction** is movement away from the midline of the body. **Adduction** is movement toward the midline. Notice how you move your arm at the shoulder for these two motions. Practice this movement first with your shoulder joint and then with your wrist joint.

- **Flexion** of a joint decreases the angle between the articulating bones, and **extension** increases the angle between the bones (Figure 16.3b). Hang your arm

Table 16.3 **Articulations of the Axial Skeleton**

Element	Joint	Type of Articulation	Movements
Skull			
Cranial and facial bones of skull	Various	Synarthroses (suture or synostosis)	None
Maxillary bone/teeth and mandible/teeth	Alveolar	Synarthrosis (gomphosis)	None
Temporal bone/mandible	Temporomandibular	Combined gliding joint and hinge diarthrosis	Elevation, depression, and lateral gliding
Vertebral Column			
Occipital bone/atlas	Atlanto-occipital	Ellipsoidal diarthrosis	Flexion/extension
Atlas/axis	Atlanto-axial	Pivot diarthrosis	Rotation
Other vertebral elements	Intervertebral (*between vertebral bodies*)	Amphiarthrosis (symphysis)	Slight movement
	Intervertebral (*between articular processes*)	Gliding diarthrosis	Slight rotation and flexion/extension
L_5/sacrum	Between L_5 body and sacral body	Amphiarthrosis (symphysis)	Slight movement
	Between inferior articular processes of L_5 and articular processes of sacrum	Gliding diarthrosis	Slight flexion/extension
Sacrum/os coxae	Sacroiliac	Gliding diarthrosis	Slight movement
Sacrum/coccyx	Sacrococcygeal	Gliding diarthrosis (*may become fused*)	Slight movement
Coccygeal bones		Synarthrosis (synostosis)	No movement
Thoracic Cage			
Bodies of T_1–T_{12} and heads of ribs	Costovertebral	Gliding diarthrosis	Slight movement
Transverse processes of T_1–T_{10}	Costovertebral	Gliding diarthrosis	Slight movement
Ribs and costal cartilages		Synarthrosis (synchondrosis)	No movement
Sternum and first costal cartilage	Sternocostal (1st)	Synarthrosis (synchondrosis)	No movement
Sternum and costal cartilages 2–7	Sternocostal (2nd–7th)	Gliding diarthrosis*	Slight movement

*Commonly converts to synchondrosis in elderly individuals.

down at your side in anatomical position. Now flex your arm by moving the elbow joint. Your hand should be up by your shoulder. Notice how close the antebrachium is to the brachium and how the angle between them has decreased. Is your flexed arm still in anatomical position? Now extend your arm to return it to anatomical position. How has the angle changed?

- **Hyperextension** moves the body beyond anatomical position. Follow Figure 16.3b and flex, extend, and hyperextend your head.

- **Circumduction** (Figure 16.3c) is movement at a ball-and-socket joint. When you repeat the motion shown in Figure 16.3c, notice how the proximal region of your arm is relatively stationary while the distal portion traces a wide circle in the air. To circumduct a leg, stick one foot out in front of you, raise it up off the floor, and again trace a circle in the air (a great movement for working the thigh muscles).

Table 16.4 Articulations of the Appendicular Skeleton

	Element	Joint	Type of Articulation	Movements
Articulations of the Pectoral Girdle and Upper Limb				
	Sternum/clavicle	Sternoclavicular	Gliding diarthrosis*	Protraction/retraction, elevation/depression, slight rotation
	Scapula/clavicle	Acromioclavicular	Gliding diarthrosis	Slight movement
	Scapula/humerus	Shoulder, or glenohumeral	Ball-and-socket diarthrosis	Flexion/extension, adduction/abduction, circumduction, rotation
	Humerus/ulna and humerus/radius	Elbow (humeroulnar and humeroradial)	Hinge diarthrosis	Flexion/extension
	Radius/ulna	Proximal radioulnar	Pivot diarthrosis	Rotation
		Distal radioulnar	Pivot diarthrosis	Pronation/supination
	Radius/carpal bones	Radiocarpal	Ellipsoidal diarthrosis	Flexion/extension, adduction/abduction, circumduction
	Carpal bone to carpal bone	Intercarpal	Gliding diarthrosis	Slight movement
	Carpal bone to metacarpal bone (I)	Carpometacarpal of thumb	Saddle diarthrosis	Flexion/extension, adduction/abduction, circumduction, opposition
	Carpal bone to metacarpal bone (II–V)	Carpometacarpal	Gliding diarthrosis	Slight flexion/extension, adduction/abduction
	Metacarpal bone to phalanx	Metacarpophalangeal	Ellipsoidal diarthrosis	Flexion/extension, adduction/abduction, circumduction
	Phalanx/phalanx	Interphalangeal	Hinge diarthrosis	Flexion/extension
Articulations of the Pelvic Girdle and Lower Limb				
	Sacrum/ilium os coxae	Sacroiliac	Gliding diarthrosis	Slight movement
	Os coxae/os coxae	Pubic symphysis	Amphiarthrosis (symphysis)	None†
	Os coxae/femur	Hip	Ball-and-socket diarthrosis	Flexion/extension, adduction/abduction, circumduction, rotation
	Femur/tibia	Knee	Complex, functions as hinge	Flexion/extension, limited rotation
	Tibia/fibula	Tibiofibular (proximal)	Gliding diarthrosis	Slight movement
		Tibiofibular (distal)	Gliding diarthrosis and amphiarthrotic syndesmosis	Slight movement
	Tibia and fibula with talus	Ankle, or talocrural	Hinge diarthrosis	Flexion/extension (dorsiflexion/plantar flexion)
	Tarsal bone to tarsal bone	Intertarsal	Gliding diarthrosis	Slight movement
	Tarsal bone to metatarsal bone	Tarsometatarsal	Gliding diatthrosis	Slight movement
	Metatarsal bone to phalanx	Metatarsophalangeal	Ellipsoidal diarthrosis	Flexion/extension, adduction/abduction
	Phalanx/phalanx	Interphalangeal	Hinge diarthrosis	Flexion/extension

*A "double gliding joint," with two joint cavities separated by an articular cartilage.
†During pregnancy, hormones weaken the symphysis and permit movement important to childbirth.

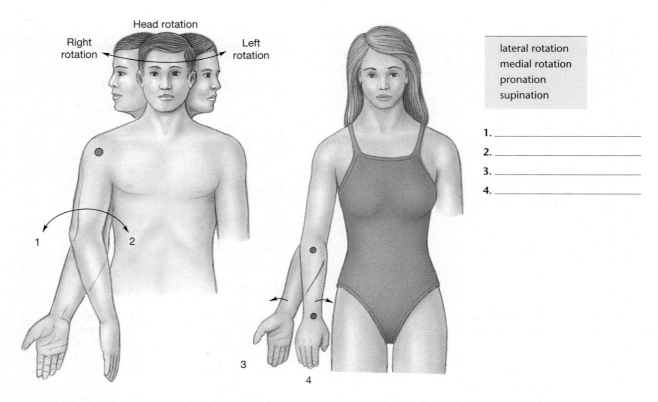

Figure 16.4 Rotational Movements

- **Lateral rotation** and **medial rotation** occur at the ball-and-socket joints in the shoulder and hip (Figure 16.4). These movements turn the rounded head of one bone in the socket of another bone. With your arm held stiff and straight in anatomical position, rotate it until the palm faces posteriorly. Your elbow joint should not move during the rotation. All movement is at the ball-and-socket joint of the pectoral girdle. When might you use this type of movement?

- **Supination** (soo-pi-NĀ-shun) moves the palm to anatomical position. **Pronation** (prō-NĀ-shun) moves the palm to face posteriorly. During these two motions, the humerus serves as a foundation for the radius to pivot around the ulna. What daily activities require you to pronate and supinate?

Study Tip | **Rotational Movements**

To see the difference between medial/lateral rotation and pronation/supination, hold both arms in anatomical position and then bend the elbow until your forearms are parallel to the floor. Now, keeping the forearm parallel to the floor, move your left hand clockwise until it hits your torso. The movement of your humerus at the shoulder when you do this is medial rotation, and when you swing your right hand away from your torso, the humerus is rotating laterally. For supination/pronation, start again with your forearm parallel to the floor and this time twist your right hand as if you were turning a doorknob. This movement is supination and pronation. ●

The following specialized motions are all illustrated in Figure 16.5:

- **Eversion** (ē-VER-zhun) is lateral movement of the ankle to move the sole outward. Moving the sole medially is **inversion**; the foot moves "in." Eversion and inversion are commonly mistaken for pronation and supination of the ankle.

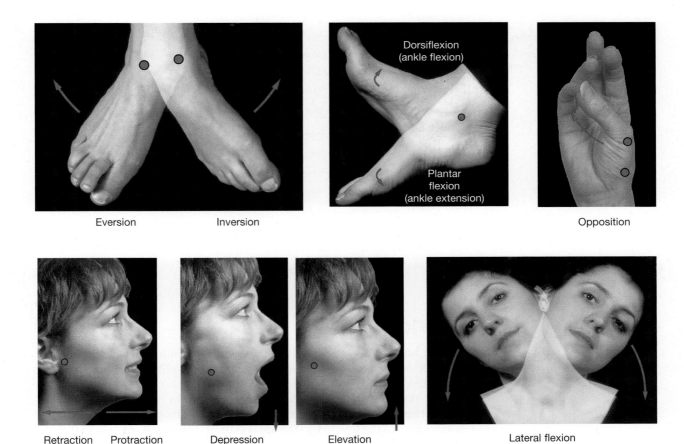

Figure 16.5 **Special Movements**

- Two other terms describing ankle movement are dorsiflexion and plantar flexion. **Dorsiflexion** permits you to walk on your heels, which means your plantae are raised up off the floor and the angle of the ankles is decreased. **Plantar flexion** moves the foot so that you can walk on your tiptoes; here the angle of the ankles is increased.
- **Opposition** is touching the thumb pad with the pad of the little finger.
- **Retraction**, which means to take back, moves structures posteriorly out of anatomical position. Move your mandible bone back, as if to demonstrate an overbite; this is retraction. Now **protract** your mandible bone by jutting your chin out as if it were a pointer. Practice retracting and protracting your shoulders, too.
- **Depression** lowers bones. This motion occurs, for instance, when you lower your mandible bone to take a bite of food. Closing your mouth is **elevation** of the mandible bone. Examine these movements in Figure 16.5. Name another joint that permits depression and elevation.
- **Lateral flexion** is the bending of the vertebral column from side to side. Most of the movement occurs in the cervical and lumbar regions.

QuickCheck Questions

4.1 How does flexion differ from extension?

4.2 How does hyperextension differ from extension?

4.3 What are pronation and supination?

4.4 How does dorsiflexion differ from plantar flexion?

| 4 | *Materials* | *Procedures* |

☐ Articulated skeleton

1. Use an articulated skeleton or your body to demonstrate each of the movements in Figures 16.3, 16.4, and 16.5.

2. Label the movements illustrated in Figures 16.3 and 16.4.

3. Give an example of each of the following movements:
   - abduction _____
   - extension _____
   - hyperextension _____
   - pronation _____
   - supination _____
   - depression _____
   - retraction _____
   - lateral rotation _____ ∎

LAB ACTIVITY 5 — The Knee Joint

The knee is a hinge joint that permits flexion and extension of the leg. This joint must be stable in order to support most of the body's weight while simultaneously adjusting to the body's frequent changes in position. To accomplish both support and mobility, the knee has many unique structural features (Figure 16.6). Most support for the knee is provided by seven bands of ligaments that encase the joint. Flexion and extension of the leg change the tension on the various ligaments so that the joint is supported throughout the range of motion. The shape of the articulating bones and the stabilizing effect of the ligaments act in concert to keep knee rotation to a minimum.

Cushions of fibrocartilage, the **lateral meniscus** (men-IS-kus; *meniskos,* a crescent) and the **medial meniscus**, pad the area between the condyles of the femur and tibia. The menisci also adjust for conformational changes in the knee when the leg is either flexed or extended. Areas where tendons move against the bones in the knee are protected with bursae. Additionally, because the patella is so close to the skin, bursae offer protection here as well.

The ligaments of the knee occur in three pairs—the collateral, popliteal, and cruciate—and a single patellar ligament.

- **Tibial** and **fibular collateral ligaments** provide medial and lateral support when a person is standing.

- Two **popliteal ligaments** extend from the head of the femur to the fibula and tibia to support the posterior of the knee.

- The **anterior** and **posterior cruciate ligaments** are inside the articular capsule. The cruciate (*cruciate,* a cross) ligaments originate on the tibial head and cross each other as they pass through the intercondylar fossa of the femur. The anterior cruciate ligament (ACL) arises on the anterior tibial head and attaches to the femur in the intercondylar fossa. The posterior cruciate ligament (PCL) is found on the posterior tibial head and crosses the ACL before attaching to the femur.

- The **patellar ligament** attaches the inferior aspect of the patella to the tibial tuberosity, adding anterior support to the knee. The large quadriceps tendon is attached to the superior margin of the patella. Cords of ligaments called the **patellar retinaculae** contribute to anterior support of the knee.

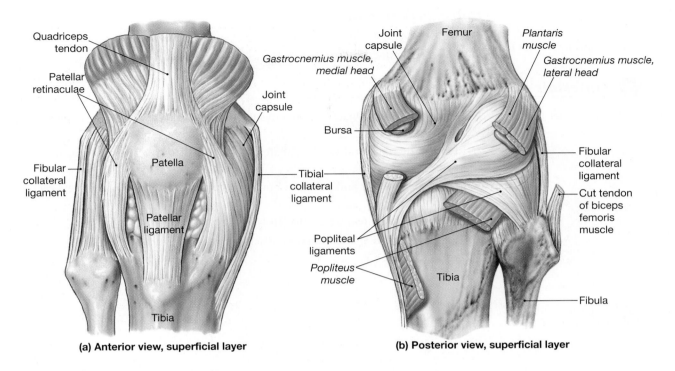

Quadriceps
tendon

Patellar
retinaculae

Fibular
collateral
ligament

Patella

Patellar
ligament

Tibia

Joint
capsule

Tibial
collateral
ligament

(a) Anterior view, superficial layer

Joint
capsule

Femur

Plantaris
muscle

Gastrocnemius muscle,
medial head

Gastrocnemius muscle,
lateral head

Bursa

Fibular
collateral
ligament

Cut tendon
of biceps
femoris
muscle

Popliteal
ligaments

Popliteus
muscle

Tibia

Fibula

(b) Posterior view, superficial layer

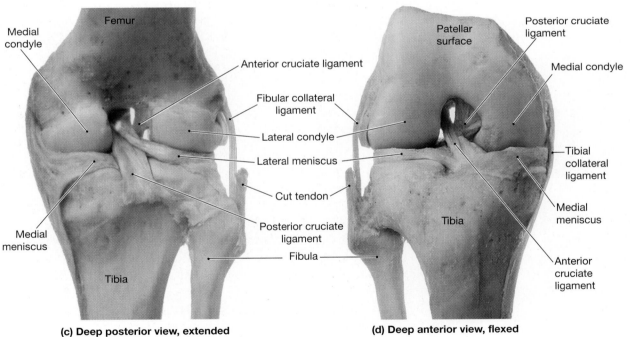

Medial
condyle

Femur

Medial
meniscus

Tibia

Anterior cruciate ligament

Fibular collateral
ligament

Lateral condyle

Lateral meniscus

Cut tendon

Posterior cruciate
ligament

Fibula

(c) Deep posterior view, extended

Patellar
surface

Posterior cruciate
ligament

Medial condyle

Tibial
collateral
ligament

Medial
meniscus

Tibia

Anterior
cruciate
ligament

(d) Deep anterior view, flexed

Figure 16.6 The Knee Joint
The right knee. Superficial **(a)** anterior and **(b)** posterior views of the extended knee joint. **(c)** Deep posterior view at
full extension. **(d)** Deep anterior view at full flexion.

QuickCheck Questions

5.1 How many ligaments are in the knee?

5.2 What structures cushion the knee?

5 Materials

□ Knee model

Procedures

1. Locate the patellar ligament on a model of the knee. Where does this ligament attach? _____

2. Identify the tibial and fibular collateral ligaments. How do they support the knee? _____

3. Locate the anterior and posterior cruciate ligaments. Where do they cross each other? _____

4. Observe the popliteal ligaments. How are they attached? _____

5. Identify the medial and lateral menisci. What is their function? _____ ■

Articulations

Name _____

Date _____

Section _____

A. Matching

Match each joint in the left column with its correct description from the right column.

_____	**1.** pivot	**A.**	forearm-to-wrist joint
_____	**2.** symphysis	**B.**	joint between parietal bones
_____	**3.** ball and socket	**C.**	rib-to-sternum joint
_____	**4.** gomphosis	**D.**	joint between vertebral bodies
_____	**5.** hinge	**E.**	femur-to-coxal bone joint
_____	**6.** suture	**F.**	phalangeal joint
_____	**7.** synostosis	**G.**	distal tibia-to-fibula joint
_____	**8.** syndesmosis	**H.**	atlas-to-axis joint
_____	**9.** condyloid	**I.**	fused frontal bones
_____	**10.** synchondrosis	**J.**	joint holding tooth in a socket

B. Matching

Match each movement in the left column with its correct description from the right column.

_____	**1.** retraction	**A.**	movement away from midline
_____	**2.** dorsiflexion	**B.**	movement to turn foot outward
_____	**3.** eversion	**C.**	palm moved to face posteriorly
_____	**4.** inversion	**D.**	palm moved to face anteriorly
_____	**5.** pronation	**E.**	movement to posterior plane
_____	**6.** plantar flexion	**F.**	movement to stand on tiptoes
_____	**7.** protraction	**G.**	movement in anterior plane
_____	**8.** supination	**H.**	movement to turn foot inward
_____	**9.** adduction	**I.**	movement to stand on heels
_____	**10.** abduction	**J.**	movement toward midline

16

C. Labeling

Label the structure of a typical synovial joint in Figure 16.7.

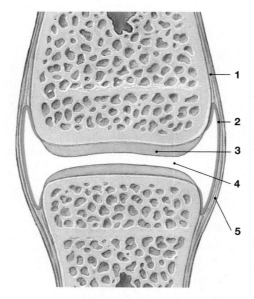

1. _____

2. _____

3. _____

4. _____

5. _____

Figure 16.7 Structure of a Synovial Joint

D. Fill in the Blanks

Describe the joints and movements involved in

1. walking
2. throwing a ball
3. turning a doorknob
4. crossing your legs while sitting
5. shaking your head "no"
6. chewing food

E. Short-Answer Questions

1. Describe the three types of functional joints.

2. What factors limit the range of movement of a joint?

3. Describe the four types of structural joints.

4. List the seven ligaments of the knee and how each supports the joint.

F. Analysis and Application

1. Describe how the articulating bones of the elbow prevent hyperextension of this joint.

2. Why do bones fuse in joints damaged by rheumatoid arthritis?

3. Which joint is unique to the hand, and how does this joint move the hand?

4. Which structural feature enables diarthrotic joints to have free movement?

5. Why is the lateral meniscus often associated with a knee injury?

Organization of Skeletal Muscles

OBJECTIVES

On completion of this exercise, you should be able to:

- Describe the basic functions of skeletal muscles.

- Describe the organization of a skeletal muscle.

- Describe the microanatomy of a muscle fiber.

- Discuss and provide examples of a lever system.

- Understand the rules that determine the names of some muscles.

Every time you move some part of your body, either consciously or unconsciously, you use muscles. Recall from Exercise 9 that there are three kinds of muscle tissue: skeletal, smooth, and cardiac. Skeletal muscles are primarily responsible for **locomotion**, movement of the body. Actions such as rolling your eyes, writing your name, and speaking are the result of highly coordinated muscle contractions. Other functions of skeletal muscle include maintenance of **posture** and **body temperature** and support of soft tissues, as with the muscles of the abdomen.

In addition to the ability to contract, muscle tissue has several other unique characteristics. Like nerve tissue, muscle tissue is **excitable** and, in response to a stimulus, produces electrical impulses called **action potentials**. Muscle tissue is **extensible** and can be stretched. When the ends of a stretched muscle are released, it recoils to its original size, like a rubber band. This property is called **elasticity**.

LAB ACTIVITY 1 Skeletal Muscle Organization

Connective Tissue Coverings

Skeletal muscles are attached to bones with **tendons.** The central, thicker part of a muscle is called the **belly.** A collagenous connective tissue layer called the **epimysium** (ep-i-MĪZ-ē-um; *epi*, on + *mys*, muscle) entirely covers each muscle (Figure 17.1). The epimysium is continuous with the fascia and deep fascia (described later) and separates the muscle from neighboring structures. The epimysium folds into the belly of the muscle as the **perimysium** (per-i-MĪZ-ē-um; *peri-*, around), and separates the muscle into bundles of muscle cells. Each individual muscle cell is called a **muscle fiber**, and the bundles of fibers are called **fascicles** (FA-sik-ulz). The parallel threadlike fibers of the fascicles can be easily seen when a muscle is teased apart with a probe. An extension of the perimysium, called the **endomysium** (en-dō-MĪZ-ē-um; *endo-*, inside), plunges deep into each fascicle and surrounds each individual muscle fiber.

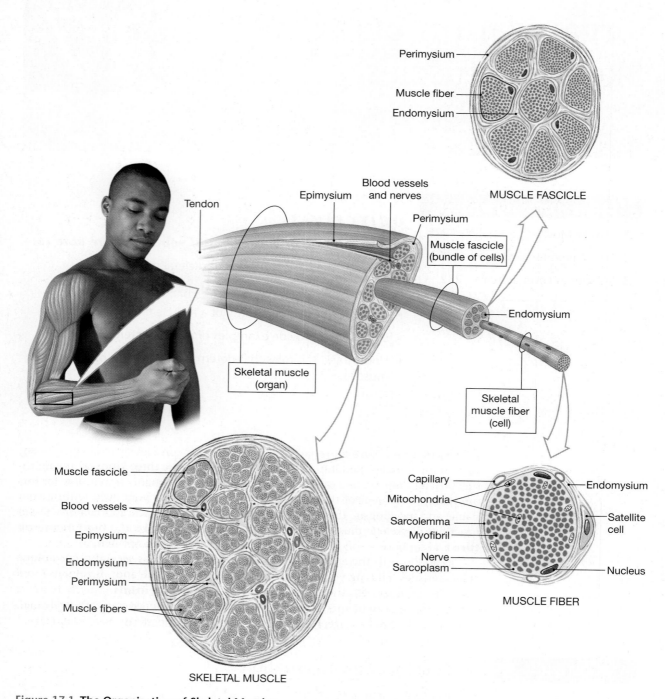

Figure 17.1 The Organization of Skeletal Muscles
A skeletal muscle consists of fascicles (bundles of muscle fibers) enclosed by the epimysium. The bundles are separated by the connective tissue fibers of the perimysium, and within each bundle the muscle fibers are surrounded by the endomysium. Each muscle fiber has many superficial nuclei, as well as mitochondria and other organelles.

The epimysium extends as a strong cord called a **tendon** that blends into the periosteum of the attached bone. When the muscle fibers contract and generate tension, they transmit this force through the connective tissue layers to the tendon, which pulls on the associated bone and produces movement.

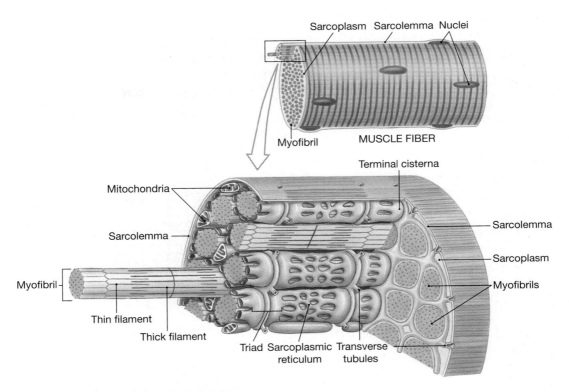

Figure 17.2 The Structure of a Skeletal Muscle Fiber
The internal organization of a muscle fiber.

Structure of a Skeletal Muscle Fiber

Each muscle fiber is a composite of many cells that fused into a single cell during embryonic development. As Figure 17.2 shows, the cell membrane of a muscle fiber is called the **sarcolemma** (sar-cō-LEM-uh; *sarkos,* flesh + *lemma,* husk) and the cytoplasm is called **sarcoplasm**. Connecting the sarcolemma to the interior of the muscle fiber are many **transverse tubules**, also called *T-tubules*. The function of these tubules is to pass contraction stimuli to deeper regions of the muscle fiber.

Inside the muscle fiber are proteins arranged in thousands of rods, called **myofibrils**, that extend the length of the fiber. Each myofibril is surrounded by a layer called the **sarcoplasmic reticulum**, where calcium ions are stored.

The branches of the sarcoplasmic reticulum fuse to form large Ca^{2+} storage chambers called **terminal cisternae** (sis-TUR-nē), which lie adjacent to the transverse tubules. A **triad** is a "sandwich" consisting of a transverse tubule plus the terminal cisterna on either side of the tubule. For a muscle to contract, calcium ions must be released from the cisternae; the transverse tubules stimulate this ion release. When a muscle relaxes, protein carriers in the sarcoplasmic reticulum transport calcium ions back into the cisternae.

Each myofibril consists of several kinds of proteins arranged in about 3,000 **thin filaments** and 1,500 **thick filaments** (Figure 17.3a). During contraction, thick and thin protein molecules interact to produce tension and shorten the muscle. The thin filaments are mostly composed of the protein **actin**, and the thick filaments are made of the protein **myosin**. The filaments are arranged in repeating patterns called **sarcomeres** (SAR-kō-mērz; *sarkos,* flesh + *meros,* part) along a myofibril. The thin filaments connect to one another at the **Z lines** on each end of the sarcomere. Each Z line is made of a protein called **actinin** (Figure 17.4a). Areas near the Z line that contain only thin filaments are **I bands**. Between I bands in a sarcomere is the **A band**, an area containing both thin and thick filaments. The edges of the A band are the **zone of overlap** where the thick

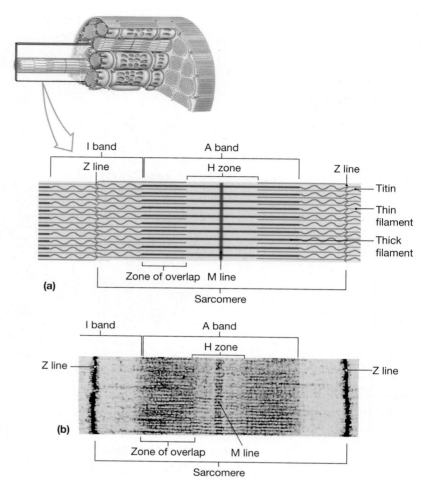

Figure 17.3 Sarcomere Structure

(a) A longitudinal section of a sarcomere. (b) A corresponding view of a sarcomere in a myofibril from a muscle fiber in the gastrocnemius muscle of the calf.

and thin filaments bind during muscle contraction. The middle region of the A band is the **H zone** and contains only thick filaments. A dense **M line** in the center of the A band attaches the thick filaments. Because the thick and thin filaments do not overlap one another completely, some areas of the sarcomere appear lighter than others. This organization results in the striated (striped) appearance of skeletal muscle tissue visible in Figure 17.3b. The molecules of thick and thin filaments interact whenever the muscle fiber contracts.

A thin filament consists of two intertwining strands of actin (Figure 17.4a). Four protein components make up the actin strands: G actin, F actin strand, nebulin, and active sites. The **G actins** are individual spherical molecules, like pearls on a necklace. Approximately 300 to 400 G actins twist together into an **F actin strand**. The G actins are held in position along the strand by the protein **nebulin**, much like the string of a necklace holds the beads in place. On each G actin molecule is an **active site** where myosin molecules bind during contraction (Figure 17.4b). Associated with the actin strands are two other proteins, **tropomyosin** (trō-pō-MĪ-ō-sin; *trope,* turning) and **troponin** (TRŌ-pō-nin). Tropomyosin follows the twisted actin strands and blocks active sites to regulate muscle contraction. Troponin holds tropomyosin in position and has binding sites for calcium ions. When calcium ions are released into the sarcoplasm, they bind to and cause troponin to change shape. This change in shape moves tropomyosin away from the binding sites, exposing the sites to myosin heads so that the interactions necessary for contraction can take place.

A thick filament is made of approximately 500 subunits of myosin (Figure 17.4d). Each subunit consists of two strands: two intertwined **tail** regions and two

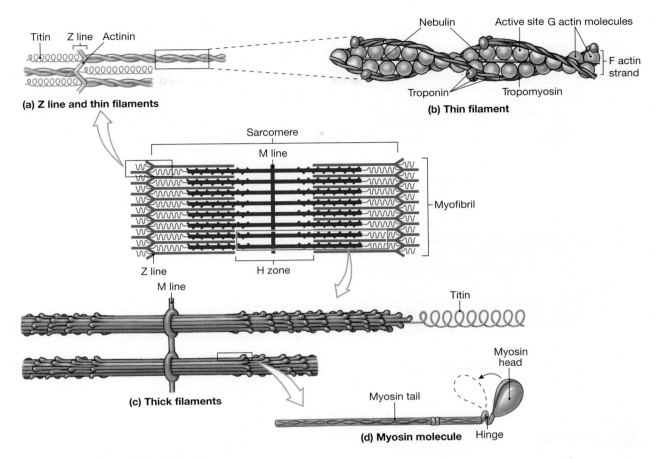

Figure 17.4 Thick and Thin Filaments

(a) The gross structure of a thin filament, showing the attachment at the Z line. (b) The organization of G actin molecules in an F actin strand, and the position of the troponin-tropomyosin complex. (c) The structure of thick filaments, showing the orientation of the myosin molecules. (d) The structure of a myosin molecule.

globular **heads**. Bundles of myosin subunits, with the tails parallel to one another and the heads projecting outward, constitute a thick filament (Figure 17.4c). A protein called **titin** attaches the thick filament to the Z line on the end of the sarcomere. Each myosin head contains a **binding site** for actin and a region that functions as an **ATPase enzyme**. This portion of the head splits an ATP molecule and absorbs the released energy to bind to a thin filament and then pivot and slide the thin filament inward.

When a muscle fiber contracts, the thin filaments are pulled deep into the sarcomere (Figure 17.5). As the thin filaments slide inward, the I band and H zone become smaller. Each myofibril consists of approximately 10,000 sarcomeres joined end to end. During contraction, the sarcomeres compress and the myofibril shortens and pulls on the sarcolemma, causing the muscle fiber to shorten as well.

QuickCheck Questions

1.1 Describe the connective tissue organization of a muscle.

1.2 What is the relationship between myofibrils and sarcomeres?

1.3 Where are calcium ions stored in a muscle fiber?

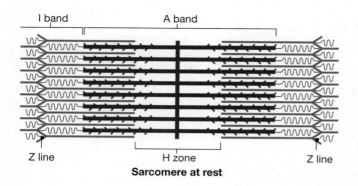

Sarcomere at rest

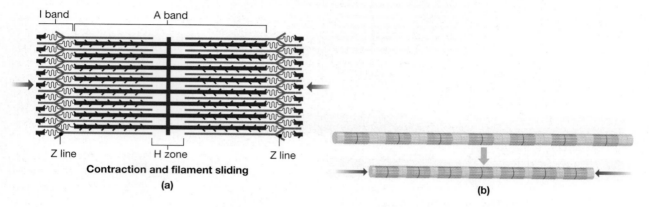

Contraction and filament sliding

(a)

(b)

Figure 17.5 Changes in the Appearance of a Sarcomere during the Contraction of a Skeletal Muscle Fiber

(a) During a contraction, the A band stays the same width, but the Z lines move closer together and the I band gets smaller. **(b)** When the ends of a myofibril are free to move, the sarcomeres shorten simultaneously and the ends of the myofibril are pulled toward its center.

1 Materials

- ☐ Muscle model
- ☐ Muscle fiber model
- ☐ Round steak or similar cut of meat
- ☐ Preserved muscle tissue
- ☐ Dissecting microscope

Procedures

1. Review the organization of muscles in Figures 17.1 to 17.5.

2. Review the histology of skeletal muscle fibers in Exercise 9.

3. Identify the connective tissue coverings of muscles on the laboratory models. If your instructor has prepared a muscle demonstration from a cut of meat, examine the meat for the various connective tissues. Are fascicles visible on the specimen?

4. Examine the muscle fiber model and identify each feature. Describe the location of the sarcoplasmic reticulum, myofibrils, sarcomeres, and filaments.

5. Draw the sarcomere and label the following structures:
   - thick and thin filaments
   - sarcomere
   - Z line
   - I band
   - A band
   - H zone
   - zone of overlap
   - M line

6. Examine a specimen of preserved muscle tissue by placing the tissue in saline solution and then teasing the muscle apart using tweezers and a probe. Notice how the fascicles appear as strands of muscle tissue. Examine the fascicles under a dissecting microscope. How are they arranged in the muscle? ■

LAB ACTIVITY **2** ## The Neuromuscular Junction

Each skeletal muscle fiber is controlled by a nerve cell called a **motor neuron**. To excite the muscle fiber, the motor neuron releases a chemical message called **acetylcholine** (as-ē-til-KŌ-lēn), abbreviated ACh. The motor neuron and the muscle fiber meet at a **neuromuscular junction** (Figure 17.6), also called a **myoneural junction**. The end of the neuron, called the **axon**, expands to form a bulbous **synaptic knob**, also called a **synaptic terminal**. In the synaptic knob are **synaptic vesicles** that contain ACh. A small gap, the **synaptic cleft**, separates the synaptic knob from a folded area of the sarcolemma called the **motor end plate**. At the motor end plate, the sarcolemma releases into the synaptic cleft the enzyme **acetylcholinesterase** (AChE), which prevents overstimulation of the muscle fiber by deactivating ACh.

When a nerve impulse, called an **action potential**, travels down a neuron and reaches the synaptic knob, the synaptic vesicles release ACh into the synaptic cleft. The ACh diffuses across the cleft and binds to ACh receptors embedded in the sarcolemma at the motor end plate. This binding of the chemical stimulus causes the sarcolemma to generate an action potential. The potential spreads across the sarcolemma and down transverse tubules, causing calcium ions to be released from the sarcoplasmic reticulum into the sarcoplasm of the muscle fiber. The calcium ions bind to troponin, which in turn moves tropomyosin and exposes the active sites on the G actins. The myosin heads attach to the active sites and ratchet the thin filaments inward, much like a tug-of-war team pulling on a rope. As thin filaments slide into the H zone, the sarcomere shortens. The additive effect of the shortening of many sarcomeres along the myofibril results in a decrease in the length of the myofibril and contraction of the muscle.

In a relaxed muscle fiber, the calcium ion concentration in the sarcoplasm is minimal. When the muscle fiber is stimulated, calcium channels in the sarcoplasmic reticulum open and calcium ions rapidly flow down the concentration gradient into the sarcoplasm. Because each myofibril is surrounded by sarcoplasmic reticulum, calcium ions are quickly and efficiently released among the thick and thin filaments of the myofibril. For the muscle to relax, calcium-ion pumps in the sarcoplasmic reticulum actively transport calcium ions out of the sarcoplasm and into the cisternae.

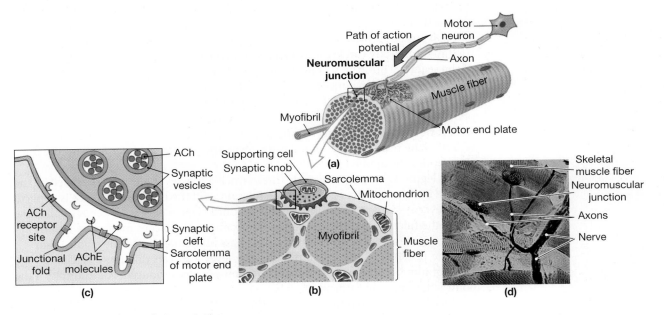

Figure 17.6 Skeletal Muscle Innervation

(a) A diagrammatic view of a neuromuscular junction. **(b)** Details of the neuromuscular junction. **(c)** Detail of the synaptic knob. **(d)** Micrograph of the neuromuscular junction.

QuickCheck Questions

2.1 What molecules are in the synaptic vesicles?

2.2 Where is the motor end plate?

2 | Materials

☐ Compound microscope

☐ Neuromuscular junction slide

Procedures

1. Review the structure of the neuromuscular junction in Figure 17.6.

2. Examine the slide of the neuromuscular junction at low and medium powers. Identify the long, dark, threadlike structures and the oval disks. Describe the appearance of the muscle fibers.

3. In the space below, sketch several muscle fibers and their neuromuscular junctions. ∎

LAB ACTIVITY | 3 | Muscles as Lever Systems

The property of contractility enables the body to move all its various parts, particularly where muscles attach to bones. The combination of muscles, bones, and joints forms a **lever system**, which is a way of increasing the amount of **work** a given force can do. If you have to lift, for example, a 100-pound object, you probably cannot exert enough force to do the lifting directly. If the object is sitting on one end of a playground seesaw, however, the same force that would not budge it when you tried to lift directly can lift the object when you apply the force to the opposite end of the seesaw. Thus, the seesaw is a lever system.

Any lever system consists of a rigid rod, the **lever** (L), that moves around a pivot point called the **fulcrum** (F). The muscle effort required to move the lever is the applied force (AF). The **resistance** (R) of the system is the weight of the lever or the weight of any object resting on the lever. In a muscle-bone lever system, the body weight is the resistance, the bone is the lever, the joint is the fulcrum, and the muscle supplies the applied force.

As Figure 17.7 shows, there are three principal classes of levers, determined by the relative positions of the resistance, the fulcrum, and the point where force is applied. Most lever systems in the human body are third class, where the force is applied between the fulcrum and the resistance. For example, the biceps moves the arm by applying a force between the resistance (the mass of the hand) and the elbow joint.

The position of the fulcrum along the lever arm can dramatically change the range of motion of a lever system and amount of work produced. For example, if the point where a muscle attaches to a bone is close to a joint, the muscle will be able to produce a great range of motion. However, this large range comes at a cost: a reduction loss in the work done by that muscle. Conversely, if another muscle of the same strength attaches to a bone farther from the joint, this second muscle can generate more work, but with a limited range of motion. This principle is called **leverage**.

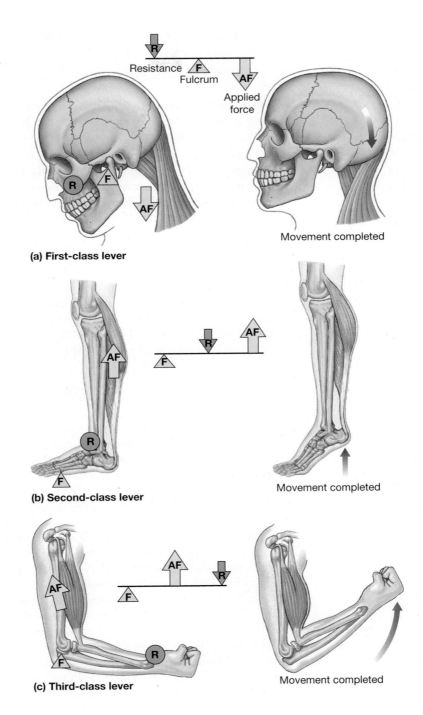

Figure 17.7 The Three Classes of Levers

(a) In a first-class lever, the applied force and the resistance are on opposite sides of the fulcrum. **(b)** In a second-class lever, the resistance lies between the applied force and the fulcrum. **(c)** In a third-class lever, the force is applied between the resistance and the fulcrum.

(a) First-class lever

Resistance   Fulcrum   Applied force

Movement completed

(b) Second-class lever

Movement completed

(c) Third-class lever

Movement completed

QuickCheck Questions

3.1 What is a fulcrum?

3.2 How are muscles attached to bone to gain more leverage?

3	**Materials**

☐ Textbook
☐ Two pencils

Procedures

1. Review the organization of lever systems in Figure 17.7. You are going to construct a simple first-class lever system to demonstrate leverage.

2. Lay the book down on the top of a desk and place one pencil parallel to and right alongside the book spine; this pencil is the fulcrum of your lever system.

Put the sharpened end of the second pencil under the spine, and rest the body of the second pencil on top of the fulcrum pencil; this second pencil is the lever.

3. Slide the fulcrum pencil away from the book until it is close to the eraser of the lever pencil.

4. Using care to prevent breaking the lever pencil, push down on it by applying a force at the eraser end. Notice how difficult it is to move the resistance (the book) with the fulcrum at the far end of the lever.

5. Slide the fulcrum pencil toward the spine of the book, to a point about midway along the lever pencil, and push down on the lever pencil as before. This time the book is easily raised. What you have demonstrated is the principle of leverage. When the fulcrum (the point of contact between the two pencils) is closer to the resistance (the book), it is easier for you to apply enough force to raise the book.

6. Next, place the pencils as they were in step 3, with the fulcrum far from the resistance, and then remove the book. Now press down on the eraser end of the lever pencil and note that, without the book, the tip moves easily.

7. Shift the fulcrum pencil toward the tip of the lever pencil. Now press down on the lever pencil. Again the tip moves easily, but this time the distance the tip moves is shorter. The force on the lever is used to move either a heavy weight over a short distance or a small weight over a greater distance. Strength is sacrificed for range of movement. Muscles systems work in a similar manner. ■

Action, Origin, and Insertion

Muscles can only generate a pulling force; they can never push. Each muscle causes a movement, called the **action**, that depends on many factors, especially the shape of the attached bones. For a muscle to produce a smooth, coordinated action, one end of it must serve as an attachment site while the other end moves the intended bone. The relatively stationary part of the muscle is called the **origin**. The opposite end of the muscle, the part that moves the bone, is called the **insertion**. During contraction, the insertion moves toward the origin to generate a pulling force and cause the muscle's action. Usually, when one muscle pulls in one direction, an **antagonistic** muscle pulls in the opposite direction to produce resistance and promote smooth movement. For example, when you flex your arm, the biceps brachii muscle is the **prime mover** and the triceps brachii is the **antagonist**. The roles of these two muscles reverse when you extend your arm.

Muscles that help the prime mover are called **synergists**. For example, the supinator muscle of the arm turns the hand so that the palm faces up. The biceps brachii acts as a synergist for this action by secondarily supinating the forearm. **Fixators** are muscles that anchor bones in place to allow other muscles to perform their actions more efficiently. For example, many muscles surrounding the scapula help fix that bone in place if the arm is required to apply a strong force. If greater flexibility is needed, the fixators relax and allow the scapula to shift its position. In this way one can have both great strength and great flexibility in the arm, although not at the same time.

Naming Muscles

Numerous methods are used to name muscles (Table 17.1). One method names muscles according to either the bones they attach to or the region of the body in which they are found. For example, the temporalis muscle is found on the temporal bone, and the rectus abdominis muscle forms the anterior muscular wall of the abdomen. Another easily identifiable muscle is the sternocleidomastoid, which originates on the sternum (sterno-) and the clavicle (cleido-) and inserts on the mastoid process of the temporal bone.

Table 17.1 **Muscle Terminology**

Terms Indicating Direction Relative to Axes of the Body	Terms Indicating Specific Regions of the Body	Terms Indicating Structural Characteristics of the Muscle	Terms Indicating Actions
Anterior (front)	Abdominis (abdomen)	**Origin**	**General**
Externus (superficial)	Anconeus (elbow)	Biceps (two heads)	Abductor
Extrinsic (outside)	Auricularis (auricle of ear)	Triceps (three heads)	Adductor
Inferioris (inferior)	Brachialis (brachium)	Quadriceps (four heads)	Depressor
Internus (deep, internal)	Capitis (head)		Extensor
Intrinsic (inside)	Carpi (wrist)	**Shape**	Flexor
Lateralis (lateral)	Cervicis (neck)	Deltoid (triangle)	Levator
Medialis/medius (medial, middle)	Cleido-/-clavius (clavicle)	Orbicularis (circle)	Pronator
Oblique	Coccygeus (coccyx)	Pectinate (comblike)	Rotator
Posterior (back)	Costalis (ribs)	Piriformis (pear-shaped)	Supinator
Profundus (deep)	Cutaneous (skin)	Platys- (flat)	Tensor
Rectus (straight, parallel)	Femoris (femur)	Pyramidal (pyramid)	
Superficialis (superficial)	Genio- (chin)	Rhomboid	**Specific**
Superioris (superior)	Glosso-/-glossal (tongue)	Serratus (serrated)	Buccinator (trumpeter)
Transversus (transverse)	Hallucis (great toe)	Splenius (bandage)	Risorius (laugher)
	Ilio- (ilium)	Teres (long and round)	Sartorius (like a tailor)
	Inguinal (groin)	Trapezius (trapezoid)	
	Lumborum (lumbar region)		
	Nasalis (nose)	**Other Striking Features**	
	Nuchal (back of neck)	Alba (white)	
	Oculo- (eye)	Brevis (short)	
	Oris (mouth)	Gracilis (slender)	
	Palpebrae (eyelid)	Lata (wide)	
	Pollicis (thumb)	Latissimus (widest)	
	Popliteus (posterior to knee)	Longissimus (longest)	
	Psoas (loin)	Longus (long)	
	Radialis (radius)	Magnus (large)	
	Scapularis (scapula)	Major (larger)	
	Temporalis (temples)	Maximus (largest)	
	Thoracis (thoracic region)	Minimus (smallest)	
	Tibialis (tibia)	Minor (smaller)	
	Ulnaris (ulna)	-tendinosus (tendinous)	
	Uro- (urinary)	Vastus (great)	

The size of a muscle is often reflected in its name. *Maximus* refers to large muscles, as in gluteus maximus, and *minimus* refers to small muscles, as in gluteus minimus. The adjective *longus* in a name implies that the muscle is long. *Brevis* refers to a short muscle.

The direction of muscle fascicles is also used to name muscles. The term *rectus* is used if the fibers are parallel to the body or limb. For example, in the rectus abdominis, the muscle fascicles are parallel to the body. *Transversus* refers to fascicles that are perpendicular to the midline, as in transversus abdominis. In the internal and external *obliques,* the fascicles cut diagonally across the body.

Many muscles have multiple origins. The triceps, for example, has three origins, and they all insert on the olecranon process of the elbow. Look for the prefixes *bi-* for two, *tri-* for three, and *quad-* for four origins.

Anatomists often conceive names based on muscle shape. The deltoid has a broad origin and inserts on a very narrow region of the humerus. This gives this muscle a triangular, or *deltoid,* shape; hence the name. The trapezius is a very large muscle covering the middle back that forms a *trapezoid* shape. The orbicularis muscles around the eye and mouth are named after their *circular* shape.

Many muscles are named based on their action. The name flexor carpi ulnaris appears complex, but it is really quite easy to understand if you examine it step-by-step. *Flexor* means the muscle flexes something. What does it flex? Look at the next two terms. *Carpi* refers to carpals, the bones of the wrist, and *ulnaris* suggests that the muscle flexes the carpi on the medial side of the wrist, where the ulna is. Therefore, the flexor carpi ulnaris is a muscle that flexes and adducts the wrist. Once you have an idea of the action, you can then determine where on the body the muscle is located. In this case, it is the forearm. The origin and insertion can sometimes be immediately determined. In our example, it is easy to guess that this muscle inserts on the carpi; the origin is a little more difficult and may require some memorization.

Try using this analysis procedure with the levator scapulae muscle.

Organization of Skeletal Muscles

Name _____

Date _____

Section _____

A. Matching

Match each term in the left column with its correct description from the right column.

_____ **1.** sarcomere

_____ **2.** epimysium

_____ **3.** perimysium

_____ **4.** endomysium

_____ **5.** myofibril

_____ **6.** striations

_____ **7.** sarcolemma

_____ **8.** transverse tubule

_____ **9.** sarcoplasmic reticulum

_____ **10.** actin

_____ **11.** myosin

_____ **12.** fascicle

A. banding patterns in muscle tissue

B. storage site for calcium ions

C. protein of thin filaments

D. group of muscle fibers

E. cylinder composed of filaments

F. protein of thick filaments

G. carries action potential deep into fiber

H. connective tissue covering fiber

I. connective tissue covering muscle

J. connective tissue covering fascicle

K. cell membrane of muscle fiber

L. repeating organization of filaments

B. Matching

Match each term in the left column with its correct description from the right column.

_____ **1.** glossal

_____ **2.** cleido

_____ **3.** scapularis

_____ **4.** abductor

_____ **5.** oris

_____ **6.** brevis

_____ **7.** adductor

_____ **8.** oculi

_____ **9.** vastus

_____ **10.** rectus

_____ **11.** tensor

_____ **12.** capitis

A. mouth

B. clavicle

C. moves away

D. great

E. tongue

F. moves toward

G. eye

H. scapula

I. tenses

J. head

K. short

L. straight

17

C. Labeling

Label the structure of a muscle fiber in Figure 17.8.

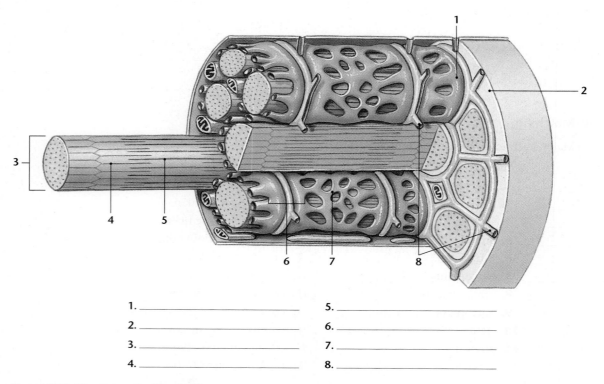

1. _____   5. _____
2. _____   6. _____
3. _____   7. _____
4. _____   8. _____

Figure 17.8 Structure of a Muscle Fiber

D. Short-Answer Questions

1. Describe the structure of a fascicle, including the connective tissue covering around and within the fascicle.

2. How does a motor neuron stimulate a muscle fiber to contract?

3. Describe the structure of a sarcomere.

E. Analysis and Application

1. Describe the lever system of the calf muscle attached to the heel of the foot. Where is the fulcrum located? Where is the lever located? What is the resistance, and what agent applies the force?

2. Give an example of a third-class muscle lever.

3. Describe the role of each thin-filament protein and each thick-filament protein in muscle contraction.

Muscles of the Head and Neck

OBJECTIVES

On completion of this exercise, you should be able to:

- Identify the origin, insertion, and action of the muscles used for facial expression and mastication.
- Identify the origin, insertion, and action of the muscles that move the eye.
- Identify the origin, insertion, and action of the muscles that move the tongue, head, and anterior neck.

The muscles of the head and neck produce a wide range of motions for making facial expressions, processing food, producing speech, and positioning the head. The names of these muscles usually indicate either the bone the muscles attach to or the structure they surround. In this exercise you will identify the major muscles used for facial expression and mastication, the muscles that move the eyes, and the muscles that position the head and neck. As you study each group, attempt to find the general location of each muscle on your body. Contract the muscle and observe the action of the muscle as your body moves.

LAB ACTIVITY 1   Muscles of Facial Expression

Occipitofrontalis

The **occipitofrontalis** consists of two muscle bellies: the **frontal belly** and the **occipital belly** (Figure 18.1). The frontal belly of the occipitofrontalis is the broad anterior muscle on the forehead that covers the frontal bone. It originates on the superior margin of the eye orbit, near the eyebrow. The muscle fibers blend into a sheet of connective tissue called the **epicranial aponeurosis** (ep-I-KRĀ-nē-ul āp-ō-nū-RŌ-sis; *epi-,* on + *kranion,* skull), which covers the epicranium, the scalp of the skull. The actions of the frontal belly include wrinkling the forehead, raising the eyebrows, and pulling the scalp forward.

The occipital belly of the occipitofrontalis covers the back of the skull. It arises on the occipital bone and the mastoid process of the temporal bone, extends superiorly, and inserts on the epicranial aponeurosis, completing the posterior part of the occipitofrontalis. This muscle draws the scalp backward, an action difficult for most people to isolate and perform.

Temporoparietalis

The **temporoparietalis** occurs on the lateral sides of the epicranium. The muscle is cut and reflected (pulled up) in Figure 18.1 to illustrate deeper muscles of the scalp.

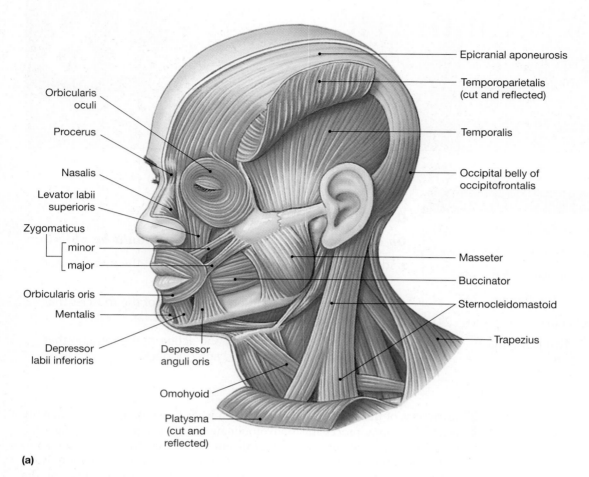

Figure 18.1A Muscles of Facial Expression
(a) Anterolateral view.

The action of the temporoparietalis is to tense the scalp and move the auricle (flap) of the ear. The origin and insertion for this muscle are on the epicranial aponeurosis.

Orbicularis Oculi

The sphincter muscle of the eye is the **orbicularis oculi** (or-bik-ū-LA-ris ok-ū-lī). It arises from the medial wall of the eye orbit, and its fibers form a band of muscle that passes around the circumference of the eye, which serves as the insertion. The muscle acts to close the eye, as during an exaggerated blink.

Corrugator Supercilii

The **corrugator supercilii** begins on the ridge between the eyebrows and the bridge of the nose and angles laterally to insert into the skin of the forehead. It acts to pull the skin downward and wrinkles the forehead into a frown. Think of a corrugated tin roof when you study this muscle. The corrugator supercilii bunches the skin of the forehead into folds, or corrugations.

Orbicularis Oris

The **orbicularis oris** is a sphincter muscle that surrounds the mouth. It originates on the bones surrounding the mouth and inserts on the lips. This muscle shapes the lips for a variety of functions, including speech, food manipulation, and facial expressions, and purses the lips together for a kiss.

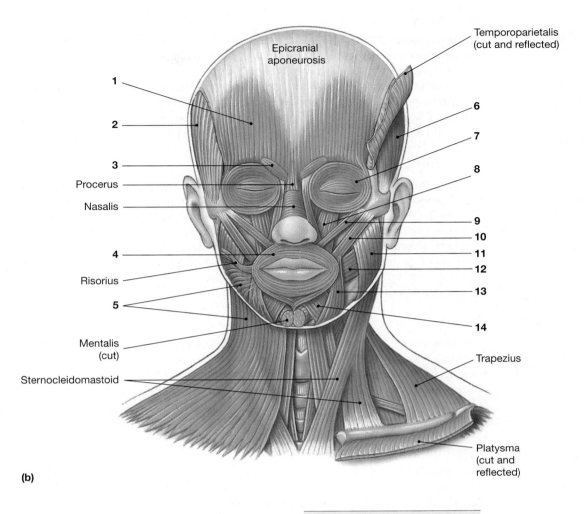

Epicranial aponeurosis

Temporoparietalis (cut and reflected)

1

2

3

Procerus

Nasalis

4

Risorius

5

Mentalis (cut)

Sternocleidomastoid

6

7

8

9

10

11

12

13

14

Trapezius

Platysma (cut and reflected)

(b)

1. _____
2. _____
3. _____
4. _____
5. _____
6. _____
7. _____
8. _____
9. _____
10. _____
11. _____
12. _____
13. _____
14. _____

buccinator
corrugator supercilii
depressor anguli oris
depressor labii inferioris
frontal belly of occipitofrontalis
levator labii superioris
masseter
orbicularis oculi
orbicularis oris
platysma
temporalis
temporoparietalis
zygomaticus major
zygomaticus minor

Figure 18.1B **(b)** Anterior view. *(Continued)*

Buccinator

The **buccinator** (BUK-si-nā-tor) is the horizontal muscle spanning between the jaws. It originates on the alveolar processes of the maxillary bone and mandible and inserts on the fibers of the orbicularis oris. When a person is chewing, the buccinator moves food across the teeth from the cheeks. In music-world slang, a "buccinator" is a trumpet player, because horn players use the buccinator to pressurize their exhalations. You use this muscle when you contract and compress your cheeks during eating, when you tense your cheeks when blowing out, and when you suck on a straw.

Depressor Labii and Levator Labii

The labii muscles insert on the lips and move them according to the muscle names. The **depressor labii** arises from the region between the symphysis and mental foramen of the mandible, inserts on the skin of the lower lip, and blends into the orbicularis oris muscle. The depressor labii pulls the lower lip down and slightly outward. The origin of the **levator labii** is immediately superior to the infraorbital foramen of the maxillary bone. Inserting on the skin at the angle of the mouth and blending into the orbicularis oris, the levator labii raises the upper lip, as when (and if) you pretend to whinny like a horse.

Risorius

The **risorius** is a narrow muscle that begins on the fascia over the parotid salivary gland and ends on the angle of the mouth. When it contracts, the risorius pulls and produces a grimace-like tensing of the mouth. Although the term *risorius* refers to a smile, the muscle is probably more associated with the expression of pain rather of pleasure. In the disease tetanus, the risorius is involved in the painful contractions that pull the corners of the mouth backward into "lockjaw."

Zygomaticus Major and Minor

The **zygomaticus major** arises from the temporal process of the zygomatic bone. The origin of the **zygomaticus minor** lies just medial to the origin of the zygomaticus major, near the maxillary suture. Both muscles insert on the skin and corners of the mouth. These muscles pull the skin of the mouth upward and laterally when you smile.

Platysma

The **platysma** (pla-TIZ-muh; *platys,* flat) is a thin, broad muscle covering the sides of the neck. It originates on the fascia covering the pectoralis and deltoid muscles and extends upward to insert on the inferior edge of the mandible. Some of the fibers of the platysma also extend into the fascia and muscles of the lower face. The platysma depresses the mandible and the soft structures of the lower face, resulting in an expression of horror and disgust. Compare this action with that of the risorius.

QuickCheck Questions

1.1 Name two facial muscles that are circular.

1.2 List the muscles associated with the epicranial aponeurosis.

---

1 *Materials*

☐ Head model
☐ Muscle chart

Procedures

1. Review the muscles of the head in Figure 18.1a and then label them all in Figure 18.1b.

2. Examine the head model and/or the muscle chart, and locate each muscle described in the preceding paragraphs.

3. Find the general location of the muscles of facial expression on your face. Practice the action of each muscle and observe how your facial expression changes. ■

LAB ACTIVITY 2   Muscles of the Eye

The **extrinsic** muscles of the eye, called **oculomotor** muscles, are those located on the outside of the eye, as shown in Figure 18.2. They insert on the sclera, which is the white, fibrous covering of the eye. (**Intrinsic** muscles are the internal muscles of the eye and are involved in focusing the eye for vision. These muscles are discussed in Exercise 36.)

Six extrinsic eye muscles control eye movements. The **superior rectus**, **inferior rectus**, **medial rectus**, and **lateral rectus** are straight muscles that move the eyeball up and down and side to side. They originate around the optic foramen within the eye orbit, insert on the sclera, and roll the eyeball as indicated by their names. The **superior** and **inferior oblique** muscles attach diagonally on the eyeball. The superior oblique has a tendon passing through a trochlea (pulley) located on the upper orbit. This muscle rolls the eye downward, and the inferior oblique rolls the eye upward.

QuickCheck Questions

2.1 Name the four rectus muscles of the eye.

2.2 Which eye muscle passes through a pulley-like structure?

2   Materials

☐ Eye model
☐ Eye muscle chart

Procedures

1. Review the muscles of the eye in Figure 18.2.

2. Examine the eye model and/or the eye muscle chart, and locate each extrinsic eye muscle.

3. Practice the action of each eye muscle by moving your eyeballs. ■

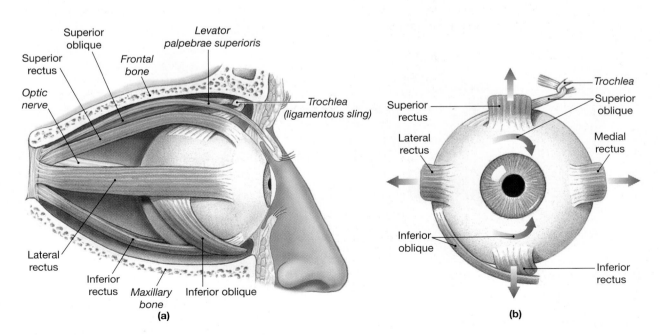

Figure 18.2 Extrinsic Eye Muscles
(**a**) Lateral view of right eye. (**b**) Anterior view of right eye.

LAB ACTIVITY 3 | Muscles of Mastication

The muscles involved in mastication (chewing) are shown in Figure 18.3. The **masseter** (MAS-se-tur) is a short, thick muscle originating on the maxillary bone and the anterior portion of the zygomatic arch. It inserts on the angle and the ramus of the mandible. The **temporalis** (tem-pō-RA-lis) covers almost the entire temporal fossa, which serves as the origin for this muscle. The temporalis inserts on the coronoid process of the mandible. When the masseter and the temporalis contract, they elevate the jaw against the teeth, producing the tremendous force used in chewing. When it acts alone, the masseter protracts the jaw slightly, which means it makes the jaw jut out in front of the facial plane. The temporalis muscle retracts the jaw.

Deep to the masseter and other cheek muscles are the **lateral** and **medial pterygoid** (TER-i-goyd; *pterygoin*, wing) muscles. They originate on the pterygoid processes of the sphenoid and insert on the medial surface of the ramus. The lateral pterygoid depresses the mandible to open the mouth, and the medial pterygoid closes the mouth. Each muscle also moves the mandible from side to side, an action called *lateral excursion*.

Study Tip | ***The Mighty Masseter***

Put your fingertips at the angle of your jaw and clench your teeth. You should feel the masseter bunch up as it forces the teeth together. ●

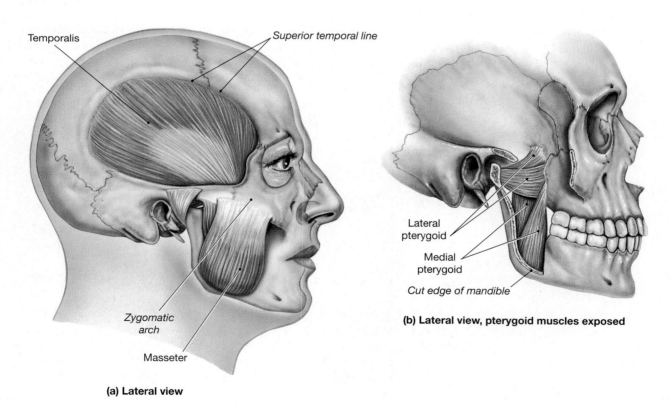

(a) Lateral view

(b) Lateral view, pterygoid muscles exposed

Figure 18.3 Muscles of Mastication
(a) The temporalis muscle passes medial to the zygomatic arch to insert on the coronoid process of the mandible. The masseter inserts on the angle and lateral surface of the mandible. **(b)** The location and orientation of the pterygoid muscles can be seen after the overlying muscles, along with a portion of the mandible, are removed.

QuickCheck Questions

3.1 To which bones do the muscles for mastication attach?

3.2 Which muscle protracts the mandible?

| **3** ***Materials*** | ***Procedures*** |

☐ Head model

☐ Muscle chart

1. Review the mastication muscles in Figures 18.1 and 18.3.

2. Examine the head model and/or the muscle chart, and locate each mastication muscle described in this activity.

3. Find the general location of the muscles of mastication on your face. Practice the action of each muscle and observe how your mandible moves. ■

LAB ACTIVITY   4   Muscles of the Tongue

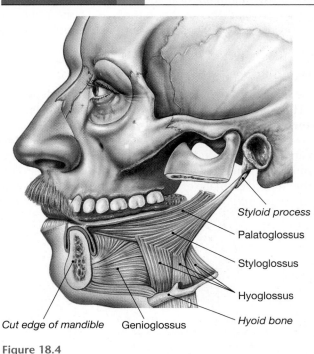

Styloid process

Palatoglossus

Styloglossus

Hyoglossus

Hyoid bone

Cut edge of mandible   Genioglossus

Figure 18.4
Muscles of the Tongue

The muscles of the tongue form the floor of the oral cavity (Figure 18.4). The root word for this muscle group is *glossus,* Greek for "tongue." Each prefix indicates the muscle's origin. The **palatoglossus** arises from the soft palate and inserts on the lateral tongue. During swallowing, the palatoglossus elevates the tongue and depresses the soft palate.

Antagonistic muscles protract and retract the tongue. Anteriorly, the **genioglossus** originates on the posterolateral area of the mandible. It depresses and protracts the tongue, as in initiating the licking of an ice cream cone. Posteriorly, the **styloglossus** and **hyoglossus** muscles arise on the styloid process and the hyoid bone, respectively. Both muscles retract the tongue. The styloglossus also raises the sides of the tongue.

QuickCheck Questions

4.1 What does the word "glossus" mean?

4.2 Where do the styloglossus and the hyoglossus muscles originate?

| **4** ***Materials*** | ***Procedures*** |

☐ Head model

☐ Muscle chart

1. Review the muscles of the tongue in Figure 18.4.

2. Examine the head model and/or the muscle chart and identify each muscle of the tongue.

3. Practice the action of each tongue muscle. The ability to curl your tongue with the styloglossus is genetically controlled by a single gene. Individuals with the dominant gene are "rollers" and can curl the tongue. Those with the recessive form of the gene are "nonrollers." Is it possible for nonrollers to learn how to roll the tongue? Are you, your parents, or your children rollers or nonrollers? ■

LAB ACTIVITY   5   Muscles of the Anterior Neck

All the muscles described in this section that carry the suffix *-hyoid* insert on the hyoid bone. This bone is suspended by muscles of the inferior and posterior mandible. It is also connected to the larynx and serves as a foundation for the tongue muscles. The muscles that insert on the hyoid bone are the **suprahyoid** and **infrahyoid** muscle groups. As the names imply, they are located either above or below the hyoid bone. The origin is different in each case, as indicated by the prefix; for example, *genio-* refers to the chin and *stylo-* refers to the styloid process of the temporal bone.

Superior Muscles of the Hyoid Bone

The suprahyoid muscles originate above the hyoid bone, and their action elevates the hyoid or depresses the mandible. The **geniohyoid** originates on the inside of the mandible near the symphysis, as shown in Figure 18.5, and acts to depress the mandible and elevate the larynx. The **stylohyoid** originates on the styloid process of the temporal bone, inserts on the hyoid bone, and acts to elevate the hyoid bone and the larynx. The **mylohyoid** originates along a broad origin on the inner border of the mandible. Along with the **digastric** muscle, the mylohyoid elevates the hyoid bone and depresses the lower jaw. The digastric has two parts: the **anterior belly** originates on the inside of the mandible near the chin, and the **posterior belly** arises on the mastoid process. The bellies insert on the hyoid bone and form a muscular swing that acts to elevate the hyoid bone or lower the mandible to open the jaw.

Inferior Muscles of the Hyoid Bone

The infrahyoid muscles arise below the hyoid bone, and their action depresses that bone and the larynx. The **omohyoid** (ō-mō-HĪ-oyd) has its origin on the clavicle and first rib. The **sternohyoid** is a thin, straplike muscle that originates on the sternal end of the clavicle. The **sternothyroid** originates on the manubrium of the sternum and terminates on the thyroid cartilage of the larynx. The omohyoid and sternohyoid muscles depress the hyoid bone. The **thyrohyoid** originates on the thyroid cartilage of the larynx, sweeps upward, and inserts on the hyoid bone to the larynx.

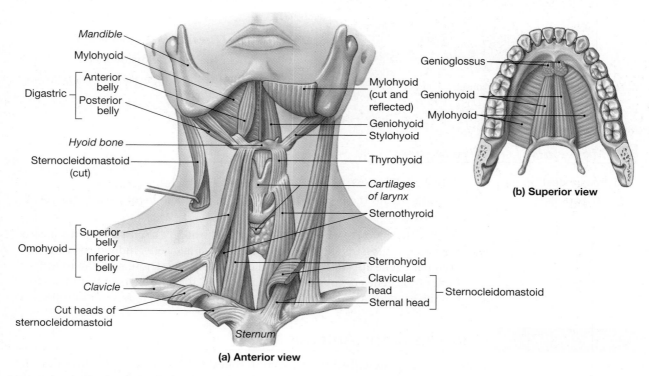

(a) Anterior view

(b) Superior view

Figure 18.5 **Muscles of the Anterior Neck**

QuickCheck Questions

5.1 What is the suffix of the name of muscles that insert on the hyoid bone?

5.2 Where is the digastric muscle located?

5 *Materials*	*Procedures*

☐ Head-torso model

☐ Muscle chart

1. Review the anterior neck muscles in Figure 18.5.

2. Locate each muscle on the head-torso model and/or the muscle chart.

3. Produce the actions of your suprahyoid and infrahyoid muscles and observe how your larynx moves. ■

LAB ACTIVITY 6

Muscles That Move the Head

The principal muscle of the anterior neck is the **sternocleidomastoid** (ster-nō-klī-dō-MAS-toyd). This long, slender muscle occurs on both sides of the neck and is named after its points of attachment on the sternum, clavicle, and mastoid process of the temporal bone (Figures 18.1 and 18.5). Because the sternocleidomastoid spans the head and sternum, the action when both sides of the muscle contract is to flex the neck and tuck the head down toward the sternum. If only one side contracts, the head is pulled toward the shoulder of that side. (There are also posterior neck muscles, but because they also move the chest, they are discussed in Exercise 19.)

QuickCheck Questions

6.1 Where does the sternocleidomastoid muscle attach?

6.2 What is the action of the sternocleidomastoid?

6 *Materials*	*Procedures*

☐ Head-torso model

☐ Muscle chart

1. Locate the sternocleidomastoid on the head-torso model and/or on the muscle chart.

2. Contract your sternocleidomastoid on one side and observe your head movement. Next, contract both sides and note how your head flexes.

3. Rotate your head until your chin almost touches your right shoulder and locate your left sternocleidomastoid just above the manubrium of the sternum. ■

Muscles of the Head and Neck

Name _____

Date _____

Section _____

A. Matching

Match each term in the left column with its correct description from the right column (O = origin, I = insertion, A = action).

_____ **1.** orbicularis oculi	**A.** moves scalp backward
_____ **2.** buccinator	**B.** elevates upper lip
_____ **3.** zygomaticus minor	**C.** rotates eye downward
_____ **4.** masseter	**D.** closes mouth
_____ **5.** frontal belly of occipitofrontalis	**E.** small muscle used in smiling
	F. large muscle used in smiling
_____ **6.** depressor labii	**G.** A: elevates jaw; O: angle of jaw
_____ **7.** temporalis	**H.** two-part neck muscle
_____ **8.** medial rectus	**I.** A: elevates larynx; O: chin; I: hyoid bone
_____ **9.** occipital belly of occipitofrontalis	**J.** tenses cheeks
	K. moves scalp forward
_____ **10.** levator labii	**L.** A: elevates jaw; O: temporal bone
_____ **11.** superior oblique	**M.** A: depresses hyoid bone; O: sternum; I: hyoid bone
_____ **12.** geniohyoid	
_____ **13.** sternohyoid	**N.** depresses lower lip
_____ **14.** platysma	**O.** moves eye medially
_____ **15.** corrugator supercilii	**P.** closes eye
_____ **16.** risorius	**Q.** wrinkles forehead
_____ **17.** mylohyoid	**R.** tenses angle of mouth laterally
_____ **18.** orbicularis oris	**S.** A: elevates hyoid bone; O: inner border of mandible
_____ **19.** zygomaticus major	
_____ **20.** digastric	**T.** thin muscle covering sides of neck, depresses jaw

18

B. Labeling

Label the muscles of the head in Figure 18.6.

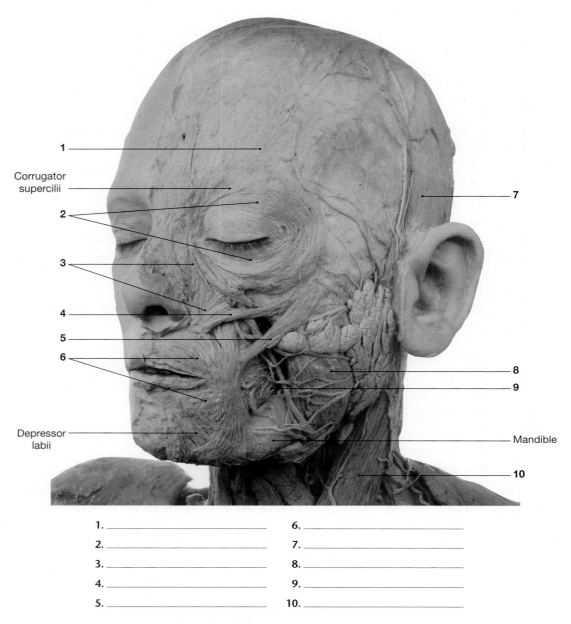

1. _____
Corrugator
supercilii _____
2. _____
3. _____
4. _____
5. _____
6. _____
Depressor
labii _____
7 _____
8 _____
9 _____
Mandible _____
10 _____

1. _____	6. _____
2. _____	7. _____
3. _____	8. _____
4. _____	9. _____
5. _____	10. _____

Figure 18.6 Cadaver Head and Neck, Lateral View
Note the superficial arteries, veins, facial nerves, parotid gland, and muscles.

C. Short-Answer Questions

1. Describe the position and action of the muscles of mastication. Which muscles oppose the action of these muscles?

2. Describe the muscles involved in smiling and grimacing.

3. What is the epicranial aponeurosis?

D. Analysis and Application

1. Describe the movement produced by each extrinsic eye muscle.

2. Explain how the muscles of the tongue and anterior neck are named.

3. Describe the actions of the digastric muscle.

Muscles of the Chest, Abdomen, Spine, and Pelvis

OBJECTIVES

On completion of this exercise, you should be able to:

- Identify the origin, insertion, and action of the major muscles of the anterior and posterior chest.

- Identify the origin, insertion, and action of the major muscles of the thoracic region.

- Identify the origin, insertion, and action of the major muscles of the abdominal region.

- Identify the origin, insertion, and action of the major muscles of the pelvis.

The torso of the body supports many muscles spread over a large area. These muscles are not as densely packed as those of the head and neck, but because they act on the limbs and the head, their actions are complex. The muscles are generally grouped based either on the torso region they cover or on the limb on which they act. The muscles of the chest (including the posterior surface of the chest) act on the arms and the head. The primary functions of the abdominal muscles are to support the abdomen, viscera, and lower back and to move the legs. Muscles of the pelvic region form the floor and walls of the pelvis. These muscles support local organs of the reproductive and digestive systems.

LAB ACTIVITY 1   Anterior Muscles of the Chest

Pectoralis Major

The largest muscle of the chest is the **pectoralis** (pek-to-RA-lis) **major** muscle (Figure 19.1), which covers most of the upper rib cage on each side of the chest. In females, the lower part of the pectoralis major is covered by the breast. You may be able to feel a portion of the pectoralis major near its origin on the sternum or clavicle while pressing your arm against your side. This muscle has several parts. The **sternal portion** arises from the costal cartilage, the sternum, and the ribs. The **clavicular portion** originates along the anterior side of the clavicle near the sternum. Both portions insert along the crest of the greater tubercle of the humerus and form the anterior wall of the axilla. The pectoralis major acts to adduct the arm and to flex and rotate the humerus at the shoulder.

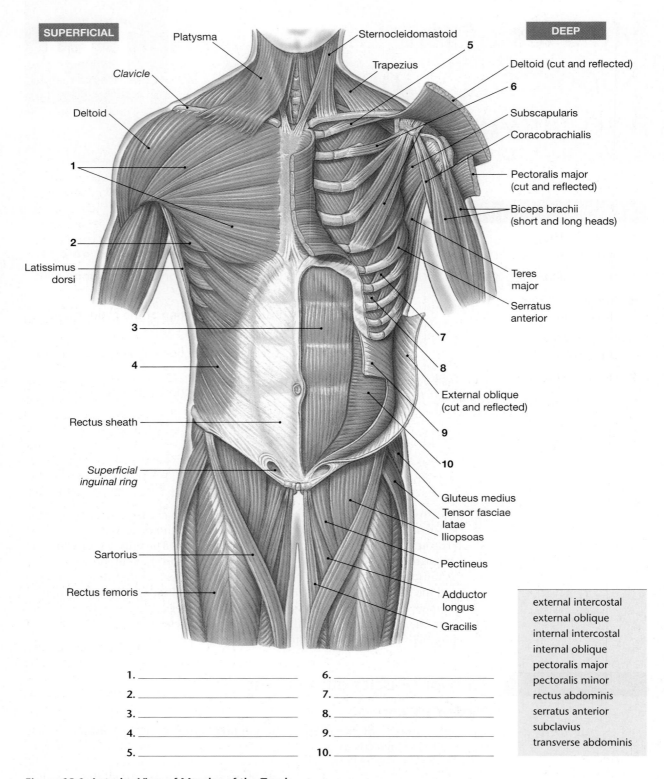

Platysma

Clavicle

Deltoid

1

2

Latissimus dorsi

3

4

Rectus sheath

Superficial inguinal ring

Sartorius

Rectus femoris

Sternocleidomastoid

Trapezius

5

Deltoid (cut and reflected)

6

Subscapularis

Coracobrachialis

Pectoralis major (cut and reflected)

Biceps brachii (short and long heads)

Teres major

Serratus anterior

7

8

External oblique (cut and reflected)

9

10

Gluteus medius

Tensor fasciae latae

Iliopsoas

Pectineus

Adductor longus

Gracilis

1. _____
2. _____
3. _____
4. _____
5. _____

6. _____
7. _____
8. _____
9. _____
10. _____

external intercostal
external oblique
internal intercostal
internal oblique
pectoralis major
pectoralis minor
rectus abdominis
serratus anterior
subclavius
transverse abdominis

Figure 19.1 Anterior View of Muscles of the Trunk

Pectoralis Minor

Tucked under the pectoralis major is the **pectoralis minor**. This muscle originates along the edges of the third, fourth, and fifth ribs and inserts on the coracoid process of the scapula. The function of this smaller muscle is very different from the function of its larger cousin: the pectoralis minor acts to pull the top of the scapula forward and depress the shoulders. It also elevates the ribs during forced inspiration, as during strenuous exercise.

Study Tip | ### *Muscle Modeling*

Your hands can be used to simulate the origin, insertion, and action for many muscles. For example, place your right hand over your left pectoralis major muscle. The heel of your palm represents the stationary origin of the muscle at the sternum, while your fingers touching the humerus represent the insertion. When you flex your fingers and pull the humerus medially, you are mimicking the major action of the muscle. ●

Subclavius

The **subclavius** (sub-KLĀ-vē-us), as the name implies, is under the clavicle. It arises from the first rib, inserts on the underside of the clavicle, and acts to depress and protract the clavicle.

Serratus Anterior

The **serratus anterior** appears as wedges on the side of the chest. This arrangement gives the muscle a sawtooth appearance similar to that of a bread knife with its *serrated* cutting edge. The muscle also looks like a hand-held fan, with the broad end of the fan resting on the ribs and the handle resting on the scapula. Straps of muscle arise on the upper edge of the first eight ribs, project back, and insert on the costal surface of the scapula near the medial border. The serratus anterior pulls the scapula forward and rotates it. It is also a *synergist,* in that it assists other muscles that move the arm. For example, when the deltoid has abducted the arm as you raise your arm above your head, the serratus anterior acts as a synergist to the trapezius (described later) in moving the arm into a vertical position.

Intercostal Muscles

The **intercostal** muscles are located between the ribs and, along with the diaphragm, change the size of the chest for breathing. The **external intercostals** span the gap between the ribs, and the **internal intercostals** are deep to the external intercostals. These muscles are difficult to palpate (feel with the hand) because they are deep to other chest muscles.

The action of the external intercostals is to elevate the ribs for inhalation. When the thoracic cage expands, the air pressure inside the lungs drops and air moves into the lungs, much like pulling open a bellows to fan a fire. Each external intercostal muscle has its origin on the lower margin of the rib immediately superior to the muscle. The insertion is on the superior border of the rib below the origin. When the external intercostal contracts, it pulls the lower rib upward and expands the thoracic cage.

The internal intercostals act to depress the ribs for forced expiration during exercise. The origin is on the upper margin of the lower ribs, and the insertion is on the upper ribs along the lower border. As they contract, the internal intercostals pull the upper rib down to squeeze air out of the lungs.

Diaphragm

The **diaphragm** is a sheet of muscle that forms the thoracic floor and separates the thoracic cavity from the abdominopelvic cavity (Figure 19.2). The diaphragm originates at many points along its edges, and the muscle fibers meet at a central tendon. When the diaphragm contracts, it pulls down on the central tendon and lowers the

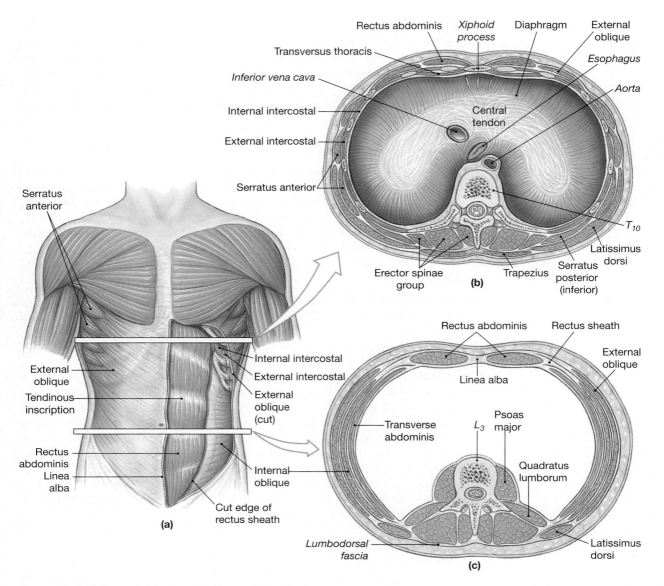

Figure 19.2 Oblique and Rectus Muscles and the Diaphragm
(a) Anterior view. (b) Sectional view at the level of the diaphragm. (c) Sectional view at the level of the umbilicus.

thoracic floor. This movement of the floor increases the volume of the thoracic cavity and results in a pressure decrease in the lungs. In other words, contracting the diaphragm to expand the thoracic cavity is the muscular process by which air is inhaled into the lungs.

QuickCheck Questions

1.1 Describe how the intercostal muscles move the thoracic cage during respiration.

1.2 Name a major and a minor chest muscle.

1 Materials

☐ Torso model
☐ Muscle chart

Procedures

1. Review the muscles of the anterior chest in Figures 19.1 and 19.2, and then label these muscles in Figure 19.1.

2. Identify each muscle of the anterior chest on the torso model and/or the muscle chart.

3. Locate the position of these muscles on your trunk. Practice the action of each chest muscle on your body. ■

LAB ACTIVITY 2 ## Abdominal Muscles

Rectus Abdominis

The **rectus abdominis** (Figure 19.2) is the vertical muscle along the midline of the abdomen between the pubic symphysis and the xiphoid process of the sternum. This muscle is divided by a midsagittal fibrous line called the **linea alba**. A well-developed rectus abdominis muscle has a washboard appearance because transverse bands of collagen called **tendinous inscriptions** separate the muscle into many segments. Bodybuilders often call the rectus abdominis the "six pack" because of the bulging segments of the muscle. Contraction of the rectus abdominis flexes the pubic symphysis and the xiphoid process toward each other, like the movement that occurs when you do sit-ups.

External and Internal Oblique Muscles

Lateral to the rectus abdominis is the **external oblique**, a thin membranous muscle that covers the side of the abdomen. The external oblique originates on the inferior borders of the lower eight ribs and inserts on the anterior portion of the iliac crest and the linea alba. The **internal oblique** lies deep to the external oblique. The internal oblique muscle fibers sweep upward from the lower abdomen and insert on the lower ribs, the xiphoid process, and the linea alba. Both the external and internal oblique muscles act with the rectus abdominis to flex the vertebral column to compress the abdomen. They also increase the pressure in the abdomen during defecation, urination, and childbirth.

Transverse Abdominis

The **transverse abdominis** is located deep to the internal oblique. It originates on the lower ribs, the iliac crest, and the lumbar vertebrae and inserts on the linea alba and the pubic symphysis. It contracts with the other abdominal muscles to compress the abdomen.

Study Tip | ***Fiber Orientation***

Find the external oblique and internal oblique on a muscle model, and notice the difference in the way the muscle fibers are oriented. The external fibers flare laterally as they are traced from bottom to top, whereas the internal fibers are directed medially. This tip is also useful in examining the external and internal intercostal muscles between the ribs. By the way, the intercostal muscles of beef and pork are the barbecue "ribs" that you might enjoy. ●

QuickCheck Questions

2.1 Describe the muscular wall of the abdomen.

2.2 Why is the rectus abdominis muscle nicknamed the "six-pack"?

2 *Materials*

☐ Torso model
☐ Muscle chart

Procedures

1. Review the muscles of the abdomen in Figures 19.1 and 19.2, and then label these muscles in Figure 19.1.

2. Identify each abdominal muscle on the torso model and/or on the muscle chart.

3. Locate the general position of these muscles on your own abdomen. ■

LAB ACTIVITY 3 Posterior Muscles of the Chest

Trapezius

The large, diamond-shaped muscle of the upper back is the **trapezius** (tra-PĒ-zē-us). It spans the gap between the scapulae and extends from the lower thoracic vertebrae to the back of the head (Figure 19.3). The trapezius has numerous origins. The upper portion arises from the **ligamentum nuchae** (li-guh-MEN-tum NOO-kē; *nucha, nape*), a mass of fibers extending from the cervical vertebrae to the occipital bone. From the ligamentum nuchae, the upper part of the trapezius inserts posteriorly on the distal third of the clavicle. The middle and lower muscle masses originate on the spinous processes of the thoracic vertebrae and insert on the acromion and spine of

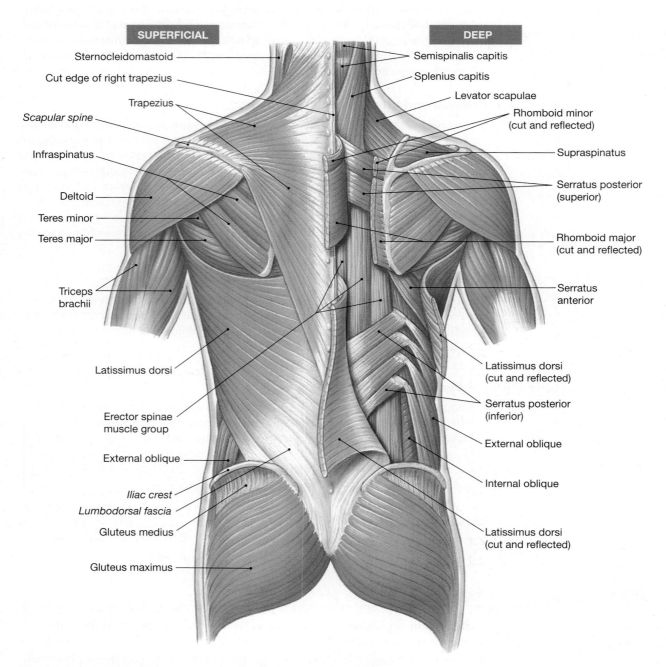

Figure 19.3 Posterior View of the Appendicular Muscles of the Trunk

the scapula. Because each part of the trapezius has a unique origin and insertion, each part also produces a unique movement. The upper trapezius elevates the clavicle and scapula by pulling in the direction of the occipital bone. The middle region adducts the scapula, and the lower fibers depress and medially rotate the scapula. Perform each of these actions separately and observe the resulting motion.

Study Tip | ***Three-in-One***

You may wish to think of the trapezius as three muscles (upper, middle, and lower) united into one and learn the action of each "portion" separately. ●

Latissimus Dorsi

The **latissimus dorsi** (la-TIS-i-mus DOR-sē) is the large muscle wrapping around the lower back. This muscle has a broad origin from the sacral and lumbar vertebrae up to the sixth thoracic vertebra. It sweeps up and inserts on the humerus, close to the attachment of the pectoralis major. It acts to adduct and extend the humerus, as when pulling a rope toward yourself.

Study Tip | ***Before or After?***

Always notice whether a muscle passes in front of or behind the bone into which it inserts. This detail determines how the bone moves. For example, the latissimus dorsi passes in front of the humerus before inserting into the intertubercular groove. This allows medial rotation of the humerus. If the muscle passed behind the humerus, the resulting action would be lateral rotation. ●

QuickCheck Questions

3.1 Describe the action of each region of the trapezius.

3.2 Where are the origin and insertion of the latissimus dorsi muscle?

3 | Materials

- ☐ Torso model
- ☐ Muscle chart

Procedures

1. Review the posterior muscles of the chest in Figure 19.3.
2. Identify each abdominal muscle on the torso model and/or the muscle chart.
3. Locate the position of these muscles on your own chest and practice their actions. ■

LAB ACTIVITY 4 Posterior Muscles of the Spine

The muscles of the spine are deep to the trapezius, latissimus dorsi, and other back muscles (Figure 19.4). Long bands of muscles from many different muscle groups stabilize and extend the vertebral column. In the lower back, the various superficial and deep muscle layers are collectively called the **erector spinae** muscles; they include the longissimus, iliocostalis, and other muscle groups. Many smaller muscles between the vertebrae—the **intertransversarii, rotatores**, and **interspinales**—help stabilize and either extend or laterally flex the vertebral column.

Scalenes

The **scalenes** muscle group of the neck consists of an **anterior scalene**, a **middle scalene**, and a **posterior scalene** (Figure 19.4b). Each originates on the transverse processes of the cervical vertebrae and inserts on the first or second rib. When the ribs are held in position, the scalenes flex the neck. If the neck is stationary, they elevate the ribs during inspiration.

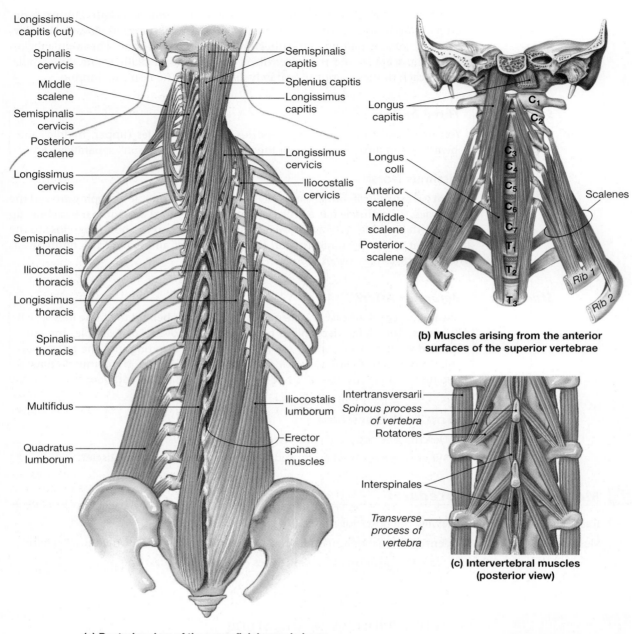

(a) Posterior view of the superficial muscle layer

(b) Muscles arising from the anterior surfaces of the superior vertebrae

(c) Intervertebral muscles (posterior view)

Figure 19.4 Muscles of the Spine

Longissimus Group

The **longissimus group** consists of three muscles: capitis, cervicis, and thoracis (Figure 19.4a). These muscles are considered superficial muscles of the spine because they are the first layer deep to the trapezius and latissimus dorsi muscles. Each longissimus muscle is named after the location of its insertion.

- The **longissimus capitis** (*capit,* head) originates on the transverse processes of the lower cervical and upper thoracic vertebrae and inserts on the temporal bone at the mastoid process. When both sides of this muscle contract, the head is extended; when only one side contracts, the head rotates and the neck flexes laterally on that side.

- The **longissimus cervicis** originates on the processes of the thoracic vertebrae and inserts on the processes of the lower cervical vertebrae. It has the same action as the capitis muscle: when both sides contract, the head is extended; contraction of one side only rotates the head and laterally flexes the neck.
- The **longissimus thoracis** originates on the processes of the lower thoracic and lumbar vertebrae and inserts on the transverse processes of the thoracic vertebrae above and on the ribs. It stabilizes the vertebrae when the vertebral column is being extended.

Spinalis Group

The **spinalis group** includes the spinalis cervicis and the spinalis thoracis. This muscle group is a superficial layer of the vertebral musculature.

- The **spinalis cervicis** has its origin on the ligamentum nuchae of the last cervical vertebra. It inserts on the spinous process of the axis and acts to extend the neck.
- The **spinalis thoracis** arises from the spinous processes of the lower thoracic and upper lumbar vertebrae and inserts on the spinous processes of the upper thoracic vertebrae. Contraction of this muscle causes the vertebral column to extend.

Splenius

Medial to the longissimus capitis, the **splenius capitis** originates on vertebrae C_1 through T_4 and inserts on the occipital bone. In addition to extending the head, the splenius capitis is a synergist to the semispinalis capitis and longissimus capitis muscles and assists those muscles in extending the head and flexing the neck laterally.

Iliocostalis Group

The **iliocostalis group** consists of three sets of superficial muscles that arise on the posterior surface of the ribs and the iliac crest. This muscle group acts to extend the spine and stabilize the thoracic vertebrae. Notice in Figure 19.4 that the longissimus thoracis and the iliocostalis muscles are difficult to differentiate in the lower back. In this region they are collectively called the erector spinae muscles.

- The **iliocostalis cervicis** has its origin on the upper borders of ribs T_1–T_7. The bands of muscle insert on the transverse processes of the lower half of the cervical vertebrae. This muscle acts to extend or laterally flex the neck and to elevate the ribs.
- The **iliocostalis thoracis** muscle arises on the lower ribs and inserts on the transverse processes of the upper thoracic vertebrae and on the last cervical vertebrae. Like the longissimus thoracis, the iliocostalis thoracis stabilizes the vertebrae when the vertebral column is being extended.
- The **iliocostalis lumborum** is the largest portion of the iliocostalis group. The origin is on the iliac and sacral crests and the spinous processes of the sacrum. The bands of muscle extend upward and insert along the angle of the lower seven ribs. This muscle acts to extend the vertebral column and depress the ribs.

Semispinalis Group

The **semispinalis group** is deep to the longissimus and iliocostalis groups.

- The **semispinalis capitis** is along the midline of the neck. It originates on the inferior cervical and upper thoracic vertebrae and inserts between the nuchal lines on the occipital bone. When both sides of this muscle contract, the head is extended; when only one side contracts, the head rotates and the neck flexes laterally on that side.

- The **semispinalis cervicis** is along the midline of the upper thoracic vertebrae. Its origin is on the transverse processes of the upper thoracic vertebrae. The insertion is on the spinous processes of cervical vertebrae C_2–C_5. This muscle extends the head or, if only one side contracts, causes lateral flexion of the neck on the opposite side.
- The **semispinalis thoracis** originates along the lower thoracic vertebrae and inserts on the spinous processes of the lower cervical and upper thoracic vertebrae (C_5 to T_4). This muscle is a synergist to the cervicis and acts to extend the head or flex the neck.

Multifidus

The **multifidus** muscle is a deep band of muscles that spans the length of the vertebral column. Each portion of the band originates on the transverse process of a vertebra and inserts on the spinous process of a vertebra that lies three or four units above the vertebra of origin. The multifidus acts with the semispinalis thoracis to extend the head or flex the neck.

Quadratus Lumborum

The **quadratus lumborum** arises on the middle portion of the iliac crest and inserts on the inferior border of the twelfth rib and the transverse processes of the lumbar vertebrae. It flexes the spine or, acting alone, bends the spine laterally toward the side of contraction.

QuickCheck Questions

4.1 Name the three muscles of the longissimus group and the action of each.

4.2 Which muscle inserts on the first and second pairs of ribs?

4 Materials

□ Torso model
□ Muscle chart

Procedures

1. Review the posterior muscles of the spine in Figure 19.4.
2. Locate each posterior spine muscle on the torso model and/or the muscle chart.
3. Locate the scalenes muscles along the neck.
4. Identify each muscle of the longissimus, iliocostalis, semispinalis, and spinalis groups. ∎

LAB ACTIVITY 5 Muscles of the Pelvic Region

The pelvic floor and wall form a bowl that holds the pelvic organs. The floor mainly consists of the **coccygeus** and the **levator ani** muscles (Figure 19.5a). The coccygeus is the more posterior of the two; it originates on the ischial spine, passes posteriorly, and inserts on the lateral and inferior borders of the sacrum. The levator ani originates on the inside edge of the pubis and the ischial spine and inserts on the coccyx.

As Figure 19.5b shows, the coccygeus and levator ani together form the muscle group called the **pelvic diaphragm**. The action of this group is to flex the coccyx and tense the pelvic floor. During pregnancy, the expanding uterus bears down on the pelvic floor, and the pelvic diaphragm supports the weight of the fetus.

The **external anal sphincter** originates on the coccyx and inserts around the anal opening. This muscle closes the anus and is consciously relaxed for defecation. Following depression and protrusion of the external anal sphincter during defecation, the levator ani elevates and retracts the anus.

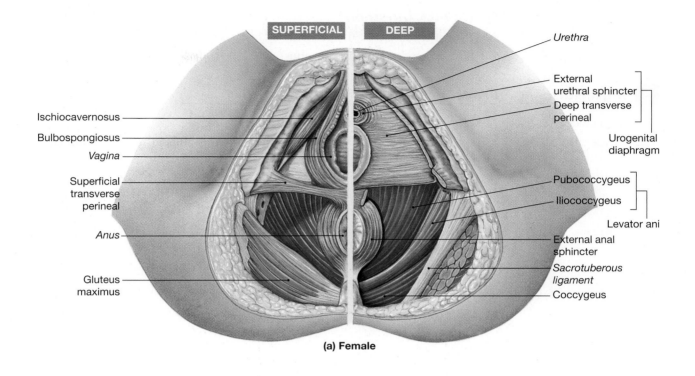

Ischiocavernosus
Bulbospongiosus
Vagina
Superficial transverse perineal
Anus
Gluteus maximus

SUPERFICIAL | DEEP

Urethra
External urethral sphincter
Deep transverse perineal
Urogenital diaphragm
Pubococcygeus
Iliococcygeus
Levator ani
External anal sphincter
Sacrotuberous ligament
Coccygeus

(a) Female

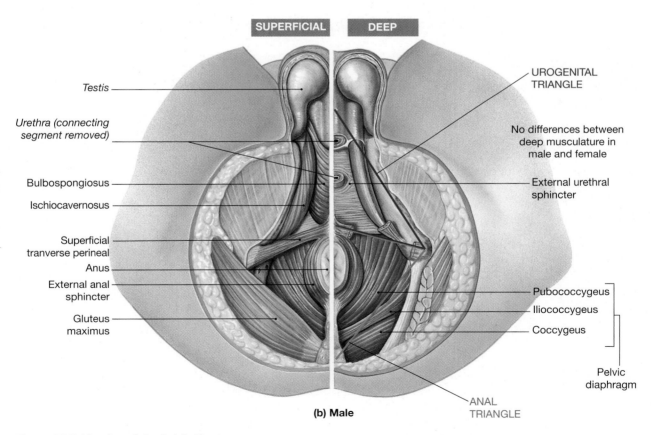

Testis
Urethra (connecting segment removed)
Bulbospongiosus
Ischiocavernosus
Superficial tranverse perineal
Anus
External anal sphincter
Gluteus maximus

SUPERFICIAL | DEEP

UROGENITAL TRIANGLE
No differences between deep musculature in male and female
External urethral sphincter
Pubococcygeus
Iliococcygeus
Coccygeus
Pelvic diaphragm
ANAL TRIANGLE

(b) Male

Figure 19.5 Muscles of the Pelvic Floor
(**a**) Female. (**b**) Male.

QuickCheck Questions

5.1 Name the muscles of the pelvic floor.

5.2 Which muscle surrounds the anus?

5 | Materials

☐ Torso model

☐ Muscle chart

Procedures

1. Review the muscles of the pelvic region in Figure 19.5.

2. Locate each muscle of the pelvic region on the torso model and/or the muscle chart. ■

Muscles of the Chest, Abdomen, Spine, and Pelvis

Name _____

Date _____

Section _____

A. Matching

Match each term in the left column with its correct description from the right column.

_____ **1.** pectoralis minor	**A.** diamond-shaped upper back muscle
_____ **2.** trapezius	**B.** superficial lateral muscle of abdomen
_____ **3.** rectus abdominis	**C.** middle lateral muscle layer of abdomen
_____ **4.** transverse abdominis	**D.** small chest muscle; depresses scapula
_____ **5.** external oblique	**E.** abdominal muscle with horizontal fibers
_____ **6.** external intercostal	**F.** vertical muscle at trunk midline; compresses abdomen
_____ **7.** serratus anterior	**G.** major muscle used during bench press
_____ **8.** latissimus dorsi	**H.** found between ribs; elevates rib cage
_____ **9.** pectoralis major	**I.** large muscle wrapping around lower back
_____ **10.** internal intercostal	**J.** fan-shaped muscle; inserts on scapula
_____ **11.** subclavius	**K.** found inferior to clavicle; depresses clavicle
_____ **12.** internal oblique	**L.** found between ribs; depresses rib cage

B. Matching

Match each term in the left column with its correct description from the right column.

_____ **1.** coccygeus	**A.** dorsal muscle that flexes lower spine
_____ **2.** quadratus lumborum	**B.** elevates anal sphincter
_____ **3.** ligamentum nuchae	**C.** posterior muscle of pelvic diaphragm
_____ **4.** levator ani	**D.** dorsal neck muscles that extend head
_____ **5.** pelvic diaphragm	**E.** posterior muscle of pelvic floor
_____ **6.** diaphragm	**F.** major muscle of inhalation
_____ **7.** longissimus	**G.** fibrous line located along midline of trunk
_____ **8.** external anal sphincter	**H.** sheet of ligament of cervical vertebrae
_____ **9.** linea alba	**I.** circular muscle in pelvic floor

19

C. Short-Answer Questions

1. Which muscle is used when you shrug your shoulders? Describe how this action is accomplished.

2. Describe the locations of the four abdominal muscles and their relationships to one another.

3. Describe the longissimus muscle group of the vertebral column.

D. Analysis and Application

1. When you blow hard through your mouth—in trying to inflate a balloon, for example—your abdominal muscles contract. Why?

2. Which muscle groups work as synergists to extend the vertebral column and head?

3. The anterior abdominal wall lacks bone. This being true, on what structure do the abdominal muscles insert?

Muscles of the Shoulder, Arm, and Hand

OBJECTIVES

On completion of this exercise, you should be able to:

- Identify the origin, insertion, and action of major muscles of the shoulder region.

- Identify the origin, insertion, and action of major muscles of the upper arm.

- Identify the origin, insertion, and action of major muscles of the forearm.

- Identify the origin, insertion, and action of the major muscles of the hand.

The function of the muscles in this region of the body is to support and move the arm, hand, and fingers. The muscles of the shoulder help to anchor the humerus to the scapula and also to stabilize the scapula on the posterior chest. Muscles of the upper arm flex and extend the arm; muscles of the forearm move the wrist and hand.

LAB ACTIVITY 1 Muscles of the Shoulder

Supraspinatus

Locate the spine of the scapula on your own back. You can use this bony protuberance as a landmark to identify several muscles. The **supraspinatus** (soo-pra-spī-NA-tus; *supra,* above) originates on the supraspinous fossa, the depression located superior to the scapular spine (Figure 20.1). The supraspinatus passes laterally under the acromion and in front of the distal end of the scapular spine to converge on a tendon that covers the superior portion of the shoulder joint. The tendon eventually inserts on the upper part of the greater tubercle of the humerus. The supraspinatus acts to abduct the arm and assists the deltoid in this action. It also keeps the head of the humerus firmly seated in the glenoid fossa.

Infraspinatus

The **infraspinatus** is located below the scapular spine. Again, use your own scapular spine to help locate this muscle on your body. The infraspinatus originates on the infraspinous fossa and inserts on the posterior surface of the greater tubercle, just inferior to the insertion of the supraspinatus. This muscle acts to laterally rotate the humerus and keeps the humeral head in the glenoid cavity by protecting the posterior side of the shoulder joint.

biceps brachii
deltoid
infraspinatus
latissimus dorsi
pectoralis major
subscapularis
suprapinatus
teres major
teres minor
triceps brachii

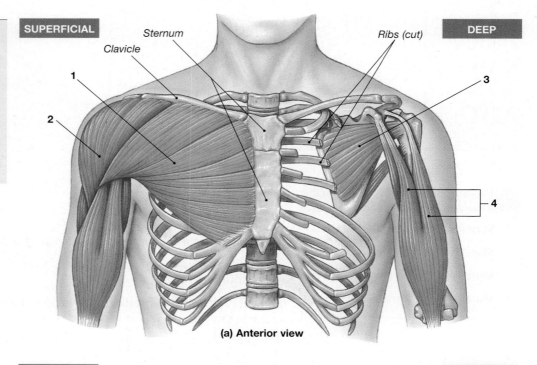

(a) Anterior view

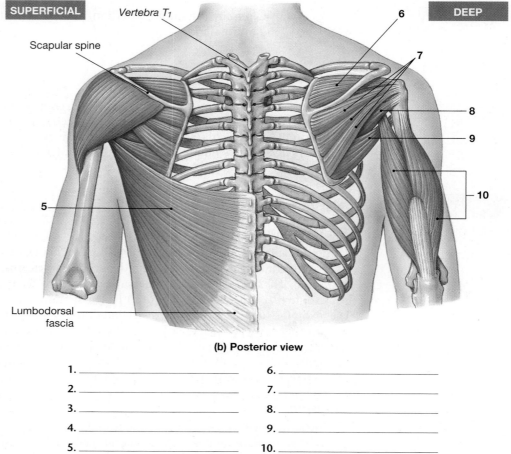

(b) Posterior view

1. _____   6. _____
2. _____   7. _____
3. _____   8. _____
4. _____   9. _____
5. _____   10. _____

Figure 20.1 Muscles That Move the Arm

270

Subscapularis

The **subscapularis**, as the name implies, is found under the scapula next to the posterior surface of the rib cage. You cannot feel this muscle on your body. It occupies most of the subscapular fossa. The muscle narrows laterally and passes in front of the humerus to insert on the upper part of the lesser tubercle. It medially rotates the humerus and so is an antagonist of the infraspinatus.

Teres Minor

Originating along the lateral border of the scapula is the **teres minor**. It is a small, flat muscle, as Figure 20.1 shows, that passes laterally and superiorly to insert on the greater tubercle of the humerus near the insertion of the infraspinatus. The teres minor rotates the humerus laterally.

The four shoulder muscles described above—supraspinatus, infraspinatus, subscapularis, and teres minor—all act to firmly seat the head of the humerus in the glenoid fossa. This fossa is a shallow depression, and without the protective function of these four muscles, the humerus would be easily dislocated from the shoulder. These muscles also make up a structure called the **rotator cuff**. Although part of the rotator cuff, the supraspinatus is not itself a rotator; rather, it is an abductor muscle.

You may be familiar with the rotator cuff if you are a baseball fan. The windup and throw of a pitcher involve circumduction of the humerus. This motion places tremendous stress on the shoulder joint and the rotator cuff—stress that causes premature degeneration of the joint. To protect the shoulder joint and muscles, bursal sacs are interspersed between the tendons of the rotator cuff muscles and the neighboring bony structures. Repeated friction on the bursae may result in an inflammation called *bursitis*.

Teres Major

The **teres major** is a thick muscle that arises on the inferior angle of the posterior surface of the scapula. The muscle converges up and laterally into a flat tendon that ends on the anterior side of the humerus. This insertion point is just medial to the intertubercular groove on the crest of the lesser tubercle. Because the teres major passes to the anterior side of the humerus, it acts to medially rotate that bone and assist the latissimus dorsi in moving the arm downward.

Rhomboid

The **rhomboid** muscles extend between the upper thoracic vertebrae and the scapula (Figure 20.2). They are deep to the trapezius, and you cannot feel them on your body. The **rhomboid major** originates along the spinous processes of the upper thoracic vertebrae and inserts on the lower medial border of the scapula. The **rhomboid minor** arises from vertebrae C_7 and T_1 and attaches to the upper border of the scapula. The rhomboid muscles act to adduct and laterally rotate the scapula.

Levator Scapulae

The **levator scapulae** muscle raises the scapula. To accomplish this action, the levator scapulae must originate superior to the scapula. It arises from the transverse processes of vertebrae C_1 through C_4 and inserts on the upper medial border of the scapula. The muscle is deep to the trapezius and therefore difficult to feel. The levator scapulae also tilts the head toward the side of contraction.

Deltoid

The very large, easily identifiable muscle that covers the shoulder is the **deltoid**. Examine Figures 20.1 and 20.2, and also locate this muscle on the upper part of your humerus, just below the point of the shoulder. It has a broad origin and several actions. As with the trapezius, it may help to consider the deltoid as having three parts. The posterior portion arises along the inferior margin of the scapular spine, the middle region

deltoid
infraspinatus
levator scapulae
rhomboid major
rhomboid minor
serratus anterior
teres major
teres minor
trapezius
triceps brachii

1. _____
2. _____
3. _____
4. _____
5. _____
6. _____
7. _____
8. _____
9. _____
10. _____

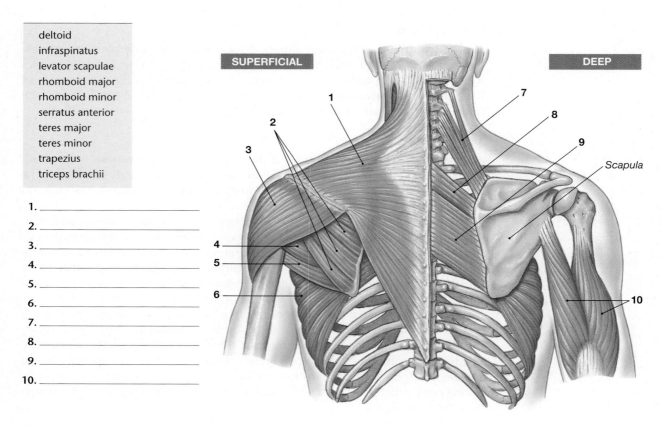

Figure 20.2 Muscles That Position the Shoulder Girdle

originates on the acromion, and the anterior portion originates on the distal end of the anterior surface of the clavicle. All parts of the deltoid converge toward the humerus and insert on the large deltoid tuberosity. The anterior portion flexes the arm, the posterior segment performs the opposite action and extends the arm, and the middle portion acts as the major abductor of the humerus.

QuickCheck Questions

1.1 Which muscles move the scapula?

1.2 Which muscle abducts the humerus?

1 Materials

☐ Torso model
☐ Arm model
☐ Muscle chart

Procedures

1. Review the muscles of the shoulder in Figures 20.1 and 20.2 and then label these muscles in these illustrations. Also label any muscles discussed in Exercise 19.

2. Locate each muscle of the shoulder on the torso model and/or the muscle chart.

3. Locate the general position of each shoulder muscle on your body. Contract each muscle and observe the action of your shoulder.

4. Observe on the arm model where the deltoid originates on the scapula and inserts on the humerus. ■

LAB ACTIVITY 2 Muscles of the Upper Arm

Biceps Brachii

The **biceps brachii** (Figure 20.3) makes up the fleshy mass of the anterior humerus when the arm is flexed. The term *biceps* refers to the presence of two origins, or "heads." The **short head** begins on the coracoid process of the scapula as a tendon that expands into the muscle belly. The **long head** arises on a roughened area on

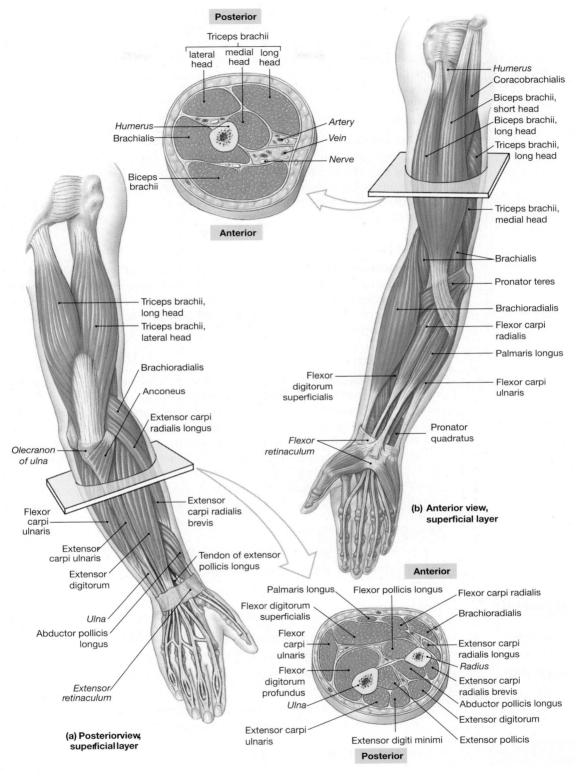

Figure 20.3 Muscles That Move the Forearm and Hand

the superior lip of the glenoid fossa of the scapula, the supraglenoid tubercle. A tendon passes over the top of the humerus into the intertubercular groove and blends into the muscle. The tendon of the long head is enclosed within a protective covering called the *intertubercular synovial sheath*. The two heads of the muscle fuse and constitute most of the mass of the anterior brachium.

The biceps brachii inserts on the radial tuberosity and is a principal flexor of the forearm. This muscle is also a supinator. When the forearm is in a prone position, the radial tuberosity faces posteriorly and the tendon of the biceps passes between the radius and ulna to insert on the radius. The contraction of the biceps pulls the radial tuberosity anteriorly and rotates the radius laterally. This rotates (supinates) the palm into a forward-facing position.

Brachialis

The **brachialis** (brā-kē-A-lis) also flexes the forearm. It is located under the distal end of the biceps brachii. You can feel a small part of it if you flex your arm and palpate the area just lateral to the tendon of the biceps. The brachialis originates on the distal humerus and inserts on the ulnar tuberosity below the coronoid process. Because the ulna is limited to a hingelike movement, the action of the brachialis is flexion of the forearm.

Coracobrachialis

The **coracobrachialis** (KOR-uh-kō-brā-kē-A-lis) is a small muscle that originates on the coracoid process of the scapula and shares a common origin with the biceps. The coracobrachialis passes inferiorly and inserts midway along the medial surface of the humerus between the origins of the brachialis and triceps muscles. It adducts and flexes the humerus.

Triceps Brachii

The **triceps brachii** is the principal antagonist to the biceps brachii and brachialis muscles. This muscle forms most of the mass of the upper posterior side of the arm. The name indicates that it has three origins, two on the scapula and one on the humerus. The **long head** begins as a flattened tendon on the scapula at the roughened area below the glenoid cavity. The **lateral head** arises from a slender origin on the posterior surface of the humerus below the head of the humerus. The **medial head** originates on a broad region covering one-third of the distal posterior humerus. The three heads merge into a common tendon that begins at about the middle of the muscle and inserts on the olecranon process of the ulna. The triceps acts to extend the forearm.

QuickCheck Questions

2.1 Which muscle causes extension of the arm?

2.2 Which muscles are antagonistic to the triceps brachii?

2 *Materials*	*Procedures*

☐ Torso model

☐ Arm model

☐ Muscle chart

1. Review and label the upper-arm muscles in Figure 20.3.
2. Complete the labeling of Figures 20.1 and 20.2.
3. Locate each muscle of the upper arm on the torso and arm models and/or the muscle chart.
4. Using Figure 20.4 as a guide, locate the general position of each upper-arm muscle on your body. Contract each muscle and observe the action of your arm. ■

LAB ACTIVITY 3 Muscles of the Forearm

Supinator

The **supinator** is found on the lateral side of the forearm deep to several muscles (Figures 20.3c and 20.5b). It arises on the lateral epicondyle of the humerus, crosses the antecubital region, and inserts on the lateral side of the radius distal to the radial head. The supinator contracts and rotates the radius into a position parallel to the ulna, resulting in supination of the forearm.

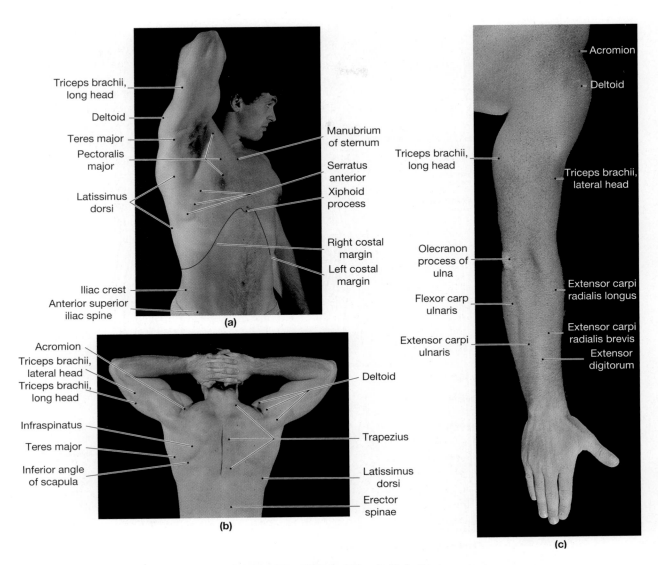

Figure 20.4 Surface Anatomy of the Neck, Shoulder, Arm, and Trunk, Male Torso
(a) Lateral view. (b) Posterior view. (c) Right arm, posterior view.

Pronator Teres and Pronator Quadratus

The **pronator teres** is the principal pronator muscle of the arm (Figures 20.3c and 20.5a). Because pronation is the opposite of supination, the origin and insertion of the pronator teres must be opposite the origin and insertion of the supinator. The pronator teres originates on the medial epicondyle of the humerus and on the proximal part of the ulna. It inserts on the lateral side of the radius about midway down the shaft of that bone. When the forearm is in a supine position, the pronator teres pulls the lateral edge of the radius toward the ulna and medially rotates the radius. This results in pronation of the forearm. Note in Figure 20.3c how the pronator teres and supinator cross over each other for antagonistic actions.

The **pronator quadratus** is found just proximal to the wrist joint, on the anterior surface of the forearm. It originates on the anterior surface of the distal end of the ulna and then passes laterally to insert on the shallow concavity on the distal end of the radius. It acts as a synergist to the pronator teres in pronating the forearm and can also cause medial rotation of the forearm as a whole.

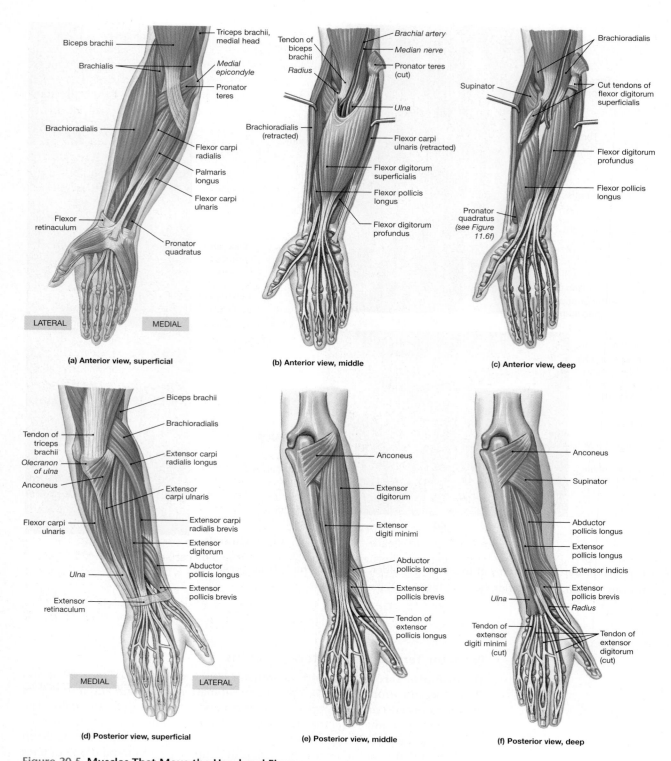

(a) Anterior view, superficial

(b) Anterior view, middle

(c) Anterior view, deep

(d) Posterior view, superficial

(e) Posterior view, middle

(f) Posterior view, deep

Figure 20.5 Muscles That Move the Hand and Fingers
Middle and deep muscle layers of the right forearm; for superficial muscles, see Figure 20.3.

Superficial Flexors of the Wrist

The superficial flexor muscles of the wrist and hand joint are located on the forearm and are shown in Figures 20.3 and 20.5. They all originate on the common flexor tendon attached to the medial epicondyle of the humerus. The first superficial flexor is the **flexor carpi radialis**, which lies just medial to the pronator teres and is the flexor muscle closest to the radius. The fibers of this muscle blend into a long tendon that inserts on the base of the second metacarpal. A smaller branch terminates on the base of the third metacarpal. The **palmaris longus** is medial to the flexor carpi radialis and on the palmar aponeurotica. The palmaris longus is continuous with the palmar fascia. Medial to the palmaris is the **flexor carpi ulnaris**. This muscle rests on the ulnar side of the forearm and inserts on the pisiform and hamate bones of the carpals and the base of the third and fourth metacarpals.

The radialis flexes and abducts the wrist. The ulnaris flexes and adducts the wrist. The palmaris longus tenses the palmar fascia and assists in wrist flexion.

Study Tip | ***Muscle Palpation***

Locate the medial epicondyle on your right humerus and place the fingers of your left hand on the muscle distal to the epicondyle. Now flex your right fingers and hand. You should feel the muscles contract near the epicondyle. Remember that the common flexor tendon originates on this epicondyle. Do the same for the lateral epicondyle, but place your fingers on the dorsal surface of the forearm and extend your right fingers and hand. The common extensor tendon arises here, and you should feel the extensor muscles contract and bulge just distal to the epicondyle in the forearm. ●

Flexor Digitorum Superficialis

The **flexor digitorum superficialis** is located deep to the superficial flexors of the hand (Figure 20.5a). It arises from several points, one of which is located on the common flexor tendon described earlier. The other origins are on the lateral surface of the coronoid process of the ulna and the anterior surface of the radius. Four tendons exit the distal end of the muscle and pass under a transverse connective tissue sheath called the **flexor retinaculum** (ret-i-NAK-ū-lum). The tendons then pass through a narrow valley bounded by carpal bones, called the *carpal tunnel*. They are covered by a synovial sheath that protects and lubricates them in the tunnel. Repeated flexing of the hand and fingers, such as extended typing or piano playing, causes the sheath to swell and compress the median nerve. Pain and numbness occur in the palm during flexion, a condition called *carpal tunnel syndrome*.

Extensors of the Hand and Fingers

The extensor muscles of the hand and fingers are located on the posterior forearm. All except one originate from a common tendon attached to the lateral epicondyle of the humerus.

The **extensor carpi ulnaris** lies next to the flexor carpi ulnaris near the ulna (Figure 20.3a). This extensor has its origin on the posterior surface of the ulna and inserts on the base of the fifth metacarpal. (As you work toward the radial side of the forearm, you will first see the extensor digiti minimi, which will not be considered here.) The action of the extensor carpi ulnaris is to extend and adduct the wrist.

The **extensor digitorum** is lateral to the extensor carpi ulnaris (Figures 20.3 and 20.5). It sends three or four tendons under a fibrous band that cuts transversely across the posterior aspect of the wrist, called the **extensor retinaculum**. The tendons insert on the dorsal surface of the base of the middle phalanges. The action of this muscle is to extend the fingers.

The **extensor carpi radialis brevis**, a short, thick muscle located just lateral to the extensor digitorum, inserts on the base of the third metacarpal. Its action is to extend and abduct the wrist. The **extensor carpi radialis longus** is the only extensor that does not originate from a tendon attached to the humerus lateral epicondyle. Instead, it arises from the humerus just proximal to the lateral epicondyle, although a few fibers do extend from the common tendon. The belly of this long muscle lies lateral to and slightly above the extensor carpi radialis brevis. It inserts on the base of the second metacarpal. The action of the extensor carpi radialis longus is also to extend and abduct the wrist. Together, the extensor carpi radialis longus and brevis oppose the actions of the forearm flexors.

Brachioradialis

The **brachioradialis** is a superficial muscle located lateral to the extensor carpi radialis longus (Figure 20.3a). You can feel the brachioradialis on the lateral side of the anterior surface of your forearm. It originates on the humerus proximal to the lateral epicondyle and superior to the origin of the extensor carpi radialis longus. The brachioradialis flexes the forearm and assists the action of the biceps and brachialis. It also aids in lateral movement of the forearm.

Study Tip

> **Tendons and Names**
>
> When you study flexors and extensors on an arm model, follow the tendons to their insertion points for clues about the names of the muscles. For example, the tendon of the flexor carpi ulnaris muscle inserts on the carpals on the ulnar side of the wrist and flexes the wrist (carpi). Because this muscle acts to flex the wrist, it must be located on the anterior surface of the forearm. ●

QuickCheck Questions

3.1 Which muscles are involved in the movement to turn a doorknob?

3.2 What is the general action of the muscles on the posterior forearm?

3 *Materials*

☐ Arm model
☐ Muscle chart

Procedures

1. Review the muscles of the forearm in Figures 20.3 and 20.5 and then complete the labeling of Figure 20.3.
2. Locate the forearm muscles on the arm model and/or the muscle chart.
3. Using Figure 20.4, locate as many of the forearm muscles on your body as possible. Contract each muscle and observe the action of your forearm. ■

LAB ACTIVITY 4 Muscles of the Hand

The muscles of the hand can be organized into two groups based on location: extrinsic muscles in the forearm and intrinsic muscles in the hand. The extrinsic muscles flex and extend the wrist and fingers, and the intrinsic muscles control fine finger and thumb movements. You can locate the muscles described here in Figure 20.6.

Muscles of the Thenar Eminence

The mass of tissue at the base of the thumb, called the **thenar eminence**, consists of several muscles. You can see these muscles making up the fleshy region on the lateral side of your palm, near the thumb. All these muscles insert on the lateral side of the base of the proximal phalanx of the thumb and originate on the flexor retinaculum, but the origins are at slightly different positions along the retinaculum and include other bones of the wrist. These differences define each muscle's individual functions.

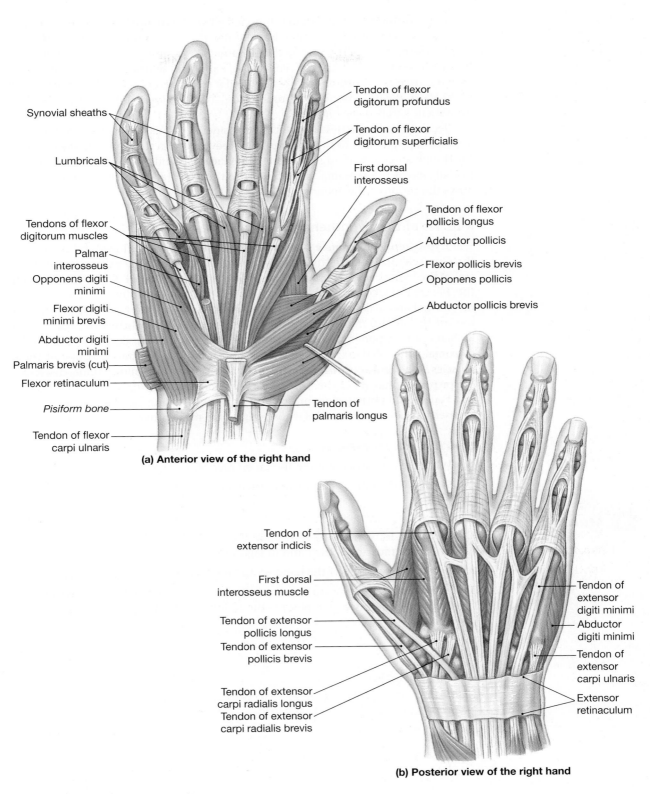

Synovial sheaths

Lumbricals

Tendons of flexor digitorum muscles

Palmar interosseus

Opponens digiti minimi

Flexor digiti minimi brevis

Abductor digiti minimi

Palmaris brevis (cut)

Flexor retinaculum

Pisiform bone

Tendon of flexor carpi ulnaris

Tendon of flexor digitorum profundus

Tendon of flexor digitorum superficialis

First dorsal interosseus

Tendon of flexor pollicis longus

Adductor pollicis

Flexor pollicis brevis

Opponens pollicis

Abductor pollicis brevis

Tendon of palmaris longus

(a) Anterior view of the right hand

Tendon of extensor indicis

First dorsal interosseus muscle

Tendon of extensor pollicis longus

Tendon of extensor pollicis brevis

Tendon of extensor carpi radialis longus

Tendon of extensor carpi radialis brevis

Tendon of extensor digiti minimi

Abductor digiti minimi

Tendon of extensor carpi ulnaris

Extensor retinaculum

(b) Posterior view of the right hand

Figure 20.6 **Muscles of the Hand**

The most medial of the thenar muscles is the **flexor pollicis brevis**. It acts to flex and adduct the thumb. Lateral to the flexor is the **abductor pollicis brevis**, which originates not only on the flexor retinaculum but also on the scaphoid and trapezium bones of the carpals. This muscle abducts the thumb. The most lateral thenar muscle is the **opponens pollicis**, and it partially originates on the trapezium of the wrist. It acts to bring the thumb across the palm toward the little finger. This movement is called *opposition*.

The **adductor pollicis** is often not considered part of the thenar eminence, as it is found just medial to the flexor pollicis brevis deep in the web of tissue between the thumb and palm. It originates on the metacarpals and carpals and inserts on the medial side of the proximal phalanx of the thumb. It adducts the thumb and opposes the action of the abductor pollicis brevis.

Muscles of the Hypothenar Eminence

The **hypothenar eminence** is found at the base of the little finger and consists of three muscles. It is the fleshy mass on the medial side of your palm. The most lateral of the hypothenar muscles is the **opponens digiti minimi**, and it originates on the flexor retinaculum and the hamate of the carpals. It inserts on the fifth metacarpal and acts in concert with the opponens pollicis to bring the little finger toward the thumb during opposition. The **flexor digiti minimi brevis**, located medial to the opponens digiti, has the same origin as this muscle but inserts on the proximal, fifth phalanx. It acts to flex the little finger. The most medial of these muscles is the **abductor digiti minimi**. It arises from the pisiform bone, inserts on the proximal phalanx of the little finger, and abducts that finger.

You should note that no muscles originate on the fingers. The phalanges serve as insertion points for muscles whose origins are more proximal.

QuickCheck Questions

4.1 What are the muscles of the thenar eminence?

4.2 What are the muscles of the hypothenar eminence?

4 *Materials*

☐ Arm or hand model
☐ Muscle chart

Procedures

1. Review the muscles of the hand in Figure 20.6.
2. Identify each muscle of the hand on the model and/or the muscle chart.
3. Use Figure 20.6 and locate as many muscles on your own hand as possible. Contract each muscle and observe the action of your hand. ■

Muscles of the Shoulder, Arm, and Hand

Name _____

Date _____

Section _____

A. Matching

Match each term in the left column with its correct description from the right column.

_____	**1.**	triceps brachii
_____	**2.**	infraspinatus
_____	**3.**	teres minor
_____	**4.**	biceps brachii
_____	**5.**	supraspinatus
_____	**6.**	brachialis
_____	**7.**	coracobrachialis
_____	**8.**	teres major
_____	**9.**	deltoid
_____	**10.**	subscapularis

A. small muscle; has common origin with biceps

B. rotator cuff; medially rotates humerus

C. originates on inferior scapula; medially rotates humerus

D. major flexor of forearm

E. major flexor and supinator of forearm

F. below scapular spine; laterally rotates humerus

G. rotator cuff muscle; assists deltoid to abduct arm

H. muscle of upper arm; flexes, extends, and abducts arm

I. originates on inferior scapula; rotator cuff muscle

J. major extensor of forearm

B. Matching

Match each term in the left column with its correct description from the right column.

_____	**1.**	opponens digiti minimi
_____	**2.**	palmaris longus
_____	**3.**	pronator teres
_____	**4.**	flexor carpi ulnaris
_____	**5.**	extensor carpi ulnaris
_____	**6.**	extensor digitorum
_____	**7.**	extensor carpi radialis
_____	**8.**	supinator
_____	**9.**	flexor retinaculum
_____	**10.**	opponens pollicis

A. tenses palmar fascia and flexes wrist

B. major pronator of arm

C. flexes and adducts wrist

D. opposes thumb

E. major supinator of forearm

F. extends and adducts wrist

G. band of connective tissue on flexor tendons

H. brings little finger toward thumb

I. extends fingers

J. extends and abducts wrist

20

C. Labeling

Label the muscles of the shoulder in Figure 20.7.

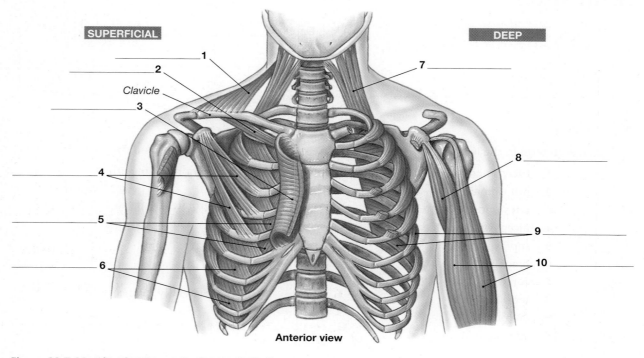

Figure 20.7 Muscles that Move the Pectoral Girdle
Anterior view.

D. Short-Answer Questions

1. Describe the muscles involved in turning the hand, as when twisting a doorknob back and forth.

2. Name the muscles responsible for flexing the arm. Which muscles are antagonists to these flexors?

3. Name a muscle for each movement of the wrist: flex, extend, abduct, and adduct.

E. Analysis and Application

1. Why would a dislocated shoulder also potentially result in injury to the rotator cuff?

2. A brace placed on your wrist to treat carpal tunnel syndrome would prevent what wrist action? What would you accomplish by limiting this action?

Muscles of the Pelvis, Leg, and Foot

OBJECTIVES

On completion of this exercise, you should be able to:

- Identify the origin, insertion, and action of the muscles of the pelvic and gluteal regions.
- Identify the origin, insertion, and action of the muscles of the anterior, medial, and posterior upper leg.
- Identify the origin, insertion, and action of the muscles of the anterior and posterior lower leg.
- Identify the origin, insertion, and action of the muscles of the foot.

The muscles of the pelvis help support the mass of the body and stabilize the pelvic girdle. Leg muscles move the thigh, knee, and foot. Flexors of the knee are on the posterior thigh, and knee extensors are anterior. Muscles that abduct the thigh are on the lateral side of the thigh, and the adductors are on the medial thigh.

LAB ACTIVITY 1 Muscles of the Pelvic and Gluteal Regions

Iliopsoas

The **iliopsoas** (il-ē-ō-sō-us) consists of two muscles, the psoas major and the iliacus (Figure 21.1). Although these muscles are located in the lower abdomen, they act on the femur and move the thigh. The **psoas** (SŌ-us) **major** originates on the body and transverse processes of vertebrae T_{12} through L_5. The muscle sweeps downward, passing between the femur and the ischial ramus, and inserts on the lesser trochanter of the femur. The **iliacus** (il-Ē-ah-kus) originates on the iliac fossa on the medial portion of the ilium and joins the tendon of the psoas. Collectively, the iliopsoas flexes the thigh, bringing the anterior surface of the upper leg toward the abdomen.

Deep Leg Rotators

The internal and external obturators and the piriformis are deep rotators of the thigh (Figure 21.1). Both the **obturator internus** and the **obturator externus** originate along the medial and lateral edges of the obturator foramen and insert on the

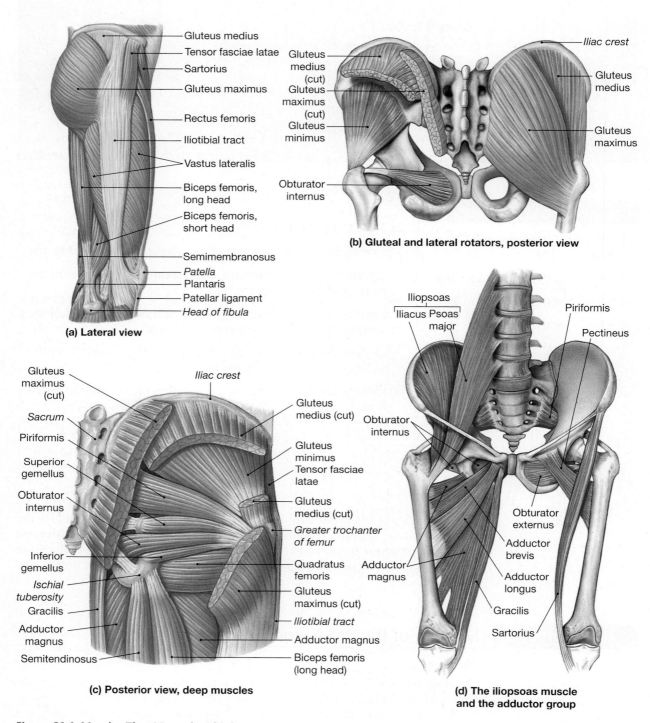

Gluteus medius
Tensor fasciae latae
Sartorius
Gluteus maximus

Rectus femoris

Iliotibial tract

Vastus lateralis

Biceps femoris, long head

Biceps femoris, short head

Semimembranosus
Patella
Plantaris
Patellar ligament
Head of fibula

(a) Lateral view

Gluteus medius (cut)
Gluteus maximus (cut)
Gluteus minimus

Obturator internus

Iliac crest

Gluteus medius

Gluteus maximus

(b) Gluteal and lateral rotators, posterior view

Gluteus maximus (cut)
Sacrum
Piriformis
Superior gemellus
Obturator internus

Inferior gemellus
Ischial tuberosity
Gracilis
Adductor magnus
Semitendinosus

Iliac crest

Gluteus medius (cut)

Gluteus minimus
Tensor fasciae latae

Gluteus medius (cut)

Greater trochanter of femur

Quadratus femoris
Gluteus maximus (cut)

Iliotibial tract

Adductor magnus
Biceps femoris (long head)

(c) Posterior view, deep muscles

Iliopsoas
Iliacus Psoas major

Obturator internus

Adductor magnus

Piriformis

Pectineus

Obturator externus

Adductor brevis

Adductor longus

Gracilis

Sartorius

(d) The iliopsoas muscle and the adductor group

Figure 21.1 Muscles That Move the Thigh

trochanteric fossa, a shallow depression on the medial side of the greater trochanter. The **piriformis** (pir-i-FOR-mis) arises from the anterior and lateral surfaces of the sacrum and inserts on the greater trochanter of the femur. Inferior to the piriformis is the **quadratus femoris** muscle. Its origin is on the lateral surface of the ischial tuberosity and inserts on the femur between the greater and lesser trochanters. The piriformis, obturators, and the quadratus femoris muscles rotate the thigh laterally. The piriformis also abducts the thigh.

Gluteal Muscles

The posterior muscles of the pelvis are three gluteal muscles that constitute the buttocks (Figure 21.1). The most superficial and prominent is the **gluteus maximus**. It is a large, fleshy muscle and is easily located as the major muscle of the buttocks. It extends and laterally rotates the thigh. It originates along the posterior of the iliac crest and passes over the sacrum and down the side of the coccyx. From this broad origin, muscle fibers pass laterally and downward to insert on a thick tendon called the **iliotibial** (il-ē-ō-TIB-ē-ul) **tract** that attaches to the lateral condyle of the tibia.

A thick muscle partially covered by the gluteus maximus, the **gluteus medius** originates on the lateral surface of the ilium and gathers laterally into a thick tendon that inserts posteriorly on the greater trochanter. The **gluteus minimus** begins on the lateral surface of the ilium, tucked under the origin of the medius. The fibers of the gluteus minimus also pass laterally to insert on the anterior surface of the greater trochanter. Both the gluteus medius and the gluteus minimus abduct and medially rotate the thigh.

Tensor Fasciae Latae

The **tensor fasciae latae** (FASH-ē-ē LĀ-tē) originates on the outer surface of the anterior superior spinous process of the ilium (Figure 21.1). The muscle fibers project downward and join the gluteus maximus on the iliotibial tract. As the name implies, the tensor fasciae latae tenses the fascia of the thigh and helps to stabilize the pelvis on the femur. The muscle also acts to abduct and medially rotate the thigh.

Study Tip | ***Deducing Action***

Rather than memorizing the action for each muscle, use the position of the muscle over the joint to determine the action. Muscles on the lateral side of the body, such as the deltoid and the tensor fasciae latae, are abductors and move the body away from the midline. Adductor muscles, such as the adductor longus, are located on the medial side of a joint and move the body toward the midline. Flexor muscles are positioned on the inner surface of a joint facing anteriorly or posteriorly. Extensor muscles are found on the outer surface of these joints. For example, the muscles over the anterior knee are on the outer surface of the knee joint and are extensor muscles. ●

QuickCheck Questions

1.1 Where are the abductors of the thigh located?

1.2 What is the iliotibial tract?

1.3 Name two muscles that rotate the thigh.

---

1 *Materials*

☐ Torso model
☐ Leg model

Procedures

1. Review the pelvic and gluteal muscles in Figure 21.1.
2. Examine the pelvis of the torso model and identify each muscle.
3. On the leg model, observe how the gluteal muscles and the tensor fasciae latae insert on the superior portion of the femur. ■

LAB ACTIVITY 2   Muscles of the Upper Leg

Muscles of the Anterior Thigh

The muscles of the anterior thigh include the quadriceps femoris muscles, which extend to lower leg, and the sartorius, which flexes the leg. The **sartorius** is a thin, ribbonlike muscle originating on the anterior superior iliac spine and passing downward, cutting obliquely across the thigh (Figure 21.2). It is the longest muscle in the human body. Tailors used to sit cross-legged while working, and because the

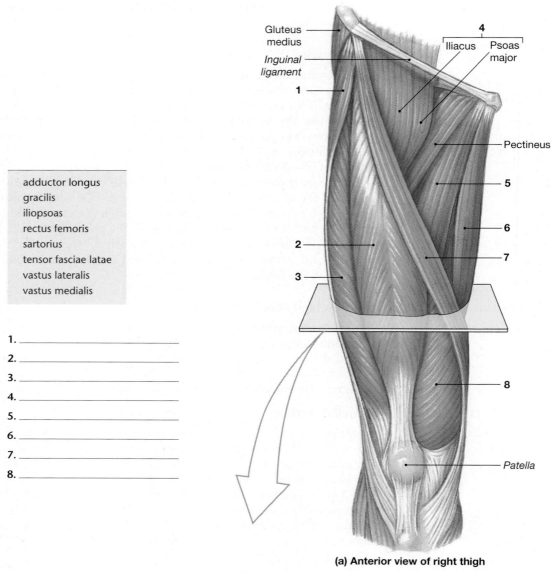

Gluteus medius

Inguinal ligament

1

2

3

4
Iliacus Psoas major

Pectineus

5

6

7

8

Patella

(a) Anterior view of right thigh

adductor longus
gracilis
iliopsoas
rectus femoris
sartorius
tensor fasciae latae
vastus lateralis
vastus medialis

1. _____
2. _____
3. _____
4. _____
5. _____
6. _____
7. _____
8. _____

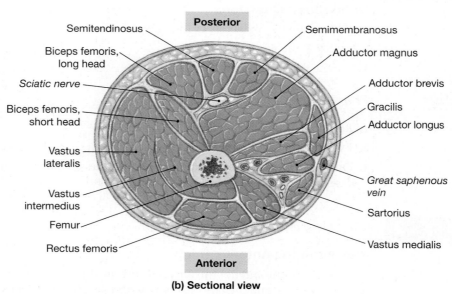

Posterior

Semitendinosus

Biceps femoris, long head

Sciatic nerve

Biceps femoris, short head

Vastus lateralis

Vastus intermedius

Femur

Rectus femoris

Semimembranosus

Adductor magnus

Adductor brevis

Gracilis

Adductor longus

Great saphenous vein

Sartorius

Vastus medialis

Anterior

(b) Sectional view

Figure 21.2 **Muscles That Move the Leg**

sartorius muscle assists in maintaining this posture, it is commonly referred to as the "tailor's muscle." It crosses the knee joint to insert on the medial surface of the tibia near the tibial tuberosity. This muscle is a flexor of the knee and thigh and a lateral rotator of the thigh.

The extensors of the lower leg are collectively called the **quadriceps femoris group**; they include the rectus femoris, vastus medialis, vastus lateralis, and vastus intermedius. They make up the bulk of the anterior mass of the thigh and are easy to locate on your leg. The **rectus femoris** (Figures 21.1a and 21.2b) is located along the midline of the anterior surface of the thigh. It originates on the iliac spine and the acetabulum. This muscle blends into the tendon of the quadriceps group and inserts on the patella. All the muscles of the quadriceps group converge on this tendon, which inserts on the tibial tuberosity. Because the rectus femoris crosses two joints, the hip and knee, it allows the hip to flex and the lower leg to extend.

The vastus muscles originate on the anterior femur and on the linea aspera of the posterior femur. They converge on the tendon of the quadriceps muscle and act together to extend the knee. Covering almost the entire medial surface of the femur is the **vastus medialis**. The **vastus lateralis** is located on the lateral side of the rectus femoris, and the **vastus intermedius** is directly under the rectus femoris.

Muscles of the Medial Thigh

The muscles on the inside of the thigh are adductors (Figure 21.3). The major ones are the gracilis, the pectineus, and the adductor muscle group. These muscles arise on the pubis bone and insert on the medial femur to adduct the thigh. Additionally, these muscles flex and medially rotate the thigh. When an athlete has an injured or "pulled" groin muscle, it is one of the adductor muscles that has been strained.

The **gracilis** (GRAS-i-lis) is the most superficial of the thigh adductors (Figures 21.1c and 21.3). It arises from the superior ramus of the pubic bone, near the symphysis, extends downward along the medial surface of the thigh, and inserts just medial to the insertion of the sartorius near the tibial tuberosity. Because it passes over the knee joint, it acts to flex the knee and adduct the thigh.

The **pectineus** (pek-TIN-ē-us) is another superficial adductor muscle of the medial thigh (Figure 21.1d). It is located between the iliacus and adductor longus muscles. It originates along the superior ramus of the pubic bone and inserts on a small ridge called the *pectineal line,* located inferior to the lesser trochanter of the femur.

The adductor muscle group includes the adductor magnus, adductor longus, and adductor brevis (Figure 21.3). The adductor muscles originate on the inferior pubis and insert on the posterior femur and are powerful adductors of the thigh. They also allow the thigh to flex and rotate medially. The **adductor magnus** is the largest of the adductor muscles. It arises on the inferior ramus of the pubis and inserts along the length of the linea aspera of the femur. It is easily observed on a leg model if the superficial muscles are removed. Superficial to the adductor magnus is the **adductor longus**. The **adductor brevis** is superior and posterior to the adductor longus.

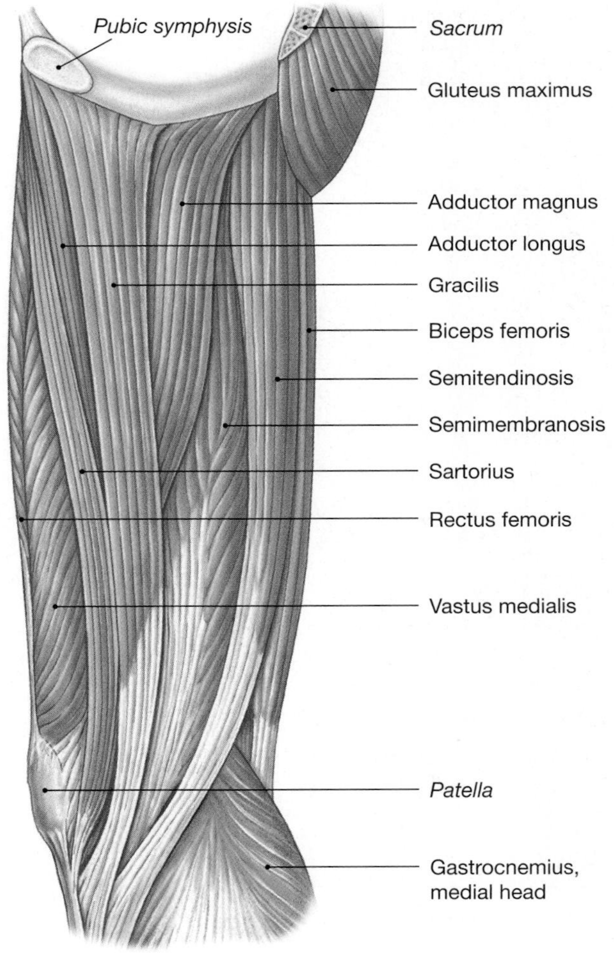

Pubic symphysis

Sacrum

Gluteus maximus

Adductor magnus

Adductor longus

Gracilis

Biceps femoris

Semitendinosis

Semimembranosis

Sartorius

Rectus femoris

Vastus medialis

Patella

Gastrocnemius, medial head

Figure 21.3
Medial Muscles of the Thigh

Muscles of the Posterior Thigh

The major muscles of the posterior thigh are collectively called the **hamstring group**. In this group are the semitendinosus, the semimembranosus, and the biceps femoris (Figure 21.4). They all have a common origin on the ischial tuberosity.

The **biceps femoris** is the lateral muscle of the posterior thigh. It has two heads and two origins, one on the ischial tuberosity and a second on the linea aspera of the femur. The two heads merge to form the belly of the muscle and insert on the lateral condyle of the tibia and the head of the fibula. Because this muscle spans the hip and the knee joints, it can flex the knee and extend the thigh.

Medial to the biceps femoris is the **semitendinosus** (sem-ē-ten-di-NŌ-sus). It is a long muscle that passes behind the knee to insert on the posteromedial surface of the tibia. The **semimembranosus** (sem-ē-mem-bra-NŌ-sus) is medial to the semi-tendinosus and inserts on the medial tibia. These muscles cross both the hip joint and the knee joint, and they both act to extend the thigh and flex the knee. The hamstrings are therefore antagonists to the quadriceps. When the thigh is flexed and drawn up toward the pelvis, the hamstrings act to extend the thigh.

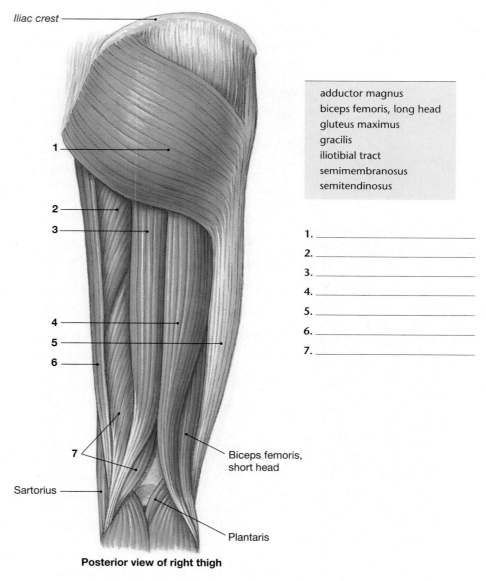

Iliac crest

adductor magnus
biceps femoris, long head
gluteus maximus
gracilis
iliotibial tract
semimembranosus
semitendinosus

1. _____
2. _____
3. _____
4. _____
5. _____
6. _____
7. _____

Biceps femoris, short head

Sartorius

Plantaris

Posterior view of right thigh

Figure 21.4 Posterior Muscles That Move the Leg

QuickCheck Questions

2.1 What are the muscles of the quadriceps femoris group?

2.2 What are the muscles of the hamstring group?

2.3 Where are the adductor muscles of the thigh located?

2 Materials

☐ Torso model
☐ Leg model

Procedures

1. Review the muscles of the thigh in Figures 21.1 to 21.4 and then label Figures 21.2 and 21.4.

2. On the torso and leg models, identify the flexors, extensors, adductors, and abductors of the leg.

3. Using Figure 21.5 as a guide, locate as many thigh muscles on your upper leg as possible. Practice the actions of the muscles and observe how your leg moves.

4. Flex your knee and feel the tendons of the semimembranosus and semitendinosus located just above the back of the knee on the medial side. Similarly, on the lateral side of the knee, just above the fibular head, the tendon of the biceps femoris can be palpated. ■

LAB ACTIVITY 3   Muscles of the Lower Leg

Muscles of the Anterior Lower Leg

The **tibialis** (tib-ē-A-lis) **anterior** is located on the anterior side of the lower leg (Figure 21.6). This muscle is easy to locate as the lateral muscle mass of the shin on the anterior edge of the tibia. The muscle has two origins, one on the anterolateral side of the tibia and the other on a membrane between the tibia and fibula. The tendon passes over the dorsal surface of the foot and inserts on the medial cuneiform and the first metatarsal. This muscle dorsiflexes and inverts the foot. Place your fingers on your tibialis and dorsiflex your foot to feel the muscle contract.

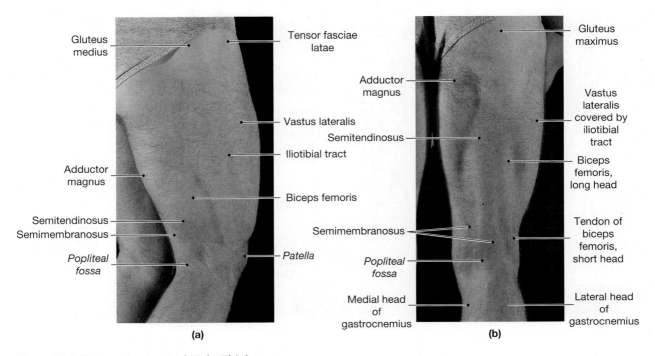

(a)   **(b)**

Figure 21.5 Surface Anatomy of Right Thigh
(a) Lateral view. **(b)** Posterior view.

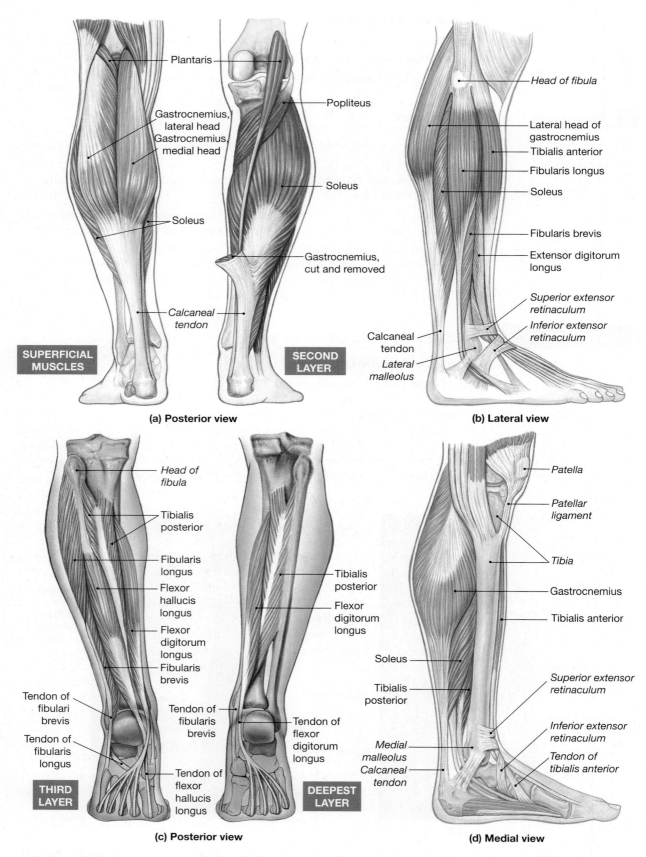

(a) Posterior view

Plantaris
Popliteus
Gastrocnemius, lateral head
Gastrocnemius, medial head
Soleus
Soleus
Gastrocnemius, cut and removed
Calcaneal tendon

SUPERFICIAL MUSCLES

SECOND LAYER

(b) Lateral view

Head of fibula
Lateral head of gastrocnemius
Tibialis anterior
Fibularis longus
Soleus
Fibularis brevis
Extensor digitorum longus
Superior extensor retinaculum
Inferior extensor retinaculum
Calcaneal tendon
Lateral malleolus

(c) Posterior view

Head of fibula
Tibialis posterior
Fibularis longus
Flexor hallucis longus
Flexor digitorum longus
Fibularis brevis
Tendon of fibulari brevis
Tendon of fibularis longus
Tendon of fibularis brevis
Tendon of flexor hallucis longus
Tibialis posterior
Flexor digitorum longus
Tendon of flexor digitorum longus

THIRD LAYER

DEEPEST LAYER

(d) Medial view

Patella
Patellar ligament
Tibia
Gastrocnemius
Tibialis anterior
Soleus
Tibialis posterior
Superior extensor retinaculum
Inferior extensor retinaculum
Tendon of tibialis anterior
Medial malleolus
Calcaneal tendon

Figure 21.6 Extrinsic Muscles That Move the Foot and Toes

Extensor Muscles of the Toes

Two extensor muscles arise on the anterior lower leg and insert on the phalanges. The **extensor hallucis longus** is lateral and deep to the tibialis anterior. It arises from the middle of the fibula and the interosseous membrane between the tibia and fibula and inserts on the dorsal surface of the distal phalanx of the great toe. This muscle extends the great toe. The tendon of this muscle becomes very prominent on the dorsal surface of the foot when this action is performed.

Lateral to the extensor hallucis longus is the **extensor digitorum longus**. It arises on the lateral condyle of the tibia and the anterior surface of the fibula. The tendon spreads into four branches that insert on the dorsal surface of the phalanges of toes 2 through 5. The extensor digitorum acts to extend toes 2 through 5. When the toes are extended, these tendons are clearly visible on the top of the foot.

Muscles of the Lateral Lower Leg

Two muscles are located on the lateral side of the lower leg. The **fibularis longus**, also called the **peroneus longus**, begins on the upper half of the shaft of the fibula, and the **fibularis brevis (peroneus brevis)** originates inferior to the origin of the longus near the middle of the fibular body (Figure 21.6). The tendons pass behind the lateral malleolus and curve under the foot. The fibularis longus inserts on the base of the first metatarsal, and the fibularis brevis inserts on the dorsal surface of the base of the fifth metatarsal. These muscles evert, or pull, the lateral edge of the foot upward. The longus also plantar flexes the foot. Practice this action with your fibularis muscles.

Muscles of the Posterior Lower Leg

The calf muscles of the posterior lower leg are the gastrocnemius and the soleus (Figure 21.6). These muscles share the calcaneal (Achilles) tendon, which inserts on the calcaneus of the foot. The **gastrocnemius** (gas-trok-NĒ-mē-us), the most superficial calf muscle, originates on the lateral and medial epicondyles of the femur. The two heads cross over the back of the knee and form a fleshy mass consisting of two bellies before blending about halfway down the lower leg. The gastrocnemius acts to plantar flex, adduct, and invert the foot. Because this muscle arises on the femur, it also flexes the knee.

The **plantaris** muscle is a short muscle of the lateral popliteal region, deep to the gastrocnemius muscle. The plantaris originates on the supracondylar ridge of the femur, and a long tendon inserts on the posterior of the calcaneus. The plantaris works with the gastrocnemius to plantar flex the foot and flex the knee.

The **popliteus** (pop-LĬ-tē-us) is deep to the plantaris muscles and crosses from its origin on the lateral condyle of the femur to insert on the posterior tibial shaft. It flexes the knee and rotates the tibia medially to unlock the knee for flexion.

Deep to the gastrocnemius is the **soleus** (SŌ-lē-us), which has an origin on the upper third of the tibia and on the fibula (Figure 21.6). The belly of the muscle passes beneath the gastrocnemius and eventually converges into the calcaneal tendon. The soleus contracts with the gastrocnemius to plantar flex, adduct, and invert the foot.

Deep to the soleus muscle of the calf is the **tibialis posterior** (Figure 21.6). The origin of this muscle is midway along the shaft of the tibia and fibula. The tendon passes medially to the calcaneus and inserts on the plantar surface of the navicular bone, the cuneiform bones, and the second, third, and fourth metatarsals. The tibialis posterior adducts, inverts, and plantar flexes the foot.

The **flexor hallucis longus** begins lateral to the origin of the tibialis posterior on the fibular shaft. Its tendon runs parallel to that of the tibialis, passes medial to the calcaneus, and inserts on the plantar surface of the distal phalanx of the great toe (Figure 21.6). It acts to flex the toe and opposes the action of the extensor hallucis longus.

The **flexor digitorum longus** muscle originates on the posterior tibia and inserts on the distal phalanges of toes 2 to 5 to flex to these toes.

QuickCheck Questions

3.1 Describe the muscles of the calf.

3.2 Which muscles move the great toe?

3.3 Describe the insertions of the muscles that plantar flex and dorsiflex the foot.

3   Materials

☐ Leg model

Procedures

1. Review the muscles of the lower leg in Figure 21.6.
2. Identify each muscle on the leg model.
3. Locate as many lower leg muscles on your own leg as possible. Practice the actions of the muscles and observe how your leg moves. ■

LAB ACTIVITY 4   Muscles of the Foot

Digitorum Brevis

The muscles that move the toes arise from the calcaneus of the foot. The **extensor digitorum brevis** is located on the dorsal surface of the foot and originates on the lateral superior surface of the calcaneus. It passes obliquely across the foot and sends out four tendons that insert into the dorsal surface of the first through fourth proximal phalanges (Figure 21.7). Because this muscle is on the top of the foot, it acts to extend the toes. One of the antagonists to this action is the **flexor digitorum brevis**, which also originates on the calcaneus, but on the plantar surface. The flexor digitorum brevis is situated in the middle of the sole of the foot. The tendons

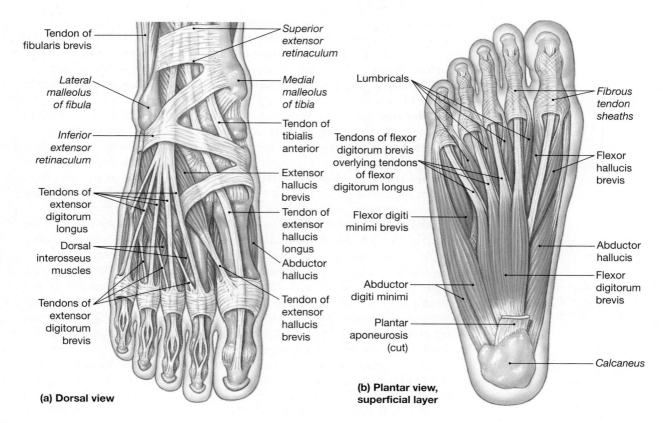

(a) Dorsal view

(b) Plantar view, superficial layer

Figure 21.7 Intrinsic Muscles of the Foot

from this muscle insert into the base of the second through fifth toes. It acts to flex the toes.

Hallucis Muscles of the Foot

The **abductor hallucis** (Figure 21.7) is found on the inner margin of the foot and originates on the plantar side of the calcaneus. It inserts on the medial side of the first phalanx of the great toe and acts to move the great toe away from the others. The **flexor hallucis brevis** originates on the plantar surface of the cuneiform and cuboid bones of the foot and splits into two heads, one medial and one lateral. Each head sends a tendon to the base of the first phalanx of the great toe, to either the lateral or the medial side. The muscle flexes the great toe and is an antagonist of the extensor hallucis longus.

Abductor Digiti Minimi

Examine the name **abductor digiti minimi** to define the action of this muscle: it abducts the little toe away from the others. The muscle is located on the outer margin of the foot and originates on the plantar and lateral surfaces of the calcaneus (Figure 21.7). It inserts on the lateral side of the proximal phalanx of the little toe.

QuickCheck Questions

4.1 What does the name "flexor hallucis brevis" mean?

4.2 Describe the action of each digitorum muscle.

4 *Materials*

☐ Leg model

Procedures

1. Review the muscles of the foot in Figure 21.7.
2. Identify each calf and foot muscle on the leg model.
3. On your own leg, locate muscles that move your foot. Practice the actions of the muscles and observe how your foot moves. ■

Muscles of the Pelvis, Leg, and Foot

Name _____

Date _____

Section _____

A. Matching

Match each term in the left column with its correct description from the right column.

_____ **1.** sartorius

_____ **2.** semitendinosus

_____ **3.** psoas major

_____ **4.** rectus femoris

_____ **5.** adductor magnus

_____ **6.** semimembranosus

_____ **7.** biceps femoris

_____ **8.** gluteus maximus

_____ **9.** vastus intermedius

_____ **10.** gracilis

_____ **11.** adductor brevis

_____ **12.** tensor fasciae latae

A. most anterior and superficial quadriceps muscle

B. quadriceps located deep to rectus femoris

C. most medial leg adductor

D. hamstring, has two heads

E. large muscle used in climbing stairs

F. short adductor of femur

G. muscle that crosses anterior thigh

H. tenses fasciae latae

I. hamstring muscle; inserts with semimembranosus

J. hamstring muscle; inserts on medial condyle of tibia

K. part of iliopsoas

L. largest adductor of femur

B. Matching

Match each term in the left column with its correct description from the right column.

_____ **1.** fibularis longus

_____ **2.** gastrocnemius

_____ **3.** flexor hallucis longus

_____ **4.** abductor digiti minimi

_____ **5.** abductor hallucis

_____ **6.** tibialis anterior

_____ **7.** extensor hallucis longus

_____ **8.** extensor digitorum longus

_____ **9.** tibialis posterior

_____ **10.** soleus

A. abducts great toe

B. on anterior lower leg; dorsiflexes foot

C. originates on anterior lower leg; extends great toe

D. long muscle that flexes great toe

E. deep calf muscle; plantar flexes foot

F. abducts little toe

G. muscle of lateral lower leg; plantar flexes foot

H. calf muscle deep to soleus

I. originates on anterior lower leg; extends toes

J. superficial calf muscle; plantar flexes and inverts foot

21

C. Labeling

Label each muscle in Figure 21.8.

1. _____
2. _____
3. _____
4. _____
5. _____
6. _____
7. _____
8. _____
9. _____
10. _____
11. _____
12. _____
13. _____
14. _____
15. _____
16. _____
17. _____
18. _____
19. _____
20. _____
21. _____
22. _____
23. _____
24. _____
25. _____

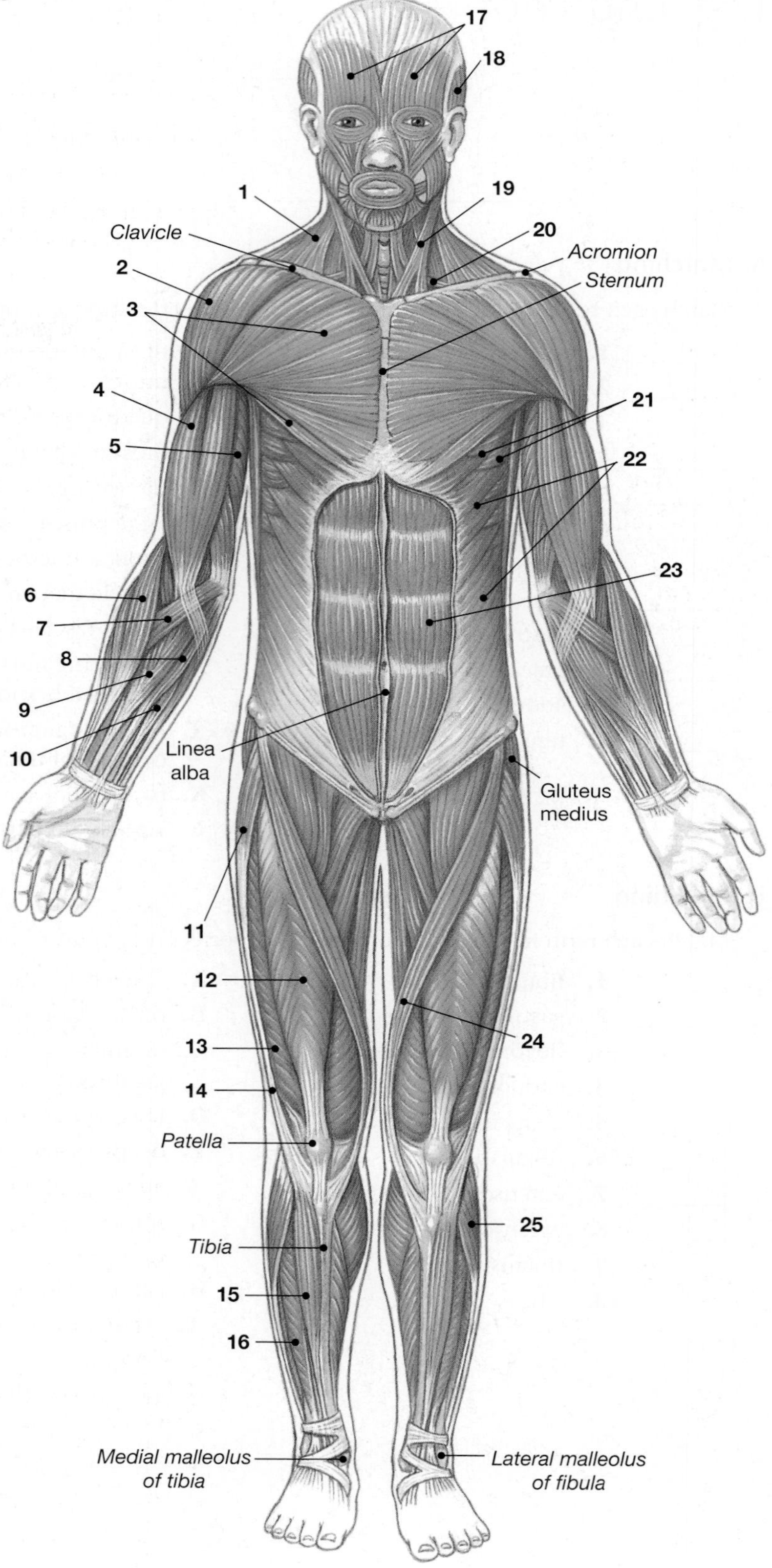

Figure 21.8 An Overview of the Major Anterior Skeletal Muscles

D. Labeling

Label each muscle in Figure 21.9.

1.
2. _____
3. _____
4. _____
5. _____
6. _____
7. _____
8. _____
9. _____
10. _____
11. _____
12. _____
13. _____
14. _____
15. _____
16. _____
17. _____
18. _____

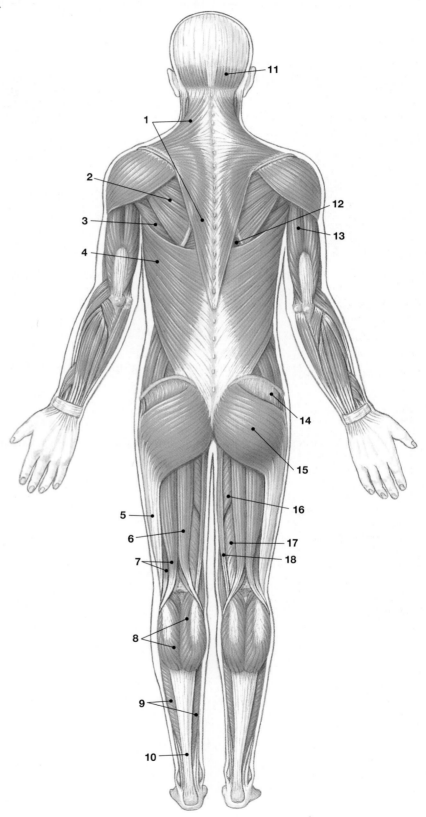

Figure 21.9 An Overview of the Major Posterior Skeletal Muscles

LABORATORY REPORT

E. Short-Answer Questions

1. Describe how the hamstring muscle group moves the leg.

2. Which muscle group is the antagonist to the muscles of the hamstring group?

3. Describe the action of the abductor and adductor muscles of the thigh.

F. Analysis and Application

1. Which leg muscles serve a function similar to the function of the arm's rotator cuff muscles?

2. How are the muscles of the leg similar to those of the arm? How are they different?

3. Describe the origin, insertion, and action of the muscles that invert and evert the foot.

Muscle Physiology

OBJECTIVES

On completion of this exercise, you should be able to:

- Relate the sliding filament theory and molecular events to muscle contraction.

- Describe the anatomy at the neuromuscular junction.

- Explain the differences among a twitch, wave summation, incomplete tetanus, and complete tetanus.

- Describe how a muscle fatigues.

- Explain the differences between isometric and isotonic contractions.

- Observe and record skeletal muscle tonus measured against a baseline activity level associated with the resting state.

- Observe and record how motor unit recruitment changes as the power of a skeletal muscle contraction increases.

- Correlate EMG "sound" intensity with level of motor unit recruitment.

- Determine maximum hand clench strength values and compare female-male strength differences.

- Record the force produced by clenched muscles, EMG, and integrated EMG when inducing fatigue.

M uscle and nerve tissues are excitable tissues that produce self-propagating electrical impulses called **action potentials**. These electrical impulses result from the movement of sodium and potassium ions through specific protein channels in the cell membrane. When a muscle fiber or a neuron is at rest, the net electrical charge inside the cell is different from the net charge outside the cell. This electrical difference is measured in millivolts (mV) and is called the **resting membrane potential**. Resting potential values differ from one type of cell to another. Typical values at the inner membrane surface are -70 mV for a neuron and -85 mV for skeletal muscles.

When a neuron stimulates a muscle, the nerve action potential causes the neuron to release specific chemicals, collectively called **neurotransmitters**, that cause an action potential and thus contraction in the muscle fiber. Sodium channels open in the sarcolemma (the "cell membrane" of the muscle fiber; Exercise 9), and sodium ions flood into the fiber, causing the sarcolemma to **depolarize**, a term used when the membrane becomes less negative. At the peak of depolarization, the sarcolemma is at +30 mV. At this millivoltage, the sodium channels close and potassium channels open. Potassium ions exit the fiber, and the fiber **repolarizes** to the resting potential. In summary, an electrical signal in the neuron causes release of a chemical signal, the neurotransmitter, which causes an electrical signal in the muscle fiber that results in contraction.

In this exercise you will investigate several types of muscle contraction. It is recommended that you review the microanatomy of skeletal muscles in Exercise 9 and the function of levers in Exercise 17 before performing these experiments.

Your laboratory may be equipped with a physiograph, an instrument that electrically stimulates muscles and records the characteristics of the contraction. Laboratory Activities 1, 2, and 3 of this exercise provide the background physiology necessary to perform such investigations. Laboratory Activities 4 and 5 utilize the Biopac Student Lab physiograph to produce and interpret human electromyographs.

LAB ACTIVITY **1** Biochemical Nature of Muscle Contraction

The **sliding filament theory** is the current model of muscle contraction. The theory uses direct laboratory observations to explain that, during contraction of a muscle fiber, the filaments slide past one another, causing the fiber to shorten. Figure 22.1 illustrates the changes in the sarcomere chain in a muscle fiber during contraction. The thin filaments, composed mostly of the protein actin, are pulled inward by the pivoting of the myosin heads of the thick filaments. As the thin filaments slide toward the middle of the sarcomere, the Z lines and I bands move closer together, the zone of overlap increases, and the width of the H zone decreases. The width of the A band does not change during the contraction.

Motor neurons communicate with skeletal muscles at neuromuscular junctions. The end of the neuron's axon swells into a synaptic knob containing synaptic vesicles full of acetylcholine (as-ē-til-KŌ-lēn) (ACh), the neurotransmitter molecule used to stimulate the skeletal muscle to contract. When a nerve action potential reaches the

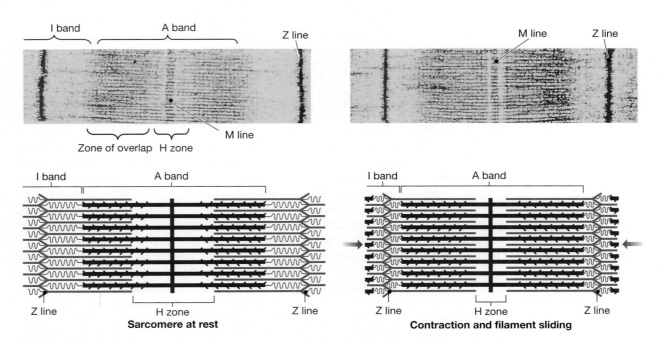

Figure 22.1 **Changes in the Appearance of a Sarcomere during Contraction of a Skeletal Muscle Fiber**

synaptic knob, the knob's membrane becomes more permeable to interstitial calcium ions present in the synaptic cleft. These ions diffuse into the knob and cause the synaptic vesicles to fuse with the knob's membrane and release ACh into the synaptic cleft. The sarcolemma at the synaptic cleft is folded into a motor end plate and contains ACh receptors for binding the neurotransmitter. When ACh binds to the receptor sites, the muscle fiber responds by producing a muscle action potential. This electrical current travels along the sarcolemma and down the transverse tubules inside the muscle fiber. This current through the fiber stimulates the release of stored calcium ions from the cisternae of the sarcoplasmic reticulum. Once calcium floods the sarcoplasm inside the muscle fiber, contraction occurs. The fiber produces an enzyme called acetylcholinesterase (AChE), which deactivates ACh and initiates muscle relaxation.

The contraction involves five chemical steps, illustrated in Figure 22.2:

- **Active-site exposure.** An active site in a muscle fiber is a spot on a thin filament where a myosin head can bind to an actin molecule. While the muscle is at rest, troponin holds tropomyosin molecules in place to cover the active sites on the actin filaments. As long as the active sites are covered, the myosin heads cannot attach to the thin filament and pull it inward. When calcium ions are released into the sarcoplasm, they bind to troponin and cause the tropomyosin to roll away from the active sites, exposing them to the myosin heads. As long as calcium ions are present in the sarcoplasm, the active sites remain exposed for myosin binding.

- **Cross-bridge attachment.** Actin and myosin molecules have a strong affinity for each other; thus, once active sites on an actin molecule are exposed, the myosin heads chemically bind to the sites. This binding of myosin to active sites is called *cross-bridge attachment.*

- **Pivoting.** While a muscle is at rest, the myosin heads split ATP and become energized by storing the energy released as the ATP converts to ADP. During pivoting, the heads use the stored energy to pull the thin filaments toward the M line of the sarcomere, causing the entire muscle fiber to contract. You can see the pivoting by comparing the angle of the myosin heads in steps 2 and 3 of Figure 22.2. After pivoting, the heads are depleted of stored energy and release the ADP and phosphate into the sarcoplasm.

- **Cross-bridge detachment.** After the myosin heads have pivoted and released their ADP, an ATP molecule binds to each head and causes it to "back out" of the active site. The heads are now ready to be reenergized for another attachment-pivot sequence.

- **Myosin reactivation.** This final step of contraction involves splitting of the ATP on the myosin heads to reenergize the heads and return them to their resting positions. The head is an ATPase and hydrolyzes ATP to ADP plus released energy.

As contraction continues, the attachment-pivoting-detachment-reactivation sequence repeats until all calcium ions are removed from the sarcoplasm and returned to the sarcoplasmic reticulum. Note that the first step, active-site exposure, occurs only at the onset of contraction. The active sites remain exposed as long as calcium ions are bonded to troponin molecules so that the tropomyosin molecules are away from the active sites.

The signal for a muscle fiber to relax is the loss of neurotransmitter stimulation on the motor end plate. The fiber secretes AChE, and this enzyme deactivates any ACh in the synaptic cleft and on the sarcolemma ACh receptors. When ACh is inactivated, the sarcolemma no longer produces an action potential. Active-transport mechanisms begin to pump calcium ions in the sarcoplasm back into the sarcoplasmic reticulum. As calcium ions leave the troponin molecules, the tropomyosin moves into place to block the active sites on actin. This blocking prevents cross-bridge attachments, the thin filaments slide outward away from the middle of each sarcomere, and the sarcomeres and muscle fiber return to their initial relaxed length. Table 22.1 summarizes the steps involved in skeletal muscle contraction.

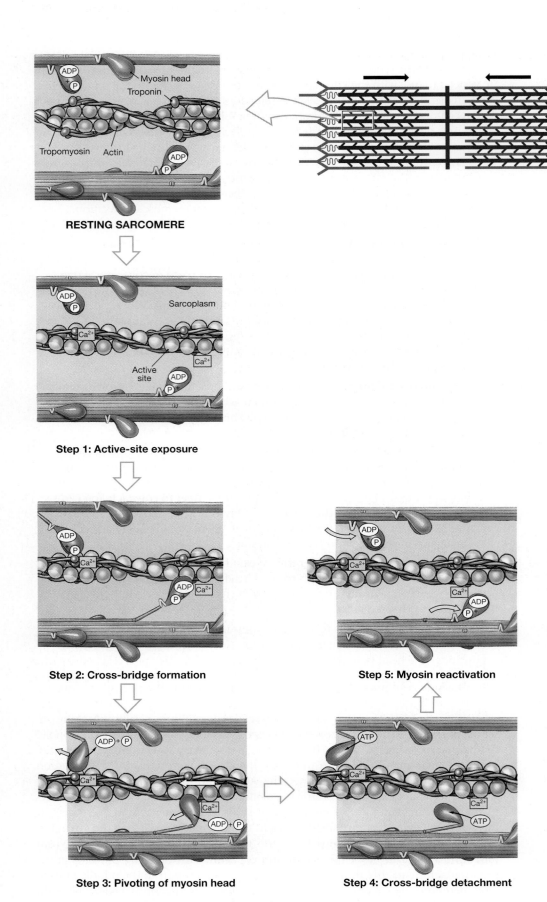

RESTING SARCOMERE

Step 1: Active-site exposure

Step 2: Cross-bridge formation

Step 3: Pivoting of myosin head

Step 4: Cross-bridge detachment

Step 5: Myosin reactivation

Figure 22.2 **Molecular Events of the Contraction Process**

Table 22.1 Steps Involved in Skeletal Muscle Contraction

Steps That Initiate a Contraction:

1. At the neuromuscular junction (NMJ), ACh released by the synaptic terminal binds to receptors on the sarcolemma.

2. The resulting change in the transmembrane potential of the muscle fiber leads to the production of an action potential that spreads across the entire surface of the muscle fiber and along the T-tubules.

3. The sarcoplasmic reticulum (SR) releases stored calcium ions, increasing the calcium concentration of the sarcoplasm in and around the sarcomeres.

4. Calcium ions bind to troponin, producing a change in the orientation of the troponin-tropomyosin complex that exposes active sites on the thin (actin) filaments. Cross-bridges form when myosin heads bind to active sites on F actin.

5. The contraction begins as repeated cycles of cross-bridge attachment, pivoting, and detachment occur, powered by the hydrolysis of ATP. These events produce filament sliding, and the muscle fiber shortens.

Steps That End a Contraction:

6. Action potential generation ceases as ACh is broken down by acetylcholinesterase (AChE).

7. The SR reabsorbs calcium ions, and the concentration of calcium ions in the sarcoplasm declines.

8. When calcium ion concentrations approach normal resting levels, the troponin-tropomyosin complex returns to its normal position. This change re-covers the active sites and prevents further cross-bridge attachment.

9. Without cross-bridge attachments, further sliding cannot take place, and the contraction ends.

10. Muscle relaxation occurs, and the muscle returns passively to its resting length.

QuickCheck Questions

1.1 Briefly summarize the sliding filament theory.

1.2 List the molecular steps of muscle contraction.

1.3 What is the role of ATP in muscle contraction?

1 Materials

- ☐ Muscle preparation (glycerinated muscle from biological supply company)
- ☐ Glycerol (supplied with muscle preparation)
- ☐ ATP solution (supplied with muscle preparation)
- ☐ Calcium ion solution (supplied with muscle preparation)
- ☐ Dissecting microscope
- ☐ Compound microscope
- ☐ Blank microscope slides
- ☐ Pipette or eye dropper
- ☐ Clean teasing needles
- ☐ Ruler calibrated in millimeters

Procedures

1. Label four clean microscope slides A, B, C, and D.

2. Place a sample of the muscle under the dissecting microscope, and use a teasing needle to gently pry the fibers apart. Separate two or three fibers and transfer this group of fibers to slide A. Repeat this teasing process three more times, placing one group of fibers on slide B, one on slide C, and the last on slide D. Add a drop of glycerol to each slide to prevent dehydration of the fibers.

3. Add a coverslip to slide A, and observe the fibers with the compound microscope at low power and at high power. Describe the appearance of these fibers here:

    slide A low _____

    slide A high _____

4. Without a coverslip, examine the fibers of slide B with the dissecting microscope. Place a millimeter ruler under the slide, measure the length of the relaxed fibers, and record your measurement here:

    slide B relaxed length _____.

5. Add a drop of ATP solution to the fibers under the dissecting microscope and observe their response. After 30 to 45 seconds, measure the length of the fibers and record your measurement here:

    slide B length after treatment _____.

6. Repeat steps 4 and 5 with slide C, this time using the calcium ion solution instead of the ATP. Record your two measurements:

    slide C relaxed length _____ length after treatment _____.

7. Repeat steps 4 and 5 with slide D, this time using both the ATP solution and the ion solution. Record your two measurements here:

    slide D relaxed length _____ length after treatment _____.

8. Dispose of the muscle preparations as indicated by your laboratory instructor. ■

LAB ACTIVITY 2 Types of Muscle Contraction

In the preceding section, muscle contraction was presented in terms of the events occurring inside a single muscle fiber. Muscle fibers do not act individually, however, because a single motor neuron controls multiple fibers. A motor neuron innervating the large muscles of the thigh, for example, stimulates more than 1,000 fibers. Any group of fibers controlled by the same neuron is called a **motor unit** and can be considered a "muscle team" that contracts together when stimulated by the neuron. The muscle fibers are said to be "on" for contraction or "off" for relaxation, a concept called the **all-or-none principle**.

The type of contraction a muscle fiber undergoes is determined by the frequency at which the fiber is stimulated by its motor neuron. If a single action potential occurs in the neuron, only a small amount of ACh will be released and the muscle fiber will twitch. A **twitch** is a single stimulation-contraction-relaxation event in the fiber. Figure 22.3 displays recordings of muscle contractions called **myograms**. The figure compares the twitching of different muscles. Each twitch has three sections: latent period, contraction phase, and relaxation phase.

The **latent period** is the time from the initial stimulation to the start of muscle contraction. During this brief period, the fiber is stimulated by ACh, releases calcium ions, exposes active sites, and attaches cross-bridges. No tension is produced during this period, and no pivoting occurs. The **contraction phase** involves shortening of the fiber and the production of muscle tension, or force.

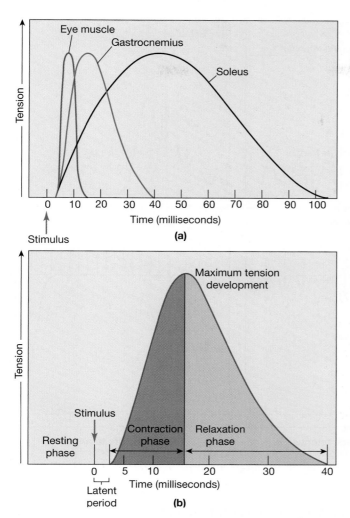

Figure 22.3 **The Twitch and the Development of Tension**
(a) A myogram showing differences in tension over time for a twitch contraction in different skeletal muscles. (b) The details of tension over time for a single twitch contraction in the gastrocnemius muscle. Notice the presence of a latent period, which corresponds to the time needed for the conduction of an action potential and the subsequent release of calcium ions by the sarcoplasmic reticulum.

During this phase, myosin heads are pivoting and cycling through the attach-pivot-detach-return sequence. As more calcium ions enter the sarcoplasm, more cross-bridges are formed, and tension increases as the thick filaments pull the thin filaments toward the center of the sarcomere. The **relaxation phase** occurs as AChE inactivates ACh and calcium ions are returned to the sarcoplasmic reticulum. The thin filaments passively slide back to their resting positions, and muscle tension decreases.

The myogram in Figure 22.4a is a recording of **treppe** (TREP-e), which is muscle contraction with complete relaxation between each stimulus. As the muscle is repeatedly stimulated, calcium ions accumulate in the cytosol, causing the first 30–50 contractions to increase in tension. The rest of Figure 22.4 illustrates the effect of increasing the frequency at which a skeletal muscle is stimulated. If the muscle is stimulated a second time before it has completely relaxed from a first stimulation, the two contractions are summed. This phenomenon, called **wave summation** (shown in Figure 22.4b), results in an increase in tension with each summation. Because the thin filaments have not returned to their resting length when the muscle is stimulated again, contractile force increases as more calcium ions are released into the sarcoplasm and more cross-bridges attach and pivot.

If the frequency of stimulation is increased further, the muscle produces peak tension with short cycles of relaxation. This type of contraction is called **incomplete tetanus** (Figure 22.4c). If the rate of stimulation is such that the relaxation phase is completely eliminated, the contraction type is **complete tetanus** (Figure 22.4d). During complete tetanus, peak tension is produced for a sustained period of time and results in a smooth, strong contraction. Most muscle work is accomplished by complete tetanus.

Clinical Application | ***Tetanus***

Surely you have had a tetanus shot. Why is the injection called a tetanus shot when tetanus is a type of muscle contraction? Often an injury introduces bacteria, *Clostridium tetani,* into the wound. The bacteria produce a toxin that binds to ACh receptors in skeletal muscles and stimulates the muscles to contract. The enzyme AChE cannot inactivate the bacterial toxin, and the muscle remains in a painful tetanic contraction for an extended period of time. Muscles for mastication are often affected by the toxin; hence the common name "lockjaw" for the symptom. As a preventative measure, a tetanus shot contains human tetanus immune globulins that prevent the *Clostridium* from surviving in the body and producing toxins. Actual treatment for the infection and toxin is usually ineffective in preventing tetanus. Is your tetanus booster shot current? ▸

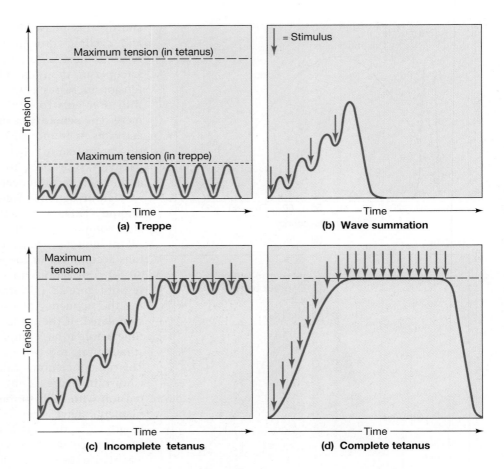

Figure 22.4 Effects of Repeated Stimulations
(a) Treppe. (b) Wave summation.
(c) Incomplete tetanus.
(d) Complete tetanus.

Because all muscle work requires complete tetanus, the type of contraction does not determine the overall tension a muscle produces. Muscle strength is varied through the number of motor units activated. The process called **recruitment** stimulates more motor units to carry or move a load placed on a muscle. As recruitment occurs, more muscle fibers are turned on and contract, and thus tension increases. Imagine holding a book in your outstretched hand. If another book is added to the load, additional motor units are recruited to increase the muscle tension to support the added weight.

Muscle fibers cannot contract indefinitely, and eventually they become **fatigued**. The force of contraction decreases as fibers lose the ability to maintain compete tetanic contractions. Fatigue is caused by a decrease in cellular energy and oxygen sources in the muscle and an accumulation of waste products. During intense muscle contraction, such as lifting a heavy object, the fibers become fatigued because of low ATP levels and a buildup of lactic acid, a by-product of anaerobic respiration. Joggers experience muscle fatigue as secondary energy reserves are depleted and damage accumulates in muscle fibers, especially in the sarcoplasmic reticulum and calcium-regulating mechanisms.

QuickCheck Questions

2.1 What are the stages of a muscle twitch?

2.2 What is the difference between incomplete tetanus and complete tetanus?

2.3 Why do skeletal muscles fatigue?

Table 22.2 **Muscle Fatigue Demonstration**

Trial	Start Time (seconds)	End Time (seconds)	Duration (seconds)
1			
2			
3			

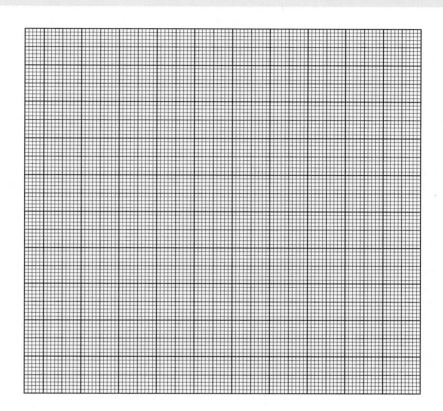

2 Materials

- ☐ Heavy object
- ☐ Stop watch

Procedures

1. Set the stop watch to zero, and record this time in Table 22.2 under Trial 1.

2. Extend your arm straight out in front of you, parallel to the floor, and load your arm by placing the heavy object in your hand. *Immediately* start the stop watch.

3. Hold the object with your arm straight for as long as possible. Once your arm starts to shake or your muscles ache, put the object down, and stop the stop watch. Record the end time for Trial 1 in Table 22.2.

4. Rest for 1 minute, and then repeat steps 1 through 3 as Trial 2.

5. Rest for another minute and then repeat steps 1 through 3 as Trial 3.

6. Calculate the total time in seconds required for each trial. Enter these values in the Duration column in Table 22.2.

7. Plot on the graph the total time until fatigue for each trial. Label the horizontal axis "Trial" and the vertical axis "Time in seconds."

8. Interpret your experimental data. Why is there a difference in the time to fatigue for each trial? ■

LAB ACTIVITY 3 Isometric and Isotonic Contractions

Two major types of complete tetanic contractions occur: isometric and isotonic (Figure 22.5). **Isometric** (*iso-,* same + *metric,* length) **contractions** occur when the muscle length is relatively constant but muscle tension changes. Because length is constant, no body movement occurs. Muscles for maintaining posture use isometric contractions to support the body weight. **Isotonic** (*tonic,* tone) **contractions** involve constant tension while the length of the muscle changes. Consider picking up a book and then flexing your arm so that you move the book up to your shoulder. Once you are holding the book, your arm muscles are "loaded" with the weight of the book. As you flex your arm, muscle tension changes minimally while muscle length varies greatly.

QuickCheck Questions

3.1 Define and give an example of an isotonic contraction.

3.2 Define and give an example of an isometric contraction.

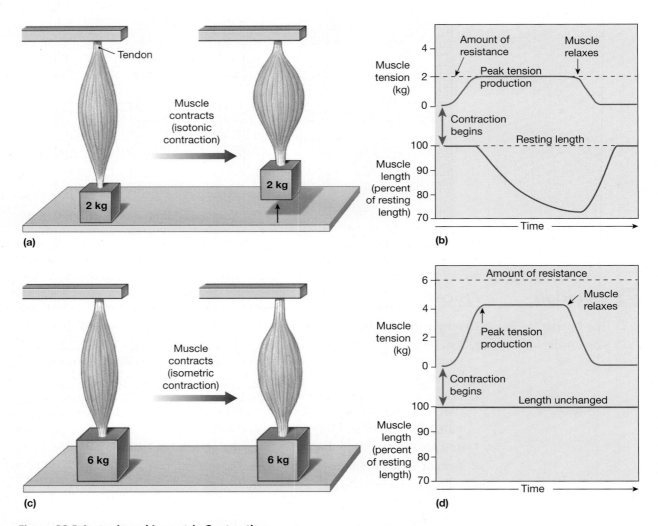

Figure 22.5 Isotonic and Isometric Contractions

(a, b) This muscle is attached to a weight less than its peak tension capabilities. On stimulation, it develops enough tension to lift the weight. Tension remains constant for the duration of the contraction, although the length of the muscle changes. This is an example of isotonic contraction. **(c, d)** The same muscle is attached to a weight that exceeds its peak tension capabilities. On stimulation, tension will rise to a peak, but the muscle as a whole cannot shorten. This is an isometric contraction.

Table 22.3 Isometric and Isotonic Contractions

	Trial 1	Trial 2	Trial 3
Tension	_____	_____	_____
Length	_____	_____	_____
Type of contraction	_____	_____	_____

3 *Materials*

Procedures

☐ Heavy object

☐ Ruler calibrated in millimeters

1. Extend your arm straight out in front of you, parallel to the floor.

2. Palpate your extended biceps brachii muscle to feel the tension of the muscle. Also notice the length of the muscle. Record your observations in the Trial 1 column in Table 22.3.

3. Load your extended arm by placing the heavy object in your hand. Palpate your biceps brachii again, and notice the degree of muscle tension and the length of the muscle. Record your observations in the Trial 2 column in Table 22.3.

4. Gently squeeze the loaded biceps brachii and repeatedly flex and extend your arm six times, moving the heavy object four to six inches each time. When you are done with this motion, record your muscle tension and length observations in the Trial 3 column in Table 22.3.

5. Describe each type of muscle contraction to complete Table 22.3. ∎

LAB ACTIVITY 4   BIOPAC: Electromyography—Standard and Integrated EMG Activity

BIOPAC

When a muscle or nerve is stimulated, it responds by producing action potentials. An action potential in a muscle fiber activates the physiological events of contraction. Because the electrical stimulation is also passively conducted to the body surface by surrounding tissues, sensors placed on the skin can detect the electrical activity produced by a muscle. The impulse produced by individual muscle fibers is minimal, but the combination of impulses from thousands of stimulated fibers produces a measurable electrical change in the overlying skin. The sensors, which are electrodes, are connected to an amplifier that passes the signals to a recorder that then produces an **electromyogram** (EMG), a graph of the muscle's electrical activity. (You are probably familiar with electrical tracings of the heart's electrical activity done in an electrocardiogram, or ECG, which will be studied in a later exercise on the heart.)

In this laboratory activity, you will use the Biopac Student Lab system to detect, record, and analyze a series of muscle impulses. The system consists of three main components: sensors (both electrodes and transducers) that detect electrical impulses and other physiological phenomena; an acquisition unit, which collects and amplifies data from the sensors; and a computer with software to record and interpret the EMG data. After applying electrodes to the skin over your forearm muscles, you will clench your fist repeatedly with increasing force, and the BIOPAC system will produce an EMG of the muscle impulses. Each time you increase the force of a fist clench, the muscles recruit additional motor units to contract and produce more tension. For more background information, you should review the physiology of muscle stimulation, contraction, and recruitment in the first three laboratory activities of this exercise.

This EMG laboratory activity is organized into four major sections. Section 1, Setup, describes where to plug in the electrode leads and how to apply the skin electrodes. Section 2, Calibration, adjusts the hardware so that it can collect accurate physiological data. Section 3, Data Recording, describes how to record the fist clench impulses once the hardware is calibrated. After you have saved the muscle data to a computer disk, Section 4, Data Analysis, instructs you how to use the software tools to interpret and evaluate the EMG.

Safety Alert

The Biopac Student Lab system is safe and easy to use, but be sure to follow the procedures as outlined in the laboratory activities. Under no circumstances should you deviate from the experimental procedures. Exercise extreme caution when using the electrodes and transducers with other equipment that also uses electrodes or transducers that may make electrical contact with you or your laboratory partner. Always assume that a current exists between any two electrodes or electrical contact points. ▲

QuickCheck Questions

4.1 What are electrodes?

4.2 What is the purpose of calibrating the BIOPAC hardware?

4 Materials

- ☐ BIOPAC electrode lead set (SS2L)
- ☐ BIOPAC disposable vinyl electrodes (EL503), 6 electrodes per subject
- ☐ BIOPAC headphones (OUT1)
- ☐ BIOPAC electrode gel (GEL1) and abrasive pad (ELPAD) or skin cleanser or alcohol prep
- ☐ Computer: Macintosh, minimum 68020; or PC Windows 98SE or better
- ☐ Biopac Student Lab software v3.0 or better
- ☐ BIOPAC acquisition unit (MP35/MP30)
- ☐ BIOPAC wall transformer (AC100A)
- ☐ BIOPAC serial cable (CBLSERA)
- ☐ BIOPAC USB adapter (USB1W)

Procedures

Section 1: Setup

1. Turn your computer ON.
2. Make sure the BIOPAC MP35/MP30 unit is OFF.
3. Plug the equipment in as shown in Figure 22.6: the electrode lead (SS2L) into CH 3 and the headphones jack (OUT1) into the back of the unit.
4. Turn the MP35/MP30 unit ON.
5. Attach three electrodes either to your own dominant forearm or, if you are working with a partner, to your partner's dominant forearm, as shown in Figure 22.7. This will be forearm 1, and the nondominant forearm will be forearm 2. For optimal signal quality, you should place the electrodes on the skin at least five minutes before starting the calibration section.
6. Attach the electrode lead set (SS2L) to the electrodes on forearm 1. Make sure the electrode lead colors match those shown in Figure 22.7. (Each pinch connector works like a small clothespin, but will latch onto the nipple of the electrode only from one side of the connector.)
7. Start the Biopac Student Lab Program on your computer.
8. Choose lesson "L01-EMG-1," and click OK.
9. Type in a filename, using a unique identifier such as you or your partner's nickname or student ID number.
10. Click OK to end the setup section.

Section 2: Calibration

This series of steps establishes the hardware's internal parameters and is critical for optimum performance.

1. On the computer screen, click on Calibrate.
2. Read the dialog box, and click OK when ready. The calibration will not begin until you click OK.
3. Wait two seconds. Then clench your forearm 1 fist as hard as possible for five or six seconds and release. The calibration will last eight seconds and will stop automatically. (You do not need to keep your fist clenched for the whole eight seconds.)

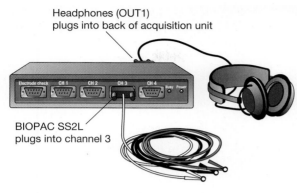

Headphones (OUT1)
plugs into back of acquisition unit

BIOPAC SS2L
plugs into channel 3

Figure 22.6 BIOPAC Cable Setup

4. Your computer screen should resemble Figure 22.8. Repeat calibration steps 1 through 3 if your screen is different. If your calibration recording did not begin with a zero baseline, this means that you clenched your fist before waiting two seconds, and you need to repeat the calibration.

Section 3: Data Recording

You will record EMG activity data for two segments: segment 1 from forearm 1 and segment 2 from forearm 2. To work efficiently, read through the rest of this activity so that you will know what to do before recording. Check the last line of the software journal, and note the total amount of time available for the recording. Stop each recording segment as soon as possible, so that you do not waste storage space on the computer.

Segment 1 (Forearm 1, Dominant).

1. On your computer, click on Record.

2. Clench your fist and hold for two seconds. Release the clench and wait two seconds. Repeat the clench-release-wait sequence while increasing the force in each sequence by equal increments so that the fourth clench uses maximum force.

3. On your computer, click on Suspend, and then review the recording on the screen. If it looks similar to Figure 22.9 (except for the black selected area at the left), go to step 5. If your recording looks different, go to step 4.

   4. Click on Redo, and repeat steps 1 through 3.

   5. Remove the electrode cable pinch connectors, peel the electrodes from your arm, and dispose of them. Use soap and water to wash the electrode gel residue from your skin. The electrodes may leave a slight ring on the skin for a few hours; this is quite normal.

Segment 2 (Forearm 2, Nondominant).

   6. Attach the remaining three electrodes to your nondominant arm (or to the nondominant arm of your partner, if you are working with one), again spacing them as shown in Figure 22.7.

   7. On your computer, click on Resume. A marker labeled "Forearm 2" will automatically be inserted when you do this.

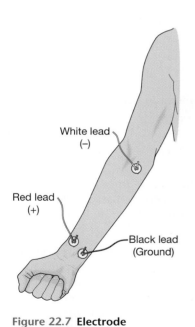

White lead
(–)

Red lead
(+)

Black lead
(Ground)

Figure 22.7 Electrode Placement and Lead Attachment

Fist is clenched for calibration procedure.

   8. Repeat a cycle of clench-release-wait, holding each clench for two seconds and waiting two seconds after release before beginning the next cycle. Increase the strength of your clench by the same amount for each cycle, with the fourth clench having the maximum force.

   9. On your computer, click on Suspend, and then review the recording on the screen. If it looks similar to Figure 22.9 (except for the black selected area at the left), go to step 11. If your recording looks different from Figure 22.9, go to step 10.

10. Click on Redo, and repeat steps 8 through 9.

11. Click on Stop. Clicking "yes" in the dialog box will end the data-recording section and will automatically save the data. Clicking "no" will bring you back to either Resume or Stop. If you want to listen to the EMG signal, go to step 12. Listening to the EMG can be a valuable tool in detecting muscle abnormalities and is performed here for general interest. If you want to end the recording, go to step 16.

Figure 22.8 Calibration EMG

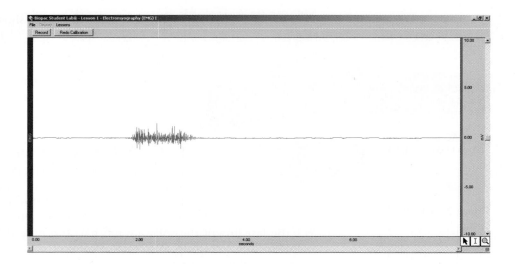

Figure 22.9 Selection of EMG Cluster for Analysis

Data for the highlighted EMG are displayed in the small menu boxes.

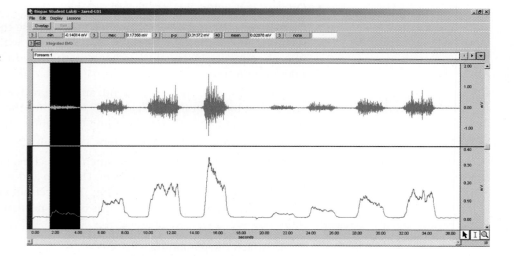

12. Put on the headphones, and click on Listen. The volume through the headphones may be very loud because of system feedback. Because the volume cannot be adjusted, you may have to position the headphones slightly off your ears for comfort.

13. Experiment by changing the clench force during the clench-release-wait cycles as you watch the screen and listen. You will hear the EMG signal through the headphones as it is being displayed on the screen. The screen will display two channels: CH 3 EMG and CH 40 Integrated EMG. The data on the screen will not be saved. The signal will run until you press Stop. If others in your laboratory group would like to hear the EMG signal, pass the headphones around before clicking Stop.

14. Click on Stop to end the listening-to-EMG portion.

15. To listen again, or to have another person listen, click Redo.

16. Click on Done. A pop-up window will appear. Make your choice, and continue as directed. If choosing the "Record from another subject" option, you will have to attach electrodes per Setup step 5 and continue the activity from Setup step 8. Remember that each person will need to use a unique filename.

Section 4: Data Analysis

1. Enter the Review Saved Data mode, and choose the correct file. Note Channel Number (CH) settings in the small menu boxes, shown in Figure 22.9 as the string of squares across the top containing the numbers, left to right, 3, 3, 3, and 40.

Channel	Displays
CH 3	Raw EMG
CH 40	Integrated EMG

2. Set up your display window for optimal viewing of the first data segment (forearm 1, dominant). Figure 22.9 shows a sample display of this segment. The following tools help you adjust the data window:

Autoscale horizontal	Horizontal (Time) scroll bar
Autoscale waveforms	Vertical (Amplitude) scroll bar
Zoom tool	Overlap button
Zoom Previous	Split button

3. Set up the measurement boxes as follows:

Channel	Measurement
CH 3	min
CH 3	max
CH 3	p-p
CH 40	mean

The measurement boxes are above the marker region in the data window. Each measurement has three sections: channel number, measurement type, and result. The first two sections are pull-down menus that are activated when you click on them. The following is a brief description of these measurements, where "selected area" is the area selected by the I-beam tool (including endpoints):

min displays the minimum value in the selected area

max displays the maximum value in the selected area

p-p finds the maximum value in the selected area and subtracts the minimum value found in the selected area

mean displays the average value in the selected area

4. Using the I-beam cursor, select an area enclosing the first EMG cluster, as, for instance, the black area on the left in Figure 22.9. Record your data in section A of Laboratory Report 2, Electromyography—Standard and Integrated EMG Activity.

5. Repeat step 4 on each successive EMG cluster.

6. Scroll to the second recording segment, which is for forearm 2 (nondominant) and begins after the first marker.

7. Repeat steps 4 and 5 for the forearm 2 data.

8. Save or print the data file. You may save the data, save notes that are in the software journal, or print the data file.

9. Exit the program.

10. Complete Laboratory Report 2, Electromyography—Standard and Integrated EMG Activity. ■

LAB ACTIVITY   5

BIOPAC: Electromyography—Motor Recruitment and Fatigue

BIOPAC

In this activity, you will examine motor unit recruitment and skeletal muscle fatigue by combining electromyography with dynamometry (*dyno-*, power + *meter*, measure). **Dynamometry** is the measurement of power, and the graphic record derived from the use of a dynamometer is called a **dynagram**. In this activity, the contraction power of clenching muscles will be determined by a **hand dynamometer** equipped with an electronic transducer for recording. For more background information, you should review the physiology of muscle stimulation, contraction, and recruitment in the first three laboratory activities of this exercise.

This recruitment and fatigue laboratory activity is organized into the same four major sections described earlier for the EMG laboratory activity: Setup, Calibration, Data Recording, and Data Analysis.

Safety Alert

The Biopac Student Lab system is safe and easy to use, but be sure to follow the procedures as outlined in the laboratory activities. Under no circumstances should you deviate from the experimental procedures. Exercise extreme caution when using the electrodes and transducers with other equipment that also uses electrodes or transducers that may make electrical contact with you or your laboratory partner. Always assume that a current exists between any two electrodes or electrical contact points. ▲

QuickCheck Questions

5.1 What does a dynamometer measure?

5.2 What is motor recruitment?

5 Materials

- ☐ BIOPAC electrode lead set (SS2L)
- ☐ BIOPAC disposable vinyl electrodes (EL503), 6 electrodes per subject
- ☐ BIOPAC headphones (OUT1)
- ☐ BIOPAC SS25L hand dynamometer
- ☐ BIOPAC electrode gel (GEL1) and abrasive pad (ELPAD) or skin cleanser or alcohol prep
- ☐ Computer: Macintosh, minimum 68020; or PC Windows 98SE or better
- ☐ Biopac Student Lab software v3.0 or better
- ☐ BIOPAC acquisition unit (MP35/MP30)
- ☐ BIOPAC wall transformer (AC100A)
- ☐ BIOPAC serial cable (CBLSERA)
- ☐ BIOPAC USB adapter (USB1W)

Procedures

Section 1: Setup

1. Turn your computer ON.
2. Make sure the BIOPAC MP35/MP30 unit is OFF.
3. Plug the equipment in as shown in Figure 22.10: the dynamometer (SS25L) into CH 1, electrode lead (SS2L) into CH3, and the headphones jack (OUT1) into the back of the unit.
4. Turn the BIOPAC MP35/MP30 unit ON.
5. Attach three electrodes either to your own dominant forearm or, if you are working with a partner, to your partner's dominant forearm, as shown in Figure 22.7. This will be forearm 1, and the nondominant forearm will be forearm 2. For optimal signal quality, you should place the electrodes on the skin at least five minutes before starting the calibration section.
6. Attach the electrode lead set (SS2L) to the electrodes on forearm 1. Make sure the electrode lead colors match those shown in Figure 22.7. (Each pinch connector works like a small clothespin, but will latch onto the nipple of the electrode only from one side of the connector.)
7. Start the Biopac Student Lab Program on your computer.
8. Choose lesson "L02-EMG-2," and click OK.
9. Type in a filename, using a unique identifier such as your or your partner's nickname or student ID number.
10. Click OK to end the setup section.

Section 2: Calibration

This series of steps establishes the hardware's internal parameters and is critical for optimum performance.

1. On the computer screen, click on Calibrate.
2. Set the dynamometer down, and click OK. To get an accurate calibration, there must be no force on the dynamometer transducer.
3. Grasp the dynamometer with the hand of forearm 1 as shown in Figure 22.11, with your hand as close to the crossbar as possible without touching the crossbar. IMPORTANT: Be sure to hold the dynamometer in the same position

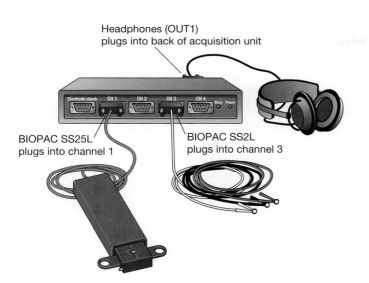

Headphones (OUT1)
plugs into back of acquisition unit

BIOPAC SS25L
plugs into channel 1

BIOPAC SS2L
plugs into channel 3

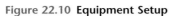

Figure 22.10 Equipment Setup

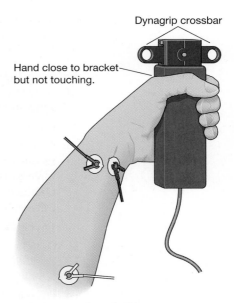

Dynagrip crossbar

Hand close to bracket
but not touching.

**Figure 22.11 Holding the Hand
Dynamometer**

for all measurements from each arm. Note your hand position here in step 3
and try to replicate it exactly in all subsequent steps in this activity.

4. Follow the instructions in the Calibrate dialog box, and click OK when ready.
The calibration will not begin until you click OK.

5. Wait two seconds. Then clench the hand dynamometer as hard as possible for
about four seconds and release.

6. Wait for the calibration to stop.

7. Your computer screen should resemble Figure 22.12. Repeat calibration steps 1
through 6 if your screen is different. If your calibration recording did not begin
with a zero baseline, this means that you clenched your fist before waiting two
seconds, and you need to repeat the calibration.

Section 3: Data Recording

You will record data for four segments: forearm 1 motor unit recruitment, forearm
1 fatigue, forearm 2 motor unit recruitment, and forearm 2 fatigue. To work efficiently, read through the rest of this activity so that you will know what to do

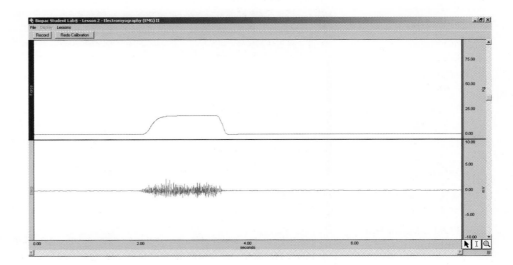

**Figure 22.12 Calibration
Recording**

before recording. Check the last line of the software journal and note the total amount of time available for the recording. Stop each recording segment as soon as possible so that you do not waste recording time (time is memory).

Segment 1 (Forearm 1 Motor Unit Recruitment).

1. Note your Assigned Increment Level from the software journal. The BIOPAC software calculates this level during your grip force calibration.

2. Click on Record.

3. Clench your fist and hold for two seconds. Release the clench and wait two seconds. Beginning with your Assigned Increment Level, increase the clench force by the assigned increment for each cycle until maximum clench force is obtained. For example, if your Assigned Increment Level is 5 kg, start at a force of 5 kg and repeat cycles at forces of 10 kg and 15 kg.

Force Calibration	Assigned Increment Level
0–25 kg	5 kg
25–50 kg	10 kg
> 50 kg	20 kg

4. On your computer, click on Suspend, and then review the recording of the screen. If it looks similar to Figure 22.13, go to step 6. If your recording looks different, go to step 5.

5. Click on Redo, and repeat steps 3 through 4.

Segment 2 (Forearm 1 Fatigue).

6. Click on Resume. A marker labeled "Continued clench at maximum force" will automatically be inserted when you do this.

7. Clench the dynamometer with your maximum force. Note this force and try to maintain it.

8. When the maximum clench force displayed on the screen has decreased by more than 50%, click on Suspend and review the data on the screen.

9. If the data are similar to Figure 22.14, go to step 11. If your screen looks different from Figure 22.14, go to step 10.

10. Click on Redo, and repeat steps 7 through 9.

11. Click on Stop. Clicking "yes" in the dialog box will end the data-recording section and will automatically save the data. Clicking "no" will bring you back to either Resume or Stop.

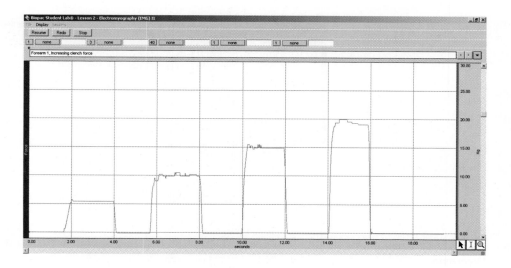

Figure 22.13 Motor Unit Recruitment

12. If you want to listen to the EMG signal, go to step 13. Listening to the EMG can be a valuable tool in detecting muscle abnormalities, and is performed here for general interest. If you want to end the recording, go to step 18.

13. Put on the headphones, and click on Listen. The volume through the headphones may be very loud because of system feedback. Because the volume cannot be adjusted, you may have to position the headphones slightly off your ears for comfort.

14. Experiment by changing the clench-release-wait cycles as you watch the screen and listen. You will hear the EMG signal through the headphones as it is being displayed on the screen. The screen will display three channels: CH 1 Force, CH 3 EMG, and CH 40 Integrated EMG. The data on the screen will not be saved. The signal will run until you press Stop. If others in your laboratory group would like to hear the EMG signal, pass the headphones around before clicking Stop.

15. Click on Stop to end the listening-to-EMG portion.

16. To listen again, or to have another person listen, click Redo.

17. Remove the electrode cable pinch connectors, peel the electrodes from your arm, and dispose of them. Use soap and water to wash the electrode gel residue from your skin. The electrodes may leave a slight ring on the skin for a few hours; this is quite normal.

Segment 3 (Forearm 2 Motor Unit Recruitment).   Attach the three remaining electrodes to your nondominant forearm. Click Forearm 2 on your screen, and repeat all the steps of Segment 1 (Forearm 1 Motor Unit Recruitment).

Segment 4 (Forearm 2 Fatigue).   Repeat all the steps of Segment 2 (Forearm 1 Fatigue).

18. Click Done. A pop-up window will appear. Make your choice, and continue as directed. If choosing the "Record from another subject" option:

   a. Attach electrodes per Setup step 5 and continue the entire lesson from Setup step 9.

   b. Each person will need to use a unique filename.

Section 4: Data Analysis

1. Enter the Review Saved Data mode, and choose the correct file. For the first part of the data analysis, choose the data file for forearm 1, saved with the filename extension "1-L02." For the second part of the data analysis, choose the data file for forearm 2, saved with the filename extension "2-L02." (BSL 3.7 generates one combined file.)

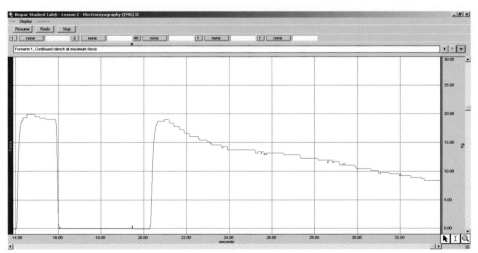

Figure 22.14 Fatigue

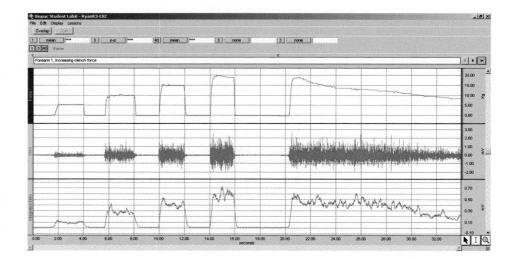

Figure 22.15 Plateaus of First Clenches

Note the Channel Number (CH) settings in the small menu boxes, shown in Figure 22.15.

Channel	Displays
CH 1	FORCE
CH 3	Raw EMG
CH 40	Integrated EMG

2. Read the software journal, and note your force increment in the Data Report.

3. Set up your display window for optimal viewing of the first data segment (forearm 1 motor unit recruitment). Figure 22.15 shows a sample display of these data. The following tools help you adjust the data window:

Autoscale horizontal   Horizontal (Time) scroll bar
Autoscale waveforms   Vertical (Amplitude) scroll bar
Zoom tool   Zoom previous

4. Set up the measurement boxes as follows:

Channel	Measurement
CH 3	min
CH 3	max
CH 3	p-p
CH 40	mean

The measurement boxes are above the marker region in the data window. Each measurement has three sections: channel number, measurement type, and result. The first two sections are pull-down menus that are activated when you click on them. The following is a brief description of these measurements, where "selected area" is the area selected by the I-beam tool (including endpoints):

min displays the minimum value in the selected area

max displays the maximum value in the selected area

p-p finds the maximum value in the selected area and subtracts the minimum value found in the selected area

mean displays the average value in the selected area

5. Using the I-beam cursor, select an area on the plateau phase of the first clench (Figure 22.15). Record your data in section A of Laboratory Report 3, Electromyography—Motor Recruitment and Fatigue.

6. Repeat step 5 on the plateau phase of each successive clench. Record your data in section A of Laboratory Report 3, Electromyography—Motor Recruitment and Fatigue.

7. Scroll to the second recording segment. This begins after the first marker and represents the continuous maximum clench.

8. Set up the measurement boxes as follows:

Channel	Measurement
CH 1	value
CH 40	Δ T

Value displays the amplitude for the channel at the point selected by the cursor. If a single point is selected, the value is for that point; if an area is selected, the value is the endpoint of the selected area. Δ **T** displays the amount of time in the selected segment (difference in time between the endpoints of the selected area).

9. Using the I-beam cursor, select a point of maximal clench force immediately following the start of segment 2 (Figure 22.15). Record your data in section A of Laboratory Report 3, Electromyography—Motor Recruitment and Fatigue.

10. Calculate 50% of the maximum clench force from step 8.

11. Find the point of 50% maximum clench force by using the I-beam cursor, and leave the cursor at this point.

12. Select the area from the point of 50% clench force back to the point of maximum clench force by using the I-beam cursor and dragging. Note the time to fatigue (CH 40 Δ T) measurement.

13. Save or print the data file. You may save the data, save notes that are in the journal, or print the data file.

14. Repeat the entire Data Analysis section, starting with step 1, for forearm 2 (Segments 3 and 4).

15. Exit the program. ■

Muscle Physiology

Name _____

Date _____

Section _____

A. Matching

Match each term in the left column with its correct description from the right column.

_____ **1.** isometric contraction
_____ **2.** complete tetanus
_____ **3.** repolarization
_____ **4.** all-or-none principle
_____ **5.** fatigue
_____ **6.** isotonic contraction
_____ **7.** depolarization
_____ **8.** treppe
_____ **9.** latent period
_____ **10.** wave summation
_____ **11.** incomplete tetanus
_____ **12.** twitch
_____ **13.** recruitment
_____ **14.** acetylcholine

A. fusion of twitches
B. tension changes more than length
C. warming-up contraction
D. shift in transmembrane potential toward 0 mV
E. contraction with no relaxation cycles
F. response to single stimulus
G. length changes more than tension
H. muscle fibers either "on" or "off"
I. contraction with rapid relaxation cycles
J. neurotransmitter
K. time prior to tension
L. reduction in contraction and performance
M. return of membrane to resting potential
N. activating more motor units

B. Short-Answer Questions

1. Describe each phase of muscle contraction:

a. expose

b. attach

c. pivot

d. detach

e. reactivate

2. Describe the events taking place at the neuromuscular junction during muscle stimulation.

3. Distinguish between isometric and isotonic contractions.

4. Explain why muscles become fatigued.

5. Describe each of the following types of muscle contraction:
   a. wave summation

   b. treppe

   c. complete tetanus

   d. twitch

   e. incomplete tetanus

Electromyography—Standard and Integrated EMG Activity

Name _____

Date _____

Section _____

A. Data and Calculations

Subject Profile

Name _____   Height _____

Age _____   Weight _____

Gender _____

1. **EMG Measurements:** Complete Table 22.4 by using the data obtained during the EMG I experiment. Refer to the computer journal for recorded data.

Table 22.4 **EMG Data**

Cluster Number	Forearm 1 (Dominant)				Forearm 2 (Nondominant)			
	Min (3 min)	Max (3 max)	p-p (3 p-p)	Mean (40 mean)	Min (3 min)	Max (3 max)	p-p (3 p-p)	Mean (40 mean)
1								
2								
3								
4								

22

2. Use the mean measurement from Table 22.4 to compute the percentage increase in EMG activity recorded between the weakest clench and the strongest clench of forearm 1.

Calculation: Answer: _____%

B. Analysis and Application

1. Does there appear to be any difference in tonus between the two forearm clench muscles?

Would you expect to see a difference? Does the subject's sex influence your expectations? Explain.

2. Compare the mean measurement for the right and left maximum clench EMG cluster. Are they the same or different? Which one suggests the greater clench strength? Explain.

3. What factors in addition to gender contribute to observed differences in clench strength?

4. Explain the source of signals detected by the EMG electrodes.

5. What does the term *motor unit recruitment* mean?

6. Define electromyography.

Electromyography—Motor Recruitment and Fatigue

Name _____

Date _____

Section _____

A. Data and Calculations

Subject Profile

Name _____   Height _____

Age _____   Weight _____

Gender _____

Dominant forearm (right or left) _____

1. Complete Table 22.5 using data from segments 1 and 3. In the "Assigned Increment Level (kg)" column, note the force increment assigned for your recording under peak 1; the increment was pasted into the software journal and should be transferred to Table 22.5 from step 2 of the Data Analysis section of Laboratory Activity 5. For subsequent peaks, add the increment (that is, 5-10-15 kg or 10-20-30 kg). You may not need nine peaks to reach max.

Table 22.5 **Data on Motor Recruitment from Segments 1 and 3**

Peak Number	Assigned Increment Level (kg)	Forearm 1 (Dominant)			Forearm 2		
		Force at Peak [CH 1] mean (kg)	Raw EMG [CH 3] p-p (mV)	Int. EMG [CH 40] mean (mV)	Force at Peak [CH 1] mean (kg)	Raw EMG [CH 3] p-p (mV)	Int. EMG [CH 40] mean (mV)
1	kg						
2	kg						
3	kg						
4	kg						
5	kg						
6	kg						
7	kg						
8	kg						
9	kg						

2. Complete Table 22.6 using Segment 2 and Segment 4 data.

Table 22.6 **Data on Fatigue from Segments 2 and 4**

Forearm 1 (Dominant)			Forearm 2		
Maximum clench force	50% of max clench force	Time to fatigue	Maximum clench force	50% of max clench force	Time to fatigue
CH 1 value	calculate	CH 40 ΔT	CH 1 value	calculate	CH 40 ΔT

B. Analysis and Application

1. Is the strength of your right arm different from the strength of your left arm?

2. Is there a difference in the absolute values of force generated by males and females in your class? What might explain any differences?

3. When you are holding an object, does the number of motor units in use remain the same? Are the same motor units used for as long as you hold the object?

4. As you fatigue, the force exerted by your muscles decreases. What physiological processes explain this decline in strength?

Organization of the Nervous System

OBJECTIVES

On completion of this exercise, you should be able to:

- Outline the organization of the nervous system.

- Describe the general functions of the central and peripheral nervous systems.

- Compare sensory receptors and effectors.

- Compare the autonomic and somatic divisions of the efferent peripheral nervous system.

- List six types of glial cells and describe a basic function of each type.

- Describe the cellular anatomy of a neuron.

- Discuss how a neuron communicates with other cells.

- Describe how learning a task influences reaction time.

- Compare reaction times for fixed-interval and pseudorandom presentations of a stimulus.

- Calculate group mean, variance, and standard deviation values for a data set.

The nervous system orchestrates body functions to maintain homeostasis. To accomplish this control, the nervous system must perform three vital tasks. First, it must detect changes in and around the body. For this task, sensory receptors monitor environmental conditions and encode information about environmental changes as electrical impulses. Second, it must process incoming sensory information and generate an appropriate motor response to adjust the activity of muscles and glands. Third, it must orchestrate and integrate all sensory and motor activities to achieve the balance of homeostasis.

LAB ACTIVITY 1   Organization of the Nervous System

The nervous system is divided into two main components (Figure 23.1): the **central nervous system (CNS)**, which consists of the **brain** and **spinal cord**, and the **peripheral nervous system (PNS)**, which communicates with the CNS by way of cranial and spinal nerves, collectively called *peripheral nerves*. To understand the meaning of the term *nerve* in anatomy, recall from Exercise 10 that cells of the nervous system are called *neurons* and that each neuron is made up of a cell body (soma), an axon, and dendrites. A **nerve**, then, is defined as a bundle of axons plus any associated blood vessels and connective tissue. The PNS is responsible for providing the

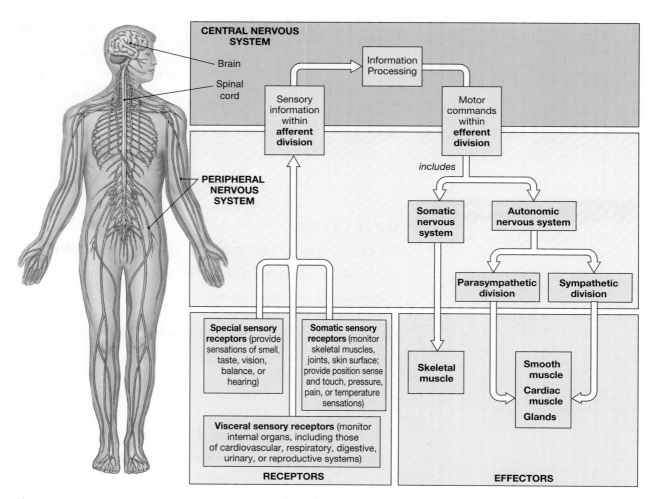

Figure 23.1 A Functional Overview of the Nervous System

CNS with information concerning changes inside the body and in the surrounding environment. Sensory information is sent along PNS nerves that join the CNS in the spinal cord or the brain. The CNS evaluates the sensory data and determines whether muscle and gland activities should be modified in response to the environmental changes. Motor commands from the CNS are then relayed to PNS nerves that carry the commands to specific muscles and glands.

The PNS is divided into afferent and efferent divisions. The **afferent division** receives sensory information from **sensory receptors**, which are the cells and organs that detect changes in the body and the surrounding environment, and then sends the sensory information to the CNS for interpretation. The CNS decides the appropriate response to the sensory information and sends motor commands to the PNS **efferent division**, which controls the activities of **effectors**; *effectors* is the general term for all the muscles and glands of the body.

The efferent division is divided into two parts. One part, the **somatic nervous system,** conducts motor responses to skeletal muscles. The other part, the **autonomic nervous system**, consists of the **sympathetic** and **parasympathetic divisions**, both of which send commands to smooth muscles, cardiac muscles, and glands.

The nerves of the PNS are divided into two groups according to the portion of the CNS with which they communicate. **Cranial nerves** are bundles of axons that communicate with the brain. Foramina of the skull provide passageways out of the skull for the cranial nerves. **Spinal nerves** join the spinal cord at intervertebral foramina and pass either into the extremities or into the body wall. There are 12 pairs of cranial nerves and 31 pairs of spinal nerves, and each pair of nerves transmits specific information between the CNS and the PNS. Functionally, all spinal nerves are "mixed" nerves, which means they carry both sensory signals and motor signals. Cranial nerves are either entirely sensory or mixed. Although a cranial or spinal nerve may transmit both sensory and motor impulses, a single neuron within the nerve transmits only one type of signal.

QuickCheck Questions

1.1 In which division of the PNS does sensory information travel?

1.2 The efferent division of the PNS controls what structures in the body?

1 Materials

☐ Figure 23.1

Procedures

Complete the following simulation of neural control by filling in the blanks, using Figure 23.1 as your guide. Imagine stepping on a tack:

1. Sensory _____ in your foot send a pain signal to the CNS via the _____ division of the _____.

2. Once information arrives in the CNS, the proper _____ response is sent to the _____ division of the PNS.

3. These voluntary signals travel along the _____ nervous system to _____, the skeletal muscles receiving the motor commands.

4. Because stepping on the tack was quite a surprise, the _____ sent impulses to the heart, smooth muscles, and glands. ■

LAB ACTIVITY 2 Histology of the Nervous System

As discussed in Exercise 10, two types of cells populate the nervous system: glial cells and neurons. Glial cells have a supportive role in protecting and maintaining nerve tissue. Neurons are the communication cells of the nervous system and are capable of propagating and transmitting electrical impulses to respond to the ever-changing needs of the body.

Glial Cells

The **glial cells**, which collectively make up a network called the **neuroglia** (noo-RŌ-glē-ah), are the most abundant cells in the nervous system. They protect, support, and anchor neurons in place. In the CNS, glial cells are involved in the production and circulation of cerebrospinal fluid. In both the CNS and the PNS, glial cells isolate and support neurons with myelin. Figure 23.2 highlights the various types of glial cells.

The CNS has four types of glial cells: astrocytes, oligodendrocytes, microglia, and ependymal cells, first described in Exercise 10. **Astrocytes** (AS-trō-sīts) have many functions. They hold neurons in place and isolate one neuron from another. They wrap footlike extensions around blood vessels, creating a blood-brain barrier that prevents certain materials from passing out of the blood and into nerve tissue (Figure 23.3).

Oligodendrocytes (o-li-gō-DEN-drō-sīts) cover axons in the CNS with a fatty myelin sheath (described later). Myelinated axons form what is called the *white matter* of the nervous system; nonmyelinated axons and cell bodies form what is called *gray matter*. **Microglia** (mī-KROG-lē-uh) are phagocytes that remove microbes and cellular debris from nerve tissue. **Ependymal** (e-PEN-dī-mul) cells line the ventricles of the brain and the central canal of the spinal cord. These cells contribute to the production of the cerebrospinal fluid that circulates in the ventricles of the brain and in the central canal of the spinal cord.

The PNS has two types of glial cells: Schwann cells, described in Exercise 10, and satellite cells. Where cell bodies cluster in groups called *ganglia*, **satellite cells** encase each neuron cell body and isolate it from the interstitial fluid to regulate the neuron's chemical environment. **Schwann cells** wrap around neurons and constitute the white matter of the PNS.

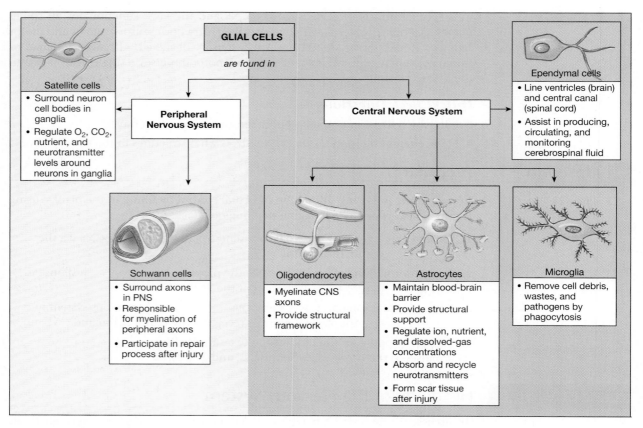

Figure 23.2 An Introduction to Glial Cells

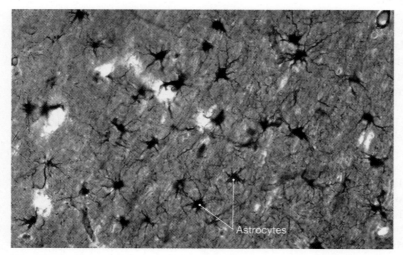

Figure 23.3
Astrocytes (400×)

Neurons

Neurons are the communicative cells of the nervous system. They generate and respond to action potentials to convey information or instructions from one cell to another. Neurons are the longest cells in the body; the ones extending from the toes into the spinal cord may be as long as three feet.

A neuron has three distinguishable features, as described in Exercise 10 and shown in Figure 23.4: dendrites, a cell body (soma), and an axon. The numerous dendrites of a neuron carry information into the large, rounded cell body, which contains the nucleus and organelles of the cell. The **perikaryon** (per-i-KAR-ē-on), which is the entire area of the cell body surrounding the nucleus, contains such organelles as mitochondria, free ribosomes, and fixed ribosomes. Also found in the perikaryon are Nissl bodies, which are groups of free ribosomes and rough endoplasmic reticulum. Nissl bodies are dark and account for the gray matter seen in gross dissection. One portion of the cell body narrows into an **axon hillock** that extends into the first part of the axon, called the **initial segment**. The single axon conducts impulses away from the cell body, either toward other neurons or toward an effector. The axon may divide into several **collateral branches** that subdivide into smaller

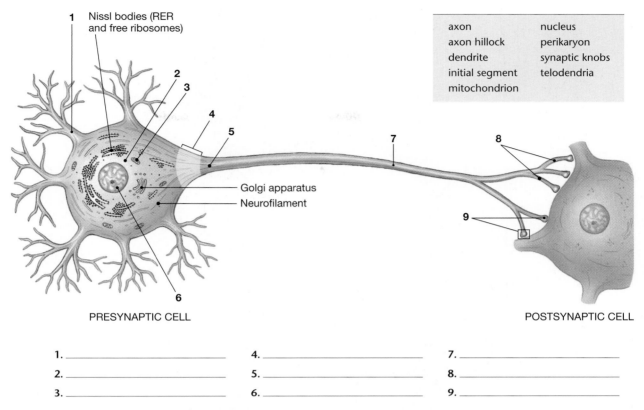

1   Nissl bodies (RER and free ribosomes)

axon	nucleus
axon hillock	perikaryon
dendrite	synaptic knobs
initial segment	telodendria
mitochondrion	

Golgi apparatus
Neurofilament

PRESYNAPTIC CELL

POSTSYNAPTIC CELL

1. _____   4. _____   7. _____

2. _____   5. _____   8. _____

3. _____   6. _____   9. _____

Figure 23.4 The Anatomy of a Multipolar Neuron
A diagrammatic view of a neuron.

branches called **telodendria** (tel-ō-DEN-drē-uh). Each telodendrium ends at a **synaptic knob** that contains chemical messengers called neurotransmitters stored in **synaptic vesicles**. When an action potential arrives at a synaptic knob, the neuron releases neurotransmitters that diffuse across the synaptic cleft and either excite or inhibit the membrane of the neighboring neuron or effector.

Neurons may synapse with other neurons; with skeletal, cardiac, or smooth muscle cells; or with gland cells. An individual neuron, however, may synapse with only one type of cell. For example, motor units in skeletal muscles are groups of muscle fibers that are all controlled by the same somatic motor neuron. The motor neuron branches, and each muscle fiber receives a synaptic knob at the neuromuscular junction. Notice in Figure 23.4 that the presynaptic neuron on the left is adjacent to a postsynaptic neuron on the right. In this cellular arrangement, the presynaptic neuron communicates through the synapse to the receiver, the postsynaptic neuron.

The axon of a neuron is usually covered with a fatty **myelin sheath,** which in the PNS is produced by Schwann cells. As a Schwann cell wraps around a small section of axon, the Schwann cell squeezes the cytoplasm out of its extensions and encases the axon in multiple layers of membrane. Notice in Figure 23.5 that between the axon's **myelinated internodes** are gaps in the sheath, called **nodes of Ranvier** (RAHN-vē-ā). The membrane of the axon, the **axolemma**, is exposed at the nodes and permits a nerve impulse to arc rapidly from node to node. The **neurilemma** (noo-ri-LEM-uh), or outer layer of the Schwann cell, covers the axolemma at the myelinated internodes.

In the CNS, myelin sheaths are formed by oligodendrocytes.

QuickCheck Questions

2.1 What are the two major types of cells in the nervous system?

2.2 What are the three main regions of a neuron?

2.3 What is a node of Ranvier?

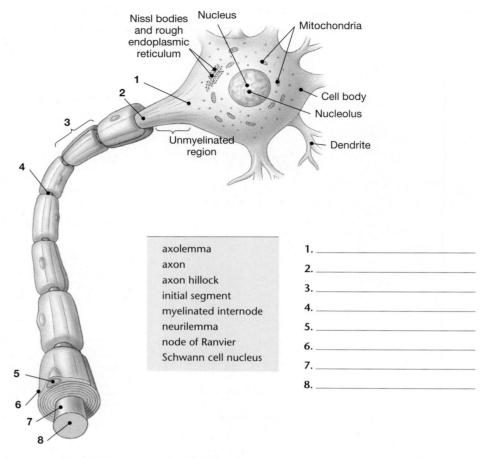

	1. _____
axolemma	2. _____
axon	3. _____
axon hillock	4. _____
initial segment	5. _____
myelinated internode	6. _____
neurilemma	7. _____
node of Ranvier	8. _____
Schwann cell nucleus	

Figure 23.5 Schwann Cells and Peripheral Axons
Myelinated neuron showing myelin sheath formed by Schwann cells.

2 *Materials*

☐ Compound microscope

☐ Prepared slides of astrocytes, large multipolar neurons, and neuromuscular junction

Procedures

1. Examine a prepared microscope slide of astrocytes, using Figure 23.3 as reference. Scan the slide at low magnification and locate a group of astrocytes.

   • Examine a single astrocyte at high magnification. Locate the nucleus and the numerous cellular extensions.

   • Draw an astrocyte in the space provided here.

2. Label the anatomy of a neuron in Figures 23.4 and 23.5.

3. Examine a prepared microscope slide of large multipolar neurons. The slide is a smear of neural tissue from the CNS and has many multipolar neurons, each with numerous dendrites and a single axon.

   • Scan the slide at low magnification and locate several neurons.

   • Select a single neuron and increase the magnification; identify the cell body and nucleus using Figure 23.6 as a reference.

   • Can you distinguish between the dendrites and the axon?

4. Use Figure 23.7 as a guide and examine a prepared microscope slide of the neuromuscular junction.

   • Scan the slide at low magnification and locate a neuron and several telodendria.

   • Increase the magnification and examine the neuromuscular junction. ■

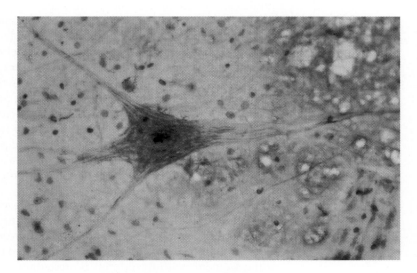

Figure 23.6 **Multipolar Motor Neuron (400×)**

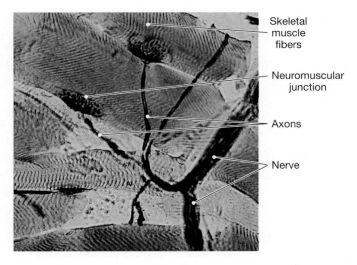

Figure 23.7 **Neuromuscular Junction**

LAB ACTIVITY 3   BIOPAC: Reaction Time and Learning

BIOPAC

The beginning of a race is a classic example of a **stimulus-response** situation, where people hear a *stimulus* (the starter's pistol) and react to it in some way (*response*). There are two key factors in stimulus-response: **reaction time** and **learning**. *Reaction time* is the delay between when the stimulus is presented and when you do something about it. *Learning* is the acquisition of knowledge or skills as a result of experience and/or instruction.

The delay between hearing the signal and responding is a function of the length of time needed for the afferent signal to reach the brain and for the brain to send an efferent signal to the muscles. With learning, the time for the various steps in the process can be shortened. As people learn what to expect, reaction time typically decreases. Reaction time varies from person to person and from situation to situation, and most people have delayed reaction times late at night and early in the morning. Longer reaction times are also a sign that people are paying less attention to the stimulus and/or are processing information.

This lesson shows how easily and quickly people learn, as demonstrated by their ability to anticipate when to press a button. The lesson uses a relatively simple variation of stimuli (pseudorandom versus fixed-interval) to determine what results in the shortest reaction times. In the **pseudorandom stimuli** segments, the computer generates a click once every 1–10 seconds. For the **fixed-interval stimuli** segments, the computer generates a click every four seconds.

With the pseudorandom presentation, the subject cannot predict when the next click will occur, and the result is that both learning and decrease in reaction time are minimal. When fixed-interval trials are performed repeatedly, average reaction time typically decreases each time new data are recorded, up to a point. Eventually, the minimal reaction time required to process information is reached, and then reaction time becomes constant.

This lesson takes a relatively simple look at reaction time and demonstrates how changing one small aspect of procedure can result in differences in reaction times. You will probably notice a difference in average reaction times between the pseudorandom and fixed-interval presentation trials, and this difference will most likely favor the segments with fixed-interval presentation stimuli. Part of the difference is probably due to the random versus nonrandom presentation of the stimuli. However, you might also notice that reaction time decreases when you change from pseudorandom presentation to fixed-interval presentation, suggesting that maybe you just got better with practice.

3 *Materials*

- ☐ BIOPAC hand switch (SS10L)
- ☐ BIOPAC headphones (OUT1)
- ☐ BIOPAC acquisition unit (MP35/MP30)
- ☐ BIOPAC wall transformer (AC100A)
- ☐ BIOPAC serial cable (CBLSERA)
- ☐ Biopac Student Lab software v.3.0 or later
- ☐ Computer: Macintosh® minimum 68020 or PC Windows® 98SE or better

Procedures

The reaction time investigation is divided into four sections: set up, calibration, recording, and data analysis. Read each section completely before attempting a recording. If you encounter a problem or need further explanation of a concept, ask your instructor.

Data collected in the recording segments must be recorded in the laboratory. You may record the data by hand or choose **Edit > Journal > Paste measurements** to paste the data into your electronic journal for future reference.

While you record a segment, markers are inserted for each response with the hand switch. In this exercise, all of the markers and labels are inserted automatically. Markers appear at the top of the computer windows as inverted triangles.

A. Setup

1. Turn your computer **ON**.

2. Make sure the acquisition unit is **OFF**.

3. **Plug the equipment in** as shown in Figure 23.8: hand switch (SS10L) in CH 1 outlet of the acquisition unit and headphones (OUT1) in back of the unit.

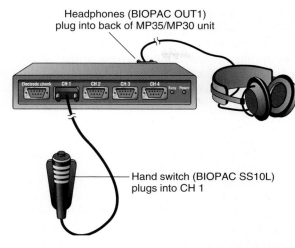

Headphones (BIOPAC OUT1)
plug into back of MP35/MP30 unit

Hand switch (BIOPAC SS10L)
plugs into CH 1

Figure 23.8
Equipment Setup

4. Turn the acquisition unit **ON**.

5. Start the Student Lab program.

6. Choose lesson **"L11-React-1,"** and click **OK**.

7. Type in a unique **filename.** No two people should have the same filename; use a unique identifier, such as your nickname or student ID number. Typing in a filename ends the setup procedure.

8. Click **OK**.

B. Calibration

1. As you prepare for the calibration recording, you should be seated and relaxed, with headphones on. Hold the hand switch with your dominant hand, so that the thumb is ready to press the button.

Note: When the Calibrate button is hit in the next step, system feedback may cause the volume through the headphones to be very loud. Because the volume cannot be adjusted, you may have to position the headphones slightly off the ears to reduce the sound.

2. Click on **Calibrate**. Before the calibration begins, a pop-up window will appear, reminding you to press the button when you hear a click. Click **OK** to begin the calibration recording.

3. Press the SS10L hand switch when you hear a click, approximately four seconds into the recording. Briefly depress the button, then release it. Do not hold the button down and do not press it more than once.

4. Wait for the calibration to end. The calibration will run for eight seconds and then stop automatically.

5. Review the data on the screen. Your screen should be similar to Figure 23.9. If your screen does look like Figure 23.9, proceed to the Data Recording Section. If your calibration screen does not resemble Figure 23.9, repeat the calibration to obtain a similar screen. Click **Redo calibration** and repeat the calibration procedure.

Two reasons for incorrect data are:

a. The baseline is not 0 millivolts.

b. The data are excessively noisy, meaning more than approximately 1mV peak-to-peak. Your data may be a little more or less noisy than the example shown in Figure 23.9.

Figure 23.9
Sample Calibration Data

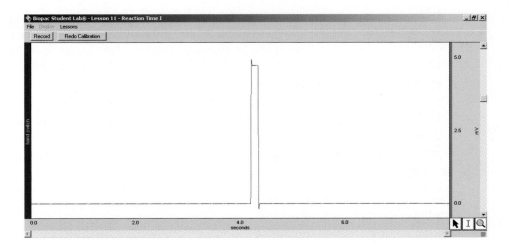

If the **Calibrate** button reappears in the window, check the connections and repeat the calibration, making sure to press the button firmly but briefly. If no signal is detected from the hand switch (flat line at 0 millivolts), the program will automatically return you to the beginning of the calibration procedure. If this happens, check the connections to the hand switch and make sure you are pressing the button firmly. Click **Redo calibration** and repeat the calibration.

C. Recording

1. Prepare for the recording. You will record four segments, each requiring you to press a button (response) as soon as possible after hearing a click (stimulus):

   a. Segments 1 and 2 present the stimuli at pseudorandom intervals every 1–10 seconds.

   b. Segments 3 and 4 present the stimuli at fixed intervals four seconds apart.

   To work efficiently, read this entire section before beginning the recording step. Check the last line of your journal and note the total amount of time available for the recording. Stop each recording segment as soon as possible so that you do not waste space on the hard drive.

2. From this point on, two persons are involved: a **subject** and a **director**. The subject should be seated and relaxed, with headphones on and eyes closed. She or he should hold the hand switch with the dominant hand, so that the thumb is ready to press the button. The director watches the screen and presses the **Record** and **Resume** buttons as required.

Note: The Student Lab software looks for only one response per stimulus. Because the software ignores responses that occur before the first click, it does not help to press the button on the SS10L numerous times before you hear the first click. If you press the button before the stimulus, or if you wait more than one second after the stimulus before pressing the button, your response will not be used in the reaction time summary.

Note: All markers are automatically inserted while recording. Do not manually insert a marker in any recording segment of this lesson.

Segment 1: Pseudorandom Trial 1

3. Once the director clicks on **Record,** a pseudorandom presentation trial will begin, with a click produced randomly every 1–10 seconds. The recording will suspend automatically after 10 clicks.

4. As soon as the subject hears a click through the headphones, he or she should press and release the SS10L button. A **marker** will automatically be inserted each time a click is generated, and an upward-pointing "pulse" will be displayed on the screen each time the button is pressed.

5. **Review** the data on the screen. After 10 clicks, the resulting graph should resemble Figure 23.10. If the subject pressed the button correctly, a pulse will be displayed after each marker. If the screen looks like Figure 23.10, go to step 6. If it does not look like Figure 23.10, click **Redo.** Three reasons for incorrect data are:

   a. The recording did not capture a pulse for each click.

   Note: The subject is allowed to miss some responses, but if more than two are missed, the recording should be redone.

   b. The pulse occurs before the marker, indicating that the subject responded prematurely.

   c. The duration of the pulse extends into the next marker, indicating that the subject held the button down too long.

Figure 23.10
Representation of Clicks

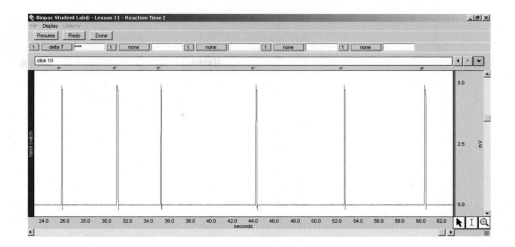

Segment 2: Pseudorandom Trial 2

6. To record a second pseudorandom presentation trial, click on **Resume**. The recording will continue from the point it last stopped, and a marker labeled "repeat pseudorandom" will be inserted when Resume is pressed.

7. As soon as the subject hears each click through the headphones, he or she should press and release the button on the SS10L. The recording will suspend automatically after 10 clicks.

8. **Review** the data on the screen. After 10 clicks, the resulting graph on your screen should resemble Figure 23.10. If it does, go to step 9. If the screen does not look like Figure 23.10, click **Redo** and repeat steps 6–8. (Data could be incorrect for the reasons listed in step 5.)

Segment 3: Fixed-Interval Trial 1

9. In this segment and in segment 4, the subject will respond to a stimulus sounded every four seconds. Click on **Resume**, and the recording will continue from the point it last stopped. A marker labeled "fixed-interval" will automatically be inserted when recording.

10. The subject should press and release the button on the SS10L at the sound of each click. The recording will suspend automatically after 10 clicks.

11. **Review** the data on the screen. After 10 clicks, the resulting graph on the screen should resemble Figure 23.10. If it does, go to step 12. If it does not, the director should click **Redo** and steps 9–11 should be repeated.

Segment 4: Fixed-Interval Trial 2

12. The director clicks on **Resume** to begin a second fixed-interval segment. The recording will continue from the point it last stopped, and a marker labeled "repeat fixed-interval" will automatically be inserted.

13. The subject should press and release the SS10L button as each click is heard. The recording will suspend automatically after 10 clicks.

14. **Review** the data on the screen. After 10 clicks, the resulting graph should resemble Figure 23.10. If it does, go to step 15.

15. If data are incorrect, the director should click **Redo** and steps 12–14 should be repeated.

16. The director clicks **Done.** A pop-up window with options will appear. Make a choice and continue as directed. If choosing the "Record from another Subject" option, remember that each subject will need to use a unique filename.

17. Unplug the hand switch and headphones.

D. Data Analysis

To compare the reaction times from the two types of presentation schedules, you can summarize the results as statistics, or measures of a population. Certain statistics are usually reported for the results of a study: mean, range, variance, and standard deviation. Mean is a measure of a central tendency. Range, variance, and standard deviation are measures of distribution—in other words, the "spread" of data. Using mean and distribution, investigators can compare the performance of groups. You will calculate your group statistics, but you will not do any formal comparisons between groups.

> The **mean** is the average of the sum of the reaction times divided by the number of subjects (*n*).
>
> The **range** is the highest score minus the lowest score. Because range is affected by extremely high and low reaction times, investigators also describe the *spread*, or distribution, of reaction times with two related statistics: variance and standard deviation.
>
> **Variance** is the average squared deviation of each number from its mean.
>
> **Standard deviation** is the square root of the variance.

1. Enter the **Review Saved Data** mode from the **Lessons** menu. After the **Done** button was pressed in the previous section, the program automatically took all 10 reaction times and calculated average reaction times for each trial and placed them in the journal (Figure 23.11). Use this journal information to fill in your data report.

2. Set up your display window for optimal viewing of the first marker and pulse of the first segment. Figure 23.12 shows the "selected area" of the first

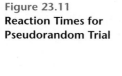

Figure 23.11
Reaction Times for Pseudorandom Trial

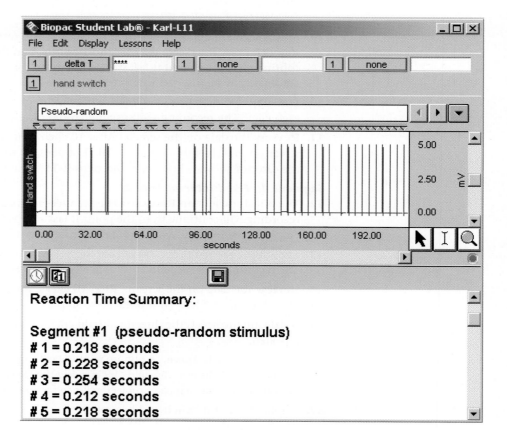

stimulus-response marker. Use the horizontal scroll bar to adjust the width of the waveforms and the vertical scroll bar to adjust their height. You may also find the Zoom tool useful for examining a segment of a graph with the X and Y axes expanded.

3. The **measurement boxes** are above the marker region in the data window. Set the boxes up as follows:

Channel	Measurement
CH1	ΔT
CH1	none
CH1	none
CH1	none

The ΔT (read as "delta time") measurement is the difference between the time at the end of the area selected by the I-beam tool and the time at the beginning of that area, including endpoints. The "none" measurement turns the measurement channel off.

You can record this and all other measurement data by hand or choose **Edit > Journal > Paste measurements** to paste the data to your journal for future reference.

4. Select an area from the first marker to the leading edge of the first pulse (Figure 23.12), and note the ΔT measurement. The marker indicates the start of the stimulus click. The leading edge of the pulse (the point where the pulse first reaches its peak) indicates when the button was first pressed. The threshold the program uses to calculate reaction time is 1.5 millivolts. The reaction time for the event shown is 0.265 second (see ΔT).

5. Look at the first reaction time result in your journal, and compare this with the ΔT measurement you found above. The two measurements should be approximately the same.

6. Repeat the preceding steps on other pulses until you are convinced that your journal readings are accurate. You can move around using the marker tools.

7. Transfer your data from the journal to Table 23.1 in section A of the Reaction Time Laboratory Report at the end of chapter 25. This step may not be

Figure 23.12
Highlighting a Stimulus-Response Pass

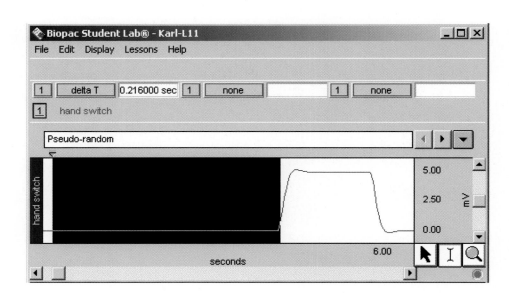

necessary if your instructor allows you to print out your journal and staple it to the report.

8. Collect data from at least four other students in your class as needed to complete Tables 23.2, 23.3, 23.4, and 23.5 of the Reaction Time Laboratory Report.

9. Either save the data or journal notes to a floppy drive, or print the data file.

10. Exit the program. ■

Organization of the Nervous System

23

Name _____

Date _____

Section _____

A. Matching

Match each structure in the left column with its correct description from the right column.

_____	**1.** dendrite	**A.** contain free and fixed ribosomes
_____	**2.** axon	**B.** main branches of axon
_____	**3.** perikaryon	**C.** outer Schwann cell membrane
_____	**4.** collateral branches	**D.** contains neurotransmitters
_____	**5.** synaptic knob	**E.** fine branches of axon
_____	**6.** axon hillock	**F.** axon region associated with Schwann cell
_____	**7.** Nissl bodies	**G.** forms blood-brain barrier
_____	**8.** telodendria	**H.** directs impulses to cell body
_____	**9.** astrocyte	**I.** region surrounding nucleus
_____	**10.** myelinated internode	**J.** also called soma
_____	**11.** neurilemma	**K.** swollen end of axon
_____	**12.** synaptic vesicles	**L.** connects cell body and axon
_____	**13.** axolemma	**M.** membrane of axon
_____	**14.** cell body	**N.** conducts impulses away from cell body

B. Drawing

Sketch a multipolar neuron, and label the dendrites, cell body, axon hillock, initial segment, axon, collateral branches, telodendria, and synaptic knob.

23

LABORATORY REPORT

C. Short-Answer Questions

1. Compare the CNS and the PNS.

2. Describe the microscopic appearance of an astrocyte.

3. Molecules of what substances are stored in synaptic knobs?

4. Name two functions of Schwann cells.

5. List the similarities of and differences between spinal nerves and cranial nerves.

6. List six types of glial cells, and indicate which are found in the CNS and which are found in the PNS.

D. Analysis and Application

1. How would an injury to the afferent neurons in the left leg affect the victim's sensory and motor functions?

2. While observing a microscopic specimen of nerve tissue from the brain, you notice an axon encased by a different cell. Describe the covering over the axon and identify the cell that has surrounded the axon.

Reaction Time

Name _____

Date _____

Section _____

A. Data and Calculations

Subject Profile

Name _____   Height _____

Age _____   Weight _____

Gender: Male / Female

1. Manual Calculation of Reaction Time
   Calculate the reaction time for the first click in Segment 1:
   ΔT = time at end of selected area—time at beginning of selected area = _____
   seconds

2. Summary of Subject's Results (from software journal)

Table 23.1 **Reaction Data**

Stimulus Number	Pseudorandom		Fixed-Interval	
	Segment 1 (1st trial)	Segment 2 (2nd trial)	Segment 3 (1st trial)	Segment 4 (2nd trial)
1				
2				
3				
4				
5				
6				
7				
8				
9				
10				
Mean				

LABORATORY REPORT

23

3. Comparison of Reaction Time and Number of Presentations
Complete *Table 23.2* with data from the first fixed-interval trial (data Segment 3), and calculate the mean for each presentation to determine if reaction times vary as a subject progresses through the series of stimulus events.

Table 23.2 Comparison of Reaction Times

Student's Name	Pseudorandom Trial 1 Data (Segment 1)			Fixed-Interval Trial 1 Data (Segment 3)		
	Stimulus 1	Stimulus 5	Stimulus 10	Stimulus 1	Stimulus 5	Stimulus 10
1.						
2.						
3.						
4.						
5.						
Mean						

4. Group Summary
Complete Table 23.3 with the means for five students, and then calculate the group mean.

Table 23.3 Reaction Time Means

Class Data Student Means	Pseudorandom Trials		Fixed-Interval Trials	
	First	Second	First	Second
1.				
2.				
3.				
4.				
5.				
Group mean				

5. Variance and Standard Deviation

$$\text{Variance} = \frac{1}{n-1}\sum_{j=1}^{n}(x_j - x)^2$$

$$\text{Standard deviation} = \sqrt{\text{variance}}$$

where

$$n = \text{number of students}$$
$$x_j = \text{mean reaction time for each student}$$
$$\bar{x} = \text{group mean (constant for all students)}$$
$$\sum_{j=1}^{n} = \text{sum of all student data}$$

Calculate the variance and standard deviation for five students with data from *Segment 2: Pseudorandom Trial 2* (Table 23.4) and from *Segment 4: Fixed-Interval Trial 2* (Table 23.5).

Table 23.4 BIOPAC Segment 2: Pseudorandom Trial 2 Data

Student	Enter Mean Reaction Time for Student (x_j)	Enter Group Mean $(\bar{x})$	Calculate Deviation $(\bar{x}_j - \bar{x})$	Calculate Deviation^2 $(x_j - \bar{x})$
1				
2				
3				
4				
5				
Sum the data for all students in the Deviation^2 column $= \sum_{j=1}^{n}(x_j - \bar{x})^2$				$=$
Variance $(\sigma^2) =$		Multiply by 0.25 $= \dfrac{1}{n-1}$		$=$
Standard Deviation $=$		square root of variance $= \sqrt{\text{variance}}$		$=$

Table 23.5 BIOPAC Segment 4: Fixed-Interval Trial 2 Data

Student	Enter Mean Reaction Time for Student (x_j)	Enter Group Mean $(\bar{x})$	Calculate Deviation $(\bar{x}_j - \bar{x})$	Calculate Deviation^2 $(x_j - \bar{x})$
1				
2				
3				
4				
5				
Sum the data for all students in the Deviation^2 column $= \sum_{j=1}^{n}(x_j - \bar{x})^2$				$=$
Variance $(\sigma^2) =$		Multiply by 0.25 $= \dfrac{1}{n-1}$		$=$
Standard Deviation $=$		square root of variance $= \sqrt{\text{variance}}$		$=$

23

B. Short-Answer Questions

1. Describe the changes in mean reaction time between the 1st and 10th stimuli presentation:

Segment 1: _____

_____

Segment 2: _____

_____

Which segment showed the greater change in mean reaction time, Segment 1 or Segment 2? _____

2. From Tables 23.2 and 23.3, estimate the minimum reaction time at which reaction time becomes constant: _____ seconds. What physiological processes occur between the time a stimulus is presented and the time the button is pressed? _____

3. From Table 23.2, which presentation schedule had the lower group mean, the pseudorandom schedule or the fixed-interval schedule? _____

4. From Tables 23.4 and 23.5, which presentation schedules seem to have less variation (lower variance and lower standard deviation), the pseudorandom schedules or the fixed-interval schedules?

_____

_____ Pseudorandom   _____ Fixed-interval

5. Based on what you see in Tables 23.4 and 23.5, state a plausible relationship between the difficulty of a task and the reaction-time statistics for the task: mean, variance, and standard deviation.

_____

_____

6. What differences would you expect between the effect of learning a task on your reaction time when you perform the task with your dominant hand and the effect of learning on reaction time when you perform the task with your nondominant hand?

_____

_____

_____

The Spinal Cord, Spinal Nerves, and Reflexes

OBJECTIVES

On completion of this exercise, you should be able to:

- Identify the major surface features of the spinal cord, including the spinal meninges.

- Identify the sectional anatomy of the spinal cord.

- Describe the organization and distribution of spinal nerves.

- List the events of a typical reflex arc.

The spinal cord connects the peripheral nervous system with the brain. In the PNS, peripheral nerves converge into spinal nerves and enter the spinal cord. Sensory neurons in these spinal nerves ascend the spinal cord and pass sensory information to the brain for interpretation. Motor neurons in the spinal nerves descend the spinal cord and carry motor commands from the brain to effectors (muscles and glands).

LAB ACTIVITY 1   Gross Anatomy of the Spinal Cord

Continuous with the medulla of the brain, the spinal cord descends approximately 45 cm (18 in.) into the spinal canal of the vertebral column and terminates between lumbar vertebrae L_1 and L_2. In young children, the spinal cord extends through most of the spine. After the age of four, the spinal cord stops lengthening, but the vertebral column continues to grow. By adulthood, therefore, the spinal cord is shorter than the spine and descends only to the level of the upper lumbar vertebrae.

As Figure 24.1 shows, the diameter of the spinal cord is not constant along its length. Two swollen regions occur where spinal nerves of the limbs join the spinal cord. The **cervical enlargement** in the neck supplies nerves to the shoulders and arms. The **lumbar enlargement** occurs near the distal end of the cord where nerves supply the pelvis and lower limbs. Inferior to the lumbar enlargement, the spinal cord narrows and terminates at the **conus medullaris**. Spinal nerves fan out from the conus medullaris in a group called the **cauda equina** (KAW-duh ek-WĪ-nuh), the "horse's tail." A thin thread of tissue, the **filum terminale**, extends past the conus medullaris to anchor the spinal cord in the sacrum.

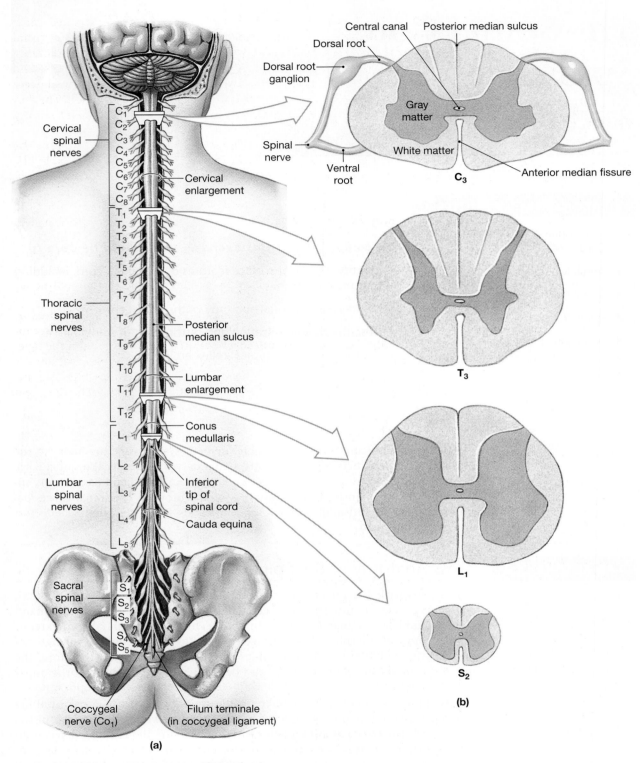

Figure 24.1 Gross Anatomy of the Adult Spinal Cord

(a) The superficial anatomy and orientation of the adult spinal cord. The numbers to the left identify the spinal nerves and indicate where the nerve roots leave the vertebral canal. The spinal cord, however, extends from the brain only to the level of vertebrae L_1–L_2; the spinal segments found at representative locations are indicated in the cross-sections. **(b)** Inferior views of cross-sections through representative segments of the spinal cord, showing the arrangement of gray matter and white matter.

The spinal cord is organized into 31 segments. Each segment has a paired spinal nerve formed by the joining of two lateral extensions, a dorsal root and a ventral root. The **dorsal root** contains sensory neurons ascending into the spinal cord from sensory receptors. The **ventral root** consists of motor neurons exiting the CNS and leading to effectors. The two roots join to form the **spinal nerve**. Each spinal nerve is therefore a *mixed nerve* and carries both sensory and motor information. The dorsal root swells at the **dorsal root ganglion**, which is where cell bodies of sensory neurons cluster.

Figure 24.2 illustrates the anatomy of the spinal cord in transverse section. The cord is divided by the deep and conspicuous **anterior median fissure** and by the shallow **posterior median sulcus**. An H-shaped area called the **gray horns** contains many glial cells and neuron cell bodies. Each horn has a specific type of neuron. The **posterior gray horns** carry sensory neurons into the spinal cord. The **lateral gray horns** contain visceral motor cell bodies, and the **anterior gray horns** carry somatic motor neurons out of the cord. Axons may cross to the opposite side of the spinal cord at the crossbars of the horns, called the **anterior** and **posterior gray commissures**. Between the commissures is a hole called the **central canal** that contains cerebrospinal fluid. The central canal is continuous with the fluid-filled ventricles of the brain.

Surrounding the gray horns are six masses of white matter: the **posterior**, **lateral**, and **anterior white columns**. The two anterior white columns are connected by the **anterior white commissure**. Within each white column the myelinated axons form distinct bundles called either **tracts** or **fascicles** that are the equivalent of nerves in the CNS. For example, in the two posterior white columns are ascending tracts of sensory neurons from receptors for touch, vibration, and pressure.

QuickCheck Questions

1.1 How is the white and gray matter of the spinal cord organized?

1.2 Which structure is useful in determining which portion of a spinal cord cross-section is the anterior region?

1.3 Why is the spinal cord shorter than the vertebral column?

1 Materials

- ☐ Spinal cord model
- ☐ Spinal cord chart
- ☐ Dissection microscope
- ☐ Compound microscope
- ☐ Prepared slide of transverse (cross) section of spinal cord

Procedures

1. Review Figures 24.1 and 24.2 and then label Figure 24.2.
2. Locate each surface feature of the spinal cord on available laboratory models and charts.
3. Locate the internal anatomy of the spinal cord on a sectional spinal cord model.
4. Examine the microscopic features of the spinal cord in transverse section:
   - View the slide at low magnification with a dissection microscope. Which features of the spinal cord enable you to distinguish anterior and posterior regions?
   - Transfer the slide to a compound microscope. Move the slide around to survey the preparation at low magnification. What structures can you use to distinguish between the posterior and anterior aspects of the spinal cord?
   - Examine the central canal and gray horns. Can you distinguish among the posterior, ventral, and anterior horns? Locate the gray commissures.
   - Examine the white columns. What is the difference between gray and white matter in the CNS?

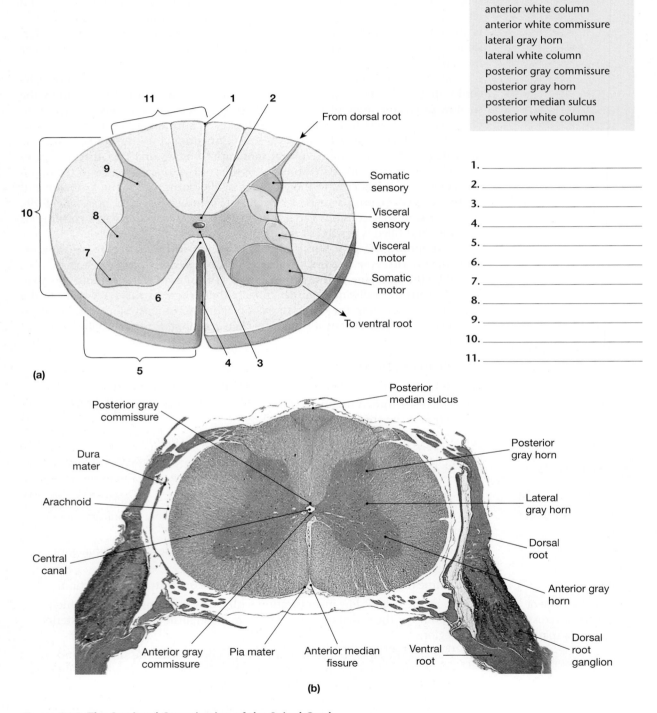

anterior gray commissure
anterior gray horn
anterior median fissure
anterior white column
anterior white commissure
lateral gray horn
lateral white column
posterior gray commissure
posterior gray horn
posterior median sulcus
posterior white column

11

1

2

From dorsal root

Somatic
sensory

Visceral
sensory

Visceral
motor

Somatic
motor

To ventral root

9

10

8

7

6

4

3

5

(a)

1. _____
2. _____
3. _____
4. _____
5. _____
6. _____
7. _____
8. _____
9. _____
10. _____
11. _____

Posterior
median sulcus

Posterior gray
commissure

Dura
mater

Arachnoid

Central
canal

Anterior gray
commissure

Pia mater

Anterior median
fissure

Ventral
root

Posterior
gray horn

Lateral
gray horn

Dorsal
root

Anterior gray
horn

Dorsal
root
ganglion

(b)

Figure 24.2 The Sectional Organization of the Spinal Cord
(a) The left half of this sectional view shows important anatomical landmarks and the major regions of white matter in the posterior white column. The right half indicates the functional organization of the gray matter in the anterior, lateral, and posterior gray horns. **(b)** A micrograph of a section through the spinal cord, showing major landmarks; compare with **(a)**.

- Draw your own version of a spinal cord cross-section in the space provided here. ∎

LAB ACTIVITY 2 ## Spinal Meninges

The spinal cord is protected within three layers of **spinal meninges** (men-IN-jēz). The outer layer, the **dura mater** (DOO-ruh MĀ-ter), is composed of tough, fibrous connective tissue (Figure 24.3). The fibrous tissue attaches to the bony walls of the spinal canal and supports the spinal cord laterally. Superficial to the dura mater is the **epidural space,** which contains adipose tissue to pad the spinal cord. The **arachnoid** (a-RAK-noyd) is the second meningeal layer. A small cavity called the **subdural space** separates the dura mater and the arachnoid. Under the arachnoid is the **subarachnoid space,** which contains cerebrospinal fluid to protect and cushion the spinal cord. The **pia mater** is the inner meningeal layer and lies directly over the nerve tissue of the spinal cord. Blood vessels supplying the spinal cord are held in place by the thin pia mater. The pia mater extends laterally on each side of the spinal cord as the **denticulate ligament**. The ligament joins the dura mater to provide lateral support to the spinal cord. Another extension of the pia mater, the filum terminale (Figure 24.1a), supports the spinal cord inferiorly.

Clinical Application | ### *Epidural Injections and Spinal Taps*

During childbirth, the expectant mother may receive an **epidural block,** a procedure that introduces anesthesia in the epidural space. A thin needle is inserted between two lumbar vertebrae, and the drug is injected into the epidural space. The anesthetic numbs only the spinal nerves of the pelvis and lower limbs and reduces the discomfort the woman feels during the powerful labor contractions of her uterus. ▶

A **spinal tap** is a procedure in which a needle is inserted into the subarachnoid space to withdraw a sample of cerebrospinal fluid. The fluid is then analyzed for the

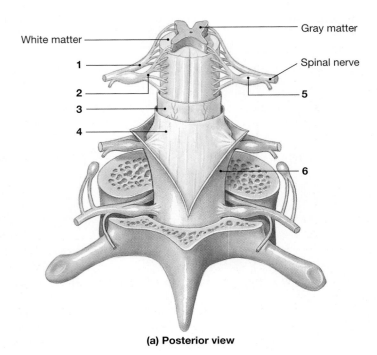

White matter

Gray matter

1

2

3

4

Spinal nerve

5

6

arachnoid
dorsal root
dorsal root ganglion
dura mater
pia mater
ventral root

1. _____

2. _____

3. _____

4. _____

5. _____

6. _____

(a) Posterior view

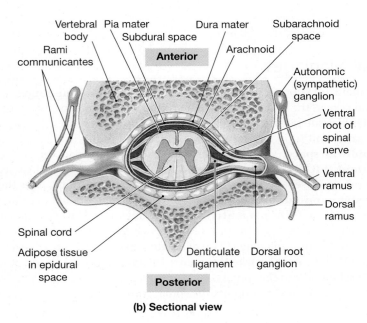

Vertebral body
Rami communicantes

Pia mater
Subdural space

Anterior

Dura mater

Arachnoid

Subarachnoid space

Autonomic (sympathetic) ganglion

Ventral root of spinal nerve

Ventral ramus

Dorsal ramus

Spinal cord

Adipose tissue in epidural space

Posterior

Denticulate ligament

Dorsal root ganglion

(b) Sectional view

Figure 24.3A and B The Spinal Cord and Spinal Meninges
(a) Posterior view of the spinal cord, showing the meningeal layers, superficial landmarks, and the distribution of gray and white matter. **(b)** Sectional view through the spinal cord and meninges, showing the peripheral distribution of the spinal nerves.

presence of microbes, wastes, and metabolites. To prevent injury to the spinal cord, the needle is inserted into the lower lumbar region inferior to the cord.

QuickCheck Questions

2.1 List the three layers of spinal meninges.

2.2 Where does cerebrospinal fluid circulate in the spinal cord?

2 *Materials*

- ☐ Spinal cord model
- ☐ Spinal cord chart
- ☐ Compound microscope
- ☐ Prepared slide of transverse (cross) section of spinal cord

Procedures

1. Review and label Figure 24.3.
2. Locate the spinal meninges on the available laboratory models and charts.
3. Examine the spinal meninges in transverse section.
   - Move the slide around to survey the preparation. Examine the outer layers surrounding the cross-sectioned spinal cord. Which of the meningeal layers can you locate?
   - Add the spinal meninges to the drawing you began in Laboratory Activity 1. ■

LAB ACTIVITY 3   Spinal Nerves

As noted in Exercise 23, two types of nerves connect PNS sensory receptors and effectors to the CNS: 12 pairs of **cranial nerves** and 31 pairs of **spinal nerves**. As their names indicate, cranial nerves connect with the brain and spinal nerves communicate with the spinal cord. As noted at the opening of this exercise, spinal nerves branch into PNS nerves, and it is spinal nerves that make up the axons of PNS sensory and motor neurons.

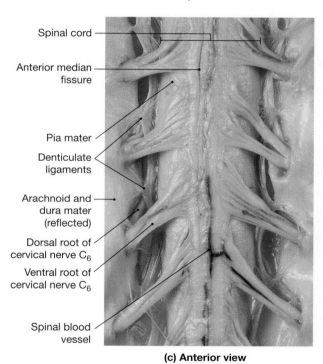

Spinal cord

Anterior median fissure

Pia mater

Denticulate ligaments

Arachnoid and dura mater (reflected)

Dorsal root of cervical nerve C_6

Ventral root of cervical nerve C_6

Spinal blood vessel

(c) Anterior view

Figure 24.3C *(Continued)* (c) Anterior view of the spinal cord and spinal nerve roots within the vertebral canal. The dura mater and arachnoid membrane have been removed; note the blood vessels in the delicate pia mater.

Figure 24.3b illustrates the distribution of neurons in a spinal nerve. The union of the dorsal and ventral roots forms a spinal nerve. The spinal nerve exits the vertebral canal by passing through an intervertebral foramen between two adjacent vertebrae. The spinal nerve then branches into a peripheral nerve with a **dorsal ramus,** which supplies the skin and muscles of the back, and a **ventral ramus**, which innervates the anterior and lateral skin and muscles. The ventral ramus has additional branches, called the **rami communicantes**, that carry autonomic neurons to the viscera. The rami communicantes lead to an **autonomic ganglion** where autonomic neurons synapse. A **white ramus** passes a **preganglionic neuron** into the autonomic ganglion, where the white ramus then synapses with a **ganglionic neuron**. A **gray ramus** carries the ganglionic neuron to whichever visceral organ the neuron innervates. The pathways of the autonomic nervous system are discussed more fully in Exercise 26.

Spinal nerves are numbered according to the vertebra located inferior to each nerve (Figure 24.4). There are 8 **cervical nerves** (C_1 through C_8), 12 **thoracic nerves** (T_1 through T_{12}), 5 **lumbar nerves** (L_1 through L_5), 5 **sacral nerves** (S_1 through S_5), and a single **coccygeal nerve**. Although there are only seven cervical vertebrae, there are eight cervical spinal nerves, the eighth being located between vertebrae C_7 and T_1. Spinal nerves are mixed nerves and contain sensory, visceral motor, and somatic motor neurons.

Groups of spinal nerves join in a network called a **plexus**. As muscles fuse during fetal development, the spinal nerves that supplied the individual muscles interconnect and create a plexus. There are four of these regions, as shown in Figure 24.4: cervical, brachial, lumbar, and sacral.

Cervical Plexus

The eight cervical nerves supply the neck, shoulder, arm, and diaphragm. The various branches of the **cervical plexus** contain nerves C_1 through C_4 and parts of C_5. This

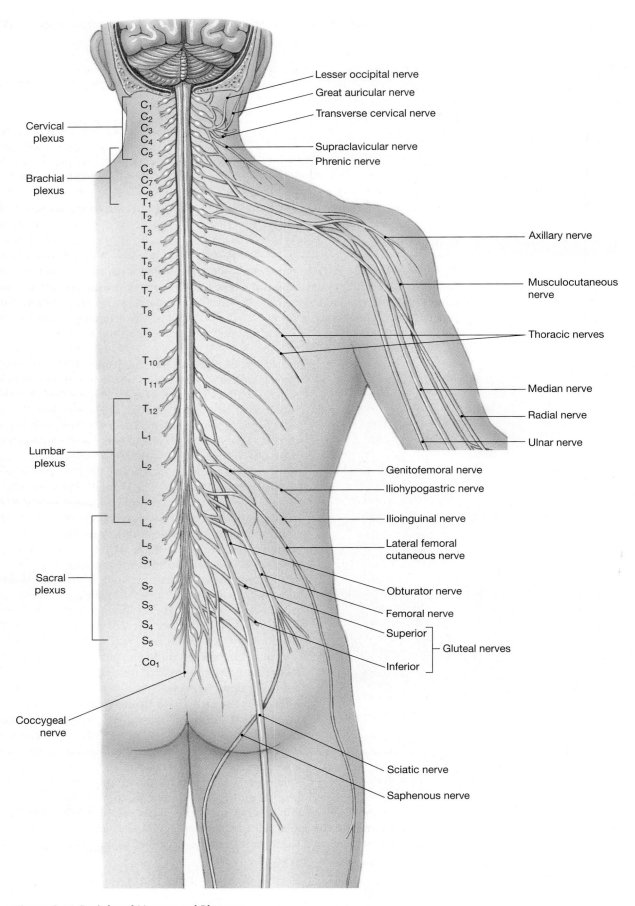

Cervical plexus

Brachial plexus

Lumbar plexus

Sacral plexus

Coccygeal nerve

C_1
C_2
C_3
C_4
C_5
C_6
C_7
C_8
T_1
T_2
T_3
T_4
T_5
T_6
T_7
T_8
T_9
T_{10}
T_{11}
T_{12}
L_1
L_2
L_3
L_4
L_5
S_1
S_2
S_3
S_4
S_5
Co_1

Lesser occipital nerve
Great auricular nerve
Transverse cervical nerve
Supraclavicular nerve
Phrenic nerve

Axillary nerve

Musculocutaneous nerve

Thoracic nerves

Median nerve
Radial nerve
Ulnar nerve

Genitofemoral nerve
Iliohypogastric nerve
Ilioinguinal nerve
Lateral femoral cutaneous nerve
Obturator nerve
Femoral nerve
Superior
Inferior — Gluteal nerves

Sciatic nerve
Saphenous nerve

Figure 24.4 Peripheral Nerves and Plexuses

An overview of selected peripheral nerves and the four major nerve plexuses: cervical, brachial, lumbar, and sacral.

plexus innervates muscles of the larynx plus the sternocleidomastoid, trapezius, and diaphragm muscles. It also innervates the skin of the neck, shoulder, and arm.

Brachial Plexus

The remaining cervical nerves are part of the **brachial plexus**, which supplies the shoulder and arm. The major branches of this plexus are the axillary, radial, musculo-cutaneous, median, and ulnar nerves. The **axillary nerve** (C_5 and C_6) supplies the deltoid and teres minor muscles and the skin of the shoulder. The **radial nerve** (C_5 through T_1) controls the extensor muscles of the arm, forearm, and digits, as well as the skin over the posterior and lateral margins of the arm. The **musculocutaneous nerve** (C_5 through C_7) supplies the flexor muscles of the upper arm and the skin of the lateral forearm. The **median nerve** (C_6 through T_1) innervates the flexor muscles of the forearm and digits, the pronator muscles, and the lateral skin of the hand. The **ulnar nerve** (C_8 and T_1) controls the flexor carpi ulnaris muscle of the forearm, other muscles of the hand, and the medial skin of the hand. Notice how overlap occurs within the brachial plexus. For example, spinal nerve C_6 innervates both flexor and extensor muscles.

Lumbar and Sacral Plexuses

The largest nerve network is called the **lumbosacral plexus**. It comprises T_{12}, L_1 through L_4, part of L_5, and S_1 through S_4. Figure 24.4 presents the distribution of nerves in this combined plexus. The major nerves of the **lumbar plexus** innervate the skin and muscles of the abdominal wall, genitalia, and thigh. The **genitofemoral nerve** supplies some of the external genitalia and the anterior and lateral skin of the thigh. The **lateral femoral cutaneous nerve** innervates the skin of the thigh from all aspects except the medial region. The **femoral nerve** controls the muscles of the anterior thigh and the adductor muscles and medial skin of the thigh.

The **sacral plexus** consists of two major nerves, the sciatic and the pudendal. The **sciatic nerve** descends the posterior leg and sends branches into the posterior thigh muscles and the musculature and skin of the lower leg. The **pudendal nerve** supplies the muscular floor of the pelvis, the perineum, and parts of the skin of the external genitalia.

Peripheral Nerves

Figure 24.5 illustrates the organization of a peripheral nerve, a branch of a spinal nerve. The nerve is compartmentalized by connective tissue in much the same way a skeletal muscle is organized. The peripheral nerve is wrapped in an outer covering called the **epineurium**. Beneath this layer is the **perineurium** that separates the axons into bundles called *fascicles*. Inside a fascicle, the **endoneurium** surrounds each axon and isolates it from neighboring axons.

QuickCheck Questions

3.1 What are the two groups of nerves in the PNS?

3.2 List the three major branches of a peripheral nerve.

3 *Materials*

□ Spinal cord laboratory model

□ Spinal cord chart

□ Compound microscope

□ Prepared slide of peripheral nerve

Procedures

1. Review Figures 24.4 and 24.5, and then label Figure 24.5.
2. Locate each nerve plexus on available laboratory models and charts.
3. Locate the spinal nerves assigned by your instructor on the laboratory models.
4. Examine the peripheral nerve slide.
   • Move the slide around to survey the preparation. Locate the epineurium. Is it continuous around the nerve?
   • Examine a single fascicle. Can you distinguish the perineurium from the epineurium?
   • Locate the individual axons inside a fascicle. Can you distinguish the myelin sheath of the axon from the endoneurium?

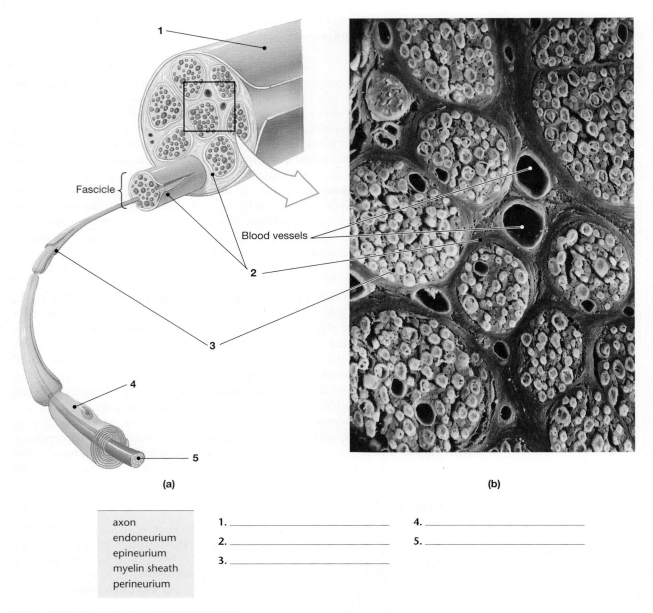

Fascicle

Blood vessels

(a)

(b)

Figure 24.5 Organization of Peripheral Nerves
(a) A typical peripheral nerve and its connective tissue wrappings. (b) A scanning electron micrograph showing the perineurium and endoneurium in great detail. (SEM × 340)

axon	1. _____ 4. _____
endoneurium	2. _____ 5. _____
epineurium	3. _____
myelin sheath	
perineurium	

• Draw and label the peripheral nerve in the space provided. ■

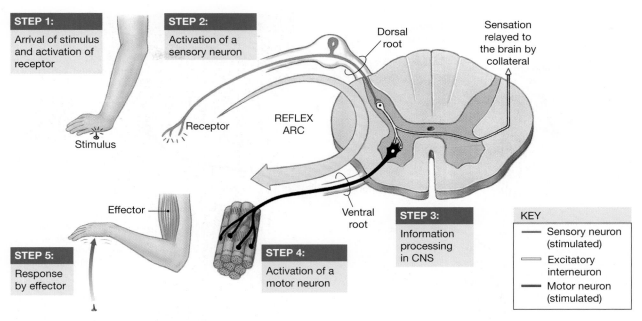

STEP 1:
Arrival of stimulus and activation of receptor

Stimulus

STEP 2:
Activation of a sensory neuron

Receptor

REFLEX ARC

Dorsal root

Sensation relayed to the brain by collateral

Effector

STEP 5:
Response by effector

Ventral root

STEP 4:
Activation of a motor neuron

STEP 3:
Information processing in CNS

KEY
— Sensory neuron (stimulated)
═ Excitatory interneuron
▬ Motor neuron (stimulated)

Figure 24.6 **Components of a Reflex Arc**

LAB ACTIVITY 4 Spinal Reflexes

Reflexes are automatic neural responses to specific stimuli. Most reflexes have a protective function. Touch something hot, and the withdrawal reflex removes your hand to prevent tissue damage. Shine a bright light into someone's eyes, and the pupils constrict to protect the retina from excessive light. Reflexes cause rapid adjustments to maintain homeostasis. The CNS does minimal processing to respond to the stimulus. The sensory and motor components of a reflex are "prewired" and initiate the reflex upon stimulation.

Figure 24.6 depicts the five steps involved in a typical reflex pathway, called a **reflex arc**. First, a receptor is activated by a stimulus. The receptor in turn activates a sensory neuron that enters the CNS, where the third step, information processing, occurs. The processing is performed at the synapse between the sensory and motor neurons. A conscious thought or recognition of the stimulus is not required to evaluate the sensory input of the reflex. The processing results in activation of a motor neuron that elicits the appropriate action, a response by the effectors. In this basic reflex arc, only two neurons are involved: one sensory and one motor. Complex reflex arcs include **interneurons** between the sensory and motor neurons.

There are many types of reflexes. **Innate** reflexes are the inborn responses of a newborn baby, such as grasping an object and suckling the breast for milk. **Cranial** reflexes have pathways in cranial nerves. **Visceral** reflexes pertain to the internal organs. **Spinal** reflexes process information in the spinal cord rather than the brain. **Somatic** reflexes involve skeletal muscles. The number of synapses in a reflex can also be used to classify reflexes. In the arc of a **monosynaptic reflex**, there is only one synapse between the sensory and motor neurons. In the arc of a **polysynaptic reflex,** there are numerous interneurons between sensory and motor neurons. The response of a polysynaptic reflex is more complex and may include both stimulation and inhibition of muscles. Reflexes are used as a diagnostic tool to evaluate the function of specific regions of the brain and spinal cord. An abnormal reflex or the lack of a reflex indicates a loss of neural function resulting from disease or injury.

You are probably familiar with the "knee jerk," a type of **stretch reflex** called the **patellar reflex,** that occurs when the tendon over the patella is hit with a rubber percussion hammer (Figure 24.7). Tapping on the patellar tendon stretches receptors called **muscle spindles** in the quadriceps muscle group of the anterior

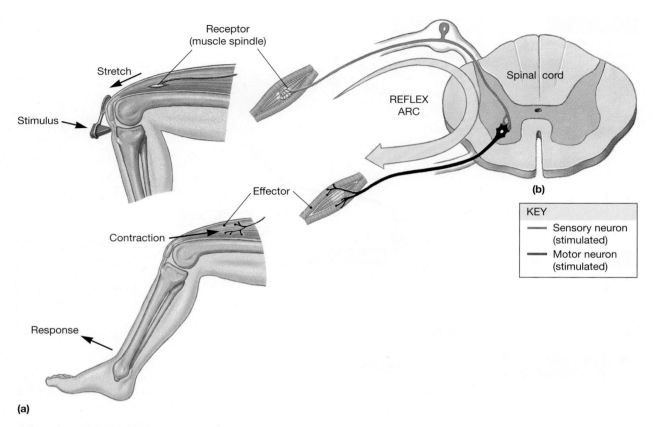

(b)

KEY

— Sensory neuron (stimulated)

— Motor neuron (stimulated)

(a)

Figure 24.7 The Patellar Reflex

(a) The patellar reflex is controlled by muscle spindles in the muscles that straighten the knee. A reflex hammer striking the muscle tendon provides the stimulus that stretches the spindle fibers. The response is an immediate increase in muscle tone and a reflexive kick. **(b)** The basic wiring of a monosynaptic reflex arc.

thigh. This stimulus evokes a rapid motor reflex to contract the quadriceps and shorten the muscles. Figure 24.8 details additional somatic reflexes.

QuickCheck Questions

4.1 What are the components of a reflex arc?

4.2 How can reflexes be used diagnostically?

4 *Materials*

☐ Reflex (percussion) hammer (with rubber head)

☐ Laboratory partner

Procedures

1. Patellar reflex (Figure 24.7):
   - Have your partner sit down and cross his or her legs at the knee.
   - On the top leg, gently tap below the patella with the percussion hammer to stimulate the tendon of the rectus femoris muscle.
   - What is the response?
   - How might this reflex help maintain upright posture?

2. Biceps reflex (Figure 24.8b):
   - This reflex tests the response of the biceps brachii muscle.
   - Have your partner rest an arm on the laboratory benchtop.
   - Place a finger over the tendon of the biceps brachii and gently tap your finger with the percussion hammer.
   - What is the response?

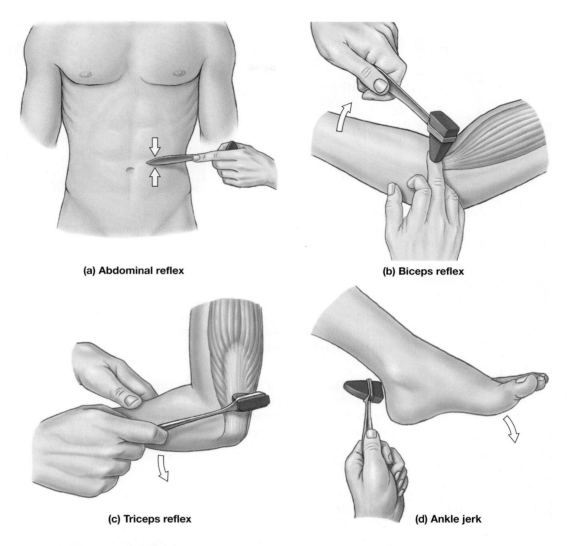

(a) Abdominal reflex

(b) Biceps reflex

(c) Triceps reflex

(d) Ankle jerk

Figure 24.8 Somatic Reflexes
(a) The abdominal reflex, an example of a superficial reflex. The biceps reflex (b), triceps reflex (c), and ankle jerk reflex (d) are examples of stretch reflexes.

3. Triceps reflex (Figure 24.8c):
   - This reflex tests the response of the triceps brachii muscle.
   - Loosely support one of your partner's forearms.
   - Gently tap the tendon of the triceps brachii at the posterior elbow.
   - What is the response?
4. Ankle calcanean reflex (Figure 24.8d):
   - This reflex tests the response of the gastrocnemius muscle when the calcanean (Achilles) tendon is stretched.
   - Have your partner sit in a chair and extend one leg forward so that the foot is off the floor.
   - Gently tap the calcanean tendon with the percussion hammer.
   - What is the response? ■

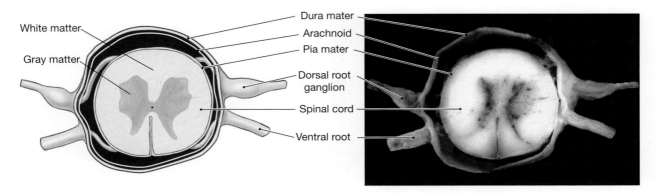

Figure 24.9 Sheep Spinal Cord Dissection
Transverse section of the sheep spinal cord detailing the spinal meninges and internal organization of the spinal cord.

LAB ACTIVITY 5 Dissection of the Spinal Cord
_____

Dissecting a preserved sheep or cow spinal cord will provide you the opportunity to examine the meningeal layers and sectional anatomy. Only a small section of the spinal cord is necessary.

 Be sure to use only a _preserved_ spinal cord for dissection. Fresh spinal cords can carry disease. Also, remember that preserved specimens have been prepared with chemicals that might irritate skin and mucous membranes. Wear safety glasses and gloves during the dissection. Disposal of the specimen requires certain safeguards, and your instructor will inform you of the disposal procedure. ▲

5 _Materials_

☐ Gloves
☐ Safety glasses
☐ Segment of preserved sheep or cow spinal cord
☐ Dissection pan
☐ Scissors
☐ Scalpel
☐ Forceps
☐ Blunt probe

Procedures

Put on gloves and safety glasses before opening the container of preserved spinal cord segments or handling one of the segments.

1. Lay the spinal cord on the dissection pan and cut a thin section about 2 cm (0.75 in.) thick. Lay this cross-section flat on the dissection pan, and observe the internal anatomy. Use Figure 24.9 as a guide to help locate the anatomy of the spinal cord.

2. Identify the gray horns, central canal, and white columns. What type of tissue is found in the gray horns? What type is found in the white columns? How can you determine the posterior margin of the cord?

3. Locate the spinal meninges by pulling the outer tissues away from the spinal cord with a forceps and blunt probe. Slip your probe between the meninges on the lateral spinal cord. Cut completely through the meninges, and gently peel them back to expose the ventral and dorsal roots. How does the dorsal root differ in appearance from the ventral root?

4. Closely examine the meninges. With your probe, separate the arachnoid from the dura mater. With a dissection pin, attempt to loosen a free edge of the pia mater. What function does each of these membranes serve?

5. Clean up your work area, wash the dissection pan and tools, and follow your instructor's directions for proper disposal of the specimen. ■

The Spinal Cord, Spinal Nerves, and Reflexes

Name _____

Date _____

Section _____

A. Matching

Match each term in the left column with its correct description from the right column.

_____	**1.** lateral gray horn	**A.** site of cerebrospinal fluid circulation
_____	**2.** posterior median sulcus	**B.** sensory branch entering spinal cord
_____	**3.** bundle of axons	**C.** surrounds axons of peripheral nerve
_____	**4.** rami communicantes	**D.** contains visceral motor cell bodies
_____	**5.** subarachnoid space	**E.** tapered end of spinal cord
_____	**6.** ventral root	**F.** fascicle
_____	**7.** dorsal ramus	**G.** shallow groove of spinal cord
_____	**8.** dorsal root ganglion	**H.** posterior branch of a spinal nerve
_____	**9.** conus medullaris	**I.** motor branch exiting spinal cord
_____	**10.** endoneurium	**J.** leads to autonomic ganglion
_____	**11.** dorsal root	**K.** contains sensory cell bodies

B. Labeling

Label the sectional anatomy of the spinal cord in Figure 24.10.

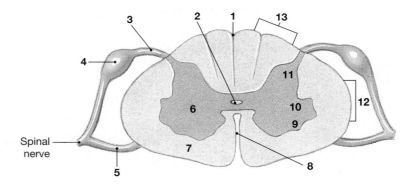

Figure 24.10 **Anatomy of the Spinal Cord**

1. _____
2. _____
3. _____
4. _____
5. _____
6. _____
7. _____
8. _____
9. _____
10. _____
11. _____
12. _____
13. _____

LABORATORY REPORT

C. Short-Answer Questions

1. Describe the organization of white and gray matter in the spinal cord.

2. Describe the spinal meninges.

3. Discuss the major nerves of the brachial plexus.

4. List the five basic steps of a reflex.

5. Compare the types of neurons in the posterior, lateral, and anterior gray horns.

D. Analysis and Application

1. Trace a sensory pathway from the flexor carpi ulnaris muscle to the spinal cord. Include the names of the peripheral nerve, the spinal nerve, and the various rami and roots involved in the pathway.

2. Starting in the spinal cord, trace a motor pathway to the adductor muscles of the thigh. Include the spinal cord root, spinal nerve, nerve plexus, and specific peripheral nerve involved in the pathway.

3. Why would an injury to a peripheral nerve cause loss of both sensory and motor functions?

4. Why does the stretch reflex cause the quadriceps femoris muscle group to contract?

Anatomy of the Brain

OBJECTIVES

On completion of this exercise, you should be able to:

- Name the three meninges that cover the brain.

- Describe the extensions of the dura mater.

- Identify the six major regions of the brain and a basic function of each.

- Identify the surface features of each region of the brain.

- Identify the 12 pairs of cranial nerves.

- Identify the anatomy of a dissected sheep brain.

The brain is one of the largest organs in the body. It weighs approximately 3 pounds and occupies the cranial cavity of the dorsal body cavity. Billions of synapses among neurons form a vast biological circuitry that no electronic computer will ever surpass. Every second the brain performs a huge number of calculations, interpretations, and visceral-activity coordinations to maintain homeostasis.

The brain is divided into six major regions: cerebrum (ser-Ē-brum or SER-ē-brum), diencephalon (dī-en-SEF-a-lon), mesencephalon, pons, medulla oblongata, and cerebellum. The medulla oblongata, pons, and mesencephalon are collectively called the **brain stem**. Some anatomists include the diencephalon as part of the brain stem. Figure 25.1 highlights each region of the brain and summarizes major functions.

The **cerebrum** is the largest portion of the brain. It is divided into right and left **cerebral hemispheres** by the **longitudinal fissure**. The hemispheres are covered with a folded **cerebral cortex** (*cortex,* bark or rind) of gray matter where neurons are not myelinated. Each small fold of the cerebral cortex is called a **gyrus** (JĪ-rus; plural *gyri*), and each small groove is called a **sulcus** (SUL-kus; plural *sulci*). Deeper grooves, such as the one separating the two hemispheres, are called **fissures**.

The **cerebellum** is under the posterior of the cerebrum. Directly anterior to the cerebellum are the components of the brain stem, the region where the brain and spinal cord meet. The slender **medulla oblongata**, or simply *medulla,* connects the spinal cord to the brain. Superior to the medulla is the swollen **pons**. Above the pons is part of the **mesencephalon**, also called the *midbrain.* Anterior to the mesencephalon is the base of the hypothalamus, a portion of the **diencephalon**.

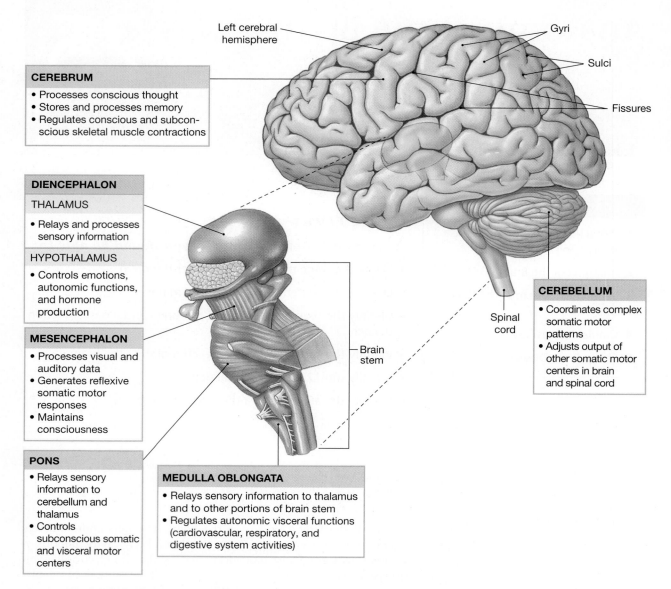

CEREBRUM
- Processes conscious thought
- Stores and processes memory
- Regulates conscious and subconscious skeletal muscle contractions

DIENCEPHALON

THALAMUS
- Relays and processes sensory information

HYPOTHALAMUS
- Controls emotions, autonomic functions, and hormone production

MESENCEPHALON
- Processes visual and auditory data
- Generates reflexive somatic motor responses
- Maintains consciousness

PONS
- Relays sensory information to cerebellum and thalamus
- Controls subconscious somatic and visceral motor centers

MEDULLA OBLONGATA
- Relays sensory information to thalamus and to other portions of brain stem
- Regulates autonomic visceral functions (cardiovascular, respiratory, and digestive system activities)

CEREBELLUM
- Coordinates complex somatic motor patterns
- Adjusts output of other somatic motor centers in brain and spinal cord

Left cerebral hemisphere

Gyri

Sulci

Fissures

Brain stem

Spinal cord

Figure 25.1 An Introduction to Brain Functions

LAB ACTIVITY 1   Protection and Support of the Brain

The brain and spinal cord are encased in tough, protective membranes called the **meninges** (men-IN-jēz; *menin*, membrane). Circulating between certain meningeal layers is **cerebrospinal fluid** (CSF), which cushions the brain and prevents it from contacting the cranial bones during a head injury, much like a car's airbag prevents a passenger from hitting the dashboard. The cranial meninges are anatomically similar to, and continuous with, the spinal meninges of the spinal cord. The cranial meninges consist of three layers: the dura mater, the arachnoid, and the pia mater (Figure 25.2).

The **dura mater** (DOO-ruh MĀ-ter; *dura*, tough + *mater*, mother), the outer meningeal covering, consists of two tissue layers. The **endosteal** portion is the outer layer and is fused with the periosteum of the flat cranial bones; the **meningeal** layer faces the arachnoid membrane. Between the inner and outer dural layers are large blood sinuses, collectively called **dural sinuses**, which drain blood from cranial veins into the jugular veins. The **superior** and **inferior sagittal sinuses** are large veins in

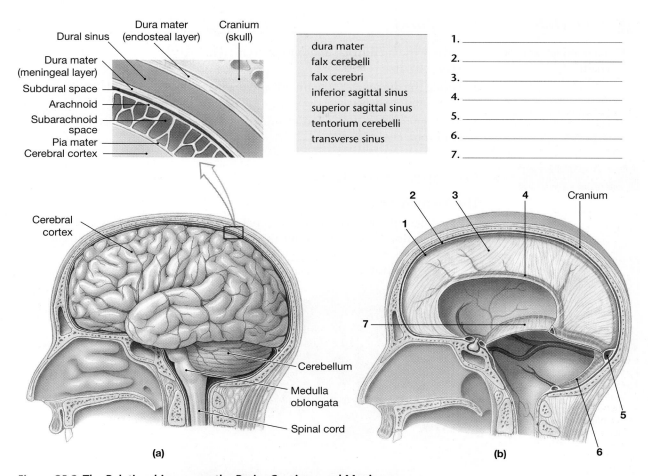

dura mater
falx cerebelli
falx cerebri
inferior sagittal sinus
superior sagittal sinus
tentorium cerebelli
transverse sinus

1. _____
2. _____
3. _____
4. _____
5. _____
6. _____
7. _____

Figure 25.2 The Relationship among the Brain, Cranium, and Meninges
(a) A lateral view of the brain, showing its position in the cranium and the
organization of the meninges. **(b)** A diagrammatic view, showing the orientation of the
falx cerebri, tentorium cerebelli, and falx cerebelli.

the dura mater between the two hemispheres of the cerebrum. The **transverse sinus**
is in the dura mater between the cerebrum and the cerebellum. Between the dura
mater and the underlying arachnoid is the **subdural space**.

Deep to the dura mater is the **arachnoid** (a-RAK-noyd; *arachno,* spider), named
after the weblike connection this membrane has with the underlying pia mater. The
arachnoid forms a smooth covering over the brain.

On the surface of the brain is the **pia** (PĒ-uh; *pia,* delicate) **mater,** which contains many blood vessels for the brain. Between the arachnoid and pia mater is the
subarachnoid space, where the CSF circulates. The chambers in the brain, called
ventricles, also contain CSF, and this fluid contains nutrients supplied by the blood.

The dura mater has extensions to further stabilize the brain. A midsagittal fold
in the dura mater forms the **falx cerebri** (FALKS ser-Ē-brē; *falx,* sickle-shaped) and
separates the right and left hemispheres of the cerebrum. Posteriorly, the dura mater
folds again as the **tentorium cerebelli** (ten-TOR-ē-um ser-e-BEL-ē; *tentorium,* a covering) and separates the cerebellum from the cerebrum. The **falx cerebelli** is a dural
fold between the hemispheres of the cerebellum.

QuickCheck Questions

1.1 What are the functions of the cranial meninges?

1.2 Between which meningeal layers does CSF circulate?

1.3 What fold separates the cerebellum from the cerebrum?

| 1 | *Materials* | **Procedures** |

☐ Brain models and/or charts

1. Review the meningeal anatomy presented in Figure 25.2, and then label Figure 25.2b.

2. Locate the dura mater, arachnoid, and pia mater on the available laboratory charts and models.

3. Examine the dura mater and identify the falx cerebri, falx cerebelli, and tentorium cerebelli. ■

| LAB ACTIVITY | 2 | Anatomy of the Brain

The Medulla Oblongata

The medulla oblongata connects the spinal cord to the brain (Figures 25.3 and 25.4). Communications descending from the brain to the spinal cord pass through the

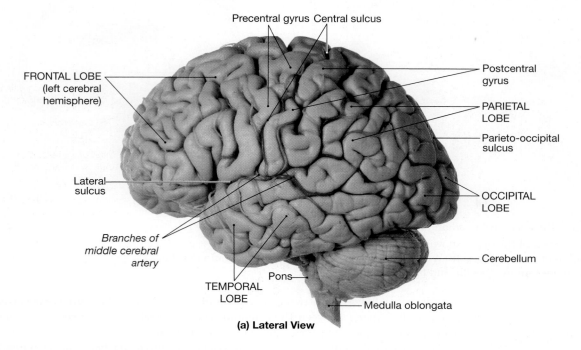

(a) **Lateral View**

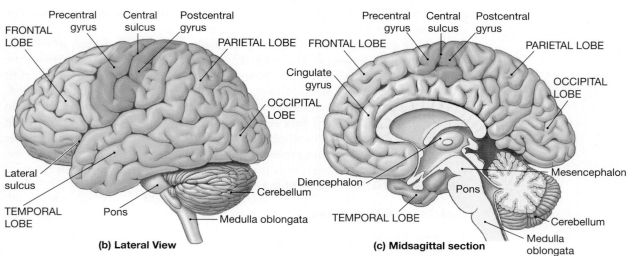

(b) **Lateral View**   (c) **Midsagittal section**

Figure 25.3 **The Brain**

(a) Photograph of lateral view. (b) Illustration of lateral view. (c) Illustration of sagittal view.

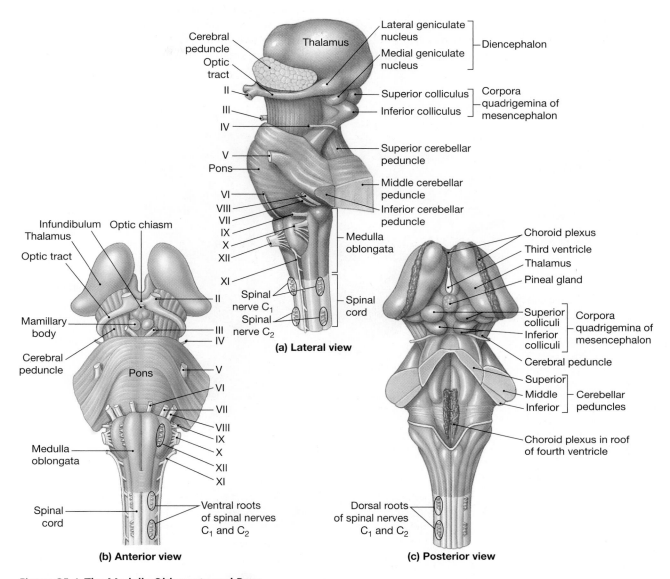

Figure 25.4 The Medulla Oblongata and Pons
(**a, b**) Nuclei and tracts in the medulla oblongata. (**c**) Nuclei and tracts in the pons.

medulla oblongata. The anterior surface of the medulla oblongata has two prominent folds called **pyramids** where motor tracts cross over, or *decussate,* to the opposite side of the body. The right and left sides of the brain have *contralateral* (opposite-side) control of the body because sensory and motor tracts decussate from one side to the other in the central nervous system. The medulla oblongata also functions as an autonomic center for visceral functions. Nuclei in this portion of the brain are vital reflex centers for the regulation of cardiovascular, respiratory, and digestive activities.

The Pons

The pons is superior to the medulla and functions as a relay station to direct sensory information to the thalamus and cerebellum. It also contains certain sensory, somatic motor, and autonomic cranial nerve nuclei.

The Mesencephalon (Midbrain)

The mesencephalon is located superior to the pons. Although covered by the cerebrum, the mesencephalon can be observed by gently pushing down on the superior surface of the cerebellum and looking below the cerebrum. The portion of the mesencephalon called the **corpora quadrigemina** (KOR-po-ra quad-ri-JEM-i-nuh) is a series of four bulges next to the nipple-like pineal gland of the diencephalon (Figures 25.4 and 25.5). The two members of the superior pair are the **superior colliculi** (kol-IK-ū-lē; *colliculus*, small hill), which function as a visual re-

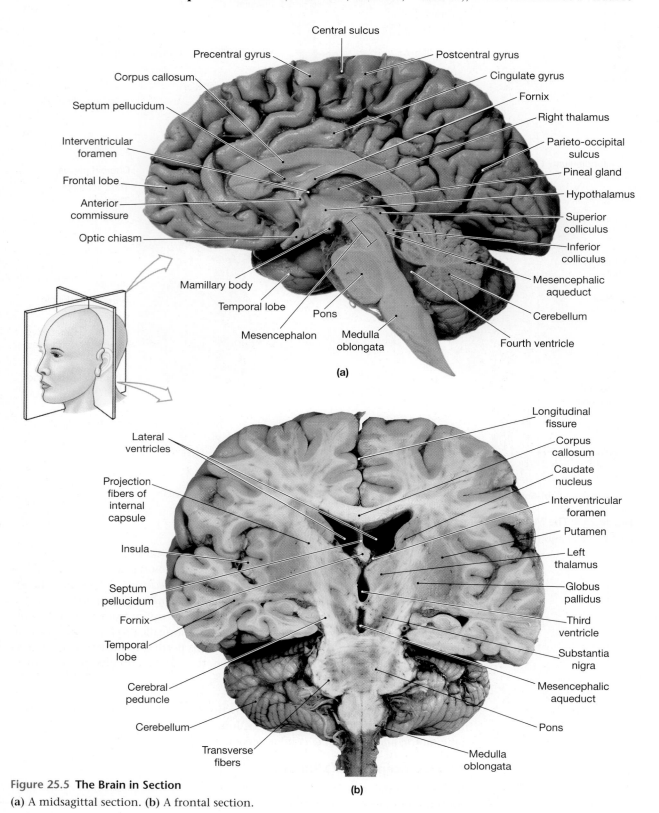

Figure 25.5 The Brain in Section
(a) A midsagittal section. (b) A frontal section.

flex center to move the eyeballs and the head to keep an object centered on the retina of the eye. The two members of the lower pair are the **inferior colliculi**, which function as an auditory reflex center to move the head to locate and follow sounds. The anterior mesencephalon between the pons and the hypothalamus consists of the **cerebral peduncles** (*peduncles;* little feet), a group of white fibers connecting the cerebral cortex with other parts of the brain.

The Cerebellum

The cerebellum (Figure 25.6) is inferior to the occipital lobe of the cerebrum. (This and the other lobes of the cerebrum are described below.) Small folds on the cerebellar cortex are called **folia** (FŌ-lē-uh; *folia,* leaves). Like the cerebrum, the cerebellum is divided into right and left **cerebellar hemispheres,** which are separated by a narrow **vermis** (VER-mis; *vermis,* worm). Each cerebellar hemisphere consists of two

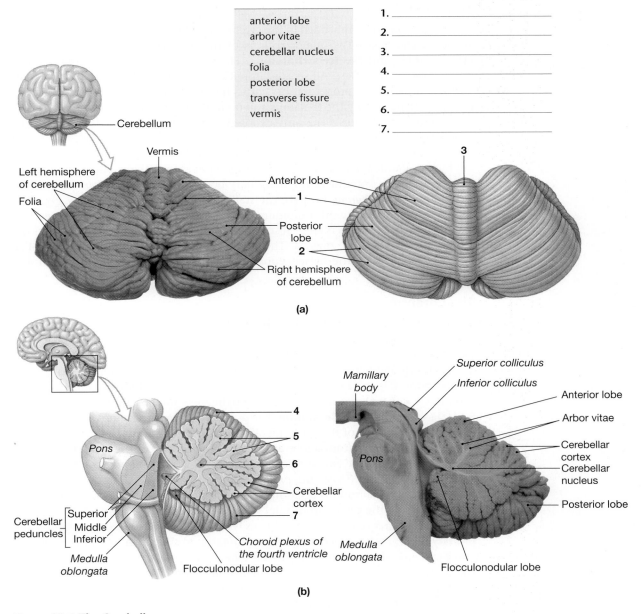

anterior lobe
arbor vitae
cerebellar nucleus
folia
posterior lobe
transverse fissure
vermis

1. _____
2. _____
3. _____
4. _____
5. _____
6. _____
7. _____

Figure 25.6 The Cerebellum
(a) Posterior, superior surface. (b) A sectional view of the cerebellum.

lobes: a smaller **anterior lobe,** which is directly inferior to the cerebrum, and a **posterior lobe**. The **transverse (primary) fissure** separates the anterior and posterior cerebellar lobes. In a sagittal section, a smaller **flocculonodular** (flok-ū-lē-NOD-ū-ler) **lobe** is visible internally where the anterior wall of the cerebellum faces the pons.

In a sagittal section, the white matter of the cerebellum is apparent. Because the tissue is highly branched, it is called the **arbor vitae** (*arbor,* tree + *vitae,* life). In the middle of the arbor vitae are the **cerebellar nuclei,** which function in the regulation of involuntary skeletal muscle contraction.

The cerebellum is primarily involved in the coordination of somatic motor functions, which means principally skeletal muscle contractions. Adjustments to postural muscles occur when impulses from the cranial nerve of the inner ear pass into the flocculonodular lobe, the part of the cerebellum where information concerning equilibrium is processed. Learned muscle patterns, such as those involved in serving a tennis ball or playing the guitar, are stored and processed in the cerebellum.

The Diencephalon

The diencephalon is embedded in the cerebrum and is visible only from the inferior aspect of the brain. Two major regions of the diencephalon are the **thalamus** (THAL-a-mus) and the **hypothalamus** (Figure 25.5). The thalamus maintains a crude sense of awareness. All sensory impulses except smell pass into the thalamus and are relayed to the proper sensory cortex for interpretation. Nonessential sensory data are filtered out by the thalamus and do not reach the sensory cortex.

The hypothalamus is the only exposed part of the diencephalon. On the exposed floor of the hypothalamus, just anterior to the mesencephalon, is a pair of rounded **mamillary** (MAM-i-lar-ē; *mammilla,* little breast) **bodies**. These bodies are hypothalamic nuclei that control eating reflexes for licking, chewing, sucking, and swallowing. Anterior to the mamillary bodies is the **infundibulum** (in-fun-DIB-ū-lum; *infundibulum,* funnel), the stalk that attaches the **pituitary gland** to the hypothalamus. The infundibulum is seen in Figure 25.4b. In sagittal section, the **intermediate mass** is an oval structure in the diencephalon that connects the right and left thalamic masses. The **pineal** (PIN-ē-ul) **gland** is the nipple-like structure superior to the mesencephalon positioned between the cerebrum and the cerebellum.

The Cerebrum

The cerebrum, or telencephalon, is the most complex part of the brain. Conscious thought, intellectual reasoning, and memory processing and storage all take place in the cerebrum.

Each cerebral hemisphere consists of five lobes, most named for the overlying cranial bone (Figure 25.7). The anterior cerebrum is the **frontal lobe**, and the prominent **central sulcus**, located approximately midposteriorly, separates the frontal lobe from the **parietal lobe**. The **occipital lobe** corresponds to the position of the occipital bone of the posterior skull. The **lateral sulcus** defines the boundary between the large frontal lobe and the **temporal lobe** of the lower lateral cerebrum. Cutting into the lateral sulcus and peeling away the temporal lobe reveals a fifth lobe, the **insula** (IN-sū-luh; *insula,* island).

Regional specializations occur in the cerebrum. The central sulcus separates the motor region of the cerebrum (frontal lobe) from the sensory region (parietal lobe). Immediately anterior to the central sulcus is the **precentral gyrus**. This gyrus is the primary motor cortex, where voluntary commands to skeletal muscles are generated. The **postcentral gyrus**, on the parietal lobe, is the primary sensory cortex, where the general sense of touch is perceived. The other four senses—sight, hearing, smell, and taste—involve the processing of complex information received from many more sensory neurons than the number involved in the sense of touch. These four senses thus require more neurons in the brain to process the sensory signals, and therefore the cerebral cortex areas devoted to processing these messages are larger than the

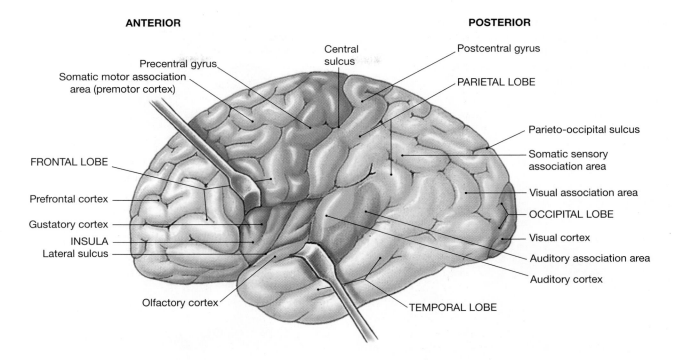

ANTERIOR

POSTERIOR

Central sulcus

Precentral gyrus

Somatic motor association area (premotor cortex)

Postcentral gyrus

PARIETAL LOBE

FRONTAL LOBE

Parieto-occipital sulcus

Somatic sensory association area

Prefrontal cortex

Visual association area

Gustatory cortex

OCCIPITAL LOBE

INSULA

Visual cortex

Lateral sulcus

Auditory association area

Auditory cortex

Olfactory cortex

TEMPORAL LOBE

Figure 25.7 The Cerebral Hemispheres

Major anatomical landmarks on the surface of the left cerebral hemisphere. Association areas are colored. To expose the insula, the lateral sulcus has been opened.

postcentral gyrus of the primary sensory cortex for touch. The occipital lobe contains the visual cortex where visual impulses from the eyes are interpreted. The temporal lobe houses the auditory cortex and the olfactory cortex.

Figure 25.7 also shows numerous *association areas,* regions that interpret sensory information from more than one sensory cortex or integrate motor commands into an appropriate response. The premotor cortex is the somatic motor association area of the anterior frontal lobe. Auditory and visual association areas occur near the corresponding sensory cortex in the occipital lobe.

The cerebral hemispheres are connected by a thick tract of white matter called the **corpus callosum**. This structure, which bridges the two hemispheres at the base of the longitudinal fissure, is easily identified as the curved white structure at the base of the cerebrum (Figure 25.5). The inferior portion of the corpus callosum is the **fornix** (FOR-niks), a white tract connecting deep structures of the limbic system, the "emotional" brain. The fornix narrows anteriorly and meets the **anterior commissure** (kom-MIS-sur), another tract of white matter connecting the cerebral hemispheres.

Frontal and Horizontal Sections of the Brain

Deep structures of the cerebrum and diencephalon are visible when the brain is sectioned on the frontal plane near the infundibulum, as in Figure 25.8. Notice in the figure how the corpus callosum transverses the brain to connect the two cerebral hemispheres. In each cerebral hemisphere, structures called either **cerebral nuclei** or **basal nuclei** are masses of gray matter involved in automating voluntary muscle contractions. Each nucleus consists of a medial **caudate nucleus** and a lateral **lentiform nucleus**. The latter is in turn made up of two parts: a **putamen** (pū-TĀ-men) and a **globus pallidus.** Between the caudate nucleus and the lentiform nucleus lies the **internal capsule**, a band of white matter that connects the cerebrum to the diencephalon, brain stem, and cerebellum.

Figure 25.8 The Basal Nuclei

(a) The relative positions of the basal nuclei in the intact brain. (b) Frontal sections. (c) A horizontal section; compare this with (d), the view in dissection.

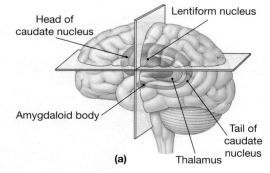

(a)

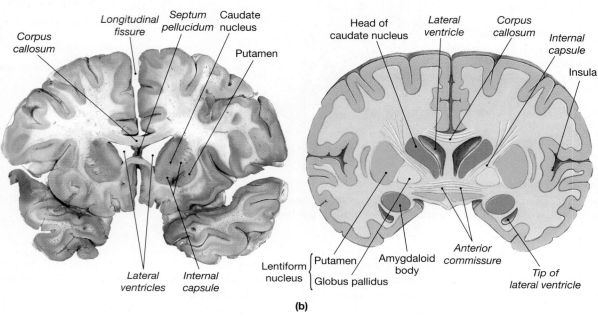

(b)

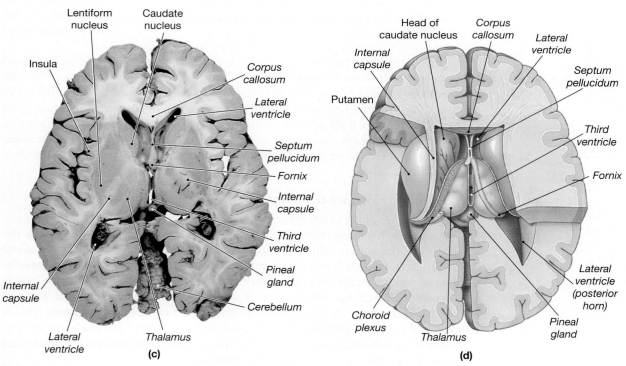

(c) (d)

Note in Figure 25.8 the organization of gray matter and white matter in the brain. The covering of the cerebrum—the cerebral cortex—is gray matter. The corpus callosum and other bridging structures of the brain consist of white matter.

QuickCheck Questions

2.1 What are the six major regions of the brain?

2.2 How are the cerebral hemispheres connected to each other?

2.3 Where is the mesencephalon?

2 Materials

- ☐ Brain models and/or charts
- ☐ Preserved and sectioned human brain (if available)

Procedures

1. Review the brain anatomy presented in Figures 25.3 through 25.8 and label Figure 25.6. Then use a model, chart, and/or preserved brain to complete procedures 2 through 5.

2. In the brain stem, identify:
   - the medulla and the two pyramids on its anterior surface.
   - the pons, superior to the medulla.
   - the cerebral peduncles on the lateral sides of the mesencephalon.
   - the corpora quadrigeminal of the mesencephalon, distinguishing between superior and inferior colliculi.

3. In the cerebellum, examine the right and left hemispheres and the vermis separating them. In each hemisphere, identify the transverse fissure and the anterior and posterior lobes. In a midsagittal section, locate the arbor vitae and the cerebellar nuclei.

4. In the diencephalon, identify:
   - the thalamus, recognizable as the lateral wall around the diencephalon.
   - the pineal gland at the posterior roof of the thalamus, just superior to the superior colliculi of the mesencephalon.
   - the wedge-shaped hypothalamus, inferior to the thalamus.
   - the infundibulum, which attaches the pituitary gland to the hypothalamus.
   - the mamillary bodies, inferior to the infundibulum.

5. Examine the cerebrum, and note how the longitudinal fissure separates it into two cerebral hemispheres. Then identify the five lobes of each hemisphere, along with the central sulcus, precentral gyri, and postcentral gyri.

6. In a sagittal section of the cerebrum, identify the corpus callosum, fornix, and anterior commissure.

7. In a frontal section and a horizontal section, locate the internal capsule, lentiform nucleus, and caudate nucleus. Distinguish between the putamen and the globus pallidus of the lentiform nucleus.

8. If a preserved brain is available, notice the difference in color between the gray matter of the cerebral cortex and the white matter of the corpus callosum, fornix, anterior commissure, and internal capsule. ■

LAB ACTIVITY 3 Ventricular System of the Brain

Deep in the brain are four chambers called *ventricles* (Figure 25.9). Two **lateral ventricles**, one in each cerebral hemisphere, extend deep into the cerebrum as horseshoe-shaped chambers. At the midline of the brain, the lateral ventricles are separated from each other by a thin membrane, the **septum pellucidum**. A brain sectioned at the midsagittal plane exposes this membrane. CSF circulates from the lateral ventricles through the **interventricular foramen** (foramen of Monro) and enters the **third ventricle**, a small chamber within the diencephalon. CSF in the

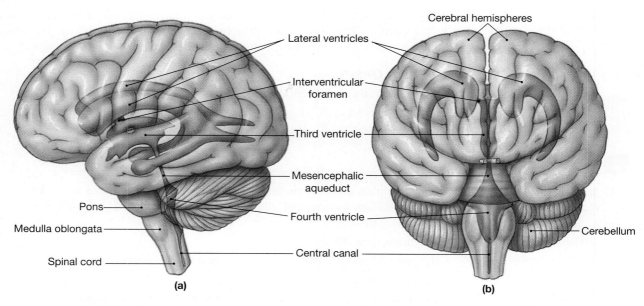

Figure 25.9 Ventricles of the Brain
The orientation and extent of the ventricles as they would appear if the brain were transparent **(a)** A lateral view. **(b)** An anterior view.

third ventricle passes through the **mesencephalic (cerebral) aqueduct** of the mesencephalon and enters the **fourth ventricle** between the brain stem and the cerebellum. In the fourth ventricle, two **lateral apertures** and a single **median aperture** direct CSF laterally to the exterior of the brain and spinal cord and into the subarachnoid space. CSF then circulates around the brain and spinal cord and is reabsorbed at **arachnoid granulations,** which project into the veins of the dural sinuses.

Each ventricle produces CSF, which then flows through the ventricles and enters the subarachnoid space surrounding the brain. The fluid is produced in each ventricle by a specialized capillary network called the **choroid plexus.** The choroid plexus of the third ventricle passes through the interventricular foramen and expands to line the floor of the lateral ventricles (Figure 25.10). The choroid plexus of the fourth ventricle lies on the posterior wall of the ventricle.

Study Tip

CSF Pathway

Although all the ventricles produce CSF, it may help you remember the complete circulatory pathway if you start with a drop of CSF in a lateral ventricle and circulate it through the third and fourth ventricles and then to the site of reabsorption. ●

Clinical Application

Hydrocephalus

The choroid plexes of an adult brain produce approximately 500 ml of cerebrospinal fluid daily. Because CSF is constantly being made, a volume equal to that produced must be removed from the central nervous system to prevent a buildup of fluid pressure in the ventricles. In an infant, if CSF production exceeds CSF reabsorption, the increase in cranial pressure expands the unfused skull, creating a condition called *hydrocephalus.* The result is an enlarged skull and possible brain damage caused by high fluid pressure on the delicate neural tissues. Surgical installation of small tubes called shunts drains the excess CSF and reduces intracranial pressure. ▶

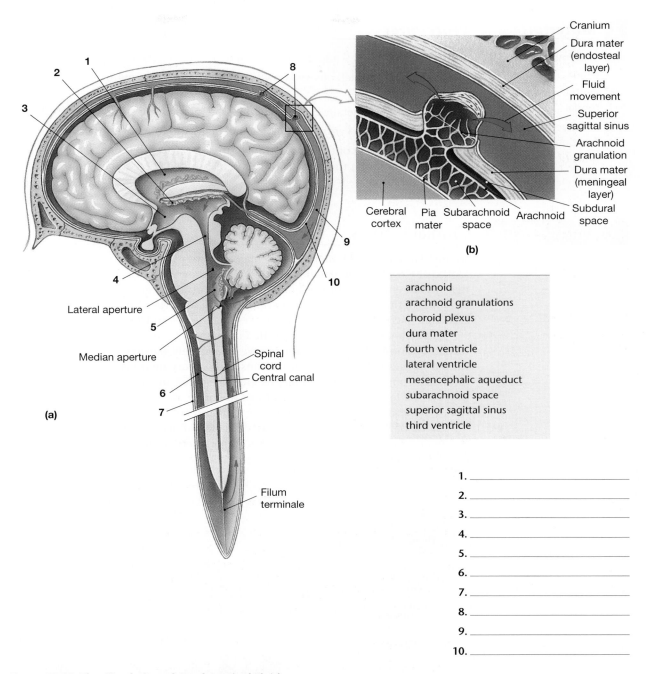

Cranium
Dura mater (endosteal layer)
Fluid movement
Superior sagittal sinus
Arachnoid granulation
Dura mater (meningeal layer)
Subdural space
Cerebral cortex   Pia mater   Subarachnoid space   Arachnoid

(b)

Lateral aperture
Median aperture
Spinal cord
Central canal
Filum terminale

(a)

arachnoid
arachnoid granulations
choroid plexus
dura mater
fourth ventricle
lateral ventricle
mesencephalic aqueduct
subarachnoid space
superior sagittal sinus
third ventricle

1. _____
2. _____
3. _____
4. _____
5. _____
6. _____
7. _____
8. _____
9. _____
10. _____

Figure 25.10 The Circulation of Cerebrospinal Fluid
(a) A sagittal section indicating the sites of formation and routes of circulation of cerebrospinal fluid (red arrows). (b) The orientation of the arachnoid granulations.

QuickCheck Questions

3.1 Where is CSF produced?

3.2 Where does CSF circulate?

3.3 How is CSF reabsorbed into the blood?

3 *Materials*

- ☐ Brain models and charts
- ☐ Ventricular system model
- ☐ Preserved and sectioned human brain (if available)

Procedures

1. Review Figures 25.9 and 25.10, and then label Figure 25.10.
2. On the brain model, observe how the lateral ventricles extend into the cerebrum.
3. If your model is detailed enough, locate the interventricular foramen.
4. Identify the third ventricle, cerebral aqueduct, and fourth ventricle.
5. Starting from one of the two lateral ventricles, trace a drop of CSF as it circulates through the brain and then is reabsorbed at an arachnoid granulation. ■

LAB ACTIVITY 4 Cranial Nerves

As noted in Exercise 23, there are two major groups of nerves: spinal and cranial. We looked at the former in detail in Exercise 24, and here we look at the latter. The cranial nerves emerge from the brain at specific locations and pass through various foramina of the skull to reach the peripheral structures they innervate. All spinal and cranial nerves occur in pairs: 12 pairs of cranial nerves and 31 pairs of spinal nerves.

Each cranial nerve pair is identified by name, roman numeral, and type of information conducted. Some cranial nerves are entirely **sensory nerves**, but most are **mixed nerves** with both motor neurons and sensory neurons. Cranial nerves that primarily conduct motor commands are considered **motor nerves,** even though they have a few sensory fibers to inform the brain about muscle tension and position. (The sense that controls muscle tension and position is called *proprioception*.)

Figure 25.11 shows the position of each cranial nerve on the inferior surface of the brain. Table 25.1 summarizes the cranial nerves and includes the foramen through which each nerve passes.

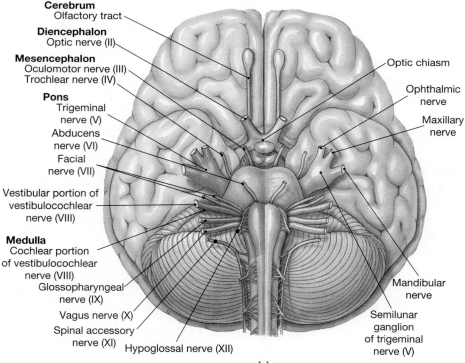

Figure 25.11 Inferior View of the Human Brain

(a)

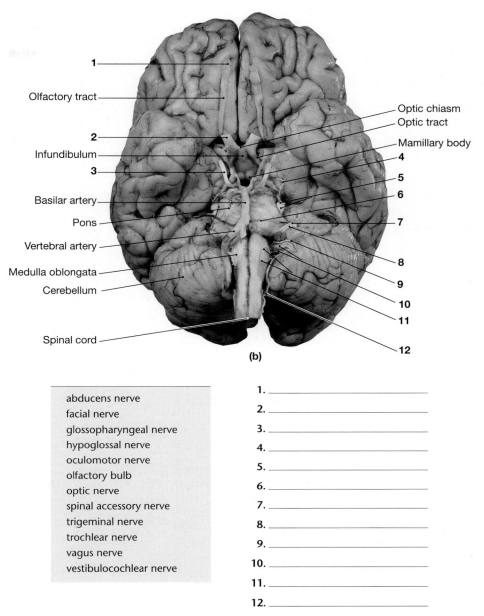

Olfactory tract

Infundibulum

Basilar artery

Pons

Vertebral artery

Medulla oblongata

Cerebellum

Spinal cord

Optic chiasm
Optic tract
Mamillary body

(b)

abducens nerve
facial nerve
glossopharyngeal nerve
hypoglossal nerve
oculomotor nerve
olfactory bulb
optic nerve
spinal accessory nerve
trigeminal nerve
trochlear nerve
vagus nerve
vestibulocochlear nerve

1. _____
2. _____
3. _____
4. _____
5. _____
6. _____
7. _____
8. _____
9. _____
10. _____
11. _____
12. _____

Figure 25.11 (*Continued*)

Olfactory Nerve (I)

The **olfactory nerve** is composed of sensory fibers for the sense of smell and is located in the roof of the nasal cavity. The fibers terminate at the olfactory bulb, which then extends as the **olfactory tract**. This tract is part of the cerebrum and for this reason is usually not called a cranial nerve.

Optic Nerve (II)

The **optic nerve** carries visual information. This nerve originates in the retina, the neural part of the eye that is sensitive to changes in the amount of light entering the eye. The nerve is easy to identify as the X at the **optic chiasm** inferior to the hypothalamus. It is at this point that some of the sensory fibers cross to the nerve on the opposite side of the brain. The optic nerve enters the thalamus, which relays the visual signal to the occipital lobe. Some of the fibers enter the superior colliculus for visual reflexes.

Table 25.1 **The Cranial Nerves**

Cranial Nerve (number)	Sensory Ganglion	Branch	Primary Function	Foramen	Innervation
Olfactory (I)			Special sensory	Cribriform plate of ethmoid bone	Olfactory epithelium
Optic (II)			Special sensory	Optic canal	Retina of eye
Oculomotor (III)			Motor	Superior orbital fissure	Inferior, medial, superior rectus, inferior oblique and levator palpebrae superioris muscles; intrinsic eye muscles
Trochlear (IV)			Motor	Superior orbital fissure	Superior oblique muscle
Trigeminal (V)	Semilunar		Mixed		Areas associated with the jaws
		Ophthalmic	Sensory	Superior orbital fissure	Orbital structures, nasal cavity, skin of forehead, upper eyelid, eyebrows, nose (part)
		Maxillary	Sensory	Foramen rotundum	Lower eyelid; superior lip, gums, and teeth; cheek, nose (part), palate and pharynx (part)
		Mandibular	Mixed	Foramen ovale	Sensory: inferior gums, teeth, lips; palate (part) and tongue (part) Motor: muscles of mastication
Abducens (VI)			Motor	Superior orbital fissure	Lateral rectus muscle
Facial (VII)	Geniculate		Mixed	Internal acoustic canal to facial canal; exits at stylomastoid foramen	Sensory: taste receptors on anterior 2/3 of tongue Motor: muscles of facial expression, lacrimal gland, submandibular gland, sublingual salivary glands
Vestibulocochlear (Acoustic) (VIII)			Special sensory	Internal acoustic canal	
		Cochlear Vestibular			Cochlea (receptors for hearing) Vestibule (receptors for motion and balance)
Glossopharyngeal (IX)	Superior (jugular) and inferior (petrosal)		Mixed	Jugular foramen	Sensory: posterior 1/3 of tongue; pharynx and palate (part); receptors for blood pressure, pH, oxygen, and carbon dioxide concentrations Motor: pharyngeal muscles and parotid salivary gland
Vagus (X)	Jugular and nodose		Mixed	Jugular foramen	Sensory: pharynx; auricle and external auditory canal; diaphragm; visceral organs in thoracic and abdominopelvic cavities Motor: palatal and pharyngeal muscles, and visceral organs in thoracic and abdominopelvic cavities
Accessory (XI)		Medullary	Motor	Jugular foramen	Skeletal muscles of palate, pharynx, and larynx (with vagus nerve)
		Spinal	Motor	Jugular foramen	Sternocleidomastoid and trapezius muscles
Hypoglossal (XII)			Motor	Hypoglossal canal	Tongue musculature

Oculomotor Nerve (III)

The **oculomotor nerve** innervates four extrinsic eyeball muscles—the superior, medial, and inferior rectus muscles and the inferior oblique muscle—and the levator palpebrae muscle of the eyelid. Autonomic motor fibers also control the intrinsic muscles of the iris and the ciliary body. The oculomotor nerve is located on the ventral mesencephalon just posterior to the optic nerve.

Trochlear Nerve (IV)

The **trochlear** (TROK-lē-ar) **nerve** supplies motor fibers to the superior oblique muscle of the eye and originates where the mesencephalon joins the pons. The root of the nerve exits the mesencephalon on the lateral surface. Because it is easily cut or twisted off during removal of the dura mater, many dissection specimens do not have this nerve intact. The superior oblique eye muscle passes through a trochlea, or "pulley"; hence the name of the nerve.

Trigeminal Nerve (V)

The **trigeminal** (trī-JEM-i-nal) **nerve** is the largest of the cranial nerves. It is located on the lateral pons near the medulla and services much of the face. In life, the nerve has three branches: *ophthalmic, maxillary,* and *mandibular.* The ophthalmic branch innervates sensory structures of the forehead, orbit, and nose. The maxillary branch contains sensory fibers for structures in the roof of the mouth, including half of the maxillary teeth. The mandibular branch carries the motor portion of the nerve to the muscles of mastication. Sensory signals from the lower lip, gum, muscles of the tongue, and one-third of the mandibular teeth are also part of the mandibular branch.

Abducens Nerve (VI)

The **abducens** (ab-DŪ-senz) **nerve** controls the lateral rectus extrinsic muscle of the eye. When this muscle contracts, the eyeball is abducted; hence the name. The nerve originates on the medulla and is positioned posterior and medial to the trigeminal nerve.

Facial Nerve (VII)

The **facial nerve** is located on the medulla oblongata, posterior and lateral to the abducens nerve. It is a mixed nerve, with sensory fibers for the anterior two-thirds of the taste buds and somatic and autonomic motor fibers. The somatic motor neurons innervate the muscles of facial expression, such as the zygomaticus muscle. Visceral motor neurons control the activity of the salivary glands, lacrimal (tear) glands, and nasal mucous glands.

Clinical Application

I'm So Happy I Could Cry?

Why do we cry? Primarily, crying is a protective mechanism that cleans the surface of the eye after some object has touched the eyeball. Why do we cry when we are sad, though, or sometimes even when we are happy? The answer is that strong emotions, such as sorrow and joy, are coordinated by the sympathetic branch of the autonomic nervous system. Sympathetic innervation regulates secretion by glands, including secretion of tears by the lacrimal glands. Thus, such events as, say, receiving a sentimental gift from a loved one activate sympathetic neurons, which in turn cause the release of tears of joy. ▶

Vestibulocochlear Nerve (VIII)

The **vestibulocochlear nerve**, also called the *auditory nerve,* is a sensory nerve of the inner ear located on the medulla oblongata near the facial nerve. The vestibulocochlear nerve has two branches. The vestibular branch gathers information regarding the sense of balance from the vestibule and semicircular canals of the inner ear. The cochlear branch conducts auditory sensations from the cochlea, the organ of hearing in the inner ear.

Glossopharyngeal Nerve (IX)

The **glossopharyngeal** (glos-ō-fah-RIN-jē-al) **nerve** is a mixed nerve of the tongue and throat. It supplies the medulla oblongata with sensory information from the posterior third of the tongue (remember, the facial nerve innervates the anterior two-thirds of the taste buds) and from the palate and pharynx. This nerve also conveys barosensory and chemosensory information from the carotid sinus and the carotid body, where blood pressure and dissolved blood gases are monitored, respectively. Motor innervation by the glossopharyngeal nerve controls the pharyngeal muscles involved in swallowing and in the activity of the salivary glands.

Vagus Nerve (X)

The **vagus** (VĀ-gus) **nerve** is a complex nerve on the medulla oblongata that has mixed sensory and motor functions. Sensory information from the pharynx, diaphragm, and most of the internal organs of the thoracic and abdominal cavities ascends along the vagus nerve and synapses with autonomic nuclei in the medulla. The motor portion controls the involuntary muscles of the respiratory, digestive, and cardiovascular systems. The vagus is the only cranial nerve to descend below the neck. It enters the ventral body cavity, but it does not pass to the thorax via the spinal cord; rather, it follows the musculature of the neck. Because this nerve regulates the activities of the organs of the thoracic and abdominal cavities, disorders of the nerve result in systemic disruption of homeostasis. Parasympathetic fibers in the vagus nerve control swallowing, digestion, heart rate, and respiratory patterns, and if this control is compromised, sympathetic stimulation goes unchecked and the organs respond as during exercise or stress. The cardiovascular and respiratory systems increase their activities, and the digestive system shuts down.

Spinal Accessory Nerve (XI)

The **spinal accessory nerve** is a motor nerve controlling the skeletal muscles involved in swallowing and the sternocleidomastoid and trapezius muscles of the neck. It is the only cranial nerve with fibers originating from both the medulla oblongata and the spinal cord. Numerous threadlike branches from these two regions unite in the spinal accessory nerve.

Hypoglossal Nerve (XII)

The **hypoglossal** (hī-pō-GLOS-al) **nerve** is positioned more to the midline than the other cranial nerves on the medulla. This motor nerve supplies motor fibers that control tongue movements for speech and swallowing.

QuickCheck Questions

4.1 What are the three major types of cranial nerves?

4.2 To which parts of the brain do the cranial nerves attach?

4 *Materials*

- ☐ Brain model and/or charts
- ☐ Isopropyl (rubbing) alcohol
- ☐ Wintergreen oil
- ☐ Eye chart
- ☐ Sugar solution
- ☐ Quinine solution
- ☐ Tuning fork
- ☐ Beaker of ice and cold probes
- ☐ Beaker of warm water and warm probes

Procedures

1. Review the cranial nerves presented in Figure 25.11, and then label Figure 25.11b.
2. Locate each cranial nerve on brain models and charts.
3. Your instructor may ask you to test the function of selected cranial nerves. Table 25.2 lists the basic tests used to access the general function of each nerve. ■

Table 25.2 Cranial Nerve Tests

Cranial Nerve	Nerve Function Test
I. Olfactory	Hold open container of rubbing alcohol under subject's nose and have subject identify odor. Repeat with open container of wintergreen oil.
II. Optic	Test subject's visual field by moving a finger back and forth in front of subject's eyes. Use eye chart to test visual acuity.
III. Oculomotor	Examine subject's pupils for equal size. Have subject follow an object with eyes.
IV. Trochlear	Tested with oculomotor nerve. Have subject roll eyes downward.
V. Trigeminal	Check motor functions of nerve by having subject move mandible in various directions. Check sensory functions with warm and cold probes on forehead, upper lip, and lower jaw.
VI. Abducens	Tested with oculomotor nerve. Have subject move eyes medially.
VII. Facial	Use sugar solution to test anterior of tongue for sweet taste reception. Observe facial muscle contractions for even muscle tone on each side of face while subject smiles, frowns, and purses lips.
VIII. Vestibulocochlear	Cochlear branch—Hold vibrating tuning fork in air next to ear, and then touch fork to mastoid process for bone-conduction test. Vestibular branch—Have subject close eyes and maintain balance.
IX. Glossopharyngeal	While subject coughs, check position of uvula on posterior of soft palate. Use quinine solution to test posterior of tongue for bitter taste reception.
X. Vagus	While subject coughs, check position of uvula on posterior of soft palate.
XI. Spinal accessory	Hold subject's shoulder while the subject rotates it to test the strength of sternocleidomastoid muscle. Hold head while subject rotates it to test trapezius strength.
XII. Hypoglossal	Observe subject protract and retract tongue from mouth, checking for even movement on two lateral edges of tongue.

LAB ACTIVITY 5 Sheep Brain Dissection

The sheep brain, like all other mammalian brains, is similar in structure and function to the human brain. One major difference between the human brain and that of other animals is the orientation of the brain stem. Humans are vertical animals and walk on two legs. The spinal cord is perpendicular to the ground, and so the brain stem must also be vertical. In four-legged animals, the spinal cord and brain stem are parallel to the ground.

All vertebrate animals—sharks, fish, amphibians, reptiles, birds, and mammals—have a brain stem for basic body functions. These animals can learn through experience, a complex neurological process that requires higher-level processing and memory storage, as occurs in the human cerebrum. Imagine the complex motor activity necessary for locomotion in these animals.

Dissecting a sheep brain will enhance your study of models and charts of the human brain. Take your time during the dissection, and follow the directions carefully. Refer to this laboratory manual and its illustrations often during the procedures.

 Safety Alert

You must—repeat, *must*—practice the highest level of laboratory safety while handling and dissecting the brain. Keep the following guidelines in mind during the dissection:

1. Wear gloves and safety glasses to protect yourself from the fixatives used to preserve the specimen.
2. Do not dispose of the fixative from your specimen. You will later store the specimen in the fixative to keep the specimen moist and to keep it from decaying.
3. Be extremely careful when using a scalpel or other sharp instrument. Always direct cutting and scissor motion away from you to prevent an accident if the instrument slips on moist tissue.
4. Before cutting a given tissue, make sure it is free from underlying and/or adjacent tissues so that they will not be accidentally severed.
5. Never discard tissue in the sink or trash. Your instructor will inform you of the proper disposal procedure. ▲

5 Materials

☐ Gloves
☐ Safety glasses
☐ Dissecting tools
☐ Dissection tray
☐ Preserved sheep brain (preferably with dura mater intact)

Procedures

I. The Meninges

If your sheep brain does not have the dura mater, skip to part II.

1. On the intact dura mater, locate the falx cerebri and the tentorium cerebelli on the overlying dorsal surface of the dura mater. How does the tissue of the falx cerebri compare with the dura mater covering the hemispheres?
2. If your specimen still has the ethmoid, a mass of bone on the anterior frontal lobe, slip a probe between the bone and the dura mater. Carefully pull the bone off the specimen, using scissors to snip away any attached dura mater. Examine the removed ethmoid and identify the **crista galli**, which is the crest of bone where the meninges attach.
3. Gently insert a probe between the dura mater and the brain to separate the two. With scissors, cut around the base of the dura mater, leaving the inferior portion intact over the cranial nerves. Use the scissors sparingly, and be careful not to cut or remove any of the cranial nerves. To detach the dura mater in one piece, grasp it at the tentorium cerebelli deep between the cerebellum and the cerebrum, and pull it straight out.
4. Open the detached dura mater and identify the falx cerebri and tentorium cerebelli. (One difference between the sheep brain and the human brain is that the sheep brain does not have a falx cerebelli.)

II. External Brain Anatomy

1. Examine the cerebrum, identifying the frontal, parietal, occipital, and temporal lobes. The insula is a deep lobe and is not visible externally.
   - How deep is the longitudinal fissure separating the right and left cerebral hemispheres?
   - Observe the gyri and sulci on the cortical surface.
   - Examine the surface between sulci for the arachnoid and pia mater.
2. View the dorsal surface of the cerebellum. Compare the size of the folia with the size of the cerebral gyri. Unlike the human brain, the sheep cerebellum is not divided medially into two lateral hemispheres.

3. To examine the dorsal anatomy of the mesencephalon, hold the sheep brain as in Figure 25.12 and gently depress the cerebellum. The mesencephalon will then be visible between the cerebrum and cerebellum.

   - Identify the medial pineal gland and the four elevated masses of the corpora quadrigemina.

   - Distinguish between the superior colliculi and the inferior colliculi. What is the function of these masses?

4. Turn the brain to view the ventral surface, as in Figure 25.13. Note how the spinal cord joins the medulla oblongata.

   - Identify the pons and the cerebral peduncles of the mesencephalon.

   - Locate the single mamillary body on the hypothalamus. (Remember that the mamillary body of the human brain is a *paired* mass.)

   - The pituitary gland has most likely been removed from your specimen; however, you can still identify the stub of the infundibulum that attaches the pituitary to the hypothalamus.

5. Using Figure 25.13 as a guide, identify as many cranial nerves on your sheep brain as possible. Nerves I through III and nerve V are usually intact and easy to identify. Your laboratory instructor may ask you to observe several sheep brains in order to study all the cranial nerves. The three branches of the trigeminal nerve were cut when the brain was removed from the sheep and therefore are not present on any specimen. The glossopharyngeal nerve may have been removed inadvertently when the specimen was being prepared; even if this nerve is present in your specimen, however, you should note that it is difficult to identify on the sheep brain.

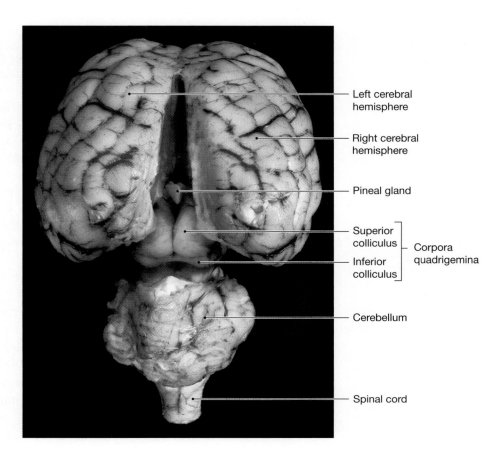

Figure 25.12 The Mesencephalon of the Sheep

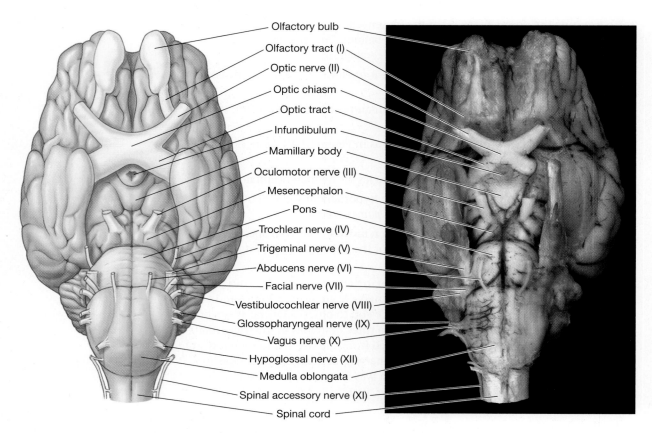

Olfactory bulb
Olfactory tract (I)
Optic nerve (II)
Optic chiasm
Optic tract
Infundibulum
Mamillary body
Oculomotor nerve (III)
Mesencephalon
Pons
Trochlear nerve (IV)
Trigeminal nerve (V)
Abducens nerve (VI)
Facial nerve (VII)
Vestibulocochlear nerve (VIII)
Glossopharyngeal nerve (IX)
Vagus nerve (X)
Hypoglossal nerve (XII)
Medulla oblongata
Spinal accessory nerve (XI)
Spinal cord

Figure 25.13 Ventral View of Sheep Brain, Showing Origins of Cranial Nerves

III. Internal Brain Anatomy

1. Lay the sheep brain on the dissecting tray, and place the blade of a large butcher knife in the anterior region of the longitudinal fissure, pointing the blade tip downward. With a single smooth motion, move the knife through the brain from anterior to posterior to separate the right and left halves along the midsagittal plane. (If only a scalpel is available, bisect the brain with as few long, smooth cuts as possible.)

2. Using Figure 25.14 as a guide, identify the following regions of the internal anatomy:

   • Locate the corpus callosum, anterior commissure, and fornix, all white tracts interconnecting regions of the cerebrum.

   • Gently slide a blunt probe between the corpus callosum and fornix and inside the lateral ventricle. How deep does the ventricle extend into the cerebrum?

   • Locate the septum pellucidum, the membrane between the fornix and corpus callosum that separates the two lateral ventricles.

   • Inside the lateral ventricle, locate the choroid plexus, which appears as a granular mass of tissue.

3. Locate the intermediate mass surrounded by the third ventricle. The lateral walls of the third ventricle are the medial margins of the thalamus.

   • Inferior to the intermediate mass is the hypothalamus.

   • Identify the infundibulum and pituitary gland if they are present in your specimen.

   • The posterior mass of the hypothalamus is the mamillary body.

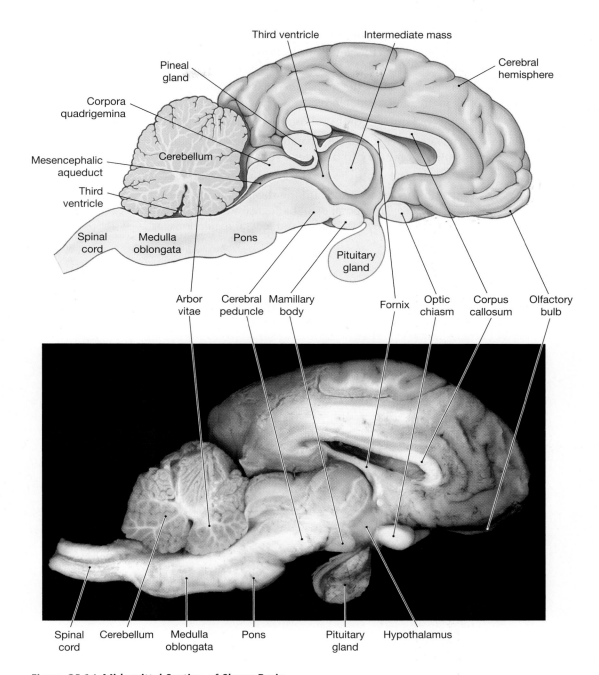

Figure 25.14 Midsagittal Section of Sheep Brain

4. Next examine the cerebellum, located superior to the medulla and pons.
   - In sagittal section the white matter of the arbor vitae is clearly visible.
   - Between the cerebellum and brain stem is the fourth ventricle. In the sheep brain, which is oriented on a horizontal axis, the fourth ventricle is inferior to the cerebellum.
5. To view deep structures of the cerebrum and diencephalon, make two frontal cuts on the brain and examine the resulting slice.
   - Put the brain halves together, and lay them on the dissection tray with the ventral surface facing up. Use a large knife to section the brain along a frontal plane through the infundibulum.

- Make another frontal cut to slice off a thin slab of brain.
- Lay the slab on the tray with the anterior surface upward, the area where you made your first cut.

6. Using Figure 25.8 of the human brain as your guide, notice the distribution of gray matter and white matter.

- Observe how the corpus callosum joins each cerebral hemisphere.
- The lateral ventricles are separated by the thin septum pellucidum. The third ventricle is inferior to the lateral ventricles.
- Lateral to the ventricles are the gray nuclei of the cerebral nuclei.

7. When finished, store or discard the sheep brain as directed by your laboratory instructor. Proper disposal of all biological waste protects the local environment and is mandated by local, state, and federal regulations. ■

Anatomy of the Brain

Name _____

Date _____

Section _____

A. Matching

Match each structure in the left column with its correct description from the right column.

_____ **1.** folia

_____ **2.** cerebrum

_____ **3.** mamillary body

_____ **4.** longitudinal fissure

_____ **5.** inferior colliculus

_____ **6.** optic chiasm

_____ **7.** falx cerebri

_____ **8.** hypothalamus

_____ **9.** central sulcus

_____ **10.** cerebral peduncles

_____ **11.** dura mater

_____ **12.** vermis

_____ **13.** subarachnoid space

_____ **14.** pons

_____ **15.** tentorium cerebelli

A. area where optic nerve crosses to opposite side of brain

B. forms floor of diencephalon

C. part of mesencephalon

D. outer meningeal layer

E. site of cerebrospinal fluid circulation

F. small folds on cerebellum

G. narrow central region of cerebellum

H. separates cerebellum and cerebrum

I. mass posterior to infundibulum

J. area of brain superior to medulla

K. divides motor and sensory cortex

L. tissue between cerebral hemispheres

M. contains five lobes

N. part of corpora quadrigemina

O. cleft between cerebral hemispheres

B. Matching

Match each structure in the left column with its correct description from the right column.

_____ **1.** third ventricle

_____ **2.** septum pellucidum

_____ **3.** thalamus

_____ **4.** corpus callosum

_____ **5.** pineal gland

_____ **6.** superior colliculus

_____ **7.** arachnoid granulation

_____ **8.** fornix

_____ **9.** mesencephalic aqueduct

_____ **10.** arbor vitae

A. part of corpora quadrigemina

B. white tract between cerebral hemispheres

C. duct through mesencephalon

D. white matter of cerebellum

E. site of cerebrospinal fluid reabsorption

F. white matter inferior to lateral ventricles

G. chamber of diencephalon

H. nipple-like gland in diencephalon

I. forms lateral walls of third ventricle

J. separates lateral ventricles

25

C. Labeling

Label Figure 25.15, which shows the inferior surface of the brain.

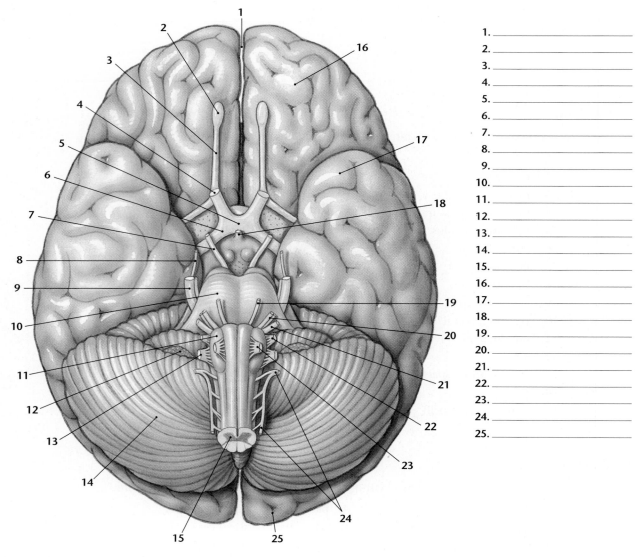

1. _____
2. _____
3. _____
4. _____
5. _____
6. _____
7. _____
8. _____
9. _____
10. _____
11. _____
12. _____
13. _____
14. _____
15. _____
16. _____
17. _____
18. _____
19. _____
20. _____
21. _____
22. _____
23. _____
24. _____
25. _____

Figure 25.15 Inferior Surface of the Brain

D. Labeling

Label Figure 25.16, a midsagittal close-up view of the human brain.

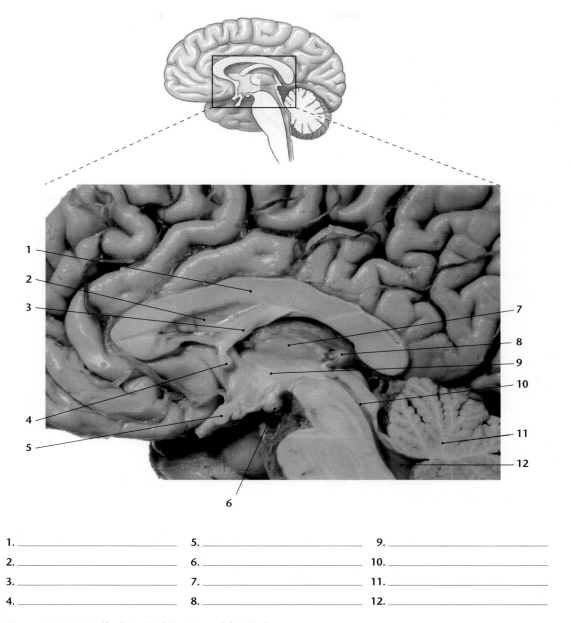

1. _____ 5. _____ 9. _____

2. _____ 6. _____ 10. _____

3. _____ 7. _____ 11. _____

4. _____ 8. _____ 12. _____

Figure 25.16 Detail of Sagittal Section of the Brain

E. Short-Answer Questions

1. List the six major regions of the brain.

2. Which cranial nerves conduct the sensory and motor impulses of the eye?

3. List the location and specific anatomy of the corpora quadrigemina.

4. Describe the extensions of the dura mater.

5. What is the function of the precentral gyrus?

6. Trace a drop of CSF from a lateral ventricle to reabsorption at an arachnoid granulation.

F. Analysis and Application

1. Imagine watching a bird fly across your line of vision. What part of your brain is active in keeping an image of the moving bird on your retina?

2. You have just eaten a medium-sized pepperoni pizza and lain down to digest it. Which cranial nerve stimulates the muscular activity of your digestive tract?

3. A child is preoccupied with a large cherry lollipop. What part of the child's brain is responsible for the licking and eating reflexes?

4. Your favorite movie has made you cry yet again. Which cranial nerve is responsible for your tears?

5. A patient is brought into the emergency room with severe whiplash. He is not breathing and has lost cardiac function. What part of the brain has most likely been damaged?

6. A woman is admitted to the hospital with Bell's palsy caused by an inflamed facial nerve. What symptoms will you, as the attending physician, observe, and how would you test her facial nerve?

Autonomic Nervous System

OBJECTIVES

On completion of this exercise, you should be able to:

- Outline a typical autonomic pathway.

- Compare the location of the preganglionic outflow from the CNS in the sympathetic and parasympathetic divisions.

- Compare the lengths of and the neurotransmitters released by each fiber in the sympathetic and parasympathetic divisions.

- Trace the sympathetic pathways into a chain ganglion, into a collateral ganglion, and into the adrenal medulla.

- Trace the parasympathetic pathways into cranial nerves III, VII, IX, and X, and into the pelvic nerves.

- Compare the sympathetic and parasympathetic responses.

All motor commands from the central nervous system (CNS) are communicated to the efferent branch of the peripheral nervous system (PNS). This PNS motor branch is divided into the somatic and autonomic motor systems. Voluntary motor signals to skeletal muscles are carried by the **somatic nervous system**. Involuntary motor instructions to smooth muscles, heart muscle, and glands, collectively called *visceral effectors,* are relayed along the **autonomic nervous system (ANS)**. If necessary, review the organization of the nervous system in Exercise 23.

The ANS is anatomically subdivided into the **sympathetic division** and the **parasympathetic division** (Figure 26.1). The parasympathetic, or **craniosacral** (krā,-nē-ō-SĀ-krul), **division** is composed of efferent neurons exiting the cranium in certain cranial nerves and efferent neurons that descend the spinal cord and exit at the sacral level. The sympathetic, or **thoracolumbar** (tho-ra-kō-LUM-bar), **division** includes efferent neurons that descend the spinal cord to the thoracic and upper lumbar levels and exit the CNS at these levels.

All autonomic motor pathways consist of two groups of neurons, **preganglionic** and **ganglionic**, so named because they synapse with each other in bulb-like PNS structures called autonomic **ganglia**. The portion of any autonomic neuron between the CNS and an autonomic ganglion is called a **preganglionic neuron**; the portion between the autonomic ganglion and the target muscle is a **ganglionic neuron** (Figure 26.2). Preganglionic neurons of the sympathetic division descend from the brain into either the thoracic region or the lumbar

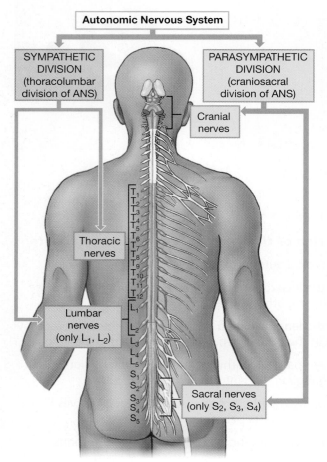

Figure 26.1
The Autonomic Nervous System

region of the spinal cord, and those of the parasympathetic division either descend from the brain to the sacral region of the spinal cord or else exit the brain via a cranial nerve. Preganglionic axons, called **preganglionic fibers**, synapse with ganglionic neurons in the autonomic ganglia. Ganglionic axons, called **ganglionic fibers**, synapse with smooth muscles, the heart, and glands.

Sympathetic and parasympathetic subdivisions of the ANS differ in three major respects:

Location of Preganglionic Exit Points from the CNS. All autonomic preganglionic neurons originate in the brain stem. Where these neurons have their cell bodies and where they exit the CNS differ between the sympathetic and parasympathetic motor pathways. The sympathetic preganglionic neurons exit the spinal cord at segments T_1 through L_2. (This is why an alternative name for the sympathetic division is the thoracolumbar division.) The parasympathetic preganglionic neurons either exit the brain in cranial nerves or descend the spinal cord and exit the CNS in the sacral region (thus the alternative name craniosacral division).

Location of Autonomic Ganglia in the PNS. All autonomic ganglia are in the PNS. Their proximity to the CNS, however, differs between sympathetic ganglia and parasympathetic ganglia. Sympathetic ganglia— the sympathetic *chain ganglia* and *collateral ganglia*—are located close to the spinal cord. This anatomical location of the ganglia results in short preganglionic neurons and long ganglionic neurons. Parasympathetic ganglia are called **intramural ganglia**. Intramural means "within the walls," which perfectly describes the location of the ganglia, embedded in the walls of the visceral effectors. Parasympathetic preganglionic neurons are long and extend from the brain or the sacral spinal segments into the intramural ganglia. Parasympathetic ganglionic neurons are short because the intramural ganglia are already in the organ the ganglionic neurons innervate. In Figure 26.2, notice both the difference in the locations of the ganglia and the difference in the preganglionic and ganglionic lengths.

Effect of Neurotransmitter on Effectors. Most involuntary effectors have **dual innervation**, which means they are innervated by both sympathetic ganglionic neurons and parasympathetic ganglionic neurons. Thus the two divisions of the ANS share the role of regulating autonomic function. Typically, one division stimulates a given effector, and the other division inhibits that same effector. The preganglionic neurons of both divisions release acetylcholine (ACh) into a ganglion, but the *ganglionic* neurons of the two divisions release different neurotransmitters to the target effector cells. During times of excitement, emotional stress, and emergencies, sympathetic ganglionic neurons release norepinephrine (NE) to effectors and cause a **fight-or-flight response** that increases overall alertness. Heart rate, blood pressure, and respiratory rate all increase, sweat glands begin to secrete, and digestive and urinary functions cease. Parasympathetic ganglionic neurons release ACh, which slows the body for normal, energy-conserving homeostasis. This **rest-and-digest response** decreases cardiovascular and respiratory activity and increases the rate at which food and wastes are processed.

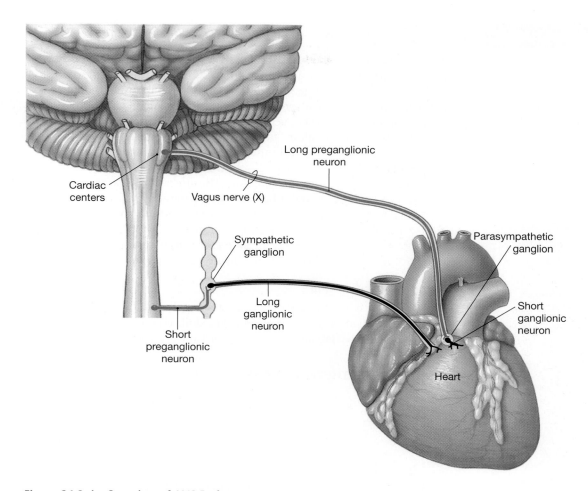

Figure 26.2 An Overview of ANS Pathways

Sympathetic pathways consist of short preganglionic neurons that release acetylcholine (ACh) in sympathetic ganglia. They synapse with long ganglionic neurons that release norepinephrine (NE) at the visceral effector. The sympathetic response is generalized as a "fight or flight" response. Parasympathetic pathways in the sacral spinal segments have long preganglionic neurons that exit the spinal cord in pelvic nerves instead of ventral roots. They release ACh in intramural ganglia in the walls of the viscera. Preganglionic neurons synapse with short ganglionic neurons that also release ACh. The general parasympathetic response is "rest and digest."

Study Tip | *ANS Function*

An easy way to remember the sympathetic fight-or-flight response is to consider how your organs must adjust their activities during an emergency situation, such as being chased by a wild animal. Sympathetic impulses increase your heart rate, blood pressure, and respiratory rate. Muscle tone increases in preparation for fighting off the animal or running for your life. As arterioles in skeletal muscles dilate, more blood flows into the muscles to support their high activity level. Sympathetic stimulation decreases the activity of your digestive tract. An emergency situation is not the time to work at digesting your lunch! With digestive actions slowed, the body shunts blood from the abdominal organs and delivers more blood to skeletal muscles. Once you are out of danger, sympathetic stimulation decreases and parasympathetic stimulation predominates to return your body to the routine "housekeeping" chores of digesting food, eliminating wastes, and conserving precious cell energy. ●

The Sympathetic (Thoracolumbar) Division

The organization of the sympathetic division of the ANS is diagrammed in Figure 26.3. Preganglionic neurons originate in the pons and the medulla of the brain stem. These autonomic motor neurons descend in the spinal cord to the thoracic and lumbar segments, where their somae are located in the lateral gray horns. Preganglionic axons exit the spinal cord in ventral roots and pass into a spinal nerve that branches into an autonomic ganglion. In the ganglion, the preganglionic neuron synapses with a ganglionic neuron that innervates a visceral effector.

In a typical sympathetic pathway, the short sympathetic preganglionic fibers release acetylcholine into the ganglion. The effect of ACh at this synapse is always excitatory to the ganglionic fiber. The long ganglionic axon then releases norepinephrine onto the effector. How the NE affects the effector depends on the type of NE receptors present in the effector's cell membrane. Generally, the sympathetic response is to prepare the body for increased activity or a crisis situation; this is the fight-or-flight response that occurs during exercise, excitement, and emergencies.

Sympathetic Ganglia

Three types of sympathetic ganglia occur: sympathetic chain ganglia, collateral ganglia, and modified ganglia in the adrenal medulla (Figure 26.3). **Sympathetic chain ganglia** are located lateral to the spinal cord and are also called **paravertebral gan-**

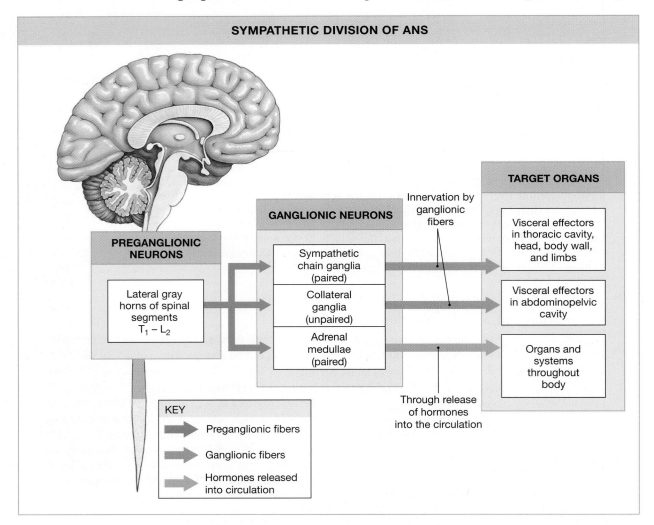

Figure 26.3 **Organization of the Sympathetic Division of the ANS**

glia. Ganglionic neurons that exit sympathetic chain ganglia innervate the visceral effectors of the thoracic cavity, head, body wall, and limbs.

Collateral ganglia are located anterior to the vertebral column and contain ganglionic neurons that lead to organs in the abdominopelvic cavity. The preganglionic fibers associated with collateral ganglia pass through the sympathetic chain ganglia without synapsing and join to form a network called the **splanchnic** (SPLANK-nik) **nerves** (Figure 26.4). This network divides and sends branches into the collateral ganglia, where the preganglionic fibers synapse with ganglionic neurons. The ganglionic fibers then synapse with abdominopelvic visceral effectors. The collateral ganglia are named after the adjacent blood vessels. The **celiac** (SĒ-lē-ak) **ganglion** supplies the liver, gallbladder, stomach, pancreas, and spleen, as Figure 26.4 shows. The **superior mesenteric ganglion** innervates the small intestine and parts of the large intestine. The **inferior mesenteric ganglion** controls most of the large intestine, the kidneys, the bladder, and the sex organs.

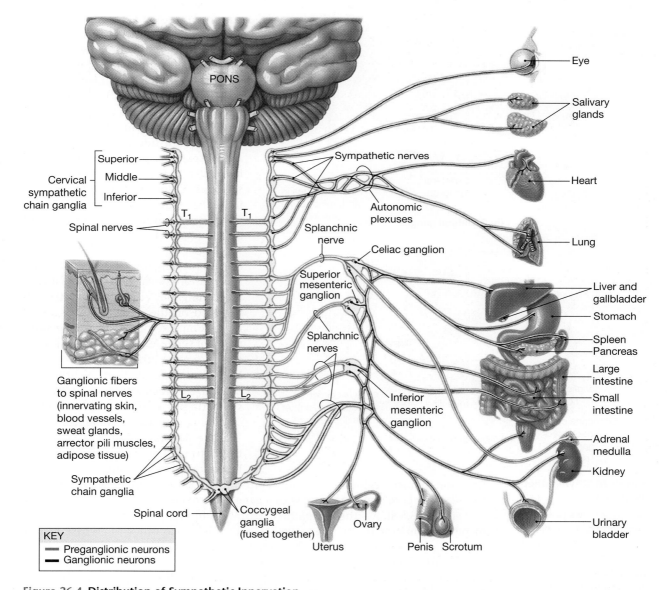

Figure 26.4 Distribution of Sympathetic Innervation

The distribution of sympathetic fibers is the same on both sides of the body. For clarity, the innervation of somatic structures is shown here on the left, and the innervation of visceral structures on the right.

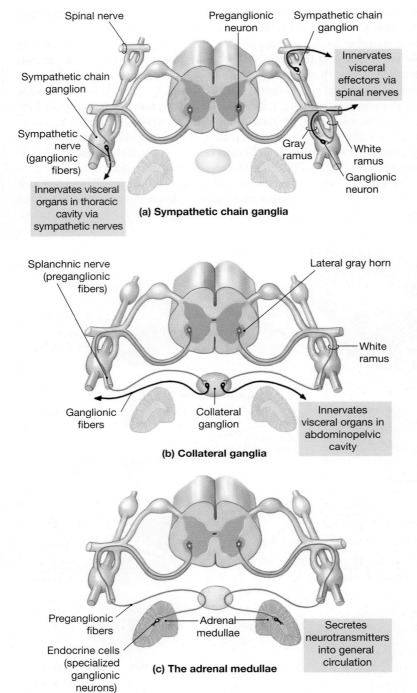

(a) Sympathetic chain ganglia

Spinal nerve

Sympathetic chain ganglion

Sympathetic chain ganglion

Preganglionic neuron

Sympathetic chain ganglion

Innervates visceral effectors via spinal nerves

Gray ramus

White ramus

Ganglionic neuron

Sympathetic nerve (ganglionic fibers)

Innervates visceral organs in thoracic cavity via sympathetic nerves

(b) Collateral ganglia

Splanchnic nerve (preganglionic fibers)

Lateral gray horn

White ramus

Ganglionic fibers

Collateral ganglion

Innervates visceral organs in abdominopelvic cavity

(c) The adrenal medullae

Preganglionic fibers

Adrenal medullae

Secretes neurotransmitters into general circulation

Endocrine cells (specialized ganglionic neurons)

Figure 26.5 Sympathetic Pathways

Superior views of sections through the thoracic spinal cord, showing the distribution of preganglionic and postganglionic fibers.

The adrenal glands are positioned on top of the kidneys. Each adrenal gland has an outer cortex layer that produces hormones and an inner **adrenal medulla** containing sympathetic ganglia and ganglionic neurons. During sympathetic stimulation, the ganglionic neurons in the medulla, like other sympathetic ganglionic neurons, release epinephrine into the bloodstream and contribute to the overall sympathetic fight-or-flight response.

The Sympathetic Pathways

Figure 26.4 outlines the general distribution of the sympathetic pathways, showing the sympathetic chain ganglia positioned lateral to the lower portion of the spinal cord. For simplicity, only somatic innervation is shown to the left of the spinal cord and only visceral innervation is shown to the right. In real life, sympathetic innervation is the same on both sides of the spinal cord. All sympathetic preganglionic neurons leave the thoracic and lumbar spinal segments, enter a spinal nerve, and pass into the sympathetic chain ganglia. Ganglionic neurons exit the ganglia and extend to the viscera.

Notice on the left of Figure 26.4 that all preganglionic fibers for somatic innervation synapse in a sympathetic chain ganglion. Visceral innervation, illustrated on the right of the figure, is more complex. A preganglionic neuron may not synapse with a ganglionic neuron in a sympathetic chain ganglion. Instead, the preganglionic neuron may enter a collateral ganglion or another sympathetic chain ganglion before synapsing with a ganglionic neuron.

The adrenal medulla is shown at the bottom right of Figure 26.4. Note how a long preganglionic neuron passes from the thoracic region of the spinal cord through several ganglia before entering the adrenal gland and synapsing with a ganglionic neuron (not shown).

Figure 26.5 shows the sympathetic pathways in more detail. The pathway utilizing the sympathetic chain ganglia passes through areas called the **white ramus** and the **gray ramus**. (Collectively, these two regions are known as the **rami communicantes.**) Once a preganglionic fiber enters a sympathetic chain ganglion via the white ramus, the fiber usually synapses with a ganglionic neuron, as shown in Figure 26.5a. The ganglionic fiber exits the sympathetic chain ganglion via either the gray ramus or an autonomic nerve. The gray ramus directs the ganglionic fiber into a spinal nerve leading to a general somatic structure, such as blood vessels supplying skeletal muscles. A ganglionic fiber in an autonomic nerve passes into the thoracic cavity to innervate the thoracic viscera.

Notice in Figure 26.5a that all the sympathetic chain ganglia on the same side of the spinal cord are interconnected. A single preganglionic neuron may enter one sympathetic chain ganglion and branch into many different chain ganglia, to synapse with up to 32 ganglionic neurons. This fanning-out of preganglionic neurons within the sympathetic chain ganglia contributes to the widespread effect that sympathetic stimulation has on the body.

Figure 26.5b details the pathway involving collateral ganglia. Note how the preganglionic axons pass through the chain ganglia and enter the splanchnic nerve (described earlier) before entering the collateral ganglia.

Sympathetic neurons supplying the adrenal gland do not synapse in a sympathetic chain ganglion or a collateral ganglion. Instead, the preganglionic fibers penetrate deep into the adrenal gland and synapse with ganglionic neurons in the adrenal medulla, as noted earlier.

Clinical Application	### Stress and the ANS

Stress stimulates the body to increase sympathetic commands from the ANS. Appetite may decrease while blood pressure and general sensitivity to stimuli may increase. The individual may become irritable and have difficulty sleeping and coping with day-to-day responsibilities. Prolonged stress can lead to disease. Coronary diseases, for example, are more common in individuals who are employed in stressful occupations or live in stressful environments. ▶

QuickCheck Questions

1.1 Why is the sympathetic division of the ANS also called the thoracolumbar division?

1.2 What is the body's general response to sympathetic stimulation?

1.3 How do the heart, lungs, and digestive tract respond to sympathetic stimulation?

1 Materials

☐ Nervous system chart
☐ Spinal cord model

Procedures

1. Review the anatomy and pathways presented in Figures 26.1 through 26.5.

2. On a chart of the nervous system, identify the lateral gray horns, ventral root, spinal nerve, white and gray rami, and 2 sympathetic chain ganglion. What is the function of the white and gray rami?

3. Locate the collateral ganglia and the adrenal gland on a spinal cord model. Where are the collateral ganglia located relative to the spinal cord?

4. On the nervous system chart, trace the following sympathetic pathways:

   a. A preganglionic fiber synapsing in a collateral ganglion.

   b. A ganglionic fiber exiting a chain ganglion and passing into a spinal nerve.

   c. A preganglionic fiber synapsing in the adrenal medulla. ■

LAB ACTIVITY 2   The Parasympathetic (Craniosacral) Division

The organization of the parasympathetic division of the ANS is diagrammed in Figure 26.6. In this division, the preganglionic neurons leave the CNS either via cranial nerves III, VII, IX, and X, or via the sacral level of the spinal cord.

In the brain, parasympathetic preganglionic neurons form autonomic nuclei that extend axons into four cranial nerves: oculomotor (III), facial (VII), glossopharyngeal (IX), and vagus (X). The preganglionic axons synapse with ganglionic neurons in one

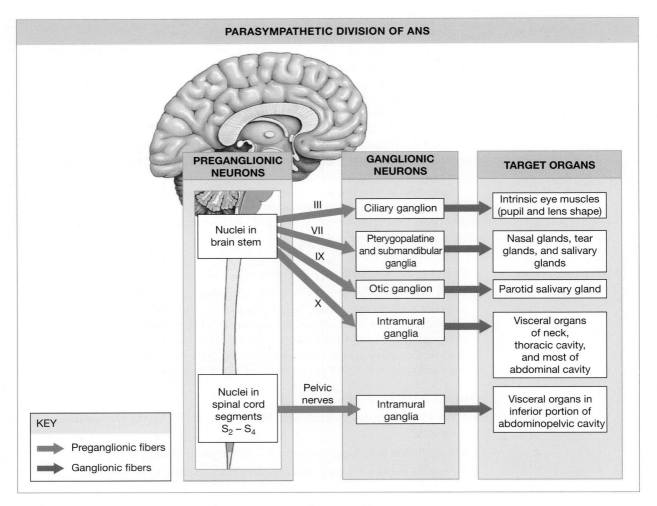

Figure 26.6 Organization of the Parasympathetic Division of the ANS

of the **parasympathetic ganglia** of the head: the **ciliary**, **pterygopalatine**, **submandibular**, or **otic ganglia**. The vagus nerve (X) contains preganglionic fibers that synapse in intramural ganglia. The target organ for each cranial nerve is included in Figures 26.6 and 26.7. If necessary, review the function of each of these cranial nerves in Exercise 25.

The sacral portion of the parasympathetic division contains preganglionic neurons in the sacral segments S_2, S_3, and S_4. The cell bodies of the sacral preganglionic neurons cluster in an autonomic nucleus located in the lateral gray horns of the spinal cord. The preganglionic fibers remain separate from spinal nerves and exit from spinal segments S_2 through S_4 as **pelvic nerves**, shown in Figure 26.7. The long preganglionic fibers extend into intramural ganglia located in the walls of the abdominal organs and then synapse with the short ganglionic neurons that control the viscera. Networks of preganglionic neurons, called the *autonomic plexuses,* occur between the vagus nerve and the pelvic nerves. In these plexuses, sympathetic preganglionic neurons and parasympathetic preganglionic neurons intermingle as they pass to their respective autonomic ganglia.

Parasympathetic preganglionic neurons release acetylcholine, which is always excitatory to a ganglionic fiber. The parasympathetic ganglionic fibers also release ACh to their visceral effectors. How the ACh affects the effectors depends on the type of ACh receptors present in the cell membrane of the effector cells. Generally, the parasympathetic response is a rest-and-digest response that slows body functions and promotes digestion and waste elimination. Table 26.1 and Figure 26.8 compare parasympathetic and sympathetic innervation.

QuickCheck Questions

2.1 Why is the parasympathetic division also called the craniosacral division?

2.2 What is the body's general response to parasympathetic stimulation?

2.3 How do the heart, lungs, and digestive tract respond to parasympathetic stimulation?

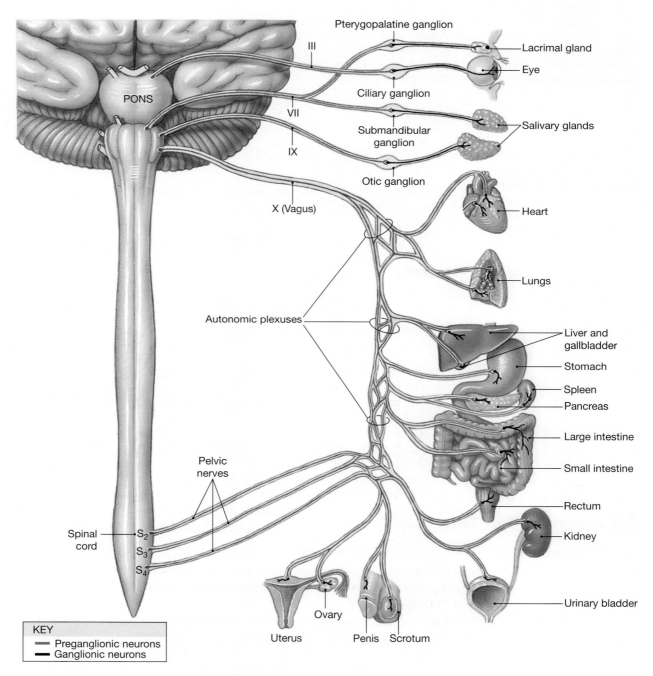

Figure 26.7 Distribution of Parasympathetic Innervation

Table 26.1 Comparison of Sympathetic and Parasympathetic Divisions of Autonomic Nervous System

Characteristic	Sympathetic Division	Parasympathetic Division
Location of CNS visceral motor neurons	Lateral gray horns of spinal segments $T_1 - L_2$	Brain stem and spinal segments $S_2 - S_4$
Location of PNS ganglia	Near vertebral column	Typically intramural
Preganglionic fibers		
Length	Relatively short	Relatively long
Neurotransmitter released	ACh	ACh
Ganglionic fibers		
Length	Relatively long	Relatively short
Neurotransmitter released	Normally NE; sometimes ACh	ACh
Neuromuscular or neuroglandular junction	Varicosities and enlarged terminal knobs that release transmitter near target cells	Junctions that release transmitter to special receptor surface
Degree of divergence from CNS to ganglion cells	Approximately 1:32	Approximately 1:6
General function(s)	Stimulates metabolism; increases alertness; prepares for emergency ("fight or flight")	Promotes relaxation, nutrient uptake, energy storage ("rest and digest")

Figure 26.8 Anatomical Differences between the Sympathetic and Parasympathetic Divisions

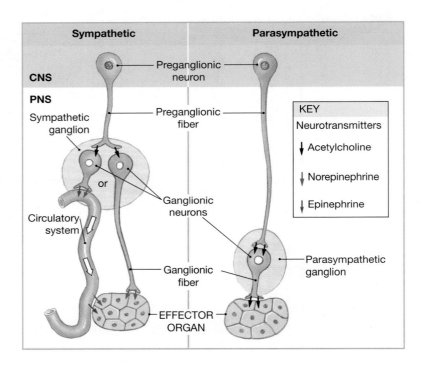

Materials

☐ Nervous system chart

Procedures

1. Review the anatomy and pathways presented in Figures 26.6 and 26.7.

2. On a chart of the nervous system, identify the oculomotor, facial, glossopharyngeal, and vagus cranial nerves. In which part of the brain are these nerves located?

3. On the nervous system chart, trace the following parasympathetic pathways:

   a. A preganglionic fiber entering a pelvic nerve and traveling to the urinary bladder.

   b. The vagus nerve from the brain to the heart.

   c. A preganglionic fiber synapsing in a ciliary ganglion. ■

Autonomic Nervous System

Name _____

Date _____

Section _____

A. Matching

Match each ANS structure in the left column with its correct description from the right column.

_____ 1. preganglionic neuron

_____ 2. gray ramus

_____ 3. adrenal medulla

_____ 4. rami communicantes

_____ 5. thoracolumbar division of ANS

_____ 6. collateral ganglia

_____ 7. intramural ganglia

_____ 8. white ramus

_____ 9. ganglionic neuron

_____ 10. craniosacral division of ANS

_____ 11. pelvic nerves

A. parasympathetic division of ANS

B. ganglia located in visceral walls

C. carries preganglionic fiber into chain ganglion

D. has cell body located in autonomic ganglion

E. releases epinephrine into blood

F. parasympathetic nerves outflowing from sacral spinal segments

G. white and gray rami

H. unpaired ganglia located anterior to spinal cord

I. has cell body located in lateral gray horn of spinal cord

J. routes ganglionic fiber into spinal nerve

K. sympathetic division of ANS

B. Drawing

In Figure 26.9, draw

1. The preganglionic and ganglionic neurons for a sympathetic pathway from the CNS to visceral effectors in the skin.

LABORATORY REPORT

2. The preganglionic and ganglionic neurons for a sympathetic pathway from the CNS to the stomach.

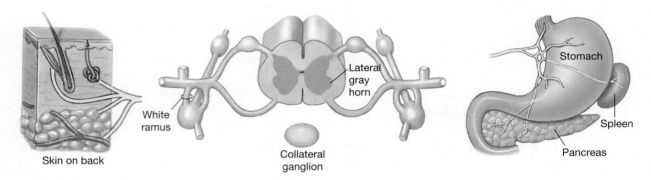

Figure 26.9 Sympathetic Pathways

C. Short-Answer Questions

1. List four responses to sympathetic stimulation and four responses to parasympathetic stimulation.

2. Discuss the anatomy of the sympathetic chain ganglia. How do fibers enter and exit these ganglia?

3. Which cranial nerves are involved in the parasympathetic division of the ANS?

4. Compare the lengths of preganglionic and ganglionic neurons in the sympathetic and parasympathetic divisions of the ANS.

D. Analysis and Application

1. Compare the effect neurotransmitters from sympathetic and parasympathetic ganglionic fibers have on smooth muscle in the digestive tract, on cardiac muscle, and on arterioles in skeletal muscles.

2. As a child, you might have been told to wait for up to an hour after eating before going swimming. Explain the rationale for this statement.

3. Compare the outflow of preganglionic neurons from the CNS in the sympathetic and parasympathetic divisions of the ANS.

General Senses

OBJECTIVES

On completion of this exercise, you should be able to:

- List the receptors for the general senses and those for the special senses.

- Discuss the distribution of cutaneous receptors.

- Describe the two-point discrimination test.

- Describe and give examples of adaptation in sensory receptors.

- Explain how referred pain can occur.

C hanges in the body's internal and external environments are detected by special cells called *sensory receptors*. Most receptors are sensitive to a specific stimulus. The taste buds of the tongue, for example, are stimulated by chemicals dissolved in saliva and not by sound waves or light rays. The human senses may be grouped into two broad categories: general senses and special senses. The **general senses,** which have simple neural pathways, are touch, temperature, pain, chemical and pressure detection, and body position (proprioception). The **special senses** have complex pathways, and the receptors for these senses are housed in specialized organs. The special senses include gustation (taste), olfaction (smell), vision, audition (hearing), and equilibrium. In this exercise you will study the receptors of the general senses.

Each sensory neuron, which may be wrapped in a sheath of connective tissue, monitors a specific region called a **receptive field.** Overlap in adjacent receptive fields enables the brain to detect where a stimulus was applied to the body. The neuron of a given receptive field is connected to a specific area of the sensory cortex. This neural connection is called a **labeled line,** and the CNS interprets sensory information entirely on the basis of the labeled line over which the information arrives.

Nerve impulses are similar to bursts of messages over a telegraph wire. The pattern of action potentials is called **sensory coding** and provides the CNS with such information as intensity, duration, variation, and movement of the stimulus.

The cerebral cortex cannot tell the difference between true and false sensations, however. For example, when you rub your eyes, you sometimes see flashes of light.

The eye-rubbing activates the optic nerve, and the sensory cortex interprets this false impulse as a visual signal. Sometimes the body projects a sensation, usually pain, to another part of the body. This phenomenon is called **referred pain.**

Every sensory receptor monitors its receptive field by two methods: tonic reception and phasic reception. **Tonic receptors** are always active; pain receptors are one example of tonic receptors. **Phasic receptors** are usually inactive and are "turned on" with stimulation. These receptors provide information on the rate of change of a stimulus. Some examples are root-hair plexuses, tactile corpuscles, and Pacinian corpuscles, which are all phasic receptors for touch.

LAB ACTIVITY 1   General-Sense Receptors

Many kinds of receptors transmit information to the CNS. Receptors for body position are called **proprioceptors** (prō-prē-ō-SEP-turz). Two types of proprioceptors are **muscle spindle receptors** in muscles, which inform the brain about muscle tension, and **Golgi tendon organs** in tendons near joints, which inform the brain about joint position. **Thermoreceptors** are sensors for changes in temperature and have wide distribution in the body, including the dermis, skeletal muscles, and the hypothalamus of the brain, our internal thermostat. Pressure receptors, or **baroreceptors,** convey signals about liquid and gas pressures. These receptors, which monitor pressure *changes,* are typically the tips of sensory-neuron dendrites in blood vessels and the lungs; the dendrite tips branch and are often referred to as **free nerve endings. Chemoreceptors** monitor changes in the concentrations of various chemicals present in body fluids.

The skin has a variety of **mechanoreceptors,** touch receptors that are bent when stimulated (Figure 27.1). **Tactile corpuscles,** also called **Meissner's** (MĪS-nerz) **corpuscles,** are nerve endings sensitive to touch and are located in the dermal papillae of the skin. **Tactile,** or **Merkel's** (MER-kelz), **disks** are embedded in the epidermis and are very responsive to touch. **Free nerve endings** are the simplest mechanisms for receptors, being nothing more than the tips of sensory-neuron dendrites. All **nociceptors** (nō-sē-SEP-turz), which are pain receptors in the epidermis, are free nerve endings. **Root-hair plexuses** are dendrites of sensory neurons wrapped around hair roots. These receptors are stimulated when an insect, for example, lands on your bare arm and moves one of the hairs there. Baroreceptors include **lamellated** (LAM-e-lā-ted; *lamella,* thin plate) **corpuscles**—also called **Pacinian** (pa-SIN-ē-an) **corpuscles**—and **Ruffini** (roo-FĒ-nē) **corpuscles.** These receptors are located deep in the dermis.

QuickCheck Questions

1.1 What is chemoreception?

1.2 What is baroreception?

1.3 What is the stimulus for mechanoreceptors?

1 *Materials*

- ☐ Compound microscope
- ☐ Microscope slide of tactile corpuscles
- ☐ Microscope slide of lamellated corpuscles

Procedures

1. Place the slide of the tactile corpuscles on the microscope stage, and focus on the specimen at low magnification. The receptor is located in the papillary region of the dermis, where the dermis folds to attach the epidermis.

2. Observe the tactile corpuscles at medium magnification.

3. Examine the slide of the lamellated corpuscle at low magnification. Note the multiple layers wrapped around the dendrites of the receptor. ■

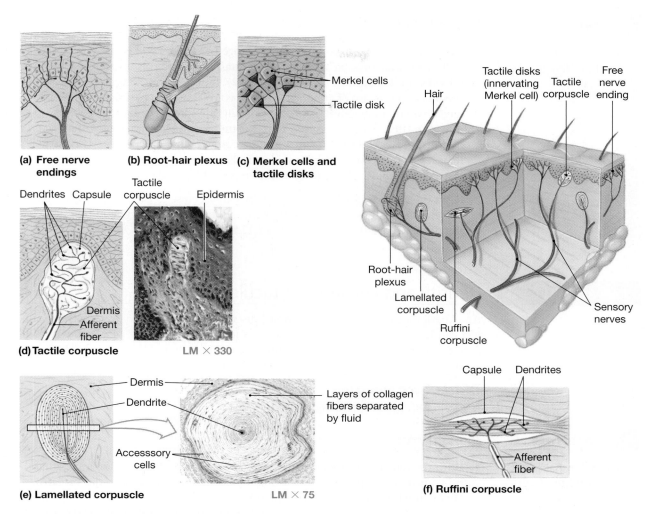

(a) Free nerve endings

(b) Root-hair plexus

(c) Merkel cells and tactile disks

Merkel cells

Tactile disk

Tactile disks (innervating Merkel cell)

Tactile corpuscle

Free nerve ending

Hair

Dendrites   Capsule

Tactile corpuscle

Epidermis

(d) Tactile corpuscle

Dermis

Afferent fiber

LM × 330

Root-hair plexus

Lamellated corpuscle

Ruffini corpuscle

Sensory nerves

Dermis

Dendrite

Accessory cells

(e) Lamellated corpuscle

Layers of collagen fibers separated by fluid

LM × 75

Capsule   Dendrites

Afferent fiber

(f) Ruffini corpuscle

Figure 27.1 **Tactile Receptors in the Skin**

LAB ACTIVITY 2   Two-Point Discrimination Test

Cutaneous receptors are not evenly distributed in the body. Some areas are densely populated with a particular receptor, whereas others have only a few or none of that receptor. This explains why your fingertips, for example, are more sensitive to touch than your scalp. The **two-point discrimination test** is used to map the distribution of touch receptors on the skin. A drawing compass with two points is used to determine the distance between cutaneous receptors. The compass points are gently pressed into the skin, and the subject decides if one or two points are felt. If the sensation is that of a single point, then only one receptor has been stimulated. By gradually increasing the distance between the points until two distinct sensations are felt, the density of the receptor population in that region can be measured.

QuickCheck Questions

2.1 What does the two-point discrimination test measure?

2.2 Are all parts of the body equally sensitive to touch?

2 Materials

☐ Drawing compass with millimeter scale

Procedures

1. Push the two points of compass as close together as they will go, read from the millimeter scale how far apart the points are, and record this distance in the space provided in the leftmost column of Table 27.1, under "Index finger." Gently place the points on the tip of an index finger of your laboratory partner, and then record whether one point or two are felt. Slightly spread the compass points, record the distance apart as read from the millimeter scale, and place them again on the same area of the fingertip; again record whether one point or two are felt. Repeat this procedure until the subject feels two distinct points. Record this distance in Table 27.1 of the laboratory report at the end of this chapter.

2. Reset the compass so that the two points are as close to each other as possible, and repeat the test on the back of the hand, back of the neck, and one side of the nose. Record the data in Table 27.1. ■

LAB ACTIVITY 3 Density and Location of Tactile Receptors

An experiment similar to the two-point discrimination test is to test whether a subject can feel a single touch from a stiff bristle of hair. In this activity, you will compare the sensitivity of two sites: the anterior and posterior of the forearm. The bristle used to measure this sensitivity is called a Von Frey hair.

QuickCheck Question

3.1 What do the Von Frey hairs measure?

3 Materials

☐ Von Frey hairs (stiff boar bristles from hair brush)
☐ Water-soluble felt-tipped marker

Procedures

1. With the felt-tipped marker, draw a small box, approximately one inch square, on your partner's posterior forearm and another box on the anterior forearm. Divide the box into 16 smaller squares.

2. *NOTE: Throughout this activity, the subject must look away and not watch as each test is run.* To test the posterior forearm's sensitivity to touch, use a Von Frey hair to gently touch the skin inside one small square. Be careful to touch only one point inside the square, and apply only enough pressure to slightly bend the bristle and stimulate the superficial tactile corpuscles. (Too much pressure will stimulate the underlying deep-touch receptors, the lamellated corpuscles.) In the "Data for posterior forearm" grid that follows, mark the corresponding small square: draw a dot if the subject felt the bristle and an x if she or he did not feel the bristle.

3. Repeat this procedure in each of the other 15 small squares you have drawn on the subject's posterior forearm. Remember that the subject must be looking away as you administer the test. When finished, the grid will contain either a dot or an x in each square.

4. Repeat steps 2 and 3 on the subject's anterior forearm, touching the skin and marking the 16 small squares of the "Data for anterior forearm" grid that follows.

5. Repeat the experiment with you as the subject and your partner administering the tests.

6. Compare the results (a) between the two regions of your forearm, (b) between the two regions of your partner's forearm, and (c) between your forearm and your partner's. Record your final count in Table 27.2 of the laboratory report at the end of this chapter. ■

Data for posterior forearm

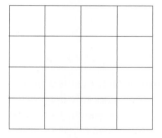

Data for anterior forearm

LAB ACTIVITY 4   Density and Location of Thermoreceptors

This experiment is a simple process of mapping the general distribution of thermoreceptors in the skin.

QuickCheck Question

4.1 Where are thermoreceptors located?

4 Materials

- ☐ Small probes (small, blunt metal rods or straightened paper clips)
- ☐ Water-soluble felt-tipped marker
- ☐ Beaker filled with ice water
- ☐ Beaker filled with 45°C water

Procedures

1. With the felt-tipped marker, draw a small box, approximately one inch square, on your partner's anterior forearm just above the wrist. Divide the box into 16 smaller squares.

2. Place a probe in the cold water for several minutes. Remove the probe from the water, dry it, and—with the subject not watching—touch the probe lightly to one of the small squares on the subject's arm. If the subject feels the cold, mark a "c" in the appropriate square of the thermoreception data grid that follows. If no cold is felt, mark an x in the square. (Write small, because during this activity you need to make two marks in each small square.)

3. Repeat in each of the other 15 squares, returning the probe to the cold water for a minute or so after each test.

4. Place a probe in the warm water for several minutes. *Important: the probe should be only warm to the touch, not hot enough to burn or cause pain.* Repeat steps 2 and 3 for each square on the subject's arm, placing an "h" in each appropriate square of the thermoreception data grid when heat is felt.

5. Have your partner repeat the experiment on your anterior wrist, and then compare the two sets of data. Record your final data in Table 27.3 of the laboratory report at the end of this chapter. ■

Data for thermoreception

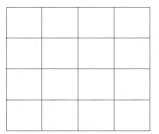

LAB ACTIVITY 5 Receptor Adaptation

Receptors display **adaptation,** which means a reduction in sensitivity to repeated stimulus. When a receptor is stimulated, it first responds strongly, but then the response declines as the stimulus is repeated. **Peripheral adaptation** is the decline in response to stimuli at receptors. This type of adaptation reduces the amount of sensory information the CNS must process. **Central adaptation** occurs in the CNS. Inhibition along a sensory pathway reduces sensory information. Phasic receptors are fast-adapting receptors, and tonic receptors are slow-adapting. In this experiment you will investigate the adaptation of thermoreceptors.

QuickCheck Question

5.1 Define the term *adaptation.*

5 Materials

- ☐ Bowl filled with ice water
- ☐ Bowl filled with 45°C water
- ☐ Bowl filled with room-temperature water

Procedures

Test 1

1. Immerse your partner's left hand in the ice water. Note in the "Initial Sensation" column of Table 27.4 of the laboratory report that the subject felt the cold. After two minutes, record in the table your partner's description of what temperature sensation is being felt in the left hand.

2. Leaving the left hand immersed, have your partner immerse her or his right hand in the ice water and describe whether the water feels colder to the left hand or to the right hand. Note this description in the "Initial Sensation" column of Table 27.4.

Test 2

Wait five minutes before proceeding to the next test, to allow blood to flow in the subject's hands and restore normal temperature sensitivity.

1. Place your partner's left hand in the ice water and his or her right hand in the 45°C water.

2. Keep hands immersed for two minutes.

3. Remove the left hand from the ice water and immerse it in the room-temperature water. Record in Table 27.4 whether your partner senses the room-temperature water as feeling hotter or colder than the ice water.

4. Now remove the right hand from the 45°C water, and immerse it in the room-temperature water. Record in Table 27.4 whether your partner senses the room-temperature water as being hotter or colder than the 45°C water. ■

LAB ACTIVITY 6 · Referred Pain

Referred pain means that the part of the body where pain is felt is different from the part of the body at which the painful stimulus is applied. An excellent example of referred pain is the pain in the forehead some people feel after quickly consuming a cold drink or a bowl of ice cream. The resulting "brain freeze," as it is commonly called, is pain "referred" from the nerves of the throat, because these nerves also innervate the forehead. Another well-known example of referred pain is the sensation felt in the left medial arm during a heart attack. The arm and heart are innervated by the same segment of the spinal cord. Interneurons in the spinal segment spread the incoming pain signals from the heart to areas innervated by that segment. In this activity, you will immerse your elbow in ice water and note any referred pain.

Figure 27.2 shows referred pain from the heart, along with three other common types. Liver and gallbladder pain is referred to the superior margin of the right shoulder. Pain from the appendix, located in the lower right quadrant, is referred to the medial area of the abdomen. Stomach, small intestine, and colon pain are also referred to the medial abdomen. The ureters are ducts in the medial aspect of the abdomen that transport urine from the kidneys to the urinary bladder; ureter pain is referred to the lateral abdomen.

QuickCheck Questions

6.1 Define the term *referred pain*.

6.2 Why does referred pain occur?

6 · Materials

☐ Bowl filled with ice water

Procedures

1. Place your elbow in the bowl of ice water, and then record in Table 27.5 of the laboratory report the location and type of your initial sensation.

2. Keeping your elbow submerged, record the location and type of sensation after 30, 60, 90, and 120 seconds. ■

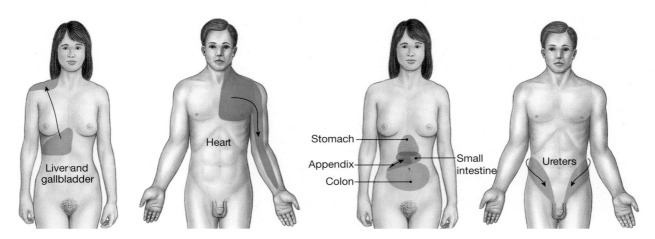

Figure 27.2 Referred Pain

<table>
<tr><td>**LAB ACTIVITY 7**</td><td># Proprioception</td></tr>
</table>

LAB ACTIVITY 7   Proprioception

Proprioception is the sense of body position. It is the sense that gives us ownership of our bodies, enabling us to walk without having to watch our feet, say, or to reach up and scratch an ear without looking in a mirror.

QuickCheck Question

7.1 Define the term *proprioception*.

7 Materials

☐ Sheet of paper
☐ Red and black felt-tipped pens

Procedures

1. Using the dominant hand and the black pen, have your partner make a small circle (1/4-inch diameter) in the middle of the sheet of paper.

2. Instruct your partner to place the tip of the pen in the middle of the circle and then remain in that position with eyes closed for a few moments.

3. Keeping the eyes closed, have your partner lift the pen two to three inches off the paper and then make a mark within the circle.

4. Repeat step 3 until 10 marks have been made on the paper.

5. Repeat the experiment with your partner using his or her nondominant hand and the red pen.

6. Record the marking accuracy results in Table 27.6 of the laboratory report. ■

General Senses

Name _____

Date _____

Section _____

A. Data Recording and Interpretation

Table 27.1 **Two-Point Discrimination Test Data**

Index Finger		Back of Hand		Back of Neck		Side of Nose	
Distance between points (mm)	1 point or 2 felt?	Distance between points (mm)	1 point or 2 felt?	Distance between (points mm)	1 point or 2 felt?	Distance between points (mm)	1 point or 2 felt?

Table 27.2 **Tactile Density Tests**

Region	Number of Touches Felt
Posterior forearm	
Anterior forearm	

Table 27.3 **Thermoreceptor Density Tests**

Temperature	Number of Touches Felt
Cold probe	
Warm probe	

LABORATORY REPORT

Table 27.4 **Adaptation Tests**

Test 1	Initial Sensation	Sensation after 2 Minutes
Left hand in ice water		
Both hands in ice water		XXXXXXXXXXXX
Test 2	Movement after 2 Minutes	Sensation in Room-Temperature Water
Left hand ice water, right hand 45°C water	Left hand to room-temperature water	Hotter or colder than ice water? (circle one)
Left hand ice water, right hand in 45°C water	Right hand to room-temperature water	Hotter or colder than 45°C water? (circle one)

Table 27.5 **Referred Pain from Elbow**

Time	Location	Type of Sensation
Upon immersion in ice water		
30 seconds after immersion		
60 seconds after immersion		
90 seconds after immersion		
120 seconds after immersion		

Table 27.6 **Proprioception**

Hand	Number of Marks within Circle
Dominant	
Nondominant	

B. Matching

Match each sense in the left column with its correct receptor from the right column.

_____ **1.** very sensitive touch receptor of epidermis

_____ **2.** baroreceptor in deep dermis

_____ **3.** touch-sensitive receptor in dermal papillae

_____ **4.** dendrite tip sensitive to pain

_____ **5.** found on hair-covered parts of body

_____ **6.** senses muscle tension; one type of proprioceptor

A. Pacinian corpuscle

B. Meissner's corpuscle

C. root-hair plexus

D. muscle spindles

E. Merkel's disk

F. free nerve ending

C. Short-Answer Questions

1. Describe the two-point discrimination test, and explain how it demonstrates receptor density in the skin.

2. Explain the concept of a receptive field.

3. What is the difference between a phasic receptor and a tonic receptor?

4. Are touch receptors phasic or tonic?

5. Based on your results from Laboratory Activity 5, which type of thermoreceptor adapts more quickly, warm or cold?

D. Analysis and Application

1. In Laboratory Activities 3 and 4, which type of receptor was most abundant: touch, cold, or warm?

2. Beth wears her hair in a tight ponytail. At first her tactile receptors are sensitive and she feels the pull from her hair, but after a while she no longer feels it. Explain the loss of sensation.

3. In Laboratory Activity 6, where did you feel the pain when you had your elbow in the cold water? Explain why pain is felt in this location.

4. Explain the differences in the proprioception marking accuracy test between your dominant and nondominant hands.

Special Senses: Olfaction and Gustation

OBJECTIVES

On completion of this exercise, you should be able to:

• Describe the location and structure of the olfactory receptors.

• Identify the microscopic features of the olfactory epithelium.

• Describe the location and structure of taste buds and papillae.

• Identify the microscopic features of taste buds.

• Explain why olfaction accentuates gustation.

Changes in the internal and external environments are detected by special cells called *sensory receptors*. Each type of receptor is sensitive to a specific stimulus. The taste buds of the tongue, for example, are stimulated by chemicals dissolved in saliva and not by sound waves or light rays. The special senses have complex pathways, and the receptors of these senses are housed in specialized organs. The special senses include gustation (taste), olfaction (smell), vision, audition (hearing), and equilibrium. In this exercise you will study the receptors of the special senses olfaction and gustation.

LAB ACTIVITY 1 Olfaction

Olfactory receptors are located in the **olfactory epithelium** lining the roof of the nasal cavity (Figure 28.1a). Most of the air we inhale passes straight through the nasal cavity and into the pharynx. Sniffing increases our sense of smell by pulling air higher into the nasal cavity, where the olfactory receptors are found.

Three types of cells occur in the olfactory epithelium. The **olfactory receptor cells** are bipolar neurons with many cilia that are sensitive to airborne molecules (Figure 28.1b). **Basal cells** are stem cells that divide and replace olfactory receptors. **Supporting cells**, also called **sustentacular cells**, provide physical support and nourishment to the receptors.

The olfactory epithelium is attached to an underlying layer of connective tissue. This layer, called the **lamina propria**, contains **olfactory (Bowman's) glands**, which secrete mucus. For a substance to be smelled, volatile molecules from the substance must first diffuse through the air from the substance to your nose. Once in the nose, the molecules must diffuse through the mucus secreted by the olfactory glands before they can stimulate the cilia of the olfactory receptors. Our sense of smell is

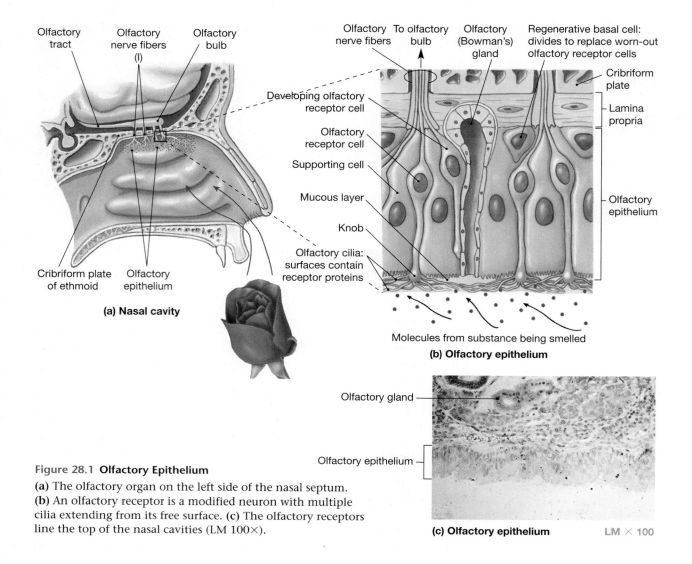

Olfactory tract
Olfactory nerve fibers (I)
Olfactory bulb
Cribriform plate of ethmoid
Olfactory epithelium

(a) Nasal cavity

Olfactory nerve fibers
To olfactory bulb
Olfactory (Bowman's) gland
Regenerative basal cell: divides to replace worn-out olfactory receptor cells
Cribriform plate
Lamina propria
Olfactory epithelium
Developing olfactory receptor cell
Olfactory receptor cell
Supporting cell
Mucous layer
Knob
Olfactory cilia: surfaces contain receptor proteins
Molecules from substance being smelled

(b) Olfactory epithelium

Olfactory gland
Olfactory epithelium

(c) Olfactory epithelium   LM × 100

Figure 28.1 Olfactory Epithelium
(a) The olfactory organ on the left side of the nasal septum.
(b) An olfactory receptor is a modified neuron with multiple cilia extending from its free surface. **(c)** The olfactory receptors line the top of the nasal cavities (LM 100×).

drastically reduced by colds and allergies because mucus production increases in the nasal cavity and blocks the diffusion of molecules to the olfactory receptors.

The **olfactory nerve**, cranial nerve I, passes through the cribriform plate of the ethmoid and enters the brain at the **olfactory bulb**. The bulb continues as the **olfactory tract** to the temporal lobe, where the olfactory cortex is located. Unlike many other senses, the olfactory pathway does not have a synapse in the thalamus. Other branches synapse in the hypothalamus and limbic system of the brain, which explains the strong emotional responses associated with olfaction.

QuickCheck Questions

1.1 Where are the olfactory receptors located?

1.2 What is the function of the olfactory glands?

1 Materials

☐ Compound microscope
☐ Prepared slide of olfactory epithelium

Procedures

1. At low magnification, focus on the olfactory epithelium. Identify the supporting cells and olfactory receptor cells, both shown in Figure 28.1.

2. Locate the lamina propria and the olfactory glands.

3. In the space provided, sketch the olfactory epithelium as viewed at medium magnification. ■

LAB ACTIVITY 2 | Olfactory Adaptation

Olfactory receptors are quick to adapt to a repeated stimulus. A few minutes after you apply cologne or perfume, for instance, you do not smell it as much as you did initially. However, if a new odor is present, the nose is immediately capable of sensing the new scent, proof that what is going on is receptor adaptation and not receptor fatigue. (If the receptors were fatigued instead of adapted, you would not be receptive to new stimuli once the receptors reached exhaustion.) Olfactory adaptation occurs along the olfactory pathway in the brain, not in the receptors. We experience olfactory adaptation when we visit, say, the house of some friends. At first, we notice that their house smells different from ours. Soon, however, we adapt to the different smell and are receptive to new smells.

In this experiment you will determine the length of time it takes for your olfactory epithelium to adapt to a particular odor. Use care when smelling the vials. Do not just put the vial right under your nose and inhale. Instead, hold the open vial about six inches in front of your nose and wave your hand over the opening to waft the odor toward your nose.

QuickCheck Questions

2.1 Where does the physiological process of olfactory adaptation occur?

2.2 When does olfactory adaptation occur?

---

2 | Materials

☐ Vial containing oil of wintergreen

☐ Vial containing isopropyl alcohol

☐ Stop watch

☐ Laboratory partner

Procedures

1. Hold the vial of wintergreen oil near your face and waft the fumes toward your nose. Ask your partner to start the stop watch.

2. Breathe through your nose to smell the wintergreen oil. Continue wafting and smelling until you no longer sense the odor. Have your partner stop the stop watch, and record the time it took for adaptation to occur in Table 28.1.

3. Immediately following loss of sensitivity to the wintergreen oil, smell the vial of alcohol. Explain how you can smell the alcohol but can no longer smell the wintergreen oil.

4. Repeat step 1–3 using the alcohol as the first vial. Is there a difference in adaptation time for the two substances?

5. Record your partner's olfactory adaptation time and the times for other classmates in Table 28.1 in the Laboratory Report at the end of this exercise. ■

LAB ACTIVITY 3 Gustation

The receptors for gustation (gus-TĀ-shun), or taste, are located in **taste buds** covering the surface of the tongue, the pharynx, and the soft palate. Taste buds are inside elevations called **papillae** (pa-PIL-lē), detailed in Figure 28.2. The base of the tongue has a number of circular papillae called **circumvallate** (sir-kum-VAL-āt) **papillae**. These papillae, arranged in an inverted "V" across the width of the tongue, contain taste buds that are sensitive to bitter substances. The tip and sides of the tongue contain buttonlike papillae called **fungiform papillae**, which are stimulated by salty, sweet, and sour substances. Approximately two-thirds of the anterior portion of the tongue is covered with **filiform papillae**, which provide a rough surface for the movement of food. Filiform papillae do not contain taste buds.

A taste bud contains up to 20 **gustatory cells**. Each cell has a small hair, or microvillus, that projects through a small taste pore. Contact with food dissolved in saliva stimulates the microvilli to produce gustatory impulses. Sensory information from the taste buds is carried to the brain by parts of three cranial nerves: the **vagus nerve** (X) serving the pharynx, the **facial nerve** (VII) serving the anterior two-thirds of the tongue, and the **glossopharyngeal nerve** (IX) serving the posterior one-third of the tongue.

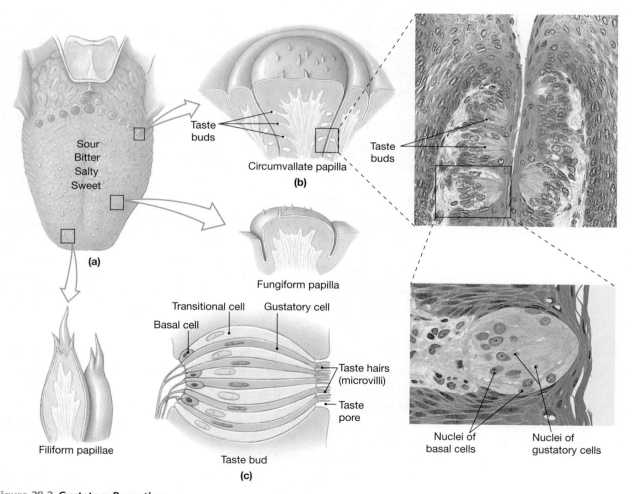

Figure 28.2 Gustatory Reception

(a) Taste receptors are located in taste buds, which form pockets in the epithelium of fungiform or circumvallate papillae. (b) Taste buds in a circumvallate papilla (LM × 3280). (c) A taste bud, showing receptor (gustatory) cells and supporting cells. The diagrammatic view shows details of the taste pore that are not visible in the micrograph (LM × 650).

3 *Materials*

☐ Compound microscope

☐ Prepared slide of taste buds

☐ PTC taste paper

Procedures

1. Observing the slide of the tongue at low and medium magnifications, use Figure 28.2 as a guide as you identify the papillae and taste buds.

2. In the space provided, sketch a few papillae with taste buds.

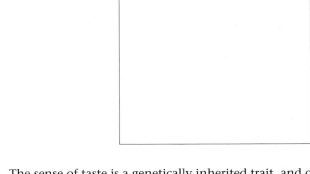

3. The sense of taste is a genetically inherited trait, and consequently two individuals can perceive the same substance in different ways. The chemical phenylthiocarbamide (PTC), for example, tastes bitter to some individuals, sweet to others, and tasteless to still others. Approximately 30 percent of the population are nontasters of PTC.

   a. Place a strip of PTC paper on your tongue, and chew it several times. Are you a taster or a nontaster?

   b. If your instructor has each student in the class record her or his taster-or-nontaster results on the chalkboard, calculate the percentage of tasters verses nontasters and complete Table 28.2 in the Laboratory Report. ■

LAB ACTIVITY **4** The Interrelationship of Olfaction and Gustation

The overall sense of taste is thousands of times more sensitive when gustatory and olfactory receptors are stimulated simultaneously. The following experiment demonstrates the effect of smell on the sense of taste. You will place small cubes of apple or onion on your laboratory partner's dry tongue and ask if he or she can detect which type of cube is present.

4 *Materials*

☐ Diced onion

☐ Diced apple

☐ Paper towels

☐ Laboratory partner

Procedures

1. Dry the surface of your partner's tongue with a clean paper towel.

2. Have your partner stand with eyes closed and nose pinched shut with the thumb and index finger.

3. Place a piece of either onion or apple on the dried tongue, and ask if the food can be identified. Record the reply, yes or no, in Table 28.3 in the Laboratory Report.

4. Have your partner, still with eyes closed and nose pinched shut, chew the piece of food, and again ask if it can be identified. Record the reply in Table 28.3.

5. Have your partner release the nose pinch, and ask one last time if the food can be identified. Record the reply in Table 28.3. ■

Special Senses: Olfaction and Gustation

Name _____

Date _____

Section _____

A. Data Recording and Interpretation

Table 28.1 Olfactory Adaptation

Student	Olfactory Adaptation Time for Wintergreen Oil (s)	Olfactory Adaptation Time for Isopropyl Alcohol (s)

Table 28.2 PTC Taste Experiment

Student	Taster	Nontaster
Percentage of class:		

Table 28.3 Gustatory/Olfactory Sensations

Food	Dry Tongue	After Chewing	Open Nostrils
Apple	_____	_____	_____
Onion	_____	_____	_____

B. Matching

Match each term in the left column with its correct description from the right column.

_____	**1.** cranial nerve for smell	**A.**	produces mucus
_____	**2.** adaptation	**B.**	sense of smell
_____	**3.** filiform papilla	**C.**	contains taste buds for bitter substances
_____	**4.** gustation	**D.**	sense of taste
_____	**5.** Bowman's gland	**E.**	loss of sensitivity due to repeated stimuli
_____	**6.** basal cell	**F.**	supportive cell of olfactory receptor
_____	**7.** circumvallate papilla	**G.**	helps manipulate food on tongue surface
_____	**8.** sustentacular cell	**H.**	cranial nerve I
_____	**9.** cranial nerve for taste	**I.**	stem cell of olfactory receptor
_____	**10.** olfaction	**J.**	cranial nerve VII

28

C. Short-Answer Questions

1. Why does a cold affect your sense of smell?

2. Where are the receptors for taste located?

3. Examine the data in Table 28.1. Did adaptation time differ greatly from one individual to another?

4. Draw the tongue, and identify where the taste buds for sweet, bitter, sour, and salty compounds are concentrated.

D. Analysis and Application

1. Describe an experiment that demonstrates how the senses of taste and smell are linked.

2. Several minutes after applying cologne, you notice that you cannot smell the fragrance. Explain this response by your olfactory receptors.

3. Does the class data in Table 28.2 clearly demonstrate which trait, taster or nontaster, is genetically dominant?

Anatomy of the Eye

OBJECTIVES

On completion of this exercise, you should be able to:

- Identify and describe the accessory structures of the eye.
- Identify and explain the actions of the six extrinsic eye muscles.
- Describe the external and internal anatomy of the eye.
- Describe the microscopic organization of the retina.
- Identify the structures of a dissected cow or sheep eye.

The eyes are complex and highly specialized sensory organs. We can see the pale light of stars and the intense, bright blue of the sky. To function in such a wide range of light conditions, the retina of the eye has two types of receptors: rods for night vision and cones for bright light and color vision. Because the level and intensity of light are always changing, the eye must regulate the size of the pupil, which allows light to enter the eye. Six oculomotor muscles surrounding the eyeball allow it to move. Four of the twelve cranial nerves control the muscular activity of the eyeball and transmit sensory signals to the brain. A sophisticated system of tear production and drainage keeps the surface of the eyeball clean and moist. In this exercise you will examine the anatomy of the eye and dissect the eye of a sheep or a cow.

LAB ACTIVITY 1 External Anatomy of the Eye

The human eyeball is a spherical organ measuring about 2.5 cm (1 in.) in diameter. Only about one-sixth of the eyeball is visible between the eyelids; the rest is recessed in the bony orbit of the skull. The slits where the upper and lower lids meet are the **lateral canthus** (KAN-thus; "corner") and **medial canthus** (Figure 29.1). A red, fleshy structure in the medial canthus, the **lacrimal caruncle** (KAR-ung-kul; "small soft mass"), contains modified sebaceous and sweat glands. Secretions from the caruncle accumulate in the medial canthus during long periods of sleep.

1. _____   9. _____

2. _____   10. _____

3. _____   11. _____

4. _____   12. _____

5. _____   13. _____

6. _____   14. _____

7. _____   15. _____

8. _____

inferior lacrimal canaliculus	lateral canthus
inferior oblique muscle	lower eyelid
inferior rectus muscle	medial canthus
lacrimal caruncle	nasolacrimal duct
lacrimal gland	superior lacrimal canaliculus
lacrimal gland ducts	superior rectus muscle
lacrimal punctum	tendon of superior
lacrimal sac	oblique muscle

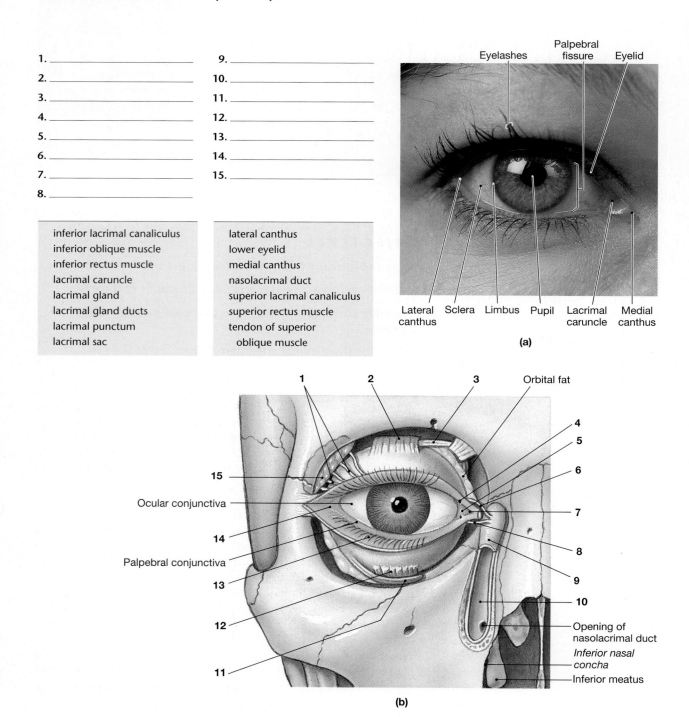

Figure 29.1 External Features and Accessory Structures of the Eye
(a) Gross and superficial anatomy of the accessory structures. (b) Details of the organization of the lacrimal apparatus.

The accessory structures of the eye are the upper eyelid, lower eyelid, eyebrow, eyelashes, lacrimal apparatus, and six extrinsic eye muscles. The eyelids, or **palpebrae** (pal-PĒ-brē), are skin-covered muscles that protect the surface of the eyeball. The **levator palpebrae superioris** muscle raises the upper eyelid and the orbicularis oculi muscle closes the eyelids. Blinking the eyelids keeps the eyeball surface lubricated and clean. A thin mucous membrane called the **palpebral conjunctiva** (kon-junk-TĪ-vuh) covers the underside of the eyelids and reflects over the anterior surface of the eyeball as the **ocular conjunctiva.** This membrane secretes mucus that reduces friction and moistens the eyeball surface. The **eyelashes** are short hairs that project from the border of each eyelid. Eyelashes and the eyebrows protect the eyeball from foreign objects, such as perspiration and dust, and partially shade the eyeball from the sun. Sebaceous **ciliary glands,** located at the base of the hair follicles, help lubricate the eyeball.

Clinical Application

Infections of the Eye

When a ciliary gland is blocked, it becomes inflamed as a **sty.** Because it is on the tip of the eyelid, the sty irritates the eyeball. **Conjunctivitis** is an inflammation of the conjunctiva that can be caused by bacteria, dust, smoke, or air pollutants; the infected eyeball usually appears red and irritated. The bacterial form of this infection is contagious and spreads easily among young children and individuals sharing such objects as tools and office equipment. ▶

The **lacrimal apparatus** (Figure 29.1) consists of the lacrimal glands, lacrimal canals, lacrimal sac, and nasolacrimal duct. The **lacrimal glands** are superior and lateral to each eyeball. Each gland contains 6 to 12 excretory **lacrimal ducts** that deliver onto the anterior surface of the eyeball a slightly alkaline solution, called either lacrimal fluid or tears, that cleans, moistens, and lubricates the surface. The lacrimal fluid also contains an antibacterial enzyme called *lysozyme* that attacks any bacteria that may be on the surface of the eyeball. The tears move medially across the eyeball surface until they enter two small openings of the medial canthus called the **lacrimal puncta.** From there the tears pass into two ducts, the **lacrimal canals,** that lead to an expanded portion of the nasolacrimal duct called the **lacrimal sac.** The **nasolacrimal duct** drains the tears into the nasal cavity.

Six extrinsic (external) muscles control the movements of the eyeball (Figure 29.2). The **superior rectus, inferior rectus, medial rectus,** and **lateral rectus** are straight muscles that move the eyeball up and down and side to side. The **superior** and **inferior oblique muscles** attach diagonally on the eyeball. The superior oblique has a tendon passing through the **trochlea** ("pulley") located on the upper orbit. The superior oblique muscle rolls the eyeball downward, and the inferior oblique rolls it upward.

QuickCheck Questions

1.1 How are lacrimal secretions drained from the surface of the eyeball?

1.2 Which extrinsic eye muscles are involved in rolling your left eyeball down and medially?

1 *Materials*

☐ Dissectible eye model
☐ Eyeball chart

Procedures

1. On the eye model and chart, locate the structures of the lacrimal apparatus, the eyelids, and other external anatomical features of the eye.

2. On the model, identify the six external muscles of the eye. Describe how each muscle moves the eye.

3. Label the structures of the eye in Figures 29.1 and 29.2. ■

LAB ACTIVITY 2   Internal Anatomy of the Eye

The eyeball is divided anatomically into three layers: the outermost **fibrous tunic,** the middle **vascular tunic** (**uvea**), and the inner **neural tunic** (**retina**), each detailed in Figure 29.3. The outermost layer is called the fibrous tunic because of the abundance of dense connective tissue. The **sclera** (SKLER-uh; "hard") is the white part of the fibrous tunic. It covers the eyeball except at the transparent **cornea**

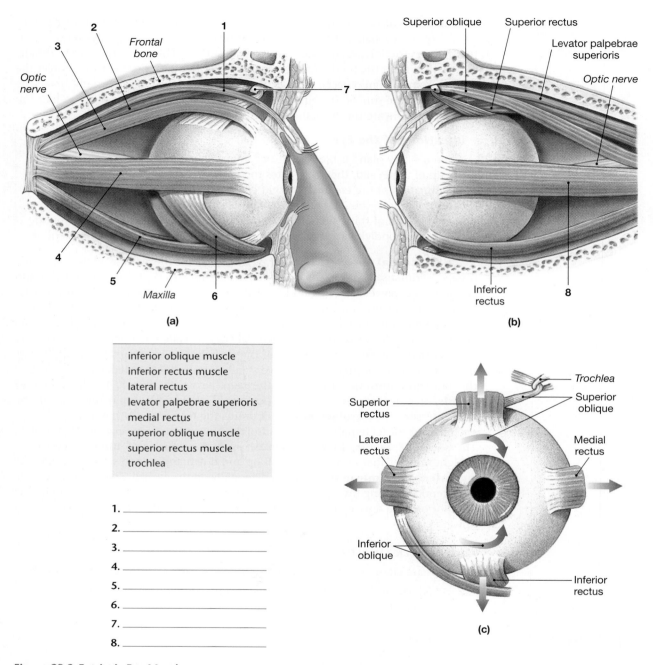

Figure 29.2 Extrinsic Eye Muscles
(a) Lateral surface, right eye. (b) Medial surface, right eye. (c) Anterior view, right eye.

(KOR-nē-uh), where light enters the eye. The sclera resists punctures and maintains the shape of the eyeball. The cornea consists primarily of many layers of densely packed collagen fibers. The **limbus** is the border between the sclera and the cornea. Around the limbus is the **scleral venous sinus** (canal of Schlemm), a small passageway that drains fluid into veins in the sclera.

The most posterior portion of the vascular tunic, the **choroid,** is highly vascularized and contains a dark pigment (melanin) that absorbs light to prevent reflection. Anteriorly, the vascular tunic is modified into the **ciliary body** and the pigmented **iris.** The ciliary body contains two structures, the **ciliary process** and

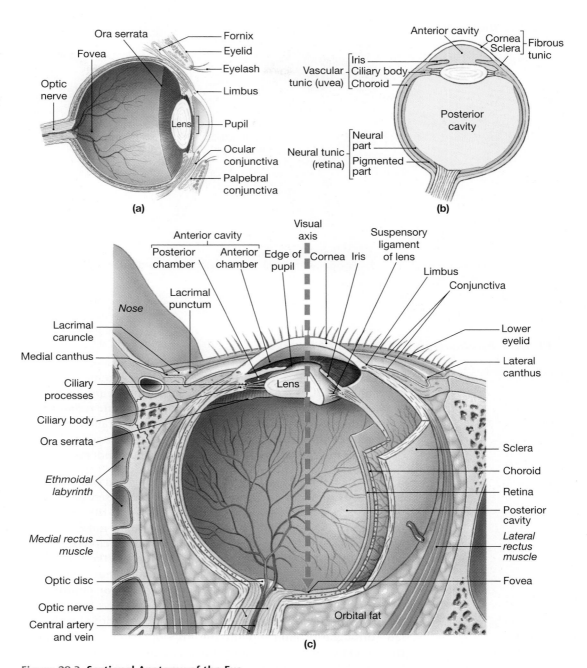

Figure 29.3 Sectional Anatomy of the Eye
(a) Major anatomical landmarks and features viewed in a sagittal section through the left eye. (b) The three layers, or tunics, of the eye. (c) A horizontal section of the right eye.

the **ciliary muscle.** The ciliary process is a series of folds with thin **suspensory ligaments** extending to the **lens.** The ciliary muscle adjusts the shape of the lens for near and far vision. The front of the iris is pigmented and has a central aperture called the **pupil. Constrictor** and **dilator muscles** of the iris change the diameter of the pupil to regulate the amount of light striking the retina (Figure 29.4). The autonomic nervous system controls both sets of muscle fibers. In bright light and for close vision, parasympathetic activation causes the constrictor muscles to narrow the pupils. In low light and for distant vision, sympathetic stimulation causes contraction of the dilator muscles to expand the pupils so that more light can enter the eye.

Figure 29.4 The Pupillary Muscles

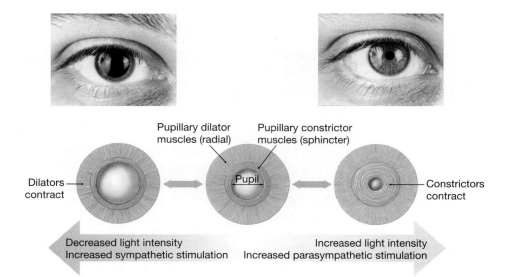

Pupillary dilator muscles (radial)

Pupillary constrictor muscles (sphincter)

Dilators contract

Pupil

Constrictors contract

Decreased light intensity
Increased sympathetic stimulation

Increased light intensity
Increased parasympathetic stimulation

Clinical Application

Diseases of the Eye

Glaucoma is a disease in which the intraocular pressure of the eye is elevated. The increased pressure damages the optic nerve and may eventually result in blindness. If blockage of the scleral venous sinus occurs, fluid accumulates in the anterior cavity and intraocular pressure rises. **Diabetic retinopathy** develops in many diabetic individuals. Diabetes causes a proliferation of blood vessels over the retina and rupture of others. These vascular changes occur gradually, but eventually the photoreceptors are damaged and blindness occurs. ▶

The neural tunic, usually referred to as the *retina* contains light-sensitive photoreceptors—the rods and cones—and two layers of sensory neurons. The anterior margin of the retina where the choroid is exposed is the **ora serrata** (Ō-ra ser-RA-tuh; "serrated mouth") and appears as a jagged edge, much like a serrated knife.

The lens divides the eyeball into an **anterior cavity,** the area between the lens and the cornea; and a **vitreous (posterior) cavity,** the area between the lens and the retina. The anterior cavity is further subdivided into an **anterior chamber** between the iris and the cornea and a **posterior chamber** between the iris and the lens. Capillaries of the ciliary processes form a watery fluid called **aqueous humor,** which is secreted into the posterior chamber. This fluid circulates through the pupil and into the anterior chamber, where it is reabsorbed into the blood by the scleral venous sinus. The aqueous humor helps to maintain the intraocular pressure of the eyeball and supply nutrients to the lens and cornea. The vitreous cavity, larger than the anterior cavity, contains the **vitreous body,** a clear, jellylike substance that holds the retina against the choroid layer and prevents the eyeball from collapsing.

QuickCheck Questions

2.1 How does aqueous humor circulate in the eyeball, and where is it reabsorbed?

2.2 How does the pupil regulate the amount of light that enters the lens?

2 *Materials*

☐ Dissectible eye model
☐ Eyeball chart

Procedures

1. On the eye model and chart, identify the three major layers of the eyeball wall.

2. Identify the sclera and cornea on the eye model. Also locate the limbus and scleral venous sinus.

3. On the model, locate the choroid, and identify the ciliary body and associated structures.

4. Identify the retina, fovea, and ora serrata on the eye model. ■

Cellular Organization of the Retina

Receptors of the retina are called **photoreceptors** and are stimulated by photons, which are particles of light. Photoreceptors are classified into two types: **rods** and **cones.** Rods are sensitive to low illumination and to motion. They are insensitive to most colors of light, and therefore we see little color at night. Cones are stimulated by moderate or bright light and respond to different wavelengths, or colors, of light. Our visual acuity is attributed to cones. The rods and cones face a **pigmented epithelium** located at the posterior of the eyeball (Figure 29.5). Light passes through the retina, reflects off the pigmented epithelium, and strikes the photoreceptors. The

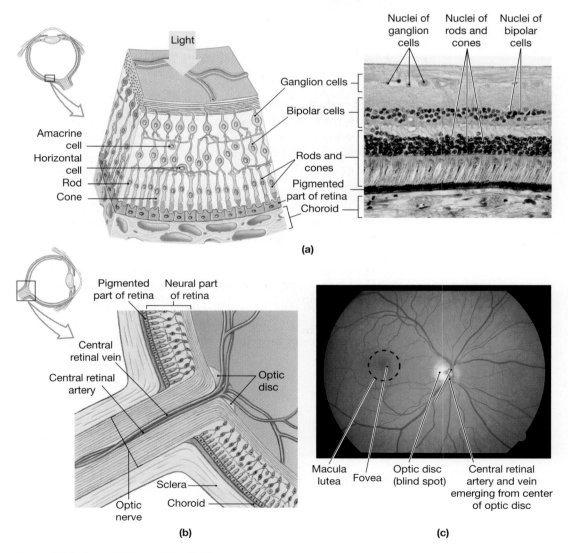

Figure 29.5 Organization of the Retina

(a) An illustration and micrograph of the cellular organization of the retina (neural tunic) (LM × 290). The photoreceptors face the choroid in the wall rather than the posterior cavity. **(b)** The optic disc in diagrammatic horizontal section. **(c)** A photograph of the retina as seen through the pupil.

photoreceptors communicate with a layer of sensory neurons called **bipolar cells.** These cells pass the visual signal on to a layer of **ganglion cells** whose axons converge at an area called the **optic disc,** where the optic nerve exits the eyeball. Horizontal cells form a network that either inhibits or facilitates communication between the photoreceptors and the bipolar cells. **Amacrine** (AM-a-krin) cells enhance communication between bipolar and ganglion cells.

The optic disc lacks photoreceptors and is a "blind spot" in your field of vision. Because the visual fields of your two eyes overlap, however, the blind spot is filled in and not noticeable. Lateral to the optic disc is an area of high cone density called the **macula lutea** (LOO-tē-uh; "yellow spot"). In the center of the macula lutea is a small depression called the **fovea** (FŌ-vē-uh, "shallow depression"). The fovea is the area of sharpest vision because of the abundance of cones. Rods are most numerous at the periphery of the retina, and we see best at night by looking out of the corners of our eyes. There are no rods in the fovea.

QuickCheck Questions

3.1 What is the optic disc and why don't you see a blind spot in your field of vision?

3.2 How are the cells of the retina organized?

3 Materials

- ☐ Compound microscope
- ☐ Prepared microscope slide of retina

Procedures

1. Focus the slide at low magnification, and use Figure 29.5 as a reference while observing the specimen. Change to medium or high magnification as you examine each layer of the retina.

2. Locate the thick choroid layer on the edge of the specimen. Next to the choroid, find the pigmented epithelium.

3. The three types of cells in the retina are clearly visible where the nuclei are grouped into three distinct bands. The photoreceptors—the rods and cones— form the dense band of nuclei next to the pigmented epithelium. Locate this layer of cells.

4. The bipolar cells form a thinner cluster of nuclei next to the photoreceptors. Identify the bipolar cells on the slide. Refer to Figure 29.5a as needed.

5. The ganglion cells of the retina have scattered nuclei and appear on the edge of the retina. Locate this layer on the slide.

6. Sketch a view of the retina at medium power in the space provided. ■

LAB ACTIVITY 4   Observation of the Retina

The retina is the only location in the body where blood vessels may be directly observed. To do so, clinicians use a lighted magnifying instrument called an **ophthalmoscope** (Figure 29.6). The instrument shines a beam of light into the eye while the examiner looks through a lens called a viewing port to observe the retina. *Important:* To protect the subject's eye, make quick observations, then move the light beam away from the eye.

QuickCheck Questions

4.1 An ophthalmoscope is used to observe what part of the eyeball?

4.2 How do you prevent damage to the eye while using an ophthalmoscope?

4 Materials

☐ Ophthalmoscope
☐ Laboratory partner

Procedures

1. Before observing the retina, familiarize yourself with the parts of the ophthalmoscope, using Figure 29.6 as a reference.

2. The examination is best performed in a darkened room. Sit face to face with your partner, the subject, who should be relaxed. Be careful not to shine the light from the ophthalmoscope into the eye for longer than one to two seconds at a time. Ask your partner to look away from the light every few seconds.

3. Hold the ophthalmoscope in your right hand to examine the subject's right retina. Begin approximately six inches from the subject's right eye and look into the ophthalmoscope with your right eyebrow against the brow rest.

4. Move the instrument closer to the subject's eye, and tilt it so that light enters the pupil at an angle. The orange-red image is the interior of the eyeball. The blood vessels should be visible as branched line as in Figure 29.5c.

5. Observe the macula, with the fovea in its center. Move closer to the subject if you cannot see the fovea. To prevent damage to the fovea, be careful not to shine the light on the fovea for longer than one second.

6. Next to the macula is the optic disc, the blind spot on the retina. Notice how blood vessels are absent from this area. ■

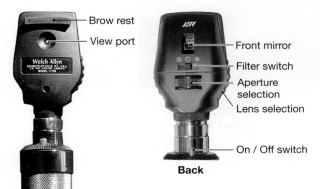

Brow rest
View port
Welch Allyn
MODEL 11770

Front mirror
Filter switch
Aperture selection
Lens selection

On / Off switch

Back

Handle

Front

Figure 29.6 **An Ophthalmoscope**

LAB ACTIVITY 5   Dissection of the Cow or Sheep Eye

The anatomy of the cow eye and sheep eye is similar to that of the human eye (Figure 29.7). Be careful while dissecting the eyeball, because the sclera is fibrous and difficult to cut. Use small strokes with the scalpel, and cut away from your fingers. Do not allow your laboratory partner to hold the eyeball while you dissect.

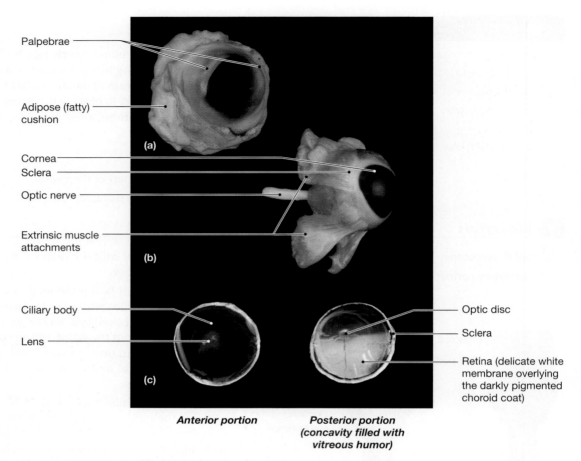

Palpebrae

Adipose (fatty) cushion

(a)

Cornea

Sclera

Optic nerve

Extrinsic muscle attachments

(b)

Ciliary body

Lens

(c)

Optic disc

Sclera

Retina (delicate white membrane overlying the darkly pigmented choroid coat)

Anterior portion

Posterior portion (concavity filled with vitreous humor)

Figure 29.7 Anatomy of the Sheep Eye
(a) Anterior view. **(b)** Lateral view. **(c)** Internal view of frontal sections.

 Safety Alert

You must—repeat, *must*—practice the highest level of laboratory safety while handling and dissecting the eyeball. Keep the following guidelines in mind during the dissection:

1. Wear gloves and safety glasses to protect yourself from the fixatives used to preserve the specimen.
2. Do not dispose of the fixative from your specimen. You will later store the specimen in the fixative to keep the specimen moist and to keep it from decaying.
3. Be extremely careful when using a scalpel or other sharp instrument. Always direct cutting and scissor motion away from you to prevent an accident if the instrument slips on moist tissue.
4. Before cutting a given tissue, make sure it is free from underlying and/or adjacent tissues so that they will not be accidentally severed.
5. Never discard tissue in the sink or trash. Your instructor will inform you of the proper disposal procedure. ▲

5 *Materials*

☐ Gloves
☐ Safety glasses
☐ Dissecting tray
☐ Dissecting tools
☐ Fresh or preserved cow or sheep eye

Procedures

1. Examine the external features of the cow or sheep eye. Depending on how the eye was removed from the animal, your specimen may have, around the eyeball, adipose tissue, portions of the extrinsic eye muscles, and the palpebrae. If so, note the amount of adipose tissue, which serves to cushion the eyeball. If your specimen lacks these structures, observe them in Figure 29.7.
   - Identify the **optic nerve** (II) exiting the eyeball at the posterior wall.
   - Examine the remnants of the **extrinsic muscles** and, if present, the **palpebrae** and eyelashes.
   - Locate the limbus, where the white sclera and the cornea join. The cornea, which is normally transparent, will be opaque if the eye has been preserved.

2. Securely hold the eyeball, and with scissors remove any adipose and extrinsic muscles from the surface, taking care not to remove the optic nerve.

3. Lay the eyeball on the dissecting tray, and with a sharp scalpel make an incision about 0.6 cm (1/4 in.) back from the cornea. Use numerous small downward strokes over the same area to penetrate the sclera.

4. Insert the scissors in the incision, and cut around the circumference of the eyeball, being sure to maintain the 0.6-cm distance back from the cornea.

5. Carefully separate the anterior and posterior cavities of the eyeball. The vitreous body should stay with the posterior cavity.

6. Examine the anterior portion of the eyeball.
   - Place a blunt probe between the lens and the ciliary processes, and carefully lift the lens up a little. The halo of delicate transparent filaments between the lens and the ciliary processes is formed by the suspensory ligaments. Notice the ciliary body where the suspensory ligaments originate and the heavily pigmented iris with the pupil in its center.
   - Locate the vitreous body in the posterior portion of the eyeball.
   - The retina is the tan-colored membrane that is easily separated from the heavily pigmented **choroid.**
   - Examine the optic disc where the retina attaches to the posterior of the eyeball.
   - The choroid has a greenish-blue membrane, the **tapetum lucidum,** which improves night vision in many animals, including sheep and cows. When headlights shine in a cow's eyes at night, this membrane reflects the light and makes the eyes glow. Humans do not have this membrane, and our night vision is not as good as that of animals that do. ■

Anatomy of the Eye

Name _____

Date _____

Section _____

A. Matching

Match each term in the left column with its correct description from the right column.

_____ **1.** transparent part of fibrous tunic

_____ **2.** thin filaments attached to lens

_____ **3.** corner of eye

_____ **4.** produces tears

_____ **5.** drains tears into nasal cavity

_____ **6.** adjusts lens shape

_____ **7.** border of sclera and cornea

_____ **8.** jellylike substance of eye

_____ **9.** depression in retina

_____ **10.** area lacking photoreceptors

_____ **11.** white region of eyeball

_____ **12.** watery fluid of anterior eye

_____ **13.** regulates light entering eye

_____ **14.** contains photoreceptors

_____ **15.** red structure in medial eye

A. limbus

B. iris

C. vitreous body

D. optic disc

E. ciliary muscle

F. caruncle

G. canthus

H. cornea

I. sclera

J. aqueous humor

K. retina

L. lacrimal gland

M. suspensory ligaments

N. fovea

O. nasolacrimal duct

B. Labeling

Label the structures of the eye in Figure 29.8.

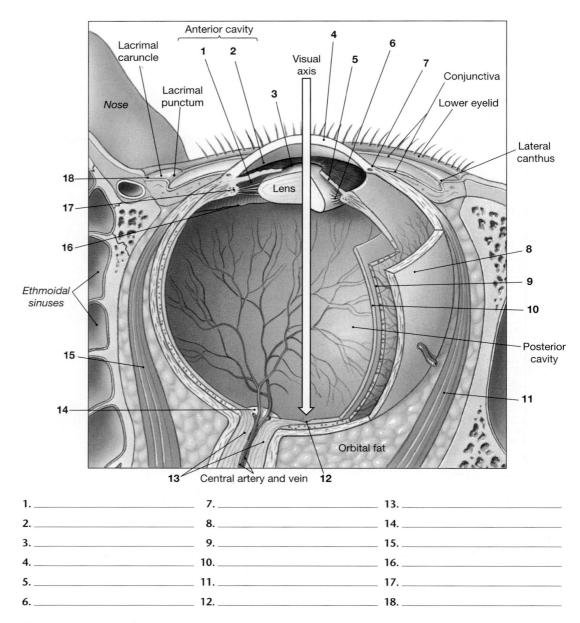

Figure 29.8 Sectional Anatomy of the Eye

A horizontal section of the right eye.

1. _____ 7. _____ 13. _____

2. _____ 8. _____ 14. _____

3. _____ 9. _____ 15. _____

4. _____ 10. _____ 16. _____

5. _____ 11. _____ 17. _____

6. _____ 12. _____ 18. _____

C. Short-Answer Questions

1. Name and describe the two major components of the fibrous tunic.

2. Describe the ciliary body of the choroid.

3. How are lacrimal secretions drained from the surface of the eye into the nasal cavity?

4. Describe the cavities of the eye and the circulation of aqueous humor.

5. How do the pupillary muscles respond to too much light entering the eye?

6. Why don't you see a blind spot in your visual field?

D. Analysis and Application

1. Which extrinsic eye muscles move the eyes right to look at an object in the far right of your visual field?

2. Macular degeneration occurs when cells in the fovea and macula are destroyed. Describe how this disease would change vision.

3. Name the structures of the eye through which light passes as the eye is stimulated, including the cellular layers of the retina.

4. Why are the pupils dilated in the eyes of someone who is excited or frightened?

Physiology of the Eye

OBJECTIVES

On completion of this exercise, you should be able to:

- Demonstrate the use of a Snellen eye chart and an astigmatism chart.

- Explain the terms myopia, hyperopia, presbyopia, and astigmatism.

- Explain why a blind spot exists in the eye and describe how it is mapped.

- Describe how to measure accommodation.

- Discuss the role of convergence in near vision.

- Describe how to record an electrooculogram.

- Compare eye movement when the eye is fixated on a stationary object with movement when the eye is tracking an object.

- Measure duration of saccades and fixation during reading.

LAB ACTIVITY 1 Visual Acuity

Sharpness of vision, or **visual acuity,** is tested with a Snellen eye chart, which consists of black letters of various sizes printed on white cardboard (Figure 30.1). A person with a visual acuity of 20/20 is considered to have normal, or **emmetropic** (em-e-TRŌ-pik), vision. If your visual acuity is 20/30, for example, you can see at 20 feet what an emmetropic eye can see at 30 feet; 20/30 vision is not as sharp as 20/20 vision.

An eye that focuses an image in front of the retina is **myopic** (mi-Ō-pik), or nearsighted, and can clearly see close objects but not distant objects. An eye that focuses an image behind the retina is **hyperopic** (HĪ-per-ō-pik), or farsighted, and can only see distant objects clearly. Corrective lenses are used to adjust for both conditions.

QuickCheck Questions

1.1 What is visual acuity, and how can it be measured?

1.2 How is the myopic eye different from the emmetropic eye?

1 Materials

- ☐ Snellen eye chart
- ☐ 25-foot tape measure
- ☐ Masking tape
- ☐ Laboratory partner

Procedures

1. Mount the eye chart on a wall at eye level. Along the floor, measure off a distance of 20 feet in front of the chart, and mark that spot on the floor with a piece of tape.

2. If you wear glasses or contact lenses, remove them before performing the vision test.

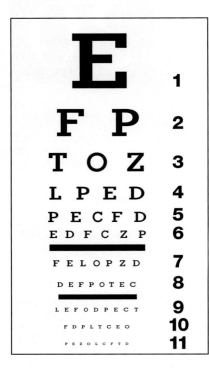

Figure 30.1 **Snellen Eye Chart**

3. Stand at the 20-foot mark, and cover your left eye with either a cupped hand or an index card. While your partner stands next to the eye chart, read a line where you can easily make out all the letters. Continue to view progressively smaller letters until your partner announces that you have not read the letters correctly. Record in Table 30.1 the visual-acuity value of this line.

4. Repeat this process with your right eye. Record your data in Table 30.1 in the Laboratory Report.

5. Now, using both eyes, read the smallest line you can see clearly. Record your data in Table 30.1.

6. If you wear glasses or contact lenses, repeat the test while wearing your corrective lenses. Record your data in Table 30.1. ■

LAB ACTIVITY 2 Astigmatism Test

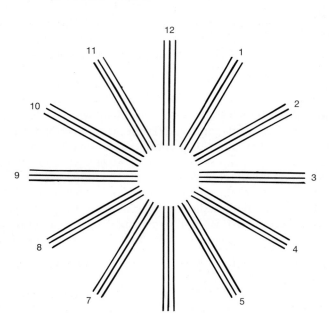

Figure 30.2 **Astigmatism Test Chart**

Astigmatism (ah-STIG-mah-tizm) is a reduction in sharpness of vision due to an irregularly shaped cornea or lens. When either of these surfaces is misshapen, it bends, or **refracts,** light rays incorrectly, resulting in blurred vision. The chart used to test for astigmatism has 12 sets of 3 lines laid out in a circular arrangement resembling a clock face (Figure 30.2).

QuickCheck Questions

2.1 What is astigmatism?

2.2 Describe the chart used to test for astigmatism.

2 *Materials*

☐ Astigmatism chart
☐ 25-foot tape measure

Procedures

1. Mount the astigmatism chart on a wall at eye level. Along the floor, measure off a distance of 20 feet in front of the chart, and mark that spot on the floor with a piece of tape.

2. If you wear glasses or contact lenses, remove them before performing this astigmatism test.

3. Stand at the 20-foot mark, and look at the white circle in the center of the chart. If all the radiating lines appear equally sharp and equally black, you have no astigmatism. If some lines appear blurred or are not consistently dark, you have astigmatism. ■

LAB ACTIVITY 3 Blind-Spot Mapping

The optic disk, or **blind spot**, is an area of the retina lacking photoreceptors. Normally you do not see your blind spot because the visual fields of your two eyes overlap and "fill in" the information missing from the blind spot.

QuickCheck Question

3.1 What is the blind spot?

3 *Materials*

□ Figure 30.3

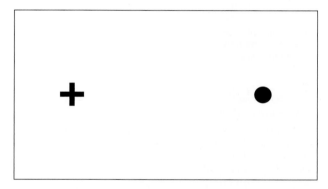

Figure 30.3 The Optic Disk
Close your left eye and stare at the cross with your right eye, keeping the cross in the center of your field of vision. Begin with the page a few inches away from your eye, and gradually increase the distance. The dot will disappear when its image falls on the blind spot, at your optic disk. To check the blind spot in your left eye, close your right eye and repeat this sequence while you stare at the dot.

Procedures

1. If you wear glasses, try the mapping procedures both with and without your glasses. Hold Figure 30.3 about two inches from your face with the cross in front of your right eye. Close your left eye, and stare at the cross with your right eye.

2. Slowly move the page away from your face. The dot disappears when its image falls on your blind spot.

3. If you have difficulty mapping your blind spot, remember not to move your eyes as the page moves. ■

LAB ACTIVITY 4 Accommodation

The process called **accommodation**, by which the eye lens changes shape to focus light on the retina, is detailed in Figure 30.4.

For objects 20 feet or further from the eye, the light rays leaving the objects and entering the eye are parallel to one another, as shown in Figure 30.4a. The ciliary muscle is relaxed, and the ciliary body is behind the lens. This causes the suspensory ligaments to pull the lens flat for proper refraction of the parallel light rays. Objects closer than 20 feet to the eye have divergent, or spreading, light rays that require more refraction to be focused on the retina. The ciliary muscle contracts, and the ciliary body shifts forward, releasing the tension on the suspensory ligaments. This release of tension causes the lens to bulge and become spherical rather than flat. The spherical lens increases refraction, and the divergent rays are bent into focus on the retina.

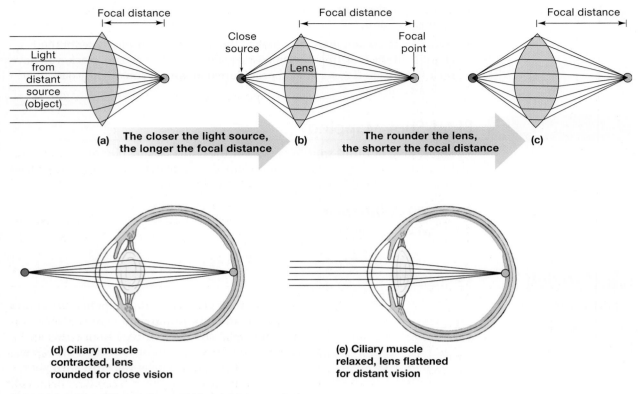

Figure 30.4 Image Formation and Visual Accommodations
A lens refracts light toward a specific point. The distance from the center of the lens to that point is the focal distance of the lens. **(a)** Light from a distant source arrives with all the light waves traveling parallel to one another. **(b)** Light from a nearby source is still diverging, or spreading out, from its source when it strikes the lens. Note the difference in focal distance after refraction. **(c)** The rounder the lens, the shorter the focal distance. **(d)** When the ciliary muscle contracts, the suspensory ligaments allow the lens to become more spherical. **(e)** For the eye to form a sharp image, the focal distance must equal the distance between the center of the lens and the retina. The lens compensates for the variations in the distance between the eye and the object in view by changing its shape. When the ciliary muscle relaxes, the ligaments pull against the margins of the lens and flatten it.

Reading and other activities requiring near vision cause eyestrain and fatigue because of the contraction of the ciliary muscle for accommodation. The lens gradually loses its elasticity as we age and causes a form of farsightedness called **presbyopia** (prez-bē-Ō-pē-uh). Many individuals have difficulty reading small type by the age of 40 and may require reading glasses to correct for the reduction in accommodation.

Accommodation is determined by measuring the closest distance from which one can see an object in sharp focus. This distance is called the **near-point** of vision. A simple test for near-point vision involves moving an object toward the eye until it becomes blurred.

QuickCheck Questions

4.1 What is accommodation?

4.2 Why does presbyopia occur with aging?

4 *Materials*

☐ Pencil
☐ Ruler

Procedures

1. Hold a pencil with the eraser up approximately two feet from your eyes and look at the ribs in the metal eraser casing.

2. Close your left eye, and slowly bring the pencil toward your open right eye.

3. Measure the distance from the eye just before the metal casing blurs. Record your measurements in Table 30.2.

4. Repeat the procedure with your left eye open.

5. Repeat the procedure with both eyes open.

6. Compare near-point distances with classmates of various ages. Record this comparative data in Table 30.2 in the Laboratory Report.

7. To focus on near objects, the eyes must rotate medially, a process called **convergence**. To observe convergence, hold the pencil at arm's length, stare at it with both eyes open, and slowly move it closer to your eyes. Which way did your eyes move? ■

LAB ACTIVITY 5 BIOPAC: Electrooculogram

BIOPAC

One of the most important functions your eyes can perform is to "fix," or "lock," on a specific object in such a way that the image is projected onto your retina at the area of greatest acuity, the fovea. Muscular control of your eyes works to keep the image on your fovea, regardless of whether the object is stationary or moving.

Two primary mechanisms are used to fixate on objects in your visual field: **voluntary fixation** allows you to direct your visual attention and lock onto the selected object, and **involuntary fixation** allows you to keep a selected object in your visual field once it has been found.

Voluntary fixation involves a conscious effort to move the eyes. You can, for instance, pick a person in a crowded room to fix your eyes on. You use this mechanism of voluntary fixation to initially select objects in your visual field; once selected, your brain "hands off" the task to involuntary fixation.

Even when you fixate on a stationary object, your eyes are not still but exhibit tiny, involuntary movements. There are three types of involuntary movements: tremors, slow drifts, and flicking. **Tremors** are a series of small tremors of the eyes at about 30–80 Hz (hertz or cycles/second). **Slow drifts** are involuntary movements that result in drifting movements of the eyes. This drift means that even if an object is stationary, the image drifts across the fovea. **Flicking** occurs as the image drifts to the edge of the fovea; as this happens, the eyeball "flicks" involuntarily so that the image is once again projected onto the fovea. Slow-drift movements and flicking movements will be in opposite directions. If the drifting movement is to the left, say, the flicking movement will be to the right.

When you wish to follow a moving object, you use large, slow movements called **tracking movements**. As you watch a bird fly across your visual field, your eyes are following an apparently smooth motion and tracking the moving bird. Although you have voluntarily directed your eyes to the bird, tracking movements are involuntary.

Saccades (sa-KĀDS) are the jumping eye motions that occurs when a person is reading or looking out the window of a moving car. Rather than a smooth tracking motion, saccades involves involuntary larger movements, to fix on a series of points in rapid succession. When this happens, your eye jumps from point to point at a rate of about three jumps per second. During saccades, the brain suppresses visual images so that you do not "see" (are not aware of) the transitional images between the fixation points. When you are reading, your eye typically spends about 10 percent of the time in saccades, moving from fixation point to fixation point, with the other 90 percent of the time fixating on words.

Eye movement can be recorded as an **electrooculogram (EOG)**, a recording of voltage changes that occur as eye position changes. Electrically, the eye is a spherical battery, with the positive terminal in front at the cornea and the negative terminal behind at the retina. The potential between the front and back of the eyeball is between 0.4 and 1.0 mV. By placing electrodes on either side of the eye, you can measure eye movement up to 70 degrees either left or right or up and down, where 0 degrees represents the eye pointed straight ahead and 90 degrees is directly lateral or vertical to the eyes. The electrodes measure the changes in potential as the cornea moves nearer or farther from the electrodes. When the eye is looking straight ahead, it is about the same distance from either electrode, and so the signal is essentially zero. When the cornea is closer to the positive electrode, that electrode records a positive difference in voltage.

QuickCheck Questions

5.1 Describe how the eyes track an object.

5.2 What is flicking?

5.3 What are saccades?

5   Materials

- ☐ Two BIOPAC electrode lead sets (SS2L)
- ☐ Six BIOPAC disposable vinyl electrodes (EL503)
- ☐ BIOPAC electrode gel (GEL1)
- ☐ Skin cleanser or rubbing alcohol
- ☐ Cotton balls
- ☐ BIOPAC acquisition unit (MP3STMP30)
- ☐ BIOPAC wall transformer (AC100A)
- ☐ BIOPAC serial cable (CBLSERA)
- ☐ Biopac Student Lab software v3.0 or better
- ☐ Computer: Macintosh®, minimum 68020 *or* PC Windows 98SE® or better
- ☐ Tape measure
- ☐ Two laboratory partners

Procedures

The EOG investigation is divided into four sections: setup, calibration, recording, and data analysis. Read each section completely before attempting a recording. If you encounter a problem or need further explanation of a concept, ask your laboratory instructor.

This experiment requires three persons. **You** will perform the test movements, one of your partners will be the **subject**, and the other partner will be the **recorder**. The recorder may either record the data by hand or choose **Edit > Journal > Paste measurements** to paste the data to the electronic journal for future reference.

Most response markers and labels are inserted automatically as a segment of the test is recorded. Markers appear at the top of the window as inverted triangles. The recorder may insert and label the marker either during or after the data are collected. To insert markers, press ESC on a Mac or F9 on a PC.

A. Setup

1. Turn the computer **ON**. The desktop should appear on the monitor. If it does not appear, ask your instructor for assistance.

2. Make sure the acquisition unit is turned **OFF**. Plug the unit into the wall and the serial cable into the unit and the computer.

3. Plug the electrode lead sets (SS2L) in as shown in Figure 30.5, with the horizontal lead set into CH 1 and the vertical lead set into CH 2. Note that on each of these two SS2L lead sets there is a pinch connector at the point where the red, white, and black leads attach to the main lead; you will use these pinch connectors in a moment when you attach the SS2L lead sets to the subject.

4. Turn the acquisition unit **ON**.

5. Have the subject remove any metal wrist or ankle bracelets. Also, *be sure the subject is not in contact with any metal objects (faucets, pipes, and so forth).*

6. Place the six electrodes (EL503) on the subject as shown in Figure 30.6. IMPORTANT: For accurate recordings, the electrodes must be horizontally and vertically aligned as described in the following steps. Before positioning each electrode, clean the subject's skin with a cotton ball dipped in skin cleanser or rubbing alcohol, and then apply a small dab of electrode gel (GEL1).

   - Attach one electrode above the right eyebrow and one below the right eye, with the two aligned vertically.

   - Attach one electrode to the lateral side of right eye and one to the lateral side of left eye, with the two aligned horizontally.

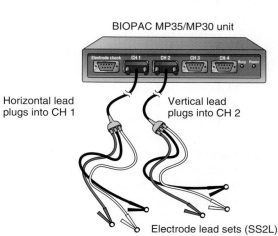

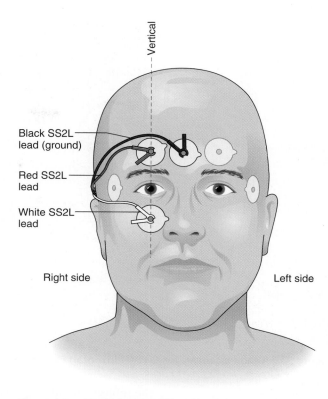

Figure 30.5 **Electrooculogram Setup**

Figure 30.6 **Lead Placement for CH2 (Vertical) Electrodes**

- Attach the fifth electrode above the nose and the sixth above the left eyebrow. These two electrodes serve as electrical grounds, and it is not critical that they be aligned.

For optimal electrode adhesion, the electrodes should be placed on the skin at least five minutes before the start of the calibration procedure. *Note:* Because these electrodes are attached near the eye, be very careful if using alcohol to clean the skin.

7. Attach the pinch connector of the vertical SS2L lead set from CH2 to the subject's shirt to relieve strain on the cable. Then attach the three leads to three of the EL503 electrodes on the subject's face, following the arrangement shown in Figure 30.6: the black lead to the electrode above the nose, the red lead to the electrode above the right eye, and the white lead to the electrode below the right eye. It is recommended that the electrode leads run behind the ears, as shown, to give proper cable strain relief.

8. Attach the pinch connector of the horizontal SS2L lead set from CH1 to the subject's shirt to relieve strain on the cable. Then attach the three leads to the other three EL503 electrodes on the subject's face, following the arrangement shown in Figure 30.7: the black lead to the electrode above the left eye, the red lead to the electrode on the lateral side of the right eye, and the white lead to the electrode to the lateral side of the left eye. Again, it is recommended that the electrode leads run behind the ears, as shown, to give proper cable strain relief.

9. Have the subject sit so that his or her eyes are in line with the center of the computer screen and he or she can see the screen easily with no head movement. Supporting the head to minimize movement is recommended.

10. Note the distance from the eyes to the computer screen; this distance will be needed later.

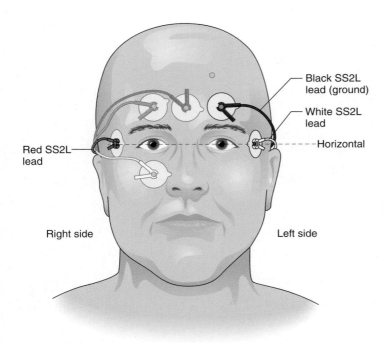

Black SS2L
lead (ground)

White SS2L
lead

Red SS2L
lead

Horizontal

Right side

Left side

Figure 30.7 Lead Placement for CH1 (Horizontal) Electrodes

11. **Start** the Student Lab program, and choose **Lesson L10-EOG-1.**

12. Type in the subject's **filename.** No two people can have the same filename, so use a unique identifier.

13. Click **OK.**

B. Calibration

The calibration procedure establishes the internal parameters for the hardware (such as gain, offset, and scaling) and is critical for optimum performance.

1. Make sure the subject is seated in the position described in "Setup" step 9. It is very important that the subject not move the head during calibration.

2. Click on **Calibrate.** A new window will be established, and a dialog box will pop up. The journal will be hidden from view during calibration.

Note: In the next step you will click on OK, and the subject should be prepared for what will happen. After you click, a dot will go in a counterclockwise rotation around the screen, and the subject will need to **track the dot with the eyes while keeping the head perfectly still.**

3. Click on **OK** to begin the calibration.

4. The **subject** should follow the dot around the screen with the eyes while keeping the head perfectly still. This calibration will continue for about 10 seconds and will stop automatically.

Check the calibration data at the end of the recording. There should be fluctuation in the data for each channel. If your data recording is similar to Figure 30.8, proceed to the Recording Lesson Data section. If your data recording is different from that shown in Figure 30.8, click on **Redo Calibration** and repeat the calibration sequence. (Possible reasons for the differences: If the subject did not follow the dot with the eyes or if the subject blinked, your recording will have large spikes. If one

Figure 30.8 Calibration Recording

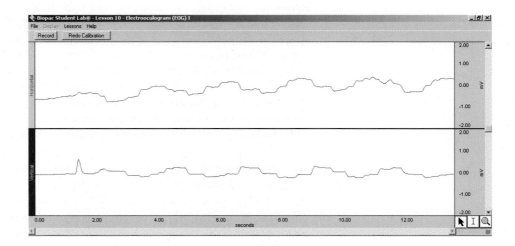

of the electrodes peeled up from the subject's face, your recording will have a too-large baseline drift.)

C. Recording Lesson Data

1. Prepare for the recording. You will record three segments: horizontal eye tracking, vertical eye tracking, and reading eye motion. In order to work efficiently, read this entire section so you will know what to do before you begin recording.

Check the last line of the journal, and note the total amount of time available for the recording. Stop each recording segment as soon as possible so that you do not waste recording space on the hard drive.

Hints for Obtaining Optimal Data:

a. The object should always be tracked with the eyes, *not the head.*

b. The subject should focus on one point of the object and should maintain that focus consistently.

c. The subject needs to sit so that head movement is minimized during recording.

d. There should be enough space near the subject so that you are able to move an object around the head at a distance of about 10 inches (25 cm).

e. When you are moving the object, try to keep it always at the same distance from the subject's head.

f. During recording, the subject should not blink. If unavoidable, the recorder should mark the blink on the recording.

g. Make sure the six electrodes stay firmly attached to the subject's face.

h. The larger the monitor, the better the eye-tracking portion of this lesson will work.

Segment 1—Horizontal Tracking

2. You and the subject should face each other in such a way that the subject is not looking at the computer screen.

3. Hold a pen in front of the subject's head at a distance of about 10 inches. Center the pen relative to the head so that the subject's eyes are looking straight ahead.

4. Instruct the subject to pick a point on the pen that is directly in her or his line of vision.

5. Click on **Record** to begin recording Segment 1 data.

6. Record for about 15 seconds.

   • Hold the pen in front of the subject for about five seconds, then move it laterally to the subject's left, then back past the center point and laterally to the subject's right, then back to center.

   From the time you start moving left till you return to center from the right, about 10 seconds should elapse. Say the movement directions aloud so that the recorder will know when to place markers to denote direction.

   • The recorder inserts a marker with each change of direction you call out: "L" for left and "R" for right. To insert markers, the recorder should press ESC on a Mac or F9 on a PC. Markers may also be entered or edited after the data are recorded.

7. Click on **Suspend** to halt the recording and review the data.

8. **Review** the data on the screen. If all went well, your data should look similar to Figure 30.9, with large deflections for the horizontal EOG (CH 1) and very little deflection for the vertical EOG (CH 2). There should be a negative peak where the subject's eyes reached their maximum displacement to the left and a positive peak where the eyes reached their maximum displacement to the right.

Figure 30.9 End of Segment 1 (Horizontal Tracking)

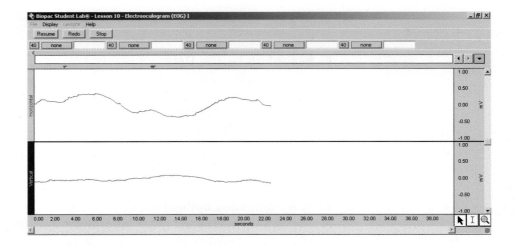

If your recording looks like that in Figure 30.9, go to step 10. If your recording does not look like Figure 30.9, go to step 9.

9. Click on **Redo**, and repeat steps 2–8. **Note:** A few blinks may be unavoidable; they will show in the data, but would not necessitate redoing the recording. The data will be **incorrect** if:

   a. Channel connections are incorrect.

   b. Lead connections are incorrect (for instance, if red lead is not connected to electrode at subject's right temple).

   c. Suspend button was pressed prematurely.

   d. Electrode peeled up, giving a large baseline drift.

   e. Subject looked away or moved the head.

Segment 2–Vertical Tracking

10. You and the subject should face each other in such a way that the subject is not looking at the computer screen.

11. Hold a pen in front of the subject's head at a distance of about 10 inches (25 cm). Center the pen relative to the head so that the subject's eyes are looking straight ahead. The subject may need to blink before resuming recording.

12. Instruct the subject to pick a point on the pen that is directly in his or her line of vision.

13. Click on **Resume**, and a marker labeled "eye tracking vertically" will automatically appear. The Segment 2 data will be recorded at the point where Segment 1 data stopped.

14. Record for about 30 seconds.

    • Hold the pen in front of the subject for about three seconds, then move it vertically up to the top edge of the subject's field of vision, then back past the center point and vertically down to the bottom edge of the field of vision, then back to center. From the time you start moving up till you return to center from the maximum down position, about 10 seconds should elapse. Say the movement directions aloud so that the recorder will know where to place markers to denote direction.

    • The recorder inserts a marker with each change of direction you call out: "U" for up and "D" for down. Markers may also be entered or edited after the data are recorded.

Figure 30.10 Segment 2 Vertical Tracking

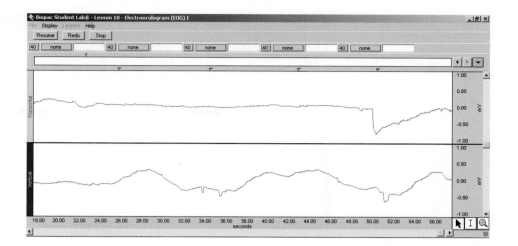

15. Click on **Suspend** to stop the recording and review the data.

16. Review the data on the screen. If all went well, your data should look similar to Figure 30.10, with large deflections for the vertical EOG (CH 2) and very little deflection for the horizontal EOG (CH 1). There should be a positive peak where the subject's eyes reached their maximum displacement upward and a negative peak where the eyes reached their maximum displacement downward.

    If your data recording is correct, go to step 18. If your recording is not correct, go to step 17.

17. Click on **Redo**, and repeat steps 10–16.

Segment 3–Reading

18. Turn to the end of the Laboratory Report, where the reading sample appears, and hold the page in front of the subject about 10 inches (25 cm) in front of his or her eyes.

19. Click on **Resume**; the recording will begin recording Segment 3 data.

20. The subject should read for about 20 seconds, reading silently to reduce any electromyogram (EMG) artifacts from facial muscle contraction. The recorder will insert a marker (ESC on Mac or F9 on PC) when the subject starts each new line of the reading sample. The recorder should watch the subject's eyes for vertical movements that indicate when the subject has finished reading one line and has begun to read the next line.

21. Click on **Suspend** to stop the recording and review the data.

22. Review the data on the screen. If all went well, your recording should look similar to Figure 30.11. If the recording is correct, go to step 24. If it is incorrect, go to step 23.

23. Click on **Redo**, and repeat steps 18–22.

24. Click on **Stop**. A dialog box comes up, asking if you are sure you want to stop the recording. Clicking "yes" will end the data-recording segment and automatically save the data. Clicking "no" will take you back to the Resume or Stop options.

Remove the SS2L pinch connectors from the subject's shirt, and peel the electrodes off the subject's face. Throw out the electrodes. Wash the electrode gel residue from the skin, using soap and water. The electrodes may leave a slight ring on the skin for a few hours. This is normal and does not indicate that anything is wrong.

If choosing the "Record from another subject" option, attach six fresh EL503 electrodes to the new subject's face as in Figure 30.6, position the subject per "Setup"

Figure 30.11 Segment 3—Reading

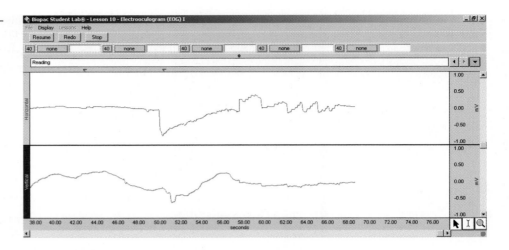

steps 5–9, and continue from "Setup" step 11. Remember that each subject will need to use a unique filename.

D. Data Analysis

1. Enter the **Review Saved Data** mode from the **Lessons menu.** A window that looks like Figure 30.12 should open. Note the CH designations: 40 for horizontal and 41 for vertical.

2. Set up your display window for optimal viewing of the first data segment, which is the one between time 0 and the first marker. The following tools help you adjust the data window:

Autoscale horizontal	Horizontal (time) scroll bar
Autoscale waveforms	Vertical (amplitude) scroll bar
Zoom tool	Zoom previous

Grids—Turn grids ON and OFF by choosing **Preferences** from the File menu.

Figure 30.12 Recording While Looking Left and Right

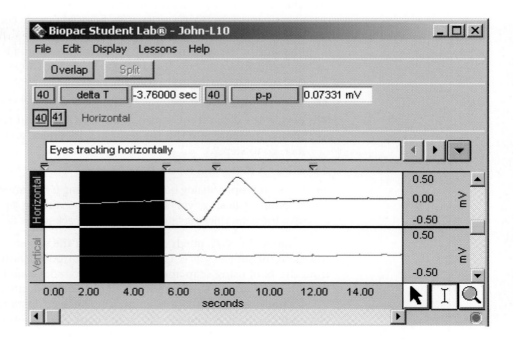

3. The measurement boxes are above the marker region in the data window. Each measurement has three sections: channel number, measurement type, and result. Channel number and measurement type are pull-down menus that are activated when you click on them. Set them up as follows:

CH 40   ΔT (difference between time at end of selected area and time at beginning of area)

CH 41   **p-p** (difference between maximum and minimum amplitude values in selected range)

CH 40   **slope** (difference in magnitude of two endpoints of selected area divided by time interval between endpoints; indicates relative speed of eye movement)

Note: The selected area is the area selected by the I-beam tool, including the endpoints.

4. Measure the amplitudes and durations for the horizontal-movement data (Figure 30.12). Remember, you can either record this and all other measurement information individually by hand, or choose **Edit** > **Journal** > **Paste measurements** to paste the data to your journal for future reference.

    When interpreting the different bumps in the data, in general, remember that:

- Large vertical bumps represent either blinks or eye movement from one line to another.
- Large horizontal bumps represent the eyes moving left to start the next line.
- Small bumps are saccades.

5. **Zoom** in on a small section of the first five seconds before the first marker of Segment 1. This section is the period of eye fixation when the eyes were staring straight ahead.

6. Find a part of the data segment with a small spike or bump, indicating flicking movement (Figure 30.13), and measure the duration and slope. Use the horizontal scroll bar to move through the recording.

7. Count the number of flicking movements in the interval from time 0 to 1 second, the interval from 1 second to 2 seconds, that from 2 seconds to

Figure 30.13 Flicking

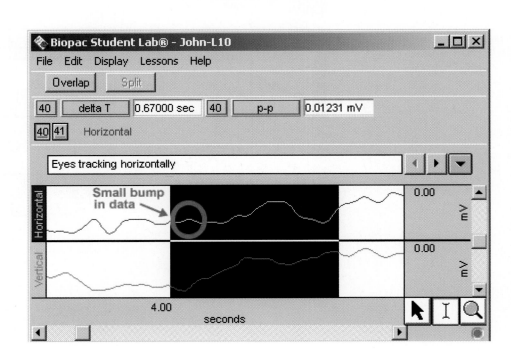

3 seconds, and that from 3 seconds to 4 seconds. Record the data in your journal.

Because slow drifting movements and flicking movements are both small, you will need to use **Zoom** and **Autoscale waveforms.** The flicking movements will occur as abrupt directional changes (steep slopes) and for shorter durations than the drifting movements. The flicking movements are difficult to measure. If it is too difficult to note them in your data, simply state that fact in the Laboratory Report.

8. Set up the measurement boxes as follows:

    CH 41   ΔT
    CH 41   p-p
    CH 41   slope

9. Take the measurements in the first five seconds of Segment 2 for flicking during fixation. Identify the data section when the subject moved his or her eyes to read the next line. The horizontal EOG will show the eyes moving left, and at the same time the vertical EOG will show the eyes moving down. Record the Segment 2 data in your journal.

10. Set up the display window to view Segment 3 data.

11. Find the saccades in the data (Figure 30.14), and record the Segment 3 data in your journal.

12. Save or print the data file. You may save the data to a floppy drive, save notes that are in the journal, or print the data file.

13. Exit the program. ■

Figure 30.14 Recording of Saccades

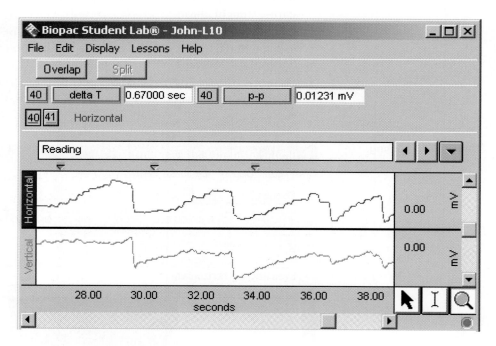

Physiology of the Eye

Name _____

Date _____

Section _____

A. Data Recording and Interpretation

Table 30.1 **Visual Acuity**

	Acuity (without glasses)	Acuity (with glasses)
Left eye	_____	_____
Right eye	_____	_____
Both eyes	_____	_____

Table 30.2 **Near-Point Determination**

Name and Age	Right Eye	Left Eye	Both Eyes
_____	_____	_____	_____
_____	_____	_____	_____
_____	_____	_____	_____

B. Matching

Match each term in the left column with its correct description from the right column.

_____ **1.** emmetropic   **A.** farsighted

_____ **2.** astigmatism   **B.** age-related farsightedness

_____ **3.** hyperopic   **C.** adjustment of lens shape

_____ **4.** myopic   **D.** irregular corneal surface

_____ **5.** presbyopia   **E.** bending of light rays

_____ **6.** visual acuity   **F.** nearsighted

_____ **7.** refraction   **G.** normal vision

_____ **8.** accommodation   **H.** sharpness of vision

C. Short-Answer Questions

1. Explain the difference between viewing a distant object and viewing one close to the eye.

2. How is the Snellen eye chart used to determine visual acuity?

LABORATORY REPORT

3. Describe how the lens changes shape to view a close object.

D. Analysis and Application

1. Why does our eyes' ability to accommodate change as we age?

2. Describe an experiment that demonstrates the presence of a blind spot on the retina.

3. Explain why the nearsighted eye cannot view distant objects in focus.

Physiology of the Eye: Electrooculogram

Name _____

Date _____

Section _____

A. Data and Calculations

Subject Profile

Name _____   Height _____

Age _____   Weight _____

Gender: Male / Female

1. Complete Table 30.3 using Segment 1 data. Be careful to be consistent with units (milliseconds versus seconds). *Note:* You need select only one example of a flicking movement.

Table 30.3 Segment 1 Data of Eye Orientation with Object Movement

Object Position → Eye Orientation → Measurement [CH #]	Stationary Object Fixation Flicking	Moving Object Tracking		
		Left	Right	Left
ΔT [CH 40]				
p-p [CH 40]				
slope [CH 40]				

Note: Velocity may be represented with a negative ("-") value because velocity vectors have a magnitude and direction.

2. Complete Table 30.4 using Segment 1 data.

Table 30.4 Flicking Movements

Time	Number of Flicking Movements
0–1 sec	
1–2 sec	
2–3 sec	
3–4 sec	

3. Complete Table 30.5 using Segment 2 data. You need select only one example of a flicking movement.

Table 30.5 Segment 2 Data of Eye Orientation with Object Movement

Eye Orientation → Measurement	Stationary Object Flicking	Moving Object Tracking		
		Up	Down	Up
ΔT [CH 41]				
p-p [CH 41]				
Slope [CH 41]				

4. Complete Table 30.6 with Segment 3 data. (You may not have seven saccades per line.)

Table 30.6 **Segment 3 Reading Data**

Measurement		First Line	Second Line
Number of saccades			
Duration of saccade:	#1		
	#2		
	#3		
	#4		
	#5		
	#6		
	#7		
Total duration of saccades/line			
Total reading time/line			
% time of saccades/total reading time			

B. Short-Answer Questions

1. For the Table 30.3 data, compare the duration (ΔT), relative changes in eye position (p-p), and speed of eye movement (slope) values for the flicking movements with these values for the tracking movements.

2. What is the stimulus for reflexive flicking movements?

3. Repeat question 1 for the data in Table 30.5.

4. Compare your Table 30.6 data results with the data of at least three other working groups in your class. What is the range of variation in percent time of saccades per total reading time?

5. Describe three types of involuntary movements when the eye is fixated on a stationary object.

6. Explain how an electrooculogram is recorded.

Reading Sample for Segment 3

Saccadic movements jump from place to place.

Alas, poor Yorick, I knew him well.

Wisdom is knowing what to do,

Skill is knowing how to do it,

and *virtue* is doing it.

David Starr Jordan

Anatomy of the Ear

OBJECTIVES

On completion of this exercise, you should be able to:

- Identify and describe the components of the external, middle, and inner ear.

- Describe the anatomy of the cochlea.

- Describe the components of the semicircular canals and the vestibule and explain their role in static and dynamic equilibrium.

The ear is divided into three regions: the **external**, or outer, **ear**; the **middle ear**; and the **inner ear**. The external ear and middle ear direct sound waves to the inner ear for hearing.

LAB ACTIVITY   1   The External Ear

The flap, or **auricle**, of the outer ear funnels sound waves into the **external auditory (acoustic) canal**, a tubular chamber that delivers sound waves to the **tympanic membrane**, commonly called the *eardrum*. The auricle has an inner foundation of elastic cartilage covered with adipose tissue and skin. The tympanic membrane is a thin sheet of fibrous connective tissue stretched across the canal and separating the external and middle regions of the ear. The canal contains wax-secreting cells in **ceruminous glands** and many hairs that prevent dust and debris from entering the ear. All these structures are represented in Figure 31.1.

QuickCheck Questions

1.1 What is the function of the auricle?

1.2 Which two regions of the ear does the tympanic membrane separate?

1 *Materials*

☐ Ear models and charts

Procedures

1. Label the three major regions of the ear in Figure 31.1.
2. Identify the auricle, external auditory canal, and tympanic membrane on the ear models and charts.
3. To appreciate how important the auricles are in directing sound into the ears, cup one hand over each ear. Did you notice a change in sound? Listen carefully for a moment to the sound, and then experiment by moving your fingers apart. Describe the change in sounds. Try again, but this time move only one finger on one hand. Did the sound change? ■

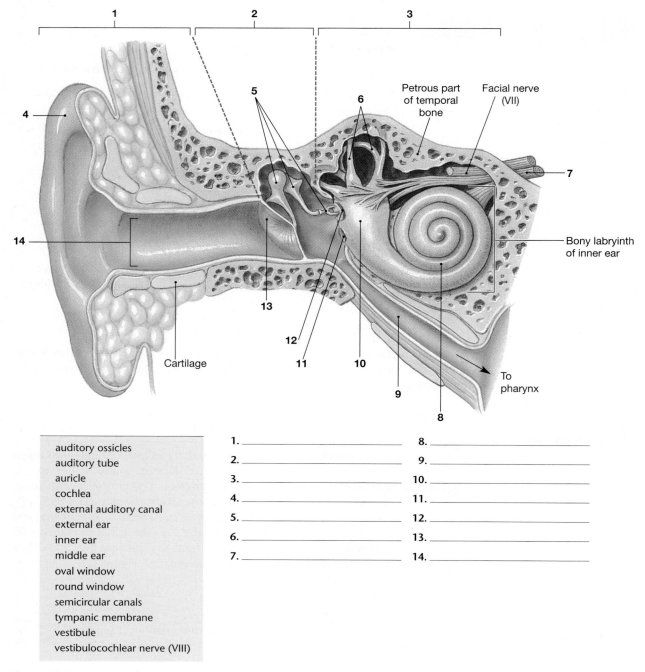

auditory ossicles
auditory tube
auricle
cochlea
external auditory canal
external ear
inner ear
middle ear
oval window
round window
semicircular canals
tympanic membrane
vestibule
vestibulocochlear nerve (VIII)

1. _____
2. _____
3. _____
4. _____
5. _____
6. _____
7. _____

8. _____
9. _____
10. _____
11. _____
12. _____
13. _____
14. _____

Figure 31.1 Anatomy of the Ear
The orientation of the external, middle, and inner ear.

LAB ACTIVITY 2   The Middle Ear

The middle ear (Figure 31.2) is a small space inside the petrous portion of the temporal bone. It is connected to the back of the upper throat (nasopharynx) by the **auditory tube**, also called the **pharyngotympanic tube** or **Eustachian tube**. This tube equalizes pressure between the external air and the cavity of the middle ear. Three small bones called **auditory ossicles** transfer vibrations from the external ear to the inner ear. The **malleus** (*malleus,* hammer) is connected to the tympanic membrane and the **incus** (*incus,* anvil), which joins to the third bone, the **stapes** (*stapes,* stirrup). Vibrations of the tympanic membrane are transferred to the malleus, which then conducts the vibrations to the incus and stapes. The stapes in turn pushes on the **oval window** of the cochlea to stimulate the auditory receptors.

The smallest skeletal muscles of the body are attached to the ossicles. The **tensor tympani** (TEN-sor tim-PAN-ē) **muscle** attaches to the malleus, and the **stapedius** (sta-PĒ-dē-us) **muscle** inserts on the stapes.

Clinical Application

Otitis Media

Infection of the middle ear is called **otitis media.** It is most common among infants and children, but occurs infrequently in adults. The infection source is typically a bacterial invasion of the throat that has migrated into the middle ear by way of the auditory tube. In children, the auditory tube is narrow and more horizontal than in adults. This orientation permits pathogens originally present in the throat to infect the middle ear. Children who frequently get middle-ear infections may have small tubes implanted through the tympanic membrane to drain fluid from the middle ear into the external auditory canal.

In severe cases, the microbes infect the air cells of the mastoid process, causing a condition known as **mastoiditis**. The passageways between the air cells become congested, and swelling occurs behind the auricle. This condition is serious, as it may spread to the brain. Powerful antibiotic therapy is necessary to treat the infection. Otherwise, a mastoidectomy may be necessary, a procedure that involves opening and draining the mastoid air cells. ▶

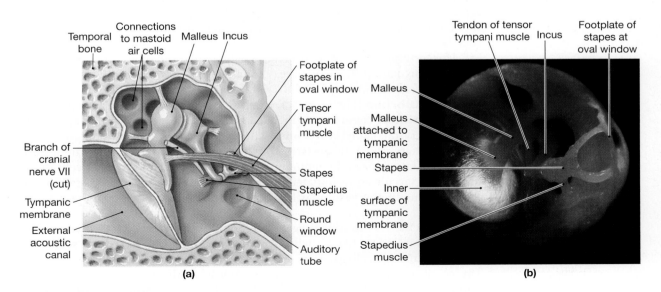

Figure 31.2 The Middle Ear
(a) The structures of the middle ear. **(b)** The tympanic membrane and auditory ossicles.

QuickCheck Questions

2.1 How does the eardrum stimulate the inner ear?

2.2 What is the function of the auditory tube?

2	*Materials*

☐ Ear models and charts

Procedures

1. Label the structures of the middle ear in Figure 31.1.

2. Identify the auditory ossicles on the ear models and charts. Notice the sequence of articulated structures: tympanic membrane, malleus, incus, stapes, oval window.

3. Identify the muscles of the middle ear: the tensor tympani and the stapedius muscles.

4. Locate the auditory tube on the ear models and charts. ■

LAB ACTIVITY 3

The Inner Ear

The inner ear consists of three compartments, each containing a specialized population of hair-cell receptors: a helical **cochlea** (KOK-lē-uh), an elongated **vestibule** (VES-ti-būl), and three **semicircular canals** that enclose receptors inside **semicircular ducts** (Figure 31.3a). Each compartment is organized into two fluid-filled chambers arranged in the form of a "pipe within a pipe." The outer "pipe," called the **bony labyrinth** (*labyrinthos;* maze of canals), is embedded in the temporal bone and contains **perilymph** (PER-i-limf). It is lined by an inner "pipe," the **membranous labyrinth**, filled with **endolymph** (en-dō-limf).

The hair-cell receptors of the inner ear are inside the membranous labyrinth. The vestibule has an **endolymphatic duct** that drains endolymph into an **endolymphatic sac**, where the liquid is reabsorbed into the blood.

The **anterior**, **lateral**, and **posterior** semicircular canals are positioned perpendicular to one another. Together they function as an organ of **dynamic balance** and work to maintain equilibrium when the body is in motion. Inside the canals the membranous labyrinth occurs as the semicircular ducts. At one end of each semicircular duct is a swollen **ampulla** (am-PUL-la) containing the **crista** (Figure 31.3b). Each crista is composed of hair cells (receptors) and supporting cells covered by a gelatinous cap called the **cupula** (KŪ-pū-la). Movement of the head causes the endolymph to either push or pull on the cupula, so that the embedded hair cells are either bent or stretched.

The vestibule is an area between the semicircular canals and the cochlea. The membranous labyrinth in this area contains two sacs, the **utricle** (Ū-tri-KUL) and the **saccule** (SAK-ūl), which contain **maculae** (MAK-ū-lē), receptors for **static equilibrium** (Figure 31.3c). Like the cristae, the maculae have hair cells and a **gelatinous covering**. Embedded in the gelatinous covering are calcium carbonate crystals called **statoconia**. The gelatinous covering and the statoconia collectively are called an **otolith**, which means "ear stone." When the head is tilted, the statoconia and membrane shift their position and the hair cells in the utricle and saccule are stimulated. Impulses from the maculae are passed to sensory neurons in the vestibular branch of the vestibulocochlear nerve (cranial nerve VIII).

The cochlea is the coiled region of the inner ear (Figure 31.4). It consists of three ducts rolled up like a shell. The **cochlear duct**, also called the **scala media**, contains hair cells that are sensitive to vibrations caused by sound waves. The cochlear duct is part of the membranous labyrinth and is filled with endolymph. Surrounding the cochlear duct are the **vestibular duct** (**scala vestibule**) and the **tympanic duct** (**scala tympani**) of the bony labyrinth.

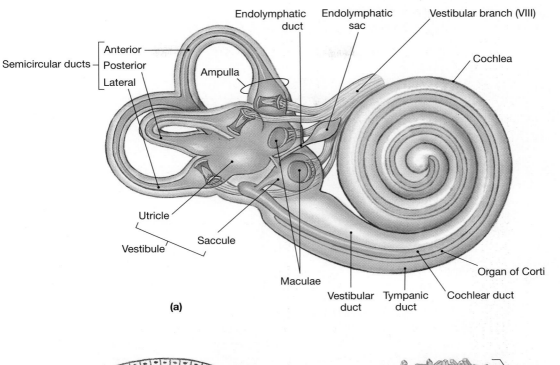

(a)

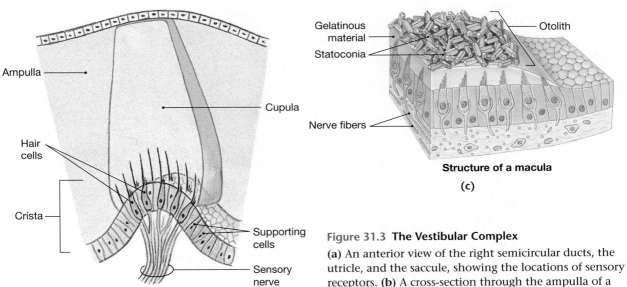

(b)

Structure of a macula

(c)

Figure 31.3 The Vestibular Complex
(a) An anterior view of the right semicircular ducts, the utricle, and the saccule, showing the locations of sensory receptors. **(b)** A cross-section through the ampulla of a semicircular duct. **(c)** The structure of a macula.

These ducts are filled with perilymph. The floor of the cochlear duct is the **basilar membrane** where the hair cells occur. The **vestibular membrane** separates the cochlear duct from the vestibular duct. These two ducts follow the helix of the cochlea, and the vestibular and tympanic ducts interconnect at the tip of the spiral.

To pass vibrations from sound waves to the inner ear, the stapes of the middle ear is connected to the vestibular duct at the oval window. When the stapes vibrates against the oval window, pressure waves are produced in the ducts of the cochlea. The waves stimulate the hair cells in the cochlear duct and then pass into the tympanic duct, where the round window stretches to dissipate the wave energy.

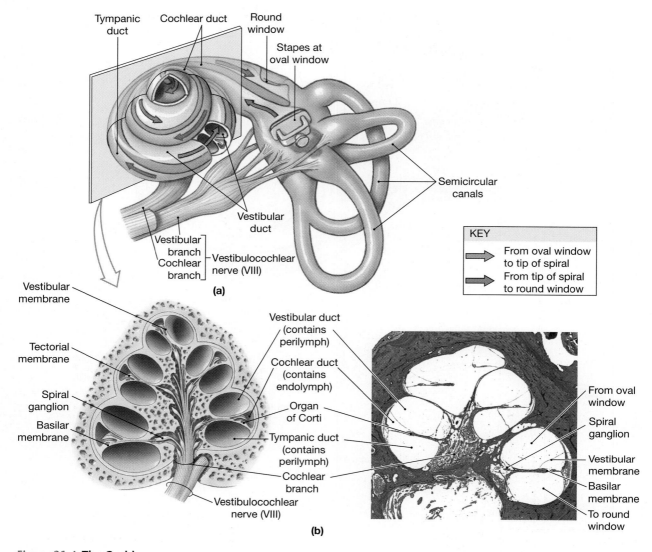

Figure 31.4 The Cochlea
(a) The structure of the cochlea. (b) Diagrammatic and sectional views of the cochlear spiral.

The cochlear duct contains the sensory receptor for hearing, called either the **spiral organ** or the **organ of Corti** (Figure 31.5). It consists of hair cells, supporting cells, and a gelatinous membrane called the **tectorial** (tek-TOR-ē-al) **membrane**. The **inner hair cells** rest on the basilar membrane near the proximal portion of the tectorial membrane, and the **outer hair cells** are at the tip of the membrane. The long stereocilia of the hair cells extend into the endolymph and contact the tectorial membrane. Sound waves cause fluid movement in the cochlea, and the hair cells are stimulated as they are pushed against the tectorial membrane. The hair cells synapse with sensory neurons in the cochlear branch of the vestibulocochlear nerve (VIII), which transmits the impulses to the auditory cortex of the brain. The **spiral ganglia** contain cell bodies of sensory neurons in the cochlear branch of nerve VIII.

QuickCheck Questions

3.1 What are the three receptors of the inner ear?

3.2 What is the function of the semicircular canals?

3.3 What is the function of the vestibule?

3.4 How is sound-wave pressure released by the cochlea?

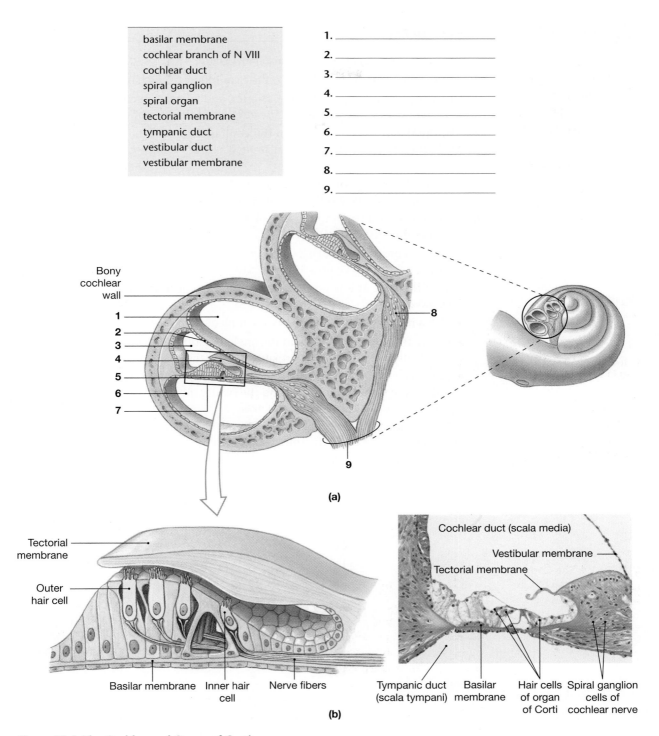

basilar membrane	1. _____
cochlear branch of N VIII	2. _____
cochlear duct	3. _____
spiral ganglion	4. _____
spiral organ	5. _____
tectorial membrane	6. _____
tympanic duct	7. _____
vestibular duct	8. _____
vestibular membrane	9. _____

Bony cochlear wall

1
2
3
4
5
6
7

8

9

(a)

Tectorial membrane

Outer hair cell

Basilar membrane   Inner hair cell   Nerve fibers

(b)

Cochlear duct (scala media)

Vestibular membrane

Tectorial membrane

Tympanic duct (scala tympani)   Basilar membrane   Hair cells of organ of Corti   Spiral ganglion cells of cochlear nerve

Figure 31.5 The Cochlea and Organ of Corti
(a) Structure of the cochlea as seen in section. (b) The three-dimensional structure of the tectorial membrane and hair cell complex of the organ of Corti.

☐ Ear models and charts

☐ Compound microscope

☐ Prepared slide of cochlea

☐ Prepared slide of crista

Procedures

1. Finish labeling Figure 31.1, and also label Figure 31.5.
2. Identify the structures of the ear on the ear models and charts.
3. Distinguish among the anterior, posterior, and lateral semicircular canals. On both a model and a chart, locate the ampulla at the base of each canal.
4. Identify the utricle and saccule of the vestibule. Identify the endolymphatic duct and sac. Note the vestibular branch of the vestibulocochlear nerve (VIII).
5. Observe the cochlea on the models and charts. Identify the vestibular, cochlear, and tympanic ducts and the vestibular and basilar membranes.
6. In the cochlear duct, examine the spiral organ. Locate the inner and outer hair cells and the tectorial membrane.
7. Examine a prepared slide of a cross-section of the cochlea and identify the structures shown in Figures 31.4 and 31.5.
   - Scan the slide at low power, and notice the repeating sections of the cochlea.
   - Pick a round section, and observe it at medium power.
   - Locate the vestibular duct and the angled vestibular membrane. Also identify the cochlear duct and the tympanic duct.
   - In the cochlear duct, identify the tectorial membrane hanging over the hair cells of the spiral organ resting on the basilar membrane. Distinguish between the inner and outer hair cells.
   - On the side where the tectorial membrane is attached, locate the oval cell bodies that constitute the spiral ganglion.
   - In the space below, sketch a cross-section of the cochlea.

8. Examine a prepared slide of the crista from a semicircular canal, and identify the structures shown in Figures 31.3 and 31.6.

- Scan the slide at low power, and notice the cochlea. Search for the crista at the base of the cochlea.
- Once you have located them, observe the crista at medium power. Identify the hair cells and the cupula.
- In the space below, sketch the section of the crista. ■

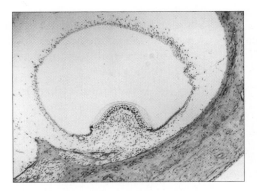

Figure 31.6 The Crista
Receptors for dynamic equilibrium are located in the ampullae of the semicircular ducts.

LAB ACTIVITY 4 Examination of the Tympanic Membrane

The tympanic membrane can be examined with an instrument called an **otoscope** (Figure 31.7). The removable tip is the **speculum**, and it is placed in the external auditory canal. Light from the instrument illuminates the tympanic membrane, which is viewed through a **magnifying lens** on the back of the otoscope.

QuickCheck Questions

4.1 What is the name of the instrument used to look at the tympanic membrane?

4.2 What is the removable tip of the instrument called?

4 Materials

- ☐ Otoscope
- ☐ Alcohol wipes
- ☐ Laboratory partner

Procedures

1. Using Figure 31.7 as reference, identify the parts of the otoscope.
2. Select the shortest but *largest-diameter* speculum that will fit into your partner's ear.
3. Either wipe the tip clean with an alcohol pad or place a new disposable cover over the speculum.
4. Turn the otoscope light on. Be sure the light beam is strong.
5. Hold the otoscope between your thumb and index finger, and either sit or stand facing one of your partner's ears. Place the tip of your extended little finger against your partner's head to support the otoscope. This finger placement is important to prevent injury by the speculum.

Disposable ear speculum

Magnifying lens

Handle

On / Off switch

Figure 31.7 Otoscope
The speculum is placed in the ear canal to examine the tympanic membrane.

6. Carefully insert the speculum into the external auditory canal while gently pulling the auricle up and posteriolaterally. Neither the otoscope nor the pulling should hurt your partner. If your partner experiences pain, stop the examination.

7. Looking into the magnifying lens, observe the walls of the external auditory canal. Note if there is any redness in the walls or any buildup of wax.

8. Manipulate the auricle and speculum until you see the tympanic membrane. A healthy membrane appears white. Also notice the vascularization of the region.

9. After the examination, either clean the speculum with a new alcohol wipe or else remove the disposable cover. Dispose of used wipes and covers in a biohazard container. ■

Anatomy of the Ear

Name _____

Date _____

Section _____

A. Matching

Match each description in the left column with its correct structure from the right column.

_____ **1.** eardrum

_____ **2.** receptors in semicircular canals

_____ **3.** coiled region of inner ear

_____ **4.** outer layer of inner ear

_____ **5.** receptor cells for hearing

_____ **6.** receptors in vestibule

_____ **7.** attachment membrane for stapes

_____ **8.** jellylike substance of crista

_____ **9.** contains endolymph

_____ **10.** membrane of spiral organ

_____ **11.** chamber inferior to organ of Corti

_____ **12.** perpendicular loops of inner ear

_____ **13.** membrane of tympanic duct

_____ **14.** membrane supporting spiral organ

_____ **15.** crystals in maculae

A. statoconia

B. basilar membrane

C. tectorial membrane

D. membranous labyrinth

E. cupula

F. bony labyrinth

G. tympanic membrane

H. cochlea

I. semicircular canals

J. round window

K. cristae

L. spiral organ

M. oval window

N. maculae

O. tympanic duct

B. Drawing

Sketch a cross-section of the cochlea. Label the ducts, membranes, and spiral organ.

LABORATORY REPORT

C. Short-Answer Questions

1. Describe the components of the external ear.

2. Describe the components of the middle ear.

3. Describe the components of the inner ear.

4. How are sound waves passed to the tympanic membrane?

5. Where in the ear are the receptors for equilibrium located?

6. Describe the receptors for hearing.

D. Analysis and Application

1. Why do children have more middle-ear infections than adults?

2. How are sound waves transmitted to the inner ear?

3. Which structures in the middle ear reduce the ear's sensitivity to sound?

4. Which parts of the inner ear are organized in a pipe-within-a-pipe arrangement?

5. If the pathway along the vestibular branch of nerve VIII has been disrupted, what symptoms would the patient display?

Physiology of the Ear

OBJECTIVES

On completion of this exercise, you should be able to:

- Explain the difference between static and dynamic equilibrium by performing various comparative tests.

- Understand the role of nystagmus in equilibrium.

- Test your range of hearing by using tuning forks.

- Compare the conduction of sound through air versus bone (Rinne test).

The inner ear serves two unique functions: balance and hearing. Imagine life without a sense of balance. We would not be able to stand, let alone play sports or enjoy a brisk walk. Balance, called **equilibrium**, feeds the brain a constant stream of information detailing the body's position relative to the ground. Although we live in a three-dimensional world, our bodies function in only two dimensions: front to back and side to side. Because our sense of the third dimension, top to bottom, is ground-based, we can get disoriented in deep water or while piloting an airplane. **Vertigo**, or motion sickness, may occur when the CNS receives conflicting sensory information from the inner ear, the eyes, and other receptors. For example, when you read in a moving car, your eyes are concentrating on the steady book, but your inner ear is responding to the motion of the car. The CNS receives the opposing sensory signals and may respond with vomiting, dizziness, sweating, and other symptoms of motion sickness.

Have you ever stood at your microwave oven and listened to your popcorn pop, waiting to push the stop button for a perfect batch? The ear is a dynamic sense organ capable of hearing multiple sound waves simultaneously. We can, for example, talk with a friend while listening to music and still hear the phone ring. The receptors for hearing, the **auditory** receptors, are located in the cochlea of the inner ear (as reviewed in Exercise 31).

LAB ACTIVITY   1   Equilibrium

As noted in Exercise 31, the receptors for equilibrium are in the membranous labyrinth of the vestibule of the inner ear. The crista in each semicircular duct is the receptor for **dynamic equilibrium** and responds to such movements as tilting of the head (Figure 32.1). When the head and body move suddenly, the otoliths, because they are heavier than the hair cells, lag behind and distort the hair cells, causing them to produce sensory impulses. The maculae, receptors for **static**

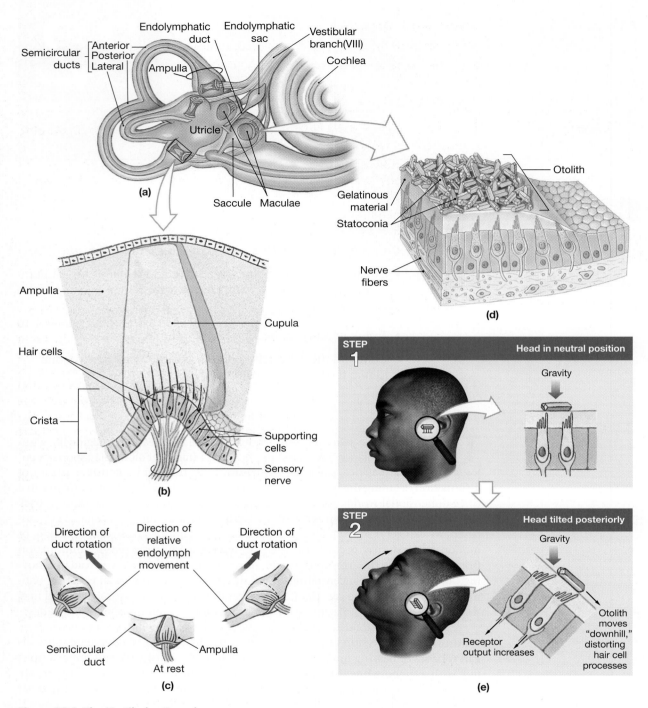

Figure 32.1 The Vestibular Complex
(a) An anterior view of the right semicircular ducts, the utricle, and the saccule, showing the locations of sensory receptors. (b) A cross-section through the ampulla of a semicircular duct. (c) Endolymph movement along the length of the duct moves the cupula and stimulates the hair cells. (d) The structure of a macula. (e) A diagrammatic view of macular function when the head is tilted back.

equilibrium, sense changes in body position relative to the direction of the pull exerted by gravity, such as when you stand on your head. Visual awareness of your surroundings enhances your sense of equilibrium as your brain continuously compares body position relative to the positions of surrounding stationary objects. A loss of visual references usually results in a loss of balance.

QuickCheck Questions

1.1 What is the function of the cristae in the semicircular ducts?

1.2 Which type of equilibrium do the maculae sense?

1 Materials

☐ Laboratory partner

Procedures

1. With your partner standing by to help if you lose your balance any time during this activity, stand on both feet in a clear area of the room.

2. With your arms at your sides and your eyes open, raise one foot and try to stand perfectly still for 45 seconds. Record your observations in Table 32.2 in the Laboratory Report.

3. While still standing on one foot, close both eyes and again try to stand perfectly still for 45 seconds. Record your observations in Table 32.2. ■

LAB ACTIVITY 2 Nystagmus

Nystagmus (nis-TAG-mus) is a reflex movement of the eyes in an attempt to maintain balance. It occurs as the visual system attempts to provide the brain with stationary references. When the head moves to the right, the eyes first move slowly to the left then quickly jump to the right to fix on some stationary object. The brain uses this object as a reference point by comparing the object's position with the body's position. This cycle of fast and slow eye movements provides the brain with brief "snapshots" of the stationary object rather than with a visual signal that is blurred because of the movement. Nystagmus can be used to evaluate how well receptors in the semicircular canals function. Immediately following any rotational motion of the head, endolymph in the semicircular ducts sloshes back and forth and continues to stimulate the receptors as if the head were still moving.

In this activity, you will look for nystagmus as a subject spins in a swivel chair. If a person spins in a chair with the head tilted forward, receptors in the lateral semicircular canals are stimulated, and the eyes move laterally. Spinning with the head leaning toward the shoulder stimulates the anterior semicircular canal and nystagmus is a vertical eye movement.

The eye movement called saccades, which is similar to nystagmus, is a tracking mechanism that occurs when a person is reading or looking out the window of a moving car. In saccades, investigated in Exercise 30, the eyes first fix on one object, then rapidly jump forward to another object.

QuickCheck Questions

2.1 Define *nystagmus*.

2.2 How does nystagmus help you maintain equilibrium?

Loss of Balance

This experiment involves spinning a subject in a chair and observing eye movements. *Do not use anyone who is prone to motion sickness.* The observer will spin the subject and record eye movement. To prevent accidents, four other individuals must hold the base of the chair steady and serve as spotters to help the subject stay in the chair during the experiment. The subject should remain in the chair for about two minutes after the experiment to regain normal equilibrium. ▲

2 *Materials*

☐ Swivel chair

☐ Four subjects

☐ Four spotters

Procedures

1. Seat the first subject in a swivel chair; have the spotters use their feet to support the chair base.

2. Instruct the subject to stare straight ahead with both eyes open.

3. Spin the chair clockwise for 10 rotations (approximately one rotation per second).

4. Quickly and carefully stop the chair, and observe the subject's eye movements. Record your observations in Table 32.3 in the Laboratory Report.

5. Keep the subject seated for approximately two minutes after spinning to regain balance.

6. Using a second subject, repeat the clockwise rotation but with the subject's head flexed. Record your observations of eye movements in Table 32.3.

7. Using a third subject, repeat the clockwise rotation but with the subject's head hyperextended. Record your observations of eye movements in Table 32.3.

8. Using a fourth subject, repeat the clockwise rotation but with the subject's head tilted toward either shoulder. Record your observations of eye movements in Table 32.3. ∎

LAB ACTIVITY 3 Hearing

We rely on the sense of hearing for communication and for an awareness of events in our immediate surroundings. Sounds are produced by vibrating objects as the vibrations cause the air around the objects first to compress and then to decompress, sending a wave of compressed and decompressed regions outward from the objects (Figure 32.2).

The number of compressed regions that pass a given point in one second is the **frequency**, or pitch, of the sound. An object vibrating rapidly produces a higher pitch than an object vibrating more slowly. The unit **hertz** (Hz) is used for the frequency of the compressed waves. (An alternative expression for this unit is cycles per second or cps.) Humans can hear sounds from about 20 to 20,000 Hz (20 to 20,000 cps). The *intensity,* or **amplitude**, of a sound is measured in **decibels** (dB). The higher, or "taller," a sound wave, the higher the amplitude of the sound and the greater its decibel rating.

Hearing occurs when sound waves enter the external auditory canal and strike the tympanic membrane, causing the membrane to vibrate at the frequency of the sound waves (Figure 32.3 and Table 32.1). The vibrations in the tympanic membrane are passed along to the auditory ossicles, causing them to vibrate. The malleus, which is connected to the tympanic membrane, vibrates and moves the incus, which moves the stapes. Vibrations of the stapes move the oval window, creating pressure waves in the perilymph of the vestibular duct. The pressure waves correspond to the sound waves that initially hit the tympanic membrane. The pressure waves pass through the cochlear duct and into the tympanic duct, causing the basilar membrane to vibrate. The pressure waves are dissipated by the stretching of the round window. If the pressure waves are not dissipated, they will cause interference with the next set of pressure waves, much like an ocean wave bouncing off a sea wall and slapping into the next incoming wave. In the ear, this interference would cause a loss of auditory acuity and loss of the ability to correctly discriminate between similar sounds.

The vibrations in the basilar membrane push specific hair cells into the tectorial membrane. The cells fire sensory impulses when their stereocilia touch the tectorial membrane and bend. The auditory information is passed over the cochlear branch of the vestibulocochlear nerve.

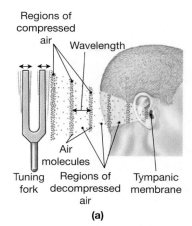

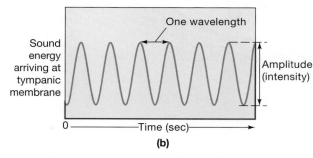

Figure 32.2 The Nature of Sound

(a) Sound waves (here, generated by a tuning fork) travel through the air as pressure waves. **(b)** A graph showing the sound energy arriving at the tympanic membrane. The distance between wave peaks is the wavelength. The number of waves arriving each second is the frequency, which we perceive as pitch. Frequencies are reported in cycles per second (cps) or hertz (Hz). The amount of energy in each wave determines the wave's amplitude, or intensity, which we perceive as the loudness of the sound.

Deep (low-frequency) sounds have long sound waves that stimulate the distal portion of the basilar membrane. High-pitch (high-frequency) sounds have short waves that stimulate the basilar membrane close to the oval window.

Deafness can be the result of many factors, not all of which are permanent. The two main categories of deafness are conduction deafness and nerve deafness. **Conduction deafness** involves damage either to the tympanic membrane or to one or more of the auditory ossicles. Proper conduction produces vibrations heard equally in both ears. If a conduction problem exists, sounds are normally heard better in the unaffected ear. Conduction tests with tuning forks, however, cause *the ear with the deafness to hear the sound louder than the normal ear.* This is due to an increased sensitivity to sounds in the ear with conduction deafness. Hearing aids are often used to correct for conduction deafness.

Nerve deafness is a result of damage to either the cochlea or the cochlear nerve. Repetitive exposure to excessively loud noises, such as music and machinery, can damage the delicate spiral organ. Nerve deafness cannot be corrected and results in a permanent loss of hearing, usually within a specific range of frequencies.

In the following tests, sound vibrations from tuning forks are conducted through the bones of the skull, bypassing normal conduction by the external and middle ear. Because the inner ear is surrounded by the temporal bone, vibrations are transmitted directly from the bone into the cochlea.

QuickCheck Questions

3.1 Where in the ear are the receptors for hearing?

3.2 What is the difference between conduction deafness and nerve deafness?

⚠ Use of Tuning Forks

To prevent damage to the tympanic membrane, *never insert a tuning fork into the external auditory canal.*

Tuning forks are designed to vibrate at a certain frequency. Gently tapping the fork on the side of your palm is sufficient to cause it to vibrate. ▲

 ## Materials

☐ Tuning forks (100 cps, 1,000 cps, and 5,000 cps)

☐ Laboratory partner

Procedures

1. A quiet environment is necessary for conducting hearing tests. The subject should use an index finger to close the ear opposite the ear being tested.

2. Weber's test:

   • With you as the subject and your partner as the examiner, have your partner strike the 100-cps tuning fork on the heel of the hand and place its base on top of your head, at the center.

   • Do you hear the vibrations from the tuning fork?

   • Are they louder in one ear than in the other?

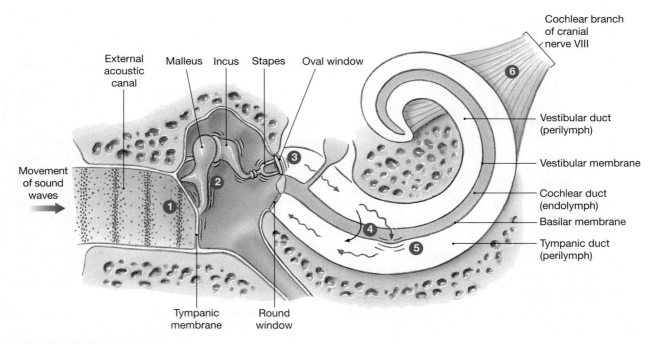

Figure 32.3 Sound and Hearing
Steps in the reception and transduction of sound energy.

Table 32.1 Steps in the Production of an Auditory Sensation (See Figure 32.3)

1. Sound waves arrive at the tympanic membrane.
2. Movement of the tympanic membrane causes displacement of the auditory ossicles (malleus, incus, and stapes).
3. Movement of the stapes at the oval window establishes pressure waves in the perilymph of the vestibular duct.
4. The pressure waves distort the basilar membrane on their way to the round window of the tympanic duct.
5. Vibration of the basilar membrane causes vibration of hair cells against the tectorial membrane.
6. Information concerning the region and intensity of stimulation is relayed to the CNS over the cochlear branch of cranial nerve VIII.

- Repeat the procedures using the 1,000-cps tuning fork.
- Do you hear one tuning fork better than the other? Record your observations in Table 32.4.

3. Rinne test:
- Have your partner sit down, and find the mastoid process behind his or her right ear.
- Strike the 1,000-cps tuning fork and place the base against the mastoid process. This bony process will conduct the sound vibrations to the inner ear.
- When your partner can no longer hear the tuning fork, quickly move it from the mastoid process and hold it close to—***but not touching***—her or his external ear. A person with normal hearing should be able to hear the fork. If conduction deafness exists in the middle ear, no sound will be heard. Record your observations in Table 32.5 in the Laboratory Report.
- Repeat this test on the left ear, and record your observations in Table 32.5.
- Repeat this test on each ear with the 5,000-cps tuning fork. Record your observations in Table 32.5. ■

Physiology of the Ear

Name _____

Date _____

Section _____

A. Experimental Observations

Record your observations in the following tables.

Table 32.2 **Equilibrium Tests**

Standing Position	Observation
On one foot, eyes open	
On one foot, eyes closed	

Table 32.3 **Nystagmus Tests**

Rotation and Head Position	Eye Movements
Clockwise, head facing straight forward	
Clockwise, head flexed	
Clockwise, head hyperextended	
Clockwise, head tilted to side	

Table 32.4 **Weber's Hearing Tests**

Frequency	Observations	
	Right Ear	Left Ear
100 cps		
1,000 cps		

Table 32.5 **Rinne Hearing Tests**

Frequency	Observations	
	Right Ear	Left Ear
1,000 cps		
5,000 cps		

32

B. Matching

Match each description in the left column with its correct structure from the right column.

_____ **1.** frequency of sound
_____ **2.** site of auditory receptors
_____ **3.** vibrates oval window
_____ **4.** transmits sound wave to auditory ossicles
_____ **5.** receptors for static equilibrium
_____ **6.** dissipates sound energy
_____ **7.** receptors for dynamic equilibrium
_____ **8.** loss of nerve function
_____ **9.** damage to tympanic membrane
_____ **10.** nystagmus
_____ **11.** amplitude
_____ **12.** vertigo

A. maculae
B. cristae
C. motion sickness
D. nerve deafness
E. eye movements during rotation
F. pitch
G. stapes
H. round window
I. tympanic membrane
J. conduction deafness
K. intensity of sound
L. cochlea

C. Short-Answer Questions

1. Describe the process of hearing.

2. Explain how sound waves striking the tympanic membrane result in movement of fluids in the inner ear.

3. Describe the receptors for dynamic and static equilibrium.

4. What is the range of sound frequencies that humans can hear?

D. Analysis and Application

1. How could the stereocilia of hair cells in the organ of Corti become damaged?

2. If conduction deafness exists, why is the sound of a tuning fork heard better in the deaf ear?

3. Explain the phenomenon of nystagmus.

The Endocrine System

OBJECTIVES

On completion of this exercise, you should be able to:

- Compare the two regulatory systems of the body: the nervous and endocrine systems.

- Identify each endocrine gland on laboratory models.

- Describe the histological appearance of each endocrine gland.

- Identify each endocrine gland when viewed microscopically.

Two systems regulate homeostasis: the nervous system and the endocrine system. These systems must coordinate their activities to maintain control of internal functions. The nervous system responds rapidly to environmental changes, sending electrical commands that can produce an immediate response in any part of the body. The duration of each electrical impulse is brief, measured in milliseconds. In contrast, the endocrine system maintains long-term control. In response to stimuli, **endocrine glands** produce regulatory molecules called **hormones** that slowly cause changes in the metabolic activities of **target cells**, which are any cells that contain membrane receptors for the hormone. Typically, the effect of the hormone is prolonged and lasts several hours.

The secretion of many hormones is regulated by negative feedback mechanisms. In **negative feedback**, a stimulus causes a response that either reduces or removes the stimulus. An excellent analogy is the operation of a central air conditioning (A/C) unit. When a room heats up, the warm temperature activates the thermostat on the wall, turning on the compressor of the A/C unit. Air flows in and cools the room and removes the stimulus (the warm temperature). Once the stimulus is completely removed, the A/C shuts off. Negative feedback is therefore a self-limiting mechanism.

An example of negative-feedback control of hormonal secretion is the regulation of insulin, a hormone from the pancreas that lowers blood sugar concentration. When blood sugar levels are high, as they are after a meal, the pancreas secretes insulin. The secreted insulin stimulates cells to increase their sugar consumption and

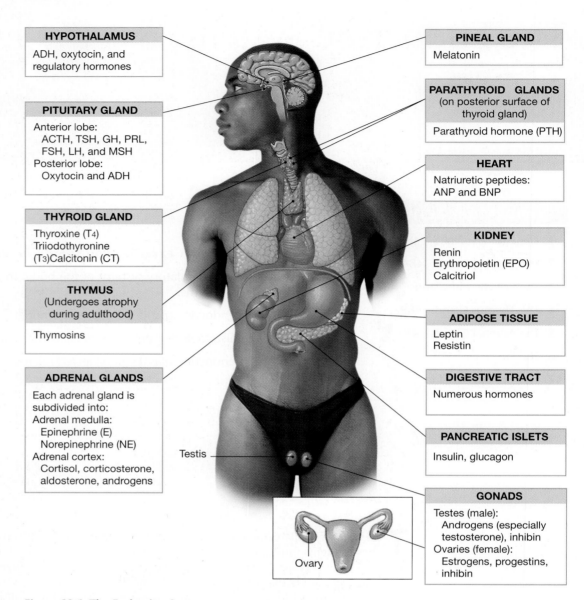

HYPOTHALAMUS
ADH, oxytocin, and regulatory hormones

PITUITARY GLAND
Anterior lobe:
 ACTH, TSH, GH, PRL, FSH, LH, and MSH
Posterior lobe:
 Oxytocin and ADH

THYROID GLAND
Thyroxine (T4)
Triiodothyronine (T3)Calcitonin (CT)

THYMUS
(Undergoes atrophy during adulthood)
Thymosins

ADRENAL GLANDS
Each adrenal gland is subdivided into:
Adrenal medulla:
 Epinephrine (E)
 Norepinephrine (NE)
Adrenal cortex:
 Cortisol, corticosterone, aldosterone, androgens

PINEAL GLAND
Melatonin

PARATHYROID GLANDS
(on posterior surface of thyroid gland)
Parathyroid hormone (PTH)

HEART
Natriuretic peptides:
ANP and BNP

KIDNEY
Renin
Erythropoietin (EPO)
Calcitriol

ADIPOSE TISSUE
Leptin
Resistin

DIGESTIVE TRACT
Numerous hormones

PANCREATIC ISLETS
Insulin, glucagon

GONADS
Testes (male):
 Androgens (especially testosterone), inhibin
Ovaries (female):
 Estrogens, progestins, inhibin

Testis

Ovary

Figure 33.1 The Endocrine System

storage, thus lowering the blood sugar concentration. As sugar levels in the blood return to normal, insulin secretion stops.

 In this exercise we will study the pituitary gland, thyroid and parathyroid glands, thymus gland, adrenal glands, pancreas, and gonads. Figure 33.1 shows the location of and hormones secreted by each endocrine gland.

LAB ACTIVITY 1 The Pituitary Gland

The **pituitary gland,** or **hypophysis** (hī-POF-i-sis), produces hormones that control the activity of other endocrine glands. It is therefore called the "master gland."

Anatomy

The pituitary is a small gland located in the sella turcica of the sphenoid of the skull, immediately inferior to the hypothalamus of the brain. Note how in Figure 33.2 the sphenoid serves as a protective cradle around the gland. A stalk, or **infundibulum**

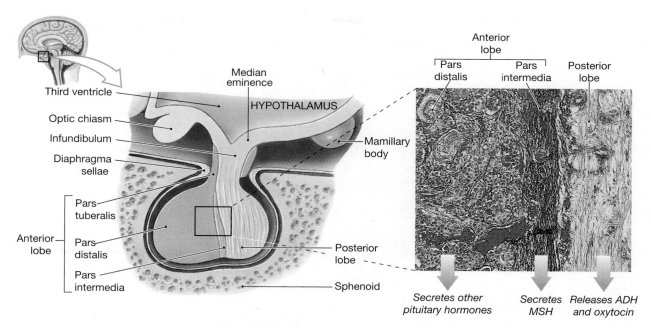

Figure 33.2 **The Anatomy and Orientation of the Pituitary Gland**

(in-fun-DIB-ū-lum), attaches the gland to the hypothalamus. **Regulatory hormones** and other molecules from the hypothalamus travel down a plexus of blood vessels in the infundibulum to reach the hypophysis.

The pituitary gland contains two lobes, anterior and posterior. The anterior lobe, the **adenohypophysis** (ad-ē-nō-hī-POF-i-sis), consists of three regions. The main portion of the gland is the **pars distalis** (dis-TAL-is); the **pars tuberalis** is a narrow portion that wraps around the infundibulum; the **pars intermedia** (in-ter-MĒ-dē-uh) is found in the interior of the gland, forming the boundary between the anterior and posterior lobes. The pars intermedia produces a single hormone, melanocyte-stimulating hormone (MSH). The pars distalis produces many hormones that target other endocrine glands, inducing them to produce and secrete their own hormones. Figure 33.3 lists the pituitary hormones and their target organs. This figure also includes the regulatory hormones from the hypothalamus that control secretions from the adenohypophysis. This interaction between the nervous system and the endocrine system integrates short-term and long-term regulation of body functions.

The posterior lobe, the **neurohypophysis** (noo-rō-hī-POF-i-sis) or **pars nervosa**, does not produce hormones. Instead, its function is to store and release antidiuretic hormone (ADH) and oxytocin (OT), which are both produced in the hypothalamus and then passed down the infundibulum to the neurohypophysis.

Histology

The pituitary gland is distinguished by the difference between the anterior and posterior lobes. The darker adenohypophysis is populated by a variety of cell types, each secreting a different hormone. This lobe contains more cells than the neurohypophysis and looks denser. Most of the cells of the anterior lobe stain reddish. The neurohypophysis consists of axons from hypothalamic neurons and appears much lighter than the anterior lobe.

QuickCheck Questions

1.1 Where is the pituitary gland located?

1.2 What is the main histological feature of the pituitary gland?

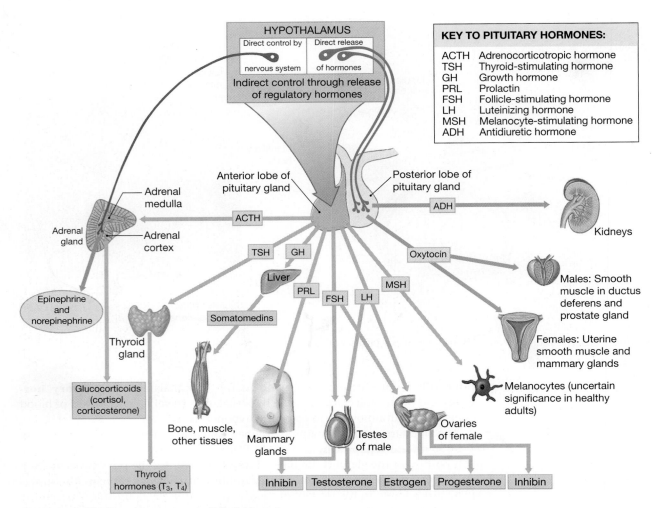

Figure 33.3 Pituitary Hormones and Their Targets

1 *Materials*

☐ Torso models
☐ Endocrine charts
☐ Dissecting and compound microscopes
☐ Prepared slide of pituitary gland

Procedures

1. Locate the pituitary gland on a torso model and an endocrine chart.

2. Use the dissecting microscope to survey the pituitary gland slide at low magnification. Can you distinguish between the two lobes of the gland?

3. Examine the slide at low and medium powers with the compound microscope. Identify the adenohypophysis and neurohypophysis, noting the different cell arrangements in each.

4. The pituitary gland regulates and influences many other glands. Return to Figure 33.1 and review the location of each endocrine gland. ■

LAB ACTIVITY 2 The Thyroid Gland

Anatomy

The **thyroid gland** is located in the anterior aspect of the neck, directly inferior to the thyroid cartilage (Adam's apple) of the larynx and just superior to the trachea (Figure 33.4a). This gland consists of two lateral lobes connected by a central mass, the **isthmus** (IS-mus). The thyroid produces two groups of hormones associated with the regulation of cellular metabolism and calcium homeostasis.

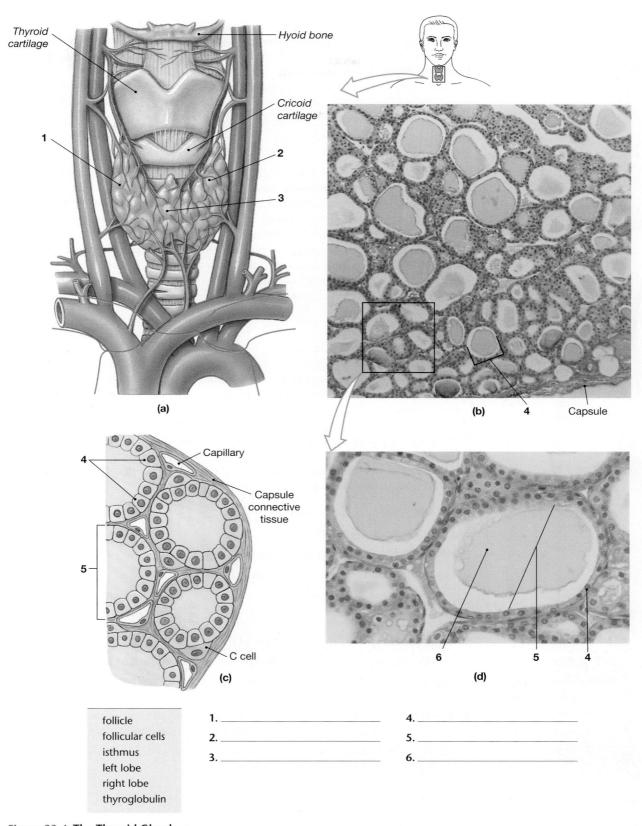

Thyroid cartilage

Hyoid bone

Cricoid cartilage

1

2

3

(a)

(b)　　4　　Capsule

4

Capillary

Capsule connective tissue

5

C cell

(c)

6　　**(d)**　　5　　4

follicle	1. _____	4. _____
follicular cells	2. _____	5. _____
isthmus	3. _____	6. _____
left lobe		
right lobe		
thyroglobulin		

Figure 33.4 The Thyroid Gland

(a) Location and gross anatomy. **(b)** Histological organization (LM × 99).
(c) Diagrammatic view of a section through the wall of the thyroid gland.
(d) Histological details, showing thyroid follicles (LM × 211).

Histology

The thyroid gland is very distinctive. It is composed of oval **follicles** embedded in connective tissue. Each follicle is composed of a single layer of cuboidal epithelium called **follicular cells**. The lumen of each follicle is filled with a **colloid**, a glycoprotein substance called **thyroglobulin** (thī-rō-GLOB-ū-lin) that stores thyroid hormones. Adjacent to the follicles are larger **C cells**, also called *parafollicular cells*.

Hormones

Follicular cells produce the hormones **thyroxine** (T_4) and **triiodothyronine** (T_3), hormones that regulate metabolic rate. These hormones are synthesized in the form of the glycoprotein thyroglobulin. It is secreted into the lumen of the follicles and stored there until needed by the body, at which time it is reabsorbed by the follicular cells and released into the bloodstream.

The C cells of the thyroid gland produce the hormone **calcitonin (CT)**, which decreases blood calcium levels. Calcitonin stimulates osteoblasts in bone tissue to store calcium in bone matrix and lower fluid calcium levels. It also inhibits osteoclasts in bone from dissolving bone matrix and releasing calcium.

Clinical Application

Hyperthyroidism

Hyperthyroidism occurs when the thyroid gland produces too much T_4 and T_3. Because these hormones increase mitochondrial ATP production and increase metabolic rate, individuals with this endocrine disorder are often thin, restless, and emotionally unstable. They fatigue easily because the cells are consuming rather than storing energy. *Graves' disease* is a form of hyperthyroidism that may result in the gland enlarging and producing a swollen goiter. *Exophthalmos* occurs as fat tissue is deposited deep in the orbit, causing the eyes to protrude. Treatment for hyperthyroidism may include partial removal of the gland or destruction of parts of it with radioactive iodine. ▶

QuickCheck Questions

2.1 Where is the thyroid gland located?

2.2 How are the thyroid cells arranged in the gland?

2 *Materials*

- ☐ Torso models
- ☐ Endocrine charts
- ☐ Dissecting and compound microscopes
- ☐ Prepared slide of thyroid gland

Procedures

1. Label the features of the thyroid gland in Figure 33.4.

2. Locate the thyroid gland on a torso model and an endocrine chart.

3. Scan the thyroid slide with the dissecting microscope. What structures are visible at this magnification?

4. Use the compound microscope to view the thyroid slide at low and medium powers. Locate a follicle, some follicular cells, thyroglobulin, and C cells.

5. In the space provided, sketch several follicles as observed at medium magnification. ■

The Parathyroid Glands

Anatomy

The **parathyroid glands** consist of two pairs of oval masses on the posterior surface of the thyroid gland. Each parathyroid mass is isolated from the underlying thyroid tissue by the parathyroid **capsule**.

Histology

The parathyroid glands are mostly composed of **chief cells**. These cells stain dark and cluster together in strings (Figure 33.5). The cells with less stain in the cytoplasm are **oxyphil cells**.

Hormones

The parathyroid glands produce **parathormone (PTH)**, also called **parathyroid hormone**, which is antagonistic to calcitonin from the thyroid gland. PTH increases the blood calcium level by stimulating osteoclasts in bone to dissolve small areas of bone matrix and release calcium ions into the blood. PTH also stimulates calcium uptake in the digestive system and reabsorption of calcium from the filtrate in the kidneys.

QuickCheck Questions

3.1 Where are the parathyroid glands located?

3.2 What two types of cells make up the parathyroid glands?

3   *Materials*

- ☐ Torso models
- ☐ Endocrine charts
- ☐ Dissecting and compound microscopes
- ☐ Prepared slide of parathyroid gland

Procedures

1. Label the parathyroid tissue in Figure 33.5.

2. Locate the parathyroid glands on a torso model and an endocrine chart.

3. Examine the parathyroid slide with the dissecting microscope. Scan the gland for thyroid follicles that may be on the slide near the parathyroid tissue.

4. Observe the parathyroid slide at low and medium powers with the compound microscope. Locate the dark-stained chief cells and the light-stained oxyphil cells.

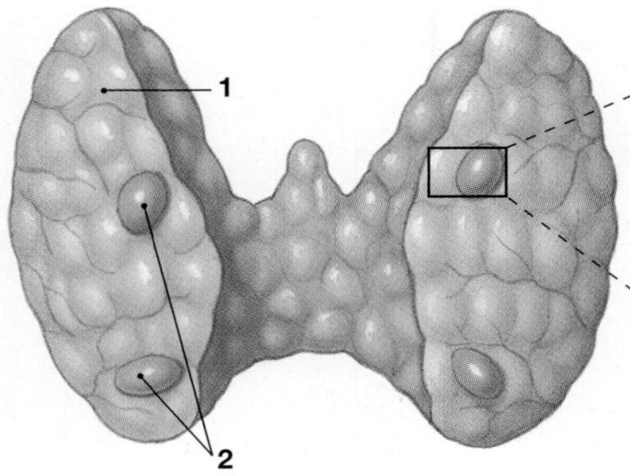

(a) Thyroid gland, posterior view

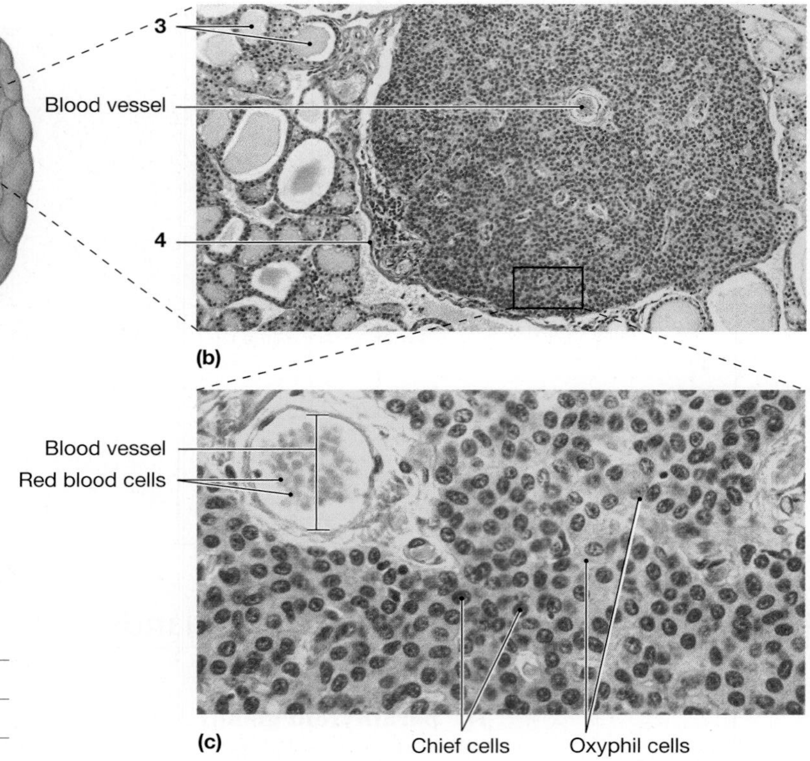

parathyroid capsule
parathyroid glands
thyroid follicles
thyroid gland

1. _____

2. _____

3. _____

4. _____

Blood vessel

(b)

Blood vessel
Red blood cells

(c)   Chief cells   Oxyphil cells

Figure 33.5 The Parathyroid Glands
(a) The location of the parathyroid glands on the posterior surfaces of the thyroid lobes.
(b) Both parathyroid and thyroid tissues (LM × 94). **(c)** Parathyroid cells (LM × 685).

5. In the space provided, sketch the parathyroid gland as observed at medium magnification. ▪

LAB ACTIVITY 4   The Thymus Gland

Anatomy

The **thymus gland** is located inferior to the thyroid gland, in the thoracic cavity posterior to the sternum (Figure 33.6). Because hormones secreted by the thymus gland facilitate development of the immune system, the gland is larger and more active in youngsters than in adults.

Histology

The thymus gland is organized into many **lobules,** separated by **septae.** Each lobule consists of a dense outer **cortex** and a light-staining central **medulla**. The cortex is

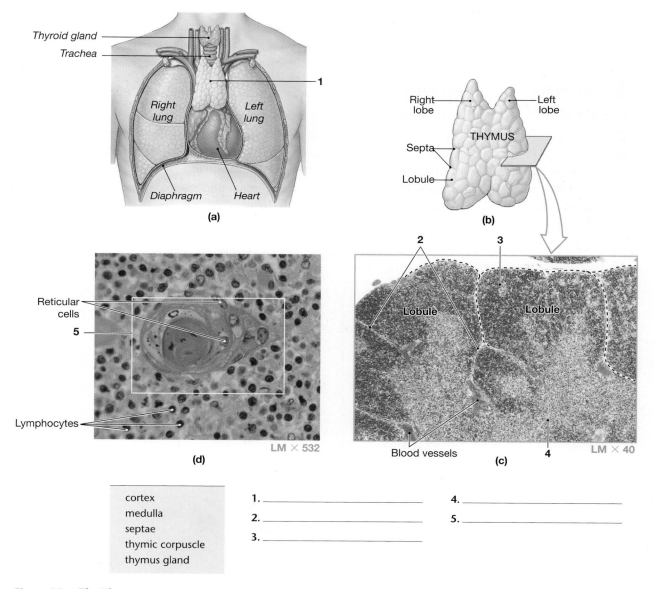

(a)

(b)

(d) LM × 532

(c) LM × 40

Reticular cells

5

Lymphocytes

Blood vessels   4

Thyroid gland

Trachea

1

Right lung

Left lung

Diaphragm   Heart

Right lobe   Left lobe

THYMUS

Septa

Lobule

2   3

Lobule   Lobule

cortex
medulla
septae
thymic corpuscle
thymus gland

1. _____
2. _____
3. _____

4. _____
5. _____

Figure 33.6 The Thymus

(a) The appearance and position of the thymus in relation to other organs in the chest. **(b)** Anatomical landmarks on the thymus. **(c)** Fibrous septa divide the tissue of the thymus into lobules resembling interconnected lymphatic nodules (LM × 40). **(d)** At higher magnification, we can see the unusual structure of Hassall's corpuscles. The small cells are lymphocytes in various stages of development (LM × 532).

populated by reticular epithelial cells that secrete the thymic hormones. The medulla is characterized by the presence of oval **thymic corpuscles** (Hassall's corpuscles), which are surrounded by cells called **lymphocytes.** Adipose and other connective tissues are abundant in an adult thymus because the function and size of the gland decrease after puberty.

Hormones

Although the epithelial cells of the thymus gland produce several hormones, the function of only one, **tymosin,** is understood. Tymosin is essential in the development and maturation of the immune system. Removal of the gland during early childhood usually results in a greater susceptibility to acute infections.

QuickCheck Questions

4.1 Where is the thymus gland located?

4.2 What is the main histological feature of the thymus gland?

4 Materials

- ☐ Torso models
- ☐ Endocrine charts
- ☐ Dissecting and compound microscopes
- ☐ Prepared slide of thymus gland

Procedures

1. Label the thymus gland features in Figure 33.6.
2. Locate the thymus gland on a torso model and an endocrine chart.
3. Scan the slide of the thymus gland with the dissecting microscope and distinguish between the cortex and the medulla.
4. Examine the thymus slide with the compound microscope at low magnification to locate a stained thymic corpuscle. Increase the magnification and examine the corpuscle. The cells surrounding the corpuscles are lymphocytes.
5. In the space provided, sketch the thymus gland as observed at medium magnification. ■

LAB ACTIVITY   5   The Adrenal Glands

Anatomy

Above each kidney is an **adrenal gland**, also called a **suprarenal gland** (soo-pra-RĒ-nal) because of its location superior to the kidney (Figure 33.7). An **adrenal capsule** protects and attaches the adrenal gland. The gland is organized into two major regions: the outer adrenal cortex and the inner adrenal medulla. The cortex is differentiated into three distinct regions, each producing specific hormones.

Histology

The adrenal **cortex** comprises three distinct layers, or zones. The **zona glomerulosa** (glō-mer-ū-LŌ-suh) is the outermost cortical region. Cells in this area are stained dark and arranged in oval clusters. The next layer, the **zona fasciculata** (fa-sik-ū-LA-tuh), is made up of larger cells organized in tight columns. These cells contain large amounts of lipid, making them appear lighter than the surrounding cortical layers (Figure 33.7). The deepest layer of the cortex, next to the medulla, is the **zona reticularis** (re-tik-ū-LAR-is). Cells in this area are small and loosely linked together in chainlike structures.

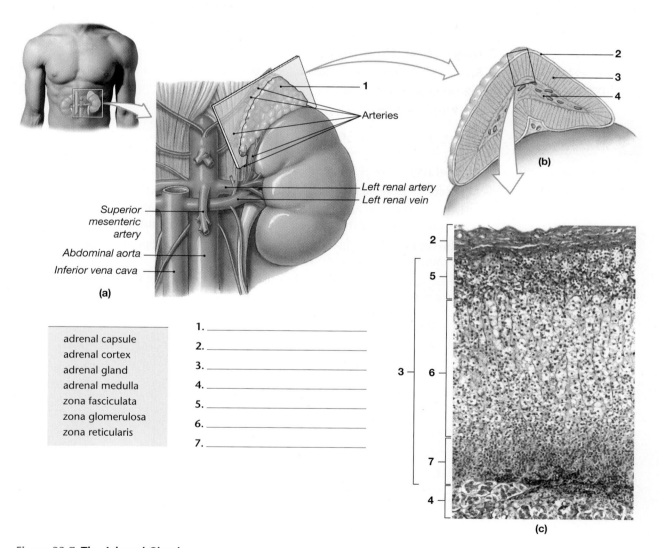

adrenal capsule
adrenal cortex
adrenal gland
adrenal medulla
zona fasciculata
zona glomerulosa
zona reticularis

1. _____
2. _____
3. _____
4. _____
5. _____
6. _____
7. _____

Figure 33.7 The Adrenal Gland

(a) Superficial view of the kidney and adrenal gland. **(b)** An adrenal gland, showing the sectional plane for part (c). **(c)** Light micrograph, with the major regions identified.

The adrenal **medulla** is the innermost part of the gland. The many blood vessels in the medulla give the tissue a dark-red coloration.

Hormones

The adrenal cortex secretes hormones collectively called *adrenocorticoids*. These hormones are lipid-based steroids. Each layer of the cortex secretes a different hormone. The zona glomerulosa secretes a group of hormones called **mineralocorticoids** that regulate, as their name implies, mineral or electrolyte concentrations of body fluids. A good example is **aldosterone**, which stimulates the kidney to reabsorb sodium from the fluid being processed into urine. The zona fasciculata produces a group of hormones called **glucocorticoids** that are involved in fighting stress, increasing glucose metabolism, and preventing inflammation. Two of the glucocorticoids, **cortisol** and **corticosterone** (kor-ti-KOS-te-rōn, are commonly used in creams to control irritating rashes and allergic responses of the skin. The zona reticularis produces **androgens**, male sex hormones. Both males and females produce small quantities of androgens in the zona reticularis.

The adrenal medulla is regulated by sympathetic neurons from the hypothalamus. In times of stress, exercise, or emotion, the hypothalamus stimulates the adrenal medulla to release its hormones, the neurotransmitters **epinephrine** (E) and **norepinephrine** (NE) into the blood, resulting in a body-wide sympathetic fight-or-flight response.

QuickCheck Questions

5.1 Where are the adrenal glands located?

5.2 What are the two major regions of the adrenal gland?

5.3 What are the three layers of the adrenal cortex?

5 Materials

- ☐ Torso models
- ☐ Endocrine charts
- ☐ Dissecting and compound microscopes
- ☐ Prepared slide of adrenal gland

Procedures

1. Label the components of the adrenal gland in Figure 33.7.
2. Locate the adrenal gland on a torso model and an endocrine chart.
3. Examine the slide of the adrenal gland with the dissecting microscope and distinguish among the capsule, adrenal cortex, and medulla.
4. Observe the adrenal gland with the compound microscope and differentiate among the three layers of the adrenal cortex.
5. In the space provided, sketch the adrenal gland, showing the details of the three cortical layers and the medulla. ■

The Pancreas

Anatomy

The **pancreas** lies between the stomach and the duodenum of the small intestine. The gland is called a *double gland* because it performs both exocrine and endocrine functions. Exocrine glands have ducts that transport secretions either to the body surface or to an exposed internal surface. Exocrine cells of the pancreas secrete digestive enzymes, buffers, and other molecules into pancreatic ducts for transport to the duodenum. Endocrine portions of the pancreas produce hormones that regulate blood sugar metabolism.

Histology

Figure 33.8 details the histology of the pancreas. This gland is composed mainly of exocrine **acinar cells,** which secrete pancreatic juice for digestive processes. These cells are dark-stained and grouped together either in small ovals or in columns. Connective tissues and pancreatic ducts are dispersed in the tissue. Embedded in the acinar cells are isolated clusters of **pancreatic islets,** or islets of Langerhans (LAN-ger-hanz). Each islet houses four types of endocrine cells: **alpha cells**, **beta cells**, **delta cells**, and **F cells**. These cells are difficult to distinguish with routine staining techniques and will not be individually examined.

Hormones

Pancreatic hormones regulate carbohydrate metabolism. Alpha cells in the pancreatic islets secrete the hormone **glucagon** (GLOO-ka-gon), which raises blood sugar concentration by catabolizing glycogen to glucose for cellular respiration. This process is called *glycogenolysis*. Beta cells secrete **insulin** (IN-suh-lin), which accelerates glucose uptake by cells and glycogenesis, the formation of glycogen. Insulin lowers blood sugar concentration by promoting the removal of sugar from the bloodstream.

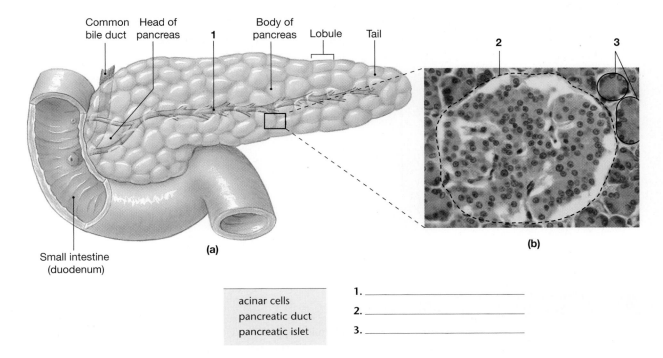

(a)

(b)

acinar cells	1. _____
pancreatic duct	2. _____
pancreatic islet	3. _____

Figure 33.8 The Endocrine Pancreas
(a) The gross anatomy of the pancreas. **(b)** A pancreatic islet surrounded by exocrine cells.

Clinical Application

Diabetes Mellitus

In **diabetes mellitus**, glucose in the blood cannot enter cells, and blood glucose levels exceed normal concentrations. In **type I diabetes**, the beta cells in the pancreas do not produce enough insulin, and cells are not stimulated to take in glucose. **Type II diabetes** occurs when the body becomes less responsive to insulin. The pancreas produces adequate amounts of insulin, but the cells are not receptive to the insulin. Diabetes is self-aggravating. Because they are glucose-starved, the alpha cells of the pancreas secrete glucagons and elevate blood glucose concentrations. ▶

QuickCheck Questions

6.1 Where is the pancreas located?

6.2 What is the exocrine function of the pancreas?

6.3 Where are the endocrine cells located in the pancreas?

6 Materials

Procedures

☐ Torso models

☐ Endocrine charts

☐ Compound microscope

☐ Prepared slide of pancreas

1. Label the histology of the pancreas in Figure 33.8.

2. Locate the pancreas on a torso model and an endocrine chart.

3. Use the compound microscope to locate the dark-stained acinar cells and the oval pancreatic ducts of the pancreas. Identify the clusters of pancreatic islets, the endocrine portion of the gland.

4. In the space provided, sketch the pancreas, labeling the pancreatic islets and the acinar cells. ■

LAB ACTIVITY 7   The Gonads

Gonads are the male and female reproductive organs that produce *gametes,* the spermatozoa and ova that fuse and start a new life. The **testes** and **ovaries** secrete hormones to regulate development of the reproductive system and maintenance of the sexually mature adult.

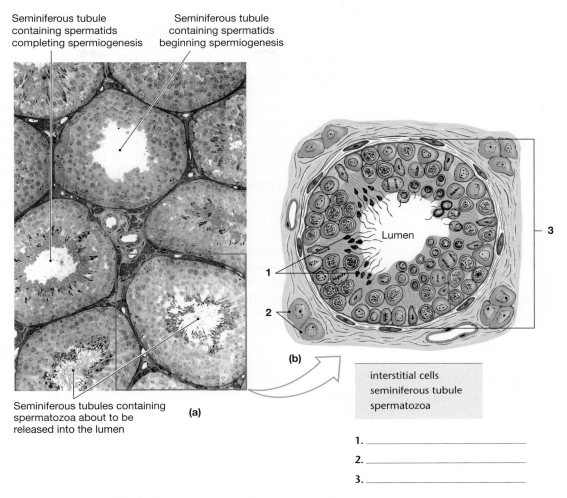

Seminiferous tubule containing spermatids completing spermiogenesis

Seminiferous tubule containing spermatids beginning spermiogenesis

Lumen

3

1

2

(b)

Seminiferous tubules containing spermatozoa about to be released into the lumen

(a)

interstitial cells
seminiferous tubule
spermatozoa

1. _____

2. _____

3. _____

Figure 33.9 The Seminiferous Tubules
(a) Transverse section through a coiled seminiferous tubule (LM × 625). **(b)** Cross-section through a single tubule.

Anatomy and Histology

Testes are the male gonads, located outside the body in the pouchlike scrotum. A testis, or testicle, is made up of coiled **seminiferous** (se-mi-NIF-e-rus) **tubules,** which produce spermatozoa. Figure 33.9 illustrates seminiferous tubules in cross-section with spermatozoa in the tubular lumen. **Interstitial cells**, located between the seminiferous tubules, are endocrine cells and secrete the male sex hormone **testosterone** (tes-TOS-ter-ōn), the hormone that produces and maintains secondary male sex characteristics, such as facial hair.

| *Clinical Application* | ***Steroid Abuse*** |

Anabolic steroids are androgens, precursors to the male sex hormone testosterone. Because testosterone stimulates muscle development and enhances the competitiveness of males, a surprising number of high school, college, and professional athletes use anabolic steroids to increase their strength, endurance, and athletic drive. The most commonly used steroid is *androstedione,* which is converted by the body to testosterone. Elevated blood testosterone levels inhibit secretion of the regulatory hormone GnRH from the hypothalamus. Decreased GnRH secretion keeps the anterior pituitary lobe from secreting FSH and LH, resulting in a decrease in sperm and testosterone production. Although steroids are used to enhance athletic performance, abuse of the steroids actually harms males and causes sterility, and a decrease in testosterone secretion.

Steroid abuse by female athletes is just as dangerous as abuse by males. Females produce small amounts of androgen in the adrenal cortex that is converted into estrogen, the main female sex hormone. Steriod use by female athletes also increases muscle mass but can cause irregular menstrual cycles and increased body hair and other secondary sex characteristics and, in some cases, baldness. In both sexes, steroid use can lead to liver failure, premature closure of epiphyseal plates, cardiovascular problems, and infertility.

The **ovaries** are the female gonads, located in the pelvic cavity. During each ovarian cycle, a small group of immature eggs, or oocytes, begins to develop an outer capsule of **follicular cells**. These **primordial follicles** develop first into **primary follicles** and then into **secondary follicles**. Eventually, one follicle becomes a **Graafian** (GRAF-ē-an)**follicle**, a fluid-filled bag containing an oocyte for release at **ovulation** (Figure 33.10). The Graafian follicle is a temporary endocrine structure

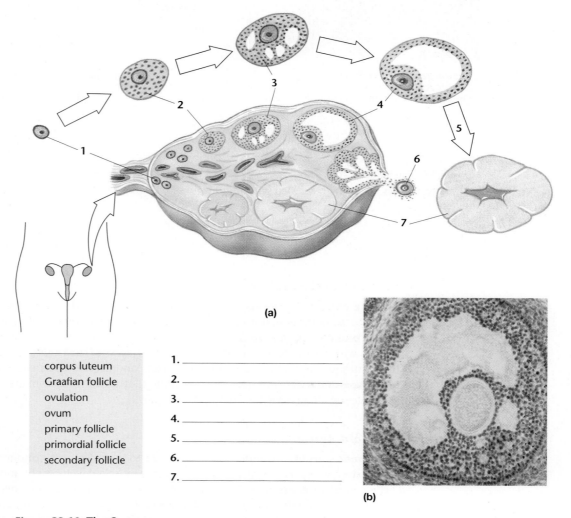

(a)

corpus luteum
Graafian follicle
ovulation
ovum
primary follicle
primordial follicle
secondary follicle

1. _____
2. _____
3. _____
4. _____
5. _____
6. _____
7. _____

(b)

Figure 33.10 The Ovary
(a) Follicular development during the ovarian cycle. **(b)** Micrograph of Graafian follicle.

and secretes the hormone **estrogen** to prepare the uterus for implantation of a fertilized egg. After ovulation, the ruptured Graafian follicle becomes the **corpus luteum** (LOO-tē-um), another temporary endocrine structure, which secretes **progesterone** (pro-JES-ter-ōn), the hormone that promotes further thickening of the uterine wall.

QuickCheck Questions

7.1 Where are the testes and ovaries located?

7.2 Where are the endocrine cells in the male gonad?

7.3 What are the endocrine structures in the ovaries?

7 Materials

- ☐ Torso models
- ☐ Endocrine charts
- ☐ Compound microscope
- ☐ Prepared slides of testis and ovary

Procedures

Testis

1. Label the structures of the testis in Figure 33.9.
2. Locate the testes on a torso model and an endocrine chart.
3. Scan the testis slide at low magnification, and identify the seminiferous tubules. Increase the magnification to locate interstitial cells between the seminiferous tubules.
4. In the space provided, sketch a cross-section of a testis, detailing the seminiferous tubules and interstitial cells.

Ovary

1. Label the ovarian structures in Figure 33.10.

2. Locate the ovaries on a torso model and an endocrine chart.

3. Scan the ovary slide at low power to locate the large Graafian follicle. Identify the developing ovum inside the follicle.

4. In the space provided, sketch the Graafian follicle. ■

The Endocrine System

Name _____

Date _____

Section _____

A. Matching

Match each endocrine structure in the left column with its correct description from the right column.

_____ **1.** thyroid follicle

_____ **2.** adrenal medulla

_____ **3.** thymic corpuscle

_____ **4.** seminiferous tubules

_____ **5.** zona glomerulosa

_____ **6.** parathyroid gland

_____ **7.** acinar cells

_____ **8.** Graafian follicle

_____ **9.** adenohypophysis

_____ **10.** master gland

_____ **11.** target cell

_____ **12.** pancreatic islets

_____ **13.** interstitial cells

_____ **14.** zona reticularis

_____ **15.** pars nervosa

_____ **16.** C cells

_____ **17.** infundibulum

_____ **18.** corpus luteum

A. contains ovum

B. four oval masses on posterior thyroid gland

C. neurohypophysis

D. produce insulin

E. cells between thyroid follicles

F. inner cortical layer of adrenal gland

G. contains hormones in colloid

H. pituitary gland

I. develops from ruptured Graafian follicle

J. innermost part of gland above kidney

K. stalk of pituitary gland

L. outer cortical layer of adrenal gland

M. produce spermatozoa

N. exocrine cells of pancreas

O. anterior pituitary gland

P. found in thymus gland

Q. produce testosterone

R. cell that responds to specific hormone

B. Drawing

Sketch a section of the thyroid gland and detail several thyroid follicles.

LABORATORY REPORT

C. Short-Answer Questions

1. Describe how negative feedback regulates the secretion of most hormones.

2. Why is the pituitary gland called the master gland of the body?

3. Describe the hormones produced by the three layers of the adrenal cortex.

4. Describe which structures are involved in the ovulation of an egg.

5. What are the endocrine functions of the pancreas?

6. If you had hyperthyroidism, how would your body respond to the hormonal imbalance caused by this condition?

D. Analysis and Application

1. What is the difference between type I and type II diabetes?

2. Although advertisements on television encourage us to eat a candy bar for quick energy, some individuals feel depressed after eating chocolate and other high-sugar snacks. What is the cause of their "sugar depression"?

3. Why is steroid use among athletes dangerous to their health?

4. Compare the histology of an adult thymus with that of an infant.

5. How is blood calcium regulated by the endocrine system?

Blood

OBJECTIVES

On completion of this exercise, you should be able to:

• List the functions of blood.

• Describe each component of blood.

• Distinguish each type of blood cell on a blood-smear slide.

• Describe the antigen-antibody reactions of the ABO and Rh blood groups.

• Safely collect a blood sample using the blood lancet puncture technique.

• Safely type a sample of blood to determine the ABO and Rh blood types.

• Correctly perform a hematocrit test.

• Correctly perform a coagulation test.

• Describe how to discard blood-contaminated wastes properly.

Blood is a liquid connective tissue that flows through blood vessels of the cardiovascular system. In response to injury, blood has the intrinsic ability to change from a liquid to a gel so as to clot and stop bleeding. Blood consists of cells and cellular pieces, collectively called the **formed elements**, carried in an extracellular fluid called blood **plasma** (PLAZ-muh). Blood has many diverse functions. It controls the chemical composition of all interstitial fluid by regulating pH and electrolyte levels. It supplies trillions of cells with life-giving oxygen, nutrients, and regulating molecules. Some of its formed elements protect the body from invasion by other life forms, such as bacteria, and others manufacture antibodies for defense against specific biological and chemical threats.

LAB ACTIVITY   1   Composition of Whole Blood

A sample of blood is approximately 55% plasma and 45% formed elements (Figure 34.1). Plasma is approximately 92% water and contains proteins to regulate the osmotic pressure of blood, proteins for the clotting process, and **antibodies**, the immune-system proteins that protect the body from invading pathogens. Electrolytes, hormones, nutrients, and some blood gases are transported in the blood plasma.

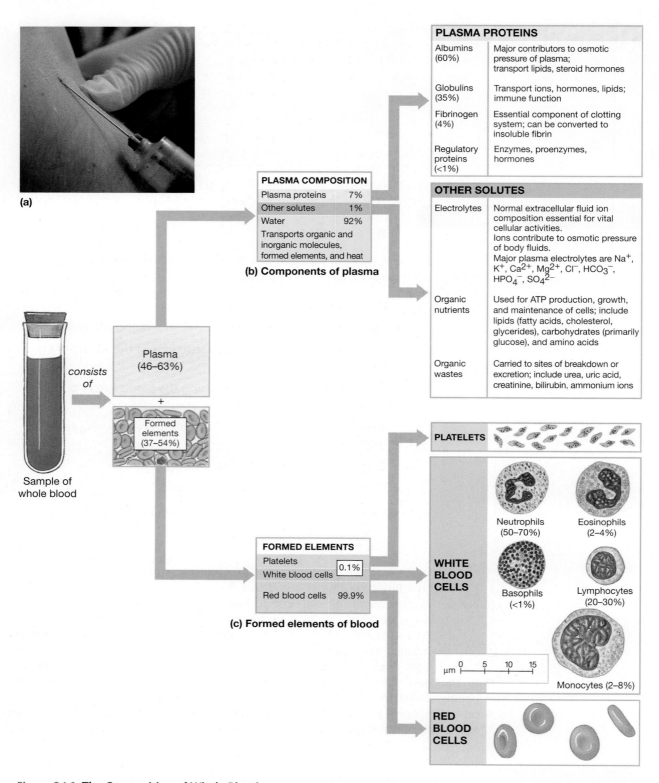

(a)

PLASMA PROTEINS

Albumins (60%)	Major contributors to osmotic pressure of plasma; transport lipids, steroid hormones
Globulins (35%)	Transport ions, hormones, lipids; immune function
Fibrinogen (4%)	Essential component of clotting system; can be converted to insoluble fibrin
Regulatory proteins (<1%)	Enzymes, proenzymes, hormones

PLASMA COMPOSITION

Plasma proteins	7%
Other solutes	1%
Water	92%

Transports organic and inorganic molecules, formed elements, and heat

(b) Components of plasma

OTHER SOLUTES

Electrolytes	Normal extracellular fluid ion composition essential for vital cellular activities. Ions contribute to osmotic pressure of body fluids. Major plasma electrolytes are Na^+, K^+, Ca^{2+}, Mg^{2+}, Cl^-, HCO_3^-, HPO_4^-, SO_4^{2-}
Organic nutrients	Used for ATP production, growth, and maintenance of cells; include lipids (fatty acids, cholesterol, glycerides), carbohydrates (primarily glucose), and amino acids
Organic wastes	Carried to sites of breakdown or excretion; include urea, uric acid, creatinine, bilirubin, ammonium ions

Plasma (46–63%)

+

Formed elements (37–54%)

Sample of whole blood

consists of

FORMED ELEMENTS

Platelets	0.1%
White blood cells	
Red blood cells	99.9%

(c) Formed elements of blood

PLATELETS

WHITE BLOOD CELLS

Neutrophils (50–70%)　Eosinophils (2–4%)

Basophils (<1%)　Lymphocytes (20–30%)

μm　0　5　10　15

Monocytes (2–8%)

RED BLOOD CELLS

Figure 34.1 The Composition of Whole Blood

(a) Drawing blood. **(b)** The composition of a typical sample of plasma. **(c)** The formed elements.

The formed elements are organized into three groups of cells: erythrocytes, leukocytes, and platelets. When stained, each group is easy to identify with a microscope. The reddish cells are erythrocytes, the cells that have visible nuclei are leukocytes, and the small cell fragments between the erythrocytes and leukocytes are platelets.

Erythrocytes (e-RITH-rō-sīts), commonly called **red blood cells** (RBCs), are red and lack a nucleus. They are the most abundant of all blood cells. Erythrocytes are biconcave discs that are noticeably thinner in the center (Figure 34.2). Also, the thinner central section of each disc is not as deeply stained as the surrounding rim. The biconcave shape gives each erythrocyte more surface area than a flat-faced disc would have, an important feature that allows rapid gas exchange between the blood and the tissues of the body. Their shape also allows erythrocytes to flex and squeeze through narrow capillaries.

The major function of RBCs is to transport blood gases. They pick up oxygen in the lungs and carry it to the tissue cells of the body. While supplying the cells with oxygen, the blood acquires carbon dioxide from the cells. The plasma and RBCs convey the carbon dioxide to the lungs for removal during exhalation. To accomplish the task of gas transport, each RBC contains millions of hemoglobin (Hb) molecules. **Hemoglobin** (HĒ-mō-glō-bin) is a complex protein molecule with four iron atoms that allow oxygen and carbon dioxide molecules to loosely bind to the Hb.

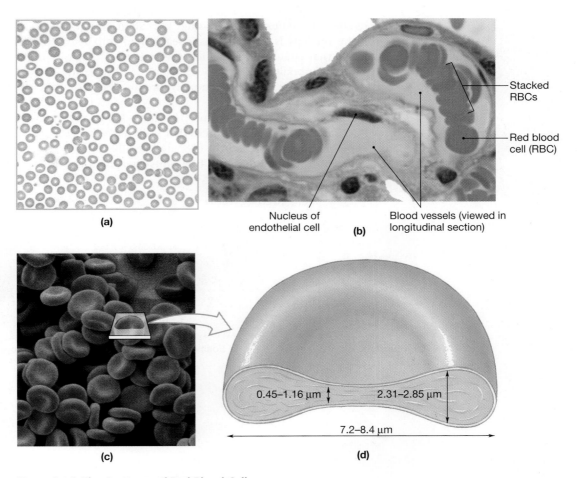

Figure 34.2 The Anatomy of Red Blood Cells
(a) When viewed in a standard blood smear, red blood cells appear as two-dimensional objects, because they are flattened against the surface of the slide. (b) When traveling through relatively narrow capillaries, RBCs may stack like dinner plates. (c) The three-dimensional structure of red blood cells. (d) A sectional view of a mature red blood cell, showing the normal ranges for its dimensions.

Unlike erythrocytes, each **leukocyte** (LOO-kō-sīt) contains a nucleus. The nucleus, which takes a very dark stain, is often branched into two or more lobes, as shown in Figure 34.3. Leukocytes lack hemoglobin and therefore do not transport blood gases. Leukocytes can pass between the endothelial cells of capillaries and enter the interstitial spaces of tissues. Most leukocytes are phagocytes and are part of the immune system. They are also called **white blood cells** (WBCs).

There are two broad classes of leukocytes: granular and agranular. The **granular leukocytes**, collectively called **granulocytes**, have granules in their cytoplasm and include the neutrophils, eosinophils, and basophils. **Agranular leukocytes**, which include the monocytes and lymphocytes, have few cytoplasmic granules.

Neutrophils (NOO-trō-filz) are the most common leukocytes and account for up to 70% of the WBC population. These granulocytes are also called **polymorphonuclear** (pol-ē-mōr-fō-NOO-klō-ar) **leukocytes** because the nuclei are complex and branch into two to five lobes. Neutrophils have many small cytoplasmic granules that stain pale purple, visible in Figure 34.3a.

Neutrophils are the first leukocytes to arrive at a wound site to begin infection control. They release cytotoxic chemicals and phagocytize bacteria. They also release hormones called **cytokines** that attract other phagocytes, such as eosinophils and monocytes, to the site of injury. Neutrophils are short-lived and survive in the blood for up to 10 hours. Active neutrophils in a wound may live only 30 minutes until they succumb to the toxins released by bacteria they have ingested.

About the same size as neutrophils, **eosinophils** (ē-ō-SIN-ō-filz) are identified by the presence of medium-sized granules that stain orange-red, as shown in Figure 34.3b. The nucleus is conspicuously segmented into two lobes. Eosinophils are phagocytes that engulf bacteria and other microbes that the immune system has coated with antibodies. They also contribute to decreasing the inflammatory response at a wound or site of infection. Approximately 3% of the circulating leukocytes are eosinophils.

Basophils (BĀ-sō-filz) constitute less than 1% of the circulating leukocytes. They have large cytoplasmic granules that stain dark blue. The granules are so large and numerous that the nucleus is obscured, as illustrated in Figure 34.3c. Smaller than neutrophils and eosinophils, basophils are sometimes difficult to locate on a

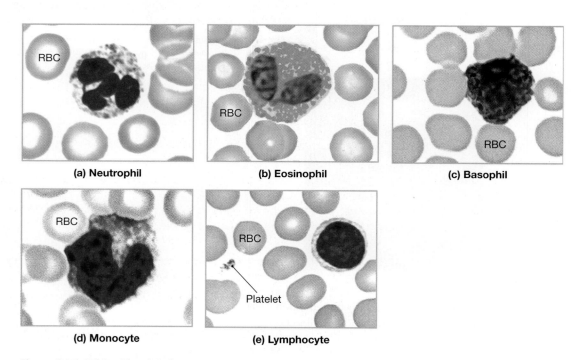

(a) Neutrophil

(b) Eosinophil

(c) Basophil

(d) Monocyte

(e) Lymphocyte

Figure 34.3 **White Blood Cells**

blood-smear slide because relatively few of them are present. They migrate to injured tissues and release histamines, which cause vasodilation, and heparin, which prevents blood from clotting. Mast cells in the tissue respond to these molecules and induce local inflammation.

Monocytes (MON-ō-sīts) are large cells with a dark stained, kidney-shaped nucleus surrounded by pale blue cytoplasm (Figure 34.3d). Approximately 2% to 8% of circulating WBCs are monocytes. On a blood-smear slide, monocytes appear roundish and may have small extensions, much like an amoeba. Even though monocytes are agranular leukocytes, materials they ingest, such as phagocytized bacteria and debris, will stain and may look like granules under the microscope.

Monocytes are wanderers. They leave the bloodstream through the capillary endothelium and patrol the body tissues in search of microbes and worn-out tissue cells. They are second to neutrophils in arriving at a wound site. When neutrophils die from phagocytizing bacteria, the monocytes phagocytize the neutrophils.

Lymphocytes (LIM-fō-sīts) are the smallest of the leukocytes and are approximately the size of a RBC (Figure 34.3e). The distinguishing feature of any lymphocyte is a large nucleus that occupies almost the entire cell, leaving room for only a small halo of pale blue cytoplasm around the edge of the cell. Lymphocytes are abundant in the bloodstream and compose 20% to 30% of all circulating leukocytes. As their name suggests, they are the main cells populating the lymphatic system. Many lymphocytes occur in lymph nodes, glands, and other lymphatic structures.

Although several types of lymphocytes exist, they cannot be individually distinguished with a light microscope. Generally, lymphocytes provide immunity from microbes and defective cells by two methods. **T-cell lymphocytes** attach to and destroy a foreign cell in a cell-mediated response by releasing cytotoxic chemicals to kill the invaders. **B-cell lymphocytes** become sensitized to a specific antigen, then manufacture and pour antibodies into the bloodstream. The antibodies attach to and help destroy foreign antigens.

Platelets (PLĀT-lets), the final type of formed element, are small cellular pieces produced from the breakdown of large megakaryocytes in the bone marrow (Figure 34.3e). Platelets lack a nucleus and other organelles. They survive in the bloodstream for a brief time and are involved in blood clotting.

QuickCheck Questions

1.1 What are the three types of formed elements in blood?

1.2 Which is the most abundant type of leukocyte?

1 Materials

☐ Compound microscope

☐ Human blood-smear slide (Wright's or Giemsa stained)

Procedures

1. With the low-power objective in position, place the blood-smear slide on the microscope stage.

2. Blood samples are thin and require careful focusing. Bring the sample into focus by slowly turning the coarse focus knob until you can clearly see cells. Now use the fine focus knob as you examine individual cells. Notice the abundance of red blood cells. The dark-stained cells are the various leukocytes. Note the small platelets between the red and white cells.

3. Scan the slide at medium magnification, and locate the different types of leukocytes. Then observe each cell at high magnification.

   • Neutrophils, the most abundant WBCs, have many lobes on the nucleus and small granules.

   • Lymphocytes are the smallest leukocytes. The nucleus is large with a thin margin of cytoplasm surrounding it.

   • A monocyte is large and irregularly shaped, with a kidney-shaped nucleus.

- An eosinophil has red-orange cytoplasmic granules and a bilobed nucleus.
- Basophils are characterized by large, dark-stained granules that obscure the nucleus.

4. Sketch each blood cell in the space provided. ◾

Erythrocyte	Neutrophil	Eosinophil
Basophil	Monocyte	Lymphocyte

Platelet

ABO and Rh Blood Groups

Your blood type is inherited from your parents' genes, and it does not change during your lifetime. Each blood type is a function of the presence or absence of specific molecules on the surface of the erythrocytes. These molecules, called either **antigens** or **agglutinogens** (a-gloo-TIN-ō-jenz), are like cellular name tags that inform your immune system that your cells belong to "self" and are not "foreign."

Table 34.1 Differences in Blood Group Distribution

Population	O	A	B	AB	RH⁺
			Percentage with Each Blood Type		
U.S. (Average)	46	40	10	4	85
African-American	49	27	20	4	95
Caucasian	45	40	11	4	85
Chinese-American	42	27	25	6	100
Filipino-American	44	22	29	6	100
Hawaiian	46	46	5	3	100
Japanese-American	31	39	21	10	100
Korean-American	32	28	30	10	100
NATIVE NORTH AMERICAN	79	16	4	<1	100
NATIVE SOUTH AMERICAN	100	0	0	0	100
AUSTRALIAN ABORIGINE	44	56	0	0	100

Each blood group also has **antibody** molecules, also called **agglutinins** (a-GLOO-ti-ninz), that are present in the blood plasma. The antibodies and antigens in an individual's blood do not interact with one another, but the antibodies do react with antigens of foreign blood cells and cause the cells to burst.

More than 50 blood groups occur in the human population. In this section, you will study the two most common, the ABO group and the Rh group. Each blood group is controlled by a different gene, and your ABO blood type does not influence your Rh blood type. Table 34.1 shows the distribution of these two blood groups in the human population.

The ABO Blood Group

There are four blood types in the **ABO blood group:** A, B, AB, and O (Figure 34.4). Two antigens, A and B, occur in different combinations that determine the ABO blood type. Type A blood has the A antigen on the cell surface, type B blood has the B antigen, type AB blood has both A and B antigens, and type O blood has neither.

The antibodies present in your blood plasma do not react with your own blood. If a different ABO blood type is introduced into your bloodstream, however, your antibodies will attach to the antigens of the foreign blood cells and cause the cells to clump, or agglutinate. Hemolysis, or bursting, then occurs to destroy the introduced blood cells.

The anti-B antibodies in type A blood do not react with the type A surface antigens, but do provide immunity against other blood types. The same is true for type B blood: the anti-A antibodies do not react with the surface antigens but do destroy other blood types. Because AB blood has neither anti-A antibodies nor anti-B antibodies, it does not react with blood of other types. People with AB blood are called *universal acceptors* because, lacking antibodies, they can accept blood of any type in a transfusion. Although the surface antigens are absent in type O blood, it has both anti-A and anti-B plasma antibodies. With no surface antigens acting as name tags, type O blood can be transfused to all blood types, and people with type O blood are called *universal donors.*

To determine blood type, the presence of antigens is detected by adding to a blood sample drops of **anti-serum** that contains either anti-A antibodies or anti-B antibodies. The antibodies in the anti-serum react with the corresponding antigens on the RBC surface. Anti-A antibodies react only with the A antigens in type A blood, anti-B antibodies react only with the B antigens in type B blood, both anti-A anti-serum and anti-B anti-serum react with type AB blood, and

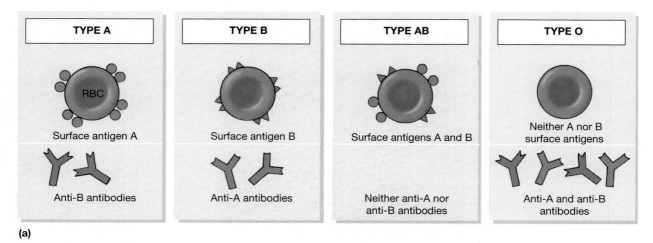

(a)

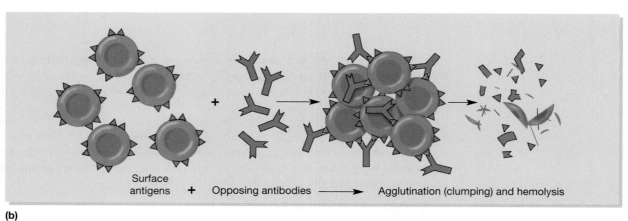

(b)

Figure 34.4 Blood Typing and Cross-Reactions
The blood type depends on the presence of surface antigens (agglutinogens) on RBC
surfaces. **(a)** The plasma contains antibodies (agglutinins) that will react with foreign
surface antigens. The relative frequencies of each blood type in the U.S. population are
indicated in Table 34.1. **(b)** In a cross-section, antibodies that encounter their target
antigens induce agglutination and hemolysis of the affected RBCs.

neither reacts with type O blood. The blood agglutinates as the antibodies react
with the antigens.

The Rh Blood Group

The **Rh blood group** is named after the Rhesus monkey, the research animal in
which the blood group was first identified. The two types of blood in this group are
Rh positive (+) and **Rh negative** (−) blood. Although this blood group is separate
from the ABO group, the two are usually used together to identify a blood type. For
example, a blood sample may be A+ or A−.

Unlike the ABO group with its two cellular antigens, the Rh group has only
one antigen, the **D antigen**, plus a single Rh antibody designated anti-D. The
D antigen is present only on RBCs that are Rh-positive; Rh-negative blood cells
lack the D antigen. Because Rh-positive blood has the D antigen, it must lack the
anti-D antibody, and this antibody is also absent in Rh-negative blood. However,
if Rh-negative blood is exposed to the D antigen in Rh-positive blood, the immune system of the Rh-negative person will produce the anti-D antibody. This becomes clinically significant in cases of pregnancy with Rh incompatibilities
between mother and fetus.

Clinical Application

Rh Factor and Hemolytic Disease of the Newborn

If an expectant mother is Rh-negative and her baby is Rh-positive, a potentially life-threatening Rh incompatibility exists for the baby. Normally, fetal blood does not mix with the mother's blood. Instead, the umbilical cord connects to the placenta, where fetal blood capillaries exchange gases, wastes, and nutrients with the mother's blood. If internal bleeding occurs, however, so that the mother is exposed to the D antigens in her baby's Rh-positive blood, she will produce anti-D antibodies. These antibodies will cross the membranes of the placenta and enter the fetal bloodstream, where they will hemolyze the fetal blood cells of this fetus and any future Rh-positive fetuses. This Rh action is called either **hemolytic disease of the newborn** or **erythroblastosis fetalis** (e-rith-rō-blas-TŌ-sis fē-TAL-is). A dosage of anti-Rh antibodies, called **RhoGam**, may be given to the mother during pregnancy and after delivery to destroy any Rh-positive fetal cells in her bloodstream. This will prevent her from developing anti-D antibodies. ▶

QuickCheck Questions

2.1 What are the two major blood groups used to identify blood type?

2.2 What surface antigens does type A blood have?

Handling Blood

1. Refer to Exercise 1, "Laboratory Safety," and review the steps for handling and disposing of blood.
2. Some infectious diseases are spread by contact with blood. Follow all instructions carefully, and protect yourself by wearing gloves and working only with your own blood.
3. Materials contaminated with blood must be disposed of properly. Your instructor will inform you of methods to dispose of lancets, slides, prep pads, and toothpicks. ▲

2 Materials

- ☐ Gloves
- ☐ Safety glasses
- ☐ Disposable sterile blood lancet
- ☐ Disposable sterile alcohol prep pad
- ☐ Disposable blood typing plate or sterile microscope slide
- ☐ Wax pencil (if using microscope slide)
- ☐ Toothpicks
- ☐ Anti-A, anti-B, and anti-D blood-typing sera
- ☐ Warming box
- ☐ Paper towels
- ☐ Biohazardous waste container
- ☐ Bleach solution in spray bottle (optional)

Your instructor may ask for a volunteer to "donate" blood to demonstrate how blood typing is done. Alternatively, many biological supply companies sell simulated blood-typing kits that contain a bloodlike solution and anti-sera. These kits contain no human or animal blood products and safely show the principles of typing human blood.

Procedures

Sample Collection

1. If you are using a slide, use the wax pencil to draw three circles across the width of the slide. Label the circles "A," "B," and "D." If you are using a typing plate, use the same labels for three of the depressions.
2. Wash both hands thoroughly with soap, and then dry them with a clean paper towel. Obtain an additional paper towel to place blood-contaminated instruments on while collecting a blood sample. Wear gloves while collecting and examining blood. If collecting a sample from yourself, wear a glove on the hand used to hold the lancet.
3. Open a sterile alcohol prep pad, and clean the tip of the index finger from which the blood will be drawn. Be sure to thoroughly disinfect the entire fingertip, including the sides. Place the used prep pad on the paper towel.
4. Open a sterile blood lancet to expose only the sharp tip. Do not use an old lancet, even if it was used on one of your own fingers. Use the sterile tip

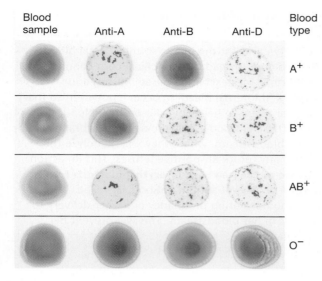

Blood sample	Anti-A	Anti-B	Anti-D	Blood type
				A$^+$
				B$^+$
				AB$^+$
				O$^-$

Figure 34.5

Blood Type Testing

Test results for blood samples from four individuals. Drops are taken from the sample at the left and mixed with solutions containing antibodies to the surface antigens A, B, AB, and D (Rh). Clumping occurs when the sample contains the corresponding surface antigen(s). The individuals' blood types are shown at right.

immediately so that there is no time for it to inadvertently become contaminated.

5. With a swift motion, jab the point of the lancet into the lateral surface of the fingertip. Place the used lancet on the paper towel until it can be disposed of in a biohazard container.

6. Gently squeeze a drop of blood either into each depression on the blood-typing plate or into the circles on the slide. If necessary, slowly "milk" the finger to work more blood out of the puncture site.

ABO and Rh Typings

1. Add a drop of anti-A serum to the sample labeled A, being very careful not to allow blood to touch (and thereby contaminate) the tip of the dropper. Repeat the process by adding a drop of anti-B serum to the B sample and a drop of anti-D to the D sample.

2. Immediately and gently mix each drop of anti-serum into the blood with a clean toothpick. To prevent cross-contamination, use a separate, clean toothpick for each sample. Place all used toothpicks on the paper towel until they can be disposed of in a biohazard container.

3. Place the slide or plate on the warming box, and agitate the blood samples by rocking the box carefully back and forth for two minutes.

4. Examine the drops for any agglutination visible with the unaided eye, and compare your blood sample with Figure 34.5. Agglutination results when the antibodies in the anti-serum react with the matching antigen on the red blood cells.

   • Type A blood agglutinates with the anti-A serum.

   • Type B blood agglutinates with the anti-B serum.

   • Type AB blood agglutinates with both anti-A and anti-B sera.

   • Type O blood does not agglutinate with either serum.

   • Rh-positive blood agglutinates with the anti-D serum.

   • Rh-negative blood does not agglutinate with the anti-D serum. *Note:* The anti-D agglutination reaction is often weaker and less easily observed than the A and B agglutination reactions. A microscope may help you observe the anti-D reaction.

5. Record your results in the first blank row of Table 34.2 in the Laboratory Report. In each cell, indicate yes or no for the presence of agglutination.

6. Collect blood-typing data from three classmates to compare agglutination responses among blood types. How does the distribution of blood types in your four-person sample compare with the distribution of types given in Table 34.1? ■

Disposal of Materials and Disinfection of Work Space

1. Dispose of all blood-contaminated materials in the appropriate biohazard box. A box for sharp objects may be available to dispose of the lancets, toothpicks, and microscope slides.

2. Your instructor may ask you to disinfect your workstation with a bleach solution. If so, wear gloves and safety glasses while wiping the surfaces clean.

3. Lastly, remove your gloves and dispose of them in the biohazard box. Remember to wash your hands after disposing of all materials. ▲

LAB ACTIVITY **3**   Packed Red Cell Volume (Hematocrit)

The **hematocrit** (he-MA-tō-krit) test, or packed cell volume (PCV), measures the volume of packed formed elements. Because RBCs far outnumber all the other formed elements, the test mainly measures their volume. The hematocrit results provide information regarding the oxygen-carrying capacity of the blood. A low hematocrit value indicates that the blood has fewer RBCs to transport oxygen. Average hematocrit values range from 40% to 54% in males and from 37% to 47% in females.

QuickCheck Questions

3.1 What does a hematocrit test measure?

3.2 What is the average hematocrit range for males? For females?

3   *Materials*

- ☐ Gloves
- ☐ Safety glasses
- ☐ Paper towels
- ☐ Disposable sterile blood lancet
- ☐ Disposable sterile alcohol prep pads
- ☐ Heparinized capillary tubes
- ☐ Seal-easy clay
- ☐ Bleach solution in spray bottle
- ☐ Microcentrifuge
- ☐ Tube reader
- ☐ Biohazardous waste disposal container

Procedures

1. Review the safety tips given in Laboratory Activity 2.

2. Follow steps 2 through 5 of Laboratory Activity 2, "Sample Collection," to obtain a blood sample.

3. *Gently* squeeze a drop of blood out of your finger. (Squeeze gently because excess pressure forces interstitial fluid into the blood, and the presence of this fluid may alter your hematocrit reading. If you are having difficulty obtaining a drop, use a clean, sterile lancet to lance your finger again in a different spot.)

4. Place a sterile heparinized capillary tube on the drop of blood. Orient the open end of the tube downward, as shown in Figure 34.6, to allow the blood to flow into the tube. Fill the tube at least two-thirds full with blood.

5. Carefully seal one end of the tube with the seal-ease clay, as shown in Figure 34.7. Do not force the delicate capillary tube into the clay, for it may break and cause you to jam glass into your hand. Instead, hold the tube the way you hold a pencil for writing, close to the tip where the blood has accumulated. Then gently turn the tube while pressing it into the clay. Leave the other end unplugged.

6. Clean any blood off the clay with the bleach solution and a paper towel.

7. Set the tube in the microcentrifuge with the clay end toward the outer margin of the chamber. Because the centrifuge spins at high speeds, the chamber must be balanced by placing tubes evenly in the chamber. Counterbalance your capillary tube by placing another sample directly across from yours. An empty tube sealed at one end with clay may be used if another student's sample is not available.

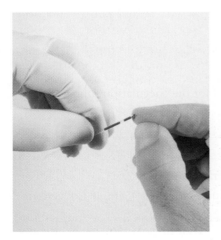

Figure 34.6
Filling Capillary Tube with Blood

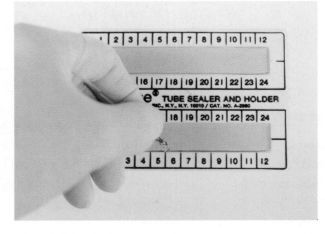

Figure 34.7
Plugging Capillary Tube with Clay

8. Screw the inner cover on with the centrifuge wrench. Do not overtighten the lid. Close the outer lid and push the latch in.

9. Set the timer to four to five minutes, and allow the centrifuge to spin. Do not attempt to open or stop the centrifuge while it is turning. Always keep loose hair and clothing away from the centrifuge.

10. After the centrifuge turns off and stops spinning, open the lid and the inner safety cover to remove the capillary tube. Your blood sample should have clear plasma at one end of the tube and packed RBCs at the other end.

11. Place the capillary tube in the tube reader. Because there are a variety of tube readers, your instructor will demonstrate how to use the reader in your laboratory.

12. Record your hematocrit measurement in Table 34.2 in the Laboratory Report. Is your hematocrit reading within the normal range?

13. Describe the appearance of your blood plasma in the space provided here:

14. Dispose of the prep pad, lancet (or lancets), and capillary tube in a biohazard container as described by your instructor. ■

LAB ACTIVITY 4 Coagulation

Blood removed from the body and allowed to sit for three to four minutes changes from a liquid to a gel. This process is called **coagulation** (cō-ag-ū-LĀ-shun), or **clotting**, and prevents excessive blood loss. Coagulation is a complex chemical chain reaction beyond the scope of this laboratory exercise. In brief, when you cut yourself, enzymes activate circulating proteins that ultimately convert the protein fibrinogen to an insoluble form called **fibrin**. The fibrin molecules join together in long threads that form a net to trap platelets and plug the wound. In the coagulation test, you will determine how fast these reactions occur in your blood.

QuickCheck Questions

4.1 What is coagulation?

4.2 Why is a nonheparinized tube used for the coagulation time test?

4 Materials

☐ Gloves
☐ Safety glasses
☐ Disposable sterile blood lancet
☐ Disposable sterile alcohol prep pad
☐ Paper towels

Procedures

Coagulation Time

1. Review the safety tips given in Laboratory Activity 2.

2. Follow steps 2 through 5 of Laboratory Activity 2, "Sample Collection," to obtain a blood sample.

3. Gently squeeze a drop of blood out of your finger. If you have difficulty obtaining a drop, use a clean, sterile lancet to lance your finger again in a different spot.

□ Nonheparinized capillary tube

□ Small metal file

□ Bleach solution in spray bottle (optional)

□ Biohazardous waste container

4. Place a sterile nonheparinized capillary tube on the drop of blood. Orient the open end of the tube downward to allow the blood to flow into the tube, as shown in Figure 34.6. Fill the tube at least two-thirds full with blood. Once the tube is prepared, note the time on a watch or clock.

5. Lay the tube on a paper towel, and after 30 seconds, break it as follows: While holding one end down as the tube lies on the towel, gently scratch the glass with the edge of a metal file, making your mark about 1 cm from the free end. Place your thumbs and index fingers on either side of the scratch, and break the tube by slowly bending it away from you.

6. Slowly separate the two broken ends of the tube, and look for a thin fibrin thread. If a thread is present, record 30 seconds as your coagulation time in Table 34.2 in the Laboratory Report.

7. If there is no fibrin, wait 30 seconds, and then break the tube again 1 cm from one end.

8. Repeat the sequence (wait 30 seconds, break the tube, look for a fibrin thread) until you see a thread. Record your coagulation time in Table 34.2.

9. How much time was required for the fibrin thread to form? ■

Disposal of Materials and Disinfection of Work Space

1. Dispose of all blood-contaminated materials in the appropriate biohazard box. A box for sharp objects may be available to dispose of the lancets, toothpicks, and microscope slides.

2. Your instructor may ask you to disinfect your workstation with a bleach solution. If so, wear gloves and safety glasses while wiping the surfaces clean.

3. Lastly, remove your gloves and dispose of them in the biohazard box. Remember to wash your hands after disposing of all materials. ▲

Blood

Name _____

Date _____

Section _____

A. Matching

Match each term in the left column with its correct description from the right column.

_____ **1.** erythrocyte	**A.** eosinophil
_____ **2.** polymorphonuclear cell	**B.** molecule on erythrocyte surface
_____ **3.** granular leukocyte	**C.** has A antigens and anti-B antibodies
_____ **4.** leukocyte	**D.** has Rh antigen
_____ **5.** antibody	**E.** neutrophil
_____ **6.** type A blood	**F.** lacks Rh antigen
_____ **7.** Rh-positive blood	**G.** red blood cell
_____ **8.** red-orange stained blood cell	**H.** contains cytoplasmic granules
_____ **9.** type B blood	**I.** reacts with a membrane molecule
_____ **10.** Rh-negative blood	**J.** has B antigens and anti-A antibodies
_____ **11.** antigen	**K.** white blood cell

B. Experimental Data and Observations

Complete Table 34.2 with data collected during the blood typing, hematocrit, and coagulation experiments.

Table 34.2 **Blood Typing Data**

Student	Anti-A Serum Reaction	Anti-B Serum Reaction	Anti-D Serum Reaction	Blood Type	Hematocrit Reading	Coagulation Time

34

C. Completion

Complete Table 34.3 by writing the function and appearance of each type of blood cell.

Table 34.3 A Summary of the Formed Elements of the Blood

Cell	Functions	Appearance
Erythrocyte		
Neutrophil		
Eosinophil		
Basophil		
Monocyte		
Lymphocyte		
Platelet		

D. Drawing

Complete each typing slide by indicating with pencil dots where agglutination occurs.

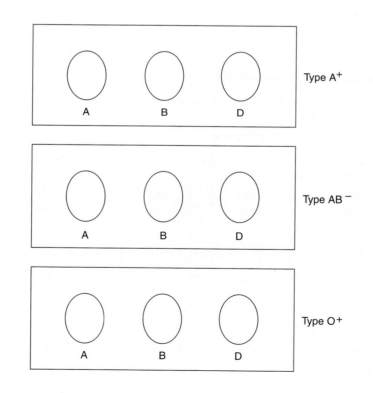

E. Short-Answer Questions

1. What is the main function of RBCs?

2. List the five types of leukocytes, and describe the function of each.

3. Describe how to do a hematocrit test. What are the average hematocrit values for males and females?

4. Describe how to type blood to detect the ABO and Rh blood groups.

5. Describe how to test the coagulation time of a blood sample.

6. How does fibrin contribute to coagulation?

F. Analysis and Application

1. Describe what would happen if type A blood were transfused into the bloodstream of someone with type B blood.

2. What happens in the blood of an Rh-negative individual who is exposed to Rh-positive blood?

3. How could you easily determine if two blood samples are compatible?

Anatomy of the Heart

OBJECTIVES

On completion of this exercise, you should be able to:

- Describe the gross external and internal anatomy of the heart.

- Identify and discuss the function of the valves of the heart.

- Identify the major blood vessels of the heart.

- Trace a drop of blood through the pulmonary circuit and the systemic circuit.

- Identify the vessels of coronary circulation.

- List the components of the conduction system of the heart.

- Describe the anatomy of a sheep heart.

Your heart beats approximately 100,000 times daily to send blood flowing into thousands of miles of blood vessels, providing nutrients and regulating substances, gases, and wastes for all body cells. All organ systems of the body depend on the cardiovascular system. Damage to the heart often results in widespread disruption of homeostasis.

For a drop of blood to complete one circuit through the body, it must be pumped by the heart twice—through the **pulmonary circuit**, which directs deoxygenated blood to the lungs; and through the **systemic circuit**, which takes oxygenated blood to the rest of the body (Figure 35.1). Each circuit delivers blood into a series of arteries, then capillaries, and finally veins that drain into the opposite side of the heart. Blood pumped to the lungs by the right side of the heart, for instance, returns to the left side.

The right side of the heart receives deoxygenated blood from the systemic circuit and pumps that blood into the pulmonary circuit. The blood moves first into the pulmonary trunk, which then branches into arteries that supply blood to the capillaries of the lungs. In the lungs, the blood becomes saturated with oxygen while releasing the carbon dioxide it picked up on its journey through the systemic circuit. Four pulmonary veins return the oxygenated blood to the left side of the heart. The heart then pumps the blood into the aorta, which branches into the major arteries of the systemic circuit. In the systemic capillaries, oxygen diffuses to the surrounding cells and carbon dioxide moves out of the cells and into the blood. The blood is once again deoxygenated and so requires two more trips through the heart.

Figure 35.1

An Overview of the Cardiovascular System

Driven by the pumping of the heart, blood flows through separate pulmonary and systemic circuits. Each circuit begins and ends at the heart and contains arteries, capillaries, and veins.

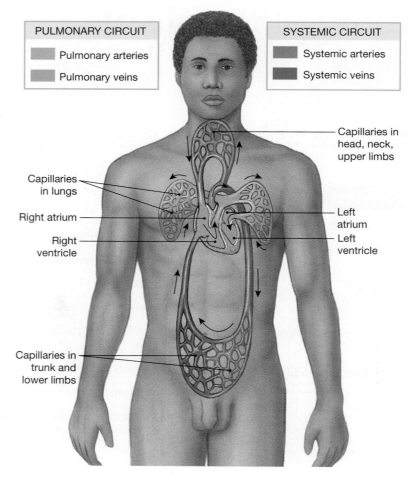

PULMONARY CIRCUIT

Pulmonary arteries

Pulmonary veins

SYSTEMIC CIRCUIT

Systemic arteries

Systemic veins

Capillaries in head, neck, upper limbs

Capillaries in lungs

Right atrium

Right ventricle

Left atrium

Left ventricle

Capillaries in trunk and lower limbs

Your anatomical studies in this exercise include the histology of cardiac muscle tissue, external and internal heart structures, blood flow through the heart, and fetal heart structures. Because all mammals have a four-chambered heart, dissection of a sheep heart will reinforce your observations of the human heart.

LAB ACTIVITY 1 Anatomy of the Heart Wall

Figure 35.2a illustrates the location of the heart within the mediastinum (mē-dē-as-TĪ-num) of the thoracic cavity. Blood vessels join the heart at the **base**, positioned medially in the mediastinum. Because the left side of the heart has more muscle mass than the right side, the **apex** at the inferior tip of the heart is more on the left side of the thoracic cavity. A line traced from your right shoulder to your left hip would pass through the axis of your heart, separating the right and left chambers.

Within the mediastinum, the heart lies inside the **pericardial** (per-i-KAR-dē-al) **cavity**. This cavity is formed by the **pericardium**, the serous membrane of the heart. Recall from your earlier studies that all serous membranes are double membranes, with a parietal layer and a visceral layer. The pericardial cavity contains **serous fluid** to reduce friction during muscular contraction.

The outer **parietal pericardium** attaches the heart in the mediastinum, and the inner **visceral pericardium**, or **epicardium**, lines the surface of the heart (Figure 35.2b). Imagine pushing your fist into an inflated balloon. The balloon around your fist represents the visceral pericardium, and the rest of the balloon

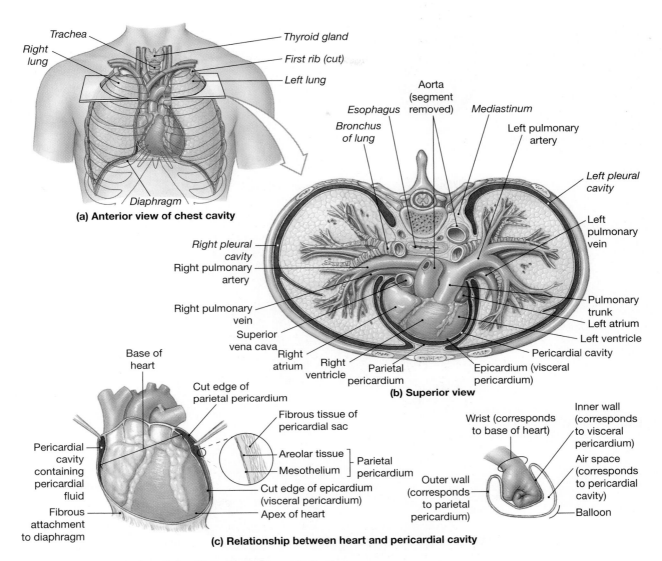

Figure 35.2 The Location of the Heart in the Thoracic Cavity
The heart is situated in the anterior part of the mediastinum, immediately posterior to the sternum. **(a)** An anterior view of the open chest cavity, showing the position of the heart and major vessels relative to the lungs. **(b)** A superior view of the heart and other organs in the mediastinum with the tissues of the lungs removed to reveal the blood vessels and airways. **(c)** The relationship between the heart and the pericardial cavity; compare with the fist-and-balloon example.

represents the parietal pericardium. The space inside the balloon is the pericardial cavity.

Both layers of the pericardium contain **mesothelium** (mes-ō-THĒ-lē-um), a simple epithelium that lines the pericardial cavity and secretes serous fluid (Figure 35.3a). **Loose connective tissue** attaches the mesothelium to the underlying structures of the pericardium. The parietal pericardium is also composed of an outer **fibrous layer** that supports the heart and vessels in the mediastinum. The fibrous layer also prevents the heart from overexpanding.

The heart wall is organized into three layers: epicardium, myocardium, and endocardium (Figure 35.3a). The epicardium is the same structure as the visceral pericardium. The **myocardium** is composed of cardiac muscle tissue and constitutes most of the heart wall. Histologically, the myocardium is composed of **striated** muscle cells

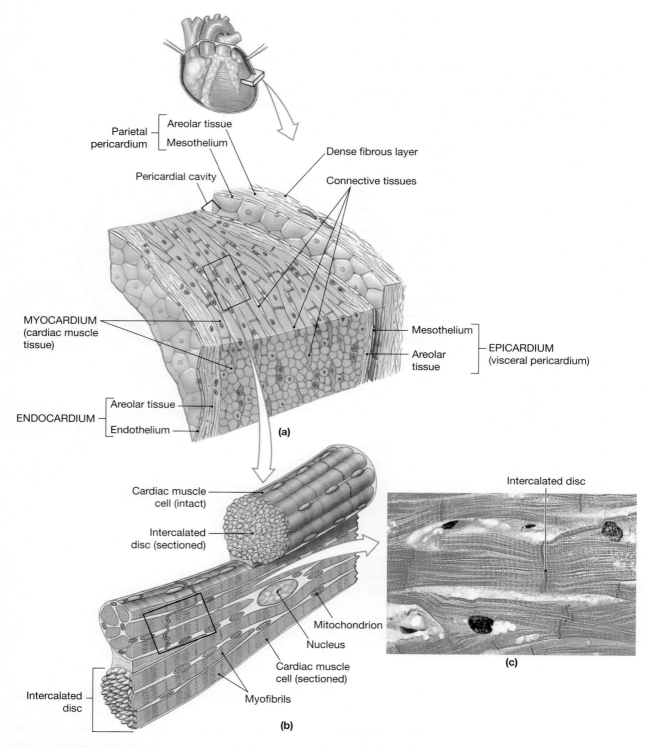

Figure 35.3 The Heart Wall
(a) A diagrammatic section through the heart wall, showing the relative positions of the epicardium, myocardium, and endocardium. **(b)** Diagrammatic and **(c)** sectional views of cardiac muscle tissue (LM × 575).

called either **cardiac muscle cells** or **cardiocytes** (Figure 35.3b). Each cardiocyte is **uninucleated** (containing a single nucleus) and branched. Cardiocytes interconnect at these branches with **intercalated** (in-TER-ka-lā-ted) **discs**. Deep to the myocardium is the **endocardium**, a thin layer of endothelial tissue lining the chambers of the heart.

QuickCheck Questions

1.1 Where is the heart located?

1.2 What are the three layers of the heart wall?

1.3 How are heart muscle cells connected?

1 *Materials*

☐ Heart models and specimens

☐ Compound microscope

☐ Prepared slide of cardiac muscle

Procedures

1. Review the anatomy presented in Figures 35.2 and 35.3.

2. Identify the layers of the heart wall on laboratory models and specimens.

3. With the microscope at low power, examine the microscopic structure of cardiac muscle, using Figure 35.3b for reference.

4. Increase the magnification to medium and locate several cardiocytes. Note the single nucleus in each cell. Intercalated discs are dark-stained lines where cardiocytes connect together. Do you see any branched cardiocytes?

5. Sketch several cardiocytes and intercalated discs in the space provided. Use both low and high magnifications for your sketches. ■

LAB ACTIVITY 2 Chambers and Vessels of the Heart

All mammalian hearts have four chambers and are anatomically divided into right and left sides, with each side having an upper and a lower chamber (Figure 35.4). The upper chambers are the **right atrium** (A-trē-um; chamber) and the **left atrium**, and the lower chambers are the **right ventricle** (VEN-tri-kul; little belly) and the **left ventricle**. The atria are receiving chambers and fill with blood returning to the heart via the veins. Blood in the atria flows into the ventricles, the pumping chambers, which squeeze their walls together to pressurize the blood and squirt it into two large arteries for distribution to the lungs via the pulmonary circuit and body tissues via the systemic circuit. In the right atrium, a structure called the **fossa ovalis** is located on the interatrial septum. This is a remnant of fetal circulation, where the foramen ovale allowed blood to bypass the fetal pulmonary circuit.

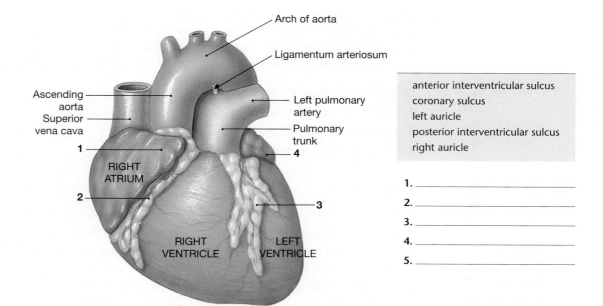

Arch of aorta

Ligamentum arteriosum

Ascending aorta

Superior vena cava

1

RIGHT ATRIUM

2

Left pulmonary artery

Pulmonary trunk

4

RIGHT VENTRICLE

LEFT VENTRICLE

3

anterior interventricular sulcus
coronary sulcus
left auricle
posterior interventricular sulcus
right auricle

1._____

2._____

3._____

4._____

5._____

(a) Anterior (sternocostal) surface

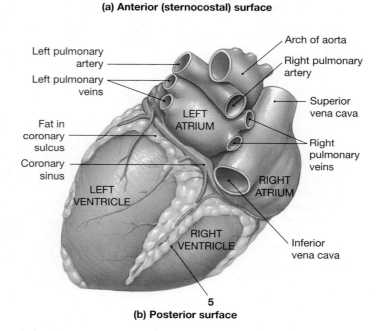

Left pulmonary artery

Left pulmonary veins

Fat in coronary sulcus

Coronary sinus

LEFT VENTRICLE

LEFT ATRIUM

Arch of aorta

Right pulmonary artery

Superior vena cava

Right pulmonary veins

RIGHT ATRIUM

RIGHT VENTRICLE

Inferior vena cava

5

(b) Posterior surface

Figure 35.4 The Superficial Anatomy of the Heart

(a) Major anatomical features on the anterior surface. **(b)** Major landmarks on the posterior surface. Coronary arteries (which supply the heart itself) are shown in red; coronary veins are shown in blue.

Each atrium is covered externally by a flap called the **auricle** (AW-ri-kul; *auris,* ear). Fat tissue and blood vessels occur along grooves in the heart wall. The **coronary sulcus** is a deep groove between the right atrium and right ventricle and extends to the posterior surface. The boundary between the right and left ventricles is marked anteriorly by the **anterior interventricular sulcus** and posteriorly by the **posterior interventricular sulcus**. Coronary blood vessels follow the sulci and branch to the myocardium.

Figure 35.5 details the internal anatomy of the heart. The wall between the atria is called the **interatrial septum**, and the ventricles are separated by the **interventricular septum**. Lining the inside of the right atrium are muscular ridges, the

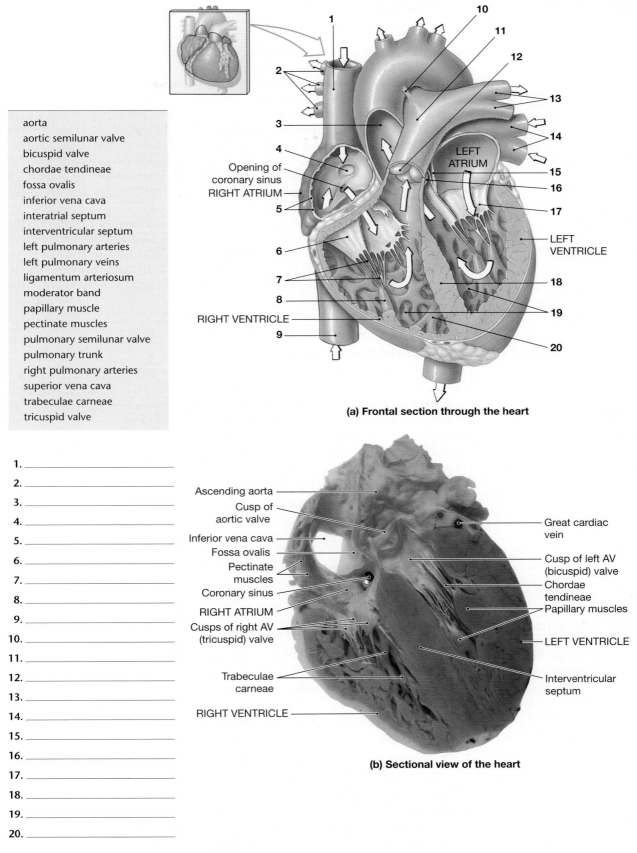

aorta
aortic semilunar valve
bicuspid valve
chordae tendineae
fossa ovalis
inferior vena cava
interatrial septum
interventricular septum
left pulmonary arteries
left pulmonary veins
ligamentum arteriosum
moderator band
papillary muscle
pectinate muscles
pulmonary semilunar valve
pulmonary trunk
right pulmonary arteries
superior vena cava
trabeculae carneae
tricuspid valve

1 10
11
12
2 13
3 14
4 LEFT
ATRIUM
Opening of 15
coronary sinus 16
RIGHT ATRIUM 17
5 LEFT
VENTRICLE
6
7 18
8 19
RIGHT VENTRICLE
9 20

(a) Frontal section through the heart

1. _____
2. _____
3. _____
4. _____
5. _____
6. _____
7. _____
8. _____
9. _____
10. _____
11. _____
12. _____
13. _____
14. _____
15. _____
16. _____
17. _____
18. _____
19. _____
20. _____

Ascending aorta
Cusp of
aortic valve
Inferior vena cava
Fossa ovalis
Pectinate
muscles
Coronary sinus
RIGHT ATRIUM
Cusps of right AV
(tricuspid) valve
Trabeculae
carneae
RIGHT VENTRICLE

Great cardiac
vein
Cusp of left AV
(bicuspid) valve
Chordae
tendineae
Papillary muscles
LEFT VENTRICLE
Interventricular
septum

(b) Sectional view of the heart

Figure 35.5 The Sectional Anatomy of the Heart

(a) A diagrammatic frontal section through the heart, showing major landmarks and
the path of blood flow (marked by arrows) through the atria, ventricles, and associated
vessels. **(b)** A sectional view of the heart.

pectinate (*pectin,* comb) **muscles**. Folds of muscle tissue called **trabeculae carneae** (tra-BEK-ū-lē CAR-nē-ē; *carneus,* fleshy) occur on the inner surface of each ventricle. The **moderator band** is a ribbon of muscle that passes conduction fibers from the interventricular septum to muscles in the right ventricle.

The right atrium receives deoxygenated blood from the **superior vena cava** (VĒ-na KĀ-vuh) and the **inferior vena cava**. From the right atrium, blood flows into the right ventricle, which then pumps the deoxygenated blood to the lungs by way of the pulmonary trunk. In the lungs, the deoxygenated blood is oxygenated as gases diffuse between the blood and the lungs. At the branch of the pulmonary trunk is the **ligamentum arteriosum**, a relic of the fetal ductus arteriosus that joined the pulmonary trunk with the aorta.

Four **pulmonary veins** drain the lungs and return the oxygenated blood to the left atrium. The oxygenated blood flows from the left atrium into the left ventricle, which then pumps the blood into the **aorta**. The aorta branches into the major systemic arteries, which transport oxygenated blood to all body tissues. In the body tissues, the blood is deoxygenated and then returns to the right atrium in one of the venae cavae to repeat the double circuit.

Study Tip | ***Anatomical Position and the Heart***

Remember anatomical position when observing right and left structures on figures and models of the heart. For example, in Figures 35.1, 35.2b, 35.4a, and 35.5 the right atrium is on the left side of the illustration. The figures depict a heart in anatomical position in front of you. Also, notice that most heart models and textbook figures are color-coded: red designates a vessel carrying oxygenated blood (arteries), and blue is used for vessels carrying deoxygenated blood (veins). Remember that arteries transport blood away from the heart and veins transport blood toward the heart. ●

QuickCheck Questions

2.1 Which chambers receive blood from veins?

2.2 Which chambers pump blood into arteries?

2.3 What structures separate the walls of the chambers?

2 Materials

☐ Heart models and specimens

Procedures

1. Review the anatomy presented in Figures 35.4 and 35.5, and then label both figures.

2. On the heart model, identify each atrium and ventricle. Which ventricle has the thicker wall?

3. Identify the pectinate muscle in the right atrium and the trabeculae carneae of the ventricles. Locate the moderator band in the inferior right ventricle.

4. Observe the external features of the heart. Note how the auricles may be used to distinguish the anterior of the heart. Trace the length of each sulcus, and notice which chambers each passes between.

5. Associated with the right atrium are the superior and inferior venae cavae. Locate these vessels on the heart model. What kind of blood do these veins drain into the heart? Where does the right atrium empty?

6. The right ventricle ejects blood into the pulmonary trunk, which branches into right and left pulmonary arteries. Which kind of blood is in these arteries, and where is it going?

7. Find the four veins that drain into the left atrium. Where is this blood coming from?

8. Identify the aorta associated with the left ventricle. From this artery arise the major systemic arteries that deliver oxygenated blood to the body. ■

LAB ACTIVITY 3   Valves of the Heart

To control and direct blood flow, the heart has two pairs of valves: atrioventricular and semilunar (Figures 35.5 and 35.6). The **atrioventricular** (AV) **valves** between atria and ventricles prevent blood from reentering the atria when the ventricles contract. The **right atrioventricular valve**, which joins the right atrium and right ventricle, has three flaps, or cusps, and is also called the **tricuspid** (trī-KUS-pid; *tri*, three + *cuspid*, flap) **valve**. The **left atrioventricular valve** between the left atrium and left ventricle has two cusps and is called either the **bicuspid valve** or the **mitral** (MĪ-tral) **valve**. The cusps of each AV valve have small cords, the **chordae tendineae** (KOR-dē TEN-di-nē-ē; tendinous cords), which are attached to **papillary** (PAP-i-ler-ē) **muscles** on the floor of the ventricles. When the ventricles contract, the AV valves are held closed by the papillary muscles pulling on the chordae tendineae.

The pair of **semilunar valves** are called the **pulmonary semilunar valve** and the **aortic semilunar valve**. The former is located at the base of the pulmonary trunk, and the latter is located at the base of the aorta. Each semilunar valve is composed of three small cusps and prevents backflow of blood into its respective ventricle when the ventricle is relaxed.

Figure 35.6 illustrates the function of the AV and semilunar valves, which generally work in opposition. When one pair of valves is open, the other pair is either closed or preparing to close. When a ventricle contracts, in a phase called *ventricular systole,* blood pressure forces the AV valve closed. To keep this valve from being inverted, and thus opened, by the force of the blood, the papillary muscles contract and pull on the chordae tendineae to secure the cusps. The increased pressure in the ventricle next forces the semilunar valve open, and blood flows into the artery. Figure 35.6a represents valve function during *ventricular diastole,* the relaxation phase of the ventricle. Blood flows from the atrium into the relaxed ventricle through the open AV valve. Note in the figure that the papillary muscles and the chordae tendineae are relaxed. The semilunar valve is closed to prevent backflow of blood from the artery into the ventricle.

Clinical Application

Mitral Valve Prolapse

A common valve problem is **mitral valve prolapse**, a condition in which the left AV valve reverses, like an umbrella in a strong wind. The papillary muscles and chordae tendineae are unable to hold the valve cusps in the closed position, and so the valve inverts. Because the opening between the atrium and ventricle is not sealed shut during ventricular contraction, blood backflows into the left atrium, and cardiac function is diminished. ▶

QuickCheck Questions

3.1 Which valve is located between the left atrium and the left ventricle?

3.2 Which valve prevents backflow of blood from the arteries attached to the right ventricle?

3.3 When a ventricle is contracting, which valve is open and which is closed?

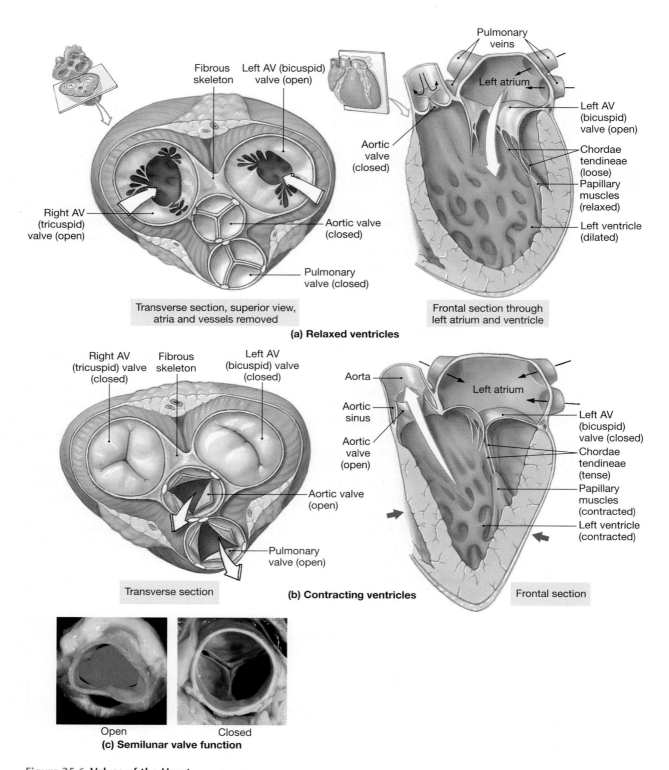

Figure 35.6 Valves of the Heart

White arrows indicate blood flow into or out of a ventricle; black arrows, blood flow into an atrium; and green arrows, ventricular contraction. **(a)** When the ventricles are relaxed, the AV valves are open and the semilunar valves are closed. The chordae tendineae are loose, and the papillary muscles are relaxed. **(b)** When the ventricles are contracting, the AV valves are closed and the semilunar valves are open. In the frontal section, notice the attachment of the left AV valve to the chordae tendineae and papillary muscles. **(c)** The aortic valve in the open (left) and closed (right) positions. The individual cusps brace one another in the closed position.

3 Materials

☐ Heart models and specimens

Procedures

1. Review the valve anatomy illustrated in Figures 35.5 and 35.6.
2. Examine the AV valves in a heart model, and distinguish between the tricuspid and bicuspid valves.
3. Observe the string-like chordae tendineae attached to the valve cuspids and to the papillary muscles extending from the myocardium.
4. At the base of the pulmonary artery and the aorta, identify the semilunar valves.
5. Starting at the superior vena cava, trace a drop of blood though the heart model, and distinguish between the pulmonary and systemic circuits. ∎

LAB ACTIVITY 4 Coronary Circulation

To produce the pressure required for blood to reach all through the vascular system, the heart can never completely rest. It has a vascular system called the **coronary circulation**, which supplies the myocardium with the oxygen necessary for muscle contraction (Figure 35.7). The right and left **coronary arteries** branch off the base of the aorta and penetrate the myocardium to the outer heart wall. The right coronary artery follows the coronary sulcus along the right atrium. It supplies blood to the right atrium, to parts of both ventricles, and to the heart's conduction system. As the right coronary artery passes along the coronary sulcus, one or more **marginal arteries** arise to supply the right ventricle. The **posterior interventricular artery** branches off the right coronary artery and supplies posterior regions of the myocardium.

The left coronary artery branches to supply blood to the left atrium, left ventricle, and interventricular septum. It divides into a **circumflex branch** and an **anterior interventricular branch** (also called the left anterior descending artery). The circumflex artery follows the left side of the heart and enters the posterior coronary sulcus with the right coronary artery. The anterior interventricular artery lies within the anterior interventricular sulcus.

Clinical Application

Anastomoses and Infarctions

Branches of the interventricular arteries connect with one another, as do smaller arteries between the right coronary and circumflex arteries. These connections, called **anastomoses**, ensure that blood flow to the myocardium remains steady. In coronary artery disease, the arteries become narrower and narrower as fatty plaque is deposited inside the vessel wall. As a result, the blood flow maintained by the anastomoses is reduced. If enough plaque accumulates in critical areas, blood flow to that part of the heart becomes inadequate, and the heart muscle has an **infarction**, a heart attack. ◗

Cardiac veins collect deoxygenated blood from the myocardium (Figure 35.7). The **great cardiac vein** follows along the anterior interventricular sulcus and curves around the left side of the heart. It drains the myocardium supplied by the anterior interventricular artery. The **small cardiac vein** drains the upper right area of the heart. The **middle cardiac vein** drains the myocardium supplied by the posterior interventricular artery. The **posterior cardiac vein** drains the myocardium supplied by the circumflex artery. The cardiac veins merge as a large **coronary sinus** situated in the posterior region of the coronary sulcus. The coronary sinus empties deoxygenated blood from the myocardium into the right atrium. As noted previously, the right atrium also receives deoxygenated blood from the venae cavae.

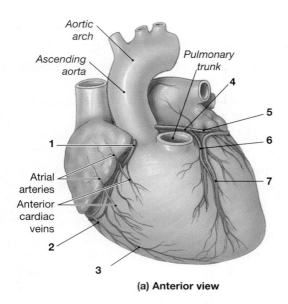

Aortic arch
Ascending aorta
Pulmonary trunk
4
5
6
7
1
Atrial arteries
Anterior cardiac veins
2
3

(a) Anterior view

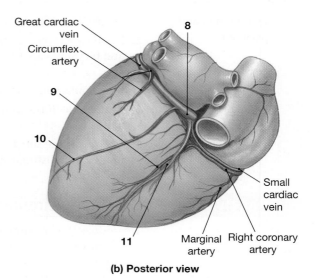

Great cardiac vein
Circumflex artery
8
9
10
Small cardiac vein
11
Marginal artery
Right coronary artery

(b) Posterior view

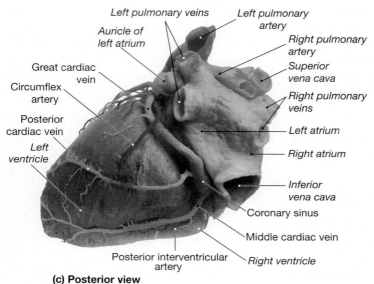

Left pulmonary veins
Left pulmonary artery
Auricle of left atrium
Right pulmonary artery
Superior vena cava
Great cardiac vein
Circumflex artery
Right pulmonary veins
Posterior cardiac vein
Left atrium
Left ventricle
Right atrium
Inferior vena cava
Coronary sinus
Middle cardiac vein
Posterior interventricular artery
Right ventricle

(c) Posterior view

anterior interventricular branch
circumflex branch
coronary sinus
great cardiac vein
left coronary artery
marginal artery
middle cardiac vein
posterior cardiac vein
posterior interventricular artery
right coronary artery
small cardiac vein

1. _____
2. _____
3. _____
4. _____
5. _____
6. _____
7. _____
8. _____
9. _____
10. _____
11. _____

Figure 35.7 Coronary Circulation
(a) Coronary vessels supplying the anterior surface of the heart. **(b)** Coronary vessels supplying the posterior surface of the heart. **(c)** A posterior view of the heart; the vessels have been injected with colored latex.

QuickCheck Questions

4.1 Where do the right and left coronary arteries arise?

4.2 Where do the cardiac veins drain?

4   *Materials*

☐ Heart models and specimens

Procedures

1. Review the coronary blood vessels in Figure 35.7, and then label the figure.
2. Follow the right coronary artery on the heart model and identify its branches, the marginal and posterior interventricular arteries. How many marginal arteries does the model show?
3. Trace the left coronary artery to its major branches, the anterior interventricular and the circumflex arteries.
4. Identify the great cardiac vein, and trace it to the coronary sinus. Where does the coronary sinus drain?
5. Just proximal to the sinus, locate the posterior cardiac vein.
6. At the distal portion of the coronary sinus, find the small cardiac vein and the middle cardiac vein. ■

LAB ACTIVITY   5   Conduction System of the Heart

Cardiac muscle tissue is unique in that it is *autorhythmic,* producing its own contraction and relaxation phases without stimulation from nerves. Nerves may increase or decrease the heart rate, but a living heart removed from the body continues to contract on its own.

Figure 35.8 details the **conduction system** of the heart. Special cells called **nodal cells** produce and conduct electrical currents to the myocardium, and it is these currents that coordinate the heart's contraction. The pacemaker of the heart is the **sinoatrial** (sī-nō-Ā-trē-al) **node** (SA node), located where the superior vena cava empties into the upper right atrium. Nodal cells in the SA node self-excite faster than nodal cells in other areas of the heart and therefore set the pace for the heart's contraction. The **atrioventricular node** (AV node) is located on the lower medial floor of the right atrium. The SA node stimulates both the atria and the AV node, and the AV node then directs the impulse toward the ventricles through the **atrioventricular bundle (bundle of His** [hiss]**)**. The atrioventricular bundle passes into the interventricular septum and branches into right and left **bundle branches**. The bundle branches divide into fine **Purkinje fibers,** which distribute the electrical impulses to the cardiocytes.

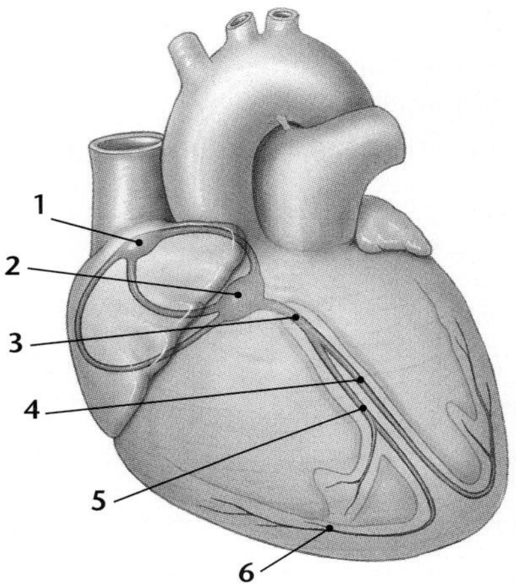

atrioventricular bundle
atrioventricular node
left bundle branch
Purkinje fiber
right bundle branch
sinoatrial node

1. _____
2. _____
3. _____
4. _____
5. _____
6. _____

Figure 35.8
The Conduction System of the Heart

QuickCheck Questions

5.1 Where is the pacemaker of the heart located?

5.2 How is the AV node connected to the ventricles?

5   *Materials*

☐ Heart models and specimens

Procedures

1. Review the conduction system in Figure 35.8, and then label the figure.
2. On the heart model, examine the sinus where the superior vena cava drains into the right atrium, and locate the SA node.
3. On the floor of the right atrium, locate the AV node. Trace the conducting path to the ventricles: AV bundle, bundle branches, and Purkinje fibers. ■

LAB ACTIVITY   6 | **Sheep Heart Dissection**

The sheep heart, like all other mammalian hearts, is similar in structure and function to the human heart. One major difference is the position of the great vessels joining the heart. In four-legged animals, the inferior vena cava has a posterior connection to the heart instead of the inferior attachment found in humans.

Dissecting a sheep heart will enhance your studies of models and charts of the human heart. Take your time while dissecting and follow the directions carefully.

Laboratory Safety

You *must* practice the highest level of laboratory safety while handling and dissecting the heart. Keep the following guidelines in mind during the dissection:

1. Wear gloves and safety glasses to protect yourself from the fixatives used to preserve the specimen.
2. Do not dispose of the fixative from your specimen. You will later store the specimen in the fixative to keep the specimen moist and to keep it from decaying.
3. Be extremely careful when using a scalpel or other sharp instrument. Always direct cutting and scissor motion away from you to prevent an accident if the instrument slips on moist tissue.
4. Before cutting a given tissue, make sure it is free from underlying and/or adjacent tissues so that they will not be accidentally severed.
5. Never discard tissue in the sink or trash. Your instructor will inform you of the proper disposal procedure. ▲

QuickCheck Questions

6.1 What type of safety equipment should you wear during the sheep heart dissection?

6.2 How should you dispose of the sheep heart and scrap tissue?

6 Materials

☐ Gloves
☐ Safety glasses
☐ Dissecting tools
☐ Dissecting tray
☐ Fresh or preserved sheep heart

Procedures

1. Put on gloves and safety glasses, and clear your workspace before obtaining your dissection specimen.
2. Wash the sheep heart with cold water to flush out preservatives and blood clots. Minimize your skin and mucous membrane exposure to the preservatives.
3. Carefully follow the instructions in this section. Cut into the heart only as instructed.

External Anatomy

1. Figure 35.9 details the external anatomy of the sheep heart. Examine the surface of the heart to see if the pericardium is present. (Often this serous membrane has been removed from preserved specimens.) Carefully use a scalpel to scrape the outer heart muscle to loosen the epicardium. Next, locate the anterior surface by orienting the heart so that the auricles face you. Under the auricles are the right and left atria. Note the base of the heart above the atria, where the large blood vessels occur. The apex is the inferior tip of the heart. Squeeze gently above the apex to locate the right and left ventricles. Locate the anterior interventricular sulcus, the fat-laden groove between the ventricles. Carefully remove some of the fat tissue to uncover some of the coronary blood vessels. Identify two other grooves—the coronary sulcus

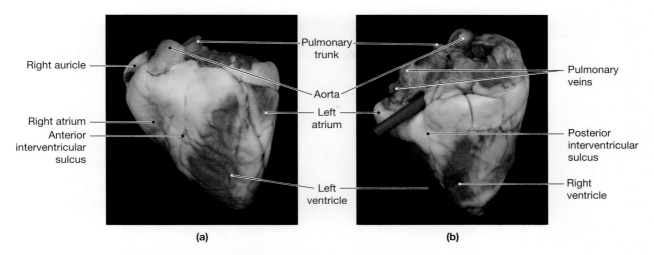

Right auricle

Right atrium

Anterior interventricular sulcus

Pulmonary trunk

Aorta

Left atrium

Left ventricle

Pulmonary veins

Posterior interventricular sulcus

Right ventricle

(a)   (b)

Figure 35.9 External Anatomy of the Sheep Heart
(a) Anterior view. (b) Posterior view.

between the right atrium and ventricle and the posterior interventricular sulcus between the ventricles on the posterior surface.

2. Identify the aorta and then the pulmonary trunk anterior to the aorta. If on your specimen the pulmonary trunk was cut long, you may be able to identify the right and left pulmonary arteries branching off the trunk. The brachiocephalic artery is the first major branch of the aorta and is often intact in preserved material.

3. Follow along the inferior margin of the right auricle to the posterior surface. The prominent vessel at the termination of the auricle is the superior vena cava. At the base of this vessel is the inferior vena cava. Next, examine the posterior aspect of the left atrium and find the four pulmonary veins. You may need to carefully remove some of the fatty tissue around the superior region of the left atrium to locate the veins.

Internal Anatomy

1. Locate the aorta at the base of the heart. Cut a frontal section passing through the aorta. Use Figure 35.10 as a reference to the internal anatomy.

2. Examine the two sides of the heart. Identify the right and left atria and the right and left ventricles. Identify the interventricular septum. Compare the myocardium of the left ventricle with that of the right ventricle. Which is thicker?

3. Note the folds of trabeculae carneae along the inner ventricular walls. Examine the right atrium for the comblike pectinate muscles lining the inner wall.

4. Locate the tricuspid and bicuspid valves. Observe the papillary muscles with chordae tendineae attached.

5. Study the wall of the left atrium for the openings of the pulmonary veins.

6. At the entrance of the aorta, locate the small cusps of the aortic semilunar valve.

7. At the base of the pulmonary trunk, locate the pulmonary semilunar valve.

8. Locate the superior and inferior venae cavae, which drain into the right atrium.

9. Upon completion of the dissection, dispose of the sheep heart as directed by your instructor and wash your hands and dissecting instruments. ■

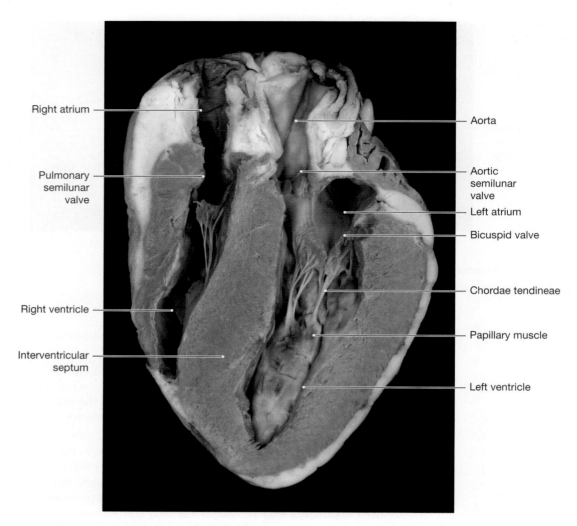

Right atrium

Pulmonary semilunar valve

Right ventricle

Interventricular septum

Aorta

Aortic semilunar valve

Left atrium

Bicuspid valve

Chordae tendineae

Papillary muscle

Left ventricle

Figure 35.10 Sagittal Section of the Sheep Heart

Anatomy of the Heart

35

Name _____

Date _____

Section _____

A. Matching

Match each heart structure in the left column with its correct description from the right column.

_____ **1.** tricuspid valve	**A.** empties into left atrium
_____ **2.** superior vena cava	**B.** left AV valve
_____ **3.** right ventricle	**C.** muscle folds of ventricles
_____ **4.** aorta	**D.** pumps blood to body tissues
_____ **5.** interventricular septum	**E.** branch off left coronary artery
_____ **6.** left ventricle	**F.** major systemic artery
_____ **7.** pulmonary veins	**G.** muscular ridges of right atrium
_____ **8.** semilunar valve	**H.** artery carrying deoxygenated blood
_____ **9.** bicuspid valve	**I.** groove on right side of heart
_____ **10.** pulmonary trunk	**J.** visceral pericardium
_____ **11.** circumflex artery	**K.** drains coronary veins into heart
_____ **12.** trabeculae carneae	**L.** wall between ventricles
_____ **13.** pectinate muscle	**M.** inferior tip of heart
_____ **14.** coronary sulcus	**N.** cardiac muscle tissue
_____ **15.** auricle	**O.** aortic or pulmonary valve
_____ **16.** coronary sinus	**P.** empties into right atrium
_____ **17.** myocardium	**Q.** attached to AV valves
_____ **18.** epicardium	**R.** right AV valve
_____ **19.** chordae tendineae	**S.** pumps blood to lungs
_____ **20.** apex	**T.** external flap of atrium

B. Short-Answer Questions

1. Describe the location of the heart in the thoracic cavity.

2. List the layers of the heart wall.

3. Trace a drop of blood through the pulmonary and systemic circuits of the heart.

4. Does the pulmonary trunk contain oxygenated blood or deoxygenated blood? Why is it an artery rather than a vein?

5. Describe how the AV valves function.

6. List the order in which an electrical impulse spreads through the heart's conduction system.

C. Analysis and Application

1. Why is the wall of the left ventricle thicker than the wall of the right ventricle?

2. Suppose a patient has mitral valve prolapse, a weakened bicuspid valve that does not close properly. How does this defect affect the flow of blood in the heart?

3. Coronary artery disease in the marginal arteries would affect which part of the myocardium?

Anatomy of the Systemic Circulation

OBJECTIVES

On completion of this exercise, you should be able to:

• Compare the histology of an artery, a capillary, and a vein.

• Describe the difference in the blood vessels serving the right and left arms.

• Describe the anatomy and importance of the circle of Willis.

• Trace a drop of blood from the ascending aorta into each abdominal organ and into the lower limbs.

• Trace a drop of blood returning to the heart from the foot.

• Discuss the unique features of the fetal circulation.

The body contains more than 60,000 miles of blood vessels to transport blood to the trillions of cells in the tissues. The basic circulatory route includes **arteries**, the vessels that distribute oxygen and nutrient-rich blood to microscopic networks of thin-walled vessels called **capillaries**. At the capillaries, nutrients, gases, wastes, and cellular products diffuse between the blood and the cells. **Veins** drain deoxygenated blood from the capillaries and direct it toward the heart, which then pumps it to the lungs to pick up oxygen and release carbon dioxide.

Smaller arteries, called **arterioles**, regulate blood flow into capillaries by relaxing or contracting smooth muscle in their walls. Contraction of the smooth muscle causes the arteriole to narrow, or **vasoconstrict**, and results in a decrease in blood flow. Relaxation opens the vessel and promotes blood flow, a process called **vasodilation**. Bands of smooth muscle called **precapillary sphincters** are positioned where each arteriole joins a **metarteriole**, a small arteriole just upstream of a capillary. The precapillary sphincters regulate blood flow through the capillary by controlling the diameter and resistance of the metarteriole. The path of least resistance in the capillary is called the **thoroughfare channel**. This channel empties into several small veins called **venules** that eventually empty into veins.

Most capillaries have several arteriole blood supplies. An **anastomosis** is where arterioles fuse together into a single vessel to supply the same capillary. Anastomoses ensure that a capillary receives blood. If one arteriole forming an anastomosis is pinched closed or damaged, the other arterioles that are part of the anastomosis will

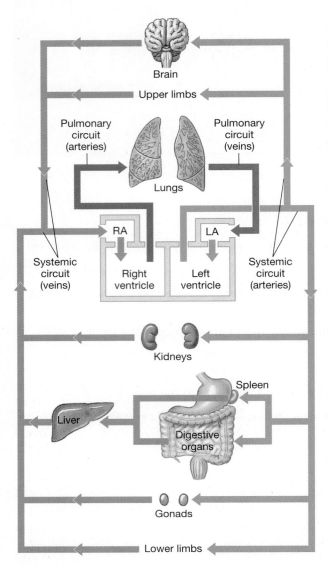

Figure 36.1
Overview of the Pattern of Circulation

Here, *RA* stands for right atrium; *LA* for left atrium.

provide blood to the shared capillary. Coronary circulation of the heart has many vital anastomoses.

Two circulatory pathways oxygenate blood and supply it to cells: the pulmonary circuit and the systemic circuit, illustrated in Figure 36.1. The **pulmonary circuit** takes deoxygenated blood from the right ventricle and transports it to the lungs via the pulmonary arteries. Oxygenated blood from the lungs is returned to the left atrium by the pulmonary veins. The **systemic circuit** supplies oxygenated blood to the organ systems. Oxygenated blood that has been carried to the left atrium via the pulmonary circuit flows into the left ventricle, which then contracts and produces enough pressure to pump the blood into the aorta (the large artery leaving the heart). Systemic arteries arising off the aorta direct blood into tissue capillaries. While the blood is in the capillaries of a given tissue, the tissue cells take oxygen and nutrients from the blood and add carbon dioxide, wastes, and regulatory molecules (enzymes and hormones) to it. Veins of the systemic circuit drain the deoxygenated blood from the capillaries and return it to the right atrium. After passing to the right ventricle, the deoxygenated blood reenters the pulmonary circuit, and the cycle repeats.

Vessels of the pulmonary circuit are presented in Exercise 35. In this exercise you will study the major arteries and veins of the systemic circuit.

Blood vessels are a continuous network of pipes, and often there is little anatomical difference along the length of a given vessel as it passes from one region of the body to another. To facilitate identification and discussion, however, anatomists assign different names to a given vessel, depending on which part of the body the vessel is passing through. The subclavian artery becomes the axillary artery, for instance, and then the brachial artery. Each name is usually related to the name of a bone or organ adjacent to the vessel; therefore, because they often run parallel to each other, arteries and veins often have the same name. For example, the large vessels in the thigh are the femoral artery and the femoral vein. As the femoral artery descends the leg, it is called the popliteal artery behind the knee and the posterior tibial artery in the calf. As the femoral vein descends the leg, it is called the popliteal vein behind the knee and the posterior tibial vein in the calf.

LAB ACTIVITY 1   Artery/Vein Comparison

The walls of arteries and veins have three layers: an outer tunica externa, a middle tunica media, and an inner tunica intima (Figure 36.2). The **tunica externa** is a connective-tissue covering that anchors the vessel to surrounding tissues. Collagen and elastic fibers give this layer strength and flexibility. The **tunica media** is a layer of smooth muscle tissue. In the tunica media of arteries (but not of veins) are elastic fibers that allow the vessels to stretch and recoil in response to blood pressure changes. In veins, collagen fibers in the tunica media provide strength. Lining the inside of the vessels is the **tunica intima**, a thin layer of simple squamous epithelium called **endothelium**.

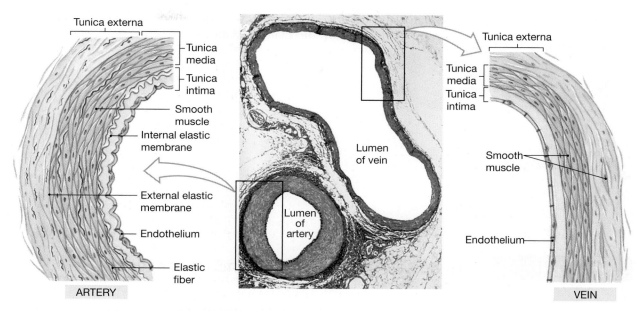

Figure 36.2 Comparison of a Typical Artery and a Typical Vein

Because blood pressure is much higher and fluctuates more in arteries than in veins, the walls of arteries are thicker than those of veins. Figure 36.2 compares an artery and a vein in cross-section. Notice how the artery cross-section is round and how the vessel has a thick tunica media with a folded tunica intima. When blood pressure increases or vasodilation occurs, the artery expands and the pleats in the tunica intima flatten as the vessel stretches.

Capillaries lack the tunica externa and tunica media. They consist of a single layer of endothelium that is continuous with the tunica intima of the artery and vein supplying and draining the capillary. Capillaries are so narrow that blood cells must line up in single file to squeeze through.

Veins have a thinner wall than arteries, and in veins the tunica intima is smooth rather than folded. The walls of a vein will collapse if the vessel is emptied of blood. Blood pressure is low in veins, and to prevent backflow, the peripheral veins have valves that function to keep blood flowing in one direction, toward the heart.

QuickCheck Questions

1.1 Describe the three layers in the wall of an artery.

1.2 How do arterial walls differ from venous walls?

1 *Materials*

☐ Compound microscope

☐ Prepared slide of artery and vein

Procedures

1. Review the structure of arteries and veins in Figure 36.2.

2. Place the artery/vein slide on the microscope stage, and locate the artery and vein at low magnification. Most slide preparations have one artery, an adjacent vein, and a nerve. The blood vessels are hollow and most likely have blood cells in the lumen. The nerve appears as a round, solid structure.

3. Increase the magnification to medium, and identify the tunica externa, tunica media, and tunica intima in the artery and vein. Examine each layer at high magnification. How does the endothelial layer of the vein differ from that of the artery?

4. Draw and label the artery and vein in the space provided. Include enough detail in your drawing to show the anatomical differences between the vessels. ■

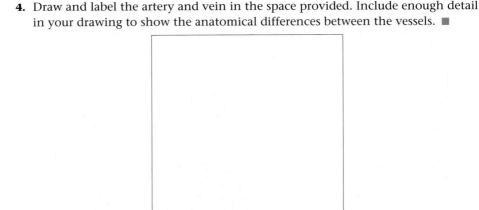

LAB ACTIVITY 2 Arteries of the Head, Neck, and Upper Limb

The aorta receives oxygenated blood from the left ventricle of the heart and distributes the blood to the major arteries that arise from the aorta and supply the head, limbs, and trunk (Figure 36.3). The aorta is curved like a question mark: it exits the base of the heart, curves upward and to the left, and then descends behind the heart to the abdominal cavity. Arteries that branch off the aortic arch serve the head, neck, and upper limb. Branches off the abdominal aorta serve the abdominal organs. The abdominal aorta enters the pelvic cavity and divides to send a branch into each lower limb.

Figure 36.4 illustrates the arteries of the chest and abdomen. The **ascending aorta** begins at the base of the heart and extends to the curved **aortic arch**, which then descends through the thorax as the **descending (thoracic) aorta**. At the point where it pierces through the diaphragm, the descending aorta becomes the **abdominal aorta**.

The first branch of the aortic arch, the **brachiocephalic** (brā-kē-ō-se-FAL-ik), or **innominate, artery**, is short and divides into the **right common carotid artery** and the **right subclavian artery**. The right common carotid artery supplies blood to the right side of the head and neck; the right subclavian artery supplies blood to the right arm. The second and third branches of the aortic arch are the **left common carotid artery** and **left subclavian artery**, respectively. Note that only the right common carotid artery and right subclavian artery are derived from the brachiocephalic artery. The left common carotid artery and left subclavian artery arise directly from the peak of the aortic arch. A **vertebral artery** branches off each subclavian artery and supplies blood to the brain and spinal cord.

The subclavian arteries supply blood to the upper limbs. Each subclavian artery passes under the clavicle, crosses the armpit as the **axillary artery**, and continues into the arm as the **brachial artery** (Figure 36.5). (Blood pressure is usually taken at the brachial artery.) At the antecubitis (elbow), the brachial artery divides into the lateral **radial artery** and the medial **ulnar artery**, each named after the bone it follows. In the palm of the hand, these arteries are interconnected by the **superficial** and **deep palmar arches**, which send small digital arteries to the fingers. Except for the brachiocephalic artery on the right side, the vascular anatomy is symmetrical in the right and left upper limbs.

The right and left common carotid arteries supply blood to the neck, face, and brain (Figure 36.6). The term *common* suggests that the vessel joins external and internal branches. Each common carotid artery ascends deep in the neck and divides at the larynx into an **external carotid artery** and an **internal carotid artery**. The base of the internal carotid swells as the **carotid sinus** and contains baroreceptors to monitor blood pressure. The external carotid artery branches to supply blood to the neck and face. The pulse in the external carotid artery can be felt by placing your fingers lateral to your thyroid cartilage (Adam's apple). The external carotid artery branches into the **facial**, **maxillary**, and **superficial temporal arteries** to serve the external structures of the head.

Figure 36.3
Overview of the Major Systemic Arteries

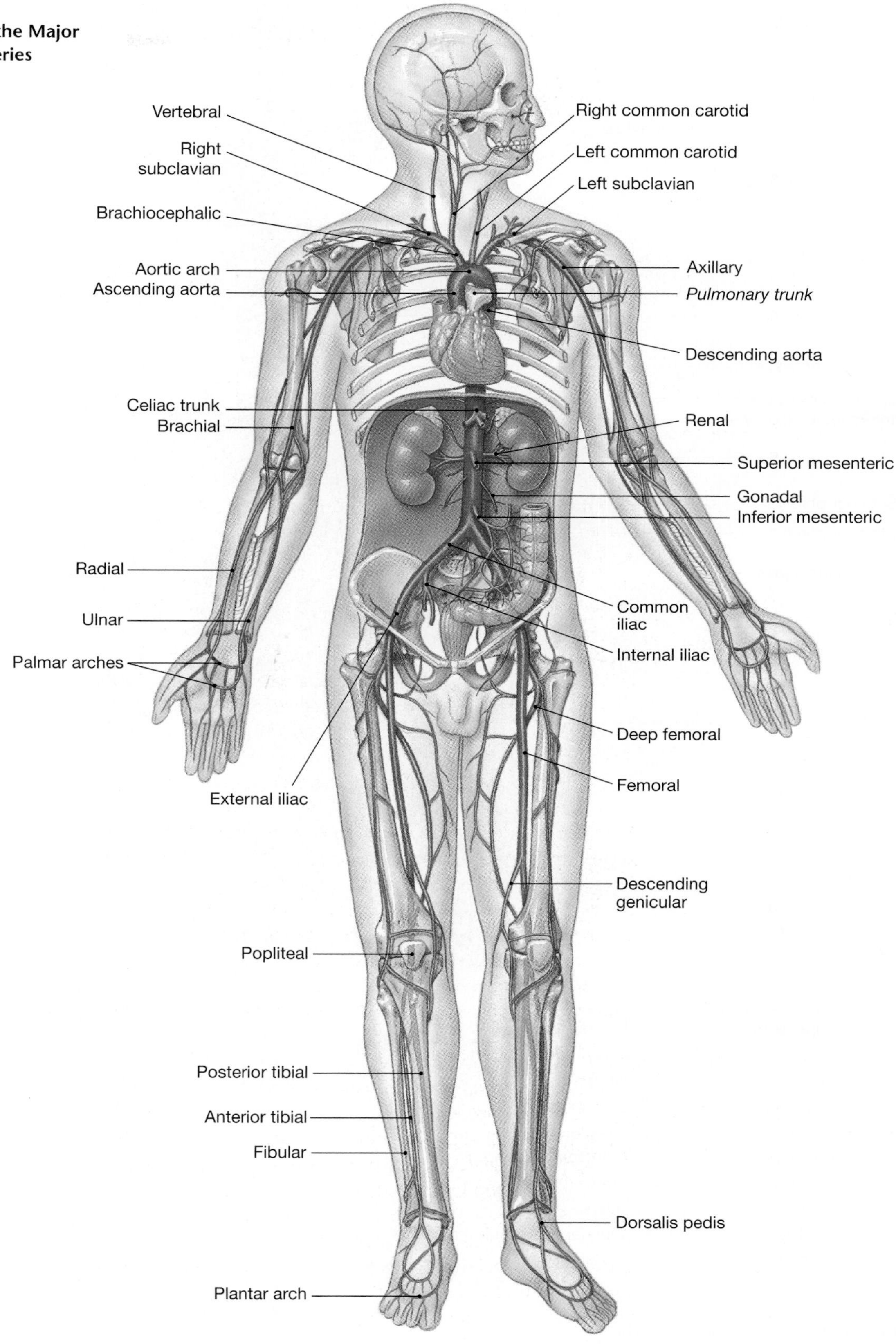

Vertebral

Right subclavian

Brachiocephalic

Aortic arch

Ascending aorta

Celiac trunk

Brachial

Radial

Ulnar

Palmar arches

External iliac

Popliteal

Posterior tibial

Anterior tibial

Fibular

Plantar arch

Right common carotid

Left common carotid

Left subclavian

Axillary

Pulmonary trunk

Descending aorta

Renal

Superior mesenteric

Gonadal

Inferior mesenteric

Common iliac

Internal iliac

Deep femoral

Femoral

Descending genicular

Dorsalis pedis

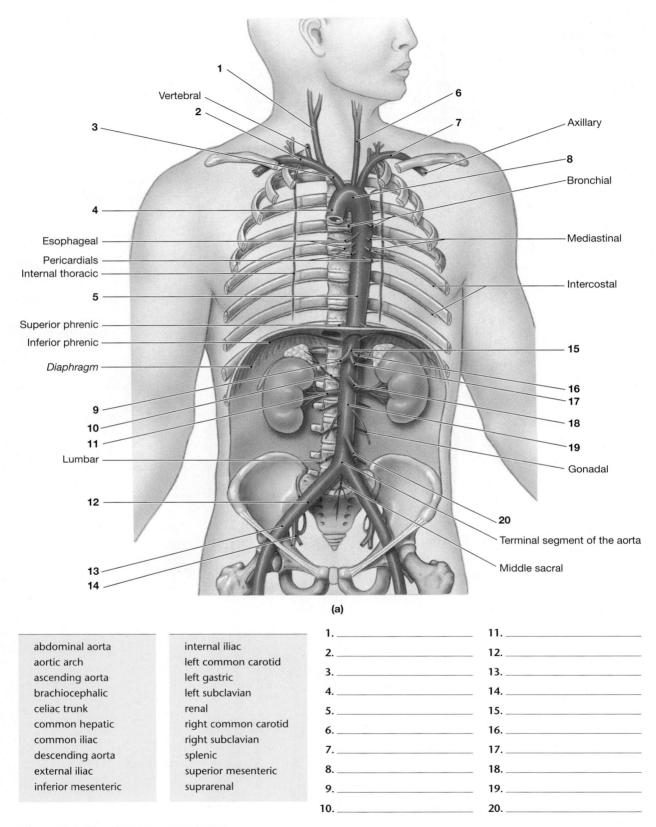

1 _____ Vertebral
2
3

6
7 Axillary
8 Bronchial

4

Esophageal
Pericardials
Internal thoracic
5

Mediastinal

Intercostal

Superior phrenic
Inferior phrenic
Diaphragm

15

9
10
11
Lumbar

16
17
18

19

Gonadal

12

20
Terminal segment of the aorta
Middle sacral

13
14

(a)

abdominal aorta	internal iliac	1. _____
aortic arch	left common carotid	2. _____
ascending aorta	left gastric	3. _____
brachiocephalic	left subclavian	4. _____
celiac trunk	renal	5. _____
common hepatic	right common carotid	6. _____
common iliac	right subclavian	7. _____
descending aorta	splenic	8. _____
external iliac	superior mesenteric	9. _____
inferior mesenteric	suprarenal	10. _____

1. _____   11. _____
2. _____   12. _____
3. _____   13. _____
4. _____   14. _____
5. _____   15. _____
6. _____   16. _____
7. _____   17. _____
8. _____   18. _____
9. _____   19. _____
10. _____ 20. _____

Figure 36.4 Major Arteries of the Trunk
A diagrammatic view, with most of the thoracic and abdominal organs removed.

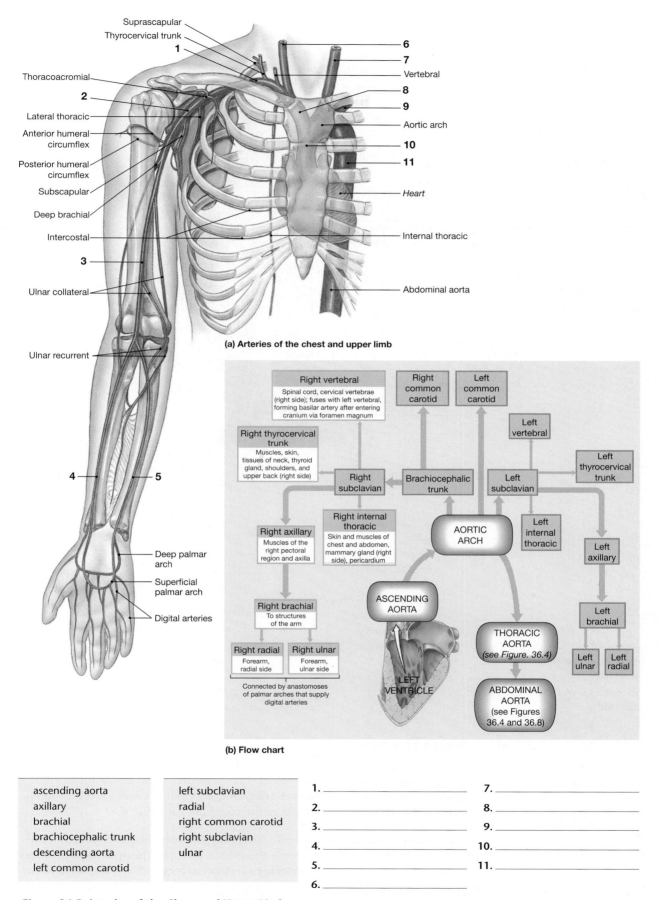

Suprascapular
Thyrocervical trunk
1
Thoracoacromial
2
Lateral thoracic
Anterior humeral circumflex
Posterior humeral circumflex
Subscapular
Deep brachial
Intercostal
3
Ulnar collateral
Ulnar recurrent
4 **5**
Deep palmar arch
Superficial palmar arch
Digital arteries

6
7
Vertebral
8
9
Aortic arch
10
11
Heart
Internal thoracic
Abdominal aorta

(a) Arteries of the chest and upper limb

Right vertebral
Spinal cord, cervical vertebrae (right side); fuses with left vertebral, forming basilar artery after entering cranium via foramen magnum

Right common carotid

Left common carotid

Left vertebral

Right thyrocervical trunk
Muscles, skin, tissues of neck, thyroid gland, shoulders, and upper back (right side)

Right subclavian

Brachiocephalic trunk

Left subclavian

Left thyrocervical trunk

Right axillary
Muscles of the right pectoral region and axilla

Right internal thoracic
Skin and muscles of chest and abdomen, mammary gland (right side), pericardium

AORTIC ARCH

Left internal thoracic

Left axillary

Right brachial
To structures of the arm

ASCENDING AORTA

LEFT VENTRICLE

THORACIC AORTA
(see Figure. 36.4)

Left brachial

Right radial
Forearm, radial side

Right ulnar
Forearm, ulnar side

Connected by anastomoses of palmar arches that supply digital arteries

ABDOMINAL AORTA
(see Figures 36.4 and 36.8)

Left ulnar

Left radial

(b) Flow chart

ascending aorta	left subclavian
axillary	radial
brachial	right common carotid
brachiocephalic trunk	right subclavian
descending aorta	ulnar
left common carotid	

1. _____   7. _____
2. _____   8. _____
3. _____   9. _____
4. _____   10. _____
5. _____   11. _____
6. _____

Figure 36.5 Arteries of the Chest and Upper Limb
(a) A diagrammatic view. **(b)** Flowchart.

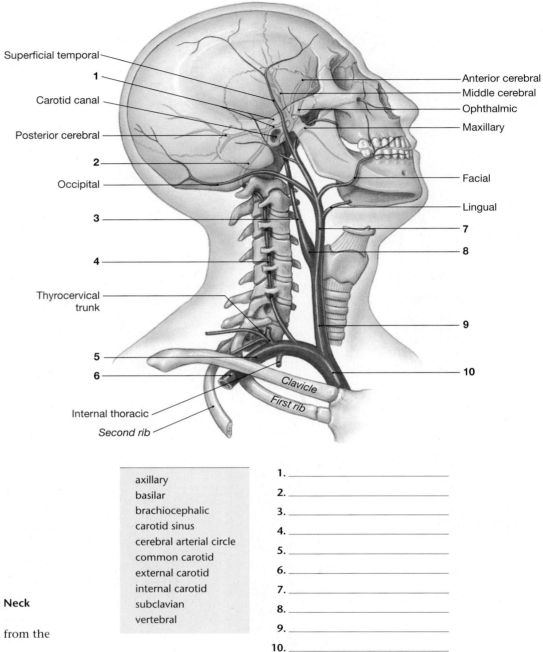

Superficial temporal

1

Carotid canal

Posterior cerebral

2

Occipital

3

4

Thyrocervical trunk

5

6

Internal thoracic

Second rib

Anterior cerebral
Middle cerebral
Ophthalmic
Maxillary

Facial

Lingual

7

8

9

10

Clavicle
First rib

Figure 36.6
Arteries of the Neck and Head
Shown as seen from the right side.

axillary
basilar
brachiocephalic
carotid sinus
cerebral arterial circle
common carotid
external carotid
internal carotid
subclavian
vertebral

1. _____
2. _____
3. _____
4. _____
5. _____
6. _____
7. _____
8. _____
9. _____
10. _____

Study Tip | **What's in a Name?**

Arteries and veins with the term *common* as part of their name always branch into an external and an internal vessel. The common carotid artery, for example, branches into an external carotid artery and an internal carotid artery. The internal and external iliac veins join as the common iliac vein. ●

The internal carotid artery ascends to the base of the brain and divides into three arteries: ophthalmic, anterior cerebral, and middle cerebral. The **ophthalmic artery** provides blood to the eyes; the **anterior cerebral artery** and **middle cerebral artery** both supply blood to the brain. These vessels are detailed in Figure 36.6.

Because of its high metabolic rate, the brain has a voracious appetite for oxygen and nutrients. A reduction in blood flow to the brain may result in permanent damage to the affected area. To ensure that the brain receives a continuous supply of blood, branches of the internal carotid arteries and other arteries interconnect, or **anastomose**, as the **cerebral arterial circle**, also called the **circle of Willis** (Figure 36.7). The right and left vertebral arteries ascend in the transverse foramina of the cervical vertebrae and enter the skull at the foramen magnum. These arteries fuse into a single **basilar artery** on the inferior surface of the brain stem. The cerebral arterial circle is formed as the basilar artery branches into left and right **posterior cerebral arteries** and left and right **posterior communicating arteries**. Each posterior communicating artery forms an anastomosis with its internal common carotid artery. On the opposite side of each common carotid, an **anterior communicating artery** branches off the carotid artery, passes anteriorly, and meets the anterior communicating artery of the other carotid to complete the cerebral arterial circle at the base of the brain.

QuickCheck Questions

2.1 How does arterial branching in the left side of the neck differ from branching in the right side?

2.2 What is an anastomosis?

2.3 Which arteries in the brain anastomose with one another?

2	*Materials*

Procedures

☐ Vascular system chart
☐ Torso model
☐ Head model
☐ Arm model

1. Review the arteries presented in Figures 36.4, 36.5, and 36.6, and then label 1 through 8 in Figure 36.4 and all of Figures 36.5 and 36.6.

2. On the torso model, examine the aortic arch, and identify the three branches arising from the superior margin of the arch.

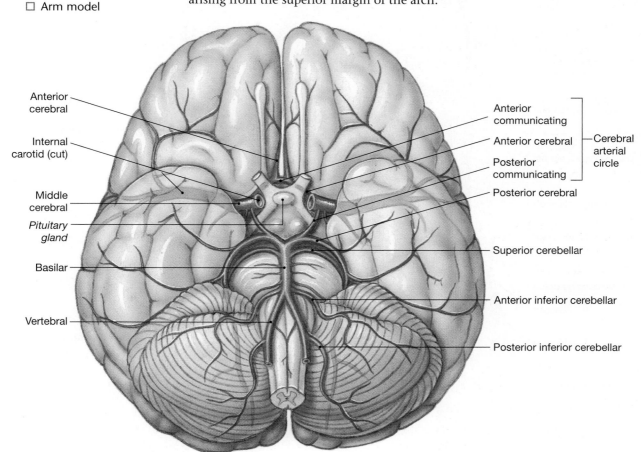

Figure 36.7 Arteries of the Brain
The major arteries on the inferior surface of the brain.

3. On the torso model, trace the arteries to the head, and note the differences between the right and left common carotid arteries.

4. On the head model, trace the arteries that converge at the cerebral arterial circle.

5. On the arm model, identify the arteries of the shoulder and arm. Note the difference in origin of the right and left subclavian arteries.

6. Locate the pulse in your radial and common carotid arteries. Use your index finger to feel the pulse. ■

LAB ACTIVITY 3   ## Arteries of the Abdominopelvic Cavity and Lower Limb

The arteries stemming from the abdominal aorta are shown in Figure 36.8. An easy way to identify the branches of the abdominal aorta is to distinguish between paired arteries, which have right and left branches, and unpaired arteries, which are not branched.

Three unpaired arteries arise from the abdominal aorta: celiac trunk, superior mesenteric artery, and inferior mesenteric artery. The **celiac** (SĒ-lē-ak) **trunk** is short

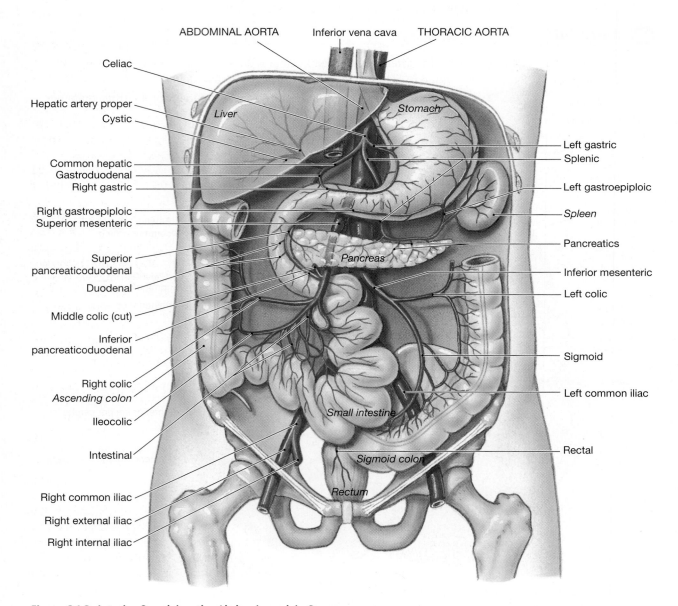

Figure 36.8 Arteries Supplying the Abdominopelvic Organs

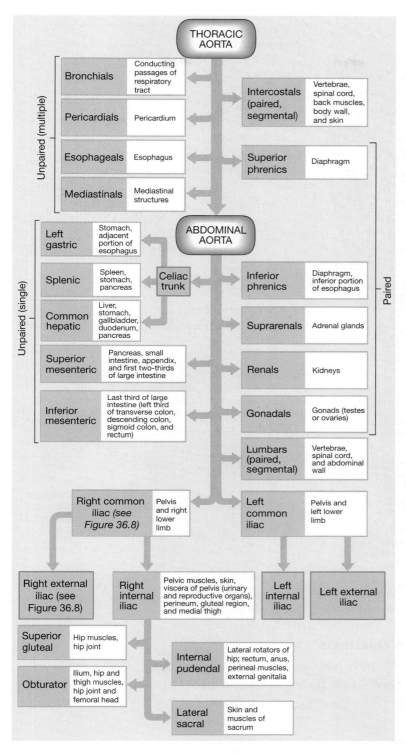

Figure 36.9
Flowchart of the Major Arteries of the Trunk

and arises inferior to the diaphragm. It splits into three arteries: common hepatic, left gastric, and splenic. The **common hepatic artery** divides to supply blood to the liver, gallbladder, and part of the stomach. The **left gastric artery** also supports blood flow to the stomach. The **splenic artery** supplies blood to the spleen, stomach, and pancreas.

Inferior to the celiac trunk is the next unpaired artery, the **superior mesenteric** (mez-en-TER-ik) **artery**. This vessel supplies blood to the large intestine, parts of the small intestine, and other abdominal organs. The third unpaired artery, the **inferior mesenteric artery**, originates before the abdominal aorta divides to enter the pelvic cavity and legs. This artery supplies parts of the large intestine and the rectum.

Figure 36.9 is a flowchart of the major arteries of the chest and abdomen.

Four major sets of paired arteries arise off the abdominal aorta. The right and left **suprarenal arteries** arise near the superior mesenteric artery and branch into the adrenal glands, located on top of the kidneys. The right and left **renal arteries**, which supply the kidneys, stem off the abdominal aorta just inferior to the suprarenal arteries. The right and left **gonadal arteries** arise near the inferior mesenteric artery and bring blood to the reproductive organs. The right and left **lumbar arteries** originate near the terminus of the abdominal aorta and service the lower body wall.

At the level of the hips, the abdominal aorta divides into the right and left **common iliac** (IL-ē-ak) **arteries** (Figure 36.8). These arteries descend through the pelvic cavity and branch into **external iliac arteries**, which enter the legs, and **internal iliac arteries**, which supply blood to the organs of the pelvic cavity. The external iliac artery pierces the abdominal wall and becomes the **femoral artery** of the thigh (Figure 36.10). A **deep femoral artery** arises to supply deep thigh muscles. The femoral artery passes though the posterior knee as the **popliteal** (pop-LIT-ē-al) **artery** and divides into the **posterior tibial artery** and the **anterior tibial artery**, each supplying blood to the lower leg. The **fibular artery**, also called the *peroneal artery*, stems laterally off the posterior tibial artery. The arteries of the lower leg branch into the foot and anastomose at the **dorsal arch** and the **plantar arch**.

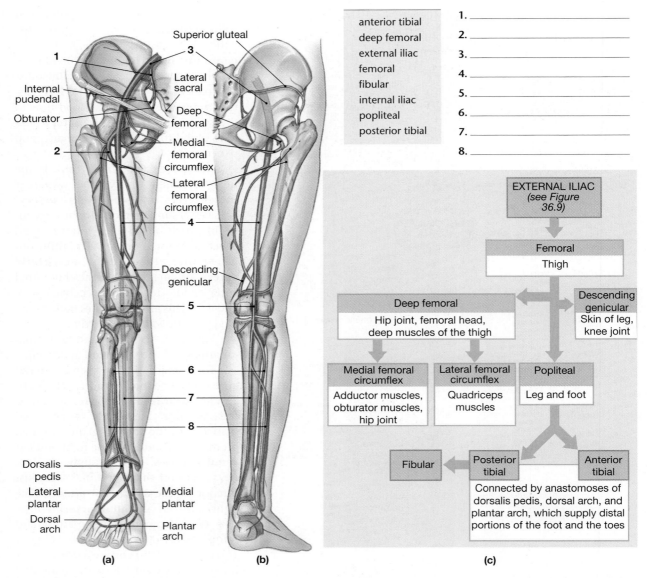

Figure 36.10 Arteries of the Lower Limb
(a) Anterior view. **(b)** Posterior view. **(c)** A flowchart of blood flow to a lower limb.

QuickCheck Questions

3.1 What are the three branches of the celiac trunk?

3.2 Which arteries supply the intestines with blood?

3.3 What does the external iliac artery become in the leg?

3	Materials	Procedures

☐ Vascular system chart
☐ Torso model
☐ Leg model

1. Review the arteries presented in Figures 36.8 and 36.9, and then label 9 through 20 in Figure 36.4 and all of Figure 36.10.

2. On the torso model, locate the celiac trunk and its three branches.

3. On the torso model, identify the superior and inferior mesenteric arteries and the other arteries stemming from the abdominal aorta.

4. On the torso model, observe how the abdominal aorta branches into the common iliac arteries.

5. On the leg model, trace the major arteries supplying the lower limb.

6. On your body, trace the location of your abdominal aorta, common iliac artery, external iliac artery, femoral artery, popliteal artery, and posterior tibial artery. ■

LAB ACTIVITY 4 Veins of the Head, Neck, and Upper Limb

Once you have learned the major systemic arteries, identification of systemic veins is easy because most arteries have a corresponding vein. This similarity in distribution can be seen by comparing Figures 36.3 and 36.11. Unlike arteries, many veins are superficial and easily seen under the skin. Systemic veins are usually painted blue on vascular models and torso models to indicate that they transport

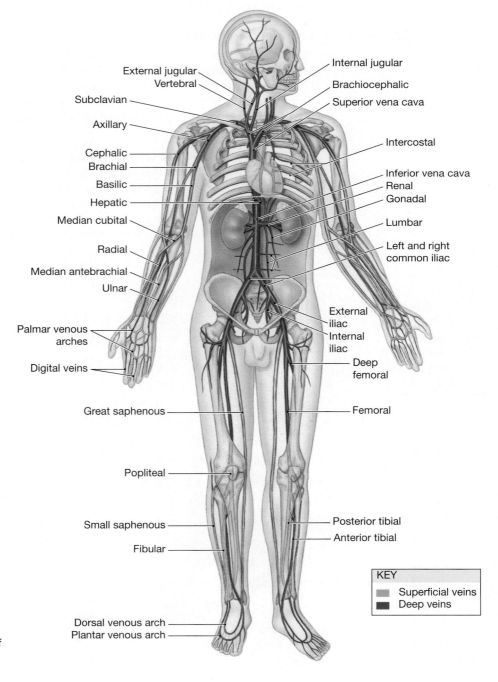

Figure 36.11 **Overview of the Major Systemic Veins**

deoxygenated blood. When identifying veins, work in the direction of blood flow, from the periphery toward the heart.

Blood in the brain drains into large veins called *sinuses* (Figure 36.12). Small-diameter veins deep inside the brain drain into progressively larger veins that empty into the **superior sagittal sinus** located in the falx cerebri separating the cerebral hemispheres. Other venous sinuses and the superior sagittal sinus drain into the **internal jugular vein**, which exits the skull via the jugular foramen, descends the neck, and empties into the **brachiocephalic vein**. Superficial veins that drain the face and scalp empty into a smaller **external jugular vein**, which also descends the neck to join the subclavian vein. The right and left brachiocephalic veins merge at the **superior vena cava** and empty deoxygenated blood into the right atrium of the heart. Blood in the right atrium enters the right ventricle, which contracts and pumps the blood to the lungs through the pulmonary circuit. The **inferior vena cava** drains veins of the abdominopelvic cavity.

Study Tip	**Brachiocephalic Veins**

The venous system has both a right and a left brachiocephalic vein, each formed by the merging of subclavian, vertebral, and internal and external jugular veins. The arterial system has a single brachiocephalic artery that branches into the right common carotid artery and right subclavian artery. On the left side of the body, the common carotid artery and subclavian artery originate directly off the aortic arch, as noted earlier. ●

Figure 36.13 illustrates the venous drainage of the arm, chest, and abdomen. The hand drains into digital veins that empty into a network of **palmar venous arches**. These vessels drain into the **cephalic vein**, which ascends along the lateral margin of the arm. The **median antebrachial vein** ascends to the elbow, is joined by the **median cubital vein** that crosses over from the cephalic vein, and becomes the **basilic vein**. The median cubital vein is often used to collect blood samples from an

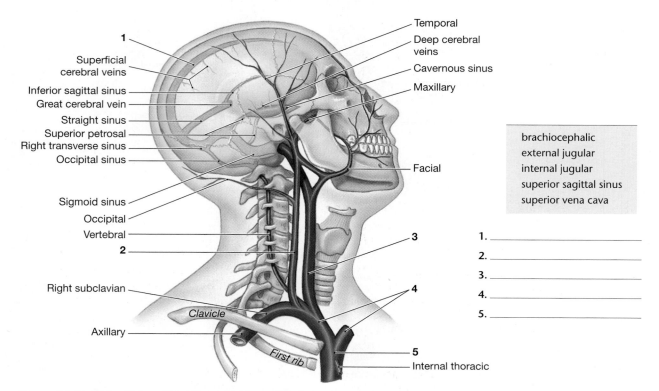

brachiocephalic
external jugular
internal jugular
superior sagittal sinus
superior vena cava

1. _____
2. _____
3. _____
4. _____
5. _____

Figure 36.12 Major Veins of the Head, Neck, and Brain
Veins draining the brain and the superficial and deep portions of the head and neck.

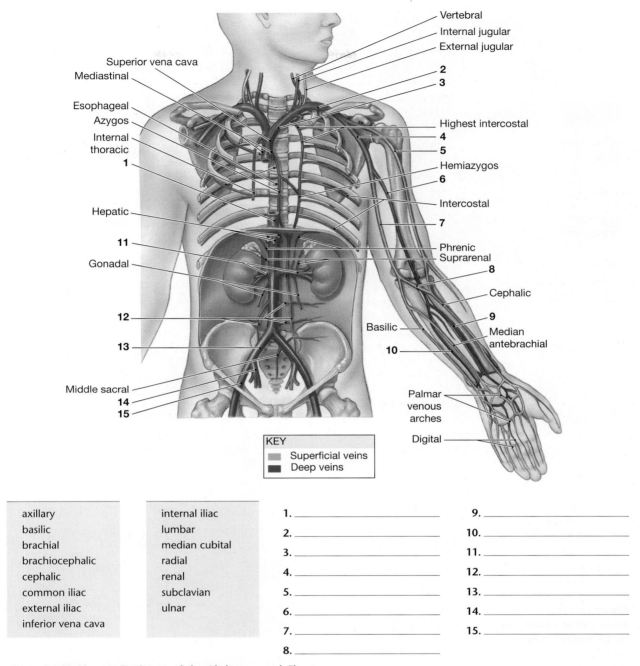

Figure 36.13 Venous Drainage of the Abdomen and Chest

axillary	internal iliac
basilic	lumbar
brachial	median cubital
brachiocephalic	radial
cephalic	renal
common iliac	subclavian
external iliac	ulnar
inferior vena cava	

1. _____   9. _____

2. _____   10. _____

3. _____   11. _____

4. _____   12. _____

5. _____   13. _____

6. _____   14. _____

7. _____   15. _____

8. _____

individual. Also in the forearm are the **radial** and **ulnar veins**, which fuse above the elbow into the **brachial vein**. The brachial and basilic veins meet at the armpit as the **axillary vein**, which joins the cephalic vein and becomes the **subclavian vein**. Subclavian veins and veins from the neck and head drain into the brachiocephalic vein, which then empties into the superior vena cava. The superior vena cava returns deoxygenated blood from the head, neck, and arms to the right atrium.

Figure 36.14 is a flowchart of the venous drainage into the venae cavae.

QuickCheck Questions

4.1 Which veins combine to form the superior vena cava?

4.2 Where is the cephalic vein?

4.3 Where is the superior sagittal sinus?

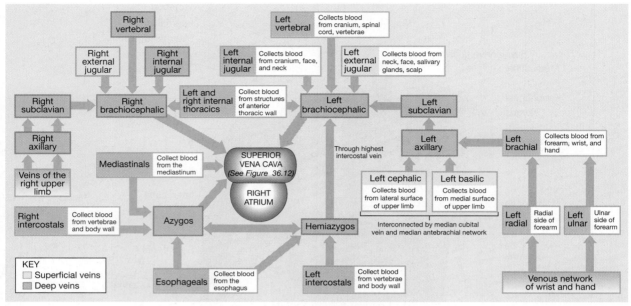

(a) Tributaries of the SVC

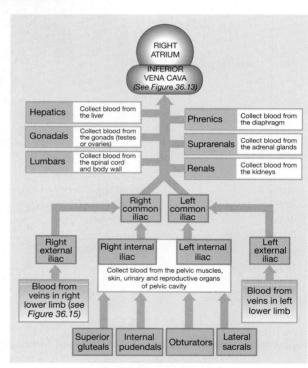

(b) Tributaries of the IVC

Figure 36.14

Flowchart of Circulation to the Venae Cavae

(a) Drainage into the Superior Vena Cava.
(b) Drainage into the Inferior Vena Cava.

4 *Materials*

☐ Vascular system chart
☐ Torso model
☐ Head model
☐ Arm model

Procedures

1. Review the head, neck, and arm veins presented in Figures 36.11 through 36.14, and then label all of Figure 36.12 and 1 through 10 in Figure 36.13.

2. On the head model or torso model, identify the superior sagittal sinus and other veins draining the head into the external and internal jugular veins.

3. On the arm model, start at the hands and name the veins draining the arm and shoulder. Notice how the superior vena cava is formed.

4. On your body, trace your cephalic vein, subclavian vein, brachiocephalic vein, and superior vena cava. ■

| **LAB ACTIVITY 5** | **Veins of the Lower Limb and Abdominopelvic Cavity** |

Veins that drain the lower limbs and abdominal organs empty into the inferior vena cava, the large vein that pierces the diaphragm and empties into the right atrium of the heart. Figure 36.15 illustrates the venous drainage of the lower limb. The foot has many anastomoses that drain into the lateral **fibular** (peroneal) **vein** and the **anterior tibial vein**, located on the medial aspect of the anterior shin. These veins, along with the **posterior tibial vein**, merge and become the **popliteal vein** of the posterior knee. The **small** (lesser) **saphenous** (sa-FĒ-nus) **vein**, which ascends from the ankle to the knee and drains blood from superficial veins, also empties into the popliteal vein. Superior to the knee, the popliteal vein becomes the **femoral vein**, which ascends along the femur to the lower pelvic girdle, where it joins the **deep femoral vein** at the external iliac vein. The **great saphenous vein** ascends from the medial side of the ankle to the upper thigh and drains into the external iliac vein. In the pelvic cavity, the **external** and **internal iliac veins** fuse to form the **common iliac vein**. The right and left common iliac veins merge and drain into the inferior vena cava.

Six major veins from the abdominal organs drain blood into the inferior vena cava (Figure 36.13). The **lumbar veins** drain the muscles of the lower body wall and the spinal cord and empty into the inferior vena cava close to the common iliac veins. A pair of **gonadal veins** empty blood from the reproductive organs into the inferior vena cava above the lumbar veins. Pairs of **renal** and **suprarenal veins** drain into the inferior vena cava next to their respective organs. Before entering the thoracic cavity to drain blood into the right atrium, the inferior vena cava collects blood from the **hepatic veins** draining the liver and the **phrenic veins** from the diaphragm.

Figure 36.16 shows the venous supply to the liver. The inferior and superior mesenteric veins drain nutrient-rich blood from the digestive tract. These veins empty into the **hepatic portal vein**, which passes the blood through the liver, where blood sugar concentration is regulated. Phagocytic cells in the liver cleanse the blood of any microbes that may have entered it through the mucous membrane of the digestive system. Blood from the hepatic arteries and hepatic portal vein mixes in the liver and is returned to the inferior vena cava by the hepatic veins.

QuickCheck Questions

5.1 How does blood drain from the leg into the inferior vena cava?

5.2 Which veins drain into the hepatic portal vein?

| **5 Materials** | **Procedures** |

☐ Vascular system model

☐ Torso model

☐ Leg model

1. Review the veins presented in Figures 36.13 through 36.16, and then label 11 through 15 in Figure 36.13 and all of Figure 36.15.

2. On the leg model, follow the draining of blood from the ankle to the inferior vena cava.

3. On the torso model, identify the veins draining the major organs. Locate where the superior and inferior mesenteric veins drain into the hepatic portal vein.

4. On your body, trace the location of the veins in your lower limb.

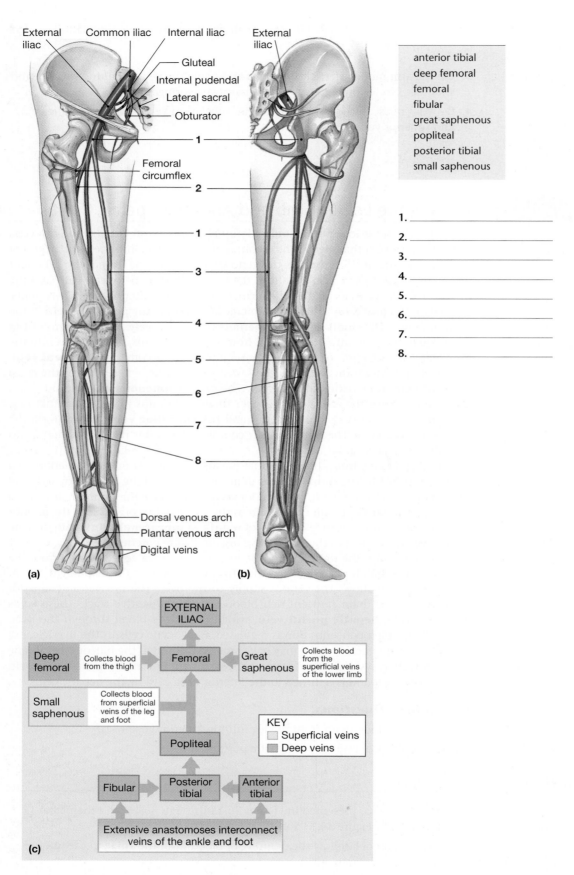

External iliac   Common iliac   Internal iliac   External iliac

Gluteal

Internal pudendal

Lateral sacral

Obturator

1

Femoral circumflex

2

1

3

4

5

6

7

8

Dorsal venous arch

Plantar venous arch

Digital veins

(a)   **(b)**

anterior tibial
deep femoral
femoral
fibular
great saphenous
popliteal
posterior tibial
small saphenous

1. _____
2. _____
3. _____
4. _____
5. _____
6. _____
7. _____
8. _____

(c)

EXTERNAL ILIAC

Deep femoral — Collects blood from the thigh → Femoral ← Great saphenous — Collects blood from the superficial veins of the lower limb

Small saphenous — Collects blood from superficial veins of the leg and foot

KEY
☐ Superficial veins
■ Deep veins

Popliteal

Fibular → Posterior tibial ← Anterior tibial

Extensive anastomoses interconnect veins of the ankle and foot

Figure 36.15 Venous Drainage from the Lower Limb
(a) Anterior view. **(b)** Posterior view. **(c)** Flowchart of venous circulation to a lower limb.

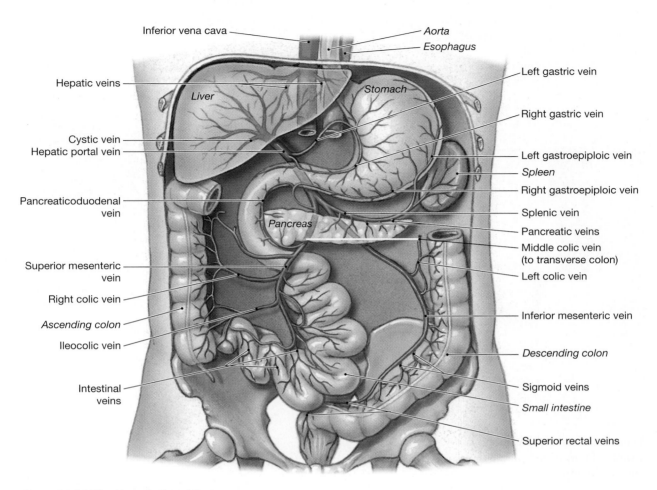

Figure 36.16 The Hepatic Portal System

5. Although you have studied the arterial and venous divisions separately, they are anatomically interconnected by capillaries. To reinforce this organization, trace blood through the following systemic routes:

   a. From the heart through the left arm and back to the heart.

   b. From the heart through the liver and back to the heart.

   c. From the heart through the right leg and back to the heart. ■

LAB ACTIVITY 6 Fetal Circulation

A fetus receives oxygen from the mother through the **placenta**, a vascular organ that connects the fetus to the wall of the mother's uterus. During development, the fetal lungs are filled with amniotic fluid, and for efficiency some of the blood is shunted away from the fetal pulmonary circuit by several structures (Figure 36.17). The **foramen ovale** is a hole in the interatrial wall. Much of the blood entering the right atrium from the inferior vena cava passes through the foramen ovale to the left atrium and avoids the right ventricle and the pulmonary circuit. Some of the blood that does enter the pulmonary trunk may bypass the lungs through a connection with the aorta, the **ductus arteriosus**. At birth, the foramen ovale closes and becomes a depression on the interatrial wall, the **fossa ovalis**. The ductus arteriosus closes and becomes the **ligamentum arteriosum**.

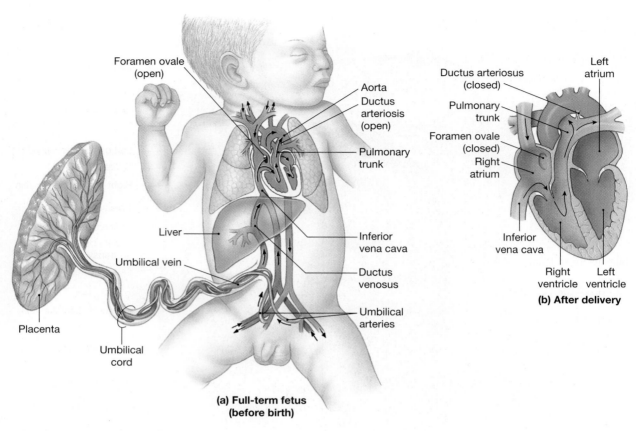

Figure 36.17 Fetal Circulation
(a) Blood flow to and from the placenta. **(b)** Blood flow through the neonatal (newborn) heart.

QuickCheck Questions

6.1 Where is the foramen ovale?

6.2 Which two structures are connected by the ductus arteriosus?

6 Materials

☐ Heart model

Procedures

1. Review the fetal structures in Figure 36.17.
2. Identify the fossa ovalis on the heart model. What was this structure in the fetus?
3. At the point where the pulmonary trunk branches, find the ligamentum arteriosum. What was this structure in the fetus, and what purpose did it serve?
4. On the heart model, trace a drop of blood through the fetal structures of the heart, starting at the right atrium. ■

Anatomy of the
Systemic Circulation

Name _____

Date _____

Section _____

A. Matching

Match each term in the left column with its correct description from the right column.

_____ **1.** artery in armpit

_____ **2.** artery having three branches

_____ **3.** vein used for taking blood samples

_____ **4.** artery on right side only

_____ **5.** long vein of leg

_____ **6.** carries deoxygenated blood to liver

_____ **7.** artery to large intestine

_____ **8.** cerebral anastomosis

_____ **9.** long vein of arm

_____ **10.** vein in knee

_____ **11.** artery to reproductive organ

_____ **12.** major artery in neck

_____ **13.** vein under clavicle

_____ **14.** found only in veins

A. subclavian

B. superior mesenteric

C. popliteal

D. cephalic

E. carotid

F. gonadal

G. valves

H. axillary

I. great saphenous

J. median cubital

K. hepatic portal vein

L. circle of Willis

M. celiac

N. brachiocephalic artery

LABORATORY REPORT

B. Labeling

Label the arteries in Figure 36.18 and the veins in Figure 36.19.

1. _____
2. _____
3. _____
4. _____
5. _____
6. _____
7. _____
8. _____
9. _____
10. _____
11. _____
12. _____
13. _____
14. _____
15. _____
16. _____
17. _____
18. _____
19. _____
20. _____
21. _____
22. _____
23. _____
24. _____
25. _____

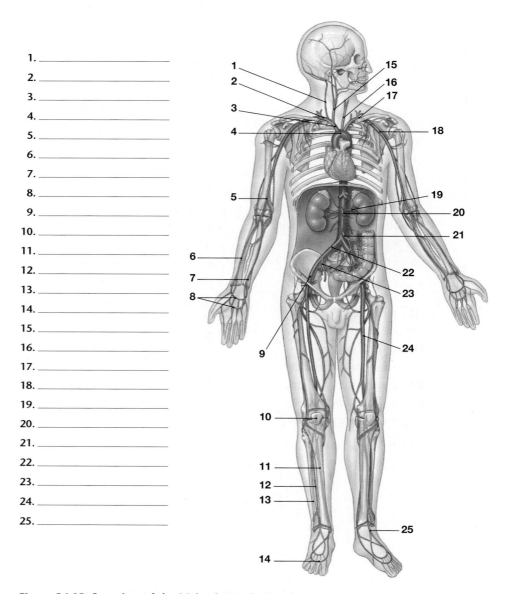

Figure 36.18 Overview of the Major Systemic Arteries

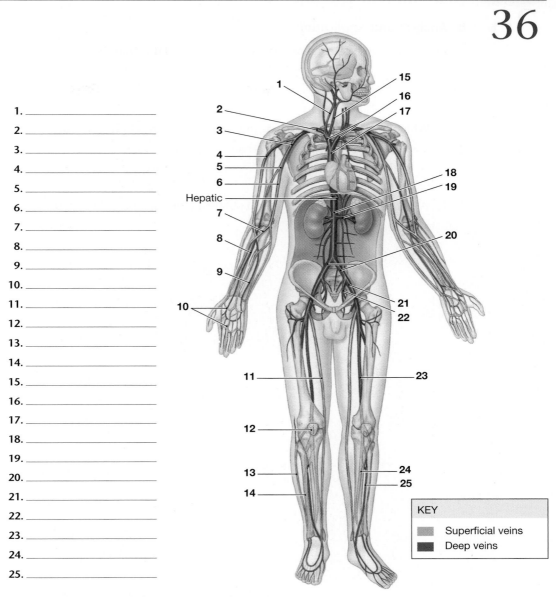

1. _____
2. _____
3. _____
4. _____
5. _____
6. _____
7. _____
8. _____
9. _____
10. _____
11. _____
12. _____
13. _____
14. _____
15. _____
16. _____
17. _____
18. _____
19. _____
20. _____
21. _____
22. _____
23. _____
24. _____
25. _____

Figure 36.19 **Overview of the Major Systemic Veins**

C. Short-Answer Questions

1. How is the anatomy of the arteries running from the aorta to the right arm different from that of the arteries running from the aorta to the left arm?

2. What is the function of valves in the peripheral veins?

3. Describe the major vessels that return deoxygenated blood to the right atrium of the heart.

36

D. Analysis and Application

1. How does the cerebral arterial circle ensure that the brain has a constant supply of blood?

2. Which vessel is normally used to measure a patient's blood pressure?

3. Which vessel is normally used to obtain a blood sample from a patient?

Cardiovascular Physiology

OBJECTIVES

On completion of this exercise you should be able to:

- Describe the pressure changes that occur during a cardiac cycle.

- Demonstrate the steps involved in blood pressure determination.

- Explain the differences in blood pressure caused by changes in body position.

- Take a pulse rate at several locations on the body.

- Read an electrocardiograph (ECG).

- Correlate electrical events as displayed on the ECG with the mechanical events of the cardiac cycle.

- Observe ECG rate and rhythm changes associated with changes in body position and breathing.

- Explain the principle of plethysmography and its usefulness in assessing changes in peripheral blood volume.

- Observe and record changes in peripheral blood volume, pulse rate, and pulse strength under a variety of experimental and physiological conditions.

- Determine the approximate speed of the pressure wave traveling between the heart and a finger.

- Describe the electrical activity associated with normal cardiac activity and how that electrical activity relates to blood flow.

To fully appreciate the cardiovascular experiments in this chapter, it is important that you understand the anatomical features of the heart and how blood circulates through its four chambers. If necessary, review these concepts in Exercise 36 before proceeding.

One complete heartbeat is called a **cardiac cycle**. During a cardiac cycle, each atrium contracts and relaxes once and each ventricle contracts and relaxes once. The contraction phase of a chamber is called **systole** (SIS-tō-lē), and the relaxation phase is termed **diastole** (dī-AS-tō-lē). The human heart averages 75 cardiac cycles, or heartbeats, per minute, with each cycle lasting 0.8 second (s). As Figure 37.1 illustrates, a cycle begins with systole of the atria, lasting 0.1 s (100 milliseconds

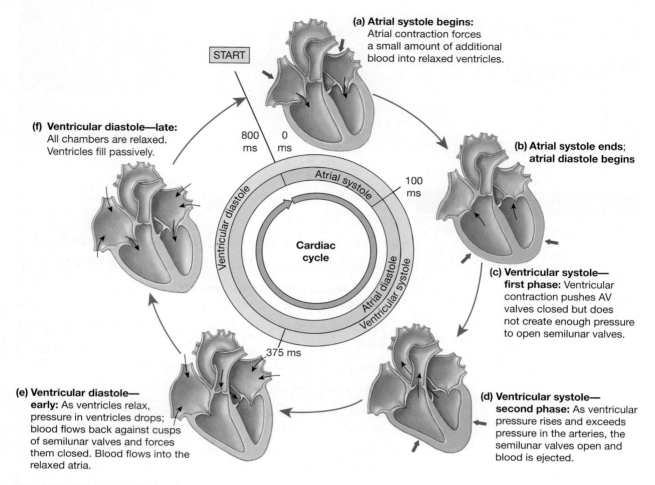

(a) **Atrial systole begins:** Atrial contraction forces a small amount of additional blood into relaxed ventricles.

START

800 ms   0 ms

100 ms

(b) **Atrial systole ends; atrial diastole begins**

(f) **Ventricular diastole—late:** All chambers are relaxed. Ventricles fill passively.

Atrial systole

Ventricular diastole

Cardiac cycle

Atrial diastole

Ventricular systole

(c) **Ventricular systole— first phase:** Ventricular contraction pushes AV valves closed but does not create enough pressure to open semilunar valves.

375 ms

(e) **Ventricular diastole— early:** As ventricles relax, pressure in ventricles drops; blood flows back against cusps of semilunar valves and forces them closed. Blood flows into the relaxed atria.

(d) **Ventricular systole— second phase:** As ventricular pressure rises and exceeds pressure in the arteries, the semilunar valves open and blood is ejected.

Figure 37.1 Phases of the Cardiac Cycle
Thin black arrows indicate blood flow, and green arrows indicate contractions.

[ms]), to fill the relaxed ventricles. Next, the ventricles enter their systolic phase and contract for 0.3 s to pump blood out of the heart. For the remaining 0.4 s of the cycle, all four chambers are in diastole and fill with blood in preparation for the next heartbeat. Most blood enters the ventricles during this resting period. Atrial systole contributes only 30% of the blood volume in the ventricle prior to ventricular systole, with fluid pressure and gravity forcing the remaining 70% that flows into the ventricles.

Each cardiac cycle is marked by an increase and a decrease in blood pressure, both in the heart and in the arteries. When a heart chamber contracts, pressure increases as a result of the squeezing together of the chamber walls. The increase in blood pressure forces blood to move either from atrium to ventricle or from ventricle to outside the heart. As a chamber relaxes, its walls move apart and pressure decreases. This drop in pressure draws blood into the chamber and refills it for the next systole.

When all four chambers are in diastole, the atrioventricular (AV) valves are open and blood flows from the atria into the ventricles. The semilunar (SL) valves are closed at this point to prevent backflow of blood from the aorta into the left ventricle and from the pulmonary artery into the right ventricle. When the left ventricle contracts, ventricular pressure increases to a point where it exceeds the pressure in the aorta that is holding the aortic SL valve shut. This difference in pressure forces blood through the valve into the aorta, and arterial blood pressure increases as a result of the increase in blood volume. When the ventricle relaxes, aortic and arterial pressures drop and the aortic SL valve closes. Similar events occur on the right side of the heart with the pulmonary SL valve.

In this exercise, you will investigate the physiology of blood pressure and the effect of posture on blood pressure. You will also listen to heart sounds and practice taking a subject's pulse.

LAB ACTIVITY 1 Listening to Heart Sounds

Listening to internal sounds of the body is called **auscultation**. A **stethoscope** is used to amplify the sounds to an audible level. The heart, lungs, and digestive tract are the most frequently auscultated systems. Auscultation provides the listener with valuable information concerning fluid accumulation in the lungs or blockages in the digestive tract. Auscultation of the heart is used as a diagnostic tool to evaluate valve function and detect the presence or absence of normal and abnormal heart sounds.

Four sounds are produced by the heart during a cardiac cycle. The first two are easily heard and are the familiar "lubb-dupp" of the heartbeat. The first heart sound (S_1), the **lubb**, is caused by the closure of the AV valves as the ventricles begin their contraction. The second sound (S_2), the **dupp**, occurs as the SL valves close at the beginning of ventricular diastole. The third and fourth sounds are faint and difficult to hear. The third sound (S_3) is produced by blood flowing into the ventricles, and the fourth (S_4) is generated by the contraction of the atria.

Figure 37.2a illustrates the landmarks for proper placement of the stethoscope **bell** (the flat metal disk that is placed against the patient's skin) to listen effectively to each heart valve. The bell has a delicate **diaphragm** that touches the skin and

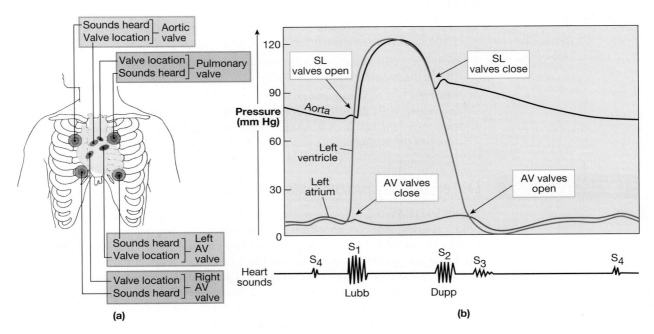

Figure 37.2 Heart Sounds

(a) Placements of a stethoscope for listening to the different sounds produced by individual valves. **(b)** The timing of heart sounds in relation to key events in the cardiac cycle.

amplifies sounds. Notice in the figure that the sites for auscultation do not overlie the anatomical location of the heart valves. This is because the soft tissue and bone overlying the heart deflect the cardiac sound waves to locations lateral to the valves. Figure 37.2b is a graphical representation of one cardiac cycle.

Clinical Application

Heart Murmur

An unusual heart sound is called a **murmur**. Not all murmurs indicate an anatomical or functional anomaly of the heart. The sound may originate from turbulent flow in a heart chamber. Some murmurs, however, are diagnostic for certain heart defects. Septal defects are holes in a wall between two chambers. A murmur is heard as blood passes through the hole from one chamber to the other. Abnormal operation of a valve also produces a murmur. In mitral valve prolapse, the left AV valve does not seal completely during ventricular systole and blood regurgitates upward into the left atrium. ❱

QuickCheck Questions

1.1 What is listening to body sounds called?

1.2 What event causes the lubb sound of the cardiac cycle?

1.3 What event causes the dupp sound?

1 Materials

☐ Stethoscope
☐ Alcohol wipes
☐ Laboratory partner

Procedures

Note: From visits to the doctor, you know that a stethoscope is usually handled only by the doctor or nurse, who first places the earpieces in her ears and then moves the stethoscope bell around all over your chest and back. Cardiac auscultation is best achieved with the stethoscope placed against bare skin, but because neither you nor your partner is a medical professional, you may both feel most comfortable in this activity if the "patient" rather than the listener holds the stethoscope bell, slipping it inside his or her own shirt and locating each valve site.

1. Clean the earpieces of the stethoscope with a sterile alcohol wipe, and dispose of the used wipe in a trash can.

2. Wear the stethoscope by placing the angled earpieces facing anterior. The angle directs the earpiece into the external ear canal.

3. Hand the bell to your partner, who should then place it on her or his chest in any one of the four bell positions shown in Figure 37.2a. Have your partner then move the bell around so that you can auscultate the AV and SL valves. Can you discriminate between the lubb and the dupp sounds? ■

LAB ACTIVITY 2 Determining Blood Pressure

Blood pressure is a measure of the force the blood exerts on the walls of the systemic arteries. Arterial pressure increases when the left ventricle contracts and pumps blood into the aorta. When the left ventricle relaxes, less blood flows into the aorta and so arterial pressure decreases until the next ventricular systole. Two pressures are therefore used to express blood pressure, a systolic pressure and a lower diastolic pressure. Average blood pressure is considered to be 120/80 mm Hg (millimeters of mercury) for a typical male and closer to 110/70 for most females. Do not be surprised when you take your blood pressure in the following exercise and discover it is not "average." Cardiovascular physiology is a dynamic mechanism, and pressures regularly change to adjust to the demands of the body.

Blood pressure is measured using an inflatable cuff called a **sphygmomanometer** (sfig-mō-ma-NOM-e-ter). Figure 37.3b demonstrates proper placement of the

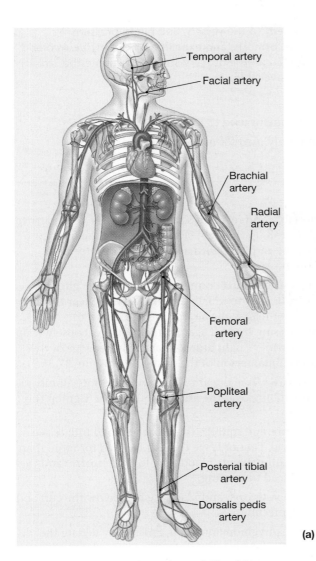

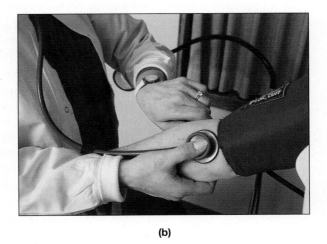

(b)

(a)

Figure 37.3 Checking the Pulse and Blood Pressure
(a) Pressure points used to check the presence and strength of the pulse. **(b)** Use of a sphygmomanometer to check arterial blood pressure.

cuff. It is wrapped around the arm just superior to the elbow and then inflated to approximately 160 mm Hg to compress and block blood flow in the brachial artery. A stethoscope is placed on the antecubital region, and pressure is gradually vented from the cuff. Once pressure in the cuff is slightly less than the pressure in the brachial artery, blood spurts through the artery and the turbulent flow makes sounds, called **Korotkoff's sounds** (kō-ROT-kofs), which are audible through the stethoscope. The pressure on the gauge when the first sound is heard is recorded as the **systolic pressure**. As more pressure is relieved from the cuff, blood flow becomes less turbulent and quieter. The sounds fade when the cuff pressure matches the **diastolic pressure** of the artery.

Clinical Application | ***High Blood Pressure and Salt Intake***

One of the first recommendations a physician gives a patient with **hypertension** is to reduce the dietary intake of salt. A high-salt diet leads to an increase in blood volume. Because the vascular system is a closed system, the additional fluid volume is "trapped" in the vessels, and blood pressure increases as a result. Think about osmosis, and explain how salt increases blood volume. Also, the next time you are in the grocery store, investigate the various salt substitutes currently available. ▶

QuickCheck Questions

2.1 What is the name of the instrument used to measure blood pressure?

2.2 Which blood vessel is commonly used for measuring blood pressure?

2 Materials

- ☐ Sphygmomanometer
- ☐ Stethoscope
- ☐ Alcohol wipes
- ☐ Cot or laboratory table
- ☐ Laboratory partner

Procedures

I. Resting Blood Pressure

1. Have your partner sit comfortably and relax for several minutes. If your partner is wearing a long-sleeved shirt, roll up the right sleeve to expose the upper brachium. Clean the stethoscope earpieces with a sterile alcohol wipe and dispose of the used pad in the trash.

2. The sphygmomanometer consists of a **cuff** connected to a **pressure gauge** by rubber tubing and a **rubber bulb** used to inflate the cuff. A **valve** near the bulb closes or opens the cuff to hold or release the air. Force all air out of the sphygmomanometer by compressing the cuff against a flat surface. Loosely wrap the deflated cuff around your partner's right arm so that the lower edge of the cuff is just superior to the antecubital region of the elbow, as in Figure 37.3b.

3. If the cuff has **orientation arrows**, line the arrows up with the antecubitis; otherwise, position the rubber tubing over the antecubital region. Tighten the cuff so it is snug against the arm.

4. Gently close the valve on the cuff, and squeeze the rubber bulb to inflate the cuff to approximately 160 mm Hg. Do not leave the cuff inflated for more than 1 minute, because the inflated cuff prevents blood flow to the forearm, and the disruption in blood flow could lead to fainting.

5. Put the stethoscope earpieces in your ears, and place the bell below the cuff and over the brachial artery at the antecubitis.

6. Carefully open the valve to the sphygmomanometer, and slowly deflate the cuff while listening for Korotkoff's sounds. When the first sound is heard, note the pressure reading on the gauge and record it in column 2 of Table 37.1.

7. Continue to vent pressure from the cuff and to listen with the stethoscope. When you hear the last faint sound, note the pressure on the gauge and record it in column 2 of Table 37.1.

8. Open the pressure valve completely, quickly finish deflating the cuff, and remove it from your partner's arm.

II. Effect of Posture on Blood Pressure

Changing posture changes the way in which gravity influences blood pressure. This is readily apparent when standing on your head. In this section your partner will lie supine (on the back) for approximately five minutes to allow for cardiovascular adjustments. You will then determine blood pressure and compare your findings with the pressures obtained in procedure I.

Table 37.1 Blood Pressure Measurements

	Resting	Supine	Standing
Measurement 1			
Measurement 2			

Table 37.2 **Effect of Exercise on Blood Pressure**

	Start	*2 min*	*4 min*	*6 min*	*8 min*	*10 min*
BP	_____	_____	_____	_____	_____	_____
MAP	_____	_____	_____	_____	_____	_____

1. Ask your partner to lie on the cot or laboratory table and relax for five minutes.

2. Wrap the sphygmomanometer around your partner's right arm, determine the supine blood pressure, and record it in the two lines of column 3 of Table 37.1.

3. Next, ask your partner to stand up and remain still for five minutes. Take the blood pressure again, and record it in the fourth column of Table 37.1.

4. What was the effect of a supine posture on the blood pressure?

III. Effect of Exercise on Blood Pressure

In this section you will determine the effect of mild exercise on blood pressure. Be sure that you have determined the resting blood pressure of your subject before-hand, as outlined in procedure I. The pressure observed in procedure I will be used as a baseline.

1. Secure the sphygmomanometer cuff around your partner's arm loose enough that it is comfortable for exercise yet is in position for taking pressure readings.

2. Have your partner jog in place for five minutes, without stopping if possible. This is not a stress test; if the subject becomes excessively winded or tired, he or she should stop immediately.

3. When the five minutes are up and your partner has stopped jogging, quickly take a blood pressure reading, and record both the systolic pressure and the diastolic pressure (write it as a fraction if you like) on the BP line of the "Start" column in Table 37.2. Have your partner stand still, and repeat the readings once every two minutes until the pressure returns to the resting values, recording each reading on the BP line of Table 37.2.

4. To compare your blood pressure readings, you must determine the **mean arterial pressure** (MAP). To calculate MAP, first determine the **pulse pressure**, which is the difference between the diastolic and systolic pressures (subtract diastolic pressure from systolic pressure). Next, add one-third of the pulse pressure to the diastolic pressure to get the MAP. Calculate the MAP for each pressure measurement, and record your data on the bottom row of Table 37.2. ∎

LAB ACTIVITY 3   Measuring the Pulse

Heart rate is usually determined by measuring the **pulse,** or **pressure wave**, in an artery. During ventricular systole, blood pressure increases and stretches the walls of arteries. When the ventricle is in diastole, blood pressure decreases and the arterial walls rebound to their relaxed diameter. This change in vessel diameter is felt as a throb—a pulse—at various **pressure points** on the body. The most commonly used pressure point is the radial artery on the lateral forearm just superior to the thumb (Figure 37.3a). Other pressure points include the common carotid artery in the neck and the popliteal artery of the posterior knee. The number of pulses in a given time interval indicates the number of cardiac cycles in that interval. As you will see in Laboratory Activity 5, arterial pulses are related to changes in the volume of blood passing a given point at a given time.

QuickCheck Questions

3.1 What does the pressure wave of a cardiac cycle represent?

3.2 Which events in the cardiac cycle cause the pulse you can feel in your anterior wrist?

3 Materials

☐ Watch or clock with second hand

☐ Laboratory partner

Procedures

1. Have your partner relax for several minutes.

2. Locate your partner's pulse in the right radial artery. Use either your index finger alone or your index and middle fingers to palpitate (feel) the pulse. Do not use your thumb for pulse measurements (because there is a pressure point in the thumb and you might not be able to distinguish your pulse from your partner's).

3. Apply light pressure to the pressure point, and count the pulse rate for 15 seconds. Multiply this number by 4 to obtain the rate per minute. Record your data in Table 37.3.

4. Repeat the pulse determination at the facial artery and the popliteal artery. Record your data in Table 37.3.

5. Is there any difference in the pulse strength at the various pressure points? ■

Table 37.3 Pulse Measurement

Pulse	Pulse Rate (beats/min)
Facial	_____
Radial	_____
Popliteal	_____

LAB ACTIVITY 4   BIOPAC: Electrocardiography

BIOPAC

During each cardiac cycle, a sequence of electrical impulses from pacemaker cells and nerves causes the heart muscle to produce electrical currents, or impulses, that result in contraction of the heart chambers. These impulses can be detected at the body surface with a series of electrodes. In this investigation, you will use the BIOPAC lead II configuration, which has a positive electrode on the left ankle, a negative electrode on the right wrist, and a ground electrode on the right ankle. A recording of the impulses is called an **electrocardiogram**, which is abbreviated either as **ECG** or as **EKG** (this latter abbreviation is an older one not used much today).

The ECG is typically printed on a standard grid, with seconds on the *x* axis and either amplitude or intensity in millivolts (mV) on the *y* axis (Figure 37.4). In the

Figure 37.4
Components of the ECG

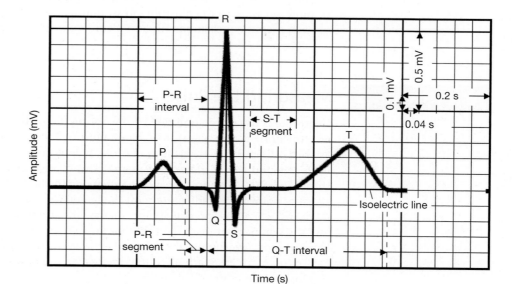

Time (s)

grid shown in Figure 37.4, each small square represents 0.04 s in the horizontal direction and 0.1 mV in the vertical direction.

The basic components of an ECG are a straight baseline, the **isoelectric line**, and waves that indicate periods of depolarization and repolarization of the heart's chambers. During **depolarization**, positively charged sodium ions enter a cell, causing the cell membrane to reverse its internal charge from negative to positive and creating an electrical current. To return to the resting negative condition, the membrane **repolarizes** by allowing positively charged potassium ions to leave the cell. The **P wave** of an ECG occurs as the atria depolarize for contraction. Atrial systole occurs approximately 0.1 s after depolarization. The P wave is followed by a large spike called the **Q-R-S complex**, caused by depolarization of the ventricles. After ventricular systole, the **T wave** results from ventricular repolarization. The ECG returns to the baseline, and the next cardiac cycle soon occurs. Atrial repolarization occurs during the Q-R-S complex and is undetected.

Within the ECG are intervals that include a wave and the return to the baseline. The **P-R interval**, which occurs between the start of the P wave and the start of the Q-R-S complex, is the time required for an impulse to travel from the SA node to the ventricular muscle. The **Q-T interval** is the cycle of ventricular depolarization and repolarization. A **segment** on an ECG is the baseline recording between any two waves. The **P-R segment** represents the time for an impulse to travel from the AV node to the ventricles. The **S-T segment** measures the delay between ventricular depolarization and repolarization. Table 37.4 presents the average length and intensity of each phase of a typical ECG.

If the electrical activity of the heart changes, the ECG will reflect the changes. For example, damage to the AV node results in an extension of the P-R segment. Irregularities in the heartbeat are called **arrhythmias** and may indicate problems in cardiac function.

The ECG laboratory activity is organized into four major procedures. The Setup section describes where to plug in the electrode leads and how to apply the skin electrodes. The Calibration section adjusts the hardware so that it can collect accurate physiological data. Once the hardware has been calibrated, the Data Recording section describes taking ECG recordings with the subject in four situations. After the ECG data has been saved to a computer disk, the Data Analysis section instructs you how to use the software tools to interpret and evaluate the ECG.

 Be sure to read the BIOPAC safety notices and carefully follow the procedures as outlined. *Under no circumstances should you deviate from the experimental procedures.* ▲

Table 37.4 **ECG Values**

Phase	Duration (second)	Amplitude (millivolt)
P wave	0.06–0.11	< 0.25
P-R interval	0.12–0.20	
P-R segment	0.08	
Q-R-S complex (R)	< 0.12	0.8–1.2
S-T segment	0.12	
Q-T interval	0.36–0.44	
T wave	0.16	< 0.5

4 *Materials*

- ☐ BIOPAC acquisition unit (MP35/30)
- ☐ BIOPAC wall transformer (AC100A)
- ☐ BIOPAC serial cable (CBLSERA)
- ☐ BIOPAC electrode lead set (SS2L)
- ☐ BIOPAC disposable vinyl electrodes (EL503), 3 electrodes per subject
- ☐ Cot or laboratory table
- ☐ Chair with armrests
- ☐ BIOPAC electrode gel (GEL1)
- ☐ Skin cleanser or soap and water
- ☐ Computer: minimum Macintosh® 68020 or PC Windows® 98SE or better
- ☐ Biopac Student Lab software V3.0 or higher

Procedures

I. Setup

1. Turn the computer ON.

2. Make sure the acquisition unit is turned OFF, and connect it to a wall outlet with the AC100A transformer.

3. Plug the CBLSERA serial cable into CH 2 of the acquisition unit (Figure 37.5).

4. Turn the acquisition unit ON.

5. Place the three EL503 electrodes on the subject as shown in Figure 37.6, using a small amount of gel (GEL1) on the skin where each electrode will be placed. The subject should not be in contact with any nearby metal objects (faucets, pipes, and so forth) and should remove any metal wrist or ankle bracelets. Attach one electrode on the medial surface of the right leg just above the ankle and one on the medial surface of the left leg just above the ankle. Place the third electrode on the right anterior forearm at the wrist. *Note:* For optimal electrode adhesion, the electrodes should be placed on the skin at least five minutes before the start of the calibration procedure.

6. Attach the three SS2L leads to the electrodes, following the color scheme shown in Figure 37.6: white lead to right wrist, black lead to right leg, red lead to left leg. The pinch connectors on the leads work like clothespins but will latch onto the electrode nipple from only one side.

7. Have the subject lie down on the cot or table and relax. Position the SS2L leads so that they are not pulling on the electrodes. Connect the clip of the CBLSERA cable to a convenient location on the subject's clothes. This will relieve cable strain.

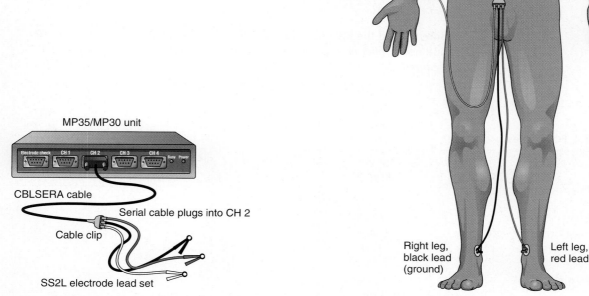

MP35/MP30 unit

CBLSERA cable

Serial cable plugs into CH 2

Cable clip

SS2L electrode lead set

Right forearm, white lead

Right leg, black lead (ground)

Left leg, red lead

Figure 37.5 BIOPAC Acquisition Unit Setup

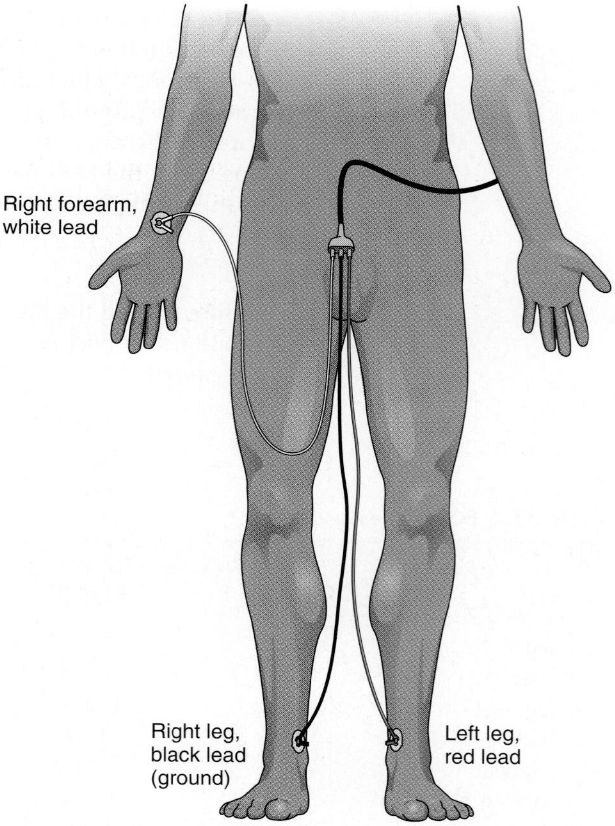

Figure 37.6 Lead II Electrode Placement

8. Start the Biopac Student Lab Program and choose Lesson L05-ECG.1.

9. Type in the subject's filename, which must be a unique identifier, then click OK. This ends the setup procedure.

II. Calibration

Calibration establishes the hardware's internal parameters (such as gain, offset, and scaling) and is critical for optimum performance. Pay close attention to the following steps.

1. Double-check that the electrodes are adhering to the skin, and make sure the subject is relaxed and lying down. If the electrodes are detaching from the skin, you will not get a good ECG signal. In addition, the electrocardiograph is very sensitive to small changes in voltage caused by contraction of skeletal muscles, and therefore the subject's arms and legs must be relaxed so that no muscle signal corrupts the ECG signal.

2. Click on the Calibrate button in the upper left corner of the Setup window. This will start the calibration.

3. Wait for the calibration to stop, which will happen automatically after eight seconds.

4. At the end of the eight-second calibration recording, the screen should resemble Figure 37.7, a greatly reduced ECG waveform with a relatively flat baseline. If your recording is correct, proceed to Data Recording. If incorrect, click on Redo Calibration.

III. Data Recording

You will make four ECG recordings of the subject: lying down, immediately after sitting up, while sitting up and breathing deeply, and after exercise. To work efficiently, read this entire section so that you will know what to do for each recording segment. The subject should remain supine and relaxed while you review the lesson. Check the last line of the journal and note the total amount of time available for the recording. Stop each recording segment as soon as possible so that you do not waste storage memory on the computer hard drive.

Hints for obtaining optimal data:

a. The subject should be relaxed and still, and should not talk or laugh during any recording segment.

b. When asked to sit up, the subject should do so in a chair, with arms relaxed on the armrest (if available).

Figure 37.7
Calibration ECG

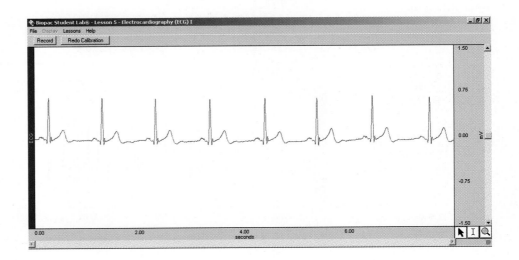

Segment 1 Data Recording—Lying Down

1. Click on Record, and record with the subject lying down motionless for 20 seconds.

2. Click on Suspend to stop the recording.

3. Review the data on the screen. If the screen looks similar to Figure 37.8, proceed to step 4. If the data are incorrect, erase the incorrect recording by clicking Redo, and repeat steps 1 and 2.

Segment 2 Data Recording—Sitting Up

4. Have the subject quickly sit up, and then click the Resume button as quickly as possible after the subject sits up. Do not click Resume while the subject is in the process of sitting up or you will record a muscle artifact. The recording will continue from the point where it last stopped, and a marker labeled "After sitting up" will automatically come up when Resume is pressed.

5. Record for 20 seconds, and then click on Suspend to stop the recording.

6. If the screen looks similar to Figure 37.9, proceed to step 7. If the data are incorrect, click on Redo and repeat steps 4 and 5.

Figure 37.8
Segment 1 ECG: Lying Down

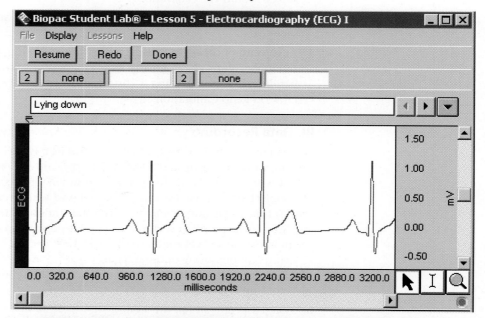

Figure 37.9
Segment 2 ECG: After Sitting Up

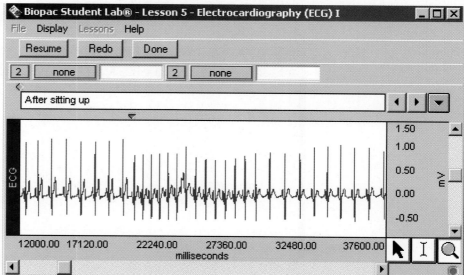

Segment 3 Data Recording—Deep Breathing

7. Have the subject move from the cot or table to the chair and then sit without moving for about two minutes, arms supported comfortably on the armrests. Then click on Resume, and at the moment you click have the subject start a series of slow, prolonged inhalations and exhalations. As you click on Resume, the recording will continue and a marker labeled "Deep Breathing" will be inserted. *Note:* It is important that the subject breathe with long, slow, deep breaths in order to minimize muscle artifacts in the recording.

8. Record for 20 seconds, and have the subject take in 5 deep breaths during the 20 seconds. Insert an "Inhale" marker at each inhalation and an "Exhale" marker at each exhalation.

9. Click on Suspend to stop the recording.

10. If the screen looks similar to Figure 37.10, proceed to step 11. *Note:* The recording may have some baseline drift, as shown in Figure 37.10. Baseline drift is fairly normal, and unless it is excessive, it does not necessitate redoing the recording. If your screen does not resemble Figure 37.10, click Redo and repeat steps 7–9.

Segment 4 Data Recording—After Exercise

11. Have the subject perform either pushups or jumping jacks for about 60 seconds to elevate the heart rate and then sit down in the chair. *Note:* You may remove the SS2L lead pinch connectors so that the subject can move about freely, but do not remove the EL503 electrodes. If you do remove the connectors, reattach them when the subject has finished exercising, following the color scheme of Figure 37.6. To capture the heart rate variation, it is important that you resume recording as quickly as possible after the subject has performed the exercise. However, it is also important that you do not click Resume while the subject is exercising, for doing so will capture motion artifacts on the recording.

12. As soon as the subject has stopped exercising and is seated, click on Resume, and record for 60 seconds. The recording will continue from the point where it last stopped, and a marker labeled "After Exercise" will be inserted.

13. Click on Suspend to stop the recording.

Figure 37.10
Segment 3 ECG: Deep Breathing

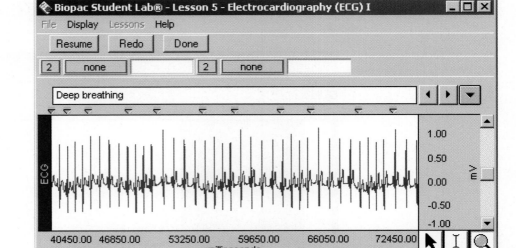

14. If the screen looks similar to Figure 37.11, proceed to step 15. If incorrect, click on Redo. *Note:* The After Exercise recording may have some baseline drift, but unless the drift is excessive, do not redo the recording.

15. Click on Done. A pop-up window with options will appear. Make your choice, and continue as directed. If choosing "Record from another subject," attach the electrodes and leads as described in the setup section, and restart the lesson to recalibrate the hardware for the new subject.

16. Remove the lead pinch connectors from the EL503 electrodes, peel off the electrodes, and throw them away (BIOPAC electrodes are not reusable). Wash the electrode gel residue from the subject's skin, using either skin cleanser or soap and water. The electrodes may leave a slight ring on the skin for a few hours. This is normal and does not indicate anything wrong.

IV. Data Analysis

In this section, you will examine ECG components and measure amplitudes and durations of the ECG components. Interpreting ECGs is a skill; it requires practice to distinguish between normal variation and those arising from medical conditions. Do not be alarmed if your ECG is different from the examples shown or from the tables and figures.

1. Enter the Review Saved Data mode from the Lessons menu. The data window that comes up should look like Figure 37.12. The channel number (CH) designation on the left of the window is CH 2 ECG Lead II.

Figure 37.11
Segment 4 ECG:
After Exercise

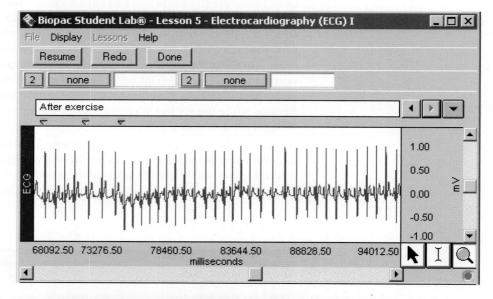

Figure 37.12 Selection of
ECG Waves

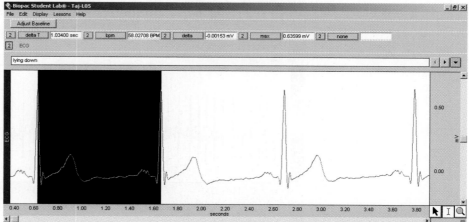

2. Set up your display window for viewing four successive beats from Segment 1 (lying down). The following tools will help you adjust the data window:

Autoscale horizontal   Horizontal (Time) Scroll Bar
Autoscale waveforms   Vertical (Amplitude) Scroll Bar
Zoom tool   Zoom Previous

Turn grids ON and OFF by choosing Display: Preferences from the File menu.

    The Adjust Baseline button allows you to position the waveform up or down in small increments so that the baseline can be exactly zero. This adjustment is not needed to get accurate amplitude measurements, but you may want to make the adjustment before making a printout or when using grids. Once the Adjust Baseline button is pressed, two more buttons, Up and Down, will be displayed. Simply click on these to move the waveform up or down.

3. The measurement boxes are above the marker region in the data window. Each measurement has three sections: channel number, measurement type, and result. The first two sections are pull-down menus that are activated when you click on them. Set up the measurement boxes as follows:

Channel	Measurement
CH 2	ΔT (delta time, difference in time between end and beginning of any area selected by I-beam tool)
CH 2	BPM (beats per minute, calculates ΔT in seconds and converts this value to minutes)
CH 2	Δ(delta amplitude, difference in amplitude between end and beginning of any area selected by I-beam tool)
CH 2	max (maximum amplitude, calculates maximum amplitude within any selected area, including endpoints)

4. Using the I-beam cursor, select and measure the area from one R wave peak to the next R wave peak as precisely as possible. Figure 37.12 shows an example of the selected area. Record your ΔT and BPM data in Table 37.5, located in the BIOPAC Electrocardiography Laboratory Report.

5. Take measurements at two other intervals in the current recording, again recording your data in Table 37.5.

6. Use the Zoom tool to zoom in on a single cardiac cycle from Segment 1 (lying down).

7. Use the I-beam cursor and measurement-box values (and refer to Figure 37.4 as necessary) to record the amplitudes and durations listed below. Record your data in Table 37.6, located in the BIOPAC Electrocardiography Laboratory Report. Figure 37.13 shows a sample setup for measuring P wave amplitude. Note that the Δ measurement shows the amplitude difference between endpoints in the selected area.

Durations		Amplitudes
P wave	Q-T interval	P wave
P-R interval	S-T segment	T wave
Q-R-S interval	T wave	Q-R-S wave

8. Repeat measurements as required for the BIOPAC Electrocardiography Laboratory Report. Follow the examples shown in Figure 37.13 to complete all the measurements required for Tables 37.7 through 37.10.

9. Save or print the data file. You may save the data to a floppy drive, save notes that are in the journal, or print the data file.

10. Exit the program. ■

Figure 37.13
Measurement of P Wave Amplitude

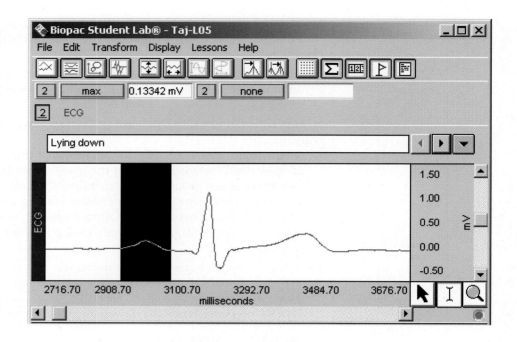

LAB ACTIVITY 5 BIOPAC: Electrocardiography and Blood Volume

BIOPAC

Each cardiac cycle pumps pressurized blood into the vascular system. This continual surge of blood from the heart creates a pressure wave, measured as the pulse, as you saw if you completed Laboratory Activity 3. A **plethysmogram** (PLĒ-thiz-mō-gram) is a recording of how the volume of blood at a given pressure point in the body changes as a pressure wave passes through that point. As the wave flows along the artery, it causes the vessel wall at any given point to first expand and then rebound to its original size.

In this activity, you will use a photoelectric transducer that passes light into the skin and measures how much light is reflected back. Blood absorbs light, and as the expansion part of a pressure wave passes through a given point along an artery, the increased blood volume at that point absorbs proportionally more light. The BIOPAC equipment converts the reflected light signals to electrical signals representing the pressure wave. Therefore, any change in the amplitude in the photoelectric transducer is directly proportional to the volume of blood in the pressure wave. In this activity you will collect ECG and blood-volume data from a subject under various physiological conditions.

The activity is organized into four major procedures. The Setup section describes where to plug in the electrode leads, where and how to apply the skin electrodes, and how to wrap the transducer on the subject's finger. The Calibration section adjusts the hardware so it can collect accurate physiological data. The Data Recording section describes taking ECG and blood-volume recordings with the subject in three positions. After the data have been saved to a computer disk, the Data Analysis section instructs you how to use the software tools to interpret and evaluate the data.

Be sure to read the BIOPAC safety notices and carefully follow the procedures as outlined. *Under no circumstances should you deviate from the experimental procedures.* ▲

5   *Materials*

- ☐ BIOPAC acquisition unit (MP35/30)
- ☐ BIOPAC wall transformer (AC100A)
- ☐ BIOPAC serial cable (CBLSERA)
- ☐ BIOPAC electrode lead set (SS2L)
- ☐ BIOPAC disposable vinyl electrodes (EL503), 3 electrodes per subject
- ☐ BIOPAC pulse plethysmograph (SS4LA or SS4L)
- ☐ Cot or laboratory table
- ☐ Chair with armrests
- ☐ Ruler or measuring tape, calibrated in centimeters
- ☐ Ice water or warm water in *plastic* bucket (*not* metal bucket)
- ☐ BIOPAC electrode gel (GEL1)
- ☐ Soft cloth
- ☐ Skin cleanser or soap and water
- ☐ Computer: minimum Macintosh® 68020 or PC Windows® 98SE or better
- ☐ Biopac Student Lab software V3.0 or higher

Procedures

I. Setup

1. Turn the computer ON.
2. Make sure the acquisition unit is turned OFF, and connect it to a wall outlet with the AC100A transformer.
3. Plug the CBLSERA serial cable into CH 1 and the SS4LA or SS4L pulse transducer into CH 2 (Figure 37.14).
4. Turn the acquisition unit ON.
5. Place the three EL503 electrodes on the subject as shown in Figure 37.6, using a small amount of gel (GEL1) on the skin where each electrode will be placed. The subject should not be in contact with any nearby metal objects (faucets, pipes, and so forth) and should remove any metal wrist or ankle bracelets. Attach one electrode on the medial surface of the right leg just above the ankle and one on the medial surface of the left leg just above the ankle. Place the third electrode on the right anterior forearm at the wrist. *Note:* For optimal electrode adhesion, the electrodes should be placed on the skin at least five minutes before the start of the calibration procedure.
6. Attach the three SS2L leads to the electrodes, following the color scheme shown in Figure 37.6: white lead to right wrist, black lead to right leg, and red lead to left leg. The pinch connectors on the leads work like clothespins but will latch onto the electrode nipple from only one side of the connector.
7. Use the piece of cloth to clean the window of the transducer sensor.
8. Position the transducer so that the sensor is on the bottom of the fingertip (the part without the fingernail) of the index finger of one of the subject's hands, and wrap the tape around the finger so that the transducer fits snugly but not so tightly that blood circulation is cut off (Figure 37.15).
9. With the ruler or measuring tape, measure two distances: from the fingertip where the sensor is attached to the subject's shoulder and from the shoulder to the middle of the sternum. Record these two distances in item 3 of Section A, Data and Calculations, in the BIOPAC Electrocardiography and Blood Volume Laboratory Report.
10. Have the subject lie down on the cot or table and relax. Position the SS2L leads so that they are not pulling on the electrodes. Connect the clip of the CBLSERA cable to a convenient location on the subject's clothes. This will relieve cable strain.
11. Start the Biopac Student Lab Program and choose Lesson L07-ECG&P-1.
12. Type in the subject's filename, which must be a unique identifier, then click OK. This ends the setup procedure.

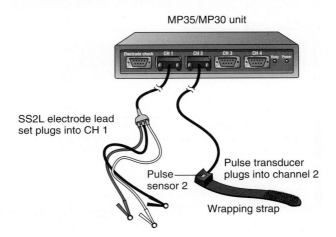

Figure 37.14 BIOPAC Acquisition Unit Setup

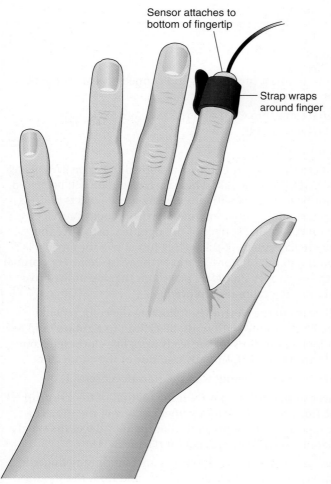

Sensor attaches to bottom of fingertip

Strap wraps around finger

Figure 37.15
Placement of Pulse Transducer

II. Calibration

1. Double-check that the electrodes are adhering to the skin, and make sure the subject is relaxed and lying down. If the electrodes are detaching from the skin, you will not get a good ECG signal. In addition, the electrocardiograph is very sensitive to small changes in voltage caused by contraction of skeletal muscles, and therefore the subject's arms and legs must be relaxed so that no muscle signal corrupts the ECG signal.

2. Click on the Calibrate button in the upper left corner of the Setup window. This will start the calibration.

3. Wait for the calibration to stop, which will happen automatically after eight seconds.

4. At the end of the eight-second calibration recording, the screen should resemble Figure 37.16: a greatly reduced ECG waveform with a relatively flat baseline in the upper band and waveforms in the pulse (blood volume) band. If your recording is correct, proceed to Data Recording. If incorrect, click on Redo Calibration.

III. Data Recording

Have the subject sit in the chair and relax, with arms on the armrests. You will record ECG on Channel 1 and changes in blood volume on Channel 2 under three conditions: arm relaxed, temperature change, and arm up. (The volume changes will be measured not directly but rather indirectly, as pressure-wave pulses in the subject's finger.) Check the last line of the journal and note the total amount of time available for the recording. Stop each recording segment as soon as possible so that you do not waste storage memory on the computer hard drive.

Hints for minimizing both baseline drift and muscle corruption of the ECG:

a. The subject should remain still and relaxed during each recording segment, because the recording from the pulse transducer is sensitive to motion and the ECG recording is sensitive to muscle artifacts.

Figure 37.16
Calibration ECG and Pulse

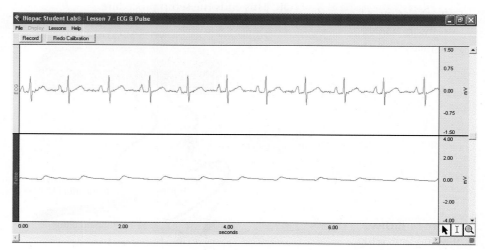

b. The subject should be quiet for each recording segment.

c. Initially, the subject's forearms should be supported on the chair's armrests.

d. Always stop recording *before* the subject prepares for the next recording segment.

e. Make sure the electrodes do not peel up from the skin.

Segment 1—Seated

1. After the subject has been sitting relaxed for several minutes, with arms on the chair armrests, click on Record, and let the hardware collect data for 15 seconds.

2. Click on Suspend to stop the recording.

3. If the screen looks similar to Figure 37.17, proceed to step 4. If the screen does not look like this, click Redo and repeat steps 1 and 2.

Segment 2—Seated with Hand in Water

4. Have the subject remain seated and place the nonrecording hand in the warm or cold water

 The container for the water cannot be metal, as a metal container could bypass the electrical isolation of the system.

5. Click on Resume, and record for 30 seconds. The recording will continue from the point where it last stopped, and a marker labeled "Seated, one hand in water" will automatically appear on the screen.

6. Click on Suspend to stop the recording.

7. If the screen looks similar to Figure 37.18, proceed to step 8. If the screen does not resemble the figure, click on Redo and repeat steps 4–6.

Segment 3—Seated with Hand Raised

8. While remaining seated, have the subject raise the nonrecording hand to extend the arm above the head and hold that position for the duration of the recording.

9. Click on Resume. The recording will continue from the point where it last stopped, and a marker labeled "Seated, arm raised above head" will automatically come up.

10. Record for 60 seconds.

11. Click on Suspend to stop the recording.

12. If the screen looks similar to Figure 37.19, proceed to step 13. If the screen does not resemble this figure, click on Redo and repeat steps 8–11.

Figure 37.17
Segment 1: Arm Relaxed

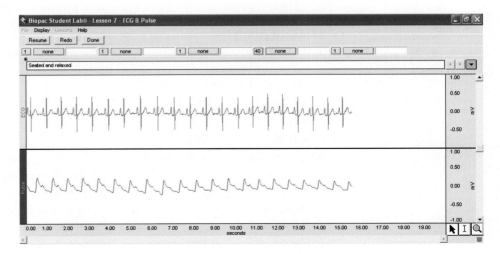

Figure 37.18
Segment 2: One Hand in Water

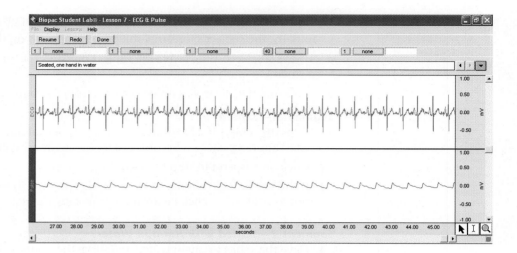

Figure 37.19
Segment 3: Arm Raised Above Head

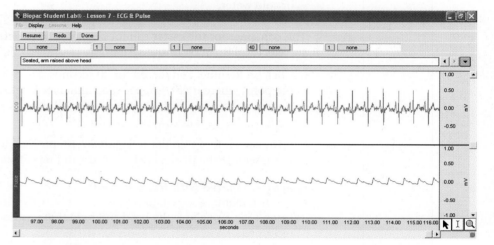

13. Click on Done. A pop-up window with four options will appear. Make your choice, and continue as directed. If choosing the "Record from another subject" option, attach fresh electrodes and the transducer as described above in the setup section. Have the new subject sit and relax, and continue from Setup step 11.

14. Remove the transducer from the subject's finger. Disconnect the electrodes from their leads, peel off the electrodes, and throw them away. Wash the electrode gel residue from the subject's skin, using either the cleanser or soap and water. The electrodes may leave a slight ring on the skin for a few hours. This is normal and does not indicate anything wrong.

IV. Data Analysis

1. Enter the Review Saved Data mode from the Lessons menu. The window that comes up should look the same as that shown in Figure 37.20. Note Channel Number (CH) designation:

Channel	Displays
CH 1	ECG
CH 40	Pulse

2. Set up your display window for optimal viewing of the entire recording. The following tools will help you adjust the window display:

Autoscale horizontal	Horizontal (Time) Scroll Bar
Autoscale waveforms	Vertical (Amplitude) Scroll Bar
Zoom tool	Zoom Previous

Grids—turn grids ON and OFF by choosing Preferences from the File menu.

Figure 37.20
Sample ECG and Pulse Data

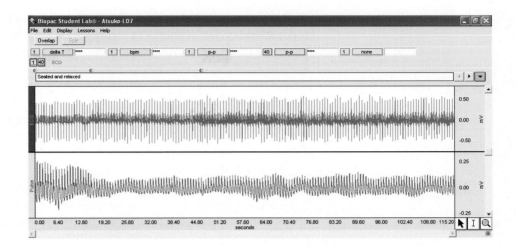

3. The measurement boxes are above the marker region in the data window. Each measurement has three sections: channel number, measurement type, and result. The first two sections are pull-down menus that are activated when you click on them. Set up the measurement boxes as follows:

Channel	Measurement
CH 1	ΔT (delta time, difference in time between end and beginning of any area selected by I-beam tool)
CH 1	BPM (beats per minute, calculates ΔT in seconds and converts this value to minutes)
CH 1	p-p (finds maximum value in selected area and subtracts minimum value found in selected area)
CH 40	p-p

4. Zoom in on a small section of the Segment 1 data. Be sure to zoom in far enough that you can easily measure the intervals between peaks for approximately four cardiac cycles.

5. Using the I-beam cursor, select the area between two successive R waves (one cardiac cycle). Try to go from one R wave peak to the adjacent R wave peak as precisely as possible (Figure 37.21).

6. Measure ΔT and BPM for the selected area, and record your data in the "R-R interval" and "heart rate" portions of Table 37.11, located in the BIOPAC Electrocardiography and Blood Volume Laboratory Report.

Figure 37.21
Measurement Between R Wave Peaks

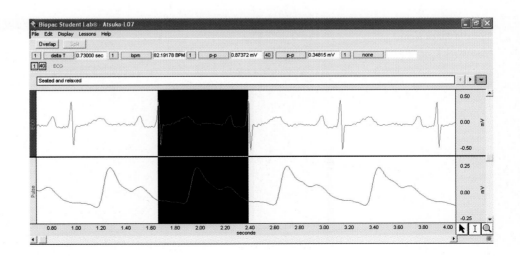

7. Using the I-beam cursor, select the area between two successive pulse peaks (one cardiac cycle), as shown in Figure 37.22. Measure ΔT and BPM for the selected area, and record your data in the "pulse interval" and "pulse rate" portions of Table 37.11.

8. Repeat the ΔT and BPM measurements for each data segment, and record your data in Table 37.11.

9. Select an individual pulse peak for each segment, and determine its amplitude, using the CH 40 p-p measurements (Figure 37.23). Record your data in Table 37.12, located in the BIOPAC Electrocardiography and Blood Volume Laboratory Report.

Important: Measure the first pulse peak after the recording is resumed. The body's homeostatic regulation of blood pressure and volume occurs quickly. The increase or decrease in your results will depend on the timing of your data relative to the speed of physiological adjustments.

10. Using the I-beam cursor, select the interval between one R wave and the adjacent pulse peak. Record the time interval (ΔT) between the two peaks in the Electrocardiography and Blood Volume Laboratory Report, Section A Data and Calculations, item 3.

11. Save or print the data file. You may save the data to a floppy drive, save notes that are in the journal, or print the data file.

12. Exit the program. ■

Figure 37.22
Measurement Between Two Pulse Peaks

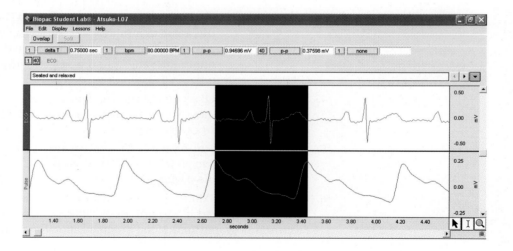

Figure 37.23
Sample Data from Segment 1

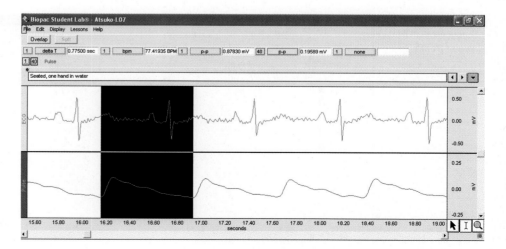

Cardiovascular Physiology

Name _____

Date _____

Section _____

A. Matching

Match each term in the left column with its correct description from the right column.

_____ **1.** cardiac cycle   **A.** unusual heart sound
_____ **2.** diastole   **B.** contraction phase of heartbeat
_____ **3.** systole   **C.** single heartbeat
_____ **4.** auscultation   **D.** listening to body sounds
_____ **5.** Korotkoff's sounds   **E.** closed during ventricular systole
_____ **6.** first heart sound   **F.** relaxation phase of heartbeat
_____ **7.** second heart sound   **G.** sounds heard in stethoscope
_____ **8.** murmur   **H.** sound due to closure of AV valves
_____ **9.** semilunar valve   **I.** closed during ventricular diastole
_____ **10.** atrioventricular valve   **J.** sound due to closure of SL valves

B. Short-Answer Questions

1. Briefly describe the events of a cardiac cycle.

2. Why does blood pressure fluctuate in arteries?

3. Describe how a sphygmomanometer is used to determine blood pressure.

4. Describe how blood pressure changes during exercise and during rest.

5. Calculate the MAP for a blood pressure of 130 mm/85 mm.

BIOPAC:
Electrocardiography

Name _____

Date _____

Section _____

A. Data and Calculations

Subject Profile

Name _____   Height _____

Age _____   Weight _____

Sex: Male / Female

1. Complete the following tables with the lesson data indicated, and calculate the mean and range as appropriate.

Segment 1—Supine, resting, regular breathing

Table 37.5

Measurement	From Channel	Cardiac Cycle 1	2	3	Mean	Range
ΔT	CH 2	_____	_____	_____	_____	_____
BPM	CH 2	_____	_____	_____	_____	_____

Table 37.6

ECG Component	Duration ΔT [CH 2] Cycle 1	Cycle 2	Cycle 3	Mean	Amplitude (mV) Δ [CH 2] Cycle 1	Cycle 2	Cycle 3	Mean
P wave	_____	_____	_____	_____	_____	_____	_____	_____
P-R interval	_____	_____	_____	_____	_____	_____	_____	_____
P-R segment	_____	_____	_____	_____	_____	_____	_____	_____
Q-R-S complex	_____	_____	_____	_____	_____	_____	_____	_____
Q-T interval	_____	_____	_____	_____	_____	_____	_____	_____
S-T segment	_____	_____	_____	_____	_____	_____	_____	_____
T wave	_____	_____	_____	_____	_____	_____	_____	_____

Table 37.7

Ventricular Readings	CH 2 ΔT Cycle 1	Cycle 2	Cycle 3	Mean
Q-T interval (corresponds to ventricular systole)	_____	_____	_____	_____
End of T wave to subsequent R wave (corresponds to ventricular diastole)	_____	_____	_____	_____

LABORATORY REPORT

Segment 2—Supine, deep breathing

Table 37.8

Rhythm	CH	Cycle 1	Cycle 2	Cycle 3	Mean
Inhalation					
ΔT	CH 2	_____	_____	_____	_____
BPM	CH 2	_____	_____	_____	_____
Exhalation					
ΔT	CH 2	_____	_____	_____	_____
BPM	CH 2	_____	_____	_____	_____

Segment 3—Sitting

Table 37.9

Heart Rate	CH	Cycle 1	Cycle 2	Cycle 3	Mean
ΔT	CH 2	_____	_____	_____	_____
BPM	CH 2	_____	_____	_____	_____

Segment 4—After exercise

Table 37.10

Ventricular Readings	CH 2 ΔT			
	Cycle 1	Cycle 2	Cycle 3	Mean
Q-T interval *(corresponds to ventricular systole)*	_____	_____	_____	_____
End of T wave to subsequent R wave *(corresponds to ventricular diastole)*	_____	_____	_____	_____

B. Data Summary and Questions

1. Is there always one P wave for every Q-R-S complex?   Yes   No
2. Describe the shape of a P wave and of a T wave:

3. Do the wave durations and amplitudes for all subjects fall within the normal ranges listed in Table 37.4?   Yes   No
4. Do the S-T segments mainly measure between −0.1 mV and 0.1 mV?   Yes   No
5. Is there any baseline drift in the recording?   Yes   No

C. Analysis and Application

1. Summarize the heart rate data in the space provided below. Explain the changes in heart rate as conditions change, and describe the physiological mechanisms causing these rate changes.

Condition	Heart Rate (BPM)	
	Mean	Range
Supine, regular breathing	_____	_____
Supine, deep breathing, inhalation	_____	_____
Supine, deep breathing, exhalation	_____	_____
Sitting, regular breathing	_____	_____
After exercise, start of recording	_____	_____
After exercise, end of recording	_____	_____

2. Duration (ΔT)

Condition	ΔT	
	Mean	Range
Supine, regular breathing		
Inhalation	_____	_____
Exhalation	_____	_____
Supine, deep breathing		
Inhalation	_____	_____
Exhalation	_____	_____

Are there differences in the cardiac cycle with the respiratory cycle?

Condition	QT Interval	
	Mean	Range
Supine, regular breathing		
Ventricular systole	_____	_____
Ventricular diastole	_____	_____
After exercise		
Ventricular systole	_____	_____
Ventricular diastole	_____	_____

What changes do you observe between the duration of systole and diastole with the subject resting and the duration of systole and diastole after the subject has exercised?

BIOPAC:
Electrocardiography
and Blood Volume

Name _____

Date _____

Section _____

A. Data and Calculations

Subject Profile

Name _____   Height _____

Age _____   Weight _____

Sex: Male / Female

1. Comparison of ECG with plethysmogram (Segments 1–3)
   Complete Table 37.11 with data from three cycles from each segment, and calculate the average.

Table 37.11

Condition	Measurement			Cycle 1	Cycle 2	Cycle 3	Average
Arm relaxed	R-R interval	ΔT	CH 1	_____	_____	_____	_____
Segment 1	Heart rate	BPM	CH 1	_____	_____	_____	_____
	Pulse interval	ΔT	CH 1	_____	_____	_____	_____
	Pulse rate	BPM	CH 1	_____	_____	_____	_____
Temp. change	R-R interval	ΔT	CH 1	_____	_____	_____	_____
Segment 2	Heart rate	BPM	CH 1	_____	_____	_____	_____
	Pulse interval	ΔT	CH 1	_____	_____	_____	_____
	Pulse rate	BPM	CH 1	_____	_____	_____	_____
Arm up	R-R interval	ΔT	CH 1	_____	_____	_____	_____
Segment 3	Heart rate	BPM	CH 1	_____	_____	_____	_____
	Pulse interval	ΔT	CH 1	_____	_____	_____	_____
	Pulse rate	BPM	CH 1	_____	_____	_____	_____

2. Blood volume changes (Segments 1–3)
   Complete Table 37.12 with data from each recording segment.

Table 37.12

Measurement	Arm resting Segment 1	Temperature change Segment 2	Arm up Segment 3
Q-R-S amplitude	_____	_____	_____
Pulse amplitude (mV)	_____	_____	_____

3. Calculation of Pulse Speed

Distance between subject's sternum and shoulder: _____ cm

Distance between subject's shoulder and fingertip: _____ cm

Total distance from sternum to fingertip: _____ cm

Segment 1 Data

Time between R wave and pulse peak: _____ s

Speed: _____ cm/s

Segment 3 Data

Time between R wave and pulse peak: _____ s

Speed: _____ cm/s

B. Short-Answer Questions

1. Are the values of heart rate and pulse rate given in Table 37.11 similar for each condition or different for each condition? Propose an explanation for any similarity or difference you observe.

2. Determine how much the Q-R-S amplitude values recorded in Table 37.12 changed as conditions changed:

   Segment 2 amplitude minus segment 1 amplitude: _____ mV

   Segment 3 amplitude minus segment 1 amplitude: _____ mV

3. Determine how much the pulse amplitude values recorded in Table 37.12 changed as arm position changed:

   Segment 2 amplitude minus segment 1 amplitude: _____ mV

   Segment 3 amplitude minus segment 1 amplitude: _____ mV

4. Does the amplitude of the Q-R-S complex change as the pulse amplitude changes? Why or why not?

5. Describe one mechanism that causes changes in blood volume to your fingertip.

6. In your calculation of pulse speed in part A, item 3, of this Laboratory Report, did you find a difference between the segment 1 speed and the segment 3 speed? If yes, explain the reason for the difference.

7. Which components of the cardiac cycle (atrial systole and diastole, ventricular systole and diastole) are discernible in the pulse tracing?

8. Would you expect the pressure-wave velocities measured in your own body to be very close to those measured in other students? Why or why not?

9. Explain any pressure-wave amplitude or frequency changes that occurred with changes in arm position.

Lymphatic System

OBJECTIVES

On completion of this exercise, you should be able to:

- List the functions of the lymphatic system.

- Describe the exchange of blood plasma, interstitial fluid, and lymph.

- Describe the structure of a lymph node.

- Explain how the lymphatic system drains into the vascular system.

- Describe the gross anatomy and basic histology of the spleen.

The lymphatic system includes the lymphatic vessels, lymph nodes, tonsils, spleen, and thymus gland (Figure 38.1). **Lymphatic vessels** transport a liquid called **lymph** from the tissue spaces to the veins of the cardiovascular system. Scattered along each lymphatic vessel are **lymph nodes** containing lymphocytes and phagocytic macrophages. The macrophages remove invading bacteria and other substances from the lymph before the lymph is returned to the blood. Although lymphocytes, one type of white blood cells, are a formed element of the blood, they are the main cells of the lymphatic system and colonize dense populations in lymph nodes and the spleen. The antigens present in invading bacteria and other foreign substances cause the body to produce antibodies. As macrophages capture these foreign antigens in the lymph, lymphocytes exposed to the antigens activate the immune system to respond to the intruding cells. B-cell lymphocytes produce antibodies that chemically combine with and destroy the antigens.

The thymus gland is involved in the development of the functional immune system in infants. In adults, this gland controls the maturation of lymphocytes. We covered the thymus gland with the endocrine system in Exercise 33.

Figure 38.2 illustrates an overview of fluid circulation in the body. Pressure in the capillaries that are part of the systemic arterial network forces fluids and solutes out of the capillaries and into the interstitial spaces. This filtrate, called either **extracellular fluid** or **interstitial fluid**, bathes the cells with nutrients, dissolved gases, hormones, and other materials. After this exchange, osmotic pressure forces most of the extracellular fluid back into the capillaries. The

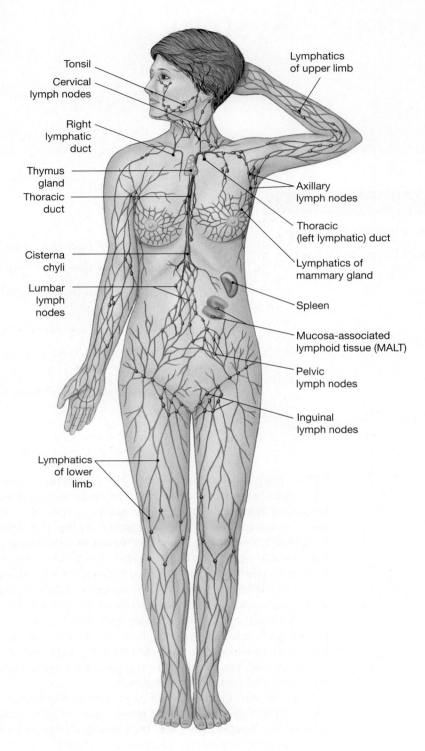

Tonsil

Cervical
lymph nodes

Right
lymphatic
duct

Thymus
gland

Thoracic
duct

Cisterna
chyli

Lumbar
lymph
nodes

Lymphatics
of lower
limb

Lymphatics
of upper limb

Axillary
lymph nodes

Thoracic
(left lymphatic) duct

Lymphatics of
mammary gland

Spleen

Mucosa-associated
lymphoid tissue (MALT)

Pelvic
lymph nodes

Inguinal
lymph nodes

**Figure 38.1 Components
of the Lymphatic System**

portion that does not reenter the capillaries enters the lymphatic system instead, slowly diffusing into the lymphatic vessels and becoming lymph. The lymph travels through the lymphatic vessels to the lymph nodes, where phagocytes remove abnormal cells and microbes from the lymph. A pair of **lymphatic ducts** joins with veins near the heart and returns the lymph to the blood. Approximately three liters of fluid per day are forced out of the capillaries and flow through lymphatic vessels as lymph.

Figure 38.2
Fluid Connective Tissues
The functional relationships between blood and lymph: Blood travels through the circulatory system, pushed by the contractions of the heart. In capillaries, hydrostatic (blood) pressure forces fluid and dissolved solutes out of the circulatory system. This fluid mixes with the extracellular fluid already in the tissue. Extracellular fluid slowly enters lymphatic vessels; now called lymph, it travels along the lymphatics and reenters the circulatory system at one of the veins that returns blood to the heart.

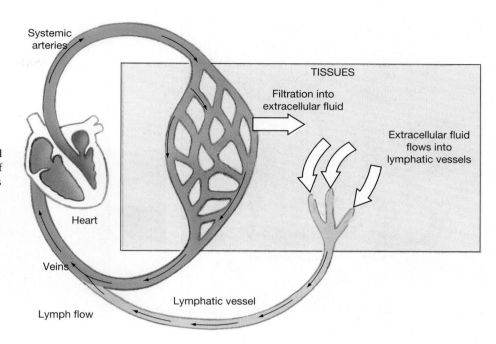

LAB ACTIVITY 1   Lymphatic Vessels

Lymphatic vessels, or simply **lymphatics,** occur next to the vessels of the vascular system, as Figure 38.3a shows. Lymphatic vessels are structurally similar to veins. The vessel wall has similar layers and valves to prevent backflow of fluid. Lymphatic pressures are very low, and the lymphatic valves are close together to keep the lymph circulating toward the body trunk. As Figure 38.3a shows, the lymphatic capillaries, which gradually expand to become the lymphatic vessels, are closed at one end, with the closed ends lying near the arterial blood capillaries. Lymph enters the lymphatic system at or near these closed ends and then moves into the lymphatic vessels, which conduct the lymph toward the body trunk and into even larger lymphatics that empty into veins near the heart.

Two large lymphatic vessels, the **thoracic duct** and the **right lymphatic duct**, return lymph to the venous circulation (Figure 38.1 and 38.4). Most of the lymph is returned to the circulation by the thoracic duct, which commences at the level of the second lumbar vertebra on the posterior abdominal wall behind the abdominal aorta. Lymphatics from the legs, pelvis, and abdomen drain into an inferior saclike portion of the duct called the **cisterna chyli** (KĪ-lē; *cistern,* storage well; *chyl,* juice). The thoracic duct pierces the diaphragm and ascends to the base of the heart, where it joins with the left subclavian vein to return the lymph to the blood. The only lymph that does not drain into the thoracic duct is from lymphatic vessels in the right arm and the right side of the chest, neck, and head. These areas drain into the right lymphatic duct near the right clavicle. This duct empties lymph into the right subclavian vein located near the base of the heart.

QuickCheck Questions

1.1 What is lymph?

1.2 Where is lymph returned to the vascular system?

1.3 What is the function of lymphocytes?

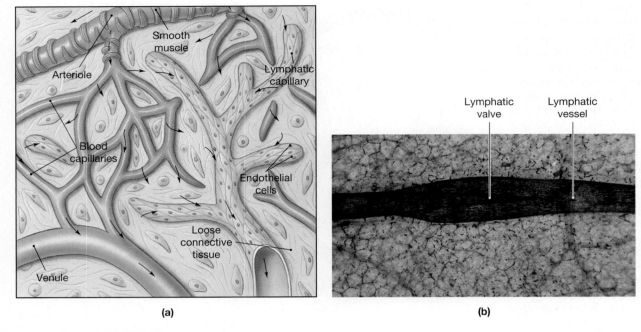

(a)

(b)

Figure 38.3 Lymphatic Capillaries
(a) A three-dimensional view of the association of blood capillaries, tissue, interstitial fluid, and lymphatic capillaries. Arrows show the direction of interstitial fluid and lymph movement. **(b)** A valve within a small lymphatic vessel (LM × 43).

1 *Materials*

☐ Torso model or lymphatic system chart

Procedures

1. Locate the thoracic duct and the cisterna chyli on a torso model or lymphatic chart, and label them in Figure 38.4. Which areas of the body drain lymph into this duct? Where does this duct return lymph to the blood?

2. Locate the right lymphatic duct on the torso model or lymphatic chart, and label the duct in Figure 38.4. Where does this vessel join the vascular system? Which regions of the body drain lymph into this duct?

3. Complete the labeling of Figure 38.4. ■

LAB ACTIVITY 2 Lymphoid Tissues and Lymph Nodes

Two major groups of lymphatic structures occur in connective tissues: **encapsulated lymph organs** and **diffuse lymphoid tissues**. The encapsulated lymph organs include lymph nodes, the thymus gland, and the spleen. Each encapsulated organ is separated from the surrounding connective tissue by a fibrous capsule. Diffuse lymphoid tissues do not have a defined boundary separating them from the connective tissue.

A *lymph node* is an oval, nodular organ that functions like a filter cartridge. As lymph passes though the node, phagocytes remove microbes, debris, and other antigens from the lymph. Lymph nodes are scattered throughout the lymphatic system, as depicted in Figure 38.1. Lymphatics from the lower limbs pass through a network of **inguinal** lymph nodes. **Pelvic** and **lumbar** nodes filter lymph from the pelvic and abdominal lymphatics. Many lymph nodes occur in the upper limbs and in the **axillary** and **cervical** regions. The breasts in women also have many lymphatic vessels and nodes. Infections often occur in a lymph node before they spread systemically. A swollen or painful lymph node suggests an increase in lymphocyte abundance and general immunological activity in response to antigens in the lymph nodes.

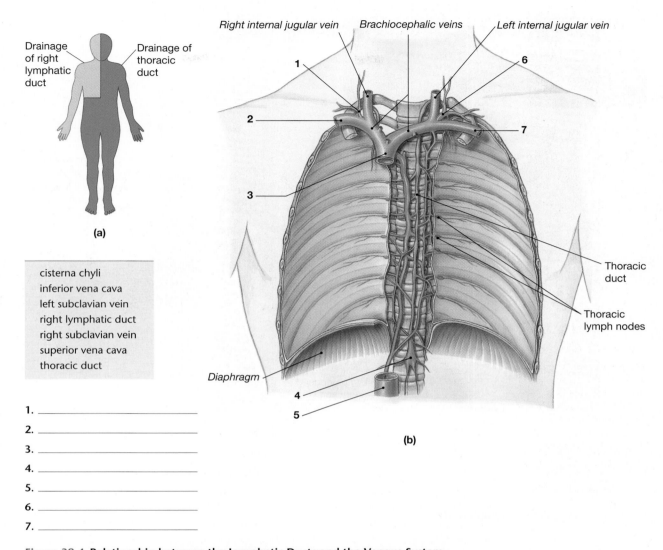

Right internal jugular vein   *Brachiocephalic veins*   *Left internal jugular vein*

Drainage
of right
lymphatic
duct

Drainage of
thoracic
duct

(a)

Thoracic
duct

Thoracic
lymph nodes

Diaphragm

(b)

cisterna chyli
inferior vena cava
left subclavian vein
right lymphatic duct
right subclavian vein
superior vena cava
thoracic duct

1. _____
2. _____
3. _____
4. _____
5. _____
6. _____
7. _____

Figure 38.4 Relationship between the Lymphatic Ducts and the Venous System
(a) The thoracic duct carries lymph originating in tissues inferior to the diaphragm and
from the left side of the upper body. The smaller right lymphatic duct services the rest of
the body. **(b)** The lymphatic vessels of the trunk. The thoracic duct empties into the left
subclavian vein. The right lymphatic duct drains into the right subclavian vein.

Figure 38.5 details the organization of a lymph node. Each node is encased in
a dense connective tissue **capsule**. Collagen fibers from the capsule extend as par-
titions called **trabeculae** into the interior of the node. The outer **cortex** contains
sinuses full of lymph fluid. Lymphocytes are produced in and surround many pale-
staining **germinal centers**. Deep to the cortex is the **medulla**, where **medullary
cords** of lymphocytes extend into sinuses. Lymph enters a node in **afferent
lymphatic vessels**. As the lymph flows through the cortical and medullary sinuses,
macrophages phagocytize abnormal cells, pathogens, and debris. Draining the
lymph node are **efferent lymphatic vessels,** which exit the node at a slit called
the **hilus**.

Lymphoid nodules, which are diffuse lymphoid tissue, are found in connective tissue under the lining of the digestive, urinary, and respiratory systems. Microbes that penetrate the exposed epithelial surface pass into lymphoid nodules and
into the lymph fluid, where lymphocytes and macrophages destroy and remove the

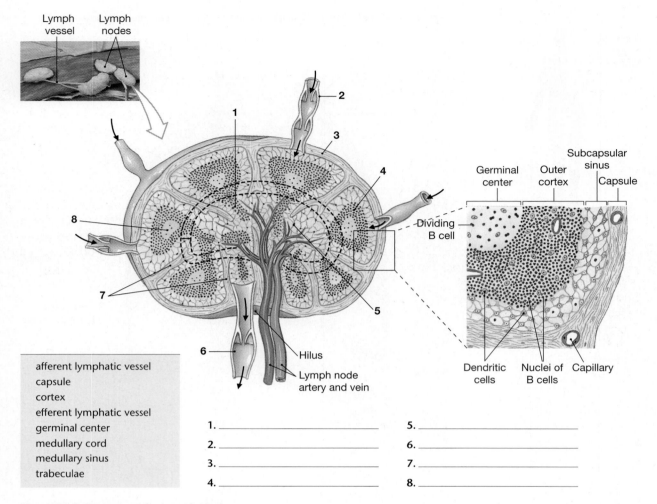

Figure 38.5 **Structure of a Lymph Node**

afferent lymphatic vessel
capsule
cortex
efferent lymphatic vessel
germinal center
medullary cord
medullary sinus
trabeculae

1. _____   5. _____

2. _____   6. _____

3. _____   7. _____

4. _____   8. _____

foreign cells from the lymph. Some nodules have a germinal center where lymphocytes are produced by cell division.

   Tonsils are lymphoid nodules in the pharynx. A pair of **lingual tonsils** occurs at the posterior base of the tongue. The **palatine tonsils** are easily viewed hanging off the posterior arches of the oral cavity. A single **pharyngeal tonsil**, or **adenoids**, is located in the upper pharynx near the opening to the nasal cavity.

Clinical Application

Tonsillitis

The lymphatic system usually has the upper hand in the immunological battle against invading bacteria and viruses. Occasionally, however, microbes manage to populate a lymphoid nodule. When tonsils are infected, they swell and become irritated. This condition is called **tonsillitis** and is treated with antibiotics to control the infection. If the problem is recurrent, the tonsils are removed in a surgical procedure called a *tonsillectomy*. Usually, the palatine tonsils are removed. If the pharyngeal tonsils are also infected or are abnormally large, they are removed, too. ◗

QuickCheck Questions

2.1 What are the names of the two types of lymphatic structures in the body?

2.2 Where are lymph nodes located?

2   *Materials*

☐ Compound
    microscope

☐ Prepared slide of
    lymph node

Procedures

1. Review the structure of a lymph node shown in Figure 38.5, and then label the figure.

2. Examine the lymph node slide at low magnification, and identify the capsule and trabeculae.

3. Change the microscope to high magnification, and examine a germinal center inside the cortex. What types of cells are produced in the germinal center? ∎

LAB ACTIVITY   3   The Spleen

The **spleen** is the largest lymphatic organ in your body. It is located lateral to the stomach along the greater curvature (Figure 38.6). The spleen is surrounded by a **capsule** that protects the underlying tissue. This underlying tissue, called **pulp,** consists of **red pulp** and **white pulp**. Branches of the **splenic artery**, called **trabecular arteries**, are surrounded by white pulp, which contains large populations

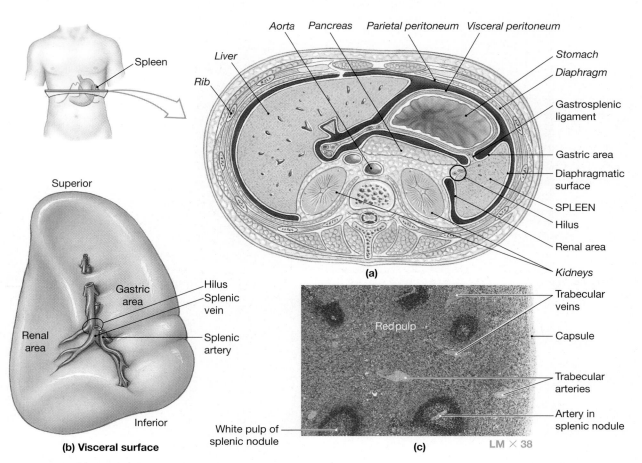

Figure 38.6 The Spleen

(a) A transverse section through the trunk, showing the typical position of the spleen within the abdominopelvic cavity. The shape of the spleen roughly conforms to the shapes of adjacent organs. **(b)** The external appearance of the intact spleen, showing major anatomical landmarks. Compare this view with that of part (a). **(c)** The histological appearance of the spleen. White pulp is dominated by lymphocytes; it appears blue because the nuclei of lymphocytes stain very dark. Red pulp contains a preponderance of red blood cells (LM × 38).

of lymphocytes. Capillaries of the trabecular arteries open into the red pulp. As blood flows through the red pulp, free and fixed phagocytes remove abnormal red blood cells and other antigens.

QuickCheck Questions

3.1 Where is the spleen located?

3.2 What tissues are in the white pulp?

3.3 Which vessels open into the red pulp?

3 Materials

- ☐ Torso model or chart showing spleen
- ☐ Compound microscope
- ☐ Prepared slide of spleen

Procedures

1. Locate the spleen on the torso model or chart. Identify the hilus, splenic artery, and splenic vein. On the visceral surface, locate the gastric area of the spleen, which is in contact with the stomach, and the renal area, which is in contact with the kidneys.

2. Examine the spleen slide at low magnification, and identify the dark-stained regions of white pulp and the lighter regions of red pulp. Examine several white pulp masses for the presence of an artery. ■

Lymphatic System

Name _____

Date _____

Section _____

A. Matching

Match each structure in the left column with its correct description from the right column.

_____	**1.** efferent vessel	**A.** empties into right subclavian vein
_____	**2.** medullary cords	**B.** empties into lymph node
_____	**3.** cisterna chyli	**C.** splenic tissue containing red blood cells
_____	**4.** right lymphatic duct	**D.** fluid in lymphatic vessels
_____	**5.** red pulp	**E.** lymphocytes deep in node
_____	**6.** lymph node	**F.** empties into left subclavian vein
_____	**7.** thoracic duct	**G.** full of macrophages and lymphocytes
_____	**8.** white pulp	**H.** drains lymph node
_____	**9.** lymph	**I.** lymphocytes surrounding trabecular artery
_____	**10.** afferent vessel	**J.** saclike region of thoracic duct

B. Short-Answer Questions

1. Describe the organization of a lymph node.

2. Discuss the major functions of the lymphatic system.

3. Explain how lymph is returned to the blood.

4. Describe the anatomy of the spleen.

38

C. Analysis and Application

1. How are blood plasma, extracellular fluid, and lymph interrelated?

2. How does the way lymph drains from the right thoracic duct differ from the way it drains from the left thoracic duct?

3. How can cancer cells spread by entering the lymphatic system?

Anatomy of the Respiratory System

OBJECTIVES

On completion of this exercise, you should be able to:

- Identify and describe the structures of the nasal cavity.

- Distinguish among the three regions of the pharynx.

- Identify and describe the cartilages and ligaments of the larynx.

- Identify the gross and microscopic structure of the trachea.

- Classify the branches of the bronchial tree.

- Identify and describe the gross and microscopic structure of the lungs.

All cells require a constant supply of oxygen (O_2) to maintain cellular respiration. A major waste product of cellular metabolism is carbon dioxide (CO_2). The respiratory system is responsible for oxygenating the blood and removing carbon dioxide from it.

The respiratory system, shown in Figure 39.1, consists of the nose, nasal cavity and sinuses, pharynx, larynx, trachea, bronchi, and lungs. The **upper respiratory system** includes the nose, nasal cavity, sinuses, and pharynx. These structures filter, warm, and moisten air before it enters the **lower respiratory system**, which comprises the larynx, trachea, bronchi, and lungs. The larynx regulates the opening into the lower respiratory system and produces speech sounds. The trachea and bronchi maintain an open airway to the lungs. In the lungs, millions of alveolar sacs exchange O_2 and CO_2 gases with the blood in pulmonary capillaries.

LAB ACTIVITY **1**   The Nose and Pharynx

The primary route for air entering the respiratory system is through two openings, the **external nares** (NĀR-ēz), or nostrils (Figure 39.2). Just inside each external naris is an expanded **vestibule** (VES-ti-byool) containing coarse hairs. The hairs help to prevent large airborne materials like dirt and insects from entering the respiratory system. A midsagittal **nasal septum** divides the nasal cavity. This bony septum is formed by union of the **perpendicular plate** of the ethmoid with the vomer. The external portion of the nose is composed of **nasal cartilages** that form the bridge and tip of the nose and part of the nasal septum. The nasal cavity is a vibrating, or resonating, chamber for the voice.

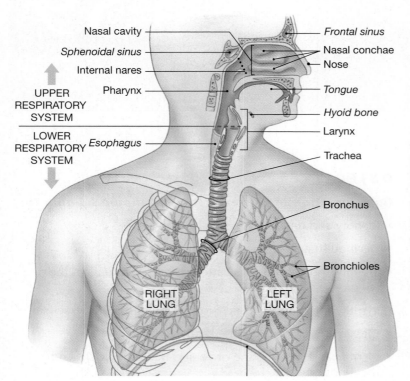

Nasal cavity
Sphenoidal sinus
Internal nares
Pharynx

UPPER
RESPIRATORY
SYSTEM
LOWER
RESPIRATORY   *Esophagus*
SYSTEM

Frontal sinus
Nasal conchae
Nose
Tongue
Hyoid bone
Larynx
Trachea
Bronchus
Bronchioles

RIGHT
LUNG
LEFT
LUNG

Figure 39.1 The Components of the Respiratory System

The conductng portion of the respiratory system is shown here. The smaller bronchioles and alveoli are not shown.

The **superior**, **middle**, and **inferior nasal conchae** are bony shelves that project from the lateral walls of the nasal cavity. The bone of each concha curls underneath and forms a tube, or **meatus**, that causes air to swirl in the nasal cavity. This turbulence moves inhaled air across the sticky epithelial lining, where dust and debris are removed. The floor of the nasal cavity is the superior portion of the **hard palate**, formed by the maxillae, palatine bones, and muscular **soft palate**. Hanging off the posterior edge of the soft palate is the conical **uvula** (Ū-vū-luh). Two openings, the **internal nares**, lead from the nasal cavity into the uppermost portion of the throat.

The throat, or **pharynx** (FAR-inks), is divided into three regions: nasopharynx, oropharynx, and laryngopharynx. The **nasopharynx** (nā-zō-FAR-inks) is located above the soft palate and serves as a passageway for airflow from the nasal cavity. A **pseudostratified ciliated columnar epithelium** lines the nasopharynx, trachea, and bronchial passageways and functions to warm, moisten, and clean inhaled air. The pharynx is connected to the oral cavity at an opening called the **fauces** (FAW-sēz). When a person is eating, food pushes past the uvula, and the soft palate raises to prevent the food from entering the nasopharynx. Located on the posterior wall of the nasopharynx is a single **pharyngeal tonsil**, also called the *adenoids*. On the lateral walls are the openings of the **auditory (pharyngotympanic) tubes**, commonly called the *eustachian tubes*. These tubes allow equalization of pressure between the air in the middle chamber of the ears and the air external to the body.

The **oropharynx**, which extends from the soft palate down to the epiglottis, contains the **palatine** and **lingual tonsils**. The **laryngopharynx** (la-rin-gō-FAR-inks) is located between the hyoid bone and the entrance to the esophagus. Because food passes through the oropharynx and laryngopharynx, these areas are lined with a **stratified squamous epithelium**.

QuickCheck Questions

1.1 Name the components of the upper respiratory system.

1.2 Name the components of the lower respiratory system.

1.3 Describe the passageways into and out of the nasal cavity.

1.4 List the three regions of the pharynx.

1 Materials

☐ Head model
☐ Respiratory system chart
☐ Hand mirror

Procedures

1. Review the gross anatomy of the nose in Figure 39.2, and then label items 1, 5, and 9–12 in this figure. Locate these structures on the head model and respiratory system chart.

2. Review the three regions of the pharynx in Figure 39.2, and then complete the labeling of this figure. Locate each region on the head model.

3. Using a hand mirror, examine the inside of your mouth. Locate your hard and soft palates, uvula, palatine tonsils, and oropharynx. ■

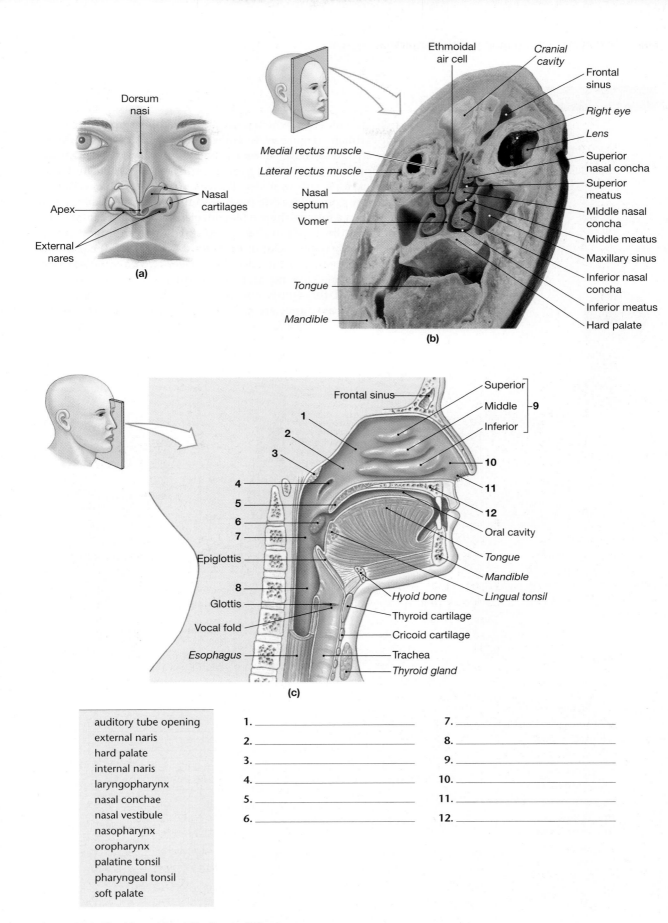

(a)

Dorsum nasi

Nasal cartilages

Apex

External nares

(b)

Ethmoidal air cell

Cranial cavity

Frontal sinus

Right eye

Lens

Superior nasal concha

Superior meatus

Middle nasal concha

Middle meatus

Maxillary sinus

Inferior nasal concha

Inferior meatus

Hard palate

Medial rectus muscle

Lateral rectus muscle

Nasal septum

Vomer

Tongue

Mandible

(c)

Frontal sinus

Superior

Middle

Inferior

9

1

2

3

4

5

6

7

8

10

11

12

Oral cavity

Tongue

Mandible

Lingual tonsil

Epiglottis

Glottis

Vocal fold

Esophagus

Hyoid bone

Thyroid cartilage

Cricoid cartilage

Trachea

Thyroid gland

auditory tube opening
external naris
hard palate
internal naris
laryngopharynx
nasal conchae
nasal vestibule
nasopharynx
palatine tonsil
pharyngeal tonsil
soft palate

1. _____
2. _____
3. _____
4. _____
5. _____
6. _____

7. _____
8. _____
9. _____
10. _____
11. _____
12. _____

Figure 39.2 The Nose, Nasal Cavity, and Pharynx

(a) The nasal cartilages and external landmarks on the nose. **(b)** The meatuses, the maxillary sinuses, and the ethmoidal air cells of the ethmoidal labyrinth. **(c)** The nasal cavity and pharynx, as seen in sagittal section with the nasal septum removed.

LAB ACTIVITY 2   The Larynx

The **larynx** (LAR-inks), or voice box, which joins the laryngopharynx with the trachea, consists of nine cartilages (Figure 39.3). Three of them—the thyroid cartilage, the epiglottis, and the cricoid cartilage—form the body of the larynx. The **thyroid cartilage**, or Adam's apple, is composed of hyaline cartilage. It is visible under the skin on the anterior neck, especially in males. The **cricoid** (KRĪ-koyd) **cartilage** is a ring of hyaline cartilage connecting the trachea to the base of the larynx. The **epiglottis** (ep-i-GLOT-is) is a tongue-shaped piece of elastic cartilage that falls over the opening, or **glottis,** of the larynx during swallowing to prevent ingested food from entering the respiratory tract. A pair of **arytenoid** (ar-i-TĒ-noyd) **cartilages** articulate with the superior border of the cricoid cartilage. Two **corniculate** (kor-NIK-ū-lāt) **cartilages** articulate with the arytenoid cartilages and are involved in the opening and closing of the glottis and in the production of sound. The **cuneiform** (kū-NĒ-i-form) **cartilages** are club-shaped cartilages anterior to the corniculate cartilages.

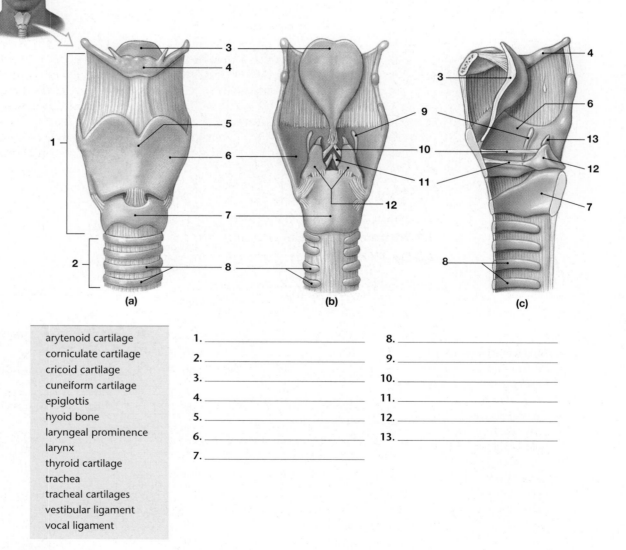

(a)   (b)   (c)

arytenoid cartilage
corniculate cartilage
cricoid cartilage
cuneiform cartilage
epiglottis
hyoid bone
laryngeal prominence
larynx
thyroid cartilage
trachea
tracheal cartilages
vestibular ligament
vocal ligament

1. _____
2. _____
3. _____
4. _____
5. _____
6. _____
7. _____

8. _____
9. _____
10. _____
11. _____
12. _____
13. _____

Figure 39.3 The Larynx

(a) Anterior view. (b) Posterior view. (c) A sagittal section through the larynx.

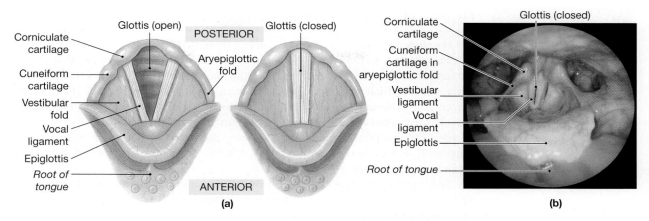

Figure 39.4 The Glottis
(a) A diagrammatic superior view of the entrance to the larynx, with the glottis open (left) and closed (right). **(b)** A fiber-optic view of the entrance to the larynx, corresponding to the right-hand image in part **(a)**. Note that the glottis is almost completely closed by the vocal ligaments.

Spanning the glottis from the thyroid cartilage are two pairs of folds (Figure 39.4). The superior ones are the **vestibular ligaments**, the false vocal cords. They prevent foreign materials from entering the glottis and tightly close the glottis during coughing and sneezing. The inferior ones are the **vocal ligaments**, also called the *true vocal cords*. To produce sounds, air is exhaled over the true vocal cords. Laryngeal muscles adjust the tension on the cords to change the pitch or frequency of the sound. Tightly stretched vocal cords produce a high-pitched sound.

QuickCheck Questions

2.1 How many pieces of cartilage are in the larynx?

2.2 What are the glottis and the epiglottis?

2.3 Describe the structures that produce the voice.

2 *Materials*

☐ Larynx model

☐ Torso model

☐ Respiratory system chart

Procedures

1. Label the gross anatomy of the larynx in Figure 39.3.

2. Examine the larynx on the larynx model, the torso model, or the respiratory system chart.

   • Locate the thyroid and cricoid cartilages. Are they continuous around the larynx?

   • Study the position of the epiglottis. How does it act like a chute to direct food into the esophagus?

   • Open the larynx model, and identify the arytenoid, corniculate, and cuneiform cartilages.

   • Locate the vestibular and vocal ligaments.

3. Put your finger on your thyroid cartilage and swallow. How does the cartilage move when you swallow? Is it possible to swallow and make a sound simultaneously?

4. While holding your thyroid cartilage, make first a high-pitched sound and then a low-pitched sound. Describe the tension in your throat muscles for each sound, and relate the muscle tension to the tension on the vocal ligaments. ■

The Trachea and Bronchial Tree

The **trachea** (TRĀ-kē-uh), or windpipe, is a tubular structure approximately 11 cm (4.25 in.) long and 2.5 cm (1 in.) in diameter (Figure 39.5). It lies anterior to the esophagus and can be felt on the front of the neck below the thyroid cartilage of the larynx. Along the length of the trachea are 15 to 20 C-shaped rings of hyaline cartilage, called the **tracheal cartilage**, that keep the airway open. The **trachealis muscle** holds the two sides of the tracheal cartilage together posteriorly. This muscle allows the esophagus diameter to increase during swallowing so that the esophagus wall presses against the adjacent trachea wall and decreases the trachea diameter momentarily.

The trachea, like the nasal cavity and the bronchial tubes, is lined with pseudostratified ciliated columnar epithelium containing many **goblet cells**. The goblet cells secrete a sticky mucus that covers the ciliated epithelium and traps particles present in the inhaled air. The **cilia** beat to move the mucus back up to the pharynx, where it is swallowed. This process of mucus movement is called the **mucus escalator**. In addition to the pseudostratified ciliated columnar epithelium, the mucosal lining of the trachea also includes the **lamina propria** of connective tissue.

The trachea divides, at a ridge called the **carina**, into the left and right **primary bronchi** (BRONG-kī). Each bronchus branches into increasingly smaller passageways to conduct air into the lungs (Figure 39.6). This branching pattern

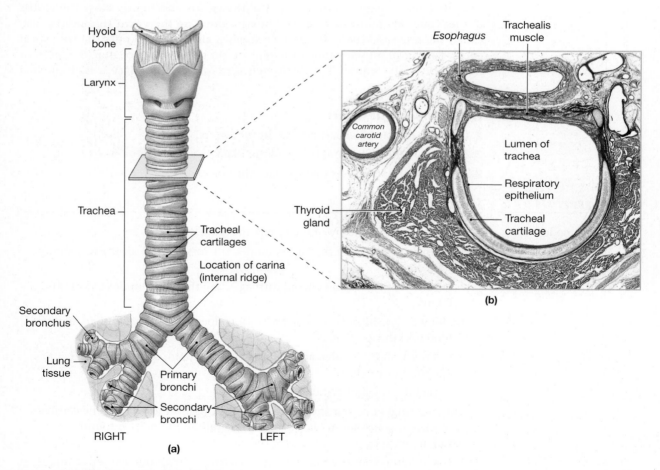

Figure 39.5 The Anatomy of the Trachea
(a) A diagrammatic anterior view. (b) Cross-sectional view (LM × 241).

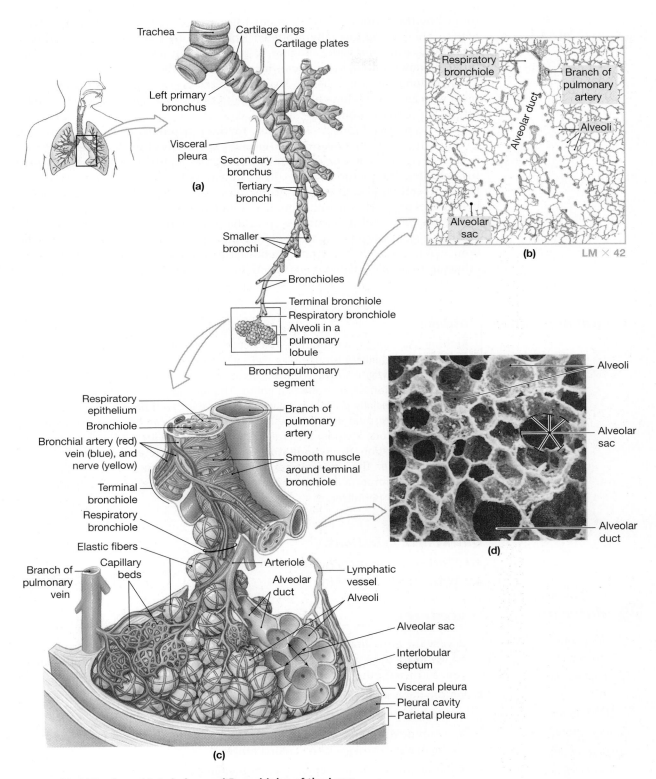

Figure 39.6 The Bronchi, Lobules, and Bronchioles of the Lung

(a) The branching pattern of bronchi in the left lung, simplified. (b) The distribution of a respiratory bronchiole supplying a portion of a lobule. (c) The structure of a single pulmonary lobule, part of a bronchopulmonary segment. (d) A SEM of the lung. Notice the open, spongy appearance of the lung tissue. Compare with (c).

formed by the divisions of the bronchial structures is called the **bronchial tree**. Objects that are accidentally aspirated (inhaled) often enter the right primary bronchus because it is wider and more vertical than the left bronchus. The primary bronchi branch into as many **secondary bronchi** as there are lobes in each lung. The right lung has three lobes, and each lobe receives a secondary bronchus to supply it with air. The left lung has two lobes, and thus two secondary bronchi branch off the left primary bronchus.

The secondary bronchi divide into **tertiary bronchi**, also called *segmental bronchi,* which enter the **bronchopulmonary segments** of the lobes (Figure 39.6). Smaller divisions called **bronchioles** enter lobules and branch into **respiratory bronchioles**. These narrow tubules divide into **alveolar ducts** that connect to clusters of **alveoli** (al-VĒ-o-lī) called **alveolar sacs**. Exchanges of gases occur across the membranes of the alveoli and pulmonary capillaries.

As the bronchial tree branches from the primary bronchi to the bronchioles, cartilage is gradually replaced with smooth muscle tissue. The epithelial lining of the bronchial tree also changes from pseudostratified ciliated columnar in the primary bronchus to simple squamous epithelium in the alveoli.

Clinical Application

Asthma

Asthma (AZ-muh) is a condition that occurs when the smooth muscle encircling the delicate bronchioles contracts and reduces the diameter of the airway. The airway is further compromised by increased mucus production and inflammation of the epithelial lining. The individual has difficulty breathing, especially during exhalation, as the narrowed passageways collapse under normal respiratory pressures. An asthma attack can be triggered by a number of factors, including allergies, chemical sensitivities, air pollution, stress, and emotion. **Bronchodilator** drugs are used to relax the smooth muscle and open the airway; other drugs reduce inflammation of the mucosa. *Albuterol* is an important bronchodilator, usually administered as an inhalant sprayed from a nebulizer.

QuickCheck Questions

3.1 What is the lining epithelium of the trachea?

3.2 What is the bronchial tree?

3   Materials

- ☐ Compound microscope
- ☐ Prepared slide of trachea
- ☐ Head model
- ☐ Torso model
- ☐ Lung model
- ☐ Respiratory system chart

Procedures

1. Review the gross anatomy of the trachea in Figure 39.5. Locate these structures on the head, torso, and lung models and the respiratory system chart. Palpate your trachea for the tracheal cartilages.

2. On the trachea slide, locate the structures labeled in Figure 39.5b.

   a. Identify the rings forming the tracheal cartilage.

   b. Examine the mucosa deep to the tracheal cartilage. The thin layer next to the hyaline cartilage is the lamina propria.

   c. Locate the respiratory epithelium at the lumen of the trachea. What type of epithelium occurs here?

   d. Sketch a section of the trachea in the space provided on page 611.

3. Study the bronchial tree on the laboratory models, and identify the primary bronchi, secondary bronchi, tertiary bronchi, bronchioles, and respiratory bronchioles. ■

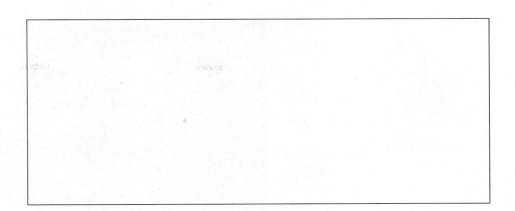

LAB ACTIVITY 4 The Lungs

The lungs are a pair of cone-shaped organs lying in the thoracic cavity (Figure 39.7). The **apex** is the conical top of each lung, and the broad inferior portion is the **base**. The **lateral surface** of each lung faces the thoracic cage, and the **medial surface** faces the mediastinum. The heart lies on a medial concavity of the left lung called the **cardiac notch**. Each lung has a **hilus**, a medial slit where the bronchial tubes, vascularization, lymphatics, and nerves reach the lung. Each lung is divided into lobes by deep fissures. The **oblique fissure** of the left lung separates the **superior** and **inferior lobes**. The right has three lobes, the **superior**, **middle**, and **inferior lobes**. The **horizontal fissure** separates the superior and middle lobes, and the **oblique fissure** separates the inferior lobe.

The mediastinum divides the thoracic cavity into two **pleural cavities**, each containing a lung isolated in its own serous membrane (Figure 39.8). Each pleural cavity is lined with a serous membrane, the **pleura**. The **parietal pleura** lines the thoracic wall, and the **visceral pleura** covers the superficial surface of the lung. The pleurae produce a slippery serous fluid that reduces friction and adhesion between the lungs and the thoracic wall during breathing.

The **alveolar walls** are constructed of simple squamous pulmonary epithelium. Scattered among the epithelium are **surfactant** (sur-FAK-tant) cells that secrete an oily coating to prevent the alveoli from sticking together after exhalation. Also in the alveolar wall are macrophages that phagocytize debris. Pulmonary capillaries cover the exterior of the alveoli, and gas exchange occurs across the thin walls. Oxygen from inhaled air diffuses through the simple pulmonary epithelium of the alveolar wall, moves across the basement membrane and the endothelium of the capillary, and enters the bloodstream. This membrane is about 0.5 mm thick and permits rapid gas exchange between the alveoli and blood.

QuickCheck Questions

4.1 How many lobes does each lung have?

4.2 Which lung has the cardiac notch?

4 Materials

- ☐ Compound microscope
- ☐ Prepared slide of lung
- ☐ Torso model
- ☐ Respiratory system chart

Procedures

1. Review the gross anatomy of the lungs in Figure 39.7, and then label that figure. Locate these structures on the torso models and on the respiratory system chart.

2. On the model, examine the right lung, and observe how the horizontal and oblique fissures divide it into three lobes. Note how the oblique fissure separates the left lung into two lobes.

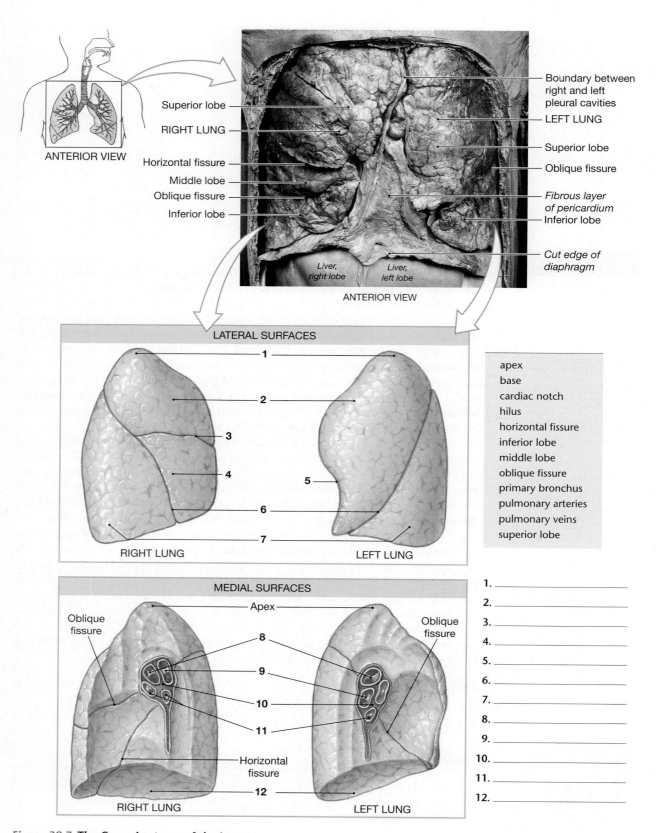

ANTERIOR VIEW

Superior lobe
RIGHT LUNG
Horizontal fissure
Middle lobe
Oblique fissure
Inferior lobe

Boundary between right and left pleural cavities
LEFT LUNG
Superior lobe
Oblique fissure
Fibrous layer of pericardium
Inferior lobe
Cut edge of diaphragm

Liver, right lobe   *Liver, left lobe*

ANTERIOR VIEW

LATERAL SURFACES

1
2
3
4
5
6
7

RIGHT LUNG   LEFT LUNG

apex
base
cardiac notch
hilus
horizontal fissure
inferior lobe
middle lobe
oblique fissure
primary bronchus
pulmonary arteries
pulmonary veins
superior lobe

MEDIAL SURFACES

Oblique fissure   Apex   Oblique fissure

8
9
10
11

Horizontal fissure

12

RIGHT LUNG   LEFT LUNG

1. _____
2. _____
3. _____
4. _____
5. _____
6. _____
7. _____
8. _____
9. _____
10. _____
11. _____
12. _____

Figure 39.7 The Gross Anatomy of the Lungs

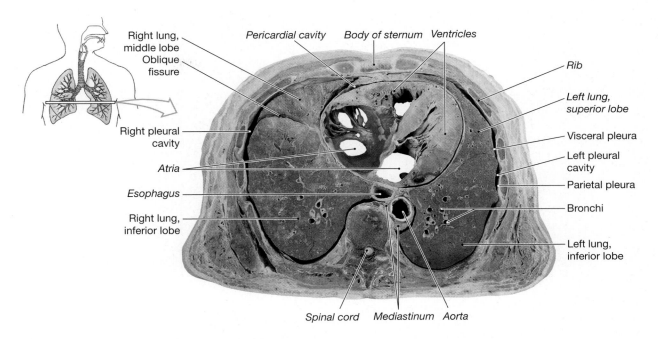

Figure 39.8 The Relationship between the Lungs and the Heart
This transverse section was taken at the level of the cardiac notch.

3. Examine the model for the parietal pleura on the thoracic wall. Where is the pleural cavity relative to the parietal pleura?

4. On the prepared slide,

   a. Identify the alveoli and an alveolar duct, using Figure 39.6 as a reference.

   b. Locate an area where the alveoli appear to have been scooped out. This passageway is an alveolar duct. Follow the duct to its end, and observe the many alveolar sacs serviced by the duct.

   c. At the opposite end of the duct, look for the thicker wall of the respiratory bronchiole and blood vessels.

   d. In the space provided below, sketch an alveolar duct and several alveolar sacs. ■

Anatomy of the Respiratory System

Name _____

Date _____

Section _____

A. Matching

Match each structure in the left column with its correct description from the right column.

_____	**1.**	C-shaped rings
_____	**2.**	internal nares
_____	**3.**	cricoid cartilage
_____	**4.**	pleurae
_____	**5.**	epiglottis
_____	**6.**	larynx
_____	**7.**	vocal ligament
_____	**8.**	cardiac notch
_____	**9.**	external nares
_____	**10.**	three lobes
_____	**11.**	thyroid cartilage
_____	**12.**	vestibular ligament

A. voice box
B. elastic cartilage flap of larynx
C. serous membrane of lungs
D. left lung
E. connects nasal cavity with throat
F. tracheal cartilage
G. true vocal cord
H. false vocal cord
I. nostrils
J. base of larynx
K. right lung
L. Adam's apple

B. Labeling

Label the anatomy presented in Figure 39.9.

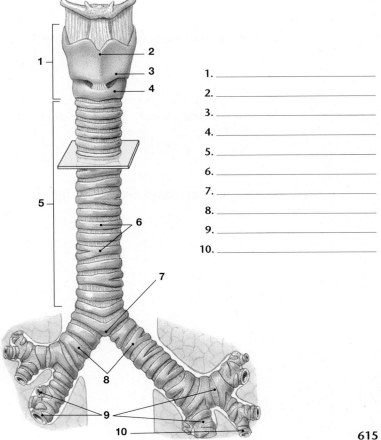

1. _____
2. _____
3. _____
4. _____
5. _____
6. _____
7. _____
8. _____
9. _____
10. _____

Figure 39.9
The Anatomy of the Trachea
A diagrammatic anterior view.

RIGHT   LEFT

LABORATORY REPORT

C. Short-Answer Questions

1. List the components of the upper and lower respiratory systems.

2. What are the functions of the superior, middle, and inferior conchae?

3. Where is the pharyngeal tonsil located?

4. Trace a breath of air from the external nares through the respiratory system to the alveolar sacs.

D. Analysis and Application

1. Where do goblet cells occur in the respiratory system, and what function do they serve?

2. Why are the oropharynx and laryngopharynx lined with stratified squamous epithelium?

3. How are speech sounds produced in the larynx?

4. Why does an asthma attack cause difficulty in breathing?

Physiology of the Respiratory System

OBJECTIVES

On completion of this exercise, you should be able to:

- Discuss pulmonary ventilation, internal respiration, and external respiration.

- Describe how the respiratory muscles move during inspiration and expiration.

- Define the various lung capacities and explain how they are measured.

- Show how to use a dry spirometer.

- Observe, record, and/or calculate selected pulmonary volumes and capacities.

- Use a pneumograph transducer and air temperature probe.

- Show how ventilation is related to temperature changes in airflow through the nostrils.

- Observe and record modifications in the rate and depth of the breathing cycle.

Respiration has three phases: pulmonary ventilation, external respiration, and internal respiration. Breathing, or **pulmonary ventilation**, is the movement of air into and out of the lungs. This movement requires coordinated contractions of the diaphragm, intercostal muscles, and abdominal muscles. **External respiration** is the exchange of gases between the lungs and the blood. Inhaled air is rich in oxygen, and this gas constantly diffuses through the alveolar wall of the lungs into the blood of the pulmonary capillaries. Simultaneously, carbon dioxide diffuses out of the blood and into the lungs, from where it is exhaled. The freshly oxygenated blood is pumped to the tissues to deliver the oxygen and take up carbon dioxide. **Internal respiration** is the exchange of gases between the blood and the tissues.

Pulmonary ventilation consists of inspiration and expiration. **Inspiration** is inhalation, the movement of oxygen-rich air into the lungs. **Expiration**, or exhalation, involves emptying the carbon dioxide-laden air from the lungs into the atmosphere. The average respiratory rate is approximately 12 breaths per minute. This rate is modified by many factors, however, such as exercise and stress, which increase the rate, and sleep and depression, which decrease it.

For pulmonary ventilation to take place, the pressure in the thoracic cavity must be different from **atmospheric pressure,** which is the pressure of the air

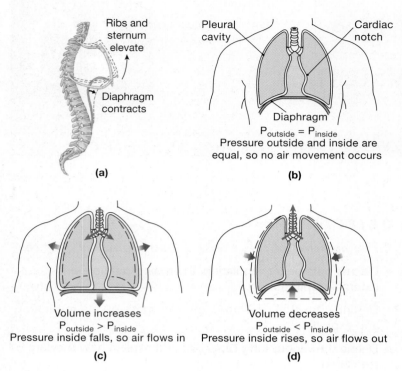

Ribs and
sternum
elevate

Diaphragm
contracts

(a)

Pleural
cavity

Cardiac
notch

Diaphragm

$P_{outside} = P_{inside}$

Pressure outside and inside are
equal, so no air movement occurs

(b)

Volume increases

$P_{outside} > P_{inside}$

Pressure inside falls, so air flows in

(c)

Volume decreases

$P_{outside} < P_{inside}$

Pressure inside rises, so air flows out

(d)

Figure 40.1

Mechanisms of Pulmonary Ventilation

(a) As the ribs are elevated or the diaphragm is depressed, the volume of the thoracic cavity increases. **(b)** An anterior view with the diaphragm at rest, with no air movement. **(c)** *Inhalation:* Elevation of the rib cage and contraction of the diaphragm increase the size of the thoracic cavity. **(d)** *Exhalation:* When the rib cage returns to its original position, the volume of the thoracic cavity decreases. Pressure rises, and air moves out of the lungs.

outside the body. Atmospheric pressure is normally 760 mm Hg, or approximately 15 pounds per square inch (psi). For inspiration to occur, the pressure in the thoracic cavity must be lower than atmospheric pressure, and for expiration the thoracic pressure must be higher than atmospheric pressure. **Pressure** is defined as how much force is applied to a given surface area, and a relationship called **Boyle's law** explains how changing the size of the thoracic cavity (and thereby changing the volume of the lungs) creates the pressure gradient necessary for breathing. The law states that the pressure of a gas in a closed container is inversely proportional to the volume of the container. Simply put, if the container is made smaller, the gas molecules exert the same amount of force on a smaller surface area and therefore the gas pressure increases.

Figure 40.1 illustrates the mechanisms of pulmonary ventilation. When the diaphragm is relaxed, it is dome-shaped. As it contracts, it lowers and flattens the floor of the thoracic cavity. This results in an increase in thoracic volume and consequently a decrease in thoracic pressure. Simultaneously, the external intercostal muscles contract and elevate the rib cage, further increasing thoracic volume and decreasing the pressure. This decrease in thoracic pressure causes a concurrent expansion of the lungs and a decrease in the pressure of the air in the lungs, the **intrapulmonic** (in-tra-PUL-mah-nik) **pressure**. Once intrapulmonic pressure falls below atmospheric pressure, air flows into the lungs.

Inspiration is an **active process** because it requires the contraction of several muscles to change pulmonary volumes and pressures. Expiration is essentially a **passive process** that occurs when the muscles just used in inspiration relax and the thoracic wall and elastic lung tissue recoil. During exercise, however, air may be actively exhaled by the combined contractions of the internal intercostal muscles and the abdominal muscles. The internal intercostal muscles depress the rib cage, and the abdominal muscles push the diaphragm higher into the thoracic cavity. Both actions decrease the thoracic volume and increase the thoracic pressure, which forces more air out during exhalation.

LAB ACTIVITY 1 Lung Volumes and Capacities

During exercise, the respiratory system must supply the muscular system with more oxygen, and therefore the respiratory rate increases, as does the volume of air inhaled and exhaled. Pulmonary volumes and capacities are generally measured when assessing health of the respiratory system, because these values change with pulmonary disease. In this section you will measure a variety of respiratory volumes.

An instrument called a **spirometer** (spī-ROM-e-ter) is used to measure respiratory volumes. The dry spirometer is a hand-held unit with disposable mouthpieces and with a small turbine that spins to measure the amount of air you exhale. As you do this activity, keep in mind that lung volumes vary according to sex, height, age, and general physical condition.

Figure 40.2 shows the volumes you will be measuring. **Tidal volume** (TV) is the amount of air one inspires and exhales during normal resting breathing. Tidal volume averages 500 ml, but additional air can be inhaled or exhaled beyond the tidal volume. The **inspiratory reserve volume** (IRV) is the amount of air that can be forcibly inspired above a normal inhalation. This volume averages 3,300 ml; if you were to gulp in air while exercising strenuously, you would take in the 500 ml of TV plus an additional 3,300 ml. The amount of air that can be forcefully exhaled after a normal exhalation is the **expiratory reserve volume** (ERV). The ERV averages 1,000 ml, which means you would expel the 500 ml of TV plus another 1,000 ml.

Vital capacity (VC) is the maximum amount of air that can be exhaled from the lungs after a maximum inhalation. This volume averages 4,800 ml in men and 3,100 ml in women and includes the combined volumes of the IRV, TV, and ERV: VC = IRV + TV + ERV.

The respiratory system always contains some air. The **residual volume** (RV) is the amount of air that cannot be forcefully exhaled from the lungs. Surfactant produced by the septal cells of the alveoli prevents the alveoli from collapsing completely during exhalation. Because the alveoli are not allowed to empty completely, they always maintain a residual volume of air, which averages 1,200 ml. **Minimal volume** is the amount of residual air—usually 30 ml to 120 ml—that stays in the lungs even if they are collapsed.

To calculate the **total lung capacity** (TLC), which averages 6,000 ml, the vital capacity is added to the residual volume: TLC = VC + RV.

Respiratory rate (RR) is the number of breaths taken per minute. RR multiplied by tidal volume gives the **minute volume** (MV), defined as the amount of air exchanged between the lungs and the environment in 1 min: MV = TV × RR.

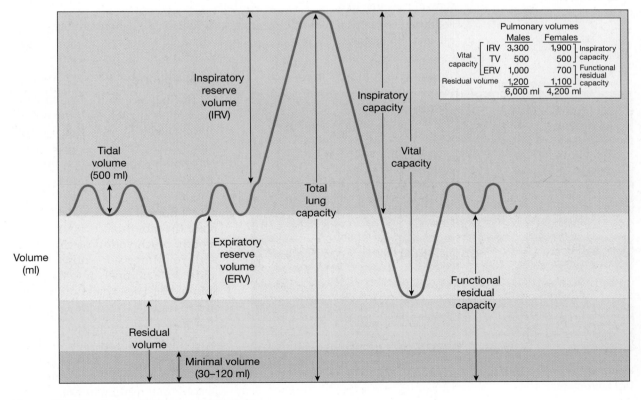

Figure 40.2 Respiratory Volumes and Capacities
The graph diagrams the relationships between respiratory volumes and capacities.

QuickCheck Questions

1.1 What instrument measures respiratory volumes?

1.2 What is vital capacity?

1.3 How is respiratory rate calculated?

Spirometry Safety

1. Do not use the dry spirometer if you have a cold or a communicable disease.
2. Always use a clean mouthpiece on the spirometer. Do not reuse a mouthpiece that has been removed from a spirometer.
3. A dry spirometer measures exhalations only. The instrument does not use an air filter, so *do not inhale* through it. *Only exhale into the instrument.*
4. Discard used mouthpieces in the designated biohazard box as indicated by your laboratory instructor. ▲

1 Materials

☐ Dry spirometer
☐ Disposable mouthpieces
☐ Noseclip (optional)
☐ Clock or watch with second hand
☐ Laboratory partner
☐ Biohazard box

Procedures

I. Setup

1. Insert a new, clean mouthpiece onto the breathing tube of the spirometer.
2. Remember: only exhale into the spirometer; the instrument cannot measure inspiratory volumes. To obtain as accurate a reading as possible, use a noseclip or your fingers to pinch your nose closed while exhaling into the spirometer.
3. Set the dial face to zero by turning the silver ring surrounding the dial face until the zero point on the scale is aligned with the point of the needle, as shown in Figure 40.3.
4. Following the steps listed next, measure each volume three times and then calculate an average.

II. Tidal Volume

1. Set the dial face so that 1,000 on the scale is aligned with the point of the needle. (You do this because tidal volume is small, and the scale on most dry spirometers is not graded before the 1,000-ml setting.)
2. Take a normal breath, quickly place the mouthpiece in your mouth, pinch your nose closed with your fingers or a noseclip, and exhale normally into the spirometer. The exhalation should not be forcible; it should be more of a sigh.
3. Record the scale reading in the tidal-volume row of Table 40.1.
4. Repeat steps 1 and 2 two times. Record each reading in the appropriate column of Table 40.1. Calculate the average of the three readings and record that average in the rightmost column of the table.

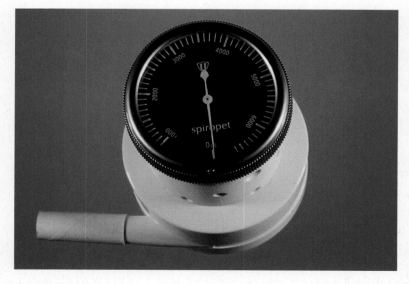

Figure 40.3
Dry Spirometer

Table 40.1 **Spirometry Data**

Volume	Reading 1	Reading 2	Reading 3	Average
Tidal volume	_____	_____	_____	_____
Expiratory reserve volume	_____	_____	_____	_____
Vital capacity	_____	_____	_____	_____
Respiratory rate	_____	_____	_____	_____
Minute volume (calculated)	_____	_____	_____	_____
Inspiratory reserve volume (calculated)	_____	_____	_____	_____

III. Expiratory Reserve Volume

1. Set the dial face to zero.
2. Exhale normally into the air (not into the mouthpiece).
3. Stop breathing for a moment, place the mouthpiece in your mouth, pinch your nose closed with your fingers or a noseclip, and forcibly exhale all the remaining air from your lungs.
4. Record the scale reading in the ERV row of Table 40.1.
5. Repeat steps 1–4 two times. Record each reading in the appropriate column of Table 40.1. Calculate the average of the three readings and record that average in the table.

IV. Vital Capacity

1. Set the dial face to zero.
2. Inhale maximally once, and then exhale maximally.
3. Inhale maximally, place the mouthpiece in your mouth, pinch your nose closed with your fingers or a noseclip, and forcibly exhale all the air from your lungs.
4. Record the scale reading in the vital-capacity row of Table 40.1.
5. Repeat steps 1–4 two times. Record each reading in the appropriate column of Table 40.1. Calculate the average of the three readings and record that average in the table.
6. Compare your vital capacity with the value listed for your sex, height, and age in Tables 40.2a and 40.2b. To use these tables, find your height across the top of the chart and your age along the left side. Follow the age row across and the height column down. The numeric value where the two intersect is the average vital capacity for all individuals of your age, height, and sex.

    Your vital capacity might differ from the average, for numerous reasons. Genetics has an influence on potential lung capacities, for instance, and lung damage from smoking or air pollution decreases vital capacity. On the positive side, cardiovascular exercises, such as swimming and jogging, increase lung volumes.

V. Respiratory Rate

1. Sit relaxed and read a textbook. Have your laboratory partner count the number of breaths you take in 20 seconds.
2. Multiply this number by 3 and record the value in the respiratory-rate row of Table 40.1.
3. Repeat steps 1 and 2 two times. Record each reading in the appropriate column of Table 40.1. Calculate the average of the three readings and record that average in the table.
4. Calculate your average minute volume by multiplying your average tidal volume by your average respiratory rate. Enter this calculated value in Table 40.1.

Table 40.2A Predicted Vital Capacities—Females

									Height in Centimeters and Inches										
Age	cm 152 in. 59.8	154 60.6	156 61.4	158 62.2	160 63.0	162 63.7	164 64.6	166 65.4	168 66.1	170 66.9	172 67.7	174 68.5	176 69.3	178 70.1	180 70.9	182 71.7	184 72.4	186 73.2	188 74.0
16	3,070	3,110	3,150	3,190	3,230	3,270	3,310	3,350	3,390	3,430	3,470	3,510	3,550	3,590	3,630	3,670	3,715	3,755	3,800
17	3,055	3,095	3,135	3,175	3,215	3,255	3,295	3,335	3,375	3,415	3,455	3,495	3,535	3,575	3,615	3,655	3,695	3,740	3,780
18	3,040	3,080	3,120	3,160	3,200	3,240	3,280	3,320	3,360	3,400	3,440	3,480	3,520	3,560	3,600	3,640	3,680	3,720	3,760
20	3,010	3,050	3,090	3,130	3,170	3,210	3,250	3,290	3,330	3,370	3,410	3,450	3,490	3,525	3,565	3,605	3,645	3,695	3,720
22	2,980	3,020	3,060	3,095	3,135	3,175	3,215	3,255	3,290	3,330	3,370	3,410	3,450	3,490	3,530	3,570	3,610	3,650	3,685
24	2,950	2,985	3,025	3,065	3,100	3,140	3,180	3,220	3,260	3,300	3,335	3,375	3,415	3,455	3,490	3,530	3,570	3,610	3,650
26	2,920	2,960	3,000	3,035	3,070	3,110	3,150	3,190	3,230	3,265	3,300	3,340	3,380	3,420	3,455	3,495	3,530	3,570	3,610
28	2,890	2,930	2,965	3,000	3,040	3,070	3,115	3,155	3,190	3,230	3,270	3,305	3,345	3,380	3,420	3,460	3,495	3,535	3,570
30	2,860	2,895	2,935	2,970	3,010	3,045	3,085	3,120	3,160	3,195	3,235	3,270	3,310	3,345	3,385	3,420	3,460	3,495	3,535
32	2,825	2,865	2,900	2,940	2,975	3,015	3,050	3,090	3,125	3,160	3,200	3,235	3,275	3,310	3,350	3,385	3,425	3,460	3,495
34	2,795	2,835	2,870	2,910	2,945	2,980	3,020	3,055	3,090	3,130	3,165	3,200	3,240	3,275	3,310	3,350	3,385	3,425	3,460
36	2,765	2,805	2,840	2,875	2,910	2,950	2,985	3,020	3,060	3,095	3,130	3,165	3,205	3,240	3,275	3,310	3,350	3,385	3,420
38	2,735	2,770	2,810	2,845	2,880	2,915	2,950	2,990	3,025	3,060	3,095	3,130	3,170	3,205	3,240	3,275	3,310	3,350	3,385
40	2,705	2,740	2,775	2,810	2,850	2,885	2,920	2,955	2,990	3,025	3,060	3,095	3,135	3,170	3,205	3,240	3,275	3,310	3,345
42	2,675	2,710	2,745	2,780	2,815	2,850	2,885	2,920	2,955	2,990	3,025	3,060	3,100	3,135	3,170	3,205	3,240	3,275	3,310
44	2,645	2,680	2,715	2,750	2,785	2,820	2,855	2,890	2,925	2,960	2,995	3,030	3,060	3,095	3,130	3,165	3,200	3,235	3,270
46	2,615	2,650	2,685	2,715	2,750	2,785	2,820	2,855	2,890	2,925	2,960	2,995	3,030	3,060	3,095	3,130	3,165	3,200	3,235
48	2,585	2,620	2,650	2,685	2,715	2,750	2,785	2,820	2,855	2,890	2,925	2,960	2,995	3,030	3,060	3,095	3,130	3,160	3,195
50	2,555	2,590	2,625	2,655	2,690	2,720	2,755	2,785	2,820	2,855	2,890	2,925	2,955	2,990	3,025	3,060	3,090	3,125	3,155
52	2,525	2,555	2,590	2,625	2,655	2,690	2,720	2,755	2,790	2,820	2,855	2,890	2,925	2,955	2,990	3,020	3,055	3,090	3,125
54	2,495	2,530	2,560	2,590	2,625	2,655	2,690	2,720	2,755	2,790	2,820	2,855	2,885	2,920	2,950	2,985	3,020	3,050	3,085
56	2,460	2,495	2,525	2,560	2,590	2,625	2,655	2,690	2,720	2,755	2,790	2,820	2,855	2,885	2,920	2,950	2,980	3,015	3,045
58	2,430	2,460	2,495	2,525	2,560	2,590	2,625	2,655	2,690	2,720	2,750	2,785	2,815	2,850	2,880	2,920	2,945	2,975	3,010
60	2,400	2,430	2,460	2,495	2,525	2,560	2,590	2,625	2,655	2,685	2,720	2,750	2,780	2,810	2,845	2,875	2,915	2,940	2,970
62	2,370	2,405	2,435	2,465	2,495	2,525	2,560	2,590	2,620	2,655	2,685	2,715	2,745	2,775	2,810	2,840	2,870	2,900	2,935
64	2,340	2,370	2,400	2,430	2,465	2,495	2,525	2,555	2,585	2,620	2,650	2,680	2,710	2,740	2,770	2,805	2,835	2,865	2,895
66	2,310	2,340	2,370	2,400	2,430	2,460	2,495	2,525	2,555	2,585	2,615	2,645	2,675	2,705	2,735	2,765	2,800	2,825	2,860
68	2,280	2,310	2,340	2,370	2,400	2,430	2,460	2,490	2,520	2,550	2,580	2,610	2,640	2,670	2,700	2,730	2,760	2,795	2,820
70	2,250	2,280	2,310	2,340	2,370	2,400	2,425	2,455	2,485	2,515	2,545	2,575	2,605	2,635	2,665	2,695	2,725	2,755	2,780
72	2,220	2,250	2,280	2,310	2,335	2,365	2,395	2,425	2,455	2,480	2,510	2,540	2,570	2,600	2,630	2,660	2,685	2,715	2,745
74	2,190	2,220	2,245	2,275	2,305	2,335	2,360	2,390	2,420	2,450	2,475	2,505	2,535	2,565	2,590	2,620	2,650	2,680	2,710

VI. Inspiratory Reserve Volume

1. Use the average values for your vital capacity, tidal volume, and expiratory reserve volume to calculate your inspiratory reserve volume: IRV = VC − (TV + ERV).
2. Enter the calculated IRV in Table 40.1. ■

LAB ACTIVITY 2 BIOPAC: Volumes and Capacities

BIOPAC

In this activity, you will use an **airflow transducer,** and a computer will convert airflow to volume. Although this is a much quicker method of obtaining lung capacity data, the disadvantage is that the recording procedure must be followed exactly for an accurate conversion from airflow to volume.

You will measure tidal volume, inspiratory reserve volume, and expiratory reserve volume and then use these data to calculate inspiratory capacity and vital capacity. The equations in Table 40.3 can be used to obtain predicted vital capacity based on sex, height, and age. For instance, the predicted vital capacity of

Table 40.2B Predicted Vital Capacities—Males

		Height in Centimeters and Inches																		
	cm	152	154	156	158	160	162	164	166	168	170	172	174	176	178	180	182	184	186	188
Age	in.	59.8	60.6	61.4	62.2	63.0	63.7	64.6	65.4	66.1	66.9	67.7	68.5	69.3	70.1	70.9	71.7	72.4	73.2	74.0
16		3,920	3,975	4,025	4,075	4,130	4,180	4,230	4,285	4,335	4,385	4,440	4,490	4,540	4,590	4,645	4,695	4,745	4,800	4,850
18		3,890	3,940	3,995	4,045	4,095	4,145	4,200	4,250	4,300	4,350	4,405	4,455	4,505	4,555	4,610	4,660	4,710	4,760	4,815
20		3,860	3,910	3,960	4,015	4,065	4,115	4,165	4,215	4,265	4,320	4,370	4,420	4,470	4,520	4,570	4,625	4,675	4,725	4,775
22		3,830	3,880	3,930	3,980	4,030	4,080	4,135	4,185	4,235	4,285	4,335	4,385	4,435	4,485	4,535	4,585	4,635	4,685	4,735
24		3,785	3,835	3,885	3,935	3,985	4,035	4,085	4,135	4,185	4,235	4,285	4,330	4,380	4,430	4,480	4,530	4,580	4,630	4,680
26		3,755	3,805	3,855	3,905	3,955	4,000	4,050	4,100	4,150	4,200	4,250	4,300	4,350	4,395	4,445	4,495	4,545	4,595	4,645
28		3,725	3,775	3,820	3,870	3,920	3,970	4,020	4,070	4,115	4,165	4,215	4,265	4,310	4,360	4,410	4,460	4,510	4,555	4,605
30		3,695	3,740	3,790	3,840	3,890	3,935	3,985	4,035	4,080	4,130	4,180	4,230	4,275	4,325	4,375	4,425	4,470	4,520	4,570
32		3,665	3,710	3,760	3,810	3,855	3,905	3,950	4,000	4,050	4,095	4,145	4,195	4,240	4,290	4,340	4,385	4,435	4,485	4,530
34		3,620	3,665	3,715	3,760	3,810	3,855	3,905	3,950	4,000	4,045	4,095	4,140	4,190	4,225	4,285	4,330	4,380	4,425	4,475
36		3,585	3,635	3,680	3,730	3,775	3,825	3,870	3,920	3,965	4,010	4,060	4,105	4,155	4,200	4,250	4,295	4,340	4,390	4,435
38		3,555	3,605	3,650	3,695	3,745	3,790	3,840	3,885	3,930	3,980	4,025	4,070	4,120	4,165	4,210	4,260	4,305	4,350	4,400
40		3,525	3,575	3,620	3,665	3,710	3,760	3,805	3,850	3,900	3,945	3,990	4,035	4,085	4,130	4,175	4,220	4,270	4,315	4,360
42		3,495	3,540	3,590	3,635	3,680	3,725	3,770	3,820	3,865	3,910	3,955	4,000	4,050	4,095	4,140	4,185	4,230	4,280	4,325
44		3,450	3,495	3,540	3,585	3,630	3,675	3,725	3,770	3,815	3,860	3,905	3,950	3,995	4,040	4,085	4,130	4,175	4,220	4,270
46		3,420	3,465	3,510	3,555	3,600	3,645	3,690	3,735	3,780	3,825	3,870	3,915	3,960	4,005	4,050	4,095	4,140	4,185	4,230
48		3,390	3,435	3,480	3,525	3,570	3,615	3,655	3,700	3,745	3,790	3,835	3,880	3,925	3,970	4,015	4,060	4,105	4,150	4,190
50		3,345	3,390	3,430	3,475	3,520	3,565	3,610	3,650	3,695	3,740	3,785	3,830	3,870	3,915	3,960	4,005	4,050	4,090	4,135
52		3,315	3,353	3,400	3,445	3,490	3,530	3,575	3,620	3,660	3,705	3,750	3,795	3,835	3,880	3,925	3,970	4,010	4,055	4,100
54		3,285	3,325	3,370	3,415	3,455	3,500	3,540	3,585	3,630	3,670	3,715	3,760	3,800	3,845	3,890	3,930	3,975	4,020	4,060
56		3,255	3,295	3,340	3,380	3,425	3,465	3,510	3,550	3,595	3,640	3,680	3,725	3,765	3,810	3,850	3,895	3,940	3,980	4,025
58		3,210	3,250	3,290	3,335	3,375	3,420	3,460	3,500	3,545	3,585	3,630	3,670	3,715	3,755	3,800	3,840	3,880	3,925	3,965
60		3,175	3,220	3,260	3,300	3,345	3,385	3,430	3,470	3,500	3,555	3,595	3,635	3,680	3,720	3,760	3,805	3,845	3,885	3,930
62		3,150	3,190	3,230	3,270	3,310	3,350	3,390	3,440	3,480	3,520	3,560	3,600	3,640	3,680	3,730	3,770	3,810	3,850	3,890
64		3,120	3,160	3,200	3,240	3,280	3,320	3,360	3,400	3,440	3,490	3,530	3,570	3,610	3,650	3,690	3,730	3,770	3,810	3,850
66		3,070	3,110	3,150	3,190	3,230	3,270	3,310	3,350	3,390	3,430	3,470	3,510	3,550	3,600	3,640	3,680	3,720	3,760	3,800
68		3,040	3,080	3,120	3,160	3,200	3,240	3,280	3,320	3,360	3,400	3,440	3,480	3,520	3,560	3,600	3,640	3,680	3,720	3,760
70		3,010	3,050	3,090	3,130	3,170	3,210	3,250	3,290	3,330	3,370	3,410	3,450	3,480	3,520	3,560	3,600	3,640	3,680	3,720
72		2,980	3,020	3,060	3,100	3,140	3,180	3,210	3,250	3,290	3,330	3,370	3,410	3,450	3,490	3,530	3,570	3,610	3,650	3,680
74		2,930	2,970	3,010	3,050	3,090	3,130	3,170	3,200	3,240	3,280	3,320	3,360	3,400	3,440	3,470	3,510	3,550	3,590	3,630

Predicted Vital Capacities (a) Female (b) Male. From Archives of Environmental Health, Volume 12, pp. 146–189, February 1966. Reprinted with permission of the Helen Dwight Reid Educational Foundation. Published by Heldref Publications, 4000 Albemarle St., N.W., Washington, D.C. 20016. Copyright © 1966.

Table 40.3 Equations for Predicted Vital Capacity

Male	$VC = 0.052H - 0.022A - 3.60$
Female	$VC = 0.041H - 0.018A - 2.69$

VC = vital capacity in liters
H = height in centimeters
A = age in years

a 19-year-old female who is 167 centimeters tall (about 5.5 ft) is $0.041(167) - 0.018(19) - 2.69 = 3.8$ liters

QuickCheck Questions

2.1 What is the tidal volume of respiration?

2.2 What is the vital capacity volume of respiration?

2 Materials

- ☐ BIOPAC acquisition unit (MP35/30)
- ☐ BIOPAC airflow transducer (SS11L or SS11LA)
- ☐ BIOPAC disposable bacteriological filters (AFT1), one per subject plus one for calibration
- ☐ BIOPAC mouthpiece, disposable (AFT2) or autoclavable (AFT8), one per subject
- ☐ BIOPAC noseclip (AFT3), one per subject
- ☐ BIOPAC calibration syringe (AFT6)
- ☐ BIOPAC wall transformer (AC100A)
- ☐ BIOPAC serial cable (CBLSERA)
- ☐ Computer: minimum Macintosh® 68020 *or* PC running Windows® 98SE
- ☐ BIOPAC Student Lab software v3.0 or higher

Procedures

This lesson has four sections: setup, calibration, recording, and data analysis. Be sure to follow the setup instructions appropriate for your type of airflow transducer (SS11L or SS11LA). The calibration step is critical for getting accurate recordings. Four segments will be recorded and then analyzed. You may record the data by hand or choose **Edit > Journal > Paste Measurements** to paste the data into your journal for future use.

Most markers and labels are automatically inserted into the data recordings. Markers appear at the top of the window as inverted triangles. This symbol indicates that you need to insert a marker and key in a marker label similar to the text in quotes. You can insert and label the marker during or after acquisition; on a Mac, press ESC; on a PC, press F9.

I. Setup

1. Turn the computer ON.
2. Make sure the acquisition unit is turned OFF. Check the serial cable (CBLSERA) connection between the acquisition unit and the computer.
3. Plug the airflow transducer (SS11L or SS11LA) into Channel 1 as shown in Figure 40.4.
4. Plug the wall transformer (AC100A) into an electrical outlet and into the acquisition unit, and then turn the unit ON.
5. Place a filter (AFT1) on the end of the calibration syringe (AFT6). The filter is required for calibration and recording because it forces the air to move smoothly through the transducer. This assembly can be left connected for future use. You need to replace the filter only if the paper inside the filter tears.
6. Insert the syringe/filter assembly into the port of the transducer head (Figure 40.5). Inside the head is the sensor that measures airflow. If using the SS11L transducer with nonremovable head, insert the assembly into the larger-diameter 120 port. If using the SS11LA transducer with removable, cleanable head, always insert the assembly on the transducer side labeled "Inlet" so that the transducer cable exits on the left, as shown in Figure 40.5.
7. Start the Biopac Student Lab program.
8. Choose Lesson 12 (L12-PUL.1).
9. Type in the subject's filename, which must be a unique identifier.
10. Click OK. You are now ready to calibrate the hardware.

II. Calibration

The calibration establishes the hardware's internal parameters and is critical for optimum performance.

1. Pull the syringe plunger all the way out, and hold the syringe/filter assembly with the barrel of the syringe perfectly horizontal (Figure 40.6). The transducer is sensitive to gravity and so must remain upright throughout the calibration and recording, in the orientation shown in Figure 40.6. *Important:* Never hold the transducer handle when using the calibration syringe, because doing so may break the syringe tip.
2. Click on **Calibrate**. The first calibration stage will run for eight seconds and then end with an alert box. *Important:* Leave the plunger extended, and hold the assembly steady and perfectly horizontal during the entire calibration. Do not touch the plunger, because any pressure at this stage will cause inaccurate results.

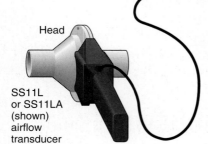

MP35/MP30 unit

Plugs into CH 1

Head

SS11L or SS11LA (shown) airflow transducer

Handle

Figure 40.4

Connecting the Airflow Transducer

Figure 40.5
**Insertion of Calibration
Syringe/Filter Assembly**

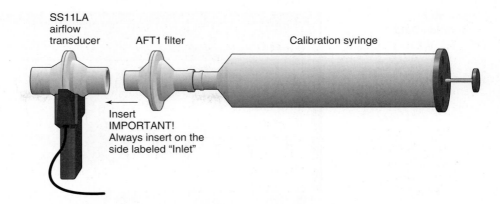

Figure 40.6
**Calibrating the
Airflow Transducer**

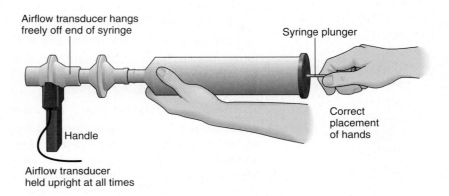

3. Prepare for the second calibration stage. Stage 1 ends with an alert box asking if you have read the directions in the journal. Read the directions in step 5 and/or the journal so that once the second calibration stage starts, you fully understand the procedures.

4. When you are ready to proceed, click **Yes.** The second calibration stage will begin and will run until you click End Calibration.

5. Cycle the syringe plunger in and out completely 5 times (10 strokes), all the while holding the syringe with your hands placed as shown in Figure 40.6. Use a rhythm of about one second per stroke with a two-second rest between strokes: take one second to push the plunger in completely, wait two seconds, take one second to pull the plunger out completely, wait two seconds. Repeat this cycle four more times.

6. Click on **End Calibration.**

7. **Check** your calibration data, which should resemble Figure 40.7. If your screen shows five downward deflections and five upward deflections, proceed to Data Recording. If your screen shows any large spikes, click on **Redo Calibration.**

III. Data Recording

1. To work efficiently, read this entire section now so that you will know what to do for each recording segment. You will be working with a partner, the subject, who should remain in a supine position and relaxed while you review the lesson. You will record airflow data for the subject for normal breathing, deep inhalation, deep exhalation, and return to normal breathing. The software will automatically calculate volumes based on the recorded airflow data.

Figure 40.7
Calibration Data

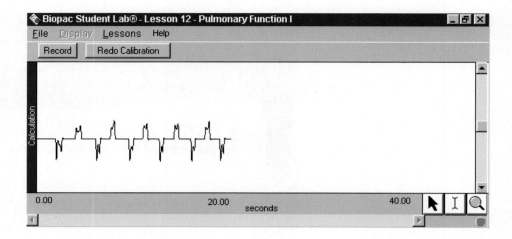

Hints for obtaining optimal data:

a. Keep the airflow transducer upright at all times.

b. If you start the recording during an inhalation, try to end during an exhalation, and vice versa. This is not absolutely critical but does increase the accuracy of the calculations.

c. The subject should not look at the screen during recording.

2. Find your transducer setup in Figure 40.8, and carefully follow the filter and mouthpiece instructions for that setup. *Important:* If your laboratory sterilizes the transducer heads after each use, make sure a clean head is installed now. Have the subject remove the filter and mouthpiece from their plastic packages. This mouthpiece will become the subject's personal one, and therefore the subject should write her or his name on the mouthpiece and filter with a permanent marker so they can be reused later.

Follow this procedure precisely to make sure the airflow transducer is sterile:

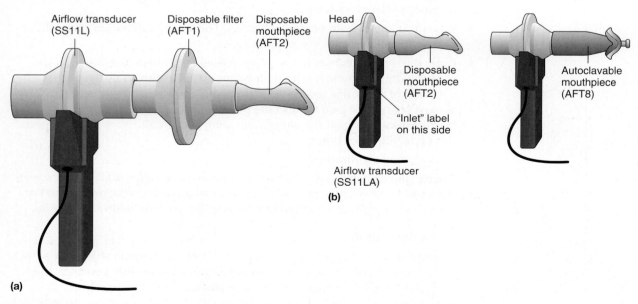

Figure 40.8
Airflow Transducer Setups
(a) SS11L (shown) or SS11LA with nonsterilized head. **(b)** SS11LA with sterilized head.

- If using the SS11L transducer with nonremovable head, insert a new filter and disposable mouthpiece (AFT1, AFT2) into the larger-diameter port on the transducer (Figure 40.8a).

- If using the SS11LA transducer and **not sterilizing** the head after each use, insert a filter and disposable mouthpiece (AFT1, AFT2) into the transducer on the side labeled "**Inlet**" (Figure 40.8a).

- If the head will be sterilized in an autoclave after use, a filter is not required for the SS11LA transducer. Insert a disposable mouthpiece (ATF2) or an autoclavable mouthpiece (AFT8) into the transducer on the side labeled "**Inlet**" (Figure 40.8b).

3. Have the subject place a noseclip on his or her nose and begin breathing through the transducer, holding the transducer upright at all times and always breathing through the side labeled "**Inlet**" (Figure 40.9).

4. Click on **Record**, and then have the subject:

    a. Breathe normally for three breaths. One breath is a complete inhale-exhale cycle.

    b. Inhale as deeply as possible.

    c. Quickly put the mouthpiece into the mouth, and exhale just to the point of normal breathing, without forcing out any additional air.

    d. Breathe normally for three breaths.

    e. With the mouthpiece in the mouth, exhale completely, forcing out as much air as possible.

    f. Breathe normally for three more cycles.

5. Click on **Stop**, ending during an exhalation if you started the recording during an inhalation, and vice versa. As soon as the **Stop** button is pressed, the Biopac Student Lab software will automatically calculate volumes based on the recorded airflow data. At the end of the calculation, both an airflow wave and a volume wave will be displayed on the screen (Figure 40.10).

6. If your screen looks similar to Figure 40.10, proceed to step 7. If your screen does not resemble Figure 40.10, click **Redo** and repeat steps 4–6. Your data would be incorrect if the subject coughed, for example, or if some exhaled air escaped from the mouthpiece.

7. Click on **Done** to exit the recording mode. Your data will automatically be saved in the "Data Files" folder on your hard drive. A pop-up window with options will appear. Make your choice and continue as directed. If choosing the "Record from another Subject" option:

    a. You will not need to recalibrate the airflow transducer. For this reason, all recordings should be completed before you proceed to data analysis.

    b. Remember to have each person use his or her own mouthpiece, bacterial filter, and noseclip.

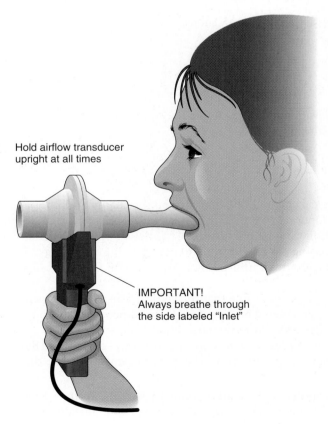

Hold airflow transducer upright at all times

IMPORTANT! Always breathe through the side labeled "Inlet"

Figure 40.9
Using the
Airflow Transducer

Figure 40.10
Sample Recording
CH0 shows volume. CH40 shows airflow.

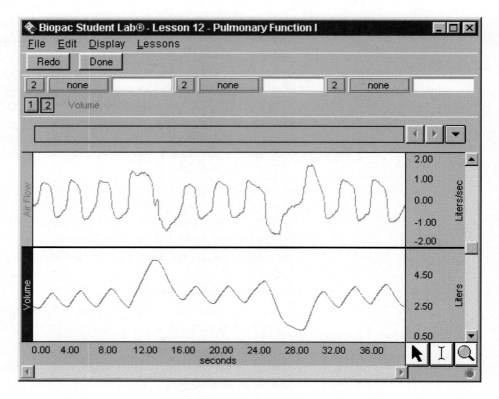

c. Repeat recording steps 1–7 for each new subject.

d. Each subject must use a unique filename.

IV. Data Analysis

The first step is to evaluate the volume data.

1. Enter the **Review Saved Data** mode, and choose the correct file. Note the channel number designations:

Channel Displays

CH 0	**Volume**
CH 40	**Airflow**

2. Turn **OFF** CH 40. To toggle a channel ON/OFF, click on the channel number box and hold down the Option key on a Mac or the Ctrl key on a PC.

   Optional: Review airflow data before turning Channel 40 off. These data do not have a lot of meaning for this lesson and may be a bit confusing at first glance, but they contain an interesting perspective on the recording.

   Looking at the airflow waveform, note that the vertical scale is in liters per second and that the wave is centered on zero. Each upward-pointing region (called a *positive peak*) of the curve corresponds to inhalation, and each downward-pointing region (a *negative peak*) corresponds to an exhalation. The deeper an inhale, the larger the positive peak; the more forceful an exhale, the larger the negative peak.

3. The measurement boxes are above the marker region in the data window. Each box has three sections: channel number, measurement type (p-p, max, min, or Δ), and result. The first two sections are pull-down menus that are activated when you click on them. Set up the boxes as follows:

Channel Displays

CH 0	**p-p** (finds maximum value in selected area and subtracts minimum value in selected area)
CH 0	**max** (displays maximum value in selected area)
CH 0	**min** (displays minimum value in selected area)
CH 0	**Δ** (delta amplitude, difference in amplitude between last and first points of selected area)

The **selected area** is the part selected by the I-beam tool and includes the endpoints.

4. Use the I-beam cursor to select the region of the first three breaths (Figure 40.11). The selected area should be from time 0 to the end of the third cycle. The p-p measurement represents TV. You can either record this and all other measurements in the data tables in the Laboratory Report or choose **Edit> Journal> Paste Measurements** to paste the data to your journal for future reference. This p-p measurement representing TV goes in Table 40.4 in the Laboratory Report.

5. Use the I-beam cursor and measurement tools to determine IRV, ERV, and VC, and record all the values in Table 40.4. To obtain the p-p measurement for VC, select the area from the start of the forced inhalation to the peak of the forced exhalation (Figure 40.12). Record the p-p value for VC in Table 40.4.

   To obtain the p-p measurement for IRV, select from the third normal-inhalation peak to the peak of the forced inhalation (Figure 40.13a). Record the p-p measurement in Table 40.4.

   To obtain the p-p measurement for ERV, locate the peaks for the series of three normal breaths taken after the forced deep inhalation but before the forced exhalation. Select from the third (downward-pointing) normal-exhalation peak to the (downward-pointing) peak of the forced exhalation (Figure 40.13b). Record the p-p measurement in Table 40.4.

   Finally, use the equations shown in Table 40.5 in the Laboratory Report and the RV value you entered there to calculate inspiratory capacity, expiratory capacity, functional residual capacity, and total lung capacity.

6. Save or print the data and journal files and exit the program. ■

Figure 40.11
Selection of First
Three Breaths

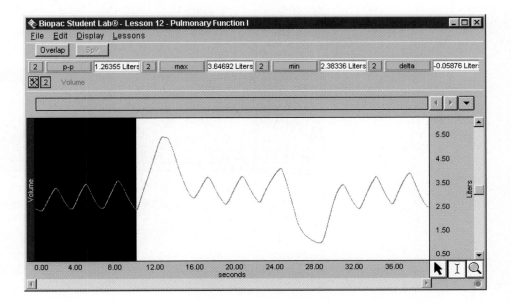

Figure 40.12
**Example of VC
from P-P Tool**

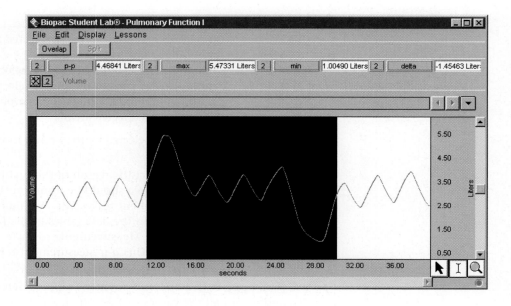

Figure 40.13
**Example of (a) IRV and
(b) ERV from P-P Tool**

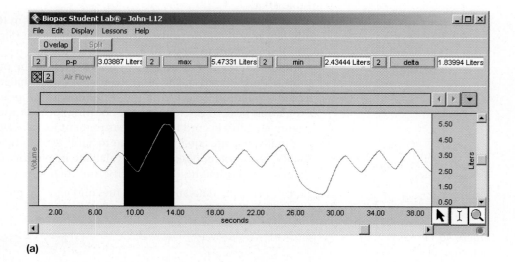

(a)

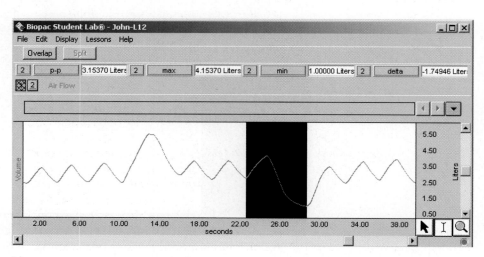

(b)

LAB ACTIVITY 3 | BIOPAC: Respiratory Rate and Depth

In this activity, you will measure ventilation by recording the rate and depth of the breathing cycle using a **pneumograph transducer**. This transducer converts changes in chest expansion and contraction to changes in voltage, which will appear as a waveform. One respiratory cycle will then be recorded as an increasing voltage (ascending segment) during inspiration and a decreasing voltage (descending segment) during expiration.

You will also record the temperature of the air flowing in and out of one nostril with a **temperature probe**. The temperature of the air passing by the probe is inversely related to the expansion or contraction of the subject's chest. During inspiration (when the chest expands), the subject breathes in air that is cool relative to body temperature. This air is then warmed in the body. During expiration (when the chest contracts), the warmer air is compressed out of the lungs and out the respiratory passages.

QuickCheck Questions

3.1 What does a pneumograph transducer measure?

3.2 What is the temperature probe used for in this investigation?

3 | Materials

- ☐ BIOPAC acquisition unit (MP35/30)
- ☐ BIOPAC pneumograph transducer (SS5LA, SS5L, or SS5LB)
- ☐ BIOPAC temperature probe (SS6L)
- ☐ BIOPAC wall transformer (AC100A)
- ☐ BIOPAC serial cable (CBLSERA)
- ☐ Single-sided (surgical) tape (TAPE1)
- ☐ Computer: minimum Macintosh® 68020 *or* PC running Windows® 98SE
- ☐ Biopac Student Lab software v3.0 or higher

Procedures

This lesson has four sections: setup, calibration, recording, and data analysis. Be sure to follow the setup instructions for your type of pneumograph transducer (SS5LA, SS5L, or SS5LB). The calibration step is critical for getting accurate recordings. Four segments will be recorded and then analyzed. You may record the data by hand or choose **Edit > Journal > Paste Measurements** to paste the data into your journal for future use.

Most markers and labels are automatically inserted into the data recordings. Markers appear at the top of the window as inverted triangles. This symbol indicates that you need to insert a marker and key in a marker label similar to the text in quotes. You can insert and label the marker during or after acquisition: on a Mac, press ESC; on a PC, press F9.

I. Setup

1. Turn the computer ON.

2. Make sure the acquisition unit is turned OFF. Plug the AC100A wall transformer into an electrical outlet and into the acquisition unit. Check the CBLSERA serial cable connection between the acquisition and the computer. Plug in the equipment as shown in Figure 40.14: the pneumograph transducer (SS5LA, SS5L, or SS5LB) into Channel 1 and the temperature probe (SS6L) into Channel 2. *Note:* Figure 40.14 shows the SS5LA model. Your laboratory might have the SS5LB or SS5L model, which both look a little different but work the same way. If using the SS5LA transducer, be very careful not to pull or yank on the rubber bowtie portion that contains the sensor element.

The temperature probe is used to measure airflow. Each inhalation brings relatively cool air across the probe, and each exhalation blows warmer air across it. The probe records these temperature changes, which are proportional to the airflow output. This indirect method is efficient when rate and relative amplitude measurements are sufficient; a direct airflow measurement requires a complex scaling procedure that we will not use.

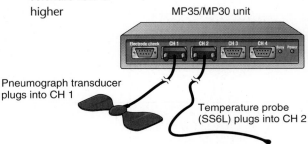

MP35/MP30 unit

Pneumograph transducer plugs into CH 1

Temperature probe (SS6L) plugs into CH 2

Figure 40.14
Connecting the Respiratory and Temperature Transducers

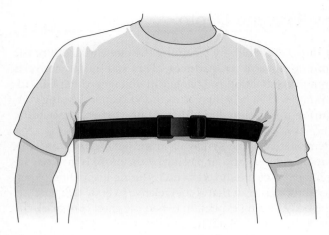

Figure 40.15
Placement of Respiratory Transducer around Chest

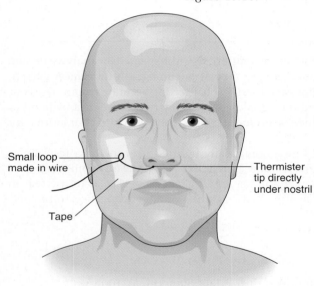

Small loop made in wire

Thermister tip directly under nostril

Tape

Figure 40.16
Placement of Temperature Transducer

3. Turn the acquisition unit ON.

4. Attach the pneumograph transducer around the subject's chest below the armpits and above the nipples (Figure 40.15). The transducer can be worn over a shirt, but the correct tension is critical: slightly tight at the point of maximal expiration. If using the SS5LA model, attach the nylon belt by threading the nylon strap through the corresponding slots on the rubber bowtie such that the strap clamps into place when tightened. If using the SS5LB or SS5L model, attach the self-sticking ends together at the correct tension.

5. Attach the temperature probe (SS6L) to the subject's face. The probe should be firmly attached so that it does not move and should be positioned below the nostril and not touching the face. It is usually best to make a small loop in the cable about two inches from the tip and tape the loop to the subject's face, as shown in Figure 40.16.

6. **Start** the Biopac Student Lab Program and choose **Lesson 8 (LO8-Resp-1)**. Type in the subject's filename, which must be a unique identifier, and then click OK.

II. Calibration

The calibration establishes the hardware's internal parameters (such as gain, offset, and scaling) and is critical for optimum performance.

1. Have the subject sit in a relaxed state and breathe normally.

2. Click the **Calibrate** button in the **Setup** window.

3. Wait two seconds. Instruct the subject to breath deeply for one cycle, then breathe normally until the calibration ends. The calibration will run for eight seconds and then stop automatically.

4. After the calibration has stopped, check your calibration data. Your screen should resemble Figure 40.17. The top channel displays data from the temperature probe and is labeled "airflow"

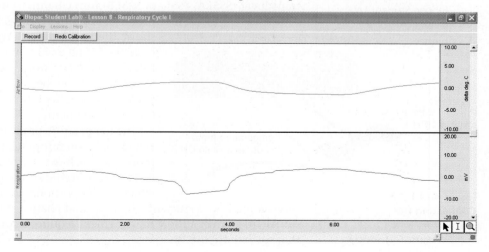

Figure 40.17
Sample Calibration Data

because the temperature at the nostril is inversely proportional to airflow in and out of the nostril.

Both recording channels should show some fluctuation. If there is no fluctuation, it is possible that either the pneumograph transducer or the temperature probe is not connected properly, and you must redo the calibration by clicking on **Redo Calibration** and repeating the sequence.

III. Recording Data

Have the subject sit down and relax. You will record four segments: normal breathing, hyperventilation, hypoventilation, and coughing. To work efficiently, read this entire section before proceeding so that you will know what to do for each segment. Check the last line of the journal and note the total amount of time available for the recording. Stop each recording segment as soon as possible so that you do not use up an excessive amount of hard drive storage space.

Hints for obtaining optimal data:

a. Subject should stop hyperventilation or hypoventilation if dizziness develops.

b. The pneumograph transducer should fit snugly around the chest prior to inspiration.

c. The temperature probe should be firmly attached so that it does not move, positioned below the nostril and not touching the face.

d. The subject should be sitting for all segments.

e. The recording should be suspended after each segment so that the subject can prepare for the next segment.

Segment 1—Normal Breathing

1. Click on **Record**.

2. Record for 15 seconds with the subject sitting in a chair, breathing normally.

3. Click on **Suspend** to halt the recording.

4. Review the data on the screen. If your screen looks similar to Figure 40.18, proceed to segment 2. If not, erase the data by clicking on **Redo**; adjust the placement of the transducer over the chest and repeat the calibration. The data would be incorrect if:

   a. The pneumograph data has plateaus instead of waveforms.

   b. The waveforms representing the temperature-probe (airflow) data are not offset from the respiration data.

   c. The temperature probe moved and is no longer directly under the nostril.

   d. The belt of the pneumograph transducer slipped.

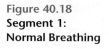
Figure 40.18
Segment 1:
Normal Breathing

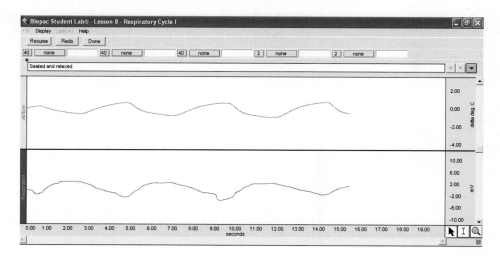

    e. The Suspend button was pressed prematurely.

    f. Any of the channels have flat data, indicating no signal. In this case, be sure the cables are all securely in their respective ports.

Segment 2—Hyperventilation and Recovery

5. Click on **Resume.** The recording will continue from the point where it last stopped, and a marker labeled "hyperventilation and recovery" will come up.

6. Have the subject first *hyperventilate* by breathing rapidly and deeply through the mouth for 30 seconds (from the 15-s position on the screen to the 45-s position) and then *recover* by breathing through the nose for 30 seconds (45-s position to 75-s position). **Record** both during hyperventilation and during recovery.

 Stop the procedure immediately if the subject starts to feel sick or excessively dizzy. ▲

7. Click on **Suspend.**

8. Review the data on the screen, using the horizontal scroll bar to look at different portions of the waveform. If your screen is similar to Figure 40.19, proceed to step 9. If your screen does not look like Figure 40.19, click **Redo** and repeat steps 5–8. The data would be incorrect for the same reasons listed in step 4.

Segment 3—Hypoventilation and Recovery *Important:* You must not begin this segment until after the subject's breathing has returned to normal.

9. Click on **Resume.** The recording will continue from the point where it last stopped, and a marker labeled "hypoventilation and recovery" will be inserted.

10. Have the subject first *hypoventilate* by breathing slowly and shallowly through the mouth for 30 seconds (75-s position on the screen to 105-s position) and then *recover* by breathing through the nose for 30 seconds (105-s position to 135-s position). **Record** both during hypoventilation and during recovery.

11. Click on **Suspend.**

12. Review the data on the screen, using the horizontal scroll bar to view different portions of the waveforms. If your screen looks similar to Figure 40.20, proceed to step 13. If your screen does not look like Figure 40.20, click **Redo** and repeat steps 9–12. The data would be incorrect for the same reasons listed in step 4.

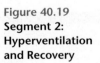

Figure 40.19 Segment 2: Hyperventilation and Recovery

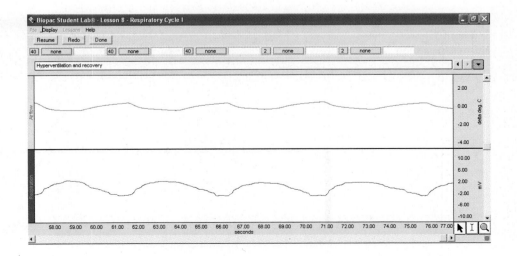

Figure 40.20
Segment 3: Hypoventilation and Recovery

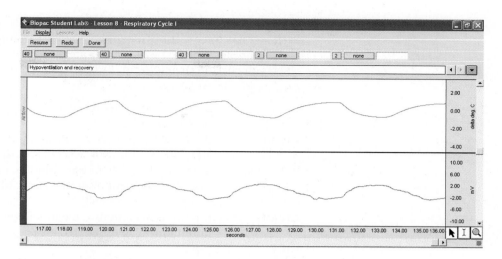

Segment 4—Coughing Followed by Reading Aloud

13. Click on **Resume**. The recording will continue from the point where it last stopped, and a marker "cough, then read aloud" will come up.

14. Have the subject cough once and then begin reading aloud a passage from the laboratory manual.

15. Click on **Suspend**.

16. Review the data on the screen, using the horizontal scroll bar to view different portions of the waveforms. If your screen looks similar to Figure 40.21, proceed to step 17. If your screen does not look like Figure 40.21, click **Redo** and repeat steps 13–16. The data would be incorrect for the same reasons listed in step 4.

17. Click on **Done**. A pop-up window with options will appear. If choosing the "Record from another subject" option, attach the pneumograph transducer and temperature probe per setup steps 4 and 5, and continue from step 1 of Segment 1—Normal Breathing.

18. Remove the pneumograph transducer and temperature probe from the subject.

IV. Data Analysis

1. Enter the **Review Saved Data** mode from the Lessons menu. Note the channel number designation:

Channel Displays

CH 2	Airflow
CH 40	Respiration

Figure 40.21
Segment 4: Cough Once, Then Read Aloud

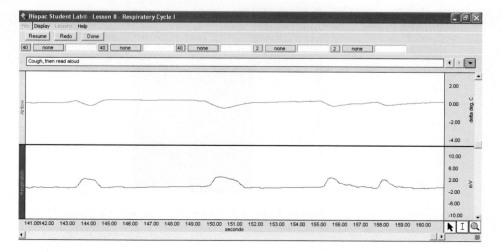

2. The measurement boxes are above the marker region in the data window. Each box has three sections: channel number, measurement type (ΔT, BPM, or p-p), and result. The first two sections are pull-down menus that are activated when you click on them. Set up the boxes as follows:

Channel	Displays
CH 40	ΔT (delta time, difference in time between end and beginning of selected area)
CH 40	BPM (beats per minute; calculates time difference between end and beginning of selected area [same as ΔT] and converts difference from seconds to minutes; because BPM uses only time measurement of selected area, BPM value is not specific to a particular channel)
CH 40	p-p (finds maximum value in selected area and subtracts minimum value in selected area)
CH 2	p-p

The **selected area** is the part selected by the I-beam tool and includes the endpoints.

3. **Zoom** in on a small section of the Segment 1 data. Zoom in far enough that you can easily measure the intervals between peaks for approximately four cycles. The following tools help you adjust the data window:

Autoscale horizontal	Horizontal (Time) Scroll Bar
Autoscale waveforms	Vertical (Amplitude) Scroll Bar
Zoom Tool	Zoom Previous

4. Select the inspiration area, as shown in Figure 40.22. The ΔT measurement gives the duration of inspiration. You can record this and all other measurements in the Laboratory Report or by choosing **Edit > Journal > Paste Measurements** to paste the data into your computer journal for future reference. This inspiration duration value goes in Table 40.6 in the Laboratory Report.

5. Select the expiration area (Figure 40.23). Here the ΔT measurement gives the duration of expiration. Either record this information in Table 40.6 or paste it into your journal.

6. Repeat steps 4 and 5 for two additional cycles of the Segment 1 data. Record the data in the appropriate columns of Table 40.6 or in your journal.

7. Select a Segment 1 area from the beginning to the end of one breathing cycle (inspiration plus expiration), as in Figure 40.24. This time interval is called the

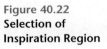

Figure 40.22
Selection of
Inspiration Region

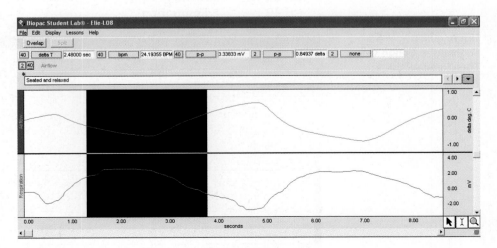

Figure 40.23
Selection of Expiration Region

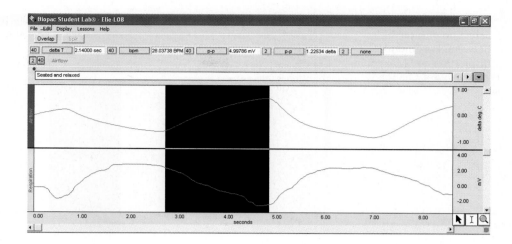

Figure 40.24
Selection of One Respiratory Cycle

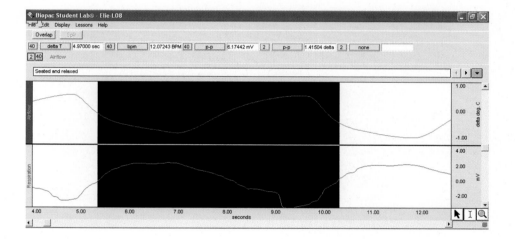

total duration. Now the ΔT measurement is the total duration, and BPM is the breathing rate of the selected area. Record your measurements either in Table 40.6 or in your journal.

8. Repeat steps 3 through 7 for data segments 2, 3, and 4. Record the data either in Table 40.7 in the Laboratory Report or in your journal. (The blacked-out cells in the Cough column of the table mean that only one cough measurement is required.)

9. Select three cycles in each of the four segments, and determine the respiration amplitude (maximum peak height) for each. The selected area should start at the middle of the descending wave in order to capture the minimum and maximum amplitudes. The p-p measurement will display the amplitude. Record the data either in Table 40.8 in the Laboratory Report or in your journal. (Again, the blacked-out cells tell you that only one cough measurement is needed.)

10. Select the interval between maximum inspiration and maximum temperature change in each of the four data segments (Figure 40.25). Record the CH 2 p-p (temperature amplitude) data and the data for ΔT between the two peaks either in Table 40.9 in the Laboratory Report or in your journal.

11. Save or print the data file and your journal notes. ■

Figure 40.25
Selection of Interval Between Maximum Inspiration and Maximum Temperature Change

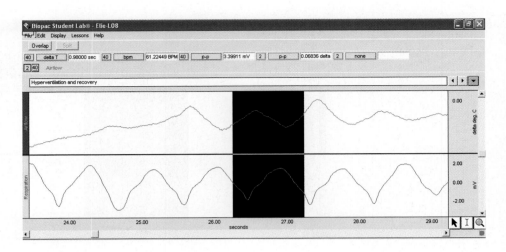

Physiology of the Respiratory System

Name _____

Date _____

Section _____

A. Matching

Match each term in the left column with its correct description from the right column.

_____ **1.** vital capacity

_____ **2.** tidal volume

_____ **3.** IRV

_____ **4.** ERV

_____ **5.** residual volume

_____ **6.** total lung capacity

_____ **7.** respiratory rate

_____ **8.** minute volume

_____ **9.** spirometer

_____ **10.** pulmonary ventilation

A. respiratory rate multiplied by tidal volume

B. volume of air that can be forcefully exhaled after normal exhalation

C. instrument used to measure respiratory volumes

D. amount of air normally inhaled or exhaled

E. inspiration and expiration

F. IRV + TV + ERV

G. number of breaths per minute

H. volume of air that can be forcefully inhaled after normal inhalation

I. vital capacity plus residual volume

J. volume of air that cannot be forcefully exhaled

B. Define

Define the following:

1. pulmonary ventilation

2. intrapulmonic pressure

3. inspiration

4. expiration

40

C. Short-Answer Questions

1. Describe how to calculate inspiratory reserve volume.

2. Use Boyle's law to explain the process of pulmonary ventilation.

3. How do external and internal respiration differ from each other?

BIOPAC: Volumes and Capacities

Name _____

Date _____

Section _____

A. Volume Measurements

Subject Profile

Name _____   Height _____

Age _____   Weight _____

Sex: Male/Female

1. Predicted Vital Capacity

Use either equation in Table 40.3 to calculate the subject's predicted vital capacity in liters.

2. Observed Volumes and Capacities

Table 40.4 **Respiratory Volume Measurements**

Type of Volume	Measurement (liters)
Tidal volume (TV)	
Inspiratory reserve volume (IRV)	
Expiratory reserve volume (ERV)	
Vital capacity (VC)	

Residual volume (RV) used: _____ liters (Default is 1.2 liters.)

Use the data you entered into Table 40.4 and the equations from the middle column of Table 40.5 to calculate inspiratory, expiratory, functional residual, and total lung capacities. Enter your results in the rightmost column of Table 40.5

Table 40.5 **Calculated Respiratory Capacities**

Capacity	Formula	Your Calculation
Inspiratory (IC)	IC = TV + IRV	
Expiratory (EC)	EC = TV + ERV	
Functional residual (FRC)	FRC = ERV + RV	
Total lung (TLC)	TLC = IRV + TV + ERV + RV	

LABORATORY REPORT

Compare the subject's lung volumes with the average volumes presented earlier in this exercise:

Average	Subject	
TV	500 ml	_____ml
IRV	3,300 ml	_____ml
ERV	1,000 ml	_____ml

3. Observed versus Predicted Vital Capacity

What is the subject's observed vital capacity as a percentage of the predicted vital capacity for her or his sex, age, and height (Table 40.2)?

_____ liters observed

$$\frac{\text{_____}}{\text{_____ liters predicted}} \times 100 = \underline{\hspace{1cm}}\%$$

Note: Vital capacities are dependent on other factors besides sex, age, and height. Therefore, 80% of predicted values are still considered normal.

B. Short-Answer Questions

1. Why does predicted vital capacity vary with height?

2. Explain how age and gender might affect lung capacity.

3. How would the volume measurements change if data were collected after vigorous exercise?

4. What is the difference between volume measurements and capacities?

5. Name the various types of pulmonary capacity studied in this exercise.

BIOPAC: Respiratory Rate and Depth

Name _____

Date _____

Section _____

A. Data and Calculations

Subject Profile

Name _____   Height _____

Age _____   Weight _____

Sex: Male/Female

1. Normal Breathing (Segment 1)
   Complete Table 40.6 with values for each cycle and calculate the means.

Table 40.6 Segment 1 Data

Rate	Measurement	Channel	Cycle 1	Cycle 2	Cycle 3	Mean
Inspiration duration	ΔT	40				
Expiration duration	ΔT	40				
Total duration	ΔT	40				
Breathing rate	BPM	40				

2. Comparison of Ventilation Rates (Segments 2–4)
   Complete Table 40.7 with measurements from CH 40 for three cycles of each segment and calculate the means.
   Note: ΔT is cycle duration, BPM is breathing rate, and Cough has only one cycle.

Table 40.7 Segments 2–4 Data

Measurement	Hyperventilation (Segment 2)		Hypoventilation (Segment 3)		Cough (Segment 4)		Read Aloud (Segment 4)	
	ΔT	BPM	ΔT	BPM	ΔT	BPM	ΔT	BPM
Cycle 1								
Cycle 2								
Cycle 3								
MeanΩ								

Note: ΔT is cycle duration, BPM is breathing rate, and Cough has only one cycle

3. Relative Ventilation Depths (Segments 1–4). Calculate the means in Table 40.8.

Table 40.8 **Ventilation Depth Comparisons**

Depth	p-p (CH 40)			Mean
	Cycle 1	Cycle 2	Cycle 3	
Normal breathing (Segment 1)				
Hyperventilation (Segment 2)				
Hypoventilation (Segment 3)				
Cough (Segment 4)				

4. Relationship between Respiratory Depth and Temperature (Segments 1–4)

Table 40.9 **Respiratory Depth and Temperature (Segments 1–4)**

Measurement	Channel	Normal Breathing (Segment 1)	Hyperventilation (Segment 2)	Hypoventilation (Segment 3)	Cough (Segment 4)	Reading Aloud (Segment 4)
p-p (temperature amplitude)	2					
ΔT (between maximum inspiration and peak temperature amplitude)	40					

B. Short-Answer Questions

1. Suppose subjects in the Respiratory Rate and Depth Laboratory Activity were told to hold their breath immediately after hyperventilating and immediately after hypoventilating. Would a subject hold her or his breath longer after hyperventilating or after hypoventilating? Explain your answer.

2. What changes occur in the body as a person hypoventilates?

3. How does the body adjust respiratory rate and depth to counteract the effect of hypoventilation?

4. In which part of the respiratory cycle is the temperature of the air being breathed highest? In which part of the cycle is the temperature lowest?

5. Explain why temperature varies during the respiratory cycle.

6. What changes in breathing occur as a person coughs?

7. What changes in breathing occur as a person reads aloud?

8. Refer to Table 40.6 data. For each cycle and for the mean values, divide 60 seconds by the total-duration time (which is in seconds) to determine the breathing rate in breaths per minute. Do the rates you calculate this way match the breathing rates you entered for each cycle in the bottom row of Table 40.6 (the rate calculated by the computer)? If your calculated rates do not match the computer's, suggest a possible explanation for the difference.

9. In your Table 40.8 data, are there differences in the relative ventilation depths from one segment to another?

10. Refer to Table 40.9. How does ventilation depth influence the temperature of air being exhaled?

Anatomy of the Digestive System

OBJECTIVES

On completion of this exercise, you should be able to:

- Identify the major layers and tissues of the digestive tract.

- Identify all digestive anatomy on laboratory models and charts.

- Describe the histological structure of the various digestive organs.

- Trace the secretion of bile from the liver to the duodenum.

- List the organs of the digestive tract and the accessory organs that empty into them.

The five major processes of digestion are (1) ingestion of food into the mouth, (2) movement of food through the digestive tract, (3) mechanical and enzymatic digestion of food, (4) absorption of nutrients into the blood, and (5) formation and elimination of indigestible material and waste. The ingested food passes through a tubular **digestive tract** extending from the mouth to the anus, and each organ of the tract has a specific function. Accessory organs occur outside the digestive tract, and these accessory organs plus the tract organs make up the **digestive system** (Figure 41.1). The accessory organs—salivary glands, liver, gallbladder, and pancreas—manufacture and secrete enzymes, hormones, and other compounds onto the inner lining of the digestive tract. Food does not pass through the accessory organs.

LAB ACTIVITY 1 Overview of the Digestive Tract

Examine the inside of your cheek with your tongue. What do you feel? The mucosal layer lining your mouth and the rest of the digestive tract is a mucous membrane that is kept wet. Glands drench the lining epithelium with enzymes, mucus, hormones, pH buffers, and other compounds to orchestrate the step-by-step breakdown of food as it passes through your mouth, pharynx, esophagus, stomach, small intestine, large intestine, and finally, rectum.

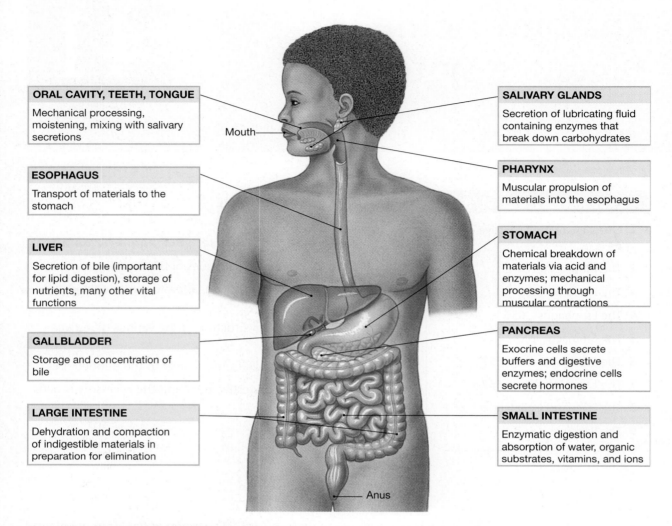

ORAL CAVITY, TEETH, TONGUE

Mechanical processing, moistening, mixing with salivary secretions

ESOPHAGUS

Transport of materials to the stomach

LIVER

Secretion of bile (important for lipid digestion), storage of nutrients, many other vital functions

GALLBLADDER

Storage and concentration of bile

LARGE INTESTINE

Dehydration and compaction of indigestible materials in preparation for elimination

SALIVARY GLANDS

Secretion of lubricating fluid containing enzymes that break down carbohydrates

PHARYNX

Muscular propulsion of materials into the esophagus

STOMACH

Chemical breakdown of materials via acid and enzymes; mechanical processing through muscular contractions

PANCREAS

Exocrine cells secrete buffers and digestive enzymes; endocrine cells secrete hormones

SMALL INTESTINE

Enzymatic digestion and absorption of water, organic substrates, vitamins, and ions

Mouth

Anus

Figure 41.1 The Components of the Digestive System
The major regions and accessory organs of the digestive tract, together with their primary functions.

Four major tissue layers compose the digestive tract: the mucosa, submucosa, muscularis externa, and serosa (Figure 41.2). Although the basic histological structure of the tract is similar along its entire length, each region has anatomical specializations reflecting that region's role in the digestive process. Figure 41.2 illustrates the anatomy of the digestive-tract lining, and Figure 41.3 shows the same thing via a micrograph of the ileum of the small intestine in transverse section.

The **mucosa** lines the digestive **lumen**, the open space within the tract. Three distinct mucosal layers can be identified: the digestive epithelium; a sheet of connective tissue called the lamina propria (LA-mi-nuh PRO-prē-uh); and a thin muscle layer, the muscularis (mus-kū-LAR-is) mucosae. The **digestive epithelium** is the only layer exposed to the lumen of the tract. From the mouth to the esophagus the digestive epithelium is **stratified squamous epithelium**, protecting the mucosa from abrasion during swallowing. The stomach and small and large intestines are lined with **simple columnar epithelium**, as food in these parts of the tract is liquid and less abrasive to the digestive epithelium. Beneath the digestive epithelium is the **lamina propria**, a layer of connective tissue that attaches the epithelium and contains blood vessels, lymphatic vessels, and nerves. A thin layer of smooth muscle, the **muscularis mucosae**, is the outermost layer of the mucosa (outermost in the sense of farthest from the lumen). The muscularis mucosae has two layers, an in-

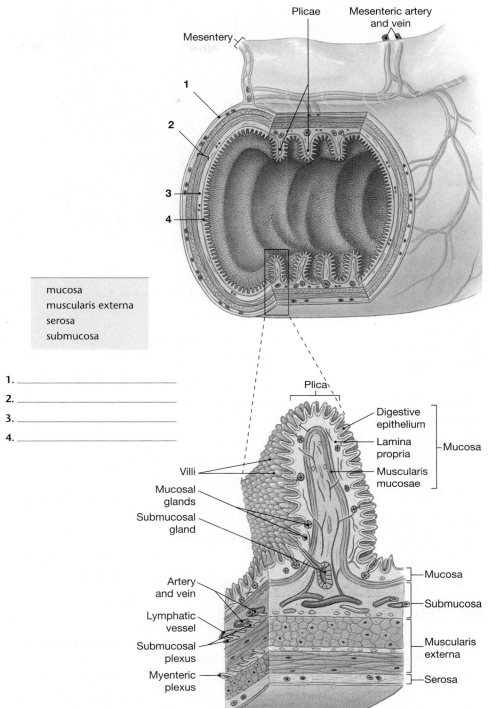

mucosa
muscularis externa
serosa
submucosa

1. _____
2. _____
3. _____
4. _____

Figure 41.2 The Structure of the Digestive Tract

A diagrammatic view of a representative portion of the digestive tract. The features illustrated are those of the small intestine.

ner **circular layer** that wraps around the tract and an outer **longitudinal layer** that extends along the length of the tract. (These two regions of the muscularis mucosae are not distinguishable in Figure 41.3.)

Superficial to the mucosa is the **submucosa**, a loose connective-tissue layer containing blood vessels, lymphatic vessels, and nerves. Superficial to the submucosa is the **muscularis externa** layer. Between the submucosa and the muscularis externa is a network of sensory and autonomic nerves, the **submucosal plexus**, which controls the tone of the muscularis mucosae.

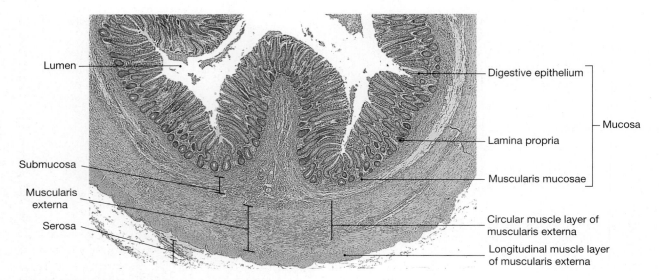

Figure 41.3 Histological Structure of the Digestive Tract
Photomicrograph of ileum showing general histological organization (LM × 160).

There are several layers of smooth muscle tissue in the muscularis externa. The inner layer is **circular muscle** that wraps around the digestive tract; when this muscle contracts, the tract gets narrower. The outer layer is **longitudinal muscle,** with fibers oriented parallel to the length of the tract. Contraction of this muscle layer shortens the tract. The layers of the muscularis externa produce waves of contraction called **peristalsis** (per-i-STAL-sis), which move materials along the digestive tract. Between the muscle layers is the **myenteric** (mī-en-TER-ik) **plexus,** nerves that control the activity of the muscularis externa.

The superficial layer of the digestive tract is the **serosa,** or **adventitia,** a loose connective-tissue covering that attaches and holds the tract in position. In the abdominal cavity, the adventitia is modified as the **peritoneum,** a serous membrane covering of the stomach and intestines.

QuickCheck Questions

1.1 What are the four major layers of the digestive tract wall?

1.2 What are the three sublayers of the mucosa?

1 Materials

☐ Compound microscope

☐ Prepared slide of ileum

☐ Intestinal wall model

Procedures

1. Review the anatomy of the digestive tract lining presented in Figure 41.2, and then label the figure.

2. Identify the major layers of the digestive tract on an intestinal wall model.

3. Identify the four major layers of the digestive tract from a cross-section of intestine.

    a. Set up the microscope, and focus on the slide using low magnification.

    b. Using Figure 41.3 as a guide, locate the lumen, the epithelium, the lamina propria, and the mucosa layer.

    c. Locate the submucosa, identified by its loose connective tissue and numerous blood and lymphatic vessels.

    d. Identify the circular and longitudinal muscles of the muscularis externa.

    e. Locate the adventitia layer.

    f. In the following space, draw a section of the microscope field, including all four main layers of the tract. ■

LAB ACTIVITY **2**   The Mouth

The mouth is the site of food ingestion, the most enjoyable part of digestion for most individuals. The mouth is formally called either the **oral cavity** or the **buccal** (BUK-al) **cavity** and is defined by the space from the lips, or **labia**, posterior to the **fauces** (FAW-sēz), the opening between the mouth and the throat (Figure 41.4). A cone-shaped **uvula** (Ū-vū-luh) is suspended from the posterior soft palate just anterior to the fauces. The lateral walls are composed of the **cheeks**, and the roof is the **hard palate** and the **soft palate** of the maxillae and palatine bones. The region between the teeth and cheeks is the **vestibule**. The floor of the mouth is muscular, mostly because of the muscles of the **tongue**. A fold of tissue, the **lingual frenulum** (FREN-ū-lum), anchors the tongue yet allows free movement for food processing and speech. Between the posterior base of the tongue and the roof of the mouth is the **palatoglossal** (pal-a-tō-GLOS-al) **arch**. At the fauces is the **palatopharyngeal arch**.

The mouth contains two structures that act as digestive-system accessory organs: the salivary glands and the teeth. Three major pairs of salivary glands, illustrated in Figure 41.5, produce the majority of the saliva, enzymes, and mucus of the oral cavity. The largest, the **parotid** (pa-ROT-id) **gland**, is in front of the ear between the skin and the masseter muscle. The **parotid duct** (Stenson's duct) pierces the masseter and enters the oral cavity to secrete saliva at the upper second molar. Saliva from this gland is a thick serous secretion and contains the enzyme salivary amylase. This enzyme starts the digestive process by splitting complex carbohydrates, such as starch, into shorter polysaccharide and disaccharide molecules.

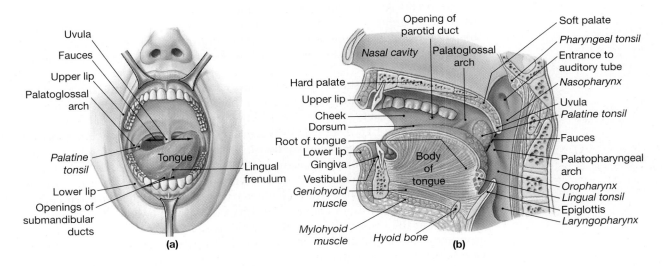

Figure 41.4 The Oral Cavity
(a) An anterior view of the oral cavity, as seen through the open mouth. **(b)** A sagittal section.

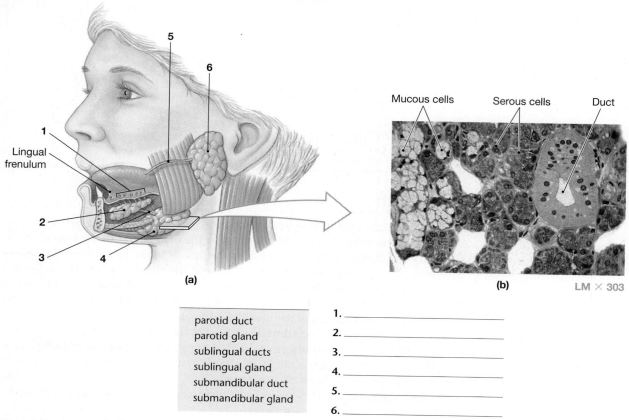

Figure 41.5 The Salivary Glands

(a) A lateral view, showing the relative positions of the salivary glands and ducts on the left side of the head. For clarity, the left ramus and body of the mandible have been removed. (b) The submandibular gland secretes a mixture of mucins, produced by mucous cells, and enzymes, produced by serous cells (LM × 303).

The **submandibular gland**, as its name implies, is under the mandible and extends from the mandibular arch posterior to the ramus. The **submandibular duct** (Wharton's duct) passes through the lingual frenulum and opens at the swelling on the central margin of this tissue. Submandibular secretions are thicker than that of the parotid glands because of the presence of mucin, a thick mucus that helps to keep food in a bolus, or ball, for swallowing. Notice the mucous cells in the micrograph in Figure 41.5.

The **sublingual** (sub-LING-gwal) **glands** are located under the tongue at the floor of the mouth. These glands secrete a thicker saliva with mucous into numerous **sublingual ducts** (ducts of Rivinus) that open along the base of the tongue.

The **teeth** are accessory digestive structures for chewing, or **mastication** (mas-ti-KĀ-shun). The **occlusal surface** is the superior area where food is ground, snipped, and torn by the tooth. Figure 41.6a details the anatomy of a typical adult tooth. The tooth is anchored in the alveolar bone of the jaw by a strong **periodontal ligament** that lines the embedded part of the tooth, the **root**. The **crown** is the portion of the tooth above the **gingiva** (JIN-ji-va), or gum. The crown and root meet at the neck, where the gingiva forms the **gingival sulcus**, a tight seal around the tooth. Although a tooth has many distinct layers, only the inner **pulp cavity** is filled with living tissue, the pulp. Supplying the pulp tissue are blood vessels, lymphatic vessels, and nerves, all of which enter the pulp cavity through the **apical foramen** at the inferior tip of the **root canal**. Surrounding the pulp

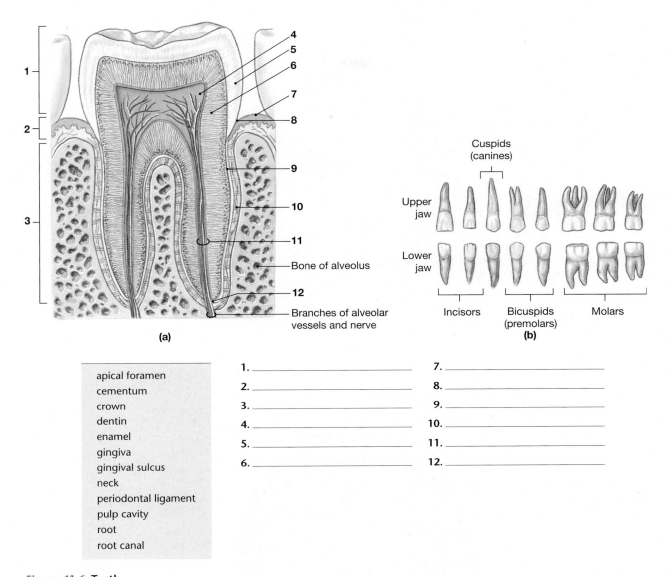

Figure 41.6 Teeth

(a) A diagrammatic section through a typical adult tooth. **(b)** The adult teeth.

apical foramen
cementum
crown
dentin
enamel
gingiva
gingival sulcus
neck
periodontal ligament
pulp cavity
root
root canal

1. _____
2. _____
3. _____
4. _____
5. _____
6. _____

7. _____
8. _____
9. _____
10. _____
11. _____
12. _____

cavity is **dentin** (DEN-tin), a hard, nonliving solid similar to bone matrix. Dentin makes up most of the structural mass of a tooth. At the root, the dentin is covered by **cementum**, a material that provides attachment for the periodontal ligament. The exposed crown is covered with **enamel**, the hardest substance produced by living organisms. Because of the hard enamel of teeth, they are often used to identify accident victims and skeletal remains that have no other identifying features.

Humans have two sets of teeth during their lifetime (Figure 41.7). The first set, the **deciduous** (de-SID-ū-us; *decidua,* to shed) **dentition,** starts to appear at about the age of six months and is replaced by the **permanent dentition** starting at around the age of six years. The deciduous teeth are commonly called the *primary teeth, milk teeth,* or *baby teeth.* There are 20 deciduous teeth, 5 in each jaw quadrant. (The mouth is divided into four quadrants: upper right, upper left, lower right, lower left.) Moving laterally from the midline of either jaw, the primary teeth are the **central incisor**, **lateral incisor**, **cuspid**, **first molar**, and **second molar** (Figure 41.7a). The permanent dentition, depicted in (Figure 41.7b, consists of 32 teeth, each quadrant containing a central incisor, lateral incisor, cuspid (canine), **first** and **second premolars** (bicuspids), and first, second, and third molars. The third molar is also called the *wisdom tooth.*

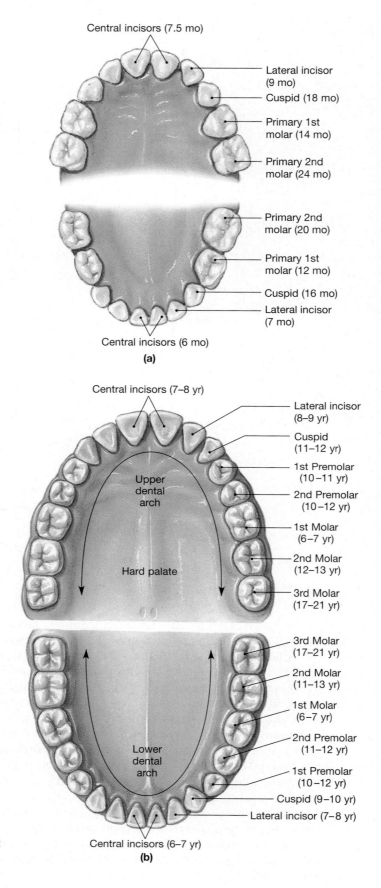

Central incisors (7.5 mo)

Lateral incisor (9 mo)

Cuspid (18 mo)

Primary 1st molar (14 mo)

Primary 2nd molar (24 mo)

Primary 2nd molar (20 mo)

Primary 1st molar (12 mo)

Cuspid (16 mo)

Lateral incisor (7 mo)

Central incisors (6 mo)

(a)

Central incisors (7–8 yr)

Lateral incisor (8–9 yr)

Cuspid (11–12 yr)

1st Premolar (10–11 yr)

2nd Premolar (10–12 yr)

1st Molar (6–7 yr)

2nd Molar (12–13 yr)

3rd Molar (17–21 yr)

Upper dental arch

Hard palate

3rd Molar (17–21 yr)

2nd Molar (11–13 yr)

1st Molar (6–7 yr)

2nd Premolar (11–12 yr)

1st Premolar (10–12 yr)

Cuspid (9–10 yr)

Lateral incisor (7–8 yr)

Lower dental arch

Central incisors (6–7 yr)

(b)

Figure 41.7
Primary and Secondary Teeth

(a) The primary teeth, with the age at eruption given in months. (b) The adult teeth, with the age at eruption given in years.

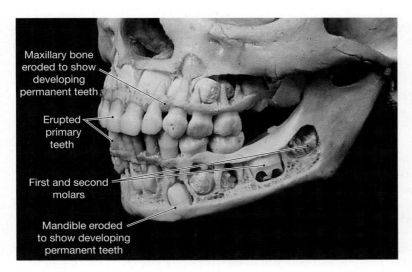

Maxillary bone eroded to show developing permanent teeth

Erupted primary teeth

First and second molars

Mandible eroded to show developing permanent teeth

Figure 41.8
Mandible and Maxillary Bones with Unerupted Teeth Exposed

Figure 41.8 is a photograph of a child's skull with the maxilla and mandible cut to reveal the permanent dentition developing at the roots of the primary dentition. Eruption of the permanent teeth helps to loosen and push out the deciduous teeth. If a permanent tooth erupts before the primary tooth is loose, however, the incoming tooth may get out of alignment with the other teeth.

QuickCheck Questions

2.1 Which two mouth structures are digestive-system accessory structures?

2.2 Where is each salivary gland located?

2.3 Describe the main layers of a tooth.

2 Materials

- ☐ Head model
- ☐ Digestive system chart
- ☐ Tooth model
- ☐ Hand mirror

Procedures

1. Review the mouth anatomy presented in Figures 41.4 and 41.5, and label Figure 41.5.

2. Identify the anatomy of the mouth on the head model and digestive system chart.

3. Identify each salivary gland and duct on the model and/or chart.

4. Use the hand mirror to locate your uvula, fauces, and palatoglossal arch. Lift your tongue and examine your submandibular duct.

5. Review the tooth anatomy in Figure 41.6 and label the figure.

6. Use the mirror to examine your teeth. Locate your incisors, cuspids, bicuspids, and molars. How many teeth do you have? Are you missing any because of extractions? Do you have any wisdom teeth? ◾

LAB ACTIVITY 3   The Pharynx

The throat, or **pharynx,** a passageway for both nutrients and air, is divided into three anatomical regions—nasopharynx, oropharynx, and laryngopharynx—as shown in Figure 41.4b. The **nasopharynx** is superior to the **oropharynx,** which is located directly posterior to the oral cavity. Muscles of the soft palate contract during swallowing and close the passageway to the nasopharynx to prevent food from entering the nasal cavity. Because food normally does not enter the nasopharynx, it can function to clean and warm inspired air and is lined with pseudostratified ciliated columnar epithelium. When you swallow a bolus of food, it passes through the fauces into the oropharynx, and inferior to the oropharynx is the **laryngopharynx**. Toward the base of this area, the pharynx branches into the larynx of the respiratory system and the esophagus leading to the stomach. The larynx has a flap called the **epiglottis** that closes the larynx so that swallowed food can enter only the esophagus and not the respiratory passageways. The lumen of the oropharynx and laryngopharynx is lined with stratified squamous epithelium to protect the walls from abrasion as swallowed food passes through this region of the digestive tract.

QuickCheck Questions

3.1 What are the three regions of the pharynx?

3.2 What is the lining epithelium of each region of the pharynx?

3 Materials

☐ Head model

☐ Digestive system chart

☐ Hand mirror

Procedures

1. Identify the anatomy of the pharynx on the head model and digestive system chart.

2. Put your finger on your Adam's apple (thyroid cartilage of the larynx) and swallow. How does your larynx move, and what is the purpose of this movement? ■

LAB ACTIVITY 4

The Esophagus

The food tube, or **esophagus,** connects the pharynx to the stomach (Figure 41.9). It is inferior to the pharynx and posterior to the trachea (windpipe). The esophagus is approximately 25 cm (10 in.) long. It pierces the diaphragm at the **esophageal hiatus** (hī-Ā-tus) to connect with the stomach in the abdominal cavity. At the stomach, the esophagus terminates in a **lower esophageal sphincter**, a muscular valve that prevents stomach contents from backwashing into the esophagus.

Clinical Application

Acid Reflux

Acid reflux, also commonly called *heartburn,* occurs when stomach acid backflows into the esophagus and irritates the mucosal lining. The term *reflux* refers to a backflow, or regurgitation, of liquid; in this case, gastric juice (which is acidic). Some individuals have a weakened lower esophageal sphincter that allows the gastric juices to reflux during gastric mixing. Recent studies indicate that acid reflux is a major cause of esophageal and pharyngeal cancer. ▶

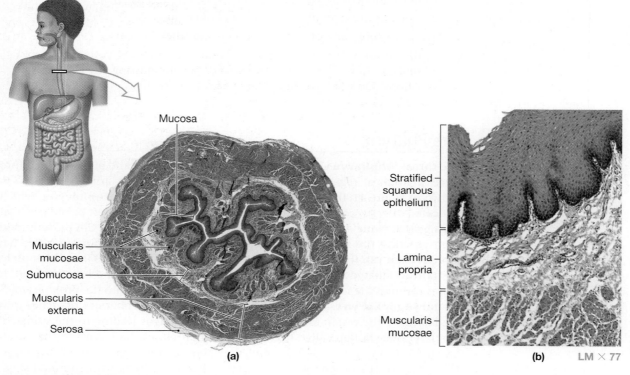

Figure 41.9 The Esophagus

(a) A transverse section through the esophagus. (b) The esophageal mucosa (LM × 77).

QuickCheck Questions

4.1 Which parts of the digestive tract does the esophagus connect?

4.2 Where is the esophageal hiatus?

4 Materials

☐ Torso model

☐ Digestive system chart

☐ Compound microscope

☐ Prepared slide of esophagus

Procedures

1. Review the structure of the esophagus in Figure 41.9.

2. Identify the anatomy of the esophagus on the torso model and digestive system chart.

3. Place the esophagus slide on the microscope, and focus on the specimen at low magnification. Observe the organization of the esophageal wall, and identify the mucosa, submucosa, muscularis externa with its inner circular and outer longitudinal layers, and serosa. Use Figure 41.9 for reference during your observations.

4. Increase the magnification and study the mucosa. Distinguish among the digestive epithelium, which is stratified squamous epithelium, the lamina propria, and the muscularis mucosae. ■

LAB ACTIVITY 5 The Stomach

Figure 41.10 depicts the gross anatomy of the **stomach**, the J-shaped organ located just inferior to the diaphragm. The four major regions of the stomach are the **cardia** (KAR-dē-uh), where the stomach connects with the esophagus; the **fundus** (FUN-dus), the upper rounded area; the **body**, the middle region; and the **pylorus** (pī-LOR-us), the narrowed distal end connected to the small intestine. The **pyloric sphincter** (also called the **pyloric valve**) controls movement of material from the stomach into the small intestine. The lateral, convex border of the stomach is the **greater curvature**, and the medial concave stomach margin is the **lesser curvature**. Extending from the greater curvature is the **greater omentum** (ō-MEN-tum), commonly referred to as the *fatty apron*. This fatty layer is part of the serosa of the stomach wall. Its functions are to protect the abdominal organs and to attach the stomach and part of the large intestine to the posterior abdominal wall. The **lesser omentum**, also part of the serosa, suspends the stomach from the liver.

Figure 41.11 shows the histology of the stomach wall. The digestive epithelium is simple columnar and folds deep into the lamina propria as **gastric pits** that extend to the base of **gastric glands**. The glands consist of numerous **parietal cells**, which secrete hydrochloric acid, and **chief cells**, which release an inactive protein-digesting enzyme called pepsinogen. The submucosa has **rugae** (ROO-gē), which are folds that enable the stomach to expand as it fills with food. Unlike what is found in other regions of the digestive tract, the muscularis externa of the stomach contains three layers of smooth muscle instead of two. The inner layer (closest to the stomach lumen) is an **oblique layer**, surrounded by a **circular layer** and then an external **longitudinal layer**. The three muscle layers contract and churn stomach contents, mixing gastric juice and liquefying the food into **chyme**. As mentioned previously, the serosa is expanded into the greater and lesser omenta.

QuickCheck Questions

5.1 What are the four major regions of the stomach?

5.2 How is the muscularis externa of the stomach unique?

5.3 Which structure of the stomach allows the organ to distend?

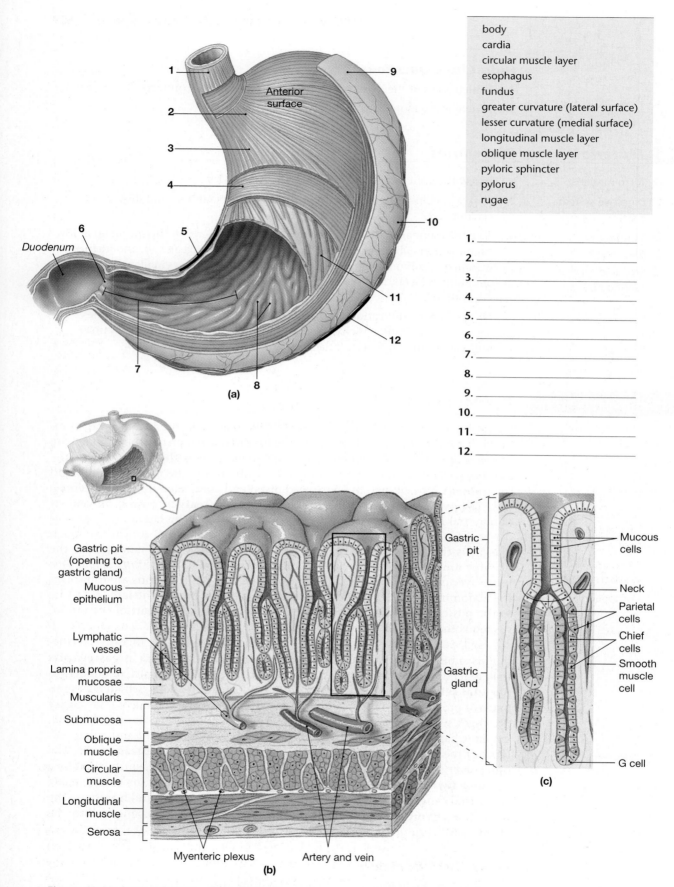

body
cardia
circular muscle layer
esophagus
fundus
greater curvature (lateral surface)
lesser curvature (medial surface)
longitudinal muscle layer
oblique muscle layer
pyloric sphincter
pylorus
rugae

1. _____
2. _____
3. _____
4. _____
5. _____
6. _____
7. _____
8. _____
9. _____
10. _____
11. _____
12. _____

Anterior surface

Duodenum

(a)

Gastric pit (opening to gastric gland)
Mucous epithelium
Lymphatic vessel
Lamina propria mucosae
Muscularis
Submucosa
Oblique muscle
Circular muscle
Longitudinal muscle
Serosa

Myenteric plexus
Artery and vein

(b)

Gastric pit
Gastric gland

Mucous cells
Neck
Parietal cells
Chief cells
Smooth muscle cell
G cell

(c)

Figure 41.10 **Gross Anatomy of the Stomach**
(a) Anterior view of the stomach showing superficial landmarks. (b) Diagrammatic view of the organization of the stomach wall. (c) A gastric gland.

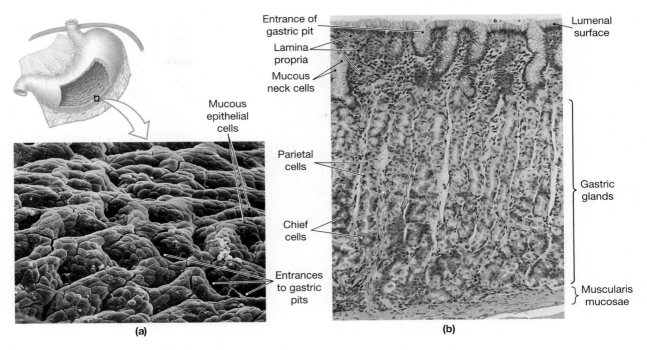

Figure 41.11 The Stomach Lining
(a) A surface view of the gastric mucosa of the full stomach, showing the entrances to the
gastric pits (SEM × 35). **(b)** A section through gastric pits and gastric glands (LM × 300).

5 *Materials*

- ☐ Torso model
- ☐ Digestive system chart
- ☐ Preserved animal stomach (optional)
- ☐ Compound microscope
- ☐ Prepared slide of stomach

Procedures

1. Review the anatomy of the stomach in Figure 41.10 and label the figure.
2. Identify the gross anatomy of the stomach on the torso model and digestive system chart.
3. If specimens are available, examine the stomach of a cat or other animal. Locate the rugae, cardia, fundus, body, pylorus, greater and lesser omenta, lower esophageal sphincter, and pyloric sphincter.
4. Place the stomach slide on the microscope stage, focus at low magnification, and observe the rugae. Note how the mucosal digestive epithelium is simple columnar.
5. Increase the magnification and locate the numerous gastric pits, which appear as invaginations along the rugae. Within the pits, distinguish between parietal cells, which are more numerous in the upper areas, and chief cells, those that have nuclei at the basal region of the cells. ∎

LAB ACTIVITY 6   The Small Intestine

The small intestine (Figure 41.12) is approximately 6.4 meters (21 ft.) long and composed of three segments: duodenum, jejunum, and ileum. Thin sheets of membrane called **mesenteries** (MEZ-en-ter-ēz) extend from the serosa to support and attach the small intestine to the posterior abdominal wall. The first 25 cm (10 in.) is the **duodenum** (doo-AH-de-num) and is attached to the distal region of the pylorus. Digestive secretions from the liver, gallbladder, and pancreas flow into ducts that merge and empty into the duodenum. This anatomy is described further in the upcoming section on the liver. The **jejunum** (je-JOO-num) is approximately 3.6 meters (12 ft.) long and is the site of most nutrient absorption. The last 2.6 meters (8 ft.) is the **ileum** (IL-ē-um), which terminates at the **ileocecal** (il-ē-ō-SĒ-kal) **valve** and empties into the cecum of the large intestine.

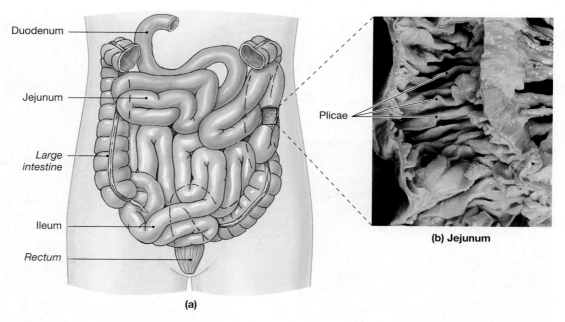

Duodenum

Jejunum

Large intestine

Ileum

Rectum

Plicae

(b) Jejunum

(a)

Figure 41.12 Regions of the Small Intestine
(a) The positions of the duodenum, jejunum, and ileum in the abdominopelvic cavity.
(b) A representative view of the jejunum.

The small intestine is the site of most digestive and absorptive activities and has specialized folds to increase the surface area for these functions (Figures 41.13 and 41.14). The submucosa and mucosa are creased together into large folds called **plicae** (PLĪ-sē). Along the plicae, the lamina propria is convoluted into small, fingerlike projections that are called **villi** and are lined with simple columnar epithelium. The epithelial cells have minute cell-membrane extensions or folds called **microvilli**.

The epithelium at the base of the villi forms a pocket of cells called **intestinal glands** (crypts of Lieberkuhn) that secrete intestinal juice rich in enzymes and pH buffers to neutralize stomach acid. Interspersed among the columnar cells are oval mucus-producing **goblet cells**. In the middle of each villus is a **lacteal** (LAK-tē-al), a lymphatic vessel that absorbs fatty acids and monoglycerides from lipid digestion. In the submucosa are scattered mucus-producing **submucosal** (Brunner's) **glands**. The submucosa of the ileum has **Peyer's patches**, large lymphatic nodules that prevent bacteria from entering the blood.

QuickCheck Questions

6.1 What are the three major regions of the small intestine?

6.2 What are plicae?

6.3 Where are the intestinal glands located?

6 Materials

- ☐ Torso model
- ☐ Digestive system chart
- ☐ Preserved animal intestines (optional)
- ☐ Compound microscope
- ☐ Prepared slide of duodenum and ileum

Procedures

1. Review the regions and organization of the small intestine in Figure 41.13, and label the figure.

2. Identify the anatomy of the small intestine on the torso model and the digestive system chart.

3. If a specimen is available, examine a segment of the small intestine of a cat or other animal.

4. Examine the duodenum slide at low magnification, and identify the mucosa, submucosa, muscularis externa, and serosa.

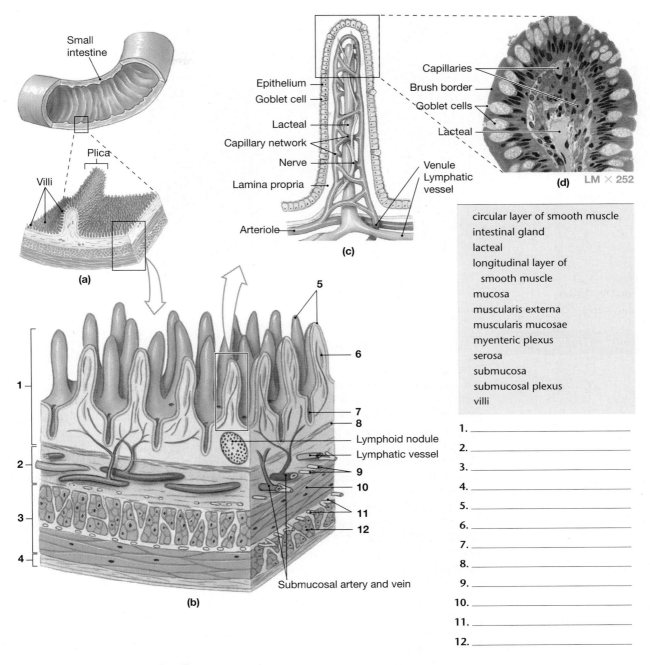

Figure 41.13 The Intestinal Wall
(a) A single plica and multiple villi. (b) The organization of the intestinal wall.
(c) Internal structures in a single villus, showing the capillary and lymphatic supplies.
(d) A villus in sectional view (LM × 252).

5. Increase the magnification to medium, and identify the simple columnar epithelium, goblet cells, lamina propria, and muscularis mucosae, using Figure 41.14 as a guide. The lacteals appear as empty ducts in the lamina propria of the villi. At the base of the villi, locate the intestinal glands.

6. Still at medium power, note the many glands in the submucosa. Trace the ducts of the glands to the mucosal surface.

7. Switch to the ileum slide, and examine it at low and medium magnifications. Locate the Peyer's patches in the submucosa. ■

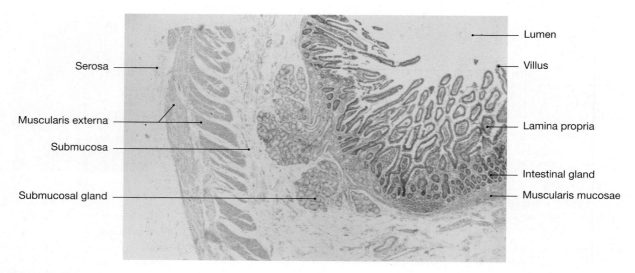

Serosa

Muscularis externa

Submucosa

Submucosal gland

Lumen

Villus

Lamina propria

Intestinal gland

Muscularis mucosae

Figure 41.14
The Duodenum
Transverse section of the human duodenum.

LAB ACTIVITY   7   The Large Intestine

The large intestine is the site of water absorption and waste compaction. It is approximately 1.5 meters (5 ft.) long and divided into two regions: colon (KŌ-lin), which makes up most of the intestine, and rectum. Figure 41.15a details the gross anatomy. The ileocecal valve regulates what enters the colon from the ileum. The first part of the colon, a pouchlike **cecum** (SĒ-kum), is located in the right lumbar region. At the medial floor of the cecum is the wormlike **appendix**. Past the cecum, the colon is divided into several regions: the **ascending colon** travels up the right side of the abdomen, bends left at the **right colic flexure**, and crosses the abdomen below the stomach as the **transverse colon**. The **left colic flexure** turns the colon downward to become the **descending colon**.

The S-shaped **sigmoid** (SIG-moyd) **colon** passes through the pelvic cavity to join the **rectum** (Figure 41.15c), which is the last 15 cm (6 in.) of the large intestine and the end of digestive tract. The opening of the rectum, the **anus,** is controlled by an **internal anal sphincter** of smooth muscle and an **external anal sphincter** of skeletal muscle. Longitudinal folds called **rectal columns** occur in the rectum where the digestive epithelium changes from simple columnar epithelium to stratified squamous epithelium.

The muscularis externa of the large intestine is modified into three separate bands of muscle, collectively called the **taenia coli** (TĒ-nē-a KŌ-lī). The muscle tone of the taenia coli constricts the colon wall into pouches called **haustra** (HAWS-truh), which permit the colon wall to expand and stretch.

QuickCheck Questions

7.1 What are the major regions of the colon?

7.2 Where is the appendix located?

7   Materials

☐ Torso model

☐ Digestive system chart

☐ Preserved animal intestines (optional)

Procedures

1. Review the anatomy of the large intestine in Figure 41.15 and label the figure.

2. Identify the gross anatomy of the large intestine on the torso model and the digestive system chart.

3. If a specimen is available, examine the colon of a cat or other animal. Locate each region of the colon, the left and right colic flexures, the taenia coli, and the haustrae. ■

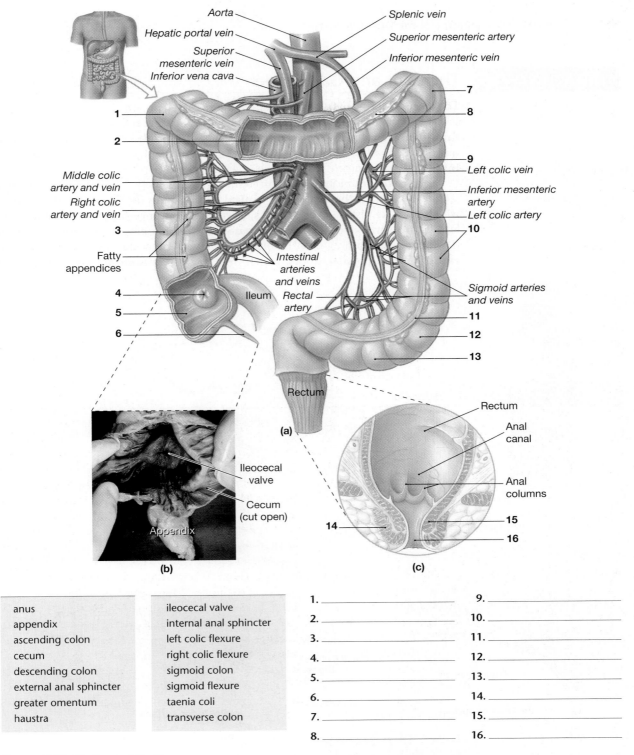

Aorta

Splenic vein

Hepatic portal vein

Superior mesenteric artery

Superior mesenteric vein

Inferior mesenteric vein

Inferior vena cava

7

1

8

2

9

Left colic vein

Middle colic artery and vein

Inferior mesenteric artery

Right colic artery and vein

Left colic artery

3

10

Fatty appendices

Intestinal arteries and veins

Ileum

Sigmoid arteries and veins

4

Rectal artery

11

5

12

6

13

Rectum

(a)

Ileocecal valve

Cecum (cut open)

Appendix

(b)

Rectum

Anal canal

Anal columns

14

15

16

(c)

anus	ileocecal valve
appendix	internal anal sphincter
ascending colon	left colic flexure
cecum	right colic flexure
descending colon	sigmoid colon
external anal sphincter	sigmoid flexure
greater omentum	taenia coli
haustra	transverse colon

1. _____ 9. _____
2. _____ 10. _____
3. _____ 11. _____
4. _____ 12. _____
5. _____ 13. _____
6. _____ 14. _____
7. _____ 15. _____
8. _____ 16. _____

Figure 41.15 **The Large Intestine**

(a) The gross anatomy and regions of the large intestine. (b) The appendix. (c) The rectum and anus.

LAB ACTIVITY 8 The Liver and Gallbladder

The **liver** is located mostly in the right abdominal region, inferior to the diaphragm. It has four major lobes (Figure 41.16). The **right** and **left lobes** are separated by the **falciform ligament,** which attaches the lobes to the abdominal wall. The square **quadrate lobe** is located on the inferior surface of the right lobe, and the **caudate lobe** is posterior, near the site of the inferior vena cava.

Each of the four lobes is organized into approximately 100,000 smaller **lobules** (Figure 41.17b). In the lobules, cells called **hepatocytes** (he-PAT-ō-sits) secrete bile, which acts like dish soap and breaks down the fat in ingested food. The bile is released into small ducts called **bile canaliculi,** which empty into **bile ductules** (DUK-tūlz) surrounding each lobule. Progressively larger ducts drain bile into the **right** and **left hepatic ducts,** which then join a **common hepatic duct**. Blood

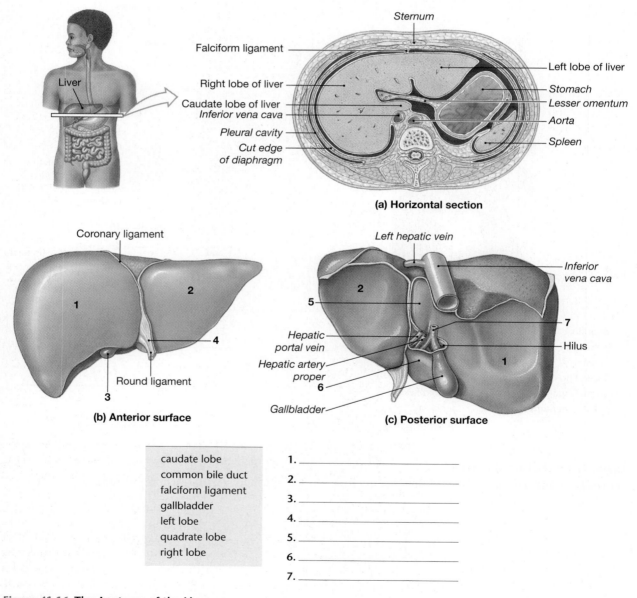

(a) Horizontal section

(b) Anterior surface

(c) Posterior surface

caudate lobe	1. _____
common bile duct	2. _____
falciform ligament	3. _____
gallbladder	4. _____
left lobe	5. _____
quadrate lobe	6. _____
right lobe	7. _____

Figure 41.16 The Anatomy of the Liver

(a) Horizontal sectional view through the superior abdomen. (b) The anterior surface of the liver. (c) The posterior surface of the liver.

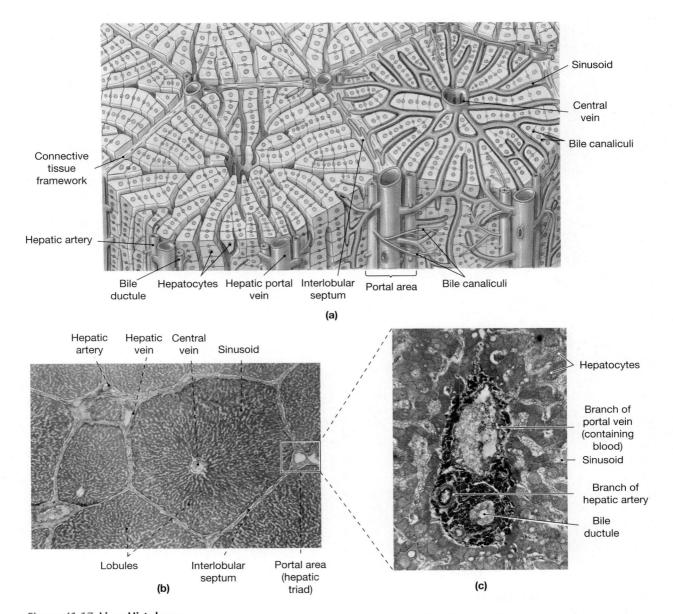

Figure 41.17 Liver Histology
(a) Diagrammatic view of lobular organization. (b) Light micrograph showing a section
through a liver lobule (LM × 38). (c) A portal area (LM × 31).

from the hepatic artery and from the hepatic portal vein drains into spaces in the
lobules called **sinusoids,** and the sinusoids then empty into a **central vein** in the
middle of each lobule. Hepatocytes lining the sinusoids phagocytize worn-out blood
cells and reprocess the hemoglobin pigments for new blood cells.

The **gallbladder** is located inferior to the right lobe of the liver. It is a small,
muscular sac that stores and concentrates bile salts used in the digestion of lipids.
Figure 41.18 details the ducts of the liver and gallbladder. The common hepatic duct
from the liver meets the **cystic duct** of the gallbladder to form the **common bile
duct**. This duct passes through the lesser omentum and joins the **pancreatic duct**
of the pancreas at the **duodenal ampulla** (am-PUL-uh). The ampulla projects into
the lumen of the duodenum at the **duodenal papilla**. A band of muscle called the
hepatopancreatic sphincter (sphincter of Oddi) regulates the flow of pancreatic
secretions and bile into the duodenum.

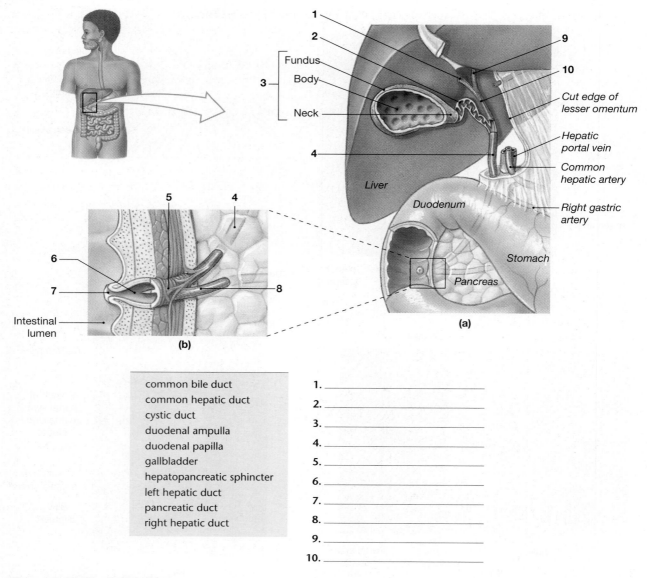

Figure 41.18 The Gallbladder

(a) A view of the inferior surface of the liver, showing the position of the gallbladder and ducts that transport bile from the liver to the gallbladder and duodenum. A portion of the lesser omentum has been cut away. (b) An interior view of the duodenum, showing the duodenal ampulla and related structures.

QuickCheck Questions

8.1 What are the names of the four lobes of the liver?

8.2 How does bile enter the small intestine?

8 *Materials*

☐ Torso model
☐ Digestive system chart
☐ Liver model

Procedures

1. Review the anatomy of the liver in Figures 41.16 and 41.17 and label Figure 41.16.

2. Review the anatomy of the gallbladder in Figure 41.18 and label the figure.

3. Identify the gross anatomy of the liver and gallbladder on the torso model, liver model, and digestive system chart.

☐ Preserved animal liver and gallbladder (optional)

☐ Compound microscope

☐ Prepared slide of liver

4. If specimens are available, examine the liver and gallbladder of a cat or other animal. Locate each liver lobe, the falciform ligament, and the hepatic, cystic, and common bile ducts.

5. Examine the liver slide at low magnification and identify the many lobules. Notice hepatocytes lining the sinusoids and the central canal of each lobule. In humans, the lobules are not well defined. Pigs and other animals have a connective tissue septum around each lobule; this septum can be seen in Figure 41.17b. ■

LAB ACTIVITY 9 The Pancreas

The **pancreas** lies between the stomach and the duodenum. The pancreas **head** is adjacent to the duodenum, the **body** is the central region, and the **tail** tapers to the distal end of the gland (Figure 41.19). The pancreas is characterized as a *double gland,* which means it has both endocrine and exocrine functions. The

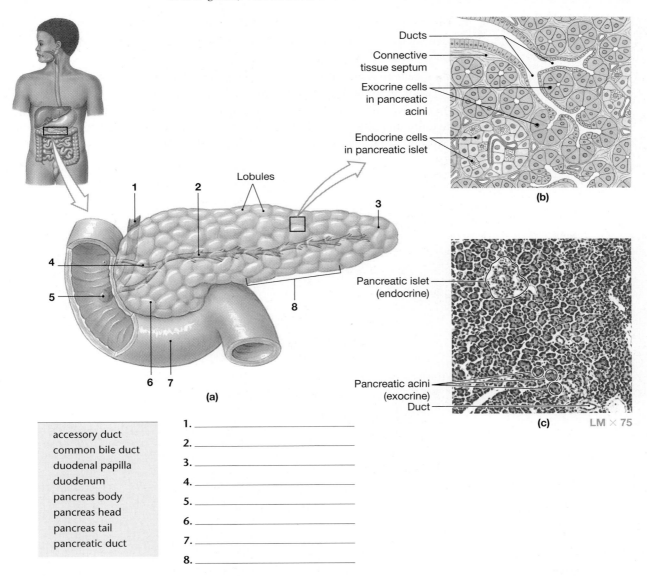

accessory duct
common bile duct
duodenal papilla
duodenum
pancreas body
pancreas head
pancreas tail
pancreatic duct

1. _____
2. _____
3. _____
4. _____
5. _____
6. _____
7. _____
8. _____

Figure 41.19 The Pancreas
(a) The gross anatomy of the pancreas. The head of the pancreas is tucked into a C-shaped curve of the duodenum that begins at the pylorus of the stomach. The cellular organization of the pancreas is shown **(b)** diagrammatically and **(c)** in a micrograph (LM × 75).

endocrine cells occur in **pancreatic islets** and secrete hormones for sugar metabolism. Most of the glandular epithelium of the pancreas has an exocrine function. These exocrine cells, called **acini** (AS-i-nī) **cells,** secrete pancreatic juice into small ducts called **acini,** located in the pancreatic glands. The acini drain into progressively larger ducts that merge as the **pancreatic duct** and, in some individuals, an **accessory duct**. The pancreatic duct joins the common bile duct at the duodenal ampulla.

QuickCheck Questions

9.1 What is the exocrine function of the pancreas?

9.2 How does the pancreatic duct connect to the duodenum?

9 Materials

☐ Torso model

☐ Digestive system chart

☐ Preserved animal pancreas (optional)

☐ Compound microscope

☐ Prepared slide of pancreas

Procedures

1. Review the anatomy of the pancreas in Figure 41.19 and label the figure.

2. Identify the anatomy of the pancreas on the torso model and the digestive system chart.

3. If a specimen is available, examine the pancreas of a cat or other animal. Locate the head, body, and tail of the organ and the pancreatic duct.

4. On the pancreas slide, observe the numerous oval pancreatic ducts at low and medium magnifications. The exocrine cells are the dark-stained acinar cells that surround groups of light-stained pancreatic islets. ■

Anatomy of the Digestive System

Name _____

Date _____

Section _____

A. Matching

Match each structure in the left column with its correct description from the right column.

_____ **1.** pyloric sphincter

_____ **2.** greater omentum

_____ **3.** incisor

_____ **4.** esophageal hiatus

_____ **5.** taenia coli

_____ **6.** haustra

_____ **7.** mesentery

_____ **8.** muscularis mucosae

_____ **9.** muscularis externa

_____ **10.** serosa

_____ **11.** fauces

_____ **12.** lingual frenulum

_____ **13.** gingiva

_____ **14.** enamel

_____ **15.** rugae

_____ **16.** plicae

_____ **17.** molar

_____ **18.** jejunum

_____ **19.** common bile duct

_____ **20.** duodenal ampulla

A. fatty apron hanging off stomach

B. muscle layer of mucosa

C. muscle folds of stomach wall

D. opening between mouth and pharynx

E. hard outer layer of tooth

F. pouches in colon wall

G. folds of intestinal wall

H. tooth used for snipping food

I. longitudinal muscle of colon

J. gum surrounding tooth

K. tooth used for grinding

L. valve between stomach and duodenum

M. swollen duct transporting bile and pancreatic juice

N. connective membrane of intestines

O. passageway for esophagus in diaphragm

P. major muscle layer superficial to submucosa

Q. middle segment of small intestine

R. duct joined by cystic duct

S. anchors tongue to floor of mouth

T. also called adventitia

B. Short-Answer Questions

1. List the three major pairs of salivary glands and the type of saliva each gland secretes.

2. List the accessory organs of the digestive system.

3. How is the wall of the stomach different from the wall of the esophagus?

4. Describe the gross anatomy of the large intestine.

5. How does the pancreas function as both an endocrine and an exocrine gland?

6. Where is the appendix located?

C. Drawing

1. Draw a longitudinal section detailing the anatomy of a typical tooth.

2. Draw a cross-section of the wall of the small intestine, showing the structures that increase the wall surface area for maximum digestion and absorption of food.

D. Analysis and Application

1. Trace a drop of bile from the point where it is produced to the point where it is released into the intestinal lumen.

2. A baby is born with esophageal atresia, an incomplete connection between the esophagus and the stomach. What will most likely happen to the infant if this defect is not corrected?

3. If the duodenal ampulla were blocked, what materials could not be released into the lumen of the small intestine?

4. List the modifications of the intestinal wall that increase surface area.

Digestive Physiology

OBJECTIVES

On completion of this exercise, you should be able to:

- Explain why enzymes are used to digest food.

- Describe a dehydration synthesis and a hydrolysis reaction.

- Describe the chemical composition of carbohydrates, lipids, and proteins.

- Discuss the function of a control group in scientific experiments.

Before nutrients can be converted to energy that is usable by the body's cells, the large organic **macromolecules** in food must be **catabolized,** or broken down, into **monomers,** the building blocks of macromolecules. The chemical reactions involved in catabolic processes are controlled by enzymes produced by the digestive system. **Enzymes** are protein **catalysts** that lower **activation energy,** which is the amount of energy required for a chemical reaction (Figure 42.1). Without enzymes, the body would have to heat up to dangerous temperatures to provide the activation energy necessary to decompose ingested food.

An enzymatic reaction involves reactants, called **substrates,** and results in a **product** (Figure 42.2). Each enzyme molecule has one or more **active sites** where substrates bind. Only substrates that are compatible with an enzyme's active sites are metabolized by the enzyme, and the enzyme is said to have **specificity** for compatible substrates. When the enzymatic chemical reaction is done, the product is released from the enzyme molecule, and the enzyme, unaltered in the reaction, can bind to other substrates and repeat the reaction.

There is a narrow range of physical conditions in which each enzyme operates at maximum efficiency. Temperature and pH are two important factors in enzymatic reactions. The protein digestion activity in this exercise provides a good example of how reaction conditions affect enzymatic reactions. The stomach produces an inactive enzyme, a **proenzyme,** called pepsinogen. When mixed with stomach acid, the pepsinogen is activated into the protein-digesting enzyme

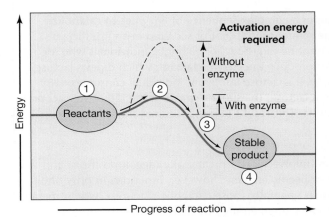

Figure 42.1
Enzymes and Activation Energy

Enzymes lower the activation energy requirements, so a reaction can occur readily, in order from 1 to 4, under conditions in the body.

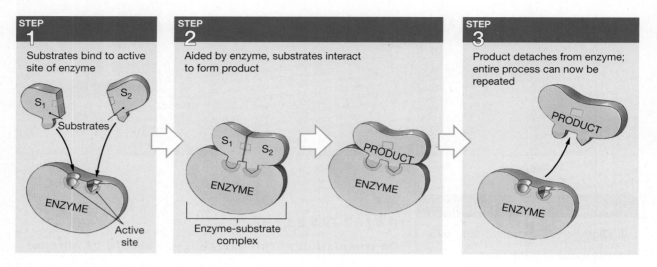

Figure 42.2 A Simplified View of Enzyme Structure and Function
Each enzyme contains a specific active site somewhere on its exposed surface.

pepsin, which begins to catabolize large protein molecules in the food. When materials from the stomach move into the small intestine, the pancreas releases sodium bicarbonate, which buffers the stomach acid mixed in with the partially digested protein and raises the pH to 7 or 8. Because pepsin is active only at an acidic pH (below 7), it becomes less active in the small intestine, while other enzymes take their turns metabolizing the protein.

The function of each enzyme is related to the shape of the enzyme molecules, much as the shape of a key determines which lock it fits. Anything that changes the molecular shape, a process called **denaturation**, renders the enzyme nonfunctional. Heat denatures some enzymes, for instance, and destroys them. This is what you do every time you cook an egg—the heat you apply denatures the protein in the egg.

Clinical Application

Lactose Intolerance

The digestive system requires a complex sequence of enzymes to catabolize food. If one enzyme in the sequence is absent or secreted in an insufficient quantity, the substrate cannot be digested. For example, individuals who are **lactose-intolerant** do not produce the enzyme lactase, which digests lactose, commonly called milk sugar. If a lactose-intolerant individual consumes dairy products, which are high in lactose, the sugar remains in the digestive tract and is slowly digested by bacteria. This results in gas, intestinal cramps, and diarrhea. ◗

Enzymes as a group are involved in both catabolic (decomposition) and anabolic (synthesis) reactions. A specific enzyme, however, functions in only one type of reaction. Digestive enzymes generally cause catabolic reactions, which metabolize ingested food into molecules small enough to cross cell membranes and supply raw materials for ATP production.

Figure 42.3 details an anabolic reaction and a catabolic reaction. Both types of reactions take place in the body as food macromolecules are digested to monomers and the monomers are then reassembled into the larger molecules the body needs. **Dehydration synthesis** reactions remove an OH group from one free monomer and a H atom from another free monomer, causing the two monomers to bond together and form a larger molecule (Figure 42.3b).

Table 42.1 summarizes the time and materials required for each activity. Use this table to help manage your laboratory time.

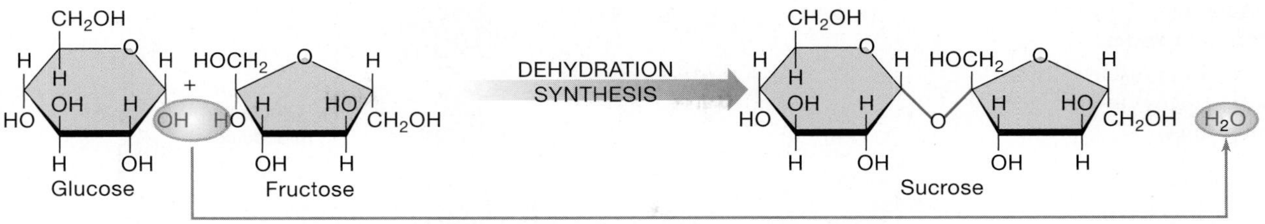

(a) During dehydration synthesis, two molecules are joined by the removal of a water molecule

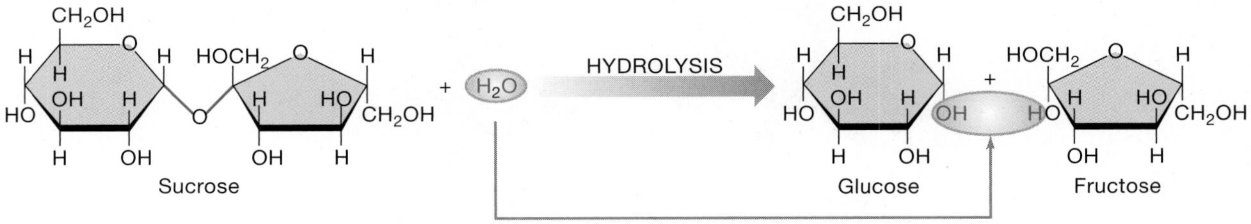

(b) Hydrolysis reverses the steps of dehydration synthesis; a complex molecule is broken down by the addition of a water molecule

Figure 42.3 The Formation and Breakdown of Complex Sugars
These reactions are performed by enzymes in the cell.

Table 42.1 Summary of Enzyme Experiments

	Carbohydrate	Lipid	Protein
Incubation time	$\frac{1}{2}$ hr	1 hr	1 hr
Number of test tubes required	6	2	2
Solutions required	Starch solution	Litmus cream	Protein solution
	Amylase	Pancreatic lipase	Pepsinogen
	Lugol's solution		0.5 M hydrochloric acid
	Benedict's reagent		Biuret's reagent

- Read through each activity from beginning to end before starting the activity.
- Wear gloves and safety glasses while pouring reagents and working near water baths.
- Report all spills and broken glass to your instructor. ▲

LAB ACTIVITY 1 Digestion of Carbohydrate

Starch and sugar molecules are classified as **carbohydrates** (kar-bō-HĪ-drātz). These molecules are composed of **saccharide** molecules. A carbohydrate composed of a single saccharide is a **monosaccharide** (mon-ō-SAK-uh-rīd), a simple sugar. Glucose and fructose are examples of monosaccharides. Monosaccharides are the monomers used to build larger sugar and starch molecules. A **disaccharide** (dī-SAK-uh-rīd) is formed when two monosaccharides bond together by a dehydration synthesis reaction, as in Figure 42.3a. Table sugar, sucrose, is a common disaccharide. Lactose, mentioned previously, is also a disaccharide.

Complex carbohydrates are **polysaccharides** (pol-ē-SAK-uh-rīdz) and consist of long chains of monosaccharides (Figure 42.4). Cells store polysaccharides as future

Figure 42.4
Polysaccharides
Liver and muscle cells store glucose in glycogen molecules.

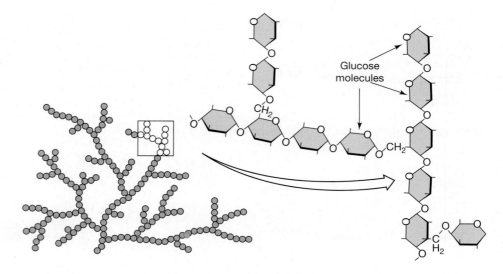

Glucose molecules

energy sources. Plant cells store them as **starch**, the molecules found in potatoes and grains; animal cells store them as **glycogen** (GLĪ-kō-jen).

In this activity, you will use the carbohydrate-digesting enzyme called *amylase* to digest the polysaccharide macromolecules in a solution of starch. Although amylase is easily obtained from your saliva, your instructor may provide amylase from a nonhuman source. Figure 42.5 summarizes the activity.

QuickCheck Questions

1.1 What is the general name for the monomers used to make carbohydrates?

1.2 What is the name of the enzyme used in the carbohydrate-digesting activity?

1.3 What does starch break down to?

Figure 42.5
Summary of Carbohydrate Digestion Procedures

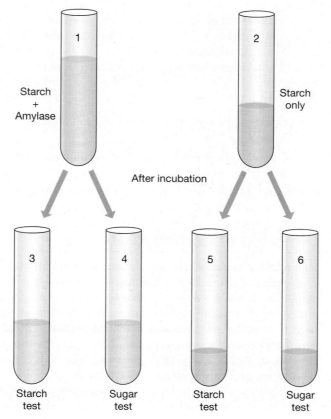

Starch + Amylase

Starch only

After incubation

Starch test   Sugar test   Starch test   Sugar test

1   *Materials*

☐ Refer to Table 42.1 for number of test tubes and solutions needed

☐ Wax pencil

☐ Test tube rack

☐ 37°C water bath (body temperature)

☐ Boiling water bath

Procedures

I. Preparation

1. Number the six test tubes with the wax pencil.

2. Add 20 ml of the starch solution to tube 1 and another 20 ml to tube 2. Be sure to shake the solution before pouring it into the tubes.

3. Add 20 ml of amylase solution to tube 1.

4. Place both tubes in the 37°C water bath, leave them there for a minimum of 30 minutes, and then remove them and place them in the test tube rack.

II. Analysis

1. Divide the solution in tube 1 equally into tubes 3 and 4, as indicated in Figure 42.5.

2. Divide the solution in tube 2 equally into tubes 5 and 6.

3. Test for the presence of starch in tubes 3 and 5 by placing one or two drops of **Lugol's solution** in each tube. A dark blue color indicates that starch is present.

4. Record your results in Table 42.2 in the Laboratory Report.

5. Test for the presence of monosaccharides in tubes 4 and 6 by placing 10 ml of **Benedict's reagent** in each tube. Mix by gently swirling the tubes, and then set them in the boiling water bath for five minutes. Do not point the tubes toward anyone because the solution could splatter and cause a burn. A color change indicates the presence of monosaccharides: from light olive to dark orange, depending on how much monosaccharide is present.

6. Record your results in Table 42.2. ■

LAB ACTIVITY   2   Digestion of Lipid

There are many classes of fats, or *lipids*. Most dietary lipids are **monoglycerides** and **triglycerides**. These lipids are constructed of one or more **fatty acid** molecules bonded to a **glycerol** molecule. Figure 42.6 shows the formation (dehydration synthesis) and decomposition (hydrolysis) of a triglyceride.

Pancreatic juice contains **pancreatic lipase**, a lipid-digesting enzyme that hydrolyzes triglycerides to monoglycerides and free fatty acids. In this activity, the released fatty acids will cause a pH change in the test tube, an indication that lipid digestion has occurred. The lipid substrate you will use is whipping cream to which a pH indicator, **litmus**, has been added. Litmus is blue in basic (alkaline) conditions and pink in acidic conditions. Figure 42.7 summarizes the lipid digestion activity.

QuickCheck Questions

2.1 What are the monomer units of lipids called?

2.2 What is the enzyme used in the lipid-digesting activity?

2   *Materials*

☐ Refer to Table 42.1 for number of test tubes and solutions needed

☐ Wax pencil

☐ 37°C water bath (body temperature)

☐ Test tube rack

Procedures

I. Preparation

1. Number the two test tubes 7 and 8. Fill each tube one-fourth full with litmus cream.

2. Add to tube 7 the same amount of pancreatic lipase as there is litmus cream in the tube.

3. Record the smell and color of the solution in each tube in the "Start" columns of Table 42.3 in the Laboratory Report.

4. Incubate the tubes for 1 hour in the 37°C water bath.

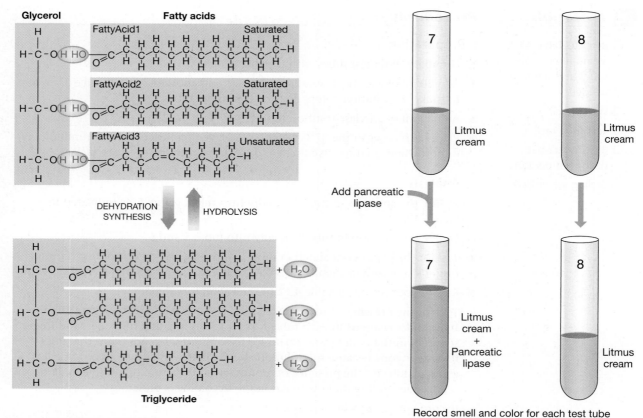

Figure 42.6 Triglyceride Formation
The formation of a triglyceride involves the attachment of fatty acids to the carbons of a glycerol molecule. In this example, a triglyceride is formed by the attachment of one unsaturated and two saturated fatty acids to a glycerol molecule.

Record smell and color for each test tube

Figure 42.7 Summary of Lipid Digestion Procedures

II. Analysis

1. After one hour, remove the tubes from the water bath and place them in the test tube rack.

2. Carefully smell the solutions by wafting fumes from each tube up to your nose. Record the smells in the "End" columns of Table 42.3.

3. Record the color of each solution in the "End" columns of Table 42.3. ∎

LAB ACTIVITY   3   Digestion of Protein

Proteins are composed of long chains of **amino acids** bonded together by **peptide bonds**. There are 20 different types of amino acids, and their position in a protein molecule determines the structure and function of the protein, much like how the location of teeth on a key determine how the key works. Ten of the amino acids are only obtained in the diet and are called essential amino acids. The other amino acids are also ingested and can be manufactured by cells. Figure 42.8 summarizes protein structure.

In this activity you will use the proenzyme pepsinogen. The proenzyme alone cannot cause digestion; it must be activated by certain chemical conditions. By doing this activity, you will not only learn about protein digestion, but will also discover the importance of the environment in which the enzyme operates. The protein substrate you will use was obtained from gelatin. Figure 42.9 summarizes the protein digestion activity.

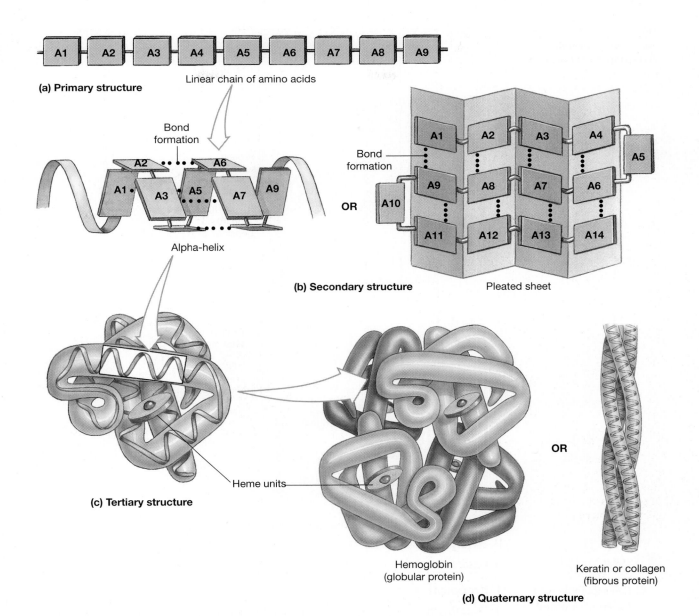

Figure 42.8 Protein Structure
(a) The primary structure of a polypeptide is the sequence of amino acids (A1, A2, A3, and so on) along its length. **(b)** Secondary structure is primarily the result of hydrogen bonding along the length of the polypeptide chain. Such bonding often produces a simple spiral (an alpha-helix) or a flattened arrangement known as a *pleated sheet.* **(c)** Tertiary structure is the coiling and folding of a polypeptide. Within the cylindrical segments of this globular protein, the polypeptide chain is arranged in an alpha-helix. **(d)** Quaternary structure develops when separate polypeptide subunits interact to form a larger molecule. A single hemoglobin molecule contains four globular subunits. In keratin and collagen, three fibrous subunits intertwine.

QuickCheck Questions

3.1 What are the monomer units of proteins called?

3.2 What is the enzyme used in the protein-digesting activity?

Figure 42.9
Summary of Protein Digestion Procedures

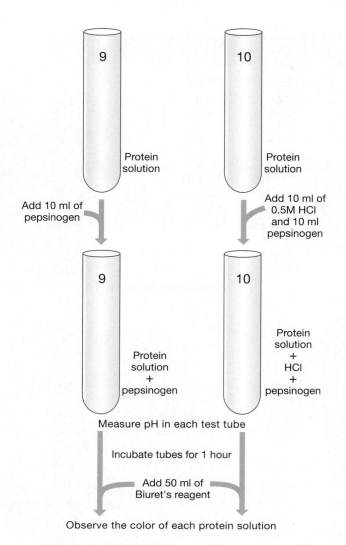

Procedures

I. Preparation

1. Number the two test tubes 9 and 10, and then add 10 ml of protein solution to each tube.

2. To tube 9, add 10 ml of the pepsinogen solution.

3. To tube 10, add 10 ml of the pepsinogen solution and 10 ml of 0.5 M hydrochloric acid.

4. Use the pH paper to measure the pH in each tube, and record your data in Table 42.4 in the Laboratory Report.

5. Incubate both tubes for 1 hour in the 37°C water bath.

II. Analysis

1. After one hour, remove the tubes from the water bath and place them in the test tube rack.

2. Test for protein digestion by adding 50 ml of **Biuret's reagent** to each tube. This solution turns pink in the presence of free amino acids and purple in the presence of protein molecules.

3. Record the color in each tube in Table 42.4. If the color is too subtle to allow you to distinguish between pink and purple, add 20 ml more of Biuret's reagent. ■

3 Materials

- ☐ Refer to Table 42.1 for number of test tubes and solutions needed
- ☐ Wax pencil
- ☐ 37°C water bath (body temperature)
- ☐ Test tube rack
- ☐ pH paper

Digestive Physiology

Name _____

Date _____

Section _____

A. Matching

Match each structure in the left column with its correct description from the right column.

_____ 1. substrate

_____ 2. product

_____ 3. amylase

_____ 4. lipase

_____ 5. pepsinogen

_____ 6. hydrochloric acid

_____ 7. Benedict's reagent

_____ 8. pepsin

_____ 9. Lugol's solution

_____ 10. Biuret's reagent

A. reagent for starch test

B. molecule that enzyme reacts with

C. enzyme that digests protein

D. reagent for sugar test

E. acid that activates stomach enzymes

F. enzyme that digests carbohydrates

G. proenzyme

H. reagent for protein test

I. substance created in chemical reaction

J. enzyme that digests triglycerides

B. Data Collection

Table 42.2 **Digestion of Starch by Amylase**

Lugol's Test for Starch		Benedict's Test for Sugar	
Tube 3	_____	Tube 4	_____
Tube 5	_____	Tube 6	_____
Conclusion	_____	Conclusion	_____

Table 42.3 **Digestion of Lipid by Pancreatic Lipase**

	Tube 7 Litmus Cream + Enzyme		Tube 8 Litmus Cream Only	
	Start	End	Start	End
Smell	_____	_____	_____	_____
Color	_____	_____	_____	_____
Conclusion	_____	_____	_____	_____

Table 42.4 **Digestion of Protein by Pepsinogen**

	Tube 9 Protein + Enzyme	Tube 10 Protein + Enzyme + HCl
pH of solution	_____	_____
Color after Biuret's Test	_____	_____
Conclusion	_____	_____

42

C. Short-Answer Questions

1. In each activity, you used a control tube that had no enzyme or other reagent added to it. What was the function of the control tubes?

2. Discuss the difference between dehydration synthesis reactions and hydrolysis reactions.

3. Describe the general chemical composition of carbohydrates, lipids, and proteins.

4. How do enzymes initiate chemical reactions?

5. List the three primary groups of nutrients.

D. Analysis and Application

1. Why were test tubes 1 and 2 in the carbohydrate laboratory activity tested for both starch *and* sugar?

2. Why did the solution in tube 7 in the lipid-digestion activity turn pink after the incubation?

3. Why was a warm-water bath used to incubate all the solutions in all three activities?

4. Why does a lactose-intolerant person have difficulty digesting milk?

Anatomy of the Urinary System

OBJECTIVES

On completion of this exercise, you should be able to:

- Identify and describe the basic anatomy of the urinary system.
- Trace the blood flow through the kidney.
- Explain the function of the kidney.
- Identify the basic components of the nephron.
- Describe the differences between the male and female urinary tracts.

All cells produce waste products through their metabolic activities. Wastes such as urea, creatinine, carbon dioxide, and excess electrolytes must be eliminated from the body to maintain homeostasis, and several organ systems carry out this task. The lungs remove carbon dioxide during exhalation, the digestive tract eliminates undigested solid wastes, and the skin removes salts and urea in sweat. The primary function of the urinary system is to control the composition, volume, and pressure of the blood by either removing or restoring blood fluid and solutes as the blood passes through the kidneys. Any excess water and solutes that accumulate during this process of continuous adjustment are considered waste and are eliminated from the body via the urinary system. These eliminated products are collectively called *urine*. The urinary system, highlighted in Figure 43.1, comprises a pair of kidneys, a pair of ureters, a urinary bladder, and a urethra.

LAB ACTIVITY 1   The Kidneys

The kidneys lie on the posterior surface of the abdomen between the waist and the eleventh and twelfth pairs of ribs. The right kidney is positioned lower than the left kidney because of the position of the liver. The kidneys, adrenal glands, and most of the ureters are retroperitoneal, meaning they are located behind the parietal peritoneum.

Figure 43.2 illustrates the sectional anatomy of the kidney. Each kidney is secured in the abdominal cavity by three layers of tissue: renal capsule, adipose capsule, and renal fascia. The **renal capsule** is a fibrous membrane that envelopes the external surface of each kidney. It protects the kidneys from trauma and infection. The **adipose capsule** is a mass of adipose tissue that envelopes the kidneys and protects them from trauma as well as helping to anchor them to the abdominal wall. The outer layer, the **renal fascia**, is composed of dense irregular connective tissue used to anchor the kidneys to the abdominal wall.

Figure 43.1
An Introduction to the Urinary System
An anterior view of the urinary system, showing the positions of the kidneys and other components.

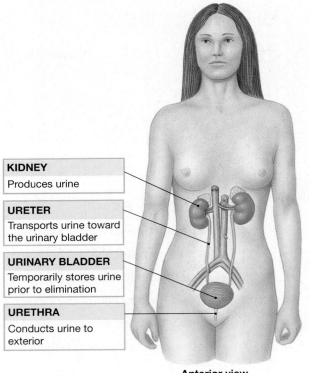

KIDNEY
Produces urine

URETER
Transports urine toward the urinary bladder

URINARY BLADDER
Temporarily stores urine prior to elimination

URETHRA
Conducts urine to exterior

Anterior view

Clinical Application | ### *Floating Kidneys*

Nephroptosis, or "floating kidneys," is the condition that results when the integrity of the adipose capsule or the renal fascia is jeopardized because of excessive weight loss: there is less adipose tissue available to secure the kidneys. This lack of support can result in the pinching or kinking of one or both ureters, preventing the normal flow of urine to the urinary bladder. ◗

A kidney is about 13 cm (5 in.) long and 2.5 cm (1 in.) thick. The medial aspect of the kidney contains a notch called the **hilus**, through which blood vessels, nerves, lymphatics, and the ureters enter and exit the kidney. The hilus also leads to a cavity in the kidney called the **renal sinus**. The **cortex** is the outer, light-red layer of the kidney. Deep to the cortex is a region called the **medulla**, which consists of triangular **renal pyramids** projecting toward the kidney center. Areas of the cortex extending between the renal pyramids are **renal columns**. At the apex of each renal pyramid is a **renal papilla** that empties urine into a small cuplike space called the **minor calyx** (KĀ-liks). Several minor calyces (KĀL-i-sēz) empty into a common space, the **major calyx**. These larger calyces merge to form the **renal pelvis**. Each kidney has a single **ureter**, a muscular tube that transports urine from the renal pelvis to the urinary bladder.

QuickCheck Questions

1.1 What is the hilus?

1.2 Where are the renal pyramids located?

1.3 Where are the renal columns located?

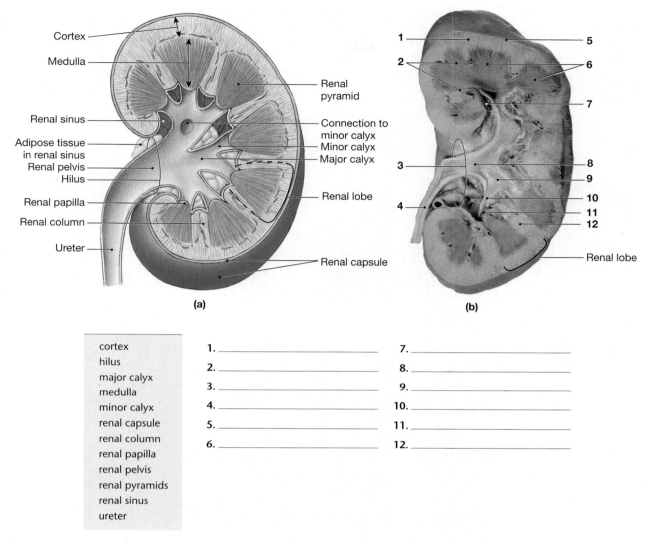

Figure 43.2 The Structure of the Kidney
(a) A diagrammatic view of a frontal section through the left kidney. (b) A frontal section of the left kidney.

1 *Materials*

☐ Kidney models
☐ Kidney charts

Procedures

1. Review the anatomy of the kidney in Figure 43.2 and label the figure.
2. Locate each structure shown in Figure 43.2 on a kidney model and/or chart. ◼

LAB ACTIVITY 2   The Nephron

The basic functional unit of the kidney is the **nephron** (NEF-ron); there are approximately 1.25 million nephrons in each kidney. In each nephron, some water, excess ions, and other waste materials are removed from the blood by filtration to produce a liquid called **filtrate**. The filtrate then circulates through a region of the nephron consisting of a series of convoluting tubules and a U-shaped section. During this passage, any substances in the filtrate still needed by the body move back into the blood. The remaining filtrate is excreted as urine. The process of urine formation is covered in more detail in Exercise 44.

Nephron anatomy is shown in Figure 43.3. Each nephron consists of two regions: a renal corpuscle and a renal tubule. The **renal corpuscle** includes a **glomerulus** (glo-MER-ū-lus) inside a double-walled capsule called **Bowman's capsule**. The glomerulus is a network of blood capillaries; as blood passes through this network, the blood pressure forces materials out of the blood and into Bowman's

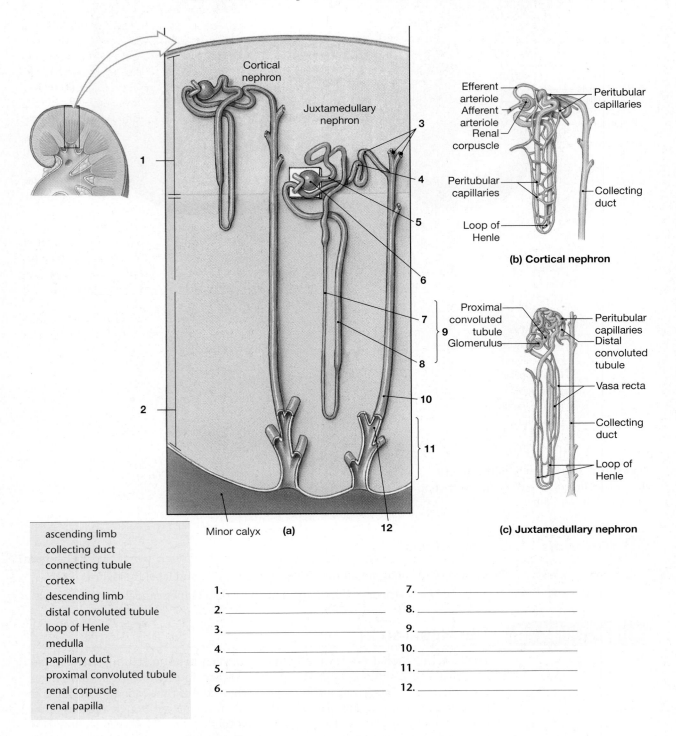

(a) General appearance and location of nephrons in the kidneys. (b) Circulation to a cortical nephron. (c) Circulation to a juxtamedullary nephron. The length of the loop of Henle is not drawn to scale.

ascending limb
collecting duct
connecting tubule
cortex
descending limb
distal convoluted tubule
loop of Henle
medulla
papillary duct
proximal convoluted tubule
renal corpuscle
renal papilla

1. _____
2. _____
3. _____
4. _____
5. _____
6. _____

7. _____
8. _____
9. _____
10. _____
11. _____
12. _____

Figure 43.3 Sectional Views of the Nephron

capsule. These materials that have left the blood are now the filtrate, which flows into the region of Bowman's capsule called the **capsular space**.

The renal corpuscle empties filtrate into the **renal tubule**, which consists of twisted and straight tubules composed mainly of cuboidal epithelium. The first segment after Bowman's capsule is a coiled tubule called the **proximal convoluted tubule**. The next portion of the tubule, the **loop of Henle** (HEN-lē), consists of a thin **descending limb** and a thick **ascending limb**. The ascending limb leads to a second convoluted section, the **distal convoluted tubule**. Surrounding the convoluted tubules are the **peritubular capillaries**, while the loop of Henle is covered with the **vasa recta capillary**. The nephron ends where the distal convoluted tubule empties into a **connecting tubule** that in turn drains into a **collecting duct**. Several nephrons drain into the same collecting duct. These ducts merge with other collecting ducts into a common **papillary duct** that drains into a minor calyx. There are between 25 and 35 papillary ducts per renal pyramid.

Two types of nephrons occur in the kidney. Approximately 85% of the nephrons are **cortical nephrons**; in these nephrons, the glomerulus and most of the renal tubule are both in the cortex. In **juxtamedullary** (juks-ta-MED-ū-lar-ē) **nephrons**, the glomerulus is at the junction of the cortex and the medulla, and the renal tubule extends deep into the medulla before turning back toward the cortex.

Bowman's capsule has an outer **parietal epithelium** (also called *capsular epithelium*) and an inner **visceral epithelium** (also called *glomerular epithelium*) that wraps around the surface of the glomerulus (Figure 43.4). Between these two layers is the capsular space. An **afferent arteriole** passes into Bowman's capsule and supplies blood to the glomerulus. An **efferent arteriole** drains the glomerulus and branches into the peritubular and vasa recta capillaries.

The visceral layer of Bowman's capsule consists of specialized cells called **podocytes**. These cells wrap extensions called **pedicels** around the endothelium of

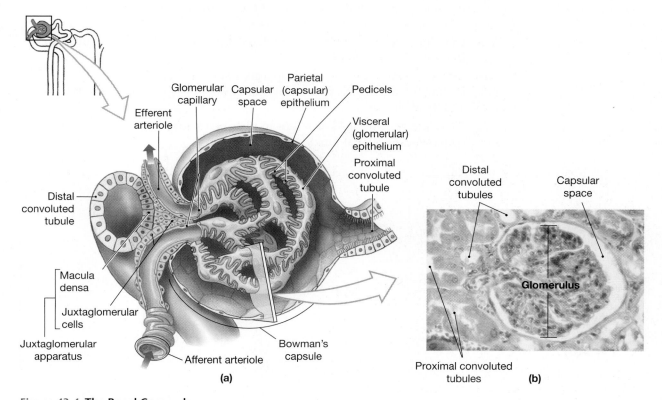

(a)

(b)

Figure 43.4 The Renal Corpuscle

(a) Important structural features of a renal corpuscle. (b) The association of proximal and distal convoluted tubules with a glomerulus (LM × 370).

the glomerulus. Small gaps between the pedicels are pores called **filtration slits**. To be filtered out of the blood passing through the glomerulus, a substance must be small enough to pass through the **capillary endothelium** and its basement membrane (lamina densa) and squeeze through the filtration slits to enter the capsular space. Any substance that can pass through these layers is removed from the blood as part of the filtrate. The filtrate therefore contains both essential materials and wastes.

The ascending limb of the loop of Henle twists back and comes into contact with the afferent arteriole that supplies its glomerulus. This point of contact is called the **juxtaglomerular apparatus** (Figure 43.4). Here the cells of the renal tubule become tall and crowded together and form the **macula densa** (MAK-ū-la DEN-sa), which monitors NaCl concentrations in this area of the renal tubule. The smooth muscle cells of the afferent arteriole secrete an enzyme called renin, a hormone that causes blood vessels to dilate. Renin also initiates the endocrine reflex that causes secretion of the hormone aldosterone, which increases the rate at which sodium ions and water are reabsorbed in the distal convoluted tubule.

QuickCheck Questions

2.1 What are the two main regions of a nephron?

2.2 What are the two kinds of nephrons?

2 Materials

☐ Kidney and nephron models

☐ Compound microscope

☐ Prepared slide of kidney

Procedures

1. Review the nephron anatomy in Figures 43.3 and 43.4, and label Figure 43.3.

2. Examine the kidney and nephron models, and locate all the structures of the nephron.

3. Mount the kidney slide on the microscope stage. Focus at low or medium magnification as needed to locate the renal cortex and, if it is present, the renal medulla. (The medulla is usually not present on most kidney slides.) Is the renal capsule visible on the outer margin of the kidney tissue?

4. Examine several renal tubules, visible as ovals on the slide. Use the micrograph in Figure 43.4 as a guide.

5. Locate a renal corpuscle, which appears as a small knot in the cortex. Distinguish among the parietal layer of Bowman's capsule, the capsular space, and the glomerulus.

6. Draw a section of the kidney slide in the space below. Label the cortex, a renal corpuscle, and a renal tubule. ■

Blood Supply to the Kidney

Every minute, approximately 25% of the total blood volume travels through the kidneys. This blood is delivered to a kidney by the **renal artery**, which branches off the abdominal aorta (Figure 43.5). Once it enters the hilus, the renal artery divides into five **segmental arteries**, which then branch into **interlobar arteries**, which pass through the renal columns. The interlobar arteries divide into **arcuate** (AR-kū-āt) **arteries**, which cross the bases of the renal pyramids and enter the renal cortex as **interlobular arteries**. These latter arteries branch into afferent arterioles, which are what form the glomerulus, the capillary tuft that sits inside the Bowman's capsule region of the nephron, as shown in Figure 43.4a.

At the end of the glomerulus opposite the afferent arteriole, an efferent arteriole exits the glomerulus and forms a bed of peritubular capillaries around the entire tubular portion of cortical nephrons—but only around the proximal and distal convoluted tubules in juxtamedullary nephrons. In the latter, the loop of Henle is surrounded not by peritubular capillaries but rather by vasa recta capillaries. Both capillary networks are involved in the reabsorption of materials from the filtrate of the renal tubules back into the blood. Both networks drain into **interlobular veins**, which then drain into **arcuate veins** along the base of the renal pyramids. **Interlobar veins** pass through the renal columns and join the **renal vein**, which drains into the inferior vena cava. Although there are segmental arteries, there are no segmental veins.

QuickCheck Questions

3.1 Which vessel branches from the abdominal aorta to supply blood to the kidney?

3.2 Where are the interlobar and interlobular arteries?

3 *Materials*

☐ Kidney and nephron models

Procedures

1. Label each blood vessel in Figure 43.5.
2. On the kidney and nephron models, trace the blood vessels that supply and drain the kidneys. ■

The Ureters, Urinary Bladder, and Urethra

The two **ureters** conduct urine, by way of gravity and peristalsis, to the urinary bladder. The ureters, detailed in Figure 43.6a, are lined with a mucus-producing epithelium that protects the ureteral walls from the acidic urine. The ureters enter the urinary bladder on the posterior surface at the top two corners of the **trigone** (TRĪ-gōn), a smooth triangular area of the urinary bladder floor.

The **urinary bladder** is a hollow, muscular organ, the function of which is temporary storage of urine. Histological details of the bladder are shown in Figure 43.6b. In the epithelium lining the mucosa, called **transitional epithelium**, the cells have many different shapes to facilitate the stretching and recoiling of the bladder wall.

In males, the bladder lies between the pubic symphysis and the rectum (Figure 43.7a). In females it is posterior to the pubic symphysis, inferior to the uterus, and superior to the vagina (Figure 43.7b). The inner wall contains folds called **rugae** that allow the bladder to expand and shrink as it fills with urine and then empties.

A single duct, the **urethra** (Figure 43.6c), drains urine from the bladder out of the body. Around the opening to the urethra are two sphincter muscles, the **internal urethral sphincter** and the **external urethral sphincter**. The **prostatic urethra**

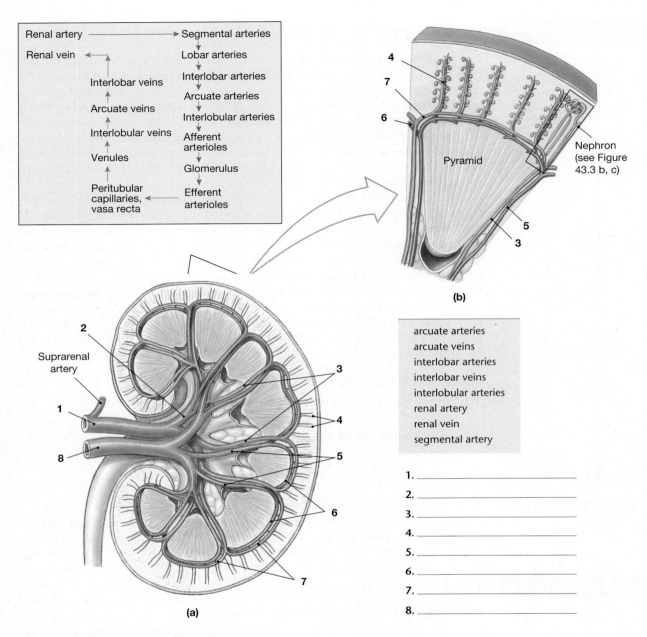

Renal artery ──────────→ Segmental arteries
                                    ↓
Renal vein ←─────────   Lobar arteries
                                    ↓
        Interlobar veins   Interlobar arteries
                                    ↓
Arcuate veins   Arcuate arteries
                                    ↓
        Interlobular veins Interlobular arteries
                                    ↓
        Venules   Afferent
                           arterioles
                                    ↓
        Peritubular   Glomerulus
        capillaries, ←─   ↓
        vasa recta   Efferent
                           arterioles

4
7
6
Pyramid
Nephron
(see Figure
43.3 b, c)
5
3

(b)

2
Suprarenal
artery
1
8
3
4
5
6
7

(a)

arcuate arteries
arcuate veins
interlobar arteries
interlobar veins
interlobular arteries
renal artery
renal vein
segmental artery

1. _____
2. _____
3. _____
4. _____
5. _____
6. _____
7. _____
8. _____

Figure 43.5 Blood Supply to the Kidneys

(a) Sectional view, showing major arteries and veins. (b) Circulation in the cortex.
(c) Circulation to a cortical nephron. (d) Circulation to a juxtamedullary nephron.

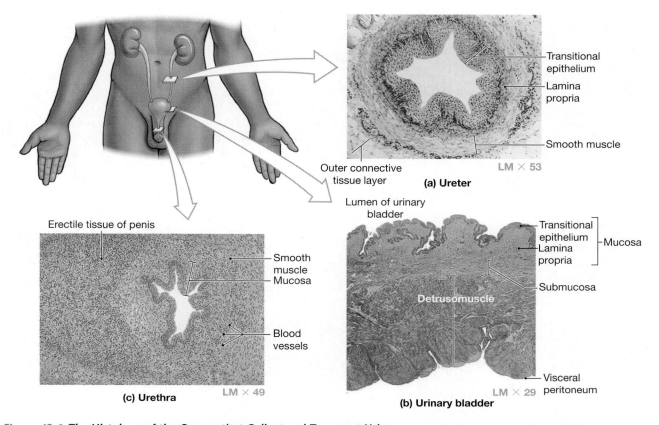

Figure 43.6 **The Histology of the Organs that Collect and Transport Urine**
(a) A transverse section through the ureter. A thick layer of smooth muscle surrounds
the lumen (LM × 53). (b) The wall of the urinary bladder (LM × 29). (c) A transverse
section through the male urethra (LM × 49).

is the portion of the duct that passes through the prostate gland located inferior to
the bladder. The internal sphincter is composed of smooth muscle and is therefore
under involuntary nerve control. The external sphincter is skeletal muscle tissue and
under voluntary control. This sphincter enables you to control the release of urine
from the urinary bladder. As the bladder fills with urine and expands, stretch receptors signal motor neurons in the spinal cord to relax the internal urethral sphincter
and contract the **detrusor** (de-TROO-sor) **muscle** of the bladder wall. This reflex automatically causes the external urethral sphincter to close. When convenient, the external urethral sphincter is relaxed and urination, or micturition, occurs.

QuickCheck Questions

4.1 Where do the ureters join the urinary bladder?

4.2 What is the trigone?

4 Materials

☐ Urinary system
models and charts

Procedures

1. Review and label Figure 43.7.

2. Locate the ureters on a urinary system model and/or chart. Trace the path urine
follows from the renal papilla to the ureter.

3. On a model, examine the wall of the urinary bladder. Identify the trigone and
the rugae. Which structures control emptying of the bladder?

4. On a model, examine the urethra. How does the male urethra differ from the
female urethra? ■

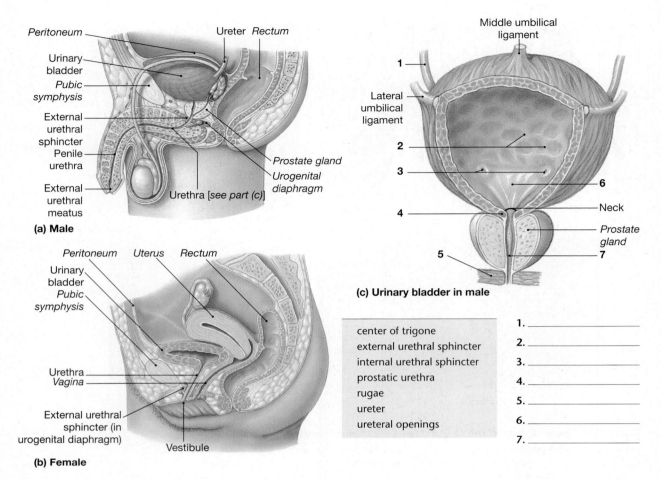

(a) Male

(b) Female

(c) Urinary bladder in male

center of trigone
external urethral sphincter
internal urethral sphincter
prostatic urethra
rugae
ureter
ureteral openings

1. _____
2. _____
3. _____
4. _____
5. _____
6. _____
7. _____

Figure 43.7 Organs for the Conduction and Storage of Urine
The ureter, urinary bladder, and urethra **(a)** in the male and **(b)** in the female.
(c) The urinary bladder of a male.

 LAB ACTIVITY 5 Dissection of the Sheep Kidney

The sheep kidney is very similar to the human kidney in both size and anatomy. Dissection of a sheep kidney will reinforce your observations of kidney models in the laboratory.

⚠ Before beginning this dissection, read again the dissection safety guidelines presented at the beginning of Laboratory Activity 5, Sheep Brain Dissection, in Exercise 25, on page 381. ▲

QuickCheck Questions

5.1 What type of safety equipment should you wear during the sheep kidney dissection?

5.2 How should you dispose of the sheep kidney and scrap tissue?

Figure 43.8

Gross Anatomy of a Sheep Kidney

A frontal section of a sheep kidney that has been injected with latex dye to highlight arteries, veins, and the urinary passageways.

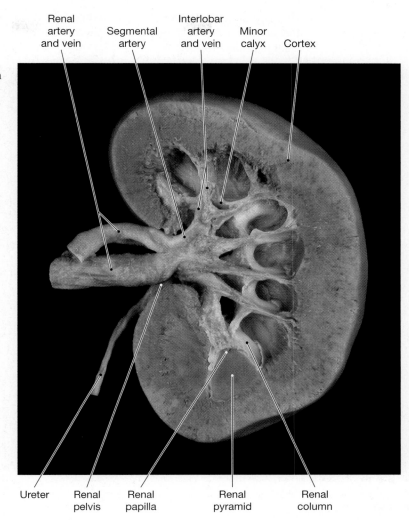

Renal artery and vein
Segmental artery
Interlobar artery and vein
Minor calyx
Cortex

Ureter
Renal pelvis
Renal papilla
Renal pyramid
Renal column

5 *Materials*

- ☐ Gloves
- ☐ Safety glasses
- ☐ Dissecting tools
- ☐ Dissecting tray
- ☐ Preserved sheep kidney

Procedures

1. Put on gloves and safety glasses, and clear your workspace before obtaining your dissection specimen.

2. Rinse the kidney with water to remove excess preservative. Minimize your skin and mucous membrane exposure to the preservatives.

3. Examine the external features of the kidney. Using Figure 43.8 as a guide, locate the hilus, a concave slit on the medial border where the renal vessels and the ureters pass through the kidney. Locate the outer renal capsule and gently lift it with a teasing needle. Below this capsule is the light-pink cortex.

4. With a scalpel, make a longitudinal cut to divide the kidney into anterior and posterior portions. A single long, smooth cut is less damaging to the internal anatomy than a sawing motion with the scalpel.

5. Distinguish between the cortex and the darker medulla, which is organized into many triangular renal pyramids. The base of each pyramid faces the cortex, and the tip narrows into a renal papilla.

6. The renal pelvis is the large, expanded end of the ureter. Extending from this area are the major calyces and then the smaller minor calyces into which the renal papillae project.

7. Upon completion of the dissection, dispose of the sheep kidney as directed by your instructor; then wash your hands and dissecting instruments. ■

Anatomy of the Urinary System

Name _____

Date _____

Section _____

A. Matching

Match each term in the left column with its correct description from the right column.

_____	**1.** renal papilla	**A.**	drains into collecting duct
_____	**2.** cortex	**B.**	located at base of pyramid
_____	**3.** Bowman's capsule	**C.**	covers surface of kidney
_____	**4.** loop of Henle	**D.**	surrounds glomerulus
_____	**5.** renal pelvis	**E.**	surrounds renal pelvis
_____	**6.** connecting tubule	**F.**	entrance for blood vessels
_____	**7.** efferent arteriole	**G.**	extends into minor calyx
_____	**8.** renal sinus	**H.**	tissue between pyramids
_____	**9.** renal pyramid	**I.**	U-shaped tubule
_____	**10.** arcuate artery	**J.**	transports urine to bladder
_____	**11.** hilus	**K.**	functional unit of kidney
_____	**12.** renal columns	**L.**	drains glomerulus
_____	**13.** renal capsule	**M.**	outer layer of kidney
_____	**14.** nephron	**N.**	medulla component
_____	**15.** ureter	**O.**	drains major calyx

B. Short-Answer Questions

1. Describe the components of the renal corpuscle.

2. Describe the different regions of a renal tubule.

3. What are two differences between cortical and juxtamedullary nephrons?

4. How does the urethra differ between males and females?

5. Where are the internal and external urethral sphincters located?

C. Analysis and Application

1. List the layers in the renal corpuscle through which filtrate must pass to enter the capsular space.

2. Trace a drop of blood from the abdominal aorta, through a kidney, and into the inferior vena cava.

3. Trace a drop of filtrate from the glomerulus to the renal papillae.

4. Trace a drop of urine from a minor calyx to the urinary bladder.

Physiology of the Urinary System

OBJECTIVES

On completion of this exercise, you should be able to:

- Define glomerular filtration, tubular reabsorption, and tubular secretion.

- Describe the physical characteristics of normal urine.

- Recognize normal and abnormal urine constituents.

- Conduct a urinalysis test.

The kidneys maintain the chemical balance of body fluids by removing metabolic wastes, excess water, and electrolytes from the blood plasma. Three physiological processes occur in the nephrons to produce urine: filtration, reabsorption, and secretion (Figure 44.1). **Filtration** occurs in the renal corpuscle as blood pressure in the glomerulus forces water, ions, and other solutes small enough to pass through the filtration slits surrounding the glomerulus out of the blood and into the capsular space. Because size is the only thing that determines what passes through the filtration slits and becomes part of the filtrate, both wastes and essential solutes are removed from the blood in the glomerulus. Any solutes and water still needed by the body reenter the blood during **reabsorption**, as cells in the renal tubule reclaim the needed materials. Movement is in both directions along the length of the renal tubule, however, and in the process called **secretion**, any unneeded blood materials that did not leave the blood in the glomerulus leave it now and become part of the filtrate. The tubular cells actively transport ions in both directions (from blood to filtrate and from filtrate to blood), often using countertransport mechanisms that result in the reabsorption of necessary ions and the secretion of unneeded ones. As the filtrate passes through the entire length of the renal tubule of the nephron, reabsorption and secretion occur over and over. Once out of the renal tubule, the filtrate is processed into urine, which drips out of the renal papillae into the minor calyxes.

As the filtrate moves through the proximal convoluted tubule (PCT), 60–70% of the water and ions and 100% of the organic nutrients, such as glucose and amino acids, are reabsorbed into the blood. The simple cuboidal epithelium in this part of the nephron has microvilli to increase the surface area for reabsorption. The loop of Henle conserves water and salt while concentrating the filtrate for modification by the distal convoluted tubule (DCT). Reabsorption in the DCT is controlled by two hormones, aldosterone and antidiuretic hormone (ADH). Most of the secretion that takes place occurs in the DCT.

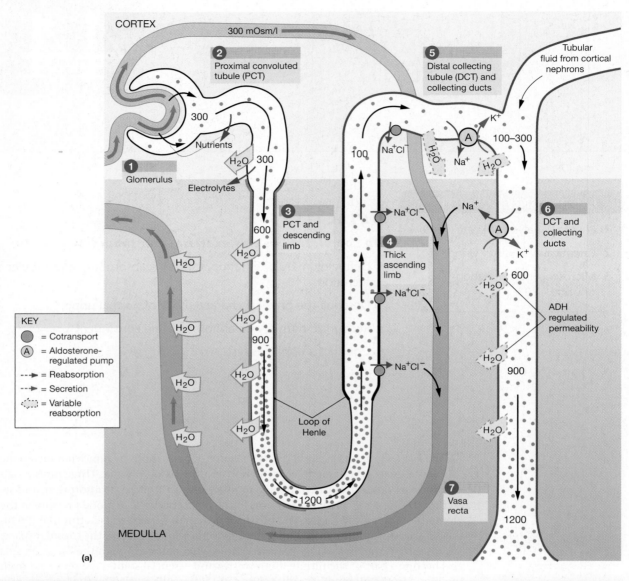

CORTEX

300 mOsm/l

2 Proximal convoluted tubule (PCT)

5 Distal collecting tubule (DCT) and collecting ducts

Tubular fluid from cortical nephrons

300

Nutrients

H_2O 300

1 Glomerulus

Electrolytes

100

Na^+Cl^-

H_2O

K^+

A

Na^+ H_2O

100–300

600

3 PCT and descending limb

Na^+Cl^-

Na^+

6 DCT and collecting ducts

H_2O

H_2O

Na^+Cl^-

A

600

H_2O

4 Thick ascending limb

K^+

ADH regulated permeability

H_2O

900

KEY
- = Cotransport
- A = Aldosterone-regulated pump
- --→ = Reabsorption
- --→ = Secretion
- = Variable reabsorption

H_2O

Na^+Cl^-

H_2O

H_2O

900

H_2O

Loop of Henle

H_2O

1200

7 Vasa recta

MEDULLA

1200

(a)

STEP 1: Glomerulus	STEP 2: Proximal convoluted tubule (PCT)	STEP 3: PCT and descending limb
The filtrate produced at the renal corpuscle has the same osmotic concentration as plasma—about 300 mOsm/ L. It has the same composition as plasma, without the plasma proteins.	In the proximal convoluted tubule (PCT), the active removal of ions and organic substrates produces a continuous osmotic flow of water out of the tubular fluid. This reduces the volume of filtrate, but keeps the solutions inside and outside the tubule isotonic.	In the PCT and descending limb of the loop of Henle, water moves into the surrounding peritublular fluids, leaving a small volume of highly concentrated tubular fluid. This reduction occurs by obligatory water reabsorption.

Figure 44.1 A Summary of Renal Function
(a) A general overview of steps and events.

The kidneys filter 25% of the body's blood each minute, producing on average 125 ml/min of filtrate. About 180 liters of filtrate are formed by the glomerulus per day, which eventually results in the production of an average daily output of 1.8 liters of urine. The composition of urine can change on a daily basis depending on one's metabolic rate and urinary output. Water accounts for about 95% of the volume of urine. The other 5% contains excess water-soluble vitamins, drugs, electrolytes, and nitrogenous wastes. Abnormal substances in urine can usually be detected by **urinalysis**, an analysis of the chemical and physical properties of urine.

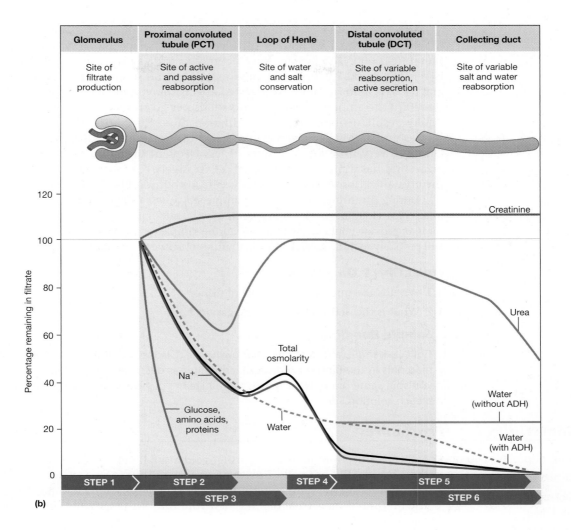

Glomerulus	Proximal convoluted tubule (PCT)	Loop of Henle	Distal convoluted tubule (DCT)	Collecting duct
Site of filtrate production	Site of active and passive reabsorption	Site of water and salt conservation	Site of variable reabsorption, active secretion	Site of variable salt and water reabsorption

(b)

STEP 4: Thick ascending limb	STEP 5: DCT and collecting ducts	STEP 6: DCT and collecting ducts	STEP 7: Vasa recta
The thick ascending limb is impermeable to water and solutes. The tubular cells actively transport Na^+ and Cl^- out of the tubule, thereby lowering the osmotic concentration of the tubular fluid. Because just Na^+ and Cl^- are removed, urea accounts for a higher proportion of the total osmotic concentration at the end of the loop.	The final adjustments in the composition of the tubular fluid occur in the DCT and the collecting system. The osmotic concentration of the tubular fluid can be adjusted through active transport (reabsorption or secretion).	The final adjustments in the volume and osmotic concentration of the tubular fluid are made by controlling the water permeabilities of the distal portions of the DCT and the collecting system. The level of exposure to ADH determines the final urine concentration.	The vasa recta absorbs the solutes and water reabsorbed by the loop of Henle and the collecting ducts. By removing these solutes and water into the main circulatory system, the vasa recta maintains the concentration gradient of the medulla.

Figure 44.1 A Summary of Renal Function *(Continued)*
(b) Specific changes in the composition and concentration of the tubular fluid as it flows along the nephron and collecting duct.

LAB ACTIVITY 1 Physical Analysis of Urine

Normal constituents of urine include water, urea, creatinine, uric acid, many electrolytes, and possibly small amounts of hormones, pigments, carbohydrates, fatty acids, mucin, and enzymes.

The average pH of urine is 6.0, and a normal range is 4.5 to 8.0. Urine pH is greatly affected by diet. Diets high in vegetable fiber result in an alkaline pH value (above 7), and high-protein diets yield an acidic pH value (below 7).

Specific gravity is the ratio of the weight of a volume of a substance to the weight of an equal volume of distilled water. The specific gravity of water is therefore 1.000. The average specific gravity for a normal urine sample is between 1.003 and 1.030. Urine contains solutes and solids that affect its specific gravity. The amount of fluids ingested affects the volume of urine excreted and therefore the amount of solutes and solids per given volume. Drinking a lot of liquids results in more frequent urination of a dilute urine that contains few solutes and solids per given volume and therefore has a low specific gravity. Drinking very little liquid results in less frequent urination of a concentrated urine that has a high specific gravity. Excessively concentrated urine results in the crystallization of solutes, usually salts, into insoluble kidney stones.

During this urinalysis you will examine *only your own urine* and will study its volume, color, cloudiness, odor, and specific gravity. Alternatively, your laboratory instructor may provide your class with a mock urine sample for analysis. This artificial sample will probably include several abnormal urine constituents for instructional purposes.

QuickCheck Questions

1.1 What is the normal range of pH of urine?

1.2 What is the definition of specific gravity?

Sample Handling

Collect, handle, and test *only your own urine.* Dispose of all urine-contaminated materials in the biohazard disposal container as described by your instructor. If you spill some urine, wear gloves as you clean your work space with a mild bleach solution. ▲

1 Materials

- ☐ 8 oz. disposable cup
- ☐ Bleach solution
- ☐ Biohazard disposal container

Procedures

1. Sample Collection: Before collecting, void a small volume of urine from your bladder. By not collecting the first few milliliters, you avoid contaminating the sample with substances such as bacteria and pus from the urethra or menstrual blood. Then void into the disposable cup until it is about one-half full.

2. Observe the **physical characteristics** of the sample and record your observations in Table 44.1 in the Laboratory Report.

   a. The color of urine varies from colorless to amber. **Urochrome** is a by-product of the breakdown of hemoglobin that gives urine its yellowish color. Color also varies because of ingested food. Vitamin supplements, certain drugs, and the amount of solutes also influence urine color. A dark-red or brown color indicates blood in the urine.

   b. Turbidity (cloudiness) is related to the amount of solids in the urine. Contributing factors include bacteria, mucus, cell casts, crystals, and epithelial cells. Observe and describe the turbidity of the urine sample. Use descriptive words such as "clear," "clouded," and "hazy."

   c. To smell the sample, place it approximately 12 in. from your face and wave your hand over the sample toward your nose. Normally, freshly voided urine has no odor, and therefore odor serves as a diagnostic tool for fresh urine. Starvation causes the body to break down fats and produce ketones, which give urine a fruity or acetone-like smell. Individuals with diabetes mellitus often produce sweet-smelling urine. (The characteristic odor associated with a urine sample that is not fresh is the result of the chemical breakdown of substances in the urine; the most characteristic odor is that of ammonia.)

   d. Use a **urinometer**, shown in Figure 44.2, to determine the specific gravity of your sample. A urinometer consists of a small glass cylinder and a urine

Read the specific gravity on the urinometer.
This specific gravity is 1.025.

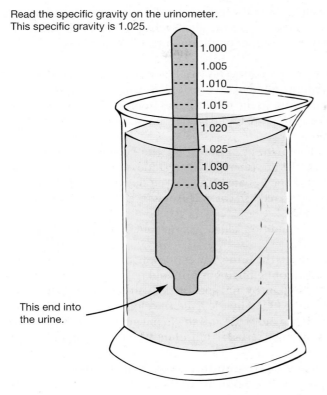

1.000
1.005
1.010
1.015
1.020
1.025
1.030
1.035

This end into
the urine.

Figure 44.2
Reading a Urinometer

hydrometer. The cylinder is used to hold the urine sample being tested. The hydrometer is a float that has been calibrated against water. Along the stem of the hydrometer is a scale used to determine the specific gravity of the sample.

- Swirl the sample in the collection cup to suspend any materials that may have settled after collection.

- Fill the glass cylinder of the urinometer at least two-thirds full with the urine sample.

- Carefully lower the urine hydrometer into the cylinder of urine. If the hydrometer does not float, add more urine to the cylinder until the hydrometer does float.

- The urine will adhere to the walls of the glass cylinder and form a trough called a **meniscus**. The scale on the hydrometer stem is read at the bottom of the meniscus, where the meniscus intersects a line on the scale. Record the specific gravity of the sample in Table 44.1.

3. Wash the glass cylinder and the hydrometer with soap and water and dry with clean paper towels. Dispose of the urine cup in the biohazard box as indicated by your instructor.

4. Wash and dry your hands before leaving the laboratory. ■

LAB ACTIVITY 2   Chemical Analysis of Urine

Certain materials in the urine suggest renal disease or injury. Excessive consumption of a substance may cause the substance to saturate the filtrate and overload the transport mechanisms of reabsorption. Because the overworked renal tubular cells cannot reclaim all of the substance, it appears in the urine.

Ketones in the urine, a condition called **ketosis** (kē-tō-sis), may be the result of starvation, diabetes mellitus, or a diet very low in carbohydrates. When the blood carbohydrate concentration is low, cells begin to catabolize fats. The products of fat catabolism are glycerol and fatty acids. Liver cells convert the fatty acids to ketones, which then diffuse out of the liver and into the blood, where they are filtered by the kidneys. In diabetes mellitus, commonly called sugar diabetes, not enough glucose enters the cells of the body. As a result, the cells use fatty acids to produce ATP. This increase in fatty acid catabolism results in the appearance of ketones in the urine.

Glucosuria is glucose in the urine and is usually an indication of diabetes mellitus. Because diabetic individuals do not have the normal cellular intake of glucose, the blood glucose concentration is abnormally high. The amount of glucose filtered out of the blood is greater than the amount the tubular cells of the nephrons can reabsorb back into the blood, and the glucose that is not reabsorbed appears in the urine. Glucosuria may also be the result of a very-high-carbohydrate meal that produces a temporary overload of glucose.

Another cause of glucosuria is stress. Production of epinephrine in response to stress results in the conversion of glycogen to glucose and its release from the liver. The elevated levels of glucose may then be secreted into the urine.

Albumin is a large protein molecule that normally cannot pass through the filtration slits of the glomerulus. A trace amount of albumin in the urine is considered normal. However, excessive albumin in the urine, a condition called **albuminuria**, suggests an increase in the permeability of the glomerular membrane. Reasons for increased permeability can be the result of physical injury, high blood pressure, disease, or bacterial toxins.

Hematuria is the presence of erythrocytes (red blood cells) in the urine and is usually an indication of bleeding caused by an inflammation or infection of the urinary tract. Causes include irritation of the renal tubules from the formation of kidney stones, trauma such as a hard blow to the kidney, blood from menstrual flow, and possible tumor formation.

When erythrocytes are hemolytized in the blood, the hemoglobin molecules break down into two chains that are filtered by the kidneys and excreted in the urine. If a large number of erythrocytes are being broken down in the circulation, the urine develops a dark-brown to reddish color. The presence of hemoglobin in the urine is called **hemoglobinuria**.

Leukocytes in the urine, a condition called **pyuria** (pī-Ū-rē-uh), indicate a urinary tract infection.

Bilirubin in large amounts in the urine, the condition known as **bilirubinuria**, is a result of the breakdown of hemoglobin from old red blood cells being removed from the circulatory system by phagocytic cells in the liver. When red blood cells are removed from the blood by the liver, the globin portion of the hemoglobin molecule is split off the molecule and the heme portion is converted to biliverdin. The biliverdin is then converted to bilirubin, which is a major pigment in bile.

Urobilinogen in the urine is called **urobilinogenuria**. Small amounts of urobilinogen in the urine are normal. It is a product of the breakdown of bilirubin by the intestines and is responsible for the normal brown color of feces. Greater than trace levels in the urine may be due to infectious hepatitis, cirrhosis, congestive heart failure, or a variety of other diseases.

Urea is produced during **deamination** (dē-am-i-NĀ-shun) reactions that remove ammonia (NH_3) from amino acids. The ammonia combines with CO_2 and forms urea (CH_4ON_2). About 4,600 mg of urea are produced daily, accounting for approximately 80% of the nitrogen waste in urine. **Creatinine** is formed from the breakdown of creatine phosphate, an energy-producing molecule found in muscle tissue. Uric acid is produced from the breakdown of the nucleic acids DNA and RNA, two molecules obtained either from foods or when body cells are destroyed.

Nitrites in the urine indicate a possible urinary tract infection.

Several inorganic ions and molecules are found in urine. Their presence is a reflection of diet and general health. Na+ and Cl− are ions from sodium chloride, the principal salt of the body. The amount of **Na+** and **Cl− ions** present in the urine varies with how much table salt is consumed in the diet. **Ammonium** (NH_4+) **ion** is a product of protein catabolism and must be removed from the blood before it reaches toxic concentrations. Many types of ions bind with sodium and form a buffer in the blood and urine to stabilize pH.

Hormones, enzymes, carbohydrates, fatty acids, pigments, and mucin all occur in small quantities in the urine.

Dip sticks are a fast, inexpensive method of determining the chemical composition of urine. These sticks hold from one to nine testing pads containing reagents that react with certain substances found in urine. Single-test sticks are usually used to determine whether glucose or ketones are present in the urine, and multiple-test sticks are used to give a more informative evaluation of the urine's chemical content.

QuickCheck Questions

2.1 What might cause glucose to be present in the urine of a person who does not have diabetes mellitus?

2.2 What are ketones, and why might they appear in the urine?

Sample Handling

Collect, handle, and test *only your own urine.* Dispose of all urine-contaminated materials in the biohazard disposal container as described by your instructor. If you spill some urine, wear gloves as you clean your work space with a mild bleach solution. ▲

2 *Materials*

☐ 8 oz. disposable cup
☐ Bleach solution
☐ Urine dip sticks
☐ Paper towels
☐ Biohazard disposal container

Procedures

1. Sample Collection: You may use the urine collected in Laboratory Activity 1. If not using the sample from **Laboratory Activity 1**, before collecting, void a small volume of urine from your bladder. By not collecting the first few milliliters, you avoid contaminating the sample with substances such as bacteria and pus from the urethra or menstrual blood. Then void into the disposable cup until it is about one-half full.

2. Review the color chart on the dip stick bottle, and note which test pads require reading at specific times.

3. Swirl your sample of urine before placing a dip stick into the urine.

4. Holding a dip stick by the end that does not have any test pads, immerse the stick in the urine so that all the test pads are wetted and then withdraw the stick. Lay it on a clean, dry paper towel to absorb any excess urine.

5. How long to wait after removing a stick from the urine varies from immediately to two minutes. Use the color chart on the side of the dip stick bottle to determine how long you must wait.

6. Record your data from the dip stick in Table 44.2 in the Laboratory Report.

7. Dispose of all used dip sticks and paper towels (and gloves if you used them) in the biohazard disposal container. Dispose of the urine cup in the biohazard box as indicated by your instructor.

8. Wash and dry your hands before leaving the laboratory. ■

LAB ACTIVITY 3 Microscopic Examination of Urine

Examination of the sediment of a centrifuged urine specimen reveals the solid components of the sample. This can be a valuable test to determine or confirm the presence of abnormal contents in the urine. A wide variety of solids can be in urine, including cells, crystals, and mucus (Figure 44.3).

Urine is generally sterile, but **microbes** can be present in a sample for several reasons, and microbes present in large numbers usually indicate infection. Microbes may contaminate a urine sample when they are present at the urethral opening and in the urethra.

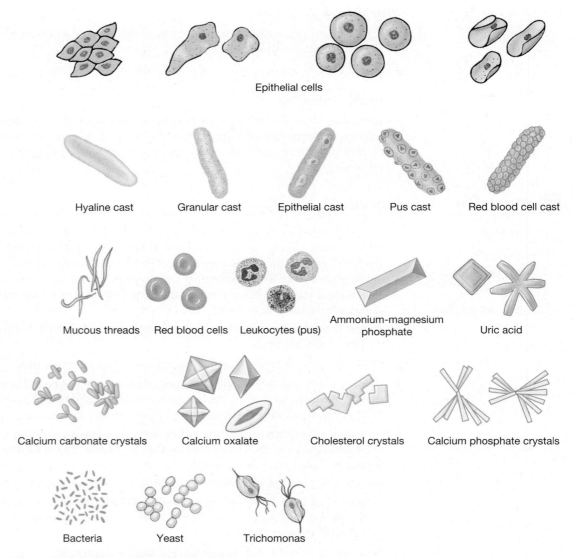

Epithelial cells

Hyaline cast   Granular cast   Epithelial cast   Pus cast   Red blood cell cast

Mucous threads   Red blood cells   Leukocytes (pus)   Ammonium-magnesium phosphate   Uric acid

Calcium carbonate crystals   Calcium oxalate   Cholesterol crystals   Calcium phosphate crystals

Bacteria   Yeast   Trichomonas

Figure 44.3 **Sediments Found in the Urine**

Clinical Application | **Kidney Stones**

Crystals can form in the urinary tract and are usually voided with the urine. This sediment consists of **casts**, which usually are small clots of blood, tissue, or crystals of mineral salts. Complete blockage of the urinary tract can occur as a result of the formation of kidney stones, or **renal calculi**, which are solid pebbles of urinary salts containing calcium, magnesium, or uric acid. Calculi can occur in the kidneys, ureters, bladder, or urethra, and may cause severe pain. If a stone completely blocks the urinary tract and will not pass out of the body, it must be removed surgically. *Lithotripsy* is a nonsurgical procedure that uses sound waves to shatter the stone into pieces small enough to pass through the tract. The patient is immersed in water, and the sound energy is directed to the area overlying the stone to destroy the calculi. ▶

QuickCheck Questions

3.1 What types of solids occur in urine?

3.2 What is one cause of renal calculi?

3 Materials

- ☐ Test tube rack
- ☐ Conical centrifuge tubes
- ☐ Wax pencil
- ☐ Centrifuge
- ☐ Pasteur pipette with bulb
- ☐ Iodine or sediment stain
- ☐ Glass slides
- ☐ Cover glasses
- ☐ Compound microscope
- ☐ Biohazard disposal container
- ☐ Bleach solution
- ☐ 10% chlorine solution

Procedures

1. Use the urine sample collected in the previous laboratory activities. Swirl the sample to suspend any solids that have settled.

2. With the wax pencil, mark two centrifuge tubes with a horizontal line two-thirds up from the bottom. Then label one tube "Sample" and the other "Blank."

3. Fill the tube labeled Sample to the mark with urine, and fill the tube labeled Blank to the mark with tap water.

4. Place the two tubes in the centrifuge opposite each other so that the centrifuge remains balanced. This step is very important to keep from damaging the centrifuge.

5. Centrifuge the sample for 8 to 10 minutes.

6. Pour off the supernatant (the liquid above the solids) from the Sample tube, and with the Pasteur pipette remove some of the sediment. Place one drop of sediment on the glass slide, add one drop of iodine or sediment stain, and place a cover glass on top of the specimen.

7. Begin viewing the stained sediment under 10× power. Look for epithelial cells, red blood cells, white blood cells, crystals, and microbes. Mucin threads and casts may also be seen in the sediment. Mucin is a complex glycoprotein secreted by unicellular exocrine glands, such as goblet cells. In water, mucin becomes mucus, a slimy coating that lubricates and protects the lining of the urinary tract. Casts are usually cylindrical and composed of proteins and dead cells. Salt crystals are also found in urine. Compare the contents of your urine sample with Figure 44.3.

8. Dispose of all used pipettes in the biohazard disposal container. Rinse test tubes in bleach solution, and then wash them with soap and water and dry with clean paper towels. Place glass slides and cover glasses in a beaker of 10% chlorine. Dispose of the urine cup in the biohazard box as indicated by your instructor.

9. Wash and dry your hands before leaving the laboratory. ■

Physiology of the Urinary System

Name _____

Date _____

Section _____

A. Matching

Match each term in the left column with its correct description from the right column.

_____ **1.** pyuria
_____ **2.** hematuria
_____ **3.** glucosuria
_____ **4.** secretion
_____ **5.** albuminuria
_____ **6.** bilirubin
_____ **7.** urochrome
_____ **8.** renal calculi
_____ **9.** deamination
_____ **10.** reabsorption
_____ **11.** ketone
_____ **12.** filtration
_____ **13.** filtrate
_____ **14.** urine
_____ **15.** urobilinogen

A. product of fat metabolism
B. yellow pigment in urine
C. product of bilirubin breakdown
D. leukocytes in urine
E. removal of materials from blood
F. fluid in minor calyx of kidney
G. removal of ammonia from amino acid
H. glucose in urine
I. fluid passing through nephron tubules
J. molecule from breakdown of hemoglobin
K. kidney stones
L. excess albumin in urine
M. addition of materials to filtrate
N. returning of materials from filtrate to blood
O. blood in urine

B. Data Collection

Table 44.1 **Physical Observations of Urine Sample**

Characteristic	Observation
Volume	_____
Color	_____
Turbidity	_____
Odor	_____
Specific gravity	_____

LABORATORY REPORT

44

Table 44.2 Chemical Evaluation of Urine Sample

	Remark
pH	
Specific gravity	
Glucose	
Ketone	
Protein	
Erythrocytes	
Bilirubin	
Other	

C. Short-Answer Questions

1. What is the normal pH range of urine?

2. What is the specific gravity of a normal sample of urine?

3. List five abnormal components of urine.

4. What substances in the urine might indicate that a person has diabetes?

5. Describe the three physiological processes of urine production.

D. Analysis and Application

1. A diabetic woman has been dieting for several months and has lost more than 25 pounds. At her annual medical checkup, a urinalysis is performed. What would you expect to find in her urine?

2. What factors might affect the odor, color, and pH of a sample of urine?

3. Mike and Fred have been hiking in the desert all afternoon. While on the trail, Fred drinks much more water than Mike. If urine samples were collected from both men, what differences in specific gravity of the samples would you expect to measure?

Anatomy of the Reproductive System

OBJECTIVES

On completion of this exercise, you should be able to:

• Describe the location of the male gonad.

• Identify the male testes, ducts, and accessory glands.

• Describe the composition of semen.

• Identify the three regions of the male urethra.

• Identify the structures of the penis.

• Identify the female ovaries, ligaments, uterine tubes, and uterus.

• Describe and recognize the three main layers of the uterine wall.

• Identify the vagina and the features of the vulva.

• Identify the structures of the mammary glands.

• Compare gamete formation in males and females.

Whereas all the other systems of the body function to support the continued life of the organism, the reproductive system functions to ensure continuation of the species. The primary sex organs, or **gonads** (GŌ-nads), of the male and female are the testes and ovaries, respectively. They function in the production of **gametes** (GAM-ēts), the spermatozoa and ova (eggs). Gonads secrete hormones that support maintenance of the male and female sex characteristics. Because these glands secrete their gametes into ducts and their hormones into the blood, they have both exocrine and endocrine functions. The ducts store the gametes, and several accessory glands in the reproductive system secrete products to protect and support the gametes.

LAB ACTIVITY   1   Male: The Testes and Spermatogenesis

The male reproductive system (Figure 45.1) consists of a pair of **testes** (TES-tēz; singular testis) located outside the abdominal cavity, conducting ducts, and the penis. The testes, which are the primary sex organ of the male, produce **spermatozoa** (sperma-tō-ZŌ-uh) and male sex hormones. The ducts transport the spermatozoa from the testes to inside the pelvic cavity, where glands add secretions to form a mixture called **semen** (SĒ-men), the liquid that is ejaculated into the female during intercourse. The

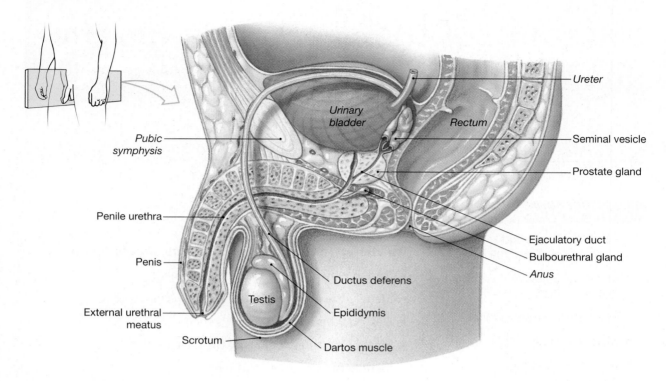

Pubic symphysis

Penile urethra

Penis

External urethral meatus

Scrotum

Urinary bladder

Rectum

Testis

Ductus deferens

Epididymis

Dartos muscle

Ureter

Seminal vesicle

Prostate gland

Ejaculatory duct

Bulbourethral gland

Anus

Figure 45.1 The Male Reproductive System
A sagittal section of the male reproductive organs.

testes are located in the **scrotum** (SKRŌ-tum), a pouch of skin hanging from the pubis region. The pouch is divided into two compartments, each containing one testis. (**Testicle** is an alternative name for testis.) The **dartos** (DAR-tōs) **muscle** forms part of the septum separating the two testes and is responsible for the wrinkling of the scrotum skin.

Each testis is about 5 cm (2 in.) in length and 2.5 cm (1 in.) in diameter. Compartments in the testis called **lobules** contain highly coiled **seminiferous** (se-mi-NIF-e-rus) **tubules** (Figure 45.2). Millions of spermatozoa are produced each day by the seminiferous tubules, and the process is called **spermatogenesis** (sper-ma-tō-JEN-e-sis). Follicle-stimulating hormone (FSH) from the anterior pituitary gland regulates spermatogenesis. Between the seminiferous tubules are small clusters of cells called **interstitial cells** that secrete **testosterone**, the male sex hormone. Testosterone is responsible for the male sex drive and for development and maintenance of the male secondary sex characteristics, such as facial hair and increased muscle and bone development.

Sexual reproduction involves the union of a female gamete and a male gamete to produce the next generation of the species. Human somatic cells (all the cells in the body except those involved in reproduction) contain 23 pairs of chromosomes, for a total of 46 individual chromosomes. Cells that contain all 46 chromosomes are **diploid** (DIP-loyd) cells. To maintain the correct number of chromosomes in the offspring created in reproduction, the gametes reduce their chromosome number by one-half and become **haploid** (HAP-loyd) cells during part of the reproduction cycle. The diploid condition is restored during fertilization, when the male haploid gamete with its 23 chromosomes joins the female haploid gamete with its 23 chromosomes join to produce a diploid cell called the **zygote**. From this new cell, an incomprehensible number of divisions ultimately shapes a new human.

Mitosis (mī-TŌ-sis) is cell division in somatic cells, where one parent cell divides to produce two identical diploid daughter cells. **Meiosis** (mī-Ō-sis) is cell

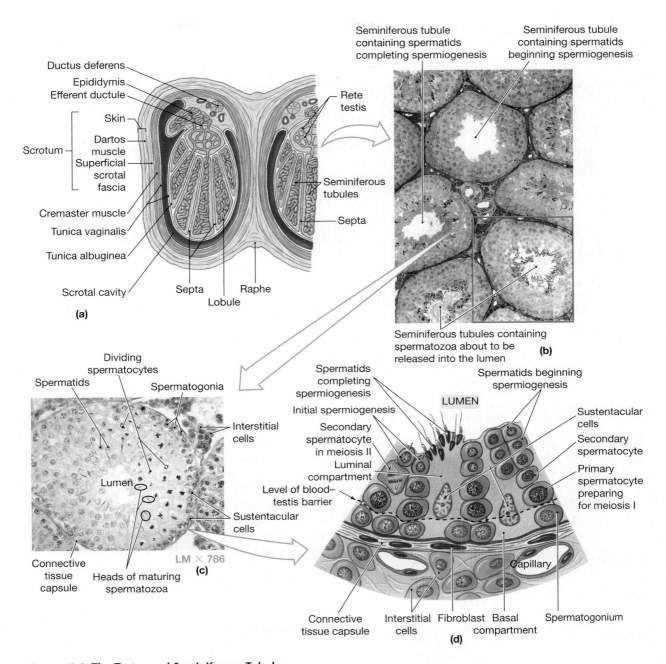

Figure 45.2 The Testes and Seminiferous Tubules
(a) A frontal section of the testes. (b) A transverse section through a coiled seminiferous tubule. (c) A cross-section through a single tubule. (d) The lining of a tubule. Sustentacular cells surround the stem cells of the tubule and support the developing spermatocytes and spermatids.

division in reproductive cells, where the products of the division are haploid gametes in the testes and the ovaries. Meiosis occurs in two cycles, meiosis I and II, and in many ways is similar to mitosis, as the comparative overview in Figure 45.3 shows.

In spermatogenesis, four haploid spermatozoa are produced from each cell that undergoes meiosis. For simplicity, Figure 45.4 illustrates meiosis in a diploid cell containing 3 chromosome pairs (6 individual chromosomes) instead of the 23 pairs found in humans. When a male reaches puberty, hormones stimulate the

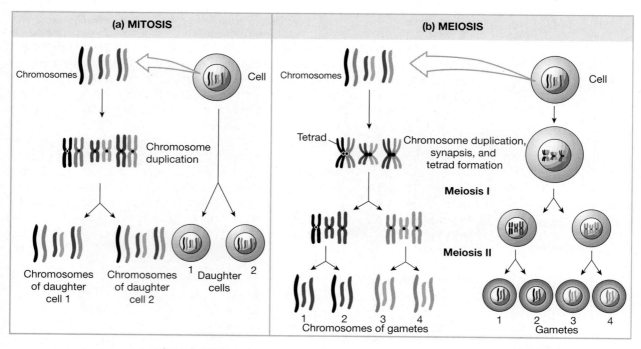

Figure 45.3 Chromosomes in Mitosis and Meiosis
(a) Steps in mitosis. **(b)** Steps in meiosis.

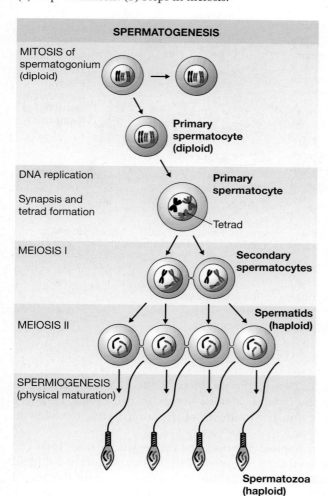

Figure 45.4 Spermatogenesis

Meiosis in the seminiferous tubules, showing the distribution of only a few chromosomes.

testes to begin spermatogenesis. **Spermatogonia** (sper-ma-tō-GŌ-nē-uh) cells in the walls of the seminiferous tubules divide by mitosis and produce more spermatogonia and diploid **primary spermatocytes** (sper-MA-tō-sits). A primary spermatocyte prepares for meiosis by duplicating its genetic material, represented by the DNA replication section of Figure 45.4. After replication, each chromosome is double-stranded and consists of two **chromatids**. Each *pair* of chromosomes, which are called **homologous chromosomes**, consists of four chromatids. The primary spermatocyte is now ready to proceed into meiosis.

Meiosis I begins with **prophase I**, as the nuclear membrane of the primary spermatocyte dissolves and the chromatids condense into chromosomes. The homologous chromosomes match into pairs in a process called **synapsis**, and the four chromatids of the pair are collectively called a **tetrad** (Figure 45.4). Because each chromosome has replicated, a total of four potential chromosomes, each still referred to as a chromatid, align in a tetrad. Because each chromatid in a tetrad belongs to the same chromosome pair, genetic information may be exchanged between chromatids. This **crossing over,** or mixing, of the genes contained in the chromatids increases the genetic variation within the population.

During **metaphase I**, the tetrads line up along the middle of the cell. Next is the critical step to reduce chromosome number to haploid. In **anaphase I** the tetrads separate and the double-stranded chromosomes move to opposite sides of the cell. This separation step is called the **reduction division** of meiosis, because haploid cells are produced. Next, in the step called **telophase I,** the cell pinches apart into two haploid secondary spermatocytes.

Meiosis II is necessary because, although the secondary spermatocytes are haploid, they have double-stranded chromosomes that must be reduced to single-stranded chromosomes. The process is similar to mitosis, with the double-stranded chromosomes lining up and separating. The two secondary spermatocytes produce four haploid **spermatids** that contain single-stranded chromosomes. The spermatids are what develop into spermatozoa. The entire process from spermatogonia to mature spermatozoa takes approximately 10 weeks.

QuickCheck Questions

1.1 Where are spermatozoa produced in the male?

1.2 What cell divides and produces the primary spermatocyte?

1.3 What is a tetrad?

1 Materials

- ☐ Male urogenital model and chart
- ☐ Meiosis models
- ☐ Compound microscope
- ☐ Prepared slide of testis

Procedures

1. Review the male anatomy in Figure 45.1.

2. After reviewing the structure of the testes in Figure 45.2, locate the scrotum, testes, and associated anatomy on the urogenital model and chart.

3. After reviewing Figures 45.3 and 45.4, identify the different cell types shown on the meiosis models.

4. Examine the testis slide, using the micrograph in Figure 45.2 for reference. Scan the slide at low magnification and observe the many seminiferous tubules. Increase the magnification to medium and locate the interstitial cells between the tubules.

5. At medium power, pick a seminiferous tubule that has distinct cells within the walls. Using Figure 45.2 as a guide, identify the spermatogonia, primary and secondary spermatocytes, and spermatids. Spermatozoa are visible in the middle of the lumen.

6. In the space provided, draw a section of a seminiferous tubule, and label the spermatogonia, primary spermatocytes, secondary spermatocytes, spermatids, and spermatozoa. ■

LAB ACTIVITY 2 Male: The Epididymis and Ductus Deferens

After spermatozoa are produced in the seminiferous tubules, they move into the **epididymis** (ep-i-DID-i-mus), a highly coiled tubule located on the posterior of the testis and visible in Figure 45.5. The spermatozoa mature in the epididymis and are stored until ejaculation out of the male reproductive system. Peristalsis of the smooth muscle of the epididymis propels the spermatozoa into the **ductus deferens** (DUK-tus DEF-e-renz), or **vas deferens**, the duct that empties into the urethra.

The ductus deferens ascends into the pelvic cavity as part of the **spermatic cord**. Other structures in the spermatic cord include blood and lymphatic vessels, nerves, and the **cremaster** (krē-MAS-ter) **muscle**, which encases the testes and raises or lowers them to maintain an optimum temperature for spermatozoa production. The ductus deferens is 46 to 50 cm (18 to 20 in.) long and is lined with pseudostratified columnar epithelium. Peristaltic waves propel spermatozoa toward the urethra. The ductus deferens passes through the **inguinal** (ing-gwi-nal) **canal** in the lower abdominal wall to enter the body cavity. This canal is a weak area and is frequently injured. An **inguinal hernia** occurs when portions of intestine protrude through the canal and slide into the scrotum. The ductus deferens continues around the posterior of the urinary bladder and widens into the **ampulla** (am-PŪL-uh) before joining the seminal vesicle at the ejaculatory duct.

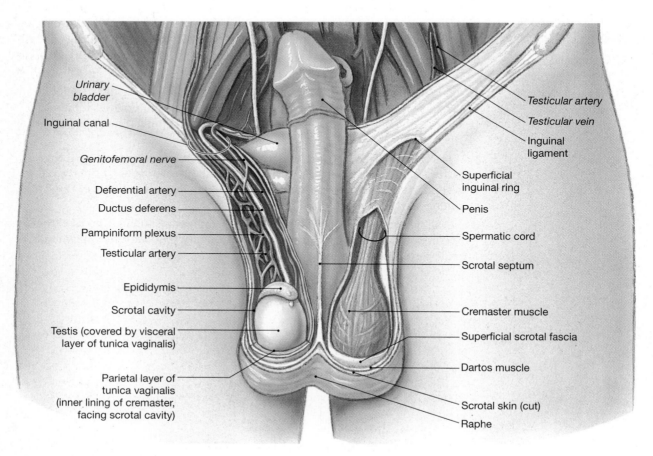

Figure 45.5 The Male Reproductive System in Anterior View

Clinical Application | ***Vasectomy***

A common method of birth control for men is a procedure called **vasectomy** (vaz-EK-to-mē). Two small incisions are made in the scrotum, and a small segment of the ductus deferens on each side is removed. A vasectomized man still produces spermatozoa, but because the duct that transports them from the epididymis to the urethra is removed, the semen that is ejaculated contains no spermatozoa. As a result, no female ovum can be fertilized. Men who have had a vasectomy still produce testosterone and have a normal sex drive. They have orgasms, and the ejaculate is approximately the same volume as in men who have not been vasectomized.

QuickCheck Questions

2.1 Where are spermatozoa stored?

2.2 Where does the ductus deferens enter the abdominal cavity?

2 *Materials*

☐ Male urogenital model and chart

Procedures

1. Review the anatomy in Figure 45.5.
2. Locate the epididymis and ductus deferens on the laboratory models.
3. Locate the spermatic cord, the cremaster muscle, and the inguinal canal. ■

LAB ACTIVITY 3

Male: The Accessory Glands

Three accessory glands—seminal vesicles, prostate gland, and bulbourethral glands—produce fluids that nourish, protect, and support the spermatozoa (Figure 45.6). The spermatozoa and fluids from the seminal vesicles, prostate gland, and bulbourethral glands mix together as semen. The average number of spermatozoa per milliliter of semen is between 50 and 150 million, and the average volume of ejaculate is between 2 and 5 ml.

The **seminal** (SEM-i-nal) **vesicles** are a pair of glands on the posterior of the urinary bladder. Each gland is approximately 15 cm (6 in.) long and merges with the ductus deferens into an **ejaculatory duct**. The seminal vesicles contribute about 60% of the total volume of semen. They secrete a viscous, alkaline fluid containing the sugar fructose. The alkaline nature of this fluid neutralizes the acidity of the male urethra and the female vagina. The fructose provides the spermatozoa with an energy source for beating of the *flagellum,* the tail on each spermatozoon that propels the cell on its way to an ovum. Seminal fluid also contains fibrinogen, which causes the semen to temporarily clot after ejaculation.

The **prostate** (PROS-tāt) gland is a single gland about 1.5 inches in diameter at the base of the urinary bladder. The ejaculatory duct passes into the prostate gland and empties into the first segment of the urethra, the **prostatic urethra**. The prostate gland secretes a milky white, slightly acidic fluid that contains clotting enzymes to coagulate the semen. These secretions contribute about 20 to 30% of the semen volume.

The prostatic urethra exits the prostate gland and passes through the floor of the pelvis as the **membranous urethra**. A pair of **bulbourethral** (bul-bō-ū-RĒ-thral) **glands**, also called *Cowper's glands,* occur on either side of the membranous urethra and add an alkaline mucus to the semen. Before ejaculation, the bulbourethral secretions neutralize the acidity of the urethra and lubricate the end of the penis for sexual intercourse. These glands contribute about 5% of the volume of semen.

QuickCheck Questions

3.1 What are the three accessory glands that contribute to the formation of semen?

3.2 Where is the membranous urethra located?

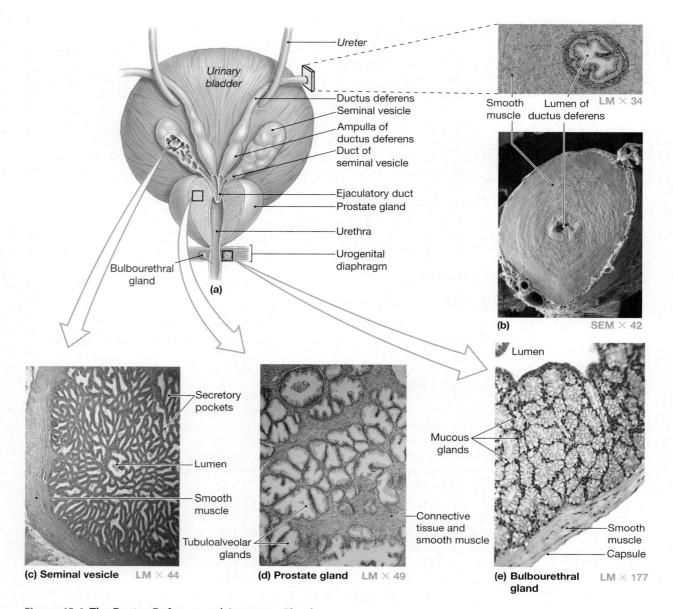

Figure 45.6 The Ductus Deferens and Accessory Glands
(a) A posterior view of the prostate gland, showing subdivisions of the ductus deferens in relation to surrounding structures. **(b)** The ductus deferens, showing the smooth muscle around the lumen (LM [top] × 34, SEM × 42) (© R. G. Kessel and R. H. Kardon, *Tissues and Organs: A Text-Atlas of Scanning Electron Microscopy*, W. H. Freeman & Co., 1979. All rights reserved). Sections of **(c)** the seminal vesicle (LM × 44), **(d)** the prostate gland (LM × 49), and **(e)** a bulbourethral gland (LM × 177).

3 Materials

☐ Male urogenital model and chart

Procedures

1. Review the anatomy in Figure 45.6.

2. On the model and/or chart, trace both ductus deferens through the inguinal canal, behind the urinary bladder, to where each unites with a seminal vesicle. Identify the swollen ampulla of the ductus deferens.

3. Identify the prostate gland, and note the ejaculatory duct that drains the ductus deferens and the seminal vesicle on each side of body. Identify the prostatic urethra that passes from the urinary bladder through the prostate gland.

4. Find the membranous urethra in the muscular pelvic floor. Identify the small bulbourethral glands on either side of the urethra. ■

Male: The Penis

The **penis**, detailed in Figure 45.7, is the male copulatory organ that delivers semen into the vagina of the female. The penis is cylindrical and has an enlarged, acorn-shaped head called the **glans**. Around its base is a margin, the **corona** (crown). On

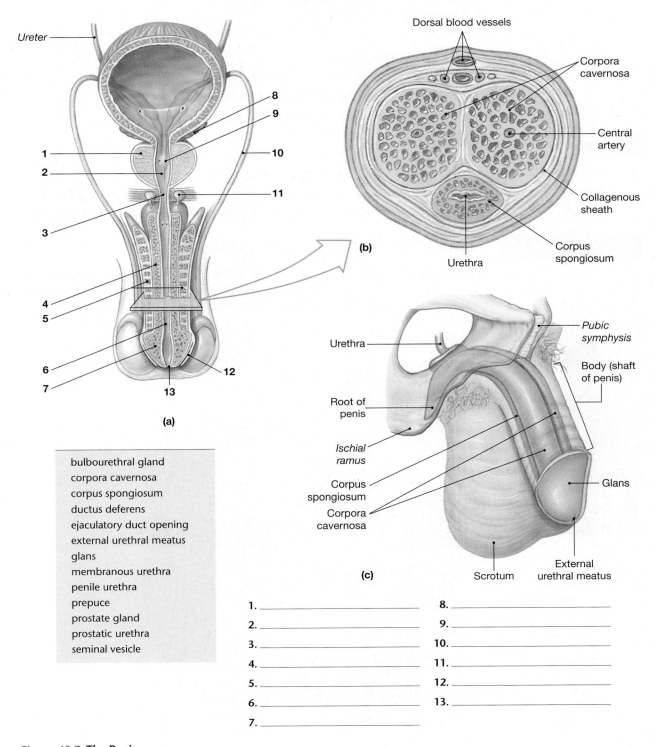

bulbourethral gland
corpora cavernosa
corpus spongiosum
ductus deferens
ejaculatory duct opening
external urethral meatus
glans
membranous urethra
penile urethra
prepuce
prostate gland
prostatic urethra
seminal vesicle

1. _____   8. _____
2. _____   9. _____
3. _____   10. _____
4. _____   11. _____
5. _____   12. _____
6. _____   13. _____
7. _____

Figure 45.7 The Penis

(a) A frontal section through the penis and associated organs. (b) A sectional view through the penis. (c) An anterior and lateral view of the penis, showing positions of the erectile tissues.

an uncircumcised penis, the glans penis is covered with a loose-fitting skin called the **prepuce** (PRĒ-pūs) or *foreskin*. **Circumcision** is surgical removal of the prepuce. The **penile urethra** transports both semen and urine through the penis and ends at the **external urethral meatus** in the tip of the glans. The **root** of the penis anchors the penis to the pelvis. The **body** consists of three cylinders of erectile tissue, a pair of dorsal **corpora cavernosa** (KOR-po-ruh ka-ver-NŌ-suh), and a single ventral **corpus spongiosum** (spon-jē-Ō-sum). During sexual arousal, the three erectile tissues become engorged with blood and cause the penis to stiffen into an **erection**.

QuickCheck Questions

4.1 What is the enlarged structure at the tip of the penis?

4.2 Which structures fill with blood during erection?

4.3 What duct transports urine and semen in the penis?

4   *Materials*

☐ Male urogenital model and chart

Procedures

1. Review the anatomy of the penis in Figure 45.7 and label the figure.

2. Identify the glans, corona, body, and root of the penis on the model and/or chart.

3. On the model, identify the corpora cavernosa and the corpus spongiosum. ■

LAB ACTIVITY   5   Female: The Ovaries and Oogenesis

The female reproductive system, highlighted in Figure 45.8, includes the ovaries, uterine tubes, uterus, vagina, external genitalia, and mammary glands. **Gynecology** is the branch of medicine that deals with the care and treatment of the female reproductive system.

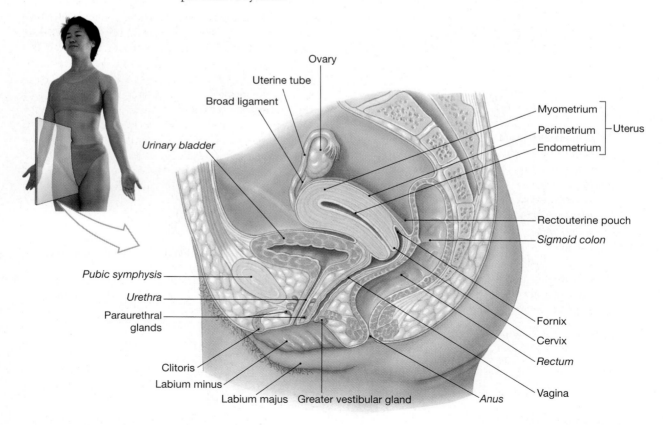

Figure 45.8 The Female Reproductive System
A sagittal section of the female reproductive organs.

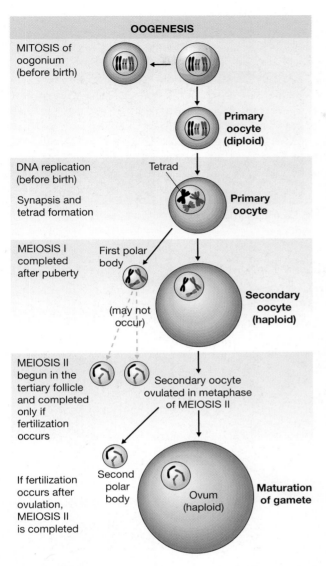

OOGENESIS

MITOSIS of oogonium (before birth)

Primary oocyte (diploid)

DNA replication (before birth)

Synapsis and tetrad formation

Tetrad

Primary oocyte

MEIOSIS I completed after puberty

First polar body

(may not occur)

Secondary oocyte (haploid)

MEIOSIS II begun in the tertiary follicle and completed only if fertilization occurs

Secondary oocyte ovulated in metaphase of MEIOSIS II

If fertilization occurs after ovulation, MEIOSIS II is completed

Second polar body

Ovum (haploid)

Maturation of gamete

Figure 45.9
Oogenesis

In oogenesis, a single primary oocyte produces an ovum and two or three nonfunctional polar bodies.

Formation of the female gamete, the **ovum** (or egg), is called **oogenesis** (ō-ō-JEN-e-sis) and occurs in the ovaries. In a female fetus being carried in its mother's uterus, meiosis I begins when **oogonia** (ō-ō-GŌ-nē-uh) divide and produce **primary oocytes** (Ō-ō-sīts), which remain suspended in this stage until the child reaches puberty. At puberty, each month, a primary oocyte divides into two **secondary oocytes**. One of the secondary oocytes is much smaller than its sister cell and is a nonfunctional cell called the **first polar body** (Figure 45.9). The remaining secondary oocyte remains suspended in meiosis II until it is ovulated. Then, if it is fertilized, this secondary oocyte finishes meiosis II to separate the double-stranded chromosomes. When this cell divides, another polar body is formed, the **second polar body**. The remaining cell is the fertilized ovum, the zygote.

Note that females produce only a single ovum by oogenesis. In males, spermatogenesis results in four spermatozoa.

Each almond-sized ovary contains from 100,000 to 200,000 undeveloped ova clustered in groupings called **egg nests**. Within the nests are **primordial follicles** consisting of oocytes, which are ova surrounded by follicular cells. Figure 45.10 details the monthly ovarian cycle, during which hormones stimulate follicular cells to proliferate and produce several **primary follicles**. These follicles increase in size, and a few become **secondary follicles**. Eventually, one secondary follicle develops into a **tertiary follicle**, also called a **Graafian** (GRAF-ē-an) **follicle**. This follicle fills with fluid and ruptures, casting out the ovum during **ovulation**. The Graafian follicle secretes **estrogen**, the hormone that stimulates rebuilding of the spongy lining of the uterus. After ovulation, the tertiary follicle becomes the **corpus luteum** (LOO-tē-um) and primarily secretes the hormone **progesterone** (prō-JES-ter-ōn), which prepares the uterus for pregnancy. If the ovum is not fertilized, the corpus luteum degenerates into the **corpus albicans** (AL-bi-kanz) and most of the rebuilt lining of the uterus is shed as the menstrual flow.

QuickCheck Questions

5.1 Where are ova produced in the female?

5.2 Which structure ruptures during ovulation to release an ovum?

5.3 What are polar bodies?

5 Materials

☐ Female reproductive system model and chart

☐ Compound microscope

☐ Prepared slide of ovary

Procedures

1. Review the female anatomy in Figure 45.8.

2. Locate the ovaries, uterine tubes, uterus, and vagina on the laboratory model and/or chart.

3. Using Figure 45.10 as a reference, scan the ovary slide at low magnification, and locate an egg nest along the periphery of the ovary.

4. Identify the primary follicles, which are larger than the primordial follicles in the nests. In the primary-follicle stage, the ovum has increased in size and is surrounded by follicular cells.

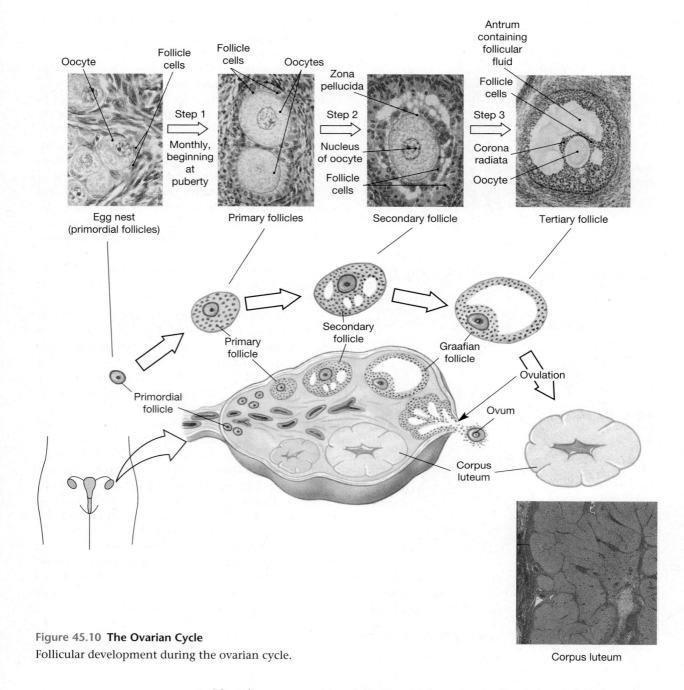

Figure 45.10 The Ovarian Cycle
Follicular development during the ovarian cycle.

Corpus luteum

5. Identify some secondary follicles, which are larger than primary follicles and have a separation between the outer and inner follicular cells.

6. Identify some tertiary follicles, which are easily distinguished by the large, fluid-filled space they contain. ■

LAB ACTIVITY 6   Female: The Uterine Tubes and Uterus

The ends of the **uterine tubes**, commonly called the *fallopian tubes,* have finger-like projections called **fimbriae** (FIM-brē-ē). These projections sweep over the surface of the ovary to capture an ovum released during ovulation and draw it into the expanded **infundibulum** (Figure 45.11). Once the ovum is inside the uterine tube, ciliated epithelium transports it toward the uterus. The tube widens

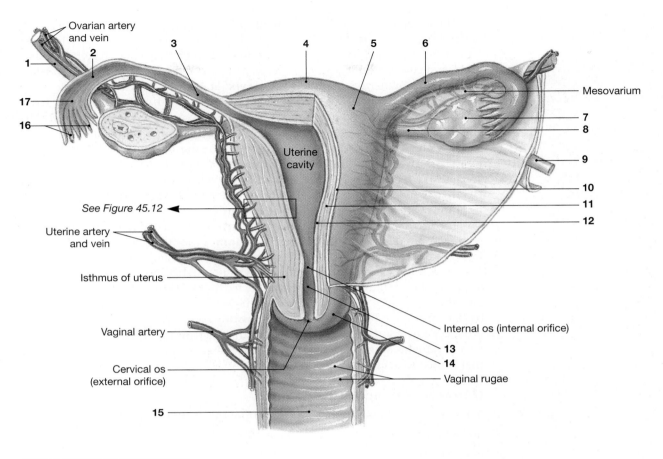

Ovarian artery
and vein

Mesovarium

Uterine
cavity

See Figure 45.12

Uterine artery
and vein

Isthmus of uterus

Vaginal artery

Internal os (internal orifice)

Cervical os
(external orifice)

Vaginal rugae

Figure 45.11 The Uterus
A posterior view with the left uterus, uterine tube, and ovary shown in section.

ampulla
cervical canal
cervix
endometrium
fimbriae
fundus
infundibulum
isthmus
myometrium
ovarian ligament
ovary
perimetrium
round ligament of uterus
suspensory ligament of ovary
uterine body
uterine tube
vagina

1. _____

2. _____

3. _____

4. _____

5. _____

6. _____

7. _____

8. _____

9. _____

10. _____

11. _____

12. _____

13. _____

14. _____

15. _____

16. _____

17. _____

midway along its length in the **ampulla** and then narrows at the **isthmus** (IS-mus) to enter the uterus. Fertilization of the ovum usually occurs in the upper third of the uterine tube.

Clinical Application | ***Tubal Ligation***

Permanent birth control for females involves removing a small segment of the uterine tubes in a process called **tubal ligation**. The female still ovulates, but the spermatozoa cannot reach the ova to fertilize them. The female still has a monthly menstrual period. ▶

The **uterus,** the pear-shaped muscular organ located between the urinary bladder and the rectum, is the site where a fertilized ovum is implanted and where the fetus develops during pregnancy. The uterus consists of three major regions: fundus, body, and cervix. The superior, dome-shaped portion of the uterus is the **fundus,** and the inferior, narrow portion is the **cervix** (SER-viks). The rest of the uterus, the part between fundus and cervix, is called the **body.** Within the uterus is a space called the **uterine cavity** that narrows at the cervix as the **cervical canal**. The **fornix** is the pouch formed where the uterus protrudes into the vagina.

A double-layered fold of peritoneum called the **mesovarium** (mes-ō-VAR-ē-um) holds the ovaries to the **broad ligament** of the uterus (Figure 45.11). The **suspensory ligaments** hold the ovaries to the wall of the pelvis, and the **ovarian ligaments** hold the ovaries to the uterus. The **round ligaments** extend laterally from the ovaries and provide posterior support.

The uterine wall consists of three main layers: perimetrium, myometrium, and endometrium (Figure 45.12). The **perimetrium** is the outer covering of the uterus. It is an extension of the visceral peritoneum and is therefore also called the *serosa*. The thick middle layer, the **myometrium** (mī-ō-MĒ-trē-um), is composed of three layers of smooth muscle and is responsible for the powerful contractions during labor. The inner layer, the **endometrium** (en-dō-MĒ-trē-um), consists of two layers, a basilar zone and a functional zone. The **basilar zone** covers the myometrium and produces a new functional zone each month. The **functional zone** is very glandular and is highly vascularized to support an implanted embryo. This is also the layer that is shed each cycle during menstruation.

QuickCheck Questions

6.1 What structure transports an ovum from the ovary to the uterus?

6.2 What are the three layers of the uterine wall?

6.3 Which layer of the uterine wall is shed during menses?

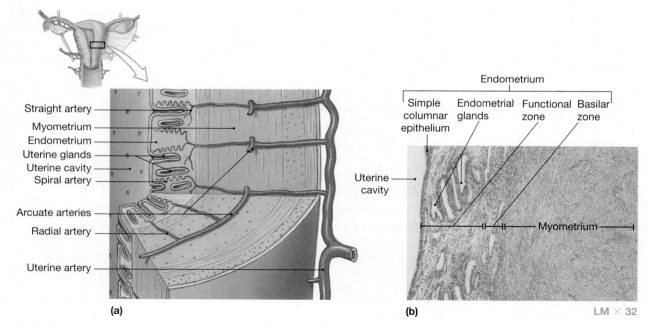

(a)

(b)   LM × 32

Figure 45.12 The Uterine Wall

(a) A sectional view of the uterine wall, showing the endometrial regions and the circulatory supply to the endometrium. **(b)** The basic structure of the endometrium (LM × 32).

6 *Materials*

☐ Female reproductive system model and chart

☐ Compound microscope

☐ Prepared slide of uterine tissue

Procedures

1. Review the uterus and associated anatomy presented in Figure 45.11, and label the figure.

2. Identify the uterine tubes, the ampulla, and the isthmus on the laboratory model and chart.

3. On the model, identify the fundus, body, and cervix.

4. Using Figure 45.12 for reference, scan the uterine-tissue slide at low and medium magnifications, and locate the thick myometrium composed of smooth muscle tissue. Identify the endometrium.

5. Draw a section of the uterine wall in the space below. ■

LAB ACTIVITY 7   Female: The Vagina and Vulva

The **vagina** is a muscular tube approximately 10 cm (4 in.) long. It is lined with stratified squamous epithelium and is the female copulatory organ, the pathway for menstrual flow, and the lower birth canal. The **vaginal orifice** is the external opening of the vagina. This opening may be partially or totally occluded by a thin fold of vascularized mucus membrane called the **hymen** (HĪ-men). On either side of the vaginal orifice are openings of the **greater vestibular glands,** glands that produce a mucous secretion that lubricates the vaginal entrance for sexual intercourse. These glands are similar to the bulbourethral glands of the male.

The **vulva** (VUL-vuh), which is the collective name for the female **external genitalia** (jen-i-TĀ-lē-uh), includes the following structures (Figure 45.13):

- The **mons pubis,** a fatty pad over the pubic symphysis, is covered with skin and pubic hair and serves as a cushion for the pubic symphysis during sexual intercourse.

- The **labia** (LĀ-bē-uh) **majora** are two fatty folds of skin extending from the mons pubis and continuing posteriorly. They are homologous to the scrotum of the male. They usually have pubic hair and contain many sudoriferous (sweat) and sebaceous (oil) glands.

- The **labia minora** (mi-NOR-uh) are two smaller parallel folds of skin containing many sebaceous glands. This pair of labia lacks hair.

- The **clitoris** (KLI-to-ris) is a small, cylindrical mass of erectile tissue similar to the penis. Like the penis, the clitoris contains a small fold of covering skin called the prepuce. The exposed portion of the clitoris is called the **glans**.

- The **vestibule** is the area between the labia minora that contains the vaginal orifice, hymen, and external urethral orifice.

- **Paraurethral glands** (Skene's glands) surround the urethra.

- The **perineum** is the diamond-shaped area between the legs from the clitoris to the anus. This area is of clinical significance because of the tremendous pressure exerted on it during childbirth. If the vagina is too narrow during childbirth, an **episiotomy** (e-pēz-ē-OT-uh-mē) is performed by making a small incision at the base of the vaginal opening toward the anus to expand the vaginal opening.

QuickCheck Questions

7.1 Where is the mons pubis located?

7.2 The vestibule is between what two sets of folds?

7.3 Which female organ has a glans?

7 **Materials**

☐ Female reproductive system model and chart

Procedures

1. Review the vulvar anatomy in Figure 45.13 and label the figure.
2. Locate the vagina and vaginal orifice on the laboratory model and/or chart. Examine the fornix, which is the point where the cervix and vagina connect.
3. Locate each component of the vulva. Note the positions of the clitoris, urethra, and vagina. ■

Figure 45.13
The Female External Genitalia

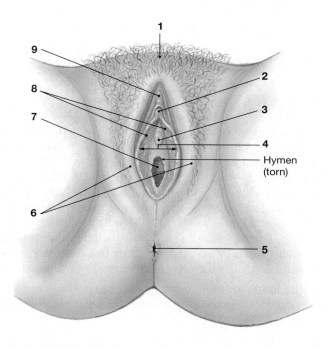

anus
glans of clitoris
labia majora
labia minora
mons pubis
prepuce of clitoris
urethral opening
vaginal orifice
vestibule

1. _____

2. _____

3. _____

4. _____

5. _____

6. _____

7. _____

8. _____

9. _____

LAB ACTIVITY   8   Female: The Mammary Glands

The **mammary glands** (Figure 45.14) are modified sweat glands that, in the process called **lactation,** produce milk to nourish a newborn infant. At puberty, the release of estrogens stimulates an increase in the size of these glands. Fat deposition is the major contributor to the size of the breast, and size does not influence the amount of milk produced. Each gland consists of 15 to 20 **lobes** separated by fat and connective tissue. Each lobe contains smaller compartments, called **lobules,** that contain milk-secreting cells called **alveoli. Lactiferous** (lak-TIF-e-rus) **ducts** drain milk from the lobules toward the **lactiferous sinuses**. These sinuses empty the milk at the raised portion of the breast called the *nipple*. A circular pigmented area called the **areola** (a-RĒ-ō-luh) surrounds the nipple.

QuickCheck Questions

8.1 What are the milk-producing cells of the breast called?

8.2 What is the areola?

8 *Materials*

☐ Breast model

Procedures

1. Review the anatomy of the breast presented in Figure 45.14.

2. On the model, trace the pathway of milk from a lobule to the surface of the nipple. ■

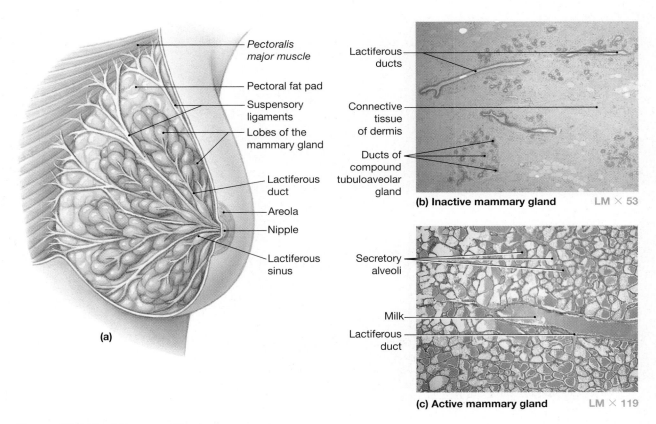

(a) The mammary gland of the left breast.

Labels (a): Pectoralis major muscle; Pectoral fat pad; Suspensory ligaments; Lobes of the mammary gland; Lactiferous duct; Areola; Nipple; Lactiferous sinus

(b) Inactive mammary gland   LM × 53

Labels (b): Lactiferous ducts; Connective tissue of dermis; Ducts of compound tubuloaveolar gland

(c) Active mammary gland   LM × 119

Labels (c): Secretory alveoli; Milk; Lactiferous duct

Figure 45.14 The Mammary Glands

(a) The mammary gland of the left breast. (b) An inactive mammary gland of a nonpregnant woman (LM × 53). (c) An active mammary gland of a nursing woman (LM × 119).

Anatomy of the Reproductive System

Name _____

Date _____

Section _____

A. Matching

Match each structure in the left column with its correct description from the right column.

_____	1. epididymis	**A.** site of spermatozoa storage
_____	2. ductus deferens	**B.** site of spermatozoa production
_____	3. bulbourethral glands	**C.** enlarged tip of penis
_____	4. glans	**D.** paired erectile cylinder
_____	5. corpora cavernosa	**E.** small glands in pelvic floor
_____	6. prepuce	**F.** first segment of urethra
_____	7. prostatic urethra	**G.** also called foreskin
_____	8. scrotum	**H.** transports spermatozoa to urethra
_____	9. membranous urethra	**I.** pouch containing testes
_____	10. seminiferous tubules	**J.** portion of urethra in pelvic floor

B. Matching

Match each structure in the left column with its correct description from the right column.

_____	1. labia minora	**A.** space between labia minora
_____	2. myometrium	**B.** flared end of uterine tube
_____	3. mons pubis	**C.** domed portion of uterus
_____	4. fundus	**D.** female external genitalia
_____	5. infundibulum	**E.** uterine protrusion into vagina
_____	6. isthmus	**F.** narrow portion of uterine tube
_____	7. vestibule	**G.** small fold lacking pubic hair
_____	8. labia majora	**H.** fatty cushion
_____	9. vulva	**I.** muscular layer of uterine wall
_____	10. cervix	**J.** large fold with pubic hair

45

C. Labeling

Label the events of meiosis in Figure 45.15.

1. _____   7. _____
2. _____   8. _____
3. _____   9. _____
4. _____   10. _____
5. _____   11. _____
6. _____   12. _____

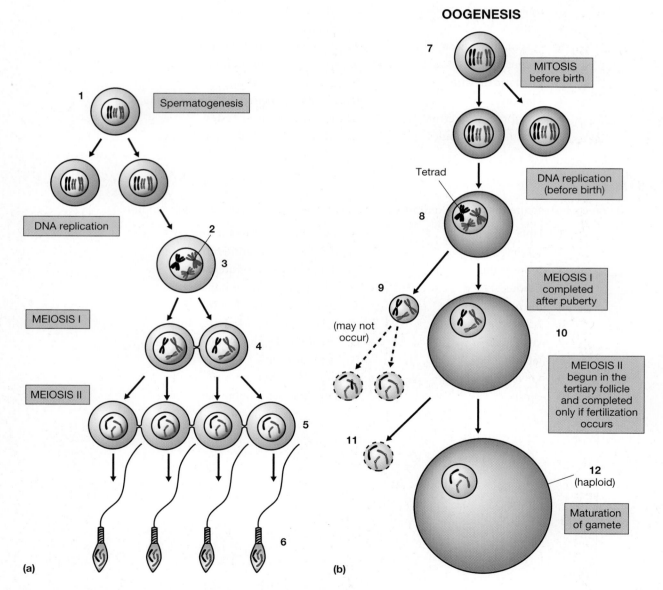

Figure 45.15 Comparison of Spermatogenesis and Oogenesis

D. Short-Answer Questions

1. List the three layers of the uterus, from superficial to deep.

2. Describe the gross anatomy of the female breast.

3. List the components of the vulva.

4. How is temperature regulated in the testes for maximal spermatozoa production?

5. Name the three regions of the male urethra.

6. What are the three accessory glands of the male reproductive system?

45

E. Analysis and Application

1. How would removal of the testes affect endocrine function and reproductive function?

2. Explain the division sequence that leads to four spermatids in male meiosis but only one ovum in female meiosis.

3. How does a vasectomy or a tubal ligation sterilize an individual?

4. How are the clitoris and the penis similar to each other?

Development

OBJECTIVES

On completion of this exercise, you should be able to:

- Describe the process of fertilization and early cleavage to the blastocyst stage.

- Describe the process of implantation and placenta formation.

- List the three germ layers and the embryonic fate of each.

- List the four extraembryonic membranes and the function of each.

- Describe the general developmental events of the first, second, and third trimesters.

- List the three stages of labor.

An intriguing topic in biology is human development. Two gametes from the parents—a spermatozoon from the father and an ovum from the mother—fuse to create a new life in the form of a single-celled zygote that has its own genetic identity and quickly develops into an **embryo**, the name given to the organism for approximately the first two months after fertilization. A phenomenal number of cell divisions and migrations occur during the early development of the embryo. Tissue layers and organ systems form in an elaborate sequence of ever-increasing complexity until a functional human is born.

In humans, the prenatal period of development occurs over a nine-month **gestation** (jes-TĀ-shun). Human gestation is divided into three trimesters, each three months in length. During the first trimester, the embryo develops cell layers that are precursors to organ systems. Specialized extraembryonic membranes, such as the amnion, support the growing embryo and fetus until birth. By the end of the second month, most organ systems have started to form, and the embryo is then called a **fetus**. The second trimester is characterized by growth in length, weight gain, and the appearance of functional organ systems. In the third trimester, increases in length and weight occur, and all organ systems either become functional or are prepared to become functional at birth. After 38 weeks of gestation, the uterus begins to rhythmically contract to deliver the fetus into the world. Although maternal changes occur during the gestation period, this exercise focuses on the development of the fetus.

Morphogenesis (mor-fō-JEN-uh-sis) is the general term for all the processes involved in the specialization of cells in the developing fetus and the migration of those cells to produce anatomical form and function. Ultimately, organ systems appear and become functional as the offspring gets closer to birth.

LAB ACTIVITY 1 First Trimester: Fertilization, Cleavage, and Blastocyst Formation

Fertilization is the act of the spermatozoon and ovum joining their haploid nuclei to produce a diploid zygote, the genetically unique cell that develops into an individual. The male ejaculates approximately 300 million spermatozoa into the female's reproductive tract during intercourse. Once exposed to the female's reproductive tract, the spermatozoa complete a process called **capacitation** (ka-pas-i-TĀ-shun), during which they increase their motility and become capable of fertilizing an ovum. Most spermatozoa do not survive the journey through the vagina and uterus, and only an estimated 100 of them reach the upper uterine tube, where fertilization occurs. Normally, only a single ovum is released from a single ovary during one ovulation cycle.

Figure 46.1 illustrates fertilization. Ovulation releases a secondary oocyte from the ovary. The oocyte is surrounded by a layer of cells called the **corona radiata** (kō-RŌ-nuh rā-dē-A-tuh). Spermatozoa must pass through the corona radiata to reach the cell membrane of the oocyte. Spermatozoa swarming around the oocyte release an enzyme called *hyaluronidase* from their acrosomal caps. The combined action of enzymes contributed by all the spermatozoa eventually creates a gap between some coronal cells, and a single spermatozoon slips into the oocyte. The membrane of the oocyte instantly undergoes chemical and electrical changes that prevent additional spermatozoa from entering the cell. The oocyte, suspended in meiosis II since ovulation, now completes meiosis while the spermatozoon prepares the paternal chromosomes for the union with the maternal chromosomes. Each set of nuclear material is called a **pronucleus**. Within 30 hours of fertilization, the male and female pronuclei come together in **amphimixis** (am-fi-MIK-sis) and undergo the first **cleavage** (KLĒ-vij), a mitotic division resulting in two cells, each called a **blastomere** (BLAS-tō-mēr). During cleavage, the existing cell mass of the ovum is subdivided by each cell division. (In other words, there is no increase in the mass of the zygote at this time.)

As the zygote slowly descends in the uterine tube toward the uterus, cleavages occur approximately every 12 hours. By the third day, the blastomeres are organized into a solid ball of nearly equal cells called a **morula** (MOR-ū-la), shown in Figure 46.2. Around day six, the morula has entered the uterus and changed into a **blastocyst** (BLAS-tō-sist), a hollow ball of cells with an internal cavity called the **blastocoele** (BLAS-tō-sel). Now the process of **differentiation**, or specialization, begins. The blastomeres making up the blastocyst are now of various sizes and have migrated into two regions. Cells on the outside compose the **trophoblast** (TRŌ-fō-blast), which will burrow into the uterine lining and eventually form part of the placenta. Cells clustered inside the blastocoele form the **inner cell mass**, which will develop into the embryo.

QuickCheck Questions

1.1 Where does fertilization normally occur?

1.2 What is a morula?

1.3 What is a blastocyst?

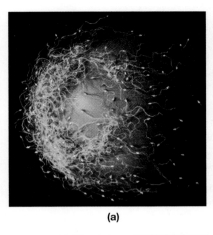

(a)

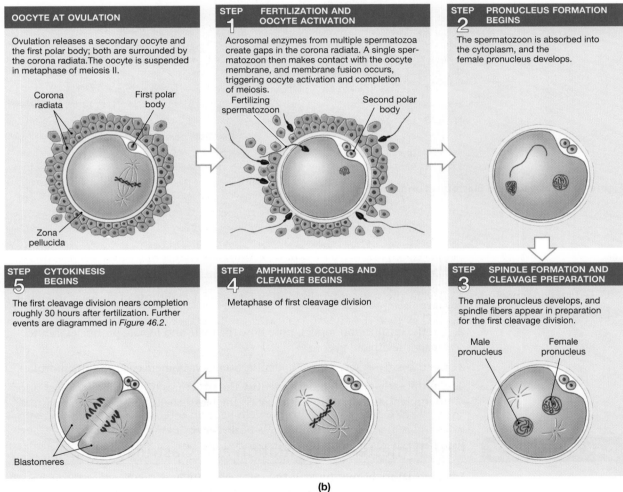

OOCYTE AT OVULATION

Ovulation releases a secondary oocyte and the first polar body; both are surrounded by the corona radiata. The oocyte is suspended in metaphase of meiosis II.

- Corona radiata
- First polar body
- Zona pellucida

STEP 1 FERTILIZATION AND OOCYTE ACTIVATION

Acrosomal enzymes from multiple spermatozoa create gaps in the corona radiata. A single spermatozoon then makes contact with the oocyte membrane, and membrane fusion occurs, triggering oocyte activation and completion of meiosis.

- Fertilizing spermatozoon
- Second polar body

STEP 2 PRONUCLEUS FORMATION BEGINS

The spermatozoon is absorbed into the cytoplasm, and the female pronucleus develops.

STEP 5 CYTOKINESIS BEGINS

The first cleavage division nears completion roughly 30 hours after fertilization. Further events are diagrammed in *Figure 46.2*.

- Blastomeres

STEP 4 AMPHIMIXIS OCCURS AND CLEAVAGE BEGINS

Metaphase of first cleavage division

STEP 3 SPINDLE FORMATION AND CLEAVAGE PREPARATION

The male pronucleus develops, and spindle fibers appear in preparation for the first cleavage division.

- Male pronucleus
- Female pronucleus

(b)

Figure 46.1 Fertilization

(a) An oocyte and numerous sperm at the time of fertilization. Notice the difference in size between the gametes. (b) Fertilization and the preparations for cleavage.

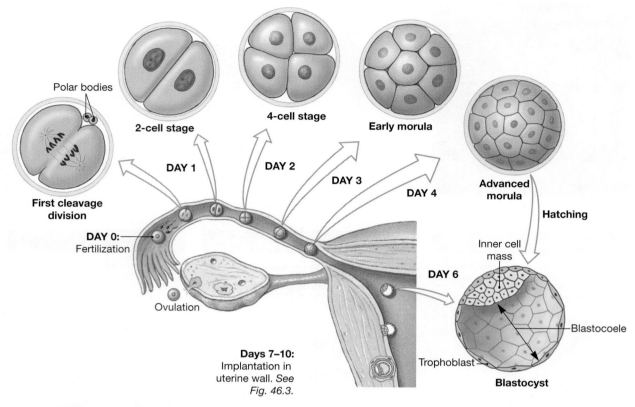

Figure 46.2 Cleavage and Blastocyst Formation

1 *Materials*

☐ Fertilization model
☐ Early-embryo model

Procedures

1. Review the steps of fertilization in Figure 46.1 and those of cleavage in Figure 46.2.

2. On the fertilization model, note how the male and female pronuclei join to create the diploid zygote.

3. On the early-embryo model, identify some blastomere cells and the morula.

4. On the early-embryo model, identify the structures of the blastocyst just prior to implantation, the amniotic cavity, the blastocoele, and the trophoblast. ■

LAB ACTIVITY 2 First Trimester: Implantation and Gastrulation

Implantation begins on day seven or eight, when the blastocyst touches the spongy uterine lining (Figure 46.3). The trophoblast near the inner cell mass faces the uterus and burrows into the functional zone of the endometrium. The plasma membranes of the trophoblast cells dissolve, and the cells mass together as a cytoplasmic layer of multiple nuclei called the **syncytial** (sin-SISH-al) **trophoblast**. The cells secrete hyaluronidase to erode a path for implantation of the blastocyst (Figure 46.3). Implantation continues until the embryo is completely covered by the functional zone of the endometrium, which occurs by the 14th day. To establish a diffusional link with the maternal circulation, the syncytial trophoblast grows extensions called **villi** into spaces in the endometrium called **lacunae**. Maternal blood from

Figure 46.3 Stages in Implantation

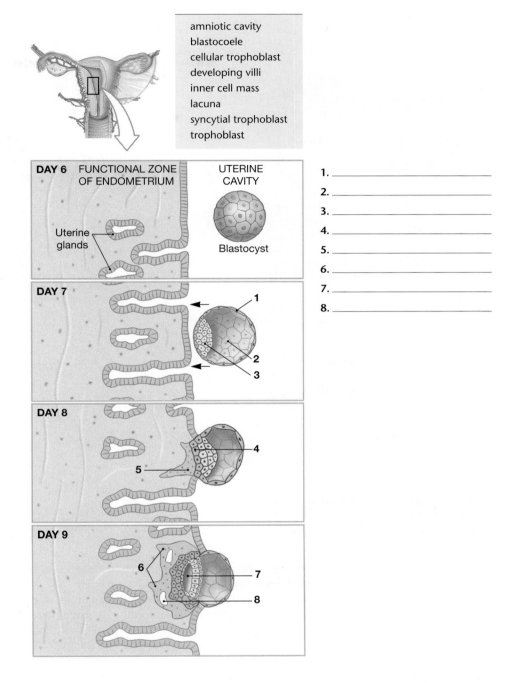

amniotic cavity
blastocoele
cellular trophoblast
developing villi
inner cell mass
lacuna
syncytial trophoblast
trophoblast

DAY 6 FUNCTIONAL ZONE OF ENDOMETRIUM

Uterine glands

UTERINE CAVITY

Blastocyst

DAY 7

1
2
3

DAY 8

4
5

DAY 9

6
7
8

1. _____
2. _____
3. _____
4. _____
5. _____
6. _____
7. _____
8. _____

the endometrium seeps into the lacunae and bathes the villi with nutrients and oxygen. These materials diffuse into the blastocyst to support the inner cell mass. Beneath the syncytial layer is the **cellular trophoblast**, which will soon help form the placenta.

On the ninth or tenth day, the middle layers of the inner cell mass drop away from the upper layer next to the cellular trophoblast. This movement of cells forms the **amniotic** (am-nē-OT-ik) **cavity**. The inner cell mass organizes into a **blastodisc** (BLAS-tō-disk), which contains two cell layers, the **epiblast** (EP-i-blast), or superficial layer, on top and the **hypoblast** (HĪ-pō-blast), or deep layer, facing the blastocoele.

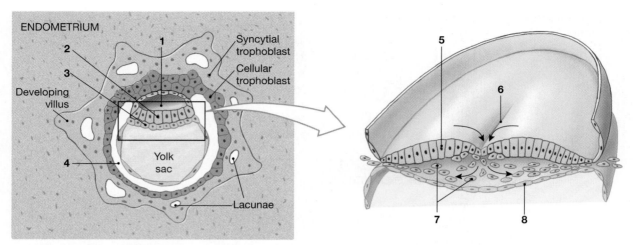

Day 10: The inner cell mass begins as two layers: a superficial layer, facing the amniotic cavity, and a deep layer, exposed to the blastocoele. Migration of cells around the amniotic cavity is the first step in the formation of the amnion. Migration of cells around the edges of the blastocoele is the first step in yolk sac formation.

Day 12: Migration of superficial cells to the interior creates a third layer. From the time this process (gastrulation) begins, the superficial layer is called *ectoderm*, the deep layer *endoderm*, and the migrating cells *mesoderm*.

amniotic cavity
blastocoele
deep layer
ectoderm
endoderm
mesoderm
primitive streak
superficial layer

1. _____   5. _____

2. _____   6. _____

3. _____   7. _____

4. _____   8. _____

Figure 46.4 The Inner Cell Mass and Gastrulation

Within the next few days, cells begin to migrate in a process called **gastrulation** (gas-troo-LĀ-shun) (Figure 46.4). Cells of the epiblast move toward the medial plane of the blastodisc to a region known as the **primitive streak**. As cells arrive at the primitive streak, infolding, or **invagination**, occurs, and cells are liberated between the epiblast and the hypoblast, producing three cell layers in the embryo. The epiblast becomes the **ectoderm**, the hypoblast is now the **endoderm**, and the cells proliferating between the two layers form the **mesoderm**. These three layers, called **germ layers**, each produce specialized tissues that contribute to the formation of the organ systems. The ectoderm forms the nervous system, as well as the skin, hair, and nails. The mesoderm contributes to the development of the skeletal and muscular systems, and the endoderm forms part of the lining of the respiratory and digestive systems.

By the end of the fourth week of development, the embryo is distinct and has a **tail fold** and a **head fold**. The dorsal and ventral surfaces and the right and left sides are well defined. The process of **organogenesis** begins as organ systems develop from the germ layers. In Figure 46.5b, a fiber-optic view of the embryo at four weeks, the heart is clearly visible and has beat since the third week of growth. **Somites** (so-MĪ-tis), embryonic precursors of skeletal muscles, appear. Elements of the nervous system are also developing. Buds for the arms and legs and small discs

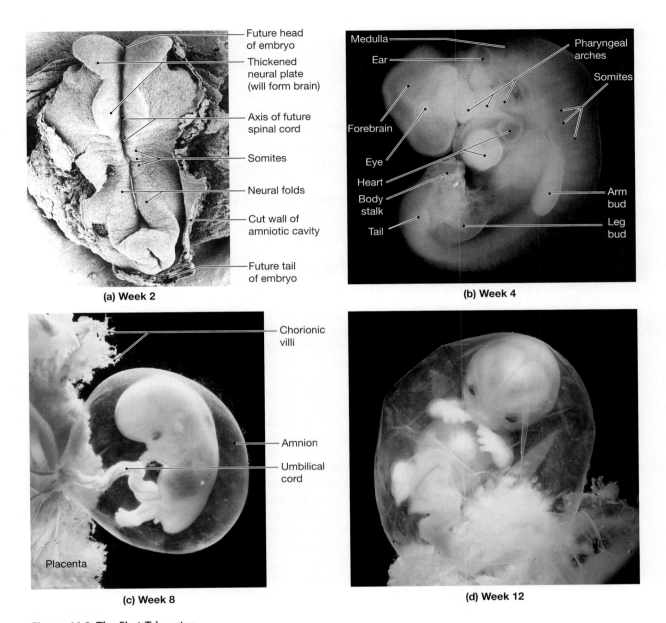

Future head of embryo

Thickened neural plate (will form brain)

Axis of future spinal cord

Somites

Neural folds

Cut wall of amniotic cavity

Future tail of embryo

(a) Week 2

Medulla

Ear

Forebrain

Eye

Heart

Body stalk

Tail

Pharyngeal arches

Somites

Arm bud

Leg bud

(b) Week 4

Chorionic villi

Amnion

Umbilical cord

Placenta

(c) Week 8

(d) Week 12

Figure 46.5 The First Trimester
(a) An SEM of the superior surface of a monkey embryo after two weeks of development. A human embryo at this stage would look essentially the same.
(b–d) Fiber-optic views of human embryos at four, eight, and twelve weeks.

for the eyes and ears are also present. By week eight, individual fingers and toes are present, and the embryo is now usually called the fetus, as noted earlier. At the end of the third month, the first trimester is completed, and every organ system has appeared in the fetus.

QuickCheck Questions

2.1 Where does implantation normally occur?

2.2 What is the syncytial trophoblast?

2.3 What are the two cellular layers of the blastodisc?

2 *Materials*

☐ 6-, 10-, and 12-day embryology models and charts

Procedures

1. Review the anatomy of the blastocyst during implantation in Figure 46.3 and label the figure.
2. On the six-day model or chart, locate the cellular and syncytial trophoblasts. The model may show the development of villi where the syncytial trophoblast has dissolved the endometrium for implantation.
3. Review the anatomy of the blastodisc in Figure 46.4 and label the figure.
4. On the 10-day model or chart, examine the blastodisc and identify the epiblast and the hypoblast.
5. On the 12-day model or chart, identify the amnion, yolk sac, ectoderm, mesoderm, and endoderm. ■

LAB ACTIVITY 3 First Trimester: Extraembryonic Membranes and the Placenta

Four extraembryonic membranes develop from the germ layers: the yolk sac, amnion, chorion, and allantois. These membranes lie outside the embryonic disc and provide protection and nourishment for the embryo/fetus. The **yolk sac** is the first membrane to appear, around the 10 day (Figure 46.6). Initially, cells from the hypoblast form a pouch under the blastodisc. Mesoderm reinforces the sac, and blood vessels appear. As the syncytial trophoblast develops more villi, the yolk sac's importance in providing nourishment for the embryo diminishes.

While the yolk sac is forming, cells in the epiblast migrate to line the inner surface of the amniotic cavity with a membrane called the **amnion** (AM-nē-on), the "water bag." As with the yolk sac, the amnion is soon reinforced with mesoderm tissue. Early embryonic growth continues, and the amnion mushrooms and envelops the entire embryo in a protective environment of **amniotic fluid**.

The **allantois** (a-LAN-tō-is) develops from the endoderm and mesoderm near the base of the yolk sac. The allantois forms part of the urinary bladder and contributes to the **body stalk**, the tissue between the embryo and the developing chorion. Blood vessels pass through the body stalk and into the villi protruding into the lacunae of the endometrium.

The outer extraembryonic membrane is the **chorion** (KOR-ē-on), formed by the cellular trophoblast and mesoderm. The chorion completely encases the embryo and the blastocoele. In the third week of growth, the chorion extends **chorionic villi** and blood vessels into the endometrial lacunae to establish the structural framework for the development of the **placenta** (pla-SENT-uh), the temporary organ through which nutrients, blood gases, and wastes are exchanged between the mother and the embryo. The embryo is connected to the placenta by the body stalk. The **yolk stalk**, where the yolk sac attaches, and the body stalk together form the **umbilical cord**. Inside the umbilical cord are two **umbilical arteries**, which transport deoxygenated blood to the placenta, and a single **umbilical vein**, which returns oxygenated blood to the embryo.

The placenta does not completely surround the embryo. By the fourth week of development, the chorionic villi have enlarged only where they face the uterine wall, and villi that face the uterine cavity become insignificant. The placenta is in contact with the area of the endometrium called the **decidua basalis** (dē-SID-ū-a ba-SA-lis), where the chorionic villi occur. The rest of the endometrium, where villi are absent, isolates the embryo from the uterine cavity and is called the **decidua capsularis** (kap-sū-LA-ris).

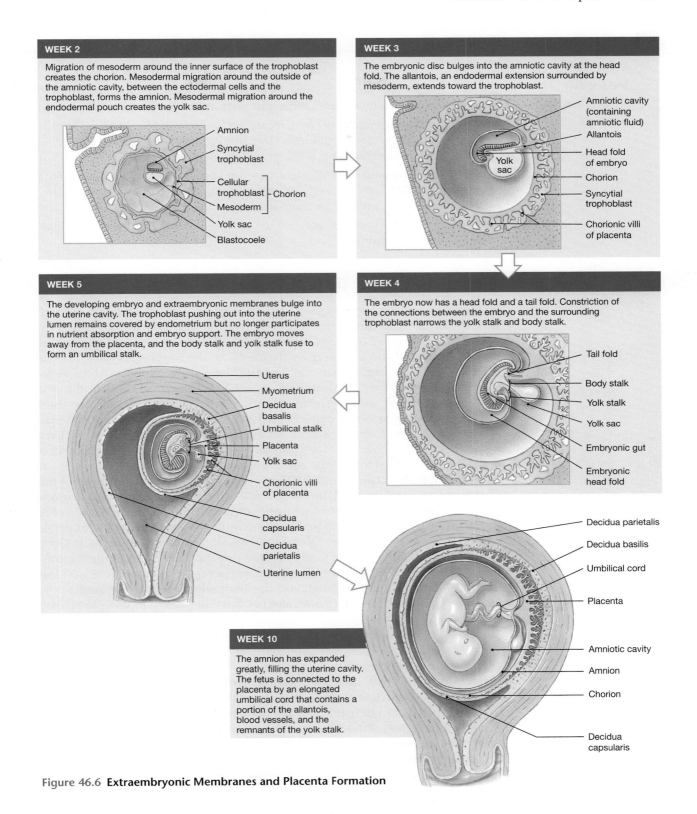

WEEK 2

Migration of mesoderm around the inner surface of the trophoblast creates the chorion. Mesodermal migration around the outside of the amniotic cavity, between the ectodermal cells and the trophoblast, forms the amnion. Mesodermal migration around the endodermal pouch creates the yolk sac.

- Amnion
- Syncytial trophoblast
- Cellular trophoblast ⎤ Chorion
- Mesoderm ⎦
- Yolk sac
- Blastocoele

WEEK 3

The embryonic disc bulges into the amniotic cavity at the head fold. The allantois, an endodermal extension surrounded by mesoderm, extends toward the trophoblast.

- Amniotic cavity (containing amniotic fluid)
- Allantois
- Head fold of embryo
- Chorion
- Syncytial trophoblast
- Chorionic villi of placenta

Yolk sac

WEEK 5

The developing embryo and extraembryonic membranes bulge into the uterine cavity. The trophoblast pushing out into the uterine lumen remains covered by endometrium but no longer participates in nutrient absorption and embryo support. The embryo moves away from the placenta, and the body stalk and yolk stalk fuse to form an umbilical stalk.

- Uterus
- Myometrium
- Decidua basalis
- Umbilical stalk
- Placenta
- Yolk sac
- Chorionic villi of placenta
- Decidua capsularis
- Decidua parietalis
- Uterine lumen

WEEK 4

The embryo now has a head fold and a tail fold. Constriction of the connections between the embryo and the surrounding trophoblast narrows the yolk stalk and body stalk.

- Tail fold
- Body stalk
- Yolk stalk
- Yolk sac
- Embryonic gut
- Embryonic head fold

WEEK 10

The amnion has expanded greatly, filling the uterine cavity. The fetus is connected to the placenta by an elongated umbilical cord that contains a portion of the allantois, blood vessels, and the remnants of the yolk stalk.

- Decidua parietalis
- Decidua basilis
- Umbilical cord
- Placenta
- Amniotic cavity
- Amnion
- Chorion
- Decidua capsularis

Figure 46.6 **Extraembryonic Membranes and Placenta Formation**

QuickCheck Questions

3.1 List the four extraembryonic membranes.

3.2 Which membrane gives rise to the placenta?

☐ Embryology models and charts

☐ Placenta model or biomount

Procedures

1. Review the extraembryonic membranes in Figures 46.5 and 46.6.

2. On an embryology model or chart, locate the yolk sac and the amnion. How does each of these membranes form, and what is the function of each?

3. On an embryology model or chart, locate the allantois and the chorion. How does each of these membranes form? Describe the chorionic villi and their significance to the embryo.

4. On the placenta model or biomount, note the appearance of the various placental surfaces. Are there any differences in appearance from one surface to another? Is the amniotic membrane attached to the placenta?

5. Examine the umbilical cord attached to the placenta, and describe the vascular anatomy in the cord. ■

LAB ACTIVITY 4 — Second and Third Trimesters and Birth

By the start of the second trimester, all major organ systems have started to form. Growth during the second trimester is fast, and the fetus doubles in size and increases its weight by 50 times. As the fetus grows, the uterus expands and displaces the other maternal abdominal organs (Figure 46.7). The fetus begins to move as its muscular system becomes functional, and articulations begin to form in the skeleton. The nervous system organizes the neural tissue that developed in the first trimester, and many sensory organs complete their formation. During the third trimester, all organ systems complete their development and become functional. The fetus responds to sensory stimuli such as a hand rubbing across the mother's swollen abdomen. The fetus gains significant weight and grows rapidly. Space for it in the uterus is limited toward the end of pregnancy.

Birth, or **parturition** (par-tū-RISH-un), involves muscular contractions of the uterine wall to expel the fetus into the outside world. Delivering the fetus is much like pulling on a turtleneck sweater. Muscle contractions must stretch the cervix of the uterus over the fetus's head, pulling the uterine wall thinner as the fetus passes into the lower birth canal, the vagina. Once true labor contractions begin, positive feedback mechanisms operate to increase the frequency and force of uterine contractions. Stretching of the uterine wall causes secretion of **oxytocin** (oks-i-TŌ-sin) from the posterior pituitary gland, and this hormone stimulates the uterine myometrium to contract. The hormone **relaxin** causes the pubic symphysis to become less rigid.

Labor is divided into three stages: dilation, expulsion, and placental (Figure 46.7). The **dilation stage** begins at the onset of true labor contractions. The cervix dilates, and the fetus begins to travel down the cervical canal. To be maximally effective at dilation, the contractions must be less than 10 minutes apart. Each contraction lasts approximately one minute and spreads from the upper cervix downward to efface, or thin, the cervix for delivery. Contractions usually rupture the amnion, and amniotic fluid flows out of the uterus and the vagina.

The **expulsion stage** occurs when the cervix is dilated completely, usually to 10 cm, and the fetus passes through the cervix and the vagina. This stage usually lasts less than two hours and results in the **delivery**, or birth, of the newborn. Once the baby is breathing independently, the umbilical cord is cut, and the baby must now rely on its own organ systems to survive.

During the **placental stage**, the uterus continues to contract to break the placenta free of the endometrium and deliver it out of the body as the **afterbirth**. This stage is usually short, and many women deliver the afterbirth within 5 to 10 minutes after the birth of the fetus.

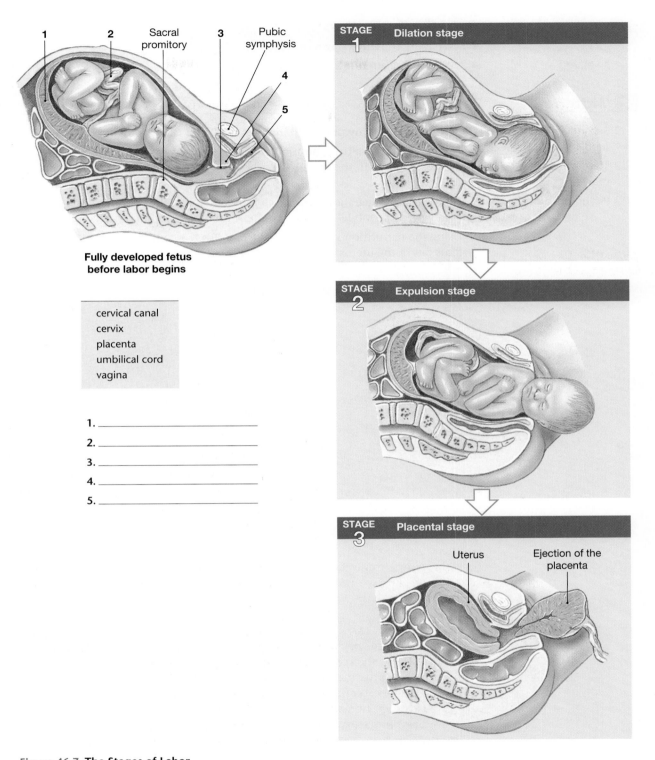

Fully developed fetus before labor begins

cervical canal
cervix
placenta
umbilical cord
vagina

1. _____
2. _____
3. _____
4. _____
5. _____

Figure 46.7 The Stages of Labor

Growth of the newborn is rapid and much development is still occurring in the baby. Neurological, skeletal, and muscular systems continue to develop. Within a few months, the infant can hold its head up, and it begins to reach out to grasp objects for examination.

QuickCheck Questions

4.1 What are the three stages of labor?

4.2 What is the afterbirth?

4 Materials

☐ Second-trimester model

☐ Third-trimester model

☐ Parturition model

Procedures

1. Review the anatomy presented in Figure 46.7 and label the figure.

2. Describe how the fetus is positioned in the uterus in the second-trimester model. If shown in the model, describe the location of the amnion and the placenta.

3. Describe how the fetus is positioned in the uterus in the third-trimester model. If shown in the model, describe the location of the amnion and the placenta.

4. Using the parturition model as an aid, describe the contractions that force the fetus out of the uterus. ∎

Development

Name _____

Date _____

Section _____

A. Matching

Match each structure in the left column with its correct description from the right column.

_____ **1.** blastomere

_____ **2.** amnion

_____ **3.** allantois

_____ **4.** morula

_____ **5.** blastocoele

_____ **6.** syncytial trophoblast

_____ **7.** chorion

_____ **8.** mesoderm

_____ **9.** endoderm

_____ **10.** epiblast

_____ **11.** parturition

_____ **12.** gastrulation

A. has villi

B. differentiates between epiblast and hypoblast

C. cavity of blastocyst

D. forms part of urinary bladder

E. becomes ectoderm

F. migration of cells

G. water bag

H. birth

I. cell produced by early cleavage

J. produces lining of respiratory tract

K. solid ball of cells

L. erodes endometrium

B. Labeling

Label the anatomy of the embryo and fetus in Figure 46.8.

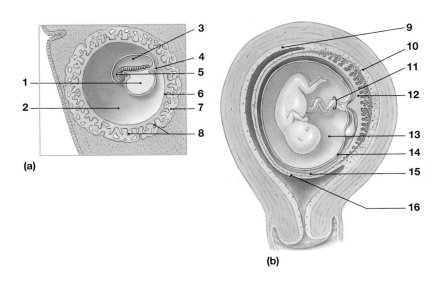

(a)

(b)

1. _____

2. _____

3. _____

4. _____

5. _____

6. _____

7. _____

8. _____

9. _____

10. _____

11. _____

12. _____

13. _____

14. _____

15. _____

16. _____

Figure 46.8 Embryonic and Fetal Development
(a) Embryo. (b) Fetus.

46

C. Short-Answer Questions

1. Describe the process of fertilization.

2. Why are so many spermatozoa required for successful fertilization?

3. Discuss the formation of the three germ layers.

4. Describe the structure of the blastocyst.

5. List the three stages of labor.

D. Analysis and Application

1. How does the amniotic cavity form?

2. What is the function of each of the four extraembryonic membranes?

3. How does a fetus obtain nutrients and gases from the maternal blood?

4. From what structures do the four major tissues groups of the body arise?

Muscles of the Cat

LABORATORY ACTIVITIES

1 Preparing the Cat for Dissection 746

2 Superficial Muscles of the Back and Shoulder   748

3 Deep Muscles of the Back and Shoulder   750

4 Superficial Muscles of the Neck, Abdomen, and Chest   752

5 Deep Muscles of the Chest and Abdomen   755

6 Muscles of the Forelimb   757

7 Muscles of the Thigh   759

8 Muscles of the Lower Hindlimb 761

OBJECTIVES

On completion of this exercise, you should be able to:

- Compare and contrast the anatomy of human muscles and cat muscles.

- Identify the muscles of the cat back and shoulder.

- Identify the muscles of the cat neck.

- Identify the muscles of the cat chest and abdomen.

- Identify the muscles of the cat forelimb and hindlimb.

Some muscles found in four-legged animals are lacking in humans, and some muscles that are fused in humans occur as separate muscles in four-legged animals. Despite these small differences, studying muscles in four-legged animals is an excellent way to learn about human muscles. This exercise on cat muscles is intended to complement your study of human muscles.

The terminology used to describe the position and location of body parts in four-legged animals differs slightly from that used for the human body because four-legged animals move forward head first, with the abdominal surface parallel to the ground. Anatomical position for a four-legged animal is all four limbs on the ground. *Superior* refers to the back (dorsal) surface, and *inferior* relates to the belly (ventral) surface. *Cephalic* means toward the front or anterior, and *caudal* refers to posterior structures.

 ### Dissection Safety Guidelines

You must—*must*—practice the highest level of laboratory safety while handling and dissecting the cat. Keep the following guidelines in mind during the dissection:

1. Wear gloves and safety glasses to protect yourself from the fixatives used to preserve the specimen.
2. Do not dispose of the fixative from your specimen. You will later store the specimen in the fixative to keep the specimen moist and to keep it from decaying.

3. Be extremely careful when using a scalpel or other sharp instrument. Always direct cutting and scissor motion away from yourself to prevent an accident if the instrument slips on moist tissue.
4. Before cutting a given tissue, make sure it is free from underlying and/or adjacent tissues so that they will not be accidentally severed.
5. Never discard tissue in the sink or trash. Your instructor will inform you of the proper disposal procedure. ▲

LAB ACTIVITY 1 Preparing the Cat for Dissection

Read this entire section and familiarize yourself completely with the exercise before proceeding. You must exercise care to prevent muscle damage as you remove the cat's skin. The degree to which the skin is attached to the underlying hypodermis varies from one part of the body to another. For instance, in the abdominopelvic region the skin is loosely attached to the body, with a layer of subcutaneous fat between the skin and the underlying muscle, whereas in the thigh the skin is held tightly to underlying muscle by tough fascia.

When it is time to remove the skin, remove it as a single intact piece and save it to wrap the body for storage. The skin wrapping keeps the body moist and keeps the muscles from drying out. If the skin does not cover the body completely, place paper towels dampened with fixative on any uncovered sections. Never rinse the cat with water; doing so will remove the preservative and promote the growth of mold. Always remoisten the body, skin, and other wrappings with fixative prior to storage. Place the cat in the plastic storage bag provided and seal the bag. Fill out a name tag, attach it to the bag, and place the bag in the assigned storage area.

1 *Materials*

- ☐ Gloves
- ☐ Safety glasses
- ☐ Dissecting tools
- ☐ Dissecting tray
- ☐ Preserved cat

Procedures

1. Place the cat dorsal side down on a dissecting tray.
2. With a scalpel, make a short, *shallow* incision on the midline of the ventral surface just anterior to the tail (line 2a in Figure D1.1a). Caution: Be sure to make your incision shallow, for the skin is very thin and too deep a cut will damage the underlying muscles. Using Figure D1.2 as a guide, insert a blunt probe or your finger to separate the skin from the underlying muscle and connective fibers. Place the *blunt-end blade* of a pair of scissors under the skin, and extend your incision anteriorly by cutting the separated skin. Continue to separate with the probe and cut with the scissors until you have an incision that extends up to the neck (line 2b in Figure D1.1a).
3. On either side of the midventral incision, use either your fingers or a blunt probe to gently separate as much skin as possible from the underlying muscle and connective fibers. If you use a scalpel, keep the blade facing the skin, and take care not to damage the musculature.
4. From the ventral surface, make several incision lines. Always, before cutting any skin, use either a blunt probe or your finger to separate the skin from the underlying tissue. The incisions to be made are:
   a. complete encirclement of the neck (line 4a in Figures D1.1a and b)
   b. on the lateral side of each forelimb from the ventral incision to the wrist (4b), completely cutting the skin around the wrists
   c. complete encirclement of the pelvis (4c)
   d. on the lateral side of each hindlimb from the ventral incision to the ankle (4d), completely cutting the skin around the ankles

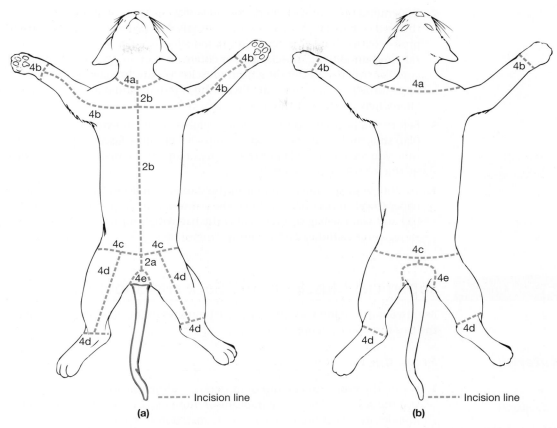

Figure D1.1 Cat Skinning Incision Lines
(a) Ventral view. (b) Dorsal view.

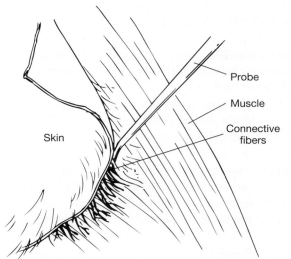

Figure D1.2
Skinning

   e. encircling the anus, the genital organs, and the tail at its base (4e). Use your fingers to loosen the anal/genital skin as much as possible. If the skin does not come off easily, free it by cutting with the scalpel.

5. Use your fingers to free the entire skin from the underlying connective tissue. Work from the ventral surface toward the dorsal surface at the posterior of the body, and then work on the ventral surface from the posterior end of the body toward the neck. Continue working from the ventral to the dorsal surface, freeing the skin from the underlying connective tissue and other attached structures. The only skin remaining on the body once you have completed this step should be the skin on the head, on the tail, and on the paws below the wrist and ankle joints.

   6. Remove all remaining skin from the side of the neck up to the ear. Be careful not to sever the external jugular vein, which is the large blue vein lying on the ventral surface of the neck. Free this vein from the underlying muscles, and clean the connective tissue from the back of the shoulder and the ventral and lateral surfaces of the neck. Do not remove the connective tissue from the midline of the back because this is the origin of the trapezius muscle group.

7. Depending on your specimen, you may observe some or all of the following: thin red or blue latex-injected blood vessels that resemble rubber bands and project between muscles and skin; in female cats, mammary glands between skin and underlying muscle; and/or cutaneous nerves, which are small, white, cordlike structures extending from muscles to skin. In male cats, leave undisturbed the skin associated with the external genitalia; you will remove it later when you dissect the reproductive system.

8. Before you begin dissecting muscles, remove as much extraneous fat, hair, platysma, and cutaneous maximus muscle and loose fascia as possible. If at any time you are confused about the nature of any of the material, check with your instructor before removing it.

9. To store your specimen, wrap it in the skin and moisten it with fixative. Use paper towels if necessary to cover the entire specimen. Return it to the storage bag and seal the bag securely. Label the bag with your name and place it in the storage area as indicated by your instructor. ■

LAB ACTIVITY 2 Superficial Muscles of the Back and Shoulder

Begin your dissection with the superficial muscles in the upper back region between the two forelimbs (Figure D1.3).

2 Materials

- ☐ Gloves
- ☐ Safety glasses
- ☐ Dissecting tools
- ☐ Dissecting tray
- ☐ Preserved cat, skin removed

Procedures

1. Locate the **trapezius group** of muscles covering the dorsal surface of the neck and the scapula. The single trapezius in humans occurs as three distinct muscles in cats. A prefix describes the insertion of each muscle:

   a. The **spinotrapezius** is the most posterior trapezius muscle. It originates on the spinous processes of posterior thoracic vertebrae and inserts on the scapular spine. It pulls the scapula dorsocaudad.

   b. The **acromiotrapezius** is a large muscle anterior to the spinotrapezius. The almost-square acromiotrapezius originates on the spinous processes of cervical and anterior thoracic vertebrae and inserts on the scapular spine. It holds the scapula in place.

   c. The **clavotrapezius** is a broad muscle anterior to the acromiotrapezius. It originates on the lambdoidal crest and axis and inserts on the clavicle. It draws the scapula cranially and dorsally.

2. The large, flat muscle posterior to the trapezius group is the **latissimus dorsi**. Its origin is on the spines of thoracic and lumbar vertebrae, and it inserts on the medial side of the humerus. The latissimus dorsi acts to pull the forelimb posteriorly and dorsally.

3. The **levator scapulae ventralis** is a flat, straplike muscle that lies on the side of the neck between the clavotrapezius and the acromiotrapezius. The occipital bone and atlas are its origin, and the vertebral border of the scapula is its insertion. This muscle, which does not occur in humans, pulls the scapula forward.

4. The **deltoid group** comprises the shoulder muscles lateral to the trapezius group. The following three cat muscles are equivalent to the single deltoid muscle in the human:

   a. The **spinodeltoid** is the most posterior of the deltoid group. It originates on the scapula ventral to the insertion of the acromiotrapezius and inserts on the proximal humerus. The action of this muscle is to flex the forelimb and rotate it laterally.

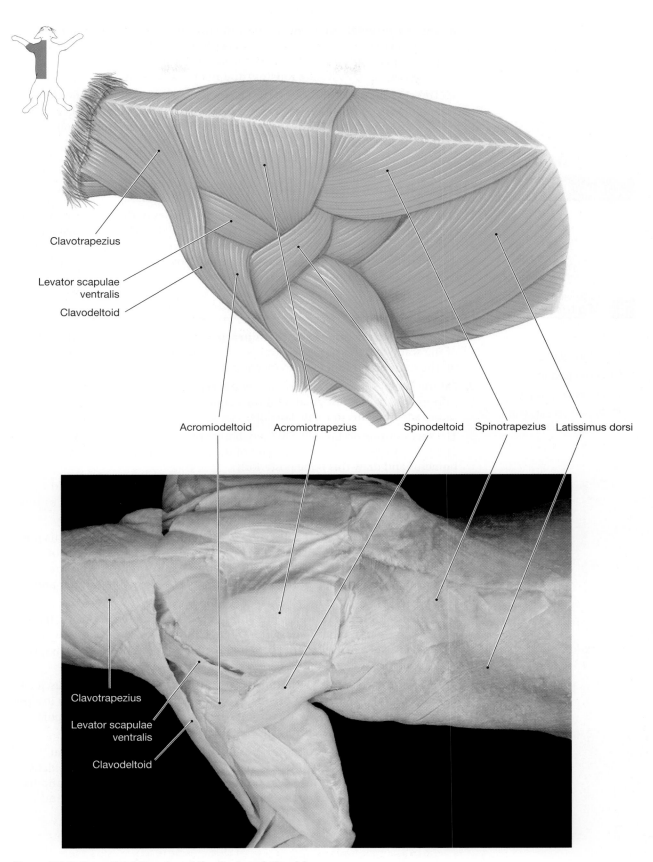

Clavotrapezius

Levator scapulae
ventralis

Clavodeltoid

Acromiodeltoid   Acromiotrapezius   Spinodeltoid   Spinotrapezius   Latissimus dorsi

Clavotrapezius

Levator scapulae
ventralis

Clavodeltoid

Figure D1.3 **Superficial Muscles of the Back and Shoulder**
Illustration and photo.

b. The **acromiodeltoid** is the middle muscle of the deltoid group. It originates on the acromion process of the scapula deep to the levator scapulae ventralis and inserts on the proximal end of the humerus. The acromiodeltoid flexes the forelimb and rotates it laterally.

c. The **clavodeltoid**, also called the clavobrachialis, originates on the clavicle and inserts on the ulna. It is a continuation of the clavotrapezius below the clavicle and extends down the forelimb from the clavicle. It functions to flex the forelimb. ∎

LAB ACTIVITY   3   Deep Muscles of the Back and Shoulder

Working on the left side of the cat, expose deeper muscles of the shoulder and back by cutting the three muscles of the trapezius group and the latissimus dorsi muscles at their insertions. Reflect these overlying muscles to expose the muscles underneath. (To *reflect* a muscle means to fold it back out of the way.) Use Figure D1.4 as a guide as you look at the following deep muscles.

3 *Materials*

- ☐ Gloves
- ☐ Safety glasses
- ☐ Dissecting tools
- ☐ Dissecting tray
- ☐ Preserved cat, skin removed

Procedures

1. Deep to the acromiotrapezius, the **supraspinatus** occupies the lateral surface of the scapula in the supraspinous fossa. It originates on the scapula and inserts on the humerus. It functions to extend the humerus.

2. On the lateral surface of the scapula, the **infraspinatus** occupies the infraspinous fossa. It originates on the scapula, inserts on the humerus, and causes the humerus to rotate laterally.

3. The **teres major** occupies the axillary border of the scapula, where it has its origin. It inserts on the proximal end of the humerus, and acts to rotate the humerus and draw this bone posteriorly.

4. The **rhomboideus group** connects the spinous processes of cervical and thoracic vertebrae with the vertebral border of the scapula. The muscles of this group hold the dorsal part of the scapula to the body.

   a. The posterior muscle of this group is the **rhomboideus major**. This fan-shaped muscle originates on the spinous processes and ligaments of posterior cervical and anterior thoracic vertebrae and inserts on the dorsal posterior angle of the scapula. It draws the scapula dorsally and anteriorly.

   b. The **rhomboideus minor** is just anterior to the rhomboideus major. Its origin is on the spines of posterior cervical and anterior thoracic vertebrae. This muscle inserts along the vertebral border of the scapula and draws the scapula forward and dorsally.

   c. The narrow, ribbonlike **rhomboideus capitis** is the anterolateral muscle of the rhomboideus group. It originates on the spinous nuchal line and inserts on the vertebral border of the scapula. It elevates and rotates the scapula. The rhomboideus capitis muscle does not occur in humans.

5. The **splenius** is a broad, flat, thin muscle that covers most of the lateral surface of the cervical and thoracic vertebrae. It is deep to the rhomboideus capitis. The splenius has its origin on the spines of thoracic vertebrae and its insertion on the superior nuchal line of the skull. It acts to both turn and raise the head. ∎

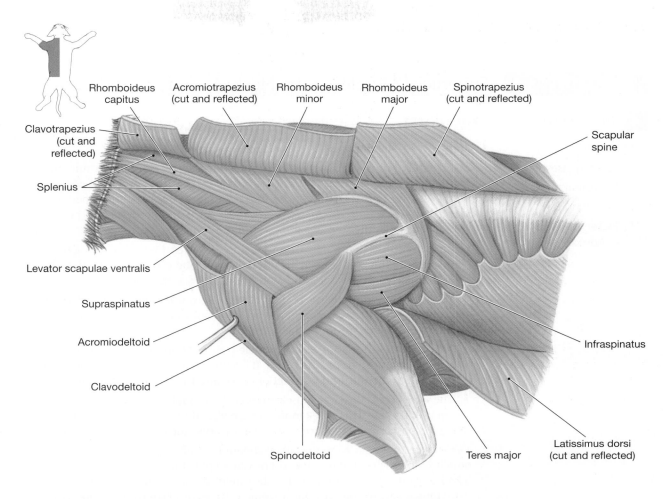

Rhomboideus capitus • Acromiotrapezius (cut and reflected) • Rhomboideus minor • Rhomboideus major • Spinotrapezius (cut and reflected)

Clavotrapezius (cut and reflected)

Splenius

Scapular spine

Levator scapulae ventralis

Supraspinatus

Acromiodeltoid

Clavodeltoid

Infraspinatus

Spinodeltoid

Teres major

Latissimus dorsi (cut and reflected)

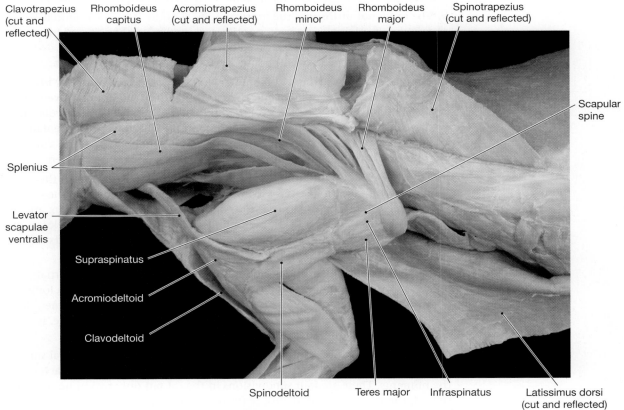

Clavotrapezius (cut and reflected)

Rhomboideus capitus

Acromiotrapezius (cut and reflected)

Rhomboideus minor

Rhomboideus major

Spinotrapezius (cut and reflected)

Splenius

Scapular spine

Levator scapulae ventralis

Supraspinatus

Acromiodeltoid

Clavodeltoid

Spinodeltoid

Teres major

Infraspinatus

Latissimus dorsi (cut and reflected)

Figure D1.4 **Deep Muscles of the Back and Shoulder—Lateral View**
Illustration and photo.

LAB ACTIVITY 4 Superficial Muscles of the Neck, Abdomen, and Chest

4 Materials

- ☐ Gloves
- ☐ Safety glasses
- ☐ Dissecting tools
- ☐ Dissecting tray
- ☐ Preserved cat, skin removed

Procedures

1. Locate the following muscles of the neck:

   a. The **sternomastoid** is a large, V-shaped muscle between the sternum and the head (Figure D1.5). This muscle originates on the manubrium of the sternum and passes obliquely around the neck to insert on the superior nuchal line and on the mastoid process. It turns and depresses the head. This muscle is the sternocleidomastoid in humans.

   b. The **sternohyoid** is a narrow muscle that lies over the larynx, along the midventral line of the neck. Its origin is the costal cartilage of the first rib, and it inserts on the hyoid bone. It acts to depress the hyoid bone.

   c. The **digastric** is a superficial muscle extending along the inner surface of the mandible. It originates on the occipital bone and mastoid process and functions as a depressor of the mandible.

   d. The **mylohyoid** is a superficial muscle running transversely in the midline and passing deep to the digastrics. It originates on the mandible and inserts on the hyoid bone, and its function is to raise the floor of the mouth.

   e. The **masseter** is the large muscle mass anteroventral to the parotid gland at the angle of the jaw. This cheek muscle originates on the zygomatic bone. It inserts on the posterolateral surface of the dentary bone and elevates the mandible.

2. Three layers of muscle form the lateral abdominal wall and insert on a fourth muscle along the ventral midline. Follow the steps below to locate these muscles. Because the three layers are very thin, be careful as you separate them. Collectively, these muscles act to compress the abdomen. They are all shown in Figure D1.6.

   a. The **external oblique** is the most superficial of the lateral abdominal muscles. It originates on posterior ribs and lumbodorsal fascia and inserts on the linea alba from the sternum to the pubis. Its fibers run from anterodorsal to posteroventral, and it acts to compress the abdomen.

   b. Cut and reflect the external oblique to view the **internal oblique**, which lies deep to the external oblique and is the second of the three lateral layers of the abdominal wall. The fibers of the internal oblique run perpendicular to those of the external oblique in a posterodorsal-to-anteroventral orientation. The internal oblique originates on the pelvis and lumbodorsal fascia and inserts on the linea alba, where it functions to compress the abdomen.

   c. Cut and reflect the internal oblique to see the third muscle layer of the abdominal wall, the **transverse abdominis**. This layer is deep to the internal oblique, has fibers that run transversely across the abdomen, and forms the deepest layer of the abdominal wall. It originates on the posterior ribs, lumbar vertebrae, and ilium and inserts on the linea alba. It acts to compress the abdomen.

   d. The **rectus abdominis** is the abdominal muscle on which the external oblique, internal oblique, and transverse abdominis all insert. It is a long, ribbonlike muscle in the midline of the ventral surface of the abdomen. It originates on the pubic symphysis and inserts on the costal cartilage. It compresses the internal organs of the abdomen.

3. The **pectoralis group** consists of the large muscles covering the ventral surface of the chest (Figure D1.6). They arise from the sternum and mostly attach to the humerus. There are four subdivisions in the cat but only two in humans. In the cat, the relatively large degree of fusion gives the pectoral muscles the appearance of a single muscle. This fusion makes the chest rather difficult to dissect, as the muscles do not separate from one another easily.

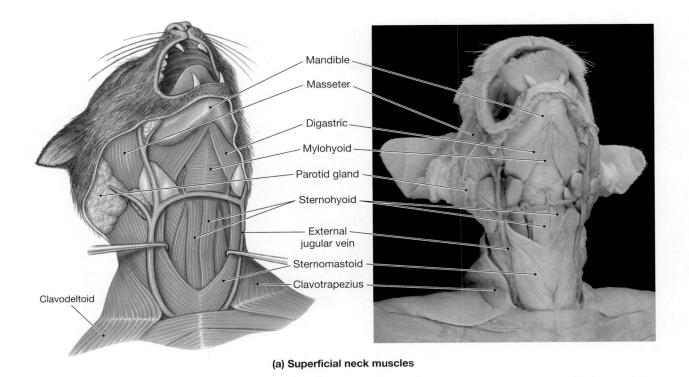

Mandible
Masseter
Digastric
Mylohyoid
Parotid gland
Sternohyoid
External
jugular vein
Sternomastoid
Clavotrapezius
Clavodeltoid

(a) Superficial neck muscles

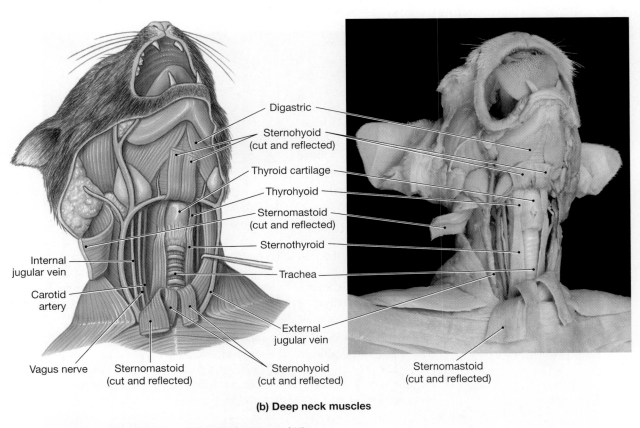

Digastric
Sternohyoid
(cut and reflected)
Thyroid cartilage
Thyrohyoid
Sternomastoid
(cut and reflected)
Sternothyroid
Trachea
External
jugular vein
Internal
jugular vein
Carotid
artery
Vagus nerve
Sternomastoid
(cut and reflected)
Sternohyoid
(cut and reflected)
Sternomastoid
(cut and reflected)

(b) Deep neck muscles

Figure D1.5 Superficial Muscles of the Neck—Ventral View
Illustrations and photos.

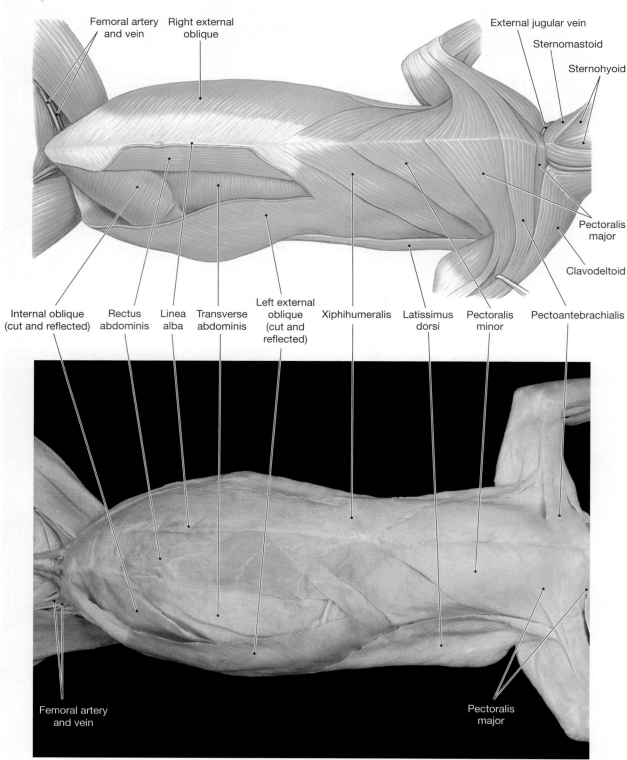

Femoral artery and vein | Right external oblique | External jugular vein | Sternomastoid | Sternohyoid | Pectoralis major | Clavodeltoid

Internal oblique (cut and reflected) | Rectus abdominis | Linea alba | Transverse abdominis | Left external oblique (cut and reflected) | Xiphihumeralis | Latissimus dorsi | Pectoralis minor | Pectoantebrachialis

Femoral artery and vein

Pectoralis major

Figure D1.6 Superficial Muscles of the Abdomen—Ventral View
Illustration and photo.

a. The **pectoantebrachialis** is the most superficial of the pectoral muscles. Its originates on the manubrium, inserts on fascia of the forearm, and adducts the forelimb. It is not found in humans.

b. The broad, triangular **pectoralis major** is posterior to the pectoantebrachialis. The origin of the pectoralis major is on the sternum, and its insertion is on the posterior humerus. It functions to adduct the arm.

c. Posterior to the pectoralis major is the **pectoralis minor**. This is the broadest and thickest muscle of the group. It extends posteriorly to the pectoralis major. It originates on the sternum, inserts near the proximal end of the humerus, and acts to adduct the arm.

d. The thin **xiphihumeralis** is posterior to the posterior edge of the pectoralis minor. The xiphihumeralis originates on the xiphoid process of the sternum and inserts by a narrow tendon on the humerus. It adducts and helps rotate the forelimb. Humans do not have a xiphihumeralis muscle. ■

LAB ACTIVITY   5   Deep Muscles of the Chest and Abdomen

5   Materials

- ☐ Gloves
- ☐ Safety glasses
- ☐ Dissecting tools
- ☐ Dissecting tray
- ☐ Preserved cat, skin removed

Procedures

1. On the lateral thoracic wall, just ventral to the scapula, a large, fan-shaped muscle, the **serratus ventralis**, originates by separate bands from the ribs (Figure D1.7). It passes ventrally to the scapula and inserts on the vertebral border of the scapula. It is homologous to the serratus anterior of humans and functions to pull the scapula forward and ventrally.

2. The **serratus dorsalis** is a serrated muscle that lies medial to the serratus ventralis. Its origin is along the mid-dorsal cervical, thoracic, and lumbar regions, and it inserts on the ribs. It acts to pull the ribs craniad and outward.

3. The **scalenus medius** is a three-part muscle on the lateral surface of the trunk. The bands of muscle unite anteriorly. The scalenus medius originates on the ribs and inserts on the transverse processes of the cervical vertebrae. It acts to flex the neck and draw the ribs anteriorly.

4. The **intercostal group** consists of two layers of muscles between the ribs that move the ribs during respiration.

    a. The **external intercostals** are deep to the external oblique. The muscle fibers of the external intercostals run obliquely from the posterior border of one rib to the anterior border of the next rib. Their origin is on the caudal border of one rib, and their insertion is on the cranial border of the next rib. They lift the ribs during inspiration.

    b. The **internal intercostals** are toward the midline and deep to the external intercostals. (They are not visible in Figure D1.7.) Bisect one external intercostal muscle to expose an internal intercostal. The fibers of the intercostals run at oblique angles: the internal from medial to lateral and the external from lateral to medial. The origin of the internal intercostals is on the superior border of the rib below, and they insert on the inferior border of the rib above. They draw adjacent ribs together and depress ribs during active expiration. ■

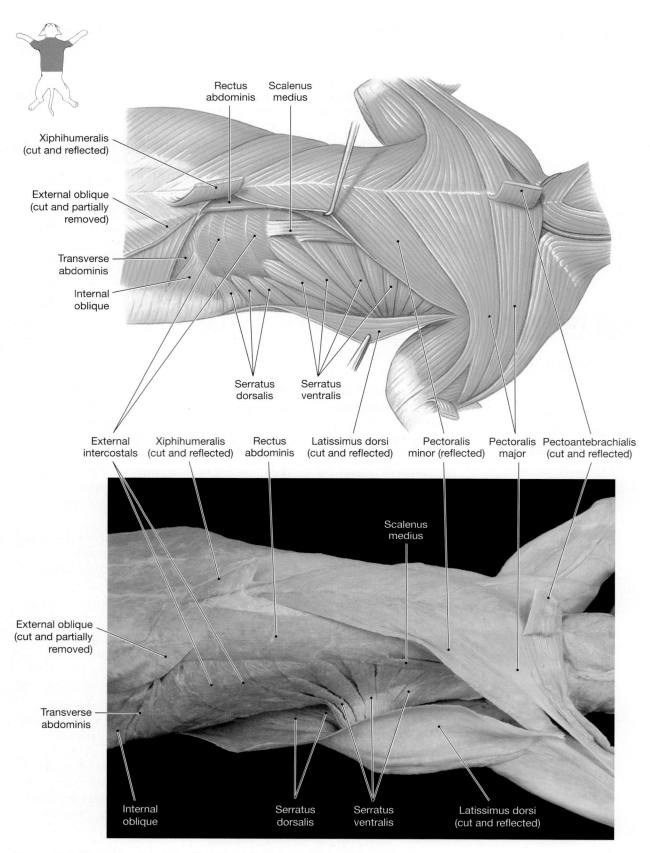

Rectus
abdominis

Scalenus
medius

Xiphihumeralis
(cut and reflected)

External oblique
(cut and partially
removed)

Transverse
abdominis

Internal
oblique

Serratus
dorsalis

Serratus
ventralis

External
intercostals

Xiphihumeralis
(cut and reflected)

Rectus
abdominis

Latissimus dorsi
(cut and reflected)

Pectoralis
minor (reflected)

Pectoralis
major

Pectoantebrachialis
(cut and reflected)

Scalenus
medius

External oblique
(cut and partially
removed)

Transverse
abdominis

Internal
oblique

Serratus
dorsalis

Serratus
ventralis

Latissimus dorsi
(cut and reflected)

Figure D1.7 **Deep Muscles of the Chest and Abdomen—Ventral View**
Illustration and photo.

LAB ACTIVITY   6   Muscles of the Forelimb

6   *Materials*   *Procedures*

- ☐ Gloves
- ☐ Safety glasses
- ☐ Dissecting tools
- ☐ Dissecting tray
- ☐ Preserved cat, skin removed

1. Observe the **epitrochlearis**, which is a broad, flat muscle covering the medial surface of the upper forelimb (Figure D1.8). This muscle appears to be an extension of the latissimus dorsi and originates on the fascia of the latissimus dorsi. It inserts on the olecranon process, where it acts to extend the forelimb. It is not found in humans.

2. The **biceps brachii** is a convex muscle interior to the pectoralis major and pectoralis minor on the ventromedial surface of the humerus. It originates on the scapula and inserts on the radial tuberosity near the proximal end of the radius. It functions as a flexor of the forelimb.

3. Observe the ribbon-like **brachioradialis** muscle on the lateral surface of the humerus (Figure D1.9). Its origin is on the mid-dorsal border of the humerus, and its insertion is on the distal end of the radius. It supinates the front paw.

4. The **extensor carpi radialis** is deep to the brachioradialis. There are two parts to this extensor muscle: the shorter, triangular **extensor carpi radialis brevis** and the deeper **extensor carpi radialis longus**. Both originate on the lateral surface of the humerus above the lateral epicondyle and insert on the bases of the second and third metacarpals. Both extensor carpi radialis muscles cause extension at the carpal joints.

5. The **pronator teres**, a narrow muscle next to the extensor carpi radialis, runs from its point of origin on the medial epicondyle of the humerus and gets smaller as it approaches insertion on the radius. It rotates the radius for pronation.

6. The **flexor carpi radialis**, found adjacent to the pronator teres, originates on the distal end of the humerus and inserts on the second and third metacarpals. It acts to flex the wrist.

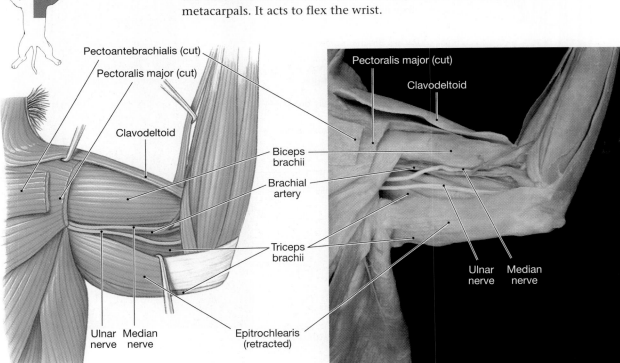

Figure D1.8 **Muscles of the Upper Forelimb—Medial View**
Illustration and photo.

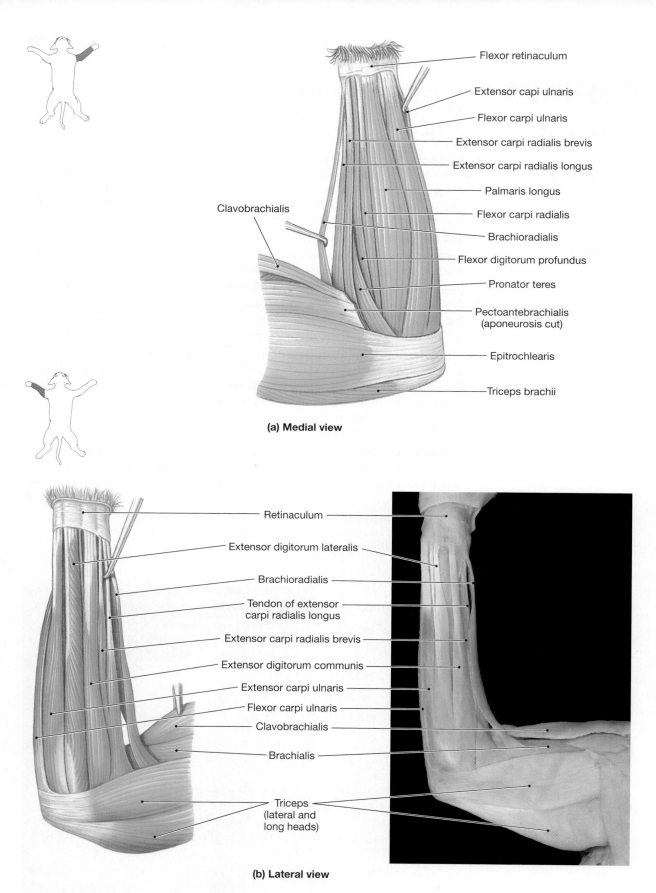

Flexor retinaculum

Extensor capi ulnaris

Flexor carpi ulnaris

Extensor carpi radialis brevis

Extensor carpi radialis longus

Palmaris longus

Flexor carpi radialis

Brachioradialis

Flexor digitorum profundus

Pronator teres

Pectoantebrachialis
(aponeurosis cut)

Epitrochlearis

Triceps brachii

Clavobrachialis

(a) Medial view

Retinaculum

Extensor digitorum lateralis

Brachioradialis

Tendon of extensor
carpi radialis longus

Extensor carpi radialis brevis

Extensor digitorum communis

Extensor carpi ulnaris

Flexor carpi ulnaris

Clavobrachialis

Brachialis

Triceps
(lateral and
long heads)

(b) Lateral view

Figure D1.9 Muscles of the Forelimb
Illustrations and photos.

7. The large, flat muscle in the center of the medial surface of the forelimb is the **palmaris longus**. Its origin is on the medial epicondyle of the humerus, and it inserts on all digits. The palmaris longus flexes the digits.

8. The flat muscle on the posterior edge of the forelimb is the **flexor carpi ulnaris**. It arises from a two-headed origin—on the medial epicondyle of the humerus and on the olecranon process—and inserts by a single tendon on the ulnar side of the carpals. It is a flexor of the wrist.

9. The **brachialis** is on the ventrolateral surface of the humerus (Figure D1.9b). This muscle originates on the lateral side of the humerus and inserts on the proximal end of the ulna. It functions to flex the forelimb.

10. The **triceps brachii** is the largest superficial muscle of the upper forelimb. It is located on the lateral and posterior surfaces of the forelimb. As the name implies, the triceps brachii has its origins on three heads. The **long head** is the large muscle mass on the posterior surface and originates on the lateral border of the scapula. The **lateral head**, which lies next to the long head on the lateral surface, originates on the deltoid ridge of the humerus. The **small medial** head lies deep to the lateral head and originates on the shaft of the humerus. All three heads have a single insertion on the olecranon process of the ulna. The function of the triceps brachii is to extend the forelimb. ■

LAB ACTIVITY 7   Muscles of the Thigh

The hindlimb of cats and other four-legged animals looks different from the human leg. Cats have long metatarsals in the arch of the foot. This makes the ankle very high, and some people might even mistake the ankle for the knee joint. Place your hand on the cat's pelvis and then slide your hand onto the thigh. Locate the distal joint, the knee, and then the ankle.

7 Materials

- ☐ Gloves
- ☐ Safety glasses
- ☐ Dissecting tools
- ☐ Dissecting tray
- ☐ Preserved cat, skin removed

Procedures

1. The **sartorius** is a wide, superficial muscle covering the anterior half of the medial aspect of the thigh (Figure D1.10). It originates on the ilium and inserts on the tibia. The sartorius adducts and rotates the femur and extends the tibia.

2. The **gracilis** is a broad muscle that covers the posterior portion of the medial aspect of the thigh. The gracilis originates on the ischium and pubic symphysis and inserts on the medial surface of the tibia. It adducts the thigh and draws it posteriorly.

3. The **tensor fasciae latae** is a triangular muscle located posterior to the sartorius (Figures D1.10b and D1.11). This muscle originates on the crest of the ilium and inserts into the fascia lata. The tensor fasciae latae extends the thigh.

4. Posterior to the tensor fasciae latae lies the **gluteus medius** (Figure D1.11b). This is the largest of the gluteus muscles, and it originates on both the ilium and the transverse processes of the last sacral and first caudal vertebrae. It inserts on the femur and acts to abduct the thigh.

   Just posterior to the gluteus medius, locate the **gluteus maximus,** a small, triangular hip muscle. It originates on the transverse processes of the last sacral and first caudal vertebrae and inserts on the proximal femur. It abducts the thigh.

5. The **adductor femoris,** deep to the gracilis, is a large muscle with an origin on the ischium and pubis. It inserts on the femur and acts to adduct the thigh.

   Anterior to the adductor femoris is the **adductor longus** muscle, visible in Figure D1.10. It is a narrow muscle that originates on the ischium and pubis and inserts on the proximal surface of the femur. It adducts the thigh.

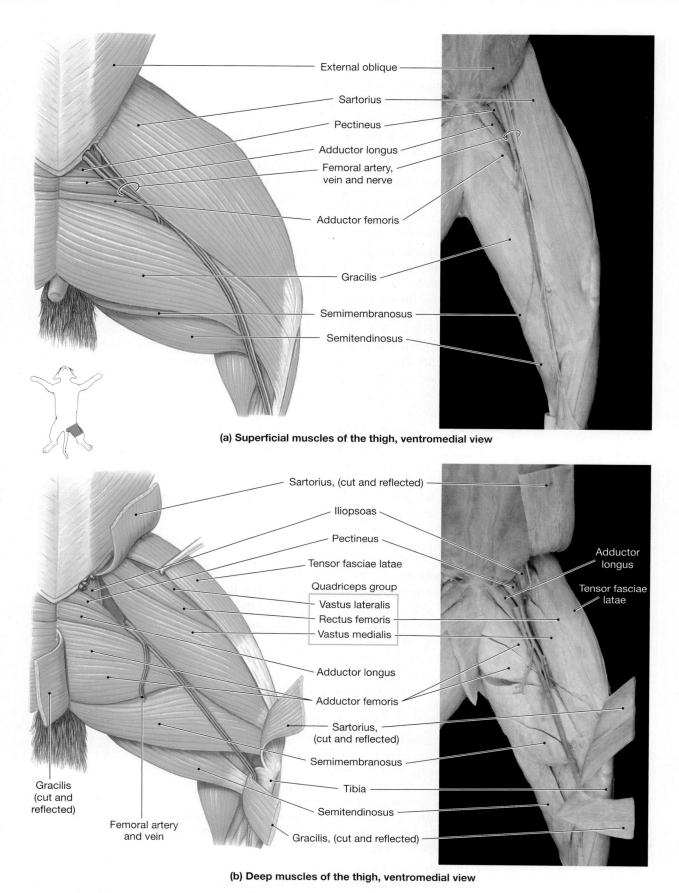

External oblique

Sartorius

Pectineus

Adductor longus

Femoral artery,
vein and nerve

Adductor femoris

Gracilis

Semimembranosus

Semitendinosus

(a) Superficial muscles of the thigh, ventromedial view

Sartorius, (cut and reflected)

Iliopsoas

Pectineus

Tensor fasciae latae

Quadriceps group

Vastus lateralis

Rectus femoris

Vastus medialis

Adductor longus

Adductor femoris

Sartorius,
(cut and reflected)

Semimembranosus

Tibia

Semitendinosus

Gracilis, (cut and reflected)

Gracilis
(cut and
reflected)

Femoral artery
and vein

Adductor
longus

Tensor fasciae
latae

(b) Deep muscles of the thigh, ventromedial view

Figure D1.10 **Muscles of the Thigh—Ventromedial View**
Illustrations and photos.

6. The **pectineus** is anterior to the adductor longus (Figure D1.10). The pectineus is a deep, small muscle posterior to both the femoral artery and the femoral vein. It originates on the anterior border of the pubis and inserts on the proximal end of the femur. It functions to adduct the thigh.

7. Locate the four large muscles that constitute the **quadriceps femoris group**. These muscles cover about one-half of the surface of the thigh. Collectively they insert into the patellar ligament and act as powerful extensors of the leg. Bisect the sartorius, and free both borders of the tensor fasciae latae. Reflect these muscles and observe that the muscles of the quadriceps femoris converge and insert on the patella.

   a. The **vastus lateralis** is the large, fleshy muscle on the anterolateral surface of the thigh. It originates along the entire length of the lateral surface of the femur.

   b. The **vastus medialis** is on the medial surface of the femur just under the sartorius. It originates on the shaft of the femur and inserts on the patellar ligament.

   c. The **rectus femoris** is a small, cylindrical muscle between the vastus medialis and the vastus lateralis muscles. In humans the rectus femoris originates on the ilium, but in cats it originates on the femur.

   d. The **vastus intermedius** (not visible in Figure D1.10) lies deep to the rectus femoris. Bisect and reflect the rectus femoris to expose the vastus intermedius. It originates on the shaft of the femur.

8. On the posterior thigh are three muscles of the **hamstring group**. These muscles span the knee joint and act to flex the leg. They are all visible in Figure D1.11, and the latter two are also visible in Figure D1.10.

   a. The **biceps femoris** is a large, broad muscle covering most of the lateral region of the thigh. It originates on the ischial tuberosity and inserts on the tibia. The biceps femoris flexes the leg and also abducts the thigh.

   b. The **semitendinosus** is visible under the posterior portion of the biceps femoris. The belly of the muscle is a uniform strap from the origin to the tendon at the insertion. The semitendinosus originates on the ischial tuberosity and inserts on the medial side of the tibia. It flexes the leg.

   c. The **semimembranosus** is a large muscle medial to the semitendinosus. It is seen best in the anteromedial view of the thigh. Transect the semitendinosus to view the semimembranosus in the posterior aspect. The semimembranosus originates on the ischium and inserts on the medial epicondyle of the femur and medial surface of the tibia. It extends the thigh. ■

LAB ACTIVITY 8 Muscles of the Lower Hindlimb

8 Materials

- ☐ Gloves
- ☐ Safety glasses
- ☐ Dissecting tools
- ☐ Dissecting tray
- ☐ Preserved cat, skin removed

Procedures

1. The **gastrocnemius**, or calf muscle, is on the posterior of the hindlimb (Figure D1.12). It has two heads of origin, medial on the knee's fascia and lateral on the distal end of the femur. The two bellies of the muscle unite at the calcaneal tendon and insert on the calcaneus. The gastrocnemius extends the foot.

2. The **flexor digitorum longus** is found between the gastrocnemius and the tibia. This muscle has two heads of origin: one on the distal end of the tibia and another on the head and shaft of the fibula. It inserts by four tendons onto the bases of the terminal phalanges. It acts as a flexor of the digits.

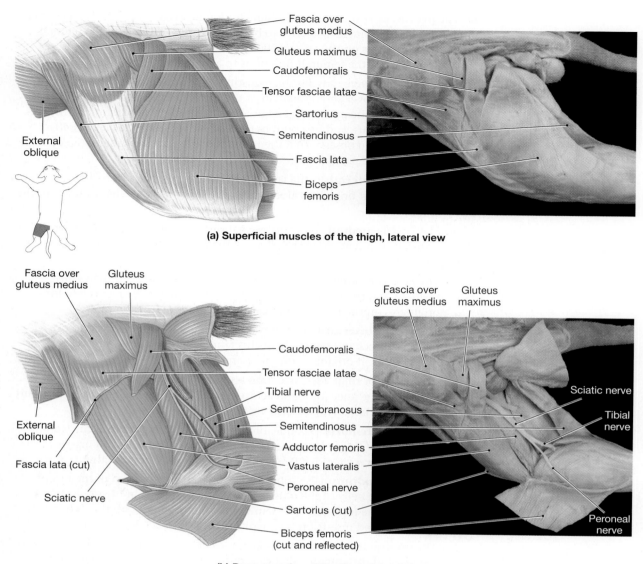

(a) Superficial muscles of the thigh, lateral view

(b) Deep muscles of the thigh, lateral view

Figure D1.11 Muscles of the Thigh—Lateral View
Illustrations and photos.

3. The **tibialis anterior** is on the anterior surface of the tibia. This muscle originates on the proximal ends of the tibia and fibula and inserts on the first metatarsal. It acts to flex the foot.

4. Deep to the gastrocnemius but visible on the lateral surface of the calf is the **soleus** (Figure D1.12b). The soleus originates on the fibula and inserts on the calcaneus. It extends the foot.

5. The **fibularis group**, also called the *peroneus muscles,* consists of three muscles deep to the soleus on the posterior and lateral surfaces.

   a. The **fibularis brevis** lies deep to the tendon of the soleus, originating from the distal portion of the fibula and inserting on the base of the fifth metatarsal. The fibularis brevis extends the foot.

   b. The **fibularis longus** is a long, thin muscle that lies on the lateral surface of the hindlimb, originating on the proximal portion of the fibula and inserting by a tendon that passes through a groove on the lateral malleolus

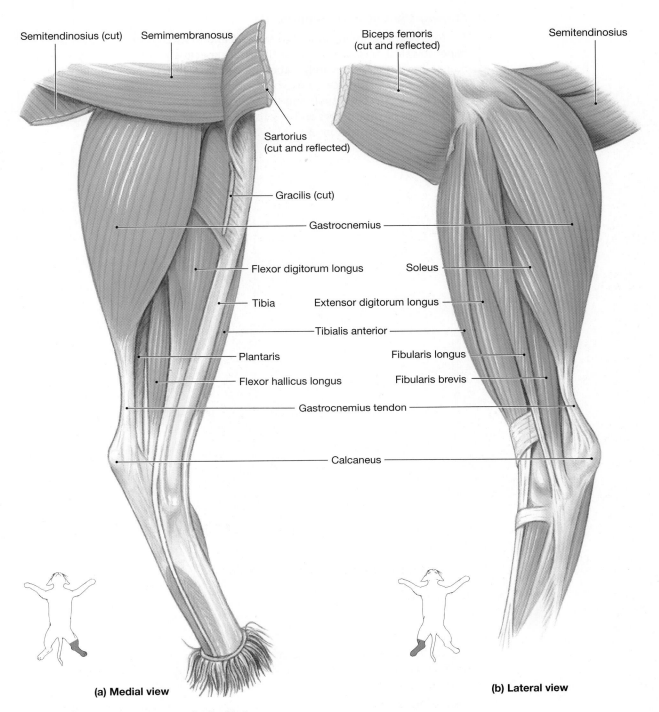

Semitendinosius (cut)   Semimembranosus

Biceps femoris
(cut and reflected)   Semitendinosius

Sartorius
(cut and reflected)

Gracilis (cut)

Gastrocnemius

Flexor digitorum longus   Soleus

Tibia   Extensor digitorum longus

Tibialis anterior

Plantaris   Fibularis longus

Flexor hallicus longus   Fibularis brevis

Gastrocnemius tendon

Calcaneus

(a) Medial view   **(b) Lateral view**

Figure D1.12 Muscles of the Left Hindlimb

and turns medially to attach to the bases of the metatarsals. This muscle acts to flex the foot.

   c. The **fibularis tertius** lies along the tendon of the fibularis longus, originating on the fibula and inserting on the base of the fifth metatarsal. This muscle extends the fifth digit and flexes the foot.

6. On the anterolateral border of the tibia, the **extensor digitorum longus** originates on the lateral epicondyle of the femur. It inserts by long tendons on each of the five digits and functions to extend the digits. ∎

Storing the Cat and Cleaning Up

To store your specimen, wrap it in the skin and moisten it with fixative. Use paper towels if necessary to cover the entire specimen. Return it to the storage bag and seal the bag securely. Label the bag with your name, and place it in the storage area as indicated by your instructor.

Wash all dissection tools and the tray, and set them aside to dry.

Dispose of your gloves and any tissues from the dissection as indicated by your laboratory instructor. ▲

Muscles of the Cat

Name _____

Date _____

Section _____

A. Matching

Match each structure in the left column with its correct description from the right column.

_____ **1.** acromiotrapezius

_____ **2.** levator scapulae ventralis

_____ **3.** spinodeltoid

_____ **4.** sternomastoid

_____ **5.** xiphihumeralis

_____ **6.** epitrochlearis

_____ **7.** pectineus

_____ **8.** vastus intermedius

_____ **9.** fibularis tertius

_____ **10.** fibularis brevis

A. adducts thigh

B. called sternocleidomastoid in humans

C. deep to rectus femoris

D. extends foot

E. flat muscle of medial upper forelimb

F. flexes and laterally rotates forelimb

G. flexes foot

H. originates on xiphoid process, adducts humerus

I. pulls scapula forward

J. square muscle, holds scapula in place

B. Short-Answer Questions

1. Describe the neck muscles of the cat.

2. Describe the flexor and extensor muscles of the cat forelimb.

3. Describe the abdominal muscles of the cat.

C. Labeling

Label the muscles in Figure D1.13.

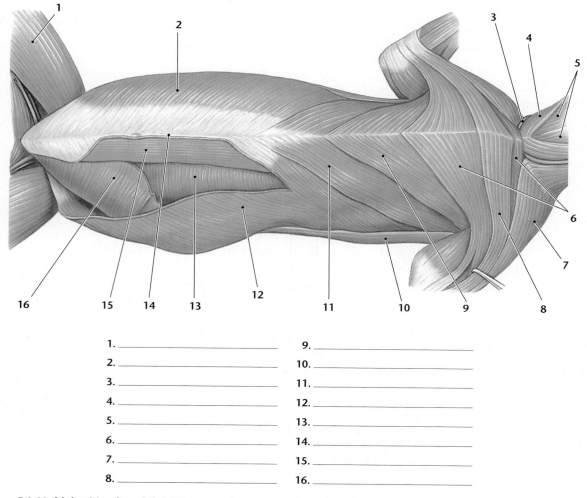

1. _____   9. _____
2. _____   10. _____
3. _____   11. _____
4. _____   12. _____
5. _____   13. _____
6. _____   14. _____
7. _____   15. _____
8. _____   16. _____

Figure D1.13 **Major Muscles of the Cat**

D. Analysis and Application

1. Explain the differences between the cat and the human deltoid and trapezius muscles.

2. How are the superficial chest muscles of the cat different from chest muscles in humans?

3. Compare cat and human muscles of the thigh and lower leg/hindlimb.

Cat Nervous System

OBJECTIVES

On completion of this exercise, you should be able to:

- Identify the major nerves of the feline brachial plexus.

- Identify the major nerves of the feline sacral plexus.

- Identify the feline spinal meninges and the dorsal and ventral roots of the spinal cord.

Cats, like humans, have pairs of spinal nerves extending laterally from the various segments of the spinal cord. Humans have 31 pairs of spinal nerves; cats have 38 to 40 pairs, depending on whether some of the distal nerves have fused (they are difficult to distinguish individually). In this exercise, you will identify the major nerves of the brachial and sacral plexuses of the feline nervous system. You will also dissect the sacral plexus and then examine the exposed spinal cord. This exercise complements the study of the human nervous system.

⚠   **Dissection Safety**

Before beginning this dissection, review the Dissection Safety Guidelines presented at the beginning of Dissection Exercise 1 on p. 745. ▲

LAB ACTIVITY **1**   Preparing the Cat

If the cat was not skinned for muscle studies, refer to Cat Dissection Exercise 1, Laboratory Activity 1: "Preparing the Cat for Dissection."

1 *Materials*

- ☐ Gloves
- ☐ Safety glasses
- ☐ Dissecting tools
- ☐ Dissecting tray
- ☐ String
- ☐ Preserved cat, skin removed

Procedures

1. Put on gloves and safety glasses and clear your workspace before obtaining your dissection specimen.

2. Secure the specimen ventral side up on the dissecting tray by spreading the limbs and tying them flat with lengths of string passing under the tray, one string for the two forelimbs and one string for the two hindlimbs.

3. Be sure to keep the specimen moist with fixative during the dissection. Keep the skin draped on areas not undergoing dissection.

4. Proceed to the next Laboratory Activity to dissect the cat. ■

LAB ACTIVITY 2   The Brachial Plexus

Dissection of the chest and forelimb reveals the brachial plexus, a network formed by the intertwining of cervical nerves C_6, C_7, and C_8 and thoracic nerve T_1. This plexus innervates muscles and other structures of the shoulder, forelimb, and thoracic wall. The nerves are delicate, so you must be careful not to damage or remove them during dissection. Use Figure D2.1 as a reference in identifying the nerves of the brachial plexus.

2 Materials

- ☐ Gloves
- ☐ Safety glasses
- ☐ Dissecting tools
- ☐ Dissecting tray
- ☐ String
- ☐ Preserved cat, skin removed

Procedures

1. Put on gloves and safety glasses, and clear your workspace before obtaining your dissection specimen.

2. Secure the specimen ventral side up on the dissecting tray by spreading the limbs and tying them flat with lengths of string passing under the tray. Use one string for the two forelimbs and one string for the two hindlimbs.

3. Reflect the left pectoralis major muscle, and observe the underlying blood vessels in the axilla. If your cat has been injected with latex paint, the arteries are injected with red paint and the veins with blue paint.

4. Use a probe and forceps to carefully remove fat and other tissue from around the vessels and nerves.

5. The largest nerve of the brachial plexus is the **radial nerve.** It lies dorsal to the axillary artery, which is the red-injected blood vessel in the axilla. The radial nerve supplies the triceps brachii muscle and other dorsal muscles of the forelimb. Trace this nerve from close to its origin near the midline to where it passes into the triceps muscle.

6. The **musculocutaneous nerve,** which is superior to the radial nerve, supplies the coracobrachial and biceps brachii muscles of the ventral forelimb and the skin of the forelimb. Trace this nerve into the musculature, and notice its two divisions.

7. Next notice the **median nerve,** which follows the red-injected brachial artery into the ventral forelimb. This nerve supplies the muscles of the ventral antebrachium of the forelimb.

8. The most posterior of the brachial plexus nerves is the **ulnar nerve.** It is often isolated from the other nerves once the surrounding fat and tissues have been removed from the plexus. Trace this nerve down the brachium to the elbow to where it supplies the muscles of the antebrachium. ∎

LAB ACTIVITY 3   The Sacral Plexus

Dissection of the hindlimb exposes the three major nerves of the sacral plexus, the neural network that supplies the muscles and structures of the hip and hindlimb. As you reflect muscles to expose the nerves, be careful not to remove or damage the nerves. Use Figure D2.2 as a reference in identifying the nerves of the sacral plexus.

3 Materials

- ☐ Gloves
- ☐ Safety glasses
- ☐ Dissecting tools
- ☐ Dissecting tray
- ☐ String
- ☐ Preserved cat, skin removed

Procedures

1. Put on gloves and safety glasses, and clear your workspace before obtaining your dissection specimen.

2. Secure the specimen ventral side up on the dissecting tray as described in Laboratory Activity 1.

3. Reflect the cut ends of the biceps femoris muscle and note the large **sciatic nerve**. This nerve supplies the muscles of the hindlimb.

4. Follow the sciatic nerve down the hindlimb to the gastrocnemius muscle, where the nerve branches into two smaller nerves.

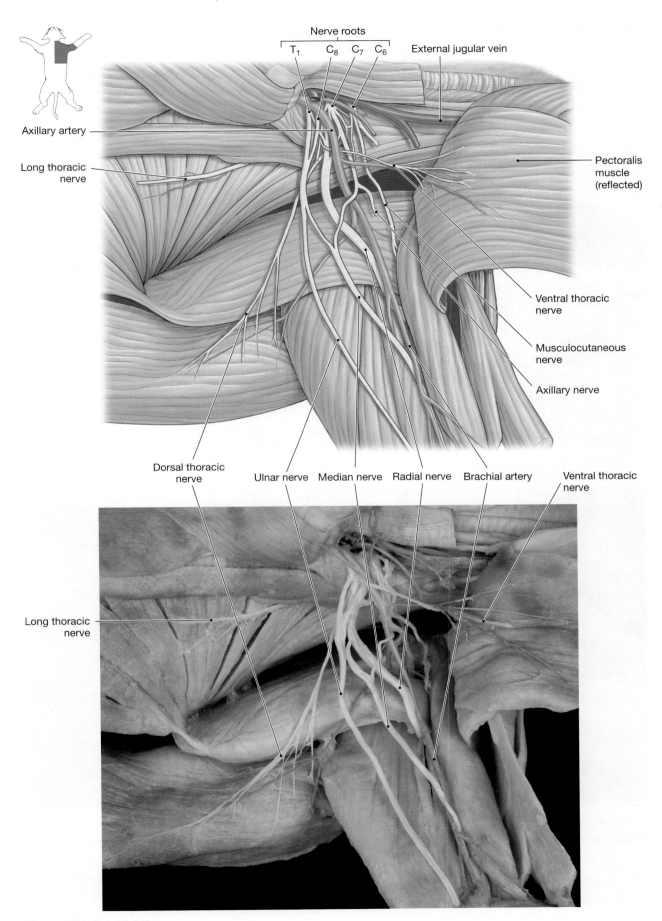

Nerve roots
T_1 C_8 C_7 C_6

External jugular vein

Axillary artery

Long thoracic nerve

Pectoralis muscle (reflected)

Ventral thoracic nerve

Musculocutaneous nerve

Axillary nerve

Dorsal thoracic nerve

Ulnar nerve

Median nerve

Radial nerve

Brachial artery

Ventral thoracic nerve

Long thoracic nerve

Figure D2.1 Brachial Plexus
Illustration and photo.

5. Identify the **tibial nerve** on the medial side of the hindlimb and the **common fibular nerve** (also called the *peroneal nerve*) on the lateral side. The tibial and common fibular nerves supply the inferior part of the hindlimb. ◼

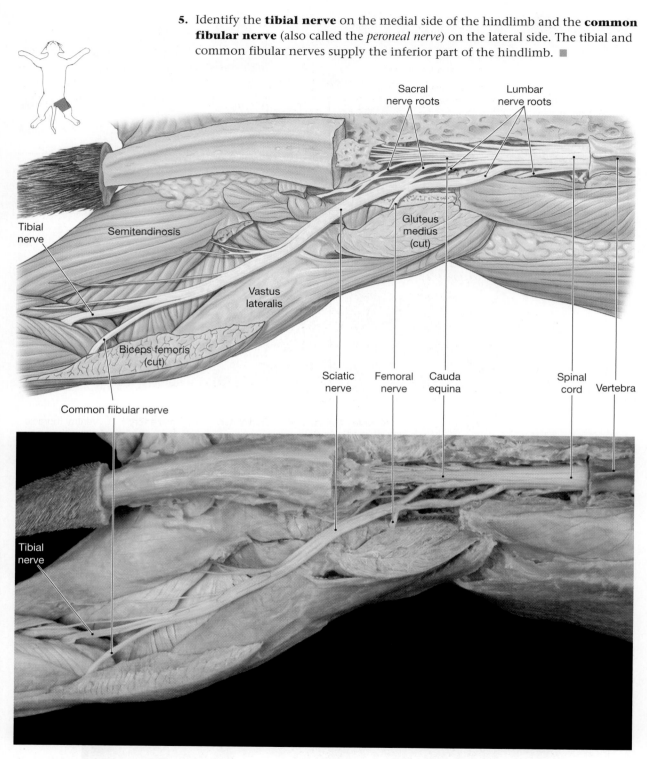

Figure D2.2 Sacral Plexus and Spinal Cord
Illustration and photo.

LAB ACTIVITY 4   The Spinal Cord

Dissection and reflection of the posterior muscles on the dorsal surface will expose the vertebral column. To save time, only a small section of the vertebral column will be dissected to study the spinal cord. Use care while using bone cutters to remove pieces of the vertebrae. Use Figure D2.2 as a reference in identifying the spinal cord.

4   *Materials*

☐ Gloves
☐ Safety glasses
☐ Dissecting tools
☐ Dissecting tray
☐ String
☐ Preserved cat, skin removed

Procedures

1. Put on gloves and safety glasses, and clear your workspace before obtaining your dissection specimen.

2. Secure the specimen ventral side down on the dissection tray, tying the limbs as described in Laboratory Activity 1.

3. Cut and reflect the large dorsal muscles that cover the vertebral column in the lumbar region of the back.

4. Use bone cutters to remove the vertebral arches of three vertebrae and thereby expose the **spinal cord**. Gently remove each piece of bone, using care not to tear or remove the membranes covering the spinal cord.

5. Examine the exposed spinal cord and identify the **dorsal roots** and **ventral roots** that form the spinal nerves.

6. Identify the outermost **dura mater** over the spinal cord.

7. Cut through the dura, and note the **arachnoid** membrane with its many fine extensions to the spinal cord.

8. Use a dissection pin to tease away a portion of the delicate **pia mater** lying on the surface of the spinal cord.

9. Note the **subarachnoid space** between the arachnoid and pia mater. Cerebrospinal fluid circulates in this space.

10. Remove a thin section of the spinal cord to view the internal organization. Identify the inner **gray horns** surrounded by the **white columns**. ■

Storing the Cat and Cleaning Up

To store your specimen, wrap it in the skin and moisten it with fixative. Use paper towels if necessary to cover the entire specimen. Return it to the storage bag and seal the bag securely. Label the bag with your name, and place it in the storage area as indicated by your instructor.

Wash all dissection tools and the tray, and set them aside to dry.

Dispose of your gloves and any tissues from the dissection as indicated by your laboratory instructor. ▲

Cat Nervous System

2

Name _____

Date _____

Section _____

A. Matching

Match each structure in the left column with its correct description from the right column.

_____	**1.** musculocutaneous nerve	**A.** serves muscles of antebrachium
_____	**2.** radial nerve	**B.** serves triceps brachii muscle
_____	**3.** tibial nerve	**C.** medial nerve of lower hindlimb
_____	**4.** median nerve	**D.** serves biceps brachii
_____	**5.** sciatic nerve	**E.** lateral nerve of lower hindlimb
_____	**6.** common fibular nerve	**F.** large nerve of sacral plexus
_____	**7.** subarachnoid space	**G.** posterior nerve of brachial plexus
_____	**8.** dura mater	**H.** site of cerebrospinal fluid circulation
_____	**9.** fibular nerve	**I.** outer membrane protecting spinal cord
_____	**10.** ulnar nerve	**J.** also known as peroneal nerve

B. Short-Answer Questions

1. Compare the brachial plexus nerves of cats and humans.

2. Describe the three major nerves of the sacral plexus.

3. What are the meningeal layers surrounding the spinal cord?

C. Analysis and Applications

1. Describe the major nerves of the forelimb.

2. How are white and gray matter organized in the spinal cord?

Cat Endocrine System

3

LABORATORY ACTIVITIES

1 Preparing the Cat for Dissection 775

2 Endocrine Glands of the Cat 776

OBJECTIVES

On completion of this exercise, you should be able to:

- Identify the main glands of the feline endocrine system.
- List the hormones produced by each gland and state the basic function of each.

In this exercise you will identify the major organs of the feline endocrine system. This exercise complements the study of the human endocrine system.

Dissection Safety

Before beginning this dissection, review the Dissection Safety Guidelines presented at the beginning of Dissection Exercise 1 on p. 745. ▲

LAB ACTIVITY   1   Preparing the Cat for Dissection

If the ventral body cavity has not been opened on your dissection specimen, complete the following procedures. Otherwise, skip to Laboratory Activity 2. Use Figure D3.1 as a reference as you dissect the ventral body cavity.

 Materials

☐ Gloves
☐ Safety glasses
☐ Dissecting tools
☐ Dissecting tray
☐ String
☐ Preserved cat, skin removed

Procedures

1. Put on gloves and safety glasses, and clear your workspace before obtaining your dissection specimen.
2. Secure the specimen ventral side up on the dissecting tray by spreading the limbs and tying them flat with lengths of string passing under the tray. Use one string for the two forelimbs and one string for the two hindlimbs.
3. Use scissors to cut a midsagittal section through the muscles of the abdomen to the sternum.
4. To avoid cutting through the bony sternum, angle your incision laterally approximately 0.5 inch, and cut the costal cartilages. Continue the parasagittal section to the base of the neck.
5. Make a lateral incision on each side of the diaphragm. Use care not to damage the diaphragm or the internal organs. Spread the thoracic walls to observe the internal organs.
6. Make a lateral section across the pubic region and angled toward the hips. Spread the abdominal walls to expose the abdominal organs. ■

LAB ACTIVITY 2 ## Endocrine Glands of the Cat

When moving internal organs aside to locate the endocrine glands, take care not to rupture the digestive tract. Because hormones are transported by the bloodstream, take note of the vascularization of the endocrine glands. Many of these vessels will be identified in a later dissection of the cardiovascular system. Use Figure D3.1 as a reference in identifying the endocrine glands.

If you are continuing from Laboratory Activity 1, begin with step 3 of the following Procedures list.

Figure D3.1
Cat Endocrine System

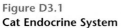

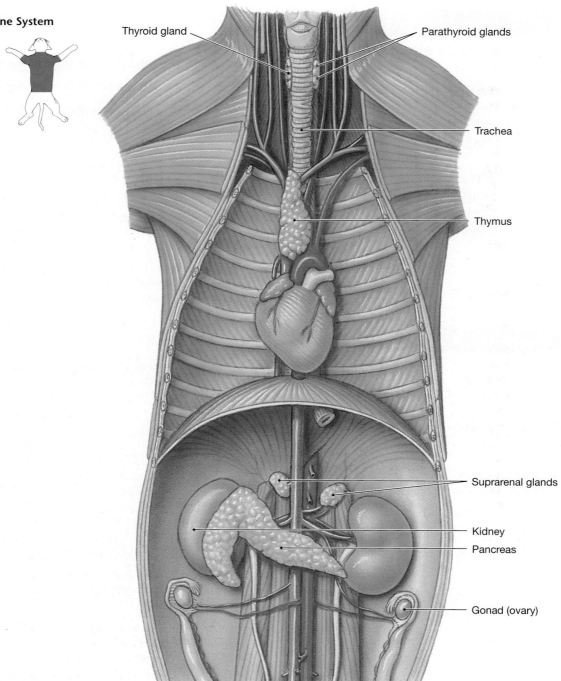

Thyroid gland

Parathyroid glands

Trachea

Thymus

Suprarenal glands

Kidney

Pancreas

Gonad (ovary)

2 *Materials*

- ☐ Gloves
- ☐ Safety glasses
- ☐ Dissecting tools
- ☐ Dissecting tray
- ☐ String
- ☐ Preserved cat, skin removed

Procedures

1. Put on gloves and safety glasses, and clear your workspace before obtaining your dissection specimen.

2. Secure the specimen ventral side up on the dissecting tray by spreading the limbs and tying them flat with lengths of string passing under the tray. Use one string for the two forelimbs and one string for the two hindlimbs.

3. Locate the windpipe, called the *trachea*, passing through the midline of the neck. Spanning the trachea on both sides is the **thyroid gland**. Note that this gland is divided into two **lateral lobes** with a small band, the **isthmus**, connecting them. The thyroid gland produces triiodothyronine (T_3) and thyroxine (T_4), hormones that increase the metabolic rate of cells. Other cells in the thyroid gland secrete the hormone calcitonin (CT), which decreases blood plasma calcium ion concentration.

4. Locate the masses of the **parathyroid gland** on the dorsal surface of each lateral lobe of the thyroid gland. The parathyroid gland secretes parathyroid hormone (PTH), which increases blood plasma calcium ion concentration.

5. Examine the superior surface of the heart and identify the **thymus**. This organ produces thymosin, a hormone that stimulates the development of the immune system. If your dissection specimen is an immature cat, the thymus should be large and conspicuous. In adult cats (and in adult humans, too), the thymus is gradually replaced by fat tissue.

6. Locate the **pancreas** lying between the stomach and small intestine. The pancreas is a "double gland" because it has both exocrine and endocrine functions. The exocrine part of the pancreas produces digestive enzymes. The endocrine part of the gland consists of islands of cells that secrete hormones that regulate the amount of sugar present in the blood. The hormone insulin decreases the amount of blood sugar by stimulating cellular intake of sugar. Glucagon has the opposite effect: it stimulates cells to release sugar into the blood.

7. Gently move the abdominal organs to one side, and locate an **adrenal gland**. In humans the adrenal glands rest on the superior margin of the kidneys; in cats the adrenals are separate from the kidneys. The adrenal glands secrete a variety of hormones. The outer **adrenal cortex** secretes three types of hormones: mineralocorticoids to regulate sodium ions, glucocorticoids to fight stress, and androgen, which is converted to sex hormones. The inner **adrenal medulla** secretes norepinephrine (NE), commonly called adrenaline, into the blood during times of excitement, stress, and exercise.

8. Locate a **kidney**, positioned dorsal to the adrenal gland. The kidneys secrete renin, an enzyme that initiates the endocrine reflex for sodium regulation. The kidneys also secrete erythropoietin (EPO), the hormone that stimulates the bone marrow to produce red blood cells.

9. The **gonads**—testes in the male and ovaries in the female—produce sex hormones. If your dissection specimen is a male, locate the **testes** in the pouchlike scrotum positioned between the hindlimbs. The testes secrete the hormone testosterone, which stimulates maturation and maintenance of the male sex organs and promotes such male characteristics as muscle development and sex drive. Testes also produce the spermatozoa that fertilize eggs during reproduction.

10. If your specimen is a female, locate the small **ovaries** in the pelvic cavity. The ovaries secrete estrogen and progesterone, hormones that prepare the uterus for gestation of an embryo. Estrogen is also responsible for development of the female sex organs, breast development, and other adult female traits. ■

Storing the Cat and Cleaning Up

To store your specimen, wrap it in the skin and moisten it with fixative. Use paper towels if necessary to cover the entire specimen. Return it to the storage bag and seal the bag securely. Label the bag with your name, and place it in the storage area as indicated by your instructor.

Wash all dissection tools and the tray, and set them aside to dry.

Dispose of your gloves and any tissues from the dissection as indicated by your laboratory instructor. ▲

Cat Endocrine System

Name _____

Date _____

Section _____

A. Matching

Match each structure in the left column with its correct description from the right column.

_____ **1.** pancreas

_____ **2.** adrenal medulla

_____ **3.** thyroid gland

_____ **4.** parathyroid gland

_____ **5.** ovary

_____ **6.** gonads

A. divided into two lobes called lateral lobes

B. secretes hormone that raises amount of calcium ions in blood

C. has both exocrine and endocrine functions

D. activated when body is under emotional stress

E. general term for sex organs

F. source of both estrogen and progesterone

B. Short-Answer Questions

1. How are hormones transported to the tissues of the body?

2. Describe the endocrine glands located in the neck.

3. Describe the location of the kidneys, and name one hormone and one enzyme produced by these organs.

C. Analysis and Application

1. How would the body respond if the thyroid gland produced an insufficient amount of thyroid hormone?

2. When male cats are neutered, the testes are removed. What kind of changes are expected in a neutered animal?

3. Suppose a cat has not eaten all day. Describe the endocrine activity of the pancreas in this animal.

4. Which endocrine gland is activated in a cat that is being chased by a dog?

Cat Circulatory System

OBJECTIVES

On completion of this exercise, you should be able to:

- Identify the major arteries and veins of the feline vascular system.

- Compare the circulatory vessels of the cat with those of the human.

In this exercise you will dissect the vascular system of the cat and identify the major arteries and veins. If your cat has been injected with latex paint, the arteries are filled with red paint and the veins with blue paint. Because this exercise complements the study of the human circulatory system, be sure to note differences between the blood vessels of the cat and those of humans.

Dissection Safety

Before beginning this dissection, review the Dissection Safety Guidelines presented at the beginning of Dissection Exercise 1 on p. 745. ▲

LAB ACTIVITY 1 Preparing the Cat for Dissection

If the thoracic and abdominal cavities have not been opened on your dissection specimen, complete the following procedures. Otherwise, proceed to Laboratory Activity 2.

1 Materials

☐ Gloves
☐ Safety glasses
☐ Dissecting tools
☐ Dissecting tray
☐ Preserved cat, skin removed

Procedures

1. Put on gloves and safety glasses, and clear your workspace before obtaining your dissection specimen.

2. Lay the specimen ventral side up on the dissecting tray and expose the thoracic and abdominal organs by making a longitudinal midline incision through the muscles of the neck, thorax, and abdominal wall.

3. Avoid cutting through the muscular diaphragm so that you can identify the vessels and structures passing between the thoracic and abdominal cavities. Your instructor will provide you with specific instructions for exposing these cavities and for isolating the blood vessels.

4. Occasionally, clotted blood fills the thoracic and abdominal cavities and must be removed. If you encounter clots, check with your instructor before proceeding. ∎

LAB ACTIVITY 2

Arteries Supplying the Head, Neck, and Thorax

Only those arteries typically injected with colored paint are listed for identification in this activity. Keep in mind that the cat has more arteries than listed here, and your instructor may assign additional vessels for you to identify. Use Figure D4.1 as a reference as you identify the vessels.

2 Materials

☐ Gloves
☐ Safety glasses
☐ Dissecting tools
☐ Dissecting tray
☐ Preserved cat, skin removed

Procedures

1. Locate the large arteries leaving the right and left ventricles of the heart:

    a. The artery connected to the right ventricle is the **pulmonary trunk**.

    b. The artery connected to the left ventricle is the **aorta**. The aorta is the main arterial blood vessel, and all major arteries except those to the lungs arise from it.

2. Trace these two large arteries, which direct blood away from the heart, to smaller arteries that direct blood to specific organs and tissues. The pulmonary trunk delivers deoxygenated blood to the lungs, and the aorta delivers oxygenated blood to the rest of the body.

    a. The aorta curves at the **aortic arch**—where vessels to the forelimbs, head, and neck arise—and then continues along the chest and abdomen on the left side of the vertebral column to the pelvis, where it divides into branches to supply the hindlimbs. The portion of the aorta anterior to the diaphragm is the **thoracic aorta**.

    b. The pulmonary trunk divides into the right and left branches of the **pulmonary artery.** The left branch of the pulmonary artery passes ventral to the aorta to reach the left lung; the right branch passes between the aortic arch and the heart to reach the right lung.

    c. The **ligamentum arteriosum** is a remnant of fetal circulation when the pulmonary trunk was connected to the aorta by a vessel called the ductus arteriosus. At birth, the ductus arteriosus closes and becomes the ligamentum arteriosum.

3. The feline aortic arch has two major branches: the **brachiocephalic artery** and the **left subclavian artery**. (In humans, there are three branches off this arch.) Via all its various branches, the brachiocephalic artery ultimately

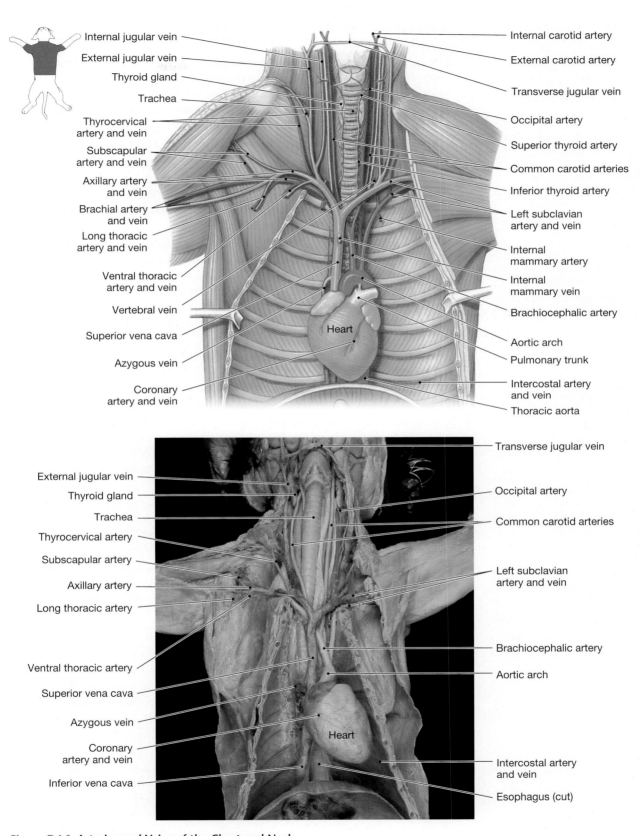

Internal jugular vein
External jugular vein
Thyroid gland
Trachea
Thyrocervical artery and vein
Subscapular artery and vein
Axillary artery and vein
Brachial artery and vein
Long thoracic artery and vein
Ventral thoracic artery and vein
Vertebral vein
Superior vena cava
Azygous vein
Coronary artery and vein

Internal carotid artery
External carotid artery
Transverse jugular vein
Occipital artery
Superior thyroid artery
Common carotid arteries
Inferior thyroid artery
Left subclavian artery and vein
Internal mammary artery
Internal mammary vein
Brachiocephalic artery
Aortic arch
Pulmonary trunk
Intercostal artery and vein
Thoracic aorta

Heart

Transverse jugular vein
External jugular vein
Thyroid gland
Trachea
Thyrocervical artery
Subscapular artery
Axillary artery
Long thoracic artery
Ventral thoracic artery
Superior vena cava
Azygous vein
Coronary artery and vein
Inferior vena cava

Occipital artery
Common carotid arteries
Left subclavian artery and vein
Brachiocephalic artery
Aortic arch
Intercostal artery and vein
Esophagus (cut)

Heart

Figure D4.1 **Arteries and Veins of the Chest and Neck**
Illustration and photo.

supplies blood to the head and the right forelimb. The left subclavian artery supplies the left forelimb.

   a. Near the level of the second rib, the brachiocephalic artery branches into the **right subclavian artery** and the **left common carotid artery**. (Note: Although the right and left subclavians have different origins, once these vessels enter their respective axillae, all the smaller vessels that branch off from each subclavian are the same in the two forelimbs.)

   b. The **right common carotid artery** branches off of the right subclavian artery.

   c. Above the larynx, each common carotid artery divides into an **internal carotid artery** and an **external carotid artery**. The two internal carotid arteries enter the skull via the foramen lacerum and join the **posterior cerebral artery** (not shown in Figure D4.1).

   d. Each external carotid artery turns medially near the posterior margin of the masseter and continues as an **internal maxillary artery**. Each internal maxillary artery gives off several branches and then branches to form the carotid plexus surrounding the maxillary branch of the trigeminal nerve near the foramen rotundum. Through various branches, each internal maxillary and each carotid plexus carries blood to the brain, eye, and other deep structures of the head.

4. Locate the left and right **superior thyroid arteries**, one on either side of the thyroid gland. Branching from its respective common carotid artery at the cranial end of the thyroid gland, each superior thyroid artery supplies blood to the thyroid gland, to superficial laryngeal muscles, and to ventral neck muscles.

5. Next locate an **internal mammary artery,** arising from the ventral surface of either subclavian at about the level of the vertebral artery (discussed next) and passing caudally to the ventral thoracic wall. Branches of both internal mammary arteries supply adjacent muscles, pericardium, mediastinum, and diaphragm.

6. Trace either **vertebral artery** as it arises from the dorsal surface of either subclavian and passes cranially through the transverse foramen of the cervical vertebrae. Each vertebral artery gives off branches to the deep neck muscles and to the spinal cord near the foramen magnum:

   a. Distal to the vertebral artery, a **costocervical artery** arises from the dorsal surface of either subclavian artery. Each costocervical artery sends branches to the deep muscles of the neck and shoulder and to the first two costal interspaces.

   b. A **thyrocervical artery** arises from the cranial aspect of either subclavian (distal to the costocervical artery) and passes cranially and laterally, supplying the muscles of the neck and chest.

7. On the aortic arch, note where 10 pairs of **intercostal arteries** are given off to the interspaces between the last 11 ribs. Also note the paired **bronchial arteries**, which arise from the thoracic aorta and supply the bronchi. Next look for the **esophageal arteries,** several small vessels of varying origin along the thoracic aorta that supply the esophagus.

8. Follow either subclavian artery from its origin, moving away from the heart. While still deep in the thoracic cavity, this vessel is called the **axillary artery** because it ultimately passes through the axilla. A right axillary artery is visible in Figure D4.1. Locate the vessels branching from either axillary artery:

   a. From the ventral surface of the axillary artery, just lateral to the first rib, the **ventral thoracic artery** arises and passes caudally to supply the medial ends of the pectoral muscles.

b. The **long thoracic artery** arises lateral to the ventral thoracic artery, passing caudally to the pectoral muscles and the latissimus dorsi.

c. A third artery branching off the subclavian artery is the **subscapular artery,** which passes laterally and dorsally between the long head of the triceps and the latissimus dorsi to supply the dorsal shoulder muscles. It gives off two branches, the **thoracodorsal artery** and the posterior **humeral circumflex artery**. ■

LAB ACTIVITY 3   Arteries Supplying the Shoulder and Forelimb (Medial View)

The axillary arteries supply blood to the brachium. Use Figure D4.2 as a reference as you identify the vessels.

3 *Materials*

- ☐ Gloves
- ☐ Safety glasses
- ☐ Dissecting tools
- ☐ Dissecting tray
- ☐ Preserved cat, skin removed

Procedures

1. Follow the axillary artery to where it enters the forelimb and is now called the **brachial artery**. Distal to the elbow, the brachial artery gives rise to the **radial artery** and the **ulnar artery**. ■

LAB ACTIVITY 4   Arteries Supplying the Abdominal Cavity

The aorta posterior to the diaphragm is called the **abdominal aorta**. As it passes through the abdomen, arteries to the digestive organs, spleen, urinary system, and reproductive system branch off of the abdominal aorta. As you identify the vessels, note the pattern of paired and unpaired vessels along the abdominal aorta. Use Figure D4.3 as a reference to identify these vessels in the cat.

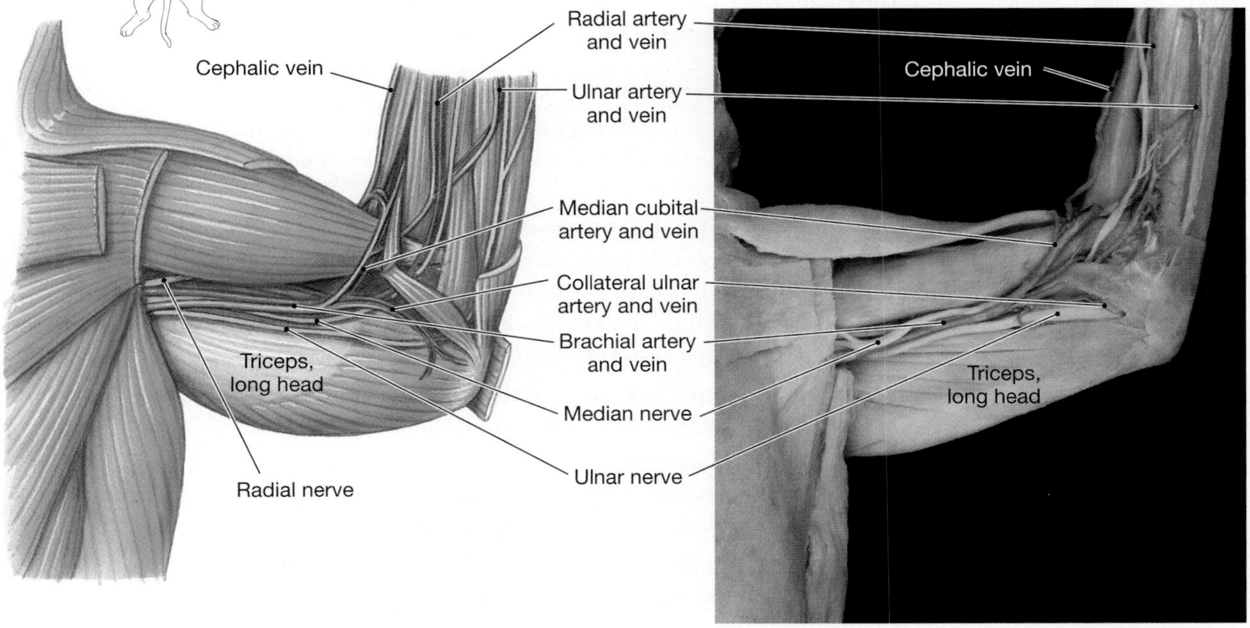

Figure D4.2 Arteries and Veins of the Forelimb
Illustration and photo.

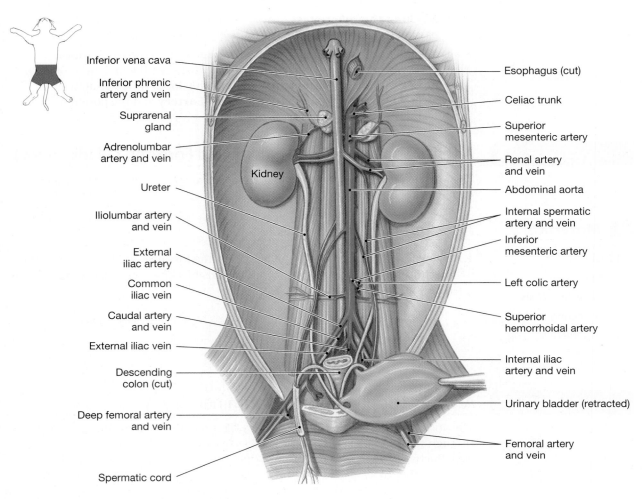

Figure D4.3 Arteries and Veins of the Abdominal Cavity
Illustration and photo.

4 ▸ *Materials*

☐ Gloves

☐ Safety glasses

☐ Dissecting tools

☐ Dissecting tray

☐ Preserved cat, skin
   removed

Procedures

1. A single vessel, the **celiac trunk**, is the first arterial branch off the abdominal aorta. Notice how it divides into three branches: the hepatic, left gastric, and splenic arteries.

   a. Along the cranial border of the gastrosplenic part of the pancreas lies the **hepatic artery**. It turns cranially near the pylorus, lying in a fibrous sheath together with the portal vein and the common bile duct. Its branches include the **cystic artery** to the gallbladder and liver and the **gastroduodenal artery** near the pylorus.

   b. Along the lesser curvature of the stomach lies the **left gastric artery**, supplying many branches to both dorsal and ventral stomach walls.

   c. The **splenic artery** is the largest branch of the celiac artery. It supplies at least two branches to the dorsal surface of the stomach and divides into anterior and posterior branches to supply these portions of the spleen.

2. Just posterior to the celiac trunk, find the unpaired **superior mesenteric artery**. It divides into numerous intestinal branches that supply the small and large intestines.

3. Next notice the paired **adrenolumbar arteries** just posterior to the superior mesenteric artery. These two arteries, which supply the suprarenal glands, pass

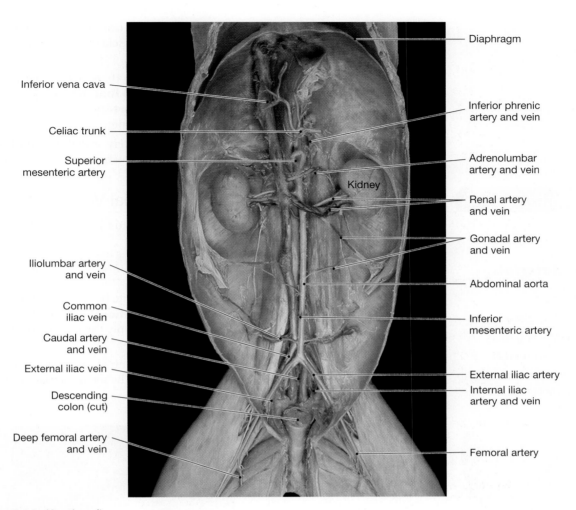

Figure D4.3 *(Continued)*

laterally along the dorsal body wall and give rise to **phrenic** and **adrenal arteries** and then supply the muscles of the dorsal body wall. The phrenic artery branches off the adrenolumbar artery and supplies the diaphragm.

4. Locate the paired **renal arteries** emerging from the abdominal aorta to supply the kidneys. In some specimens, each renal artery gives rise to an adrenal artery. Often double renal arteries supply each kidney.

5. Locate the paired **gonadal arteries** as they arise from the aorta near the caudal ends of the kidneys.

   a. In females, the gonadal arteries are called the **ovarian arteries**. They pass laterally in the broad ligament to supply the ovaries. Each artery gives a branch to the cranial end of the corresponding uterine horn.

   b. In the male, the gonadal arteries are called the **spermatic arteries**. They lie on the surface of the iliopsoas muscle, passing caudally to the internal inguinal ring and through the inguinal canal to the testes.

6. Next find the **lumbar arteries,** seven pairs of arteries arising from the dorsal surface of the aorta and supplying the muscles of the dorsal abdominal wall.

7. The unpaired **inferior mesenteric artery** arises from the abdominal aorta near the last lumbar vertebra. Close to its origin, notice how this vessel divides into the **left colic artery**, which passes anteriorly to supply the descending colon, and the **superior hemorrhoidal artery**, which passes posteriorly to supply the rectum.

8. Locate the paired **iliolumbar arteries** as they arise near the inferior mesenteric artery and pass laterally across the iliopsoas muscle to supply the muscles of the dorsal abdominal wall.

9. Near the sacrum, the abdominal aorta branches into three vessels that you should search for next. The **right** and **left external iliac arteries** lead toward the hindlimbs, and the single **internal iliac artery** serves the tail. Unlike humans, cats do not have common iliac arteries.

10. Last, look for the first branch of the internal iliac artery, called the **umbilical artery**, and the **caudal (medial sacral) artery**, which passes into the tail. ■

LAB ACTIVITY 5 Arteries Supplying the Hindlimb (Medial View)

The external iliac arteries enter the thigh to supply blood to the hindlimb. Use Figure D4.4 as a reference to locate these vessels in the cat.

5 Materials

- ☐ Gloves
- ☐ Safety glasses
- ☐ Dissecting tools
- ☐ Dissecting tray
- ☐ Preserved cat, skin removed

Procedures

1. Follow either one of the external iliac arteries, and note how, just prior to leaving the abdominal cavity, it gives off the **deep femoral artery**, which passes between the iliopsoas and the pectineus to supply the muscles of the thigh (Figure D4.3).

2. Each external iliac artery continues outside the abdominal cavity as a **femoral artery** (Figure D4.4). Note a femoral artery lying on the medial surface of the

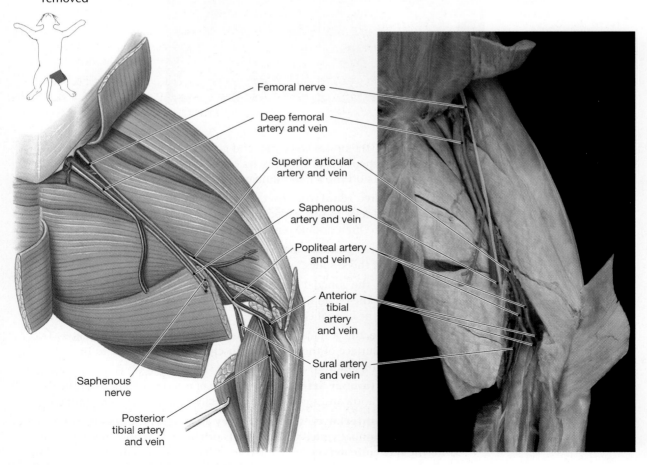

Femoral nerve

Deep femoral artery and vein

Superior articular artery and vein

Saphenous artery and vein

Popliteal artery and vein

Anterior tibial artery and vein

Sural artery and vein

Saphenous nerve

Posterior tibial artery and vein

Figure D4.4 Arteries and Veins of the Hindlimb
Illustration and photo.

thigh. At the posterior knee, this artery is called the **popliteal artery** and divides into the **anterior** and **posterior tibial arteries**. ■

Veins Draining the Head, Neck, and Thorax

Veins usually follow the pattern of the corresponding arteries. Trace the veins in the direction of venous blood flow, which is away from the extremities toward the heart. Refer to Figure D4.1 for the veins of the head, neck, and thorax.

6 *Materials*

- ☐ Gloves
- ☐ Safety glasses
- ☐ Dissecting tools
- ☐ Dissecting tray
- ☐ Preserved cat, skin removed

Procedures

1. Gently pull the heart away from a lung and examine the exposed lung root. Each lobe of the lung has a **pulmonary vein** (which usually is not injected with colored paint) that passes oxygenated blood toward the dorsal side of the heart to enter the left atrium.

2. Find the two major veins that deliver deoxygenated blood to the right atrium of the heart: the precava and the postcava veins. (In humans, these veins are called the superior vena cava and the inferior vena cava.) The **precava** is the large vessel that drains blood from the head, neck, and forelimbs to the right atrium. Its principal tributaries include the internal and external jugular veins, the subscapular veins, and the axillary veins. The **postcava** is the large vessel that returns blood from the abdomen, hindlimbs, and pelvis to the right atrium. It drains blood from numerous vessels of the abdomen.

3. Observe the smaller vessels that feed into the precava:
   a. The paired **internal thoracic (mammary) veins** unite at a small stem that joins the precava. The internal thoracic veins lie on either side of the body midline and drain the ventral chest wall.
   b. Push the heart toward the left lung to see the **azygous vein** arching over the root of the right lung and joining the precava near the right atrium. The **intercostal veins** drain blood from the intercostal muscles into the azygous vein.

4. Locate the **axillary vein,** the major vessel draining the forelimb in the armpit. The **subscapular vein** is the largest tributary to the axillary vein. A small vessel emptying into the axillary vein on its ventral surface is the **ventral thoracic vein**. It is located near the subscapular vein. The **long thoracic vein** is another small vessel emptying into the axillary vein. It is distal to the ventral thoracic vein.

5. Draining each side of the head is the **external jugular vein**, which merges with the subclavian vein. Each external jugular vein joins with a subclavian vein to form the brachiocephalic vein. The **internal jugular vein** drains the brain and spinal cord and joins the external jugular vein near its union with the brachiocephalic vein. Just superior to the hyoid bone, find the **transverse jugular vein** as it connects the left and right external jugular veins. At the shoulder, the external jugular vein receives the large **transverse scapular vein** that drains the dorsal surface of the scapula.

6. The subclavian veins drain into a pair of **brachiocephalic veins** that unite as the precava.

7. Notice the **costocervical** and **vertebral veins**, which form a common stem and dorsally connect with the brachiocephalic vein. The vertebral vein in some specimens empties into the precava vein. ■

LAB ACTIVITY 7 Veins Draining the Forelimb (Medial View)

The veins of the forelimb drain into the axillary vein and the transverse scapular vein. Refer to Figure D4.2 for the veins of the forelimb.

7 Materials

- ☐ Gloves
- ☐ Safety glasses
- ☐ Dissecting tools
- ☐ Dissecting tray
- ☐ Preserved cat, skin removed

Procedures

1. In the forelimb, locate the **radial vein** on the lateral side and the **ulnar vein** medially.

2. Near the elbow, these veins drain into the **brachial vein**, which ascends the forelimb and becomes the axillary vein.

3. The **cephalic vein** ascends the forelimb and joins the transverse scapular vein, which drains into the external jugular vein. In humans, the cephalic vein joins the axillary vein. ■

LAB ACTIVITY 8 Veins Draining the Abdominal Cavity

The postcava vein receives blood from the major veins draining the abdomen, pelvis, and hindlimbs. For ease of identification, the order in which we describe the veins emptying into the postcava is the order in which they appear as we move from the diaphragm to the tail and hindlimb. Refer to Figure D4.3 for the veins of the abdominal cavity.

8 Materials

- ☐ Gloves
- ☐ Safety glasses
- ☐ Dissecting tools
- ☐ Dissecting tray
- ☐ Preserved cat, skin removed

Procedures

1. Along the dorsal pelvis wall, a pair of common iliac veins join together as the postcava vein. Ventral to the postcava is the major artery of the abdomen, the abdominal aorta. The large **internal iliac veins** enter the common iliac vein in the pelvic cavity and drain the rectum, bladder, and internal reproductive organs. Distal to the joining of the internal iliac vein with the common iliac vein, the **external iliac vein** is a continuation of the femoral vein from the hindlimb. The **caudal vein** drains the tail and empties into the origin of the postcava.

2. Working toward the heart, identify in sequence along the veins draining into the postcava the iliolumbar veins and the **lumbar veins** that drain the abdominal muscles. Next are the gonadal veins that drain the reproductive organs. In males, these vessels drain the testes and are called the **internal spermatic veins**. Usually, the left internal spermatic vein empties into the renal vein. In females, the gonadal veins drain the ovaries and are called the **ovarian veins**. These vessels empty into the postcava vein. Note the paired **renal veins** as they drain the kidneys into the postcava. Occasionally, double renal veins drain each kidney. Also identify the **adrenolumbar veins**, which drain the adrenal glands.

3. The **hepatic veins** drain blood from the liver into the postcava. The postcava is in close contact with the liver, and as a result the hepatic veins are difficult to locate. Remove some liver tissue around the postcava, and try to expose the hepatic veins. Locate the **hepatic portal vein**, which carries blood from the intestines and other organs to the liver (Figure D4.5). Note that this vein has several veins emptying into it:

   a. The **superior mesenteric vein** is the large vein draining the small and large intestines and the pancreas. It is the largest branch of the hepatic portal vein.

   b. The **inferior mesenteric vein** follows along the inferior mesenteric artery. It drains part of the large intestine.

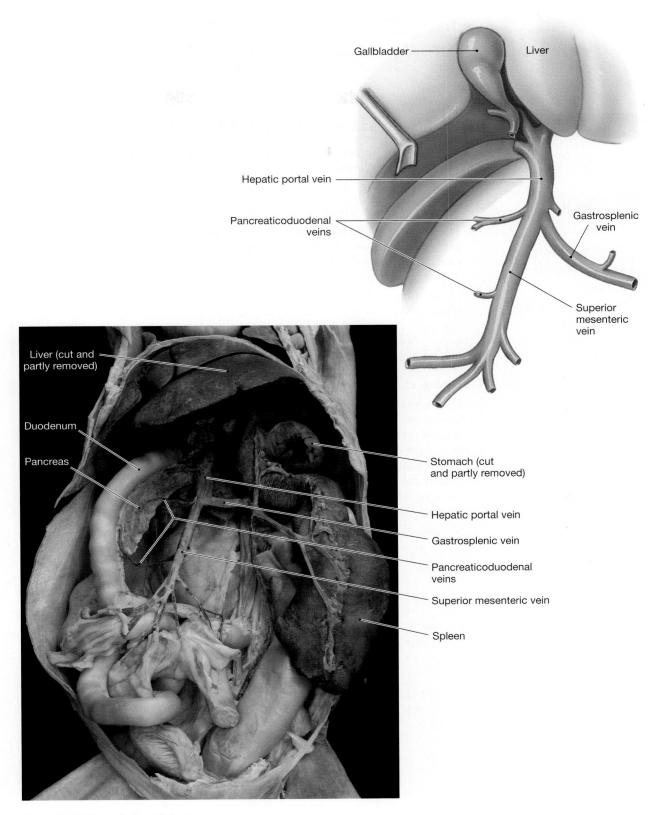

Figure D4.5 Hepatic Portal System

(a) Illustration showing the hepatic portal system in isolation.
(b) Photo showing its location in the ventral body cavity.

   c. The **gastrosplenic vein**, which drains the stomach and spleen, lies on the dorsal side of the stomach.

4. Next find the **inferior phrenic veins** as they run from the diaphragm into the postcava vein just before the postcava pierces the diaphragm. ■

LAB ACTIVITY 9   # Veins Draining the Hindlimb (Medial View)

The veins of the hindlimb basically correspond to the arteries in the region. Refer to Figure D4.4 for the veins of the hindlimb.

9 *Materials*

- ☐ Gloves
- ☐ Safety glasses
- ☐ Dissecting tools
- ☐ Dissecting tray
- ☐ Preserved cat, skin removed

Procedures

1. Find the medial and lateral branches of the **popliteal vein**, which drain the foot and calf region, uniting in the popliteal region to form the **saphenous vein**, which drains into the femoral vein. These vessels lie superficial to the muscles of the hindlimb.

2. Locate the superficial vein on the anterior surface of the thigh that is the **femoral vein**. As it enters the abdominal cavity, this vein becomes the external iliac vein. The **deep femoral vein** is a medial branch emptying into the femoral vein near the pelvis. ■

Storing the Cat and Cleaning Up

To store your specimen, wrap it in the skin and moisten it with fixative. Use paper towels if necessary to cover the entire specimen. Return it to the storage bag and seal the bag securely. Label the bag with your name and place it in the storage area as indicated by your instructor.

Wash all dissection tools and the tray, and set them aside to dry.

Dispose of your gloves and any tissues from the dissection into a biohazard box or as indicated by your laboratory instructor. Wipe your work area clean and wash your hands. ▲

Cat Circulatory System

Name _____

Date _____

Section _____

A. Matching

Match each description in the left column with its correct definition from the right column.

_____ **1.** superior vena cava in cats

_____ **2.** major branch off right subclavian artery

_____ **3.** drains blood from intercostals

_____ **4.** inferior vena cava in cats

_____ **5.** first major branch of aortic arch

_____ **6.** vein supplying blood to liver

_____ **7.** major branch off subscapular artery

_____ **8.** major branch off brachiocephalic artery

_____ **9.** vein that unites as small stem on precava

_____ **10.** artery of the tail

A. right common carotid artery

B. brachiocephalic artery

C. internal thoracic vein

D. internal iliac artery

E. precava

F. azygous vein

G. hepatic portal vein

H. postcava

I. left common carotid artery

J. thoracodorsal artery

B. Labeling

Identify the vessels in Figure D4.6.

4

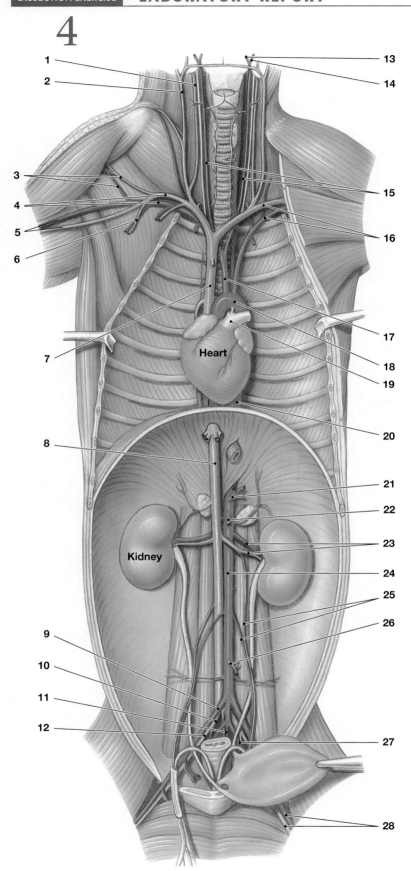

1

2

3

4

5

6

7

8

9

10

11

12

13

14

15

16

17

18

19

20

21

22

23

24

25

26

27

28

Heart

Kidney

1. _____

2. _____

3. _____

4. _____

5. _____

6. _____

7. _____

8. _____

9. _____

10. _____

11. _____

12. _____

13. _____

14. _____

15. _____

16. _____

17. _____

18. _____

19. _____

20. _____

21. _____

22. _____

23. _____

24. _____

25. _____

26. _____

27. _____

28. _____

Figure D4.6 Major Arteries and Veins of the Cat

C. Short-Answer Questions

1. Trace the feline vascular route from the abdominal aorta to the intestines and into the postcava.

2. Give an example of a feline artery and its corresponding vein that are next to each other and have the same regional name.

3. Trace a drop of blood in veins from the cat's lower hindlimb to the heart.

4. List the vessels that transport blood to and from the lungs.

5. Why is the aorta called the major artery of systemic circulation?

6. Trace a drop of oxygenated blood from the heart to the right lower forelimb.

4

D. Analysis and Application

1. How is the branching off the aortic arch in cats different from this branching in humans?

2. Name three blood vessels found in cats but not in humans.

3. Describe the differences between the origins of the left and right common carotid arteries in cats and humans.

4. How is the branching off the abdominal aorta at the pelvis in cats different from this branching in humans?

Cat Lymphatic System

OBJECTIVES

On completion of this exercise, you should be able to:

- Identify where the main feline lymphatic ducts empty into the vascular system.

- Identify lymph nodes in the feline jaw and tonsils in the mouth.

- Locate the feline spleen.

The lymphatic system protects the body against infection by producing lymphocyte blood cells, which manufacture antibodies to destroy microbes that have invaded the body. Another protective task of the lymphatic system is to collect liquid that has been forced out of blood capillaries and into extracellular spaces. This liquid, called **lymph**, is carried into the lymph nodes, where phagocytic cells remove debris and microbes from the lymph before the liquid passes out of the nodes and into the venous bloodstream. Because of this cleansing role, a node may occasionally itself become infected.

Major organs of the lymphatic system include the thymus gland, spleen, tonsils, lymph nodes, and lymphatic nodules. The nodules are similar to nodes but smaller and more scattered in the tissues of the digestive and other systems. Lymphatic vessels are long tubes similar to blood vessels except with a much lower fluid pressure. Movements of the body squeeze on the lymphatic vessels and push the contained lymph toward the heart, to the location near where the lymph is returned to the circulation.

This exercise complements the study of the human lymphatic system.

Dissection Safety

Before beginning this dissection, review the Dissection Safety Guidelines presented at the beginning of Dissection Exercise 1 on p. 745. ▲

LAB ACTIVITY   1   Preparing the Cat for Dissection

If the ventral body cavity has not been opened on your dissection specimen, complete the following procedures. Otherwise, skip to Laboratory Activity 2. Use Figure D5.1 as a reference as you dissect the ventral body cavity.

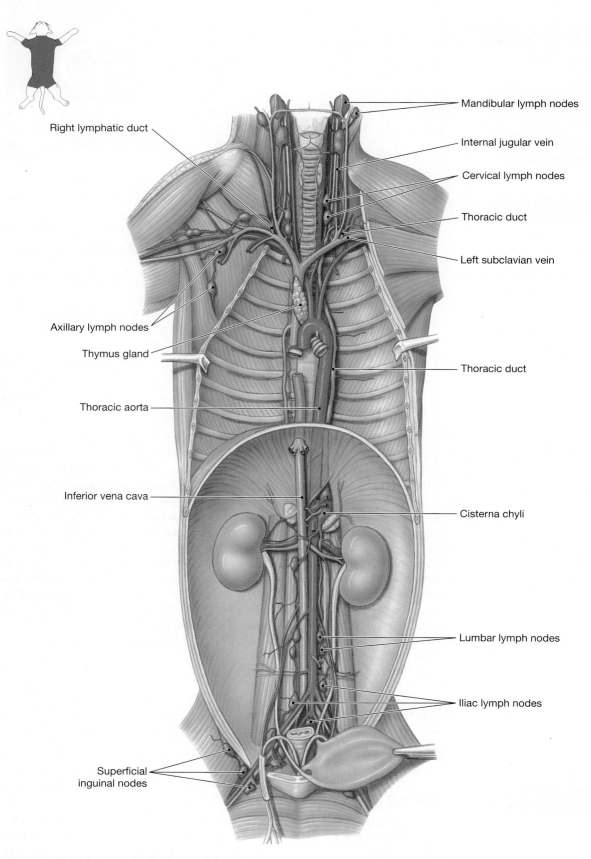

Right lymphatic duct

Axillary lymph nodes

Thymus gland

Thoracic aorta

Inferior vena cava

Superficial
inguinal nodes

Mandibular lymph nodes

Internal jugular vein

Cervical lymph nodes

Thoracic duct

Left subclavian vein

Thoracic duct

Cisterna chyli

Lumbar lymph nodes

Iliac lymph nodes

Figure D5.1 Major Lymphatic Ducts of the Cat

1 *Materials*

☐ Gloves
☐ Safety glasses
☐ Dissecting tools
☐ Dissecting tray
☐ String
☐ Preserved cat, skin removed

Procedures

1. Put on gloves and safety glasses, and clear your workspace before obtaining your dissection specimen.

2. Secure the specimen ventral side up on the dissecting tray by spreading the limbs and tying them flat with lengths of string passing under the tray. Use one string for the two forelimbs and one string for the two hindlimbs.

3. If the ventral body cavity has not been opened, use scissors to cut a midsagittal section through the muscles of the abdomen to the sternum.

4. To avoid cutting through the bony sternum, angle your incision laterally approximately 0.5 inch, and cut the costal cartilages. Continue the parasagittal section to the base of the neck.

5. Make a lateral incision on each side of the diaphragm. Use care not to damage the diaphragm or the internal organs. Spread the thoracic walls to observe the internal organs.

6. Make a lateral section across the pubic region and angled toward the hips. Spread the abdominal walls to expose the abdominal organs. ■

LAB ACTIVITY **2**   The Cat Lymphatic System

Lymphatic vessels are thin and difficult to locate. The larger lymphatic ducts near the subclavian veins may be colored with some blue latex paint that leaked in when nearby veins were injected. Use Figure D5.1 as a reference during your dissection and observations.

2 *Materials*

☐ Gloves
☐ Safety glasses
☐ Dissecting tools
☐ Dissecting tray
☐ Preserved cat, skin removed

Procedures

1. Reflect the muscles and wall of the chest to expose the thoracic cavity, and move the organs to one side. Locate the thin, brown **thoracic duct** lying along the dorsal surface of the descending aorta. This duct receives lymph from both hindlimbs, the abdomen, the left forelimb, and the left side of the chest, neck, and head. Examine the duct, and notice how the internal valves cause the duct to expand over the valve and appear beaded. Trace the duct anteriorly to where it empties into the external jugular vein near the left subclavian vein as the **left lymphatic duct**, an alternative name for the thoracic duct. In humans, the left lymphatic (thoracic) duct empties not into the external jugular vein, but rather into the left subclavian vein.

2. Trace the thoracic duct from the thorax to the abdomen. Posterior to the diaphragm, the thoracic duct is dilated into a sac called the **cisterna chyli**, where other lymphatic ducts from the hindlimbs, pelvis, and abdomen drain.

3. Return to the thorax, and locate where the left lymphatic (thoracic) duct empties into the external jugular vein. Now move to the right side of the thorax, and examine this area closely. The **right lymphatic duct** drains into the external jugular vein where the vein empties into the right subclavian vein. The right lymphatic duct drains lymph from the right forelimb and from the right side of the chest, neck, and head. Remember from step 1 that the left lymphatic (thoracic) duct drains lymph from the hindlimbs, abdomen, left forelimb, and left side of the chest, head, and neck. Review Figure 38.4 to see more detail on this uneven drainage of the right and left lymphatic vessels in humans.

4. On either side of the jaw, between the mandible and ear, identify the brown kidney-bean-shaped **lymph nodes**. Phagocytes in the lymph nodes remove debris and pathogens from the lymph. Although lymph nodes are distributed throughout the body, they are typically small and difficult to locate. The

most prominent nodes are in the inguinal and axillary regions and in the neck and jaw.

5. Open the mouth and examine the roof, called the *palate*. Posteriorly, the palate stops where the mouth joins the pharynx (throat). Note the folds of tissue forming an arch between the mouth and pharynx. Along the lateral wall of the arches are a pair of small and round **palatine tonsils**. As in humans, the feline tonsils are lymphatic organs, and their enclosed lymphocytes help fight infection.

6. On the superior surface of the heart locate the **thymus gland**. It is important in the development of the immune system. It is larger in immature cats (and in humans, too) and gradually replaced by fat in adults.

7. Next locate the **spleen**, the red, flat organ on the left side just posterior to the stomach. This lymphatic organ removes worn-out red blood cells from circulation and assists in recycling the iron from hemoglobin. Antigens in the blood stimulate the spleen to activate the immune system. ■

Storing the Cat and Cleaning Up

To store your specimen, wrap it in the skin and moisten it with fixative. Use paper towels if necessary to cover the entire specimen. Return it to the storage bag and seal the bag securely. Label the bag with your name, and place it in the storage area as indicated by your instructor.

Wash all dissection tools and the tray, and set them aside to dry.

Dispose of your gloves and any tissues from the dissection as indicated by your laboratory instructor. ▲

Cat Lymphatic System

Name _____

Date _____

Section _____

A. Matching

Match each structure in the left column with its correct description from the right column.

_____ **1.** spleen	**A.** drains lymph from part of one side of body only
_____ **2.** thoracic duct	**B.** small mass that filters lymph
_____ **3.** lymph node	**C.** expanded posterior of main lymphatic duct
_____ **4.** right lymphatic duct	**D.** gland over heart
_____ **5.** cisterna chyli	**E.** drains lymph from hindlimbs and from part of one side of body
_____ **6.** thymus	**F.** removes worn blood cells from circulation

B. Short-Answer Questions

1. Describe the location and function of the spleen.

2. What is the function of lymph nodes?

3. List the organs of the lymphatic system.

5

C. Analysis and Application

1. Compare the way the feline thoracic duct drains into the venous system with how this duct drains in the venous system in humans.

2. Compare the return of lymph on the right and left sides of the body.

3. Describe how liquid circulates from the blood, into lymphatic vessels, and returned to the blood.

Cat Respiratory System

LABORATORY ACTIVITIES

1 Preparing the Cat for Dissection 803

2 Nasal Cavity and Pharynx 804

3 Larynx and Trachea 805

4 Bronchi and Lungs 805

OBJECTIVES

On completion of this exercise, you should be able to:

- Identify the structures of the feline nasal cavity.

- Identify the three regions of the feline pharynx.

- Identify the cartilages and folds of the feline larynx.

- Identify the structures of the feline trachea and bronchi.

- Identify and describe the gross anatomy of the feline lung.

- Compare and contrast the respiratory anatomy of cats and humans.

The main function of the respiratory system is to convert deoxygenated blood into oxygenated blood. In this exercise you will identify the major organs of the feline respiratory system. As this exercise is designed to accompany the exercise on the human respiratory system, be sure to note differences between the respiratory anatomy of cats and humans.

This dissection will cover the nose, pharynx, larynx, trachea, bronchi, and lungs.

    Dissection Safety

Before beginning this dissection, review the Dissection Safety Guidelines presented at the beginning of Dissection Exercise 1 on p. 745. ▲

LAB ACTIVITY 1 Preparing the Cat for Dissection

If the ventral body cavity has not been opened on your dissection specimen, complete the following procedures. Otherwise, skip to Laboratory Activity 2. Use Figure D6.2 as a reference as you dissect the ventral body cavity.

1 *Materials*

☐ Gloves
☐ Safety glasses
☐ Dissecting tools
☐ Dissecting tray
☐ String
☐ Preserved cat, skin removed

Procedures

1. Put on gloves and safety glasses, and clear your workspace before obtaining your dissection specimen.
2. Secure the specimen ventral side up on the dissecting tray by spreading the limbs and tying them flat with lengths of string passing under the tray. Use one string for the two forelimbs and one string for the two hindlimbs.
3. Use scissors to cut a midsagittal section through the muscles of the abdomen to the sternum.
4. To avoid cutting through the bony sternum, angle your incision laterally approximately 0.5 inch, and cut the costal cartilages. Continue the parasagittal section to the base of the neck.
5. Make a lateral incision on each side of the diaphragm. Use care not to damage the diaphragm or the internal organs. Spread the thoracic walls to observe the internal organs. ■

LAB ACTIVITY 2 Nasal Cavity and Pharynx

Air enters the nasal cavity through the two nostrils, called the **external nares** (Figure D6.1). Openings called the **internal nares** (also called **choanae**) connect the posterior of the nasal cavity with the **pharynx** (the throat), which is the cavity dorsal to the soft palate of the mouth. As in humans, the cat pharynx has three regions: the anterior **nasopharynx** behind the nose, the **oropharynx** behind the mouth, and the **laryngopharynx,** where the pharynx divides into the esophagus and larynx. Air must pass through an opening called the **glottis** to enter the larynx.

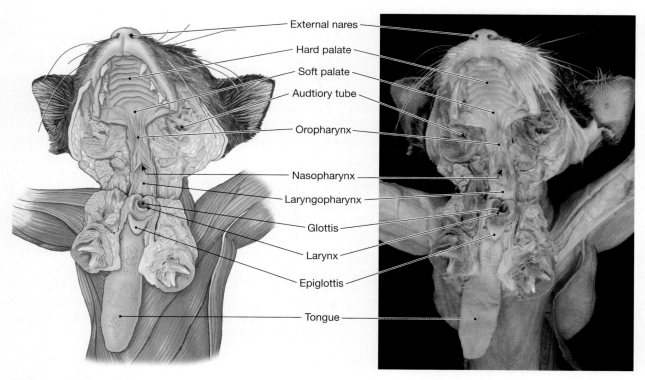

External nares
Hard palate
Soft palate
Audtiory tube
Oropharynx
Nasopharynx
Laryngopharynx
Glottis
Larynx
Epiglottis
Tongue

Figure D6.1 Oral Cavity of the Cat
Illustration and photo.

2 Materials

- ☐ Gloves
- ☐ Safety glasses
- ☐ Dissecting tools
- ☐ Dissecting tray
- ☐ Pipette or plastic tubing
- ☐ Preserved cat, skin removed

Procedures

1. Use bone cutters to cut through the angle of the mandible on one side only. Pull the lower jaw toward the uncut side, leaving the salivary glands intact on the uncut side. If necessary, secure the jaw with a pin so that it will not interfere with the rest of the dissection.

2. Carefully use a scalpel to cut through soft tissues, such as connective tissue and muscle, until you reach the pharynx.

3. Observe the internal nares and the three regions of the pharynx, and locate the glottis. ■

LAB ACTIVITY 3 Larynx and Trachea

The feline larynx has five cartilages, whereas the human larynx has nine cartilages (Figure D6.2). The **thyroid cartilage** is the large prominent ventral cartilage deep to the ventral neck muscles. Caudal to the thyroid cartilage is the **cricoid cartilage**, which is the only complete ring of cartilage in the respiratory tract. The paired **arytenoid cartilages** occupy the dorsal surface of the larynx anterior to the cricoid cartilage. A flap of cartilage called the **epiglottis** covers the glottis during swallowing and keeps food from entering the lower respiratory tract. Inside the larynx are vocal cords for production of sound. The anterior **vestibular ligaments**, commonly called the false vocal cords, protect the posterior **vocal ligaments** (true vocal cords), which vibrate and produce sounds.

Posteriorly, the larynx is continuous with the **trachea**, which is kept open by the C-shaped pieces of hyaline cartilage called the **tracheal rings**. On the dorsal side of the trachea is the food tube, the **esophagus**. Laterally, the common carotid arteries, internal jugular veins, and vagus nerve pass through the neck.

3 Materials

- ☐ Gloves
- ☐ Safety glasses
- ☐ Dissecting tools
- ☐ Dissecting tray
- ☐ Pipette or plastic tubing
- ☐ Preserved cat, skin removed

Procedures

1. Cut completely through the neck muscles to the body of a cervical vertebra. Carefully cut through any remaining connective tissue that may still be securing the larynx. Be careful not to cut the common carotid arteries or vagus nerves.

2. Locate the five cartilages of the larynx.

3. Identify the trachea and the tracheal rings. Also locate the esophagus, and on the lateral neck the common carotid arteries, internal jugular veins, and vagus nerve.

4. Expose and remove the larynx. Make a median incision on the dorsal surface of the larynx. Open the larynx to expose the elastic vocal cords between the thyroid and arytenoid cartilages. Identify the vestibular ligaments located anteriorly and the posterior pair of vocal ligaments. ■

LAB ACTIVITY 4 Bronchi and Lungs

The trachea divides (bifurcates) into left and right **primary bronchi**. The bronchi penetrate the lungs at a slit-like **hilus** and branch repeatedly to supply the alveoli with air. The lungs of the cat have more lobes than the lungs of humans. The feline left lung has three lobes and the right lung has four: three main lobes and a fourth smaller mediastinal lobe (not shown in Figure D6.2). At the site of attachment, other structures such as the pulmonary artery and pulmonary veins enter and exit the lungs.

Each lung is enclosed in a serous membrane called the *pleura*. This membrane consists of the glistening **visceral pleura** on the lung surface and the **parietal pleura** lining the thoracic wall. Between these layers is a small space called the **pleural cavity**, which contains **pleural fluid** secreted by the pleura.

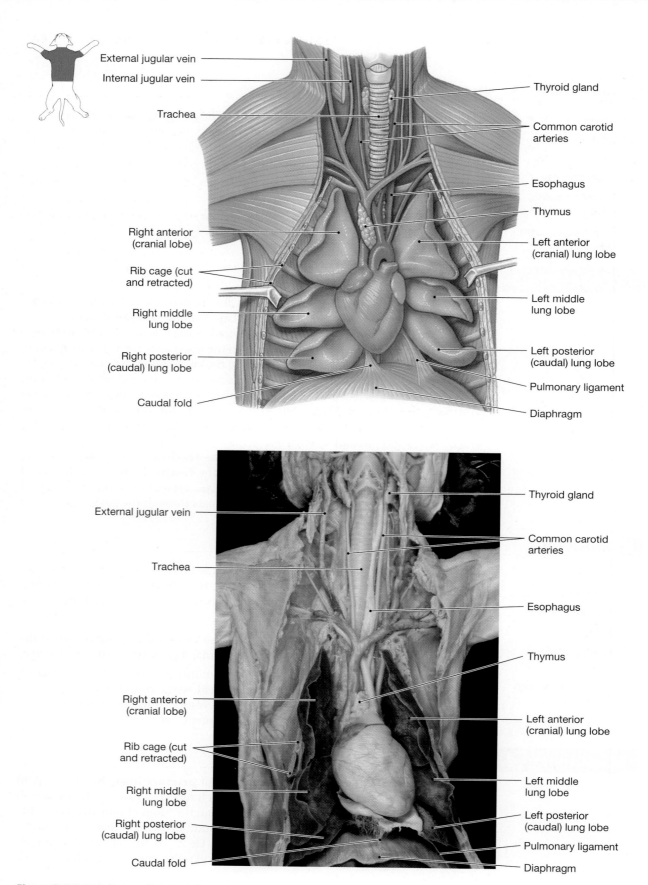

External jugular vein
Internal jugular vein
Trachea
Thyroid gland
Common carotid arteries
Esophagus
Thymus
Right anterior (cranial lobe)
Left anterior (cranial) lung lobe
Rib cage (cut and retracted)
Left middle lung lobe
Right middle lung lobe
Right posterior (caudal) lung lobe
Left posterior (caudal) lung lobe
Pulmonary ligament
Caudal fold
Diaphragm

External jugular vein
Trachea
Thyroid gland
Common carotid arteries
Esophagus
Thymus
Right anterior (cranial lobe)
Left anterior (cranial) lung lobe
Rib cage (cut and retracted)
Left middle lung lobe
Right middle lung lobe
Right posterior (caudal) lung lobe
Left posterior (caudal) lung lobe
Pulmonary ligament
Caudal fold
Diaphragm

Figure D6.2 Respiratory System of the Cat
Illustration and photo. Note: Right accessory lobe posterior to the heart not shown.

806

4 *Materials*

- ☐ Gloves
- ☐ Safety glasses
- ☐ Dissecting tools
- ☐ Dissecting tray
- ☐ Pipette or plastic tubing
- ☐ Preserved cat, skin removed

Procedures

1. Trace the trachea to where it bifurcates into left and right primary bronchi. Note the hilus, where the bronchi enter the lungs. Examine this area for pulmonary arteries and pulmonary veins.

2. Using Figure D6.2 as a guide, identify the four lobes of the right lung and the three lobes of the left lung.

3. Next locate the pleura surrounding each lung. Distinguish between the glossy visceral pleura on the lung surface and the parietal pleura on the thoracic wall.

4. The diaphragm is the sheet of muscle that divides the thoracic cavity from the abdominal cavity and is one of the major muscles involved in respiration. Locate the **phrenic nerve** that controls the diaphragm. This nerve should be clearly visible as a white "thread" along the heart.

5. Place a clean pipette or a piece of plastic tubing into the cat's mouth and push it into the laryngopharynx. Attempt to inflate the cat's lungs by gently exhaling into the tube. Observe the expansion of the lungs as they fill with air.

6. Remove a section of lung from a lobe. Observe the cut edge of the specimen and notice the spongy appearance. Are blood vessels visible? ■

Storing the Cat and Cleaning Up

To store your specimen, wrap it in the skin and moisten it with fixative. Use paper towels if necessary to cover the entire specimen. Return it to the storage bag and seal the bag securely. Label the bag with your name, and place it in the storage area as indicated by your instructor.

Wash all dissection tools and the tray, and set them aside to dry.

Dispose of your gloves and any tissues from the dissection as indicated by your laboratory instructor. ▲

Cat Respiratory System

6

Name _____

Date _____

Section _____

A. Matching

Match each short definition in the left column with its correct answer from the right column.

_____	**1.** lung with three lobes	**A.** right lung
_____	**2.** internal nares	**B.** thyroid cartilage
_____	**3.** nostrils	**C.** visceral pleura
_____	**4.** lung with four lobes	**D.** diaphragm
_____	**5.** complete ring of cartilage	**E.** tracheal ring
_____	**6.** largest cartilage of larynx	**F.** choanae
_____	**7.** innervated by phrenic nerve	**G.** airway of larynx
_____	**8.** C-ring of cartilage	**H.** left lung
_____	**9.** membrane on lung surface	**I.** cricoid cartilage
_____	**10.** membrane against thoracic wall	**J.** entrance into lung
		K. parietal pleura
_____	**11.** glottis	**L.** external nares
_____	**12.** hilus	

B. Short-Answer Questions

1. What are the three regions of the pharynx?

2. List the respiratory structures that air passes through from the external nares to the lungs.

3. Describe the similarities between the feline larynx and the human larynx.

4. Which structures produce vocal sounds?

C. Analysis and Application

1. Compare the gross anatomy of the lungs of cats and humans.

2. Knowing that the feline lungs have more lobes than the lungs in humans, speculate on how the feline secondary bronchi compare with human secondary bronchi.

3. Describe the cartilaginous structures of the respiratory system.

Cat Digestive System

OBJECTIVES

On completion of this exercise, you should be able to:

- Identify the structures of the feline mouth, stomach, and small and large intestines.

- Identify the gross anatomy of the feline liver, gallbladder, and pancreas.

- Compare and contrast the digestive system of cats and humans.

In this exercise you will be looking at the major organs and structures of the feline digestive system. Because the feline digestive system is very similar to that of the human, this dissection complements the study of the human digestive system. There are some differences, however, and these are described in the laboratory activities.

Dissection Safety

Before beginning this dissection, review the Dissection Safety Guidelines presented at the beginning of Dissection Exercise 1 on p. 745. ▲

LAB ACTIVITY **1** Preparing the Cat for Dissection

If the ventral body cavity has not been opened on your dissection specimen, complete the following procedures. Otherwise, skip to Laboratory Activity 2.

1 Materials

☐ Gloves
☐ Safety glasses
☐ Dissecting tools
☐ Dissecting tray
☐ String
☐ Preserved cat, skin removed

Procedures

1. Put on gloves and safety glasses, and clear your workspace before obtaining your dissection specimen.

2. Secure the specimen ventral side up on the dissecting tray by spreading the limbs and tying them flat with lengths of string passing under the tray. Use one string for the two forelimbs and one string for the two hindlimbs.

3. Make a longitudinal midline incision through the muscles of the neck, the bones of the sternum, and the muscles of the thorax and the abdominal wall.

4. Make a lateral incision on the cranial side of the diaphragm and another on the caudal side of the diaphragm. Use care not to damage the diaphragm or the internal organs. Spread the thoracic walls to observe the internal organs. ■

LAB ACTIVITY **2**   The Oral Cavity, Salivary Glands, Pharynx, and Esophagus

The mouth is called the **oral cavity**. The **vestibule** is the space between the teeth and lips in the front section of the oral cavity and between the teeth and cheeks on the sides of the oral cavity. The roof of the oral cavity consists of the bony **hard palate** and the posterior **soft palate**. Salivary glands in the head secrete saliva into the oral cavity. The pharynx connects the mouth to the esophagus, which in turn delivers food to the stomach.

The dentition in cats is different from that in humans, in that cats have 30 teeth whereas humans have 32. The cat dental formula is

$$\frac{3\text{-}1\text{-}3\text{-}1}{3\text{-}1\text{-}2\text{-}1}$$

The sequence of numbers from left to right represents the types and numbers of teeth: incisors–canines–premolars–molars. The numbers in the top row represent teeth on one side of the upper jaw, and those in the bottom row represent teeth on one side of the lower jaw. Thus cats have 3 incisors, 1 canine, 3 premolars, and 1 molar on each side of the upper jaw, for a total of 16 upper teeth. The lower jaw has a total of 14 teeth.

The dental formula for humans is

$$\frac{2\text{-}1\text{-}2\text{-}3}{2\text{-}1\text{-}2\text{-}3}$$

2 *Materials*

☐ Gloves
☐ Safety glasses
☐ Dissecting tools
☐ Dissecting tray
☐ Preserved cat, skin removed

Procedures

1. Observe the raised **papillae** on the surface of the **tongue**. Taste buds are located between the papillae. Note the spiny filiform papillae in the front and middle of the tongue. These function as combs when the cat grooms itself by licking its fur. Cats have more filiform papillae than humans. Lift the tongue, and identify the inferior **lingual frenulum**, the structure that attaches the tongue to the floor of the oral cavity (Figure D7.1).

2. Use the feline dental formula given earlier to identify the different teeth. Observe if any teeth are missing or damaged.

3. At the posterior of the oral cavity, locate the pharynx, the three regions of which were studied in Dissection Exercise 6. Recall that these regions are the nasopharynx dorsal to the soft palate, the oropharynx posterior to the oral cavity, and the laryngopharynx around the epiglottis and the opening to the esophagus.

4. Unless already done in a previous dissection exercise, remove the skin from one side of the head. Carefully remove the connective tissue between the jaw and the ear. Observe the small, dark, kidney-bean-shaped **lymph nodes** and the oatmeal-colored, textured **salivary glands**. Locate the large **parotid gland** inferior to the ear on the surface of the masseter muscle. The **parotid duct** passes over this muscle and enters the oral cavity. The **submandibular gland** is inferior to the parotid gland. The **sublingual gland** is anterior to the submandibular gland. Ducts of both glands open onto the floor of the mouth, but typically only the **submandibular duct** can be traced.

5. Return to the pharynx and identify the opening into the esophagus posterior to the epiglottis of the larynx. The esophagus connects the laryngopharynx to the stomach. Reflect the organs of the thoracic cavity and trace the esophagus through the diaphragm into the abdominal cavity, where it connects with the stomach. ■

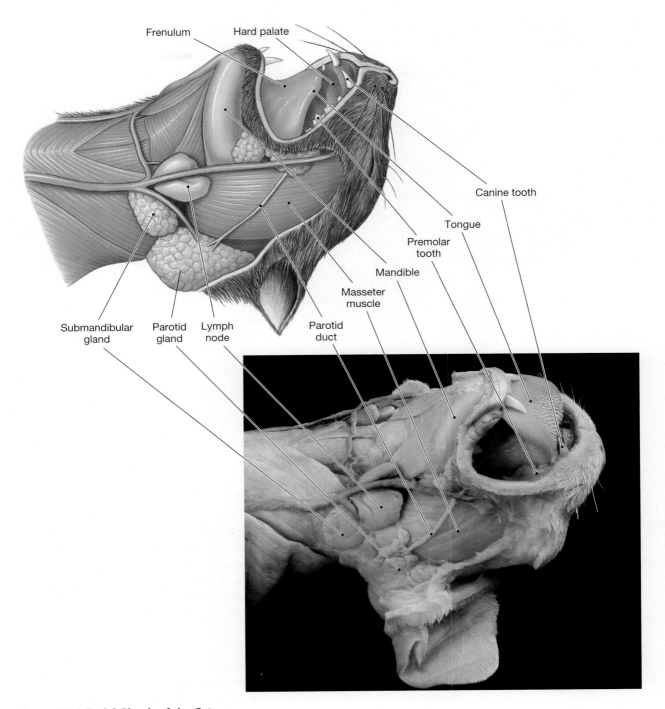

Figure D7.1 Facial Glands of the Cat
Illustration and photo.

LAB ACTIVITY 3   The Abdominal Cavity, Stomach, and Spleen

The stomach (Figure D7.2) has four major regions: the **cardia**, at the entrance of the esophagus; the **fundus**, which is the dome-shaped pouch rising above the esophagus; the **body**, the main portion of the stomach; and the **pylorus**, the posterior region of the stomach. The pylorus ends at the **pyloric sphincter**, the location where

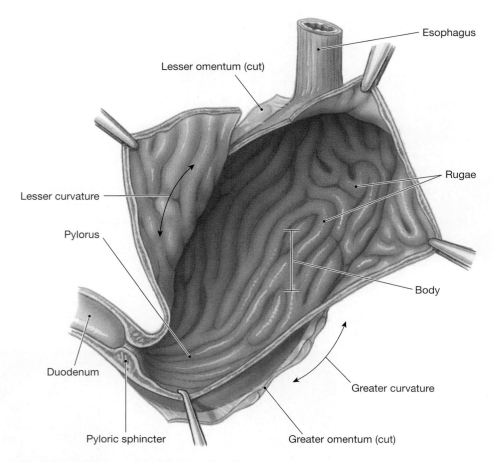

Figure D7.2 Internal Structure of the Stomach of the Cat

the digestive tube continues as the duodenum. The lateral margin of the stomach is convex and is called the **greater curvature**. The medial margin is concave and is called the **lesser curvature**.

The abdominal organs are protected by a fatty extension of the peritoneum from the greater curvature called the **greater omentum**. The **lesser omentum** is a peritoneal sheet of tissue on the lesser curvature that suspends the stomach from the liver.

3 *Materials*

☐ Gloves
☐ Safety glasses
☐ Dissecting tools
☐ Dissecting tray
☐ Preserved cat, skin removed

Procedures

1. Reflect the greater omentum to expose the abdominal organs. Note the attachment of the greater omentum to the stomach and the dorsal wall. Remove the greater omentum and discard the fatty tissue in the biohazard box, or as indicated by your instructor.

2. Locate the stomach and identify its four regions and the greater and lesser curvatures.

3. Make an incision through the stomach wall, running your scalpel along the greater curvature and continuing about two inches past the pylorus and into the duodenum. Open the stomach and observe the pyloric sphincter. Large folds of the stomach mucosa, called **rugae**, are visible in the empty stomach.

4. Posterior to the stomach, in the abdominal cavity, observe a large, dark-brown organ, the **spleen**. ■

The Small and Large Intestines

The small intestine has three regions (Figure D7.3). The first six inches is the C-shaped **duodenum**. It receives chyme from the stomach and secretions from the gallbladder and pancreas. The **jejunum** comprises the bulk of the remaining length of the small intestine. The **ileum** is the last region of the small intestine and joins with the large intestine.

The large intestine is also divided into three regions. The first, following the terminus of the small intestine, is the **cecum**, which is wider than the rest of the

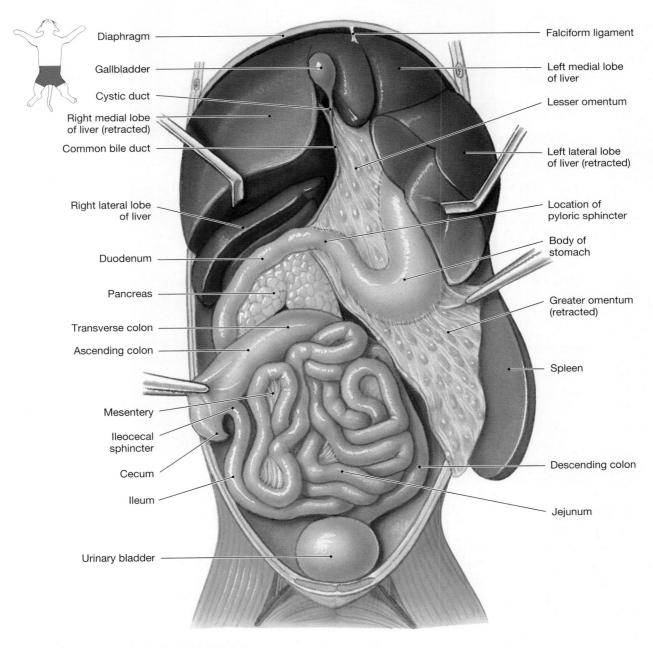

Figure D7.3 Digestive System of the Cat
Illustration.

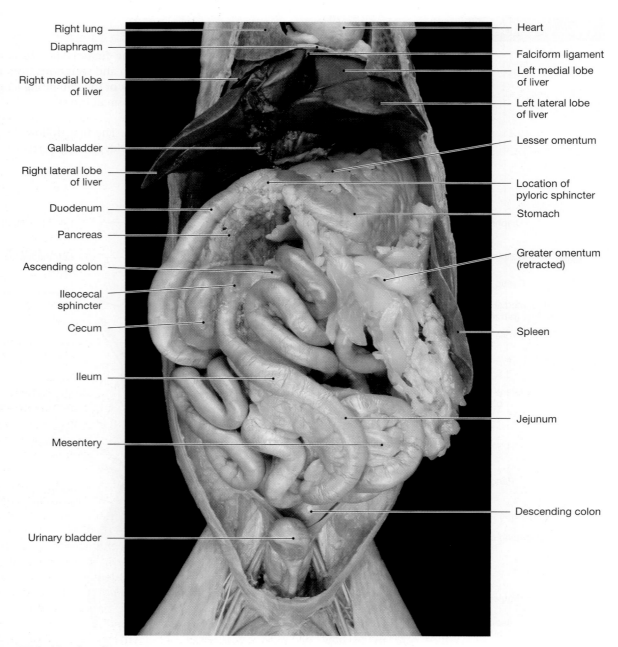

Right lung

Diaphragm

Right medial lobe
of liver

Gallbladder

Right lateral lobe
of liver

Duodenum

Pancreas

Ascending colon

Ileocecal
sphincter

Cecum

Ileum

Mesentery

Urinary bladder

Heart

Falciform ligament

Left medial lobe
of liver

Left lateral lobe
of liver

Lesser omentum

Location of
pyloric sphincter

Stomach

Greater omentum
(retracted)

Spleen

Jejunum

Descending colon

Figure D7.3 *(Continued)*
Photo. Caudate lobe of liver not shown.

large intestine and noticeably pouch-shaped. At this location is one difference between the feline and human digestive tracts: in humans, the appendix is attached to the cecum, but cats have no appendix. The greatest portion of the large intestine is the **colon**, which runs upward from the cecum, across the abdominal cavity, and then downward, terminating in the third region of the large intestine, the **rectum**.

The intestines are surrounded by the peritoneum. Sheets of peritoneum, called **mesentery**, extend between the loops of intestines. The **mesocolon** is the mesentery of the large intestine.

4 Materials

☐ Gloves
☐ Safety glasses
☐ Dissecting tools
☐ Dissecting tray
☐ Hand lens
☐ Preserved cat, skin removed

Procedures

1. Identify the three portions of the small intestine: the duodenum, the jejunum, and the ileum. Rub your fingers around on the ileum at the point where it joins the large intestine to feel the **ileocecal sphincter**, the valve that controls the flow of chyme from the ileum into the cecum.

2. Extend the cut at the pylorus to several inches along the duodenum. Reflect the cut segment of the small intestine and secure it open with dissecting pins. Use a hand lens to observe the numerous **villi**, the duodenal **ampulla**, and the opening of the duct.

3. To view the large intestine, pull the loops of the small intestine to the cat's left and let them drape out of the body cavity.

4. Take note of the three parts of the colon. The **ascending colon** lies on the right side of the abdominal cavity and begins just superior to the cecum. The **transverse colon** extends across the abdominal cavity, and the **descending colon** runs on the left side of the posterior abdominal wall.

5. Next locate the rectum, which ends at the **anus**.

6. Examine the peritoneum that supports all three regions of the colon and attaches them to the posterior body wall. Here the peritoneum is called the **mesocolon**. ■

LAB ACTIVITY 5 The Liver, Gallbladder, and Pancreas

The liver is the largest organ in the abdominal cavity and is located posterior to the diaphragm (Figure D7.3). The liver is divided into five lobes: **right** and **left medial**, **right** and **left lateral**, and **caudate (posterior)**. The liver in humans has only four lobes: right, left, caudate, and quadrate. The **falciform ligament** is a delicate membrane that attaches the liver superiorly to the diaphragm and abdominal wall. The gallbladder is a dark-green sac within a fossa in the right medial liver lobe. The liver produces bile, a substance that emulsifies lipids into small drops for digestion. The common hepatic duct transports bile from the liver. The cystic duct from the gallbladder merges with the common hepatic duct as the common bile duct, which empties bile into the duodenum.

Posterior to the stomach and within the curvature of the duodenum lies the **pancreas**, the major glandular organ of the digestive system. In the cat, the pancreas has two regions, **head** and **tail**. The region within the duodenum is the head, and the portion passing along the posterior surface of the stomach is the tail. In humans the pancreas has a broad middle portion called the *body*. The **pancreatic duct** (duct of Wirsung) transports pancreatic juice, rich in enzymes and buffers, to the duodenum. The common bile duct and the pancreatic duct join in the intestinal wall at the duodenal ampulla. Bile and pancreatic juice enter the duodenum from the ampulla.

5 Materials

☐ Gloves
☐ Safety glasses
☐ Dissecting tools
☐ Dissecting tray
☐ Preserved cat, skin removed

Procedures

1. Observe the large, brown liver posterior to the diaphragm and distinguish between the five lobes: right and left medial, right and left lateral, and caudate. Identify the gallbladder and the falciform ligament.

2. Tease the connective tissue away from the common hepatic duct, cystic duct, and common bile duct. Trace the common bile duct to its terminus at the duodenal wall.

3. Examine the head and tail of the pancreas. Expose the pancreatic duct and ampulla by using a teasing needle probe to scrape away the pancreatic tissue of the head portion. Trace the duct to the ampulla. The pancreatic and common bile ducts are adjacent to each other. ■

Storing the Cat and Cleaning Up

To store your specimen, wrap it in the skin and moisten it with fixative. Use paper towels if necessary to cover the entire specimen. Return it to the storage bag and seal the bag securely. Label the bag with your name, and place it in the storage area as indicated by your instructor.

Wash all dissection tools and the tray, and set them aside to dry.

Dispose of your gloves and any tissues from the dissection as indicated by your laboratory instructor. ▲

Cat Digestive System

Name _____

Date _____

Section _____

A. Matching

Match each structure in the left column with its correct description from the right column.

_____ **1.** greater omentum

_____ **2.** lingual frenulum

_____ **3.** pyloric sphincter

_____ **4.** hard palate

_____ **5.** cecum

_____ **6.** lesser omentum

_____ **7.** soft palate

_____ **8.** liver

_____ **9.** duodenal ampulla

_____ **10.** pancreas

_____ **11.** rugae

_____ **12.** villi

A. site where bile empties into small intestine

B. muscular roof of mouth

C. valve of stomach

D. anchors tongue to floor of mouth

E. bony roof of mouth

F. fatty sheet protecting abdominal organs

G. pouch region of large intestine

H. folds of stomach

I. suspends stomach from liver

J. folds of small intestine

K. glandular organ near duodenum

L. soft organ with five lobes

B. Short-Answer Questions

1. Name the four parts of the peritoneum and describe their position in the feline abdominal cavity.

2. Trace a bite of food through the digestive tract from the mouth to the anus.

3. Describe the duodenal ampulla and the ducts that empty into it.

4. List the salivary glands in the feline.

7

C. Analysis and Application

1. Compare the dentition of cats and humans.

2. Compare the gross anatomy of the liver in cats and humans.

3. How does the cecum of the cat differ from that of humans?

Cat Urinary System

8

OBJECTIVES

On completion of this exercise, you should be able to:

- Locate and describe the gross anatomy of the feline urinary system.
- Identify the structures of the feline kidney.
- Identify the feline ureter, urinary bladder, and urethra.

The urinary system of the cat is similar to that of humans. In your dissection of the feline urinary system, trace the pathway of urine from its site of formation in the kidney through its passage via the urinary bladder and the urethra to the exterior of the body.

Dissection Safety

Before beginning this dissection, review the Dissection Safety Guidelines presented at the beginning of Dissection Exercise 1 on p. 745. ▲

LAB ACTIVITY 1 Preparing the Cat for Dissection

If the ventral body cavity has not been opened on your dissection specimen, complete the following procedures. Otherwise, skip to Laboratory Activity 2. Use Figure D8.1 as a reference as you dissect the ventral body cavity.

1 Materials

- ☐ Gloves
- ☐ Safety glasses
- ☐ Dissecting tools
- ☐ Dissecting tray
- ☐ String
- ☐ Preserved cat, skin removed

Procedures

1. Put on gloves and safety glasses and clear your workspace before obtaining your dissection specimen.
2. Secure the specimen ventral side up on the dissecting tray by spreading the limbs and tying them flat with lengths of string passing under the tray. Use one string for the two forelimbs and one string for the two hindlimbs.
3. If the ventral body cavity has not been opened, use scissors to cut a midsagittal section through the muscles of the abdomen to the sternum.
4. To avoid cutting through the bony sternum, angle your incision laterally approximately 0.5 inch and cut the costal cartilages. Continue the parasagittal section to the base of the neck.

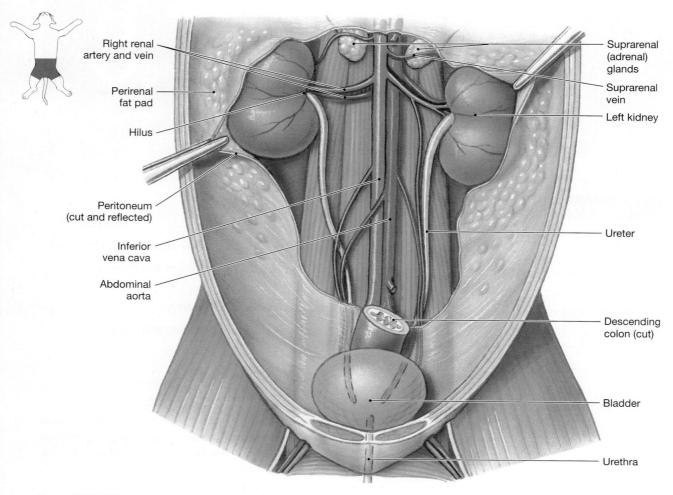

Figure D8.1 Urinary System of the Cat
(a) Illustration.

5. Make a lateral incision on each side of the diaphragm. Use care not to damage the diaphragm or the internal organs. Spread the thoracic walls to observe the internal organs.

6. Make a lateral section across the pubic region and angle toward the hips. Spread the abdominal walls to expose the abdominal organs. ∎

LAB ACTIVITY 2 External Anatomy of the Kidney

Use Figure D8.1 as a guide during your dissection. Take care in handling and repositioning organs.

2 Materials

☐ Gloves
☐ Safety glasses
☐ Dissecting tools and tray
☐ Preserved cat, skin removed

Procedures

1. Reflect the abdominal viscera to one side of the abdominal cavity, and locate the large, bean-shaped kidneys. As in humans, the kidneys are **retroperitoneal** (outside the peritoneal cavity). Each kidney is padded by **perirenal fat** that constitutes the **adipose capsule**. Remove the fat to expose the kidney. Deep to the adipose capsule, the kidney is encased in a fibrous sac called the **renal capsule.**

2. Locate the **adrenal glands**, superior to the kidneys and close to the aorta. Identify the **suprarenal arteries** (in red if your specimen has been injected with latex paint) and the **suprarenal veins** (injected blue).

Figure D8.1
(Continued)
(b) Photo.

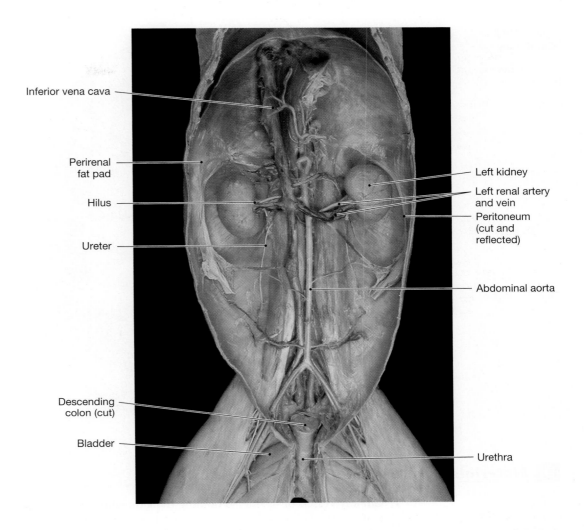

Inferior vena cava

Perirenal
fat pad

Hilus

Ureter

Left kidney

Left renal artery
and vein

Peritoneum
(cut and
reflected)

Abdominal aorta

Descending
colon (cut)

Bladder

Urethra

3. Finish exposing the kidney by removing the surrounding peritoneum and then carefully opening the renal capsule with scissors.

4. Identify the three structures that pass through the **hilus**, the concave medial surface of the kidney. These three structures are the **renal artery** (injected red), the **renal vein** (injected blue), and the cream-colored tube known as the **ureter**.

5. Follow the ureter as it descends posteriorly along the dorsal body wall to drain urine into the **urinary bladder**. Examine the bladder and locate the **suspensory ligaments** that attach the bladder to the lateral and ventral walls of the abdominal cavity. The ligaments are not visible in Figure D8.1.

6. Distinguish the various regions of the bladder, starting with the **fundus**, which is the main egg-shaped region. Then pull the bladder anteriorly to observe the region where the fundus narrows into the **neck** and continues as the **urethra**, the tube through which urine passes to the exterior of the body.

7. Note where the urethra terminates. If your specimen is male, follow the urethra as it passes into the penis. If your specimen is female, notice how the urethra and the vagina empty into a common **urogenital sinus**. ■

LAB ACTIVITY 3   Internal Anatomy of the Kidney

Use Figure D8.2 as a guide during your dissection. Take care in handling and repositioning organs.

Figure D8.2
Cat Kidney

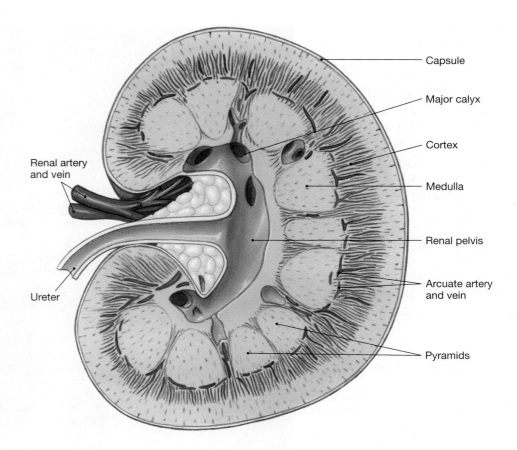

Capsule

Major calyx

Cortex

Medulla

Renal pelvis

Arcuate artery
and vein

Pyramids

Renal artery
and vein

Ureter

3 Materials

- ☐ Gloves
- ☐ Safety glasses
- ☐ Dissecting tools and
  tray
- ☐ Preserved cat, skin
  removed

Procedures

1. Cut any vessels or ligaments holding the kidney in the abdominal cavity. Remove the kidney from the cavity and place it in the dissecting tray. Make a frontal (coronal) section through the kidney so that the section passes through the middle of the hilus. Separate the two halves of the kidney.

2. Locate the outer, lighter **cortex** and the inner, darker **medulla**. The medulla region contains numerous triangular **renal pyramids**, with each two adjacent pyramids separated by a **renal column**.

3. The hollow interior of the kidney—the part not occupied by the renal pyramids and columns—is called the **renal sinus.** Observe the expanded terminus of the ureter, the **renal pelvis**, entering this region from the hilus side of the kidney.

4. The renal pelvis enters the kidney and branches into several **major calyces** (singular *calyx*) that in turn branch into many **minor calyces**. A minor calyx surrounds the tip of a pyramid where a wedge shaped renal **papilla** projects into the calyx. Urine drips out of the papilla and into the minor calyx, the major calyx, and the renal sinus. From there, it migrates out the kidney into the ureter. ∎

Storing the Cat and Cleaning Up

To store your specimen, wrap it in the skin and moisten it with fixative. Use paper towels if necessary to cover the entire specimen. Return it to the storage bag and seal the bag securely. Label the bag with your name, and place it in the storage area as indicated by your instructor.

Wash all dissection tools and the tray, and set them aside to dry.

Dispose of your gloves and any tissues from the dissection as indicated by your laboratory instructor. ▲

Cat Urinary System

Name _____

Date _____

Section _____

A. Matching

Match each term in the left column with its correct description from the right column.

_____	**1.** renal papilla	**A.** fibrous covering of kidney
_____	**2.** cortex	**B.** space within kidney containing renal pelvis
_____	**3.** renal pelvis	
_____	**4.** renal sinus	**C.** concave region of kidney surface
_____	**5.** renal pyramid	**D.** extends into minor calyx
_____	**6.** urogenital sinus	**E.** drains into renal pelvis
_____	**7.** renal column	**F.** tissue between adjacent pyramids
_____	**8.** renal capsule	**G.** transports urine to bladder
_____	**9.** ureter	**H.** outer layer of kidney
_____	**10.** hilus	**I.** drains into renal papilla
_____	**11.** minor calyx	**J.** constitutes the medulla
_____	**12.** major calyx	**K.** drains major calyces
		L. site where female urethra empties

B. Short-Answer Questions

1. Describe the location of the kidneys.

2. The kidneys are retroperitoneal. How are they protected?

3. Trace a drop of urine from the renal papilla to its exit from the body.

4. Describe the blood supply to and drainage of the kidneys.

8

C. Analysis and Application

1. How is the urinary system of the female cat different from that of the female human?

2. Describe how the feline urinary bladder is supported.

Cat Reproductive System

OBJECTIVES

On completion of this exercise, you should be able to:

- Identify the penis, prepuce, and scrotum of the male cat.
- Identify the testes, ducts, accessory glands, and urethra of the male cat.
- Identify the structures of the feline female reproductive tract.
- Identify the ovaries, ligaments, uterine tubes, and uterus of the female cat.
- Identify the vagina and the features of the vulva.

The function of the reproductive system is to produce the next generation of offspring. The reproductive systems of cats and humans are, in general, very similar, although there are differences in the uterus and in where the urethra empties. Because of the similarities, this exercise complements the study of the human reproductive system.

Be sure to observe the organs of both male and female cats. If your class does not have both a female and a male cat for each dissection team, your instructor will arrange for you to observe from time to time as some other team dissects a cat of the sex opposite that of your dissection specimen.

⚠   ***Dissection Safety***

Before beginning this dissection, review the Dissection Safety Guidelines presented at the beginning of Dissection Exercise 1 on p. 745. ▲

LAB ACTIVITY   1   Preparing the Cat for Dissection

If the ventral body cavity has not been opened on your dissection specimen, complete the following procedures. Otherwise, skip to Laboratory Activity 2 if your dissection specimen is male and to Laboratory Activity 3 if your specimen is female.

1 Materials

- ☐ Gloves
- ☐ Safety glasses
- ☐ Dissecting tools
- ☐ Dissecting tray
- ☐ String
- ☐ Preserved cat, skin removed

Procedures

1. Put on gloves and safety glasses, and clear your workspace before obtaining your dissection specimen.

2. Secure the specimen ventral side up on the dissecting tray by spreading the limbs and tying them flat with lengths of string passing under the tray. Use one string for the two forelimbs and one string for the two hindlimbs.

3. Make a longitudinal midline incision through the muscles of the neck, the bones of the sternum, and the muscles of the thorax and the abdominal wall.

4. Make a lateral incision on the cranial side of the diaphragm and another on the caudal side of the diaphragm. Use care not to damage the diaphragm or the internal organs. Spread the thoracic walls to observe the internal organs. ■

LAB ACTIVITY 2 — # The Reproductive System of the Male Cat

The feline male reproductive tract is very similar to its counterpart in human males. As in all other mammals, the feline **testes**, the sacs that produce spermatozoa, are outside the body cavity and housed inside a covering called the **scrotum**. Ventral to the scrotum is the **penis**, the tubular shaft through which the urethra passes. Although in the human the penis contains no bone, the feline penis has a small bone called the **os penis** near one side of the urethra. Refer to Figure D9.1 during your dissection.

2 Materials

- ☐ Gloves
- ☐ Safety glasses
- ☐ Dissecting tools and tray
- ☐ Preserved cat, skin removed

Procedures

1. Identify the scrotum ventral to the anus and the penis ventral to the scrotum. Carefully make an incision through the scrotum and expose the testes. The testes are covered by a peritoneal capsule called the **tunica vaginalis**.

2. On the lateral surface of each testis, locate the comma-shaped **epididymis**, where spermatozoa are stored.

3. Locate the **spermatic cord**, which is covered with connective tissue and consists of the **spermatic artery**, **spermatic vein**, **spermatic nerve**, and **ductus deferens** (vas deferens). The ductus deferens carries spermatozoa from the epididymis to the urethra for transport out of the body.

4. Trace the spermatic cord through the **inguinal canal** and **inguinal ring** into the abdominal cavity.

5. Free the connective tissue around the components of the spermatic cord. Trace the ductus deferens into the pelvic cavity, noting how this tube loops over the ureter and passes posterior to the bladder.

6. To observe the remaining reproductive structures, use bone cutters to cut through the pubic symphysis at the midline of the pelvic bone. Cut carefully, because the urethra is immediately dorsal to the bone. After cutting, split the pubic bone apart by spreading the thighs. This action exposes the structures within the pelvic cavity. Tease and remove any excess connective tissue.

7. Locate the **prostate gland**, a large, hard mass of tissue surrounding the urethra. Trace the ductus deferens from the prostate to its merging with the spermatic cord structures. Here note one difference between the feline and human male systems: the seminal vesicles present in humans are absent in cats.

8. Trace the **urethra** to the proximal end of the penis. The urethra consists of three parts: the **prostatic urethra** passing through the prostate, the **membranous urethra** passing between the prostate gland and the penis, and the **penile urethra** (also called the spongy urethra) passing through the penis.

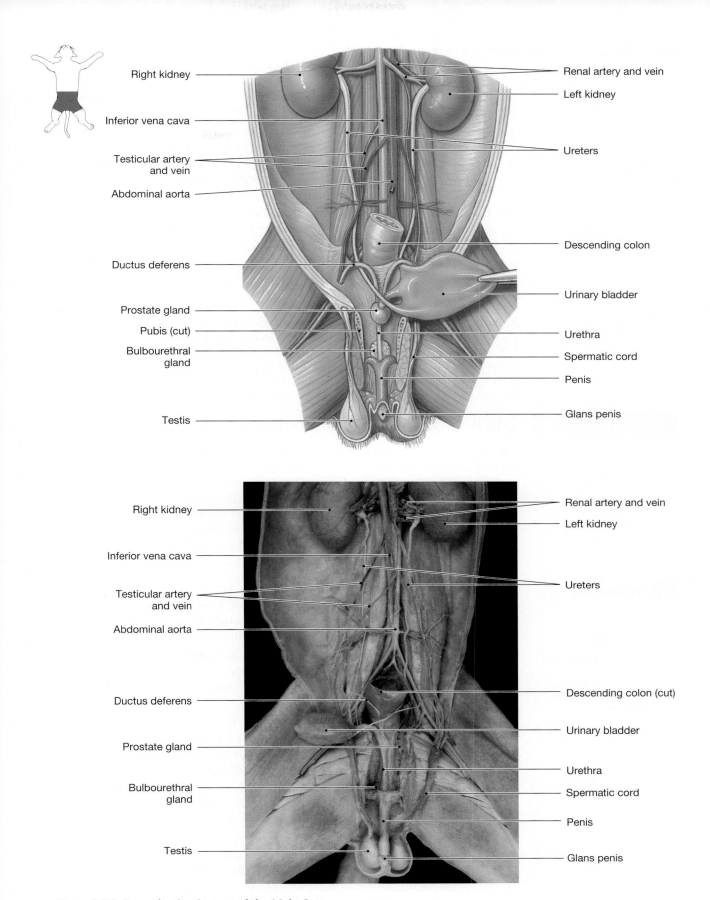

Figure D9.1 Reproductive System of the Male Cat
Illustration and photo.

9. Note the **bulbourethral glands** located on either side of the membranous urethra.

10. Identify the **prepuce**, a fold of skin covering the expanded tip of the penis, the **glans**. Make a transverse section of the penis and locate the penile urethra. Identify the cylindrical erectile tissues of the penis: the **corpus spongiosum** around the urethra and the paired **corpora cavernosa** on the dorsal side.

11. Locate the os penis, also called the **baculum** (*bacul,* a rod, staff), the small bone in the glans penis. This bone stiffens the tip of the penis. ■

Disposal of the Cat

Because this is the last dissection exercise, you will probably be disposing of the cat.

1. To dispose of your specimen, first pour any excess fixative from the storage bag into a chemical collection container provided by your instructor.

2. Place the cat and the bag in the biohazard box as indicted by your instructor. ▲

LAB ACTIVITY 3 ## The Reproductive System of the Female Cat

The reproductive systems of female cats and female humans are similar, but there are several important differences. Because cats gestate litters of multiple twins (same mother but possibly different fathers), the feline uterus is branched into right and left horns. Humans typically gestate and give birth to a single offspring, and the uterus is not branched. Another difference between female cats and humans is that the feline urethra and vagina join as a common reproductive and urinary passageway. In humans, females have separate urethral and vaginal openings. Refer to Figure D9.2 during your dissection.

3 Materials

- ☐ Gloves
- ☐ Safety glasses
- ☐ Dissecting tools and tray
- ☐ Preserved cat, skin removed

Procedures

1. Reflect the abdominal viscera to one side and locate the paired, oval **ovaries**, lying on the dorsal body wall lateral to the kidneys.

2. On the surface of the ovaries, find the small, coiled **uterine tubes**, also called **oviducts**. Their funnel-like **infundibulum**, with finger-like tips called *fimbriae,* curves around the ovary and partially covers it to catch ova released during ovulation.

3. Note that, unlike the pear-shaped uterus of the human, the uterus of the cat is Y-shaped (bicornate) and consists of two large **uterine horns** joining a single **uterine body**. Each uterine tube leads into a uterine horn. The horns are where the fertilized ova are implanted for gestation of the offspring.

4. Identify the **broad ligament** that aids in anchoring the uterine horn to the body wall. This ligament is a peritoneal fold with three parts: the **mesovarian** suspends the ovary, the **mesosalpinx** is the peritoneum around the uterine horns, and the **mesometrium** supports the uterine body and horns.

5. To observe the remaining female reproductive organs, use bone cutters to section the midline of the pubic symphysis. Cut carefully, because the urethra and vagina are immediately dorsal to the bone. After cutting, split the pubic bone apart by spreading the thighs. This action exposes the structures within the pelvic cavity. Tease and remove any excess connective tissue.

6. Return to the uterine body and follow it caudally into the pelvic cavity, where it is continuous with the **vagina**.

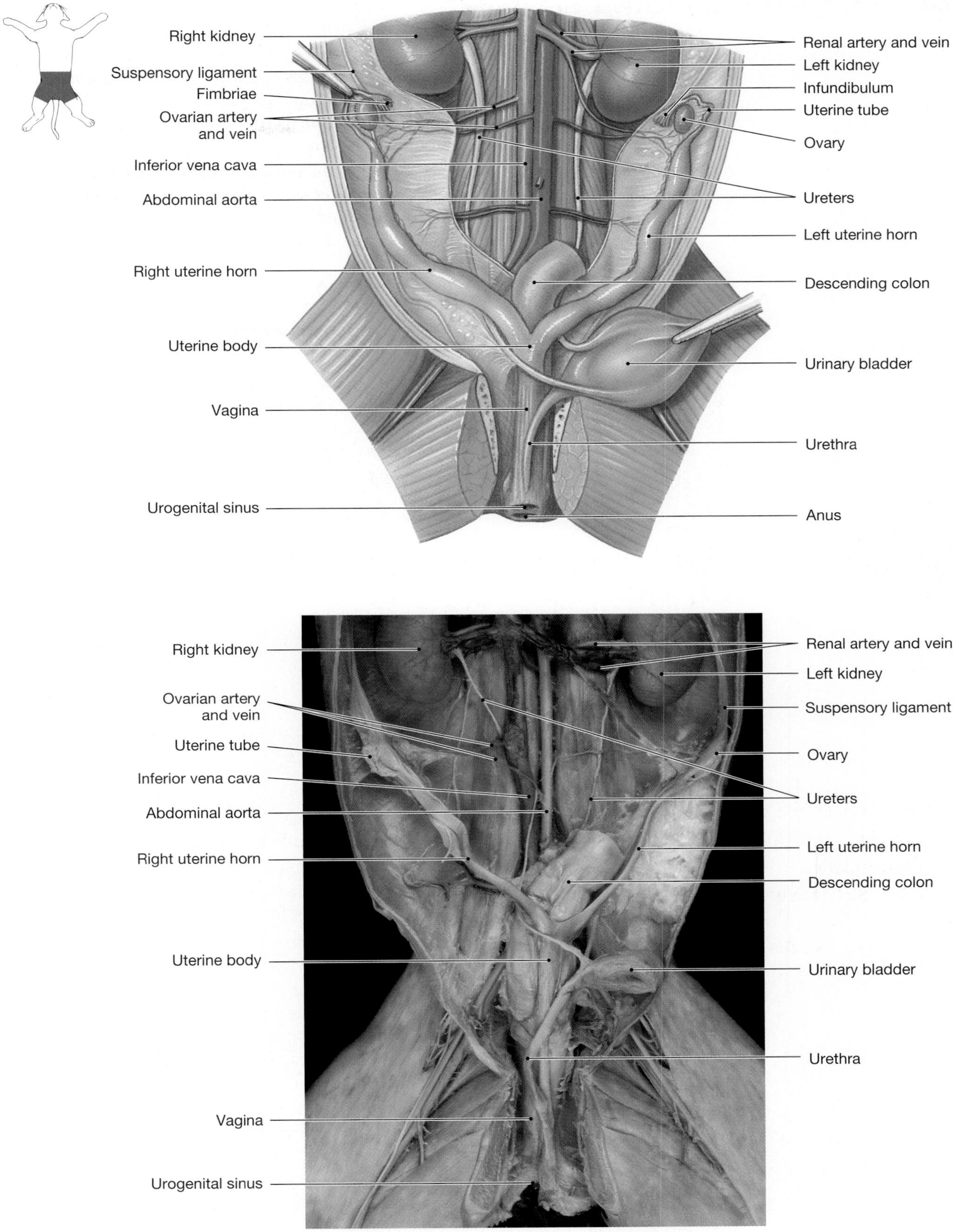
Right kidney
Suspensory ligament
Fimbriae
Ovarian artery and vein
Inferior vena cava
Abdominal aorta
Right uterine horn
Uterine body
Vagina
Urogenital sinus
Renal artery and vein
Left kidney
Infundibulum
Uterine tube
Ovary
Ureters
Left uterine horn
Descending colon
Urinary bladder
Urethra
Anus
Right kidney
Ovarian artery and vein
Uterine tube
Inferior vena cava
Abdominal aorta
Right uterine horn
Uterine body
Vagina
Urogenital sinus
Renal artery and vein
Left kidney
Suspensory ligament
Ovary
Ureters
Left uterine horn
Descending colon
Urinary bladder
Urethra

Figure D9.2 Reproductive System of the Female Cat
Illustration and photo.

7. Locate the **urethra** that emerges from the urinary bladder. The vagina is dorsal to the urethra. At the posterior end of the urethra, the vagina and urethra unite at the **urethral orifice** to form the **urogenital sinus** (vestibule), the common passage for the urinary and reproductive systems.

8. Lastly, note the urogenital sinus opening to the outside at the **urogenital aperture**. This opening is bordered by folds of skin called the **labia majora**. Together the urogenital aperture and the labia majora are considered external genitalia, and the collective name for them is the **vulva**. ■

Disposal of the Cat

Because this is the last dissection exercise, you will probably be disposing of the cat.

1. To dispose of your specimen, first pour any excess fixative from the storage bag into a chemical collection container provided by your instructor.

2. Place the cat and the bag in the biohazard box as indicted by your instructor. ▲

Cat Reproductive System

Name _____

Date _____

Section _____

A. Matching

Match each structure in the left column with its correct description from the right column.

_____	**1.**	broad ligament
_____	**2.**	bulbourethral glands
_____	**3.**	corpora cavernosa
_____	**4.**	ductus deferens
_____	**5.**	epididymis
_____	**6.**	infundibulum
_____	**7.**	labia majora
_____	**8.**	scrotum
_____	**9.**	urogenital sinus
_____	**10.**	uterine horn
_____	**11.**	seminiferous tubule
_____	**12.**	vulva

A. common urinary and reproductive passageway

B. folds around urogenital aperture

C. major branch of uterus

D. paired erectile cylinders

E. pouch containing testes

F. site of spermatozoa production

G. site of spermatozoa storage

H. small glands in pelvic floor

I. supports uterine horns

J. female external genitalia

K. transports spermatozoa to urethra

L. receives ova released during ovulation

B. Short-Answer Questions

1. Trace the route of spermatozoa through the male reproductive system.

2. Describe the three parts of the broad ligament in the female cat.

3. Trace the route of an egg from the site of ovulation to where it will implant if it is fertilized.

9

C. Analysis and Application

1. How is the feline female reproductive tract different from that of the human female?

2. How is the cat uterus different from the human uterus?

3. Describe the ways in which the feline male reproductive system differs from that of the human male.

4. Compare the location of the gonads of males and females.

Appendix

Weights and Measures

Table 1 **The U.S. System of Measurement**

Physical Property	Unit	Relationship to Other U.S. Units	Relationship to Household Units
Length	inch. (in.)	1 in. = 0.083 ft	
	foot (ft)	1 ft = 12 in. = 0.33 yd	
	yard (yd)	1 yd = 36 in. = 3 ft	
	mile (mi)	1 mi = 5,280 ft = 1,760 yd	
Volume	fluidram (fl dr)	1 fl dr = 0.125 fl oz	
	fluid ounce (fl oz)	1 fl oz = 8 fl dr = 0.0625 pt	= 6 teaspoons (tsp) = 2 tablespoons (tbsp)
	pint (pt)	1 pt = 128 fl dr = 16 fl oz = 0.5 qt	= 32 tbsp = 2 cups (c)
	quart (qt)	1 qt = 256 fl dr = 32 fl oz = 2 pt = 0.25 gal	= 4 c
	gallon (gal)	1 gal = 128 fl oz = 8 pt = 4 qt	
Mass	grain (gr)	1 gr = 0.002 oz	
	dram (dr)	1 dr = 27.3 gr = 0.063 oz	
	ounce (oz)	1 oz = 437.5 gr = 16 dr	
	pound (lb)	1 lb = 7000 gr = 256 dr = 16 oz	
	ton (t)	1 t = 2000 lb	

Table 2 The Metric System of Measurement

Physical Property	Unit	Relationship to Standard Metric Units	Conversion to U.S. Units	
Length	nanometer (nm)	1 nm = 0.000000001 m (10^{-9})	= 3.94×10^{-8} in.	25,400,000 nm = 1 in.
	micrometer (μm)	1 μm = 0.000001 m (10^{-6})	= 3.94×10^{-5} in.	25,400 mm = 1 in.
	millimeter (mm)	1 mm = 0.001 m (10^{-3})	= 0.0394 in.	25.4 mm = 1 in.
	centimeter (cm)	1 cm = 0.01 m (10^{-2})	= 0.394 in.	2.54 cm = 1 in.
	decimeter (dm)	1 dm = 0.1 m (10^{-1})	= 3.94 in.	0.25 dm = 1 in.
	meter (m)	standard unit of length	= 39.4 in.	0.0254 m = 1 in.
			= 3.28 ft	0.3048 m = 1 ft
			= 1.093 yd	0.914 m = 1 yd
	kilometer (km)	1 km = 1000 m	= 3280 ft	
			= 1093 yd	
			= 0.62 mi	1.609 km = 1 mi
Volume	microliter (μl)	1 μl = 0.000001 l (10^{-6}) = 1 cubic millimeter (mm^3)		
	milliliter (ml)	1 ml = 0.001 l (10^{-3}) = 1 cubic centimeter (cm^3 or cc)	= 0.0338 fl oz	5 ml = 1 tsp 15 ml = 1 tbsp 30 ml = 1 fl oz
	centiliter (cl)	1 cl = 0.01 l (10^{-2})	= 0.338 fl oz	2.95 cl = 1 fl oz
	deciliter (dl)	1 dl = 0.1 l (10^{-1})	= 3.38 fl oz	0.295 dl = 1 fl oz
	liter (l)	standard unit of volume	= 33.8 fl oz	0.0295 l = 1 fl oz
			= 2.11 pt	0.473 l = 1 pt
			= 1.06 qt	0.946 l = 1 qt
Mass	picogram (pg)	1 pg = 0.000000000001 g (10^{-12})		
	nanogram (ng)	1 ng = 0.000000001 g (10^{-9})	= 0.000000015 gr	66,666,666 mg = 1 gr
	microgram (μg)	1 μg = 0.000001 g (10^{-6})	= 0.000015 gr	66,666 mg = 1 gr
	milligram (mg)	1 mg = 0.001 g (10^{-3})	= 0.015 gr	66.7 mg = 1 gr
	centigram (cg)	1 cg = 0.01 g (10^{-2})	= 0.15 gr	6.67 cg = 1 gr
	decigram (dg)	1 dg = 0.1 g (10^{-1})	= 1.5 gr	0.667 dg = 1 gr
	gram (g)	standard unit of mass	= 0.035 oz	28.4 g = 1 oz
			= 0.0022 lb	454 g = 1 lb
	dekagram (dag)	1 dag = 10 g		
	hectogram (hg)	1 hg = 100 g		
	kilogram (kg)	1 kg = 1000 g	= 2.2 lb	0.454 kg = 1 lb
	metric ton (kt)	1 mt = 1000 kg	= 1.1 t	
			= 2205 lb	0.907 kt = 1 t

Temperature	Centigrade	Fahrenheit
Freezing point of pure water	0°	32°
Normal body temperature	36.8°	98.6°
Boiling point of pure water	100°	212°
Conversion	°C → °F: °F = (1.8 × °C) + 32	°F → °C: °C = (°F − 32) × 0.56

Photo Credits

837

Researchers, Inc. **25.11b** Ralph T. Hutchings **25.12, 25.13, 25.14** Shawn Miller, Organ and Animal Dissector, and Mark Nielsen, Organ and Animal Dissection Photographer/Pearson Education/ Benjamin Cummings Publishing Company **25.16** Ralph T. Hutchings

Exercise 27
27.1d Frederic H. Martini

Exercise 28
28.1c Michael G. Wood **28.2b** Pearson Education/PH College **28.2c** G. W. Willis/Terraphotographics/ Visuals Unlimited

Exercise 29
29.1a Ralph T. Hutchings **29.5a** Ed Reschke/Peter Arnold, Inc. **29.5c** Customer Medical Stock Photo, Inc. **29.6, 29.7** Shawn Miller, Organ and Animal Dissector, and Mark Nielsen, Organ and Animal Dissection Photographer/Pearson Education/Benjamin Cummings Publishing Company

Exercise 31
31.2b Lennart Nilsson/Albert Bonniers Forlag AB **31.4b** Michael J. Timmons **31.5b** Ward's Natural Science Establishment, Inc. **31.6** Michael G. Wood **31.7** William Ober, Md/ Medical and Scientific Illustration

Exercise 32
32.2b Manfred Kage/ Peter Arnold, Inc.

Exercise 33
33.2b Manfred Kage/ Peter Arnold, Inc. **33.4b, 33.4d, 33.5b, 33.5c, 33.6c 33.6d** Frederic H. Martini **33.7, 33.8b** Ward's Natural Science Establishment, Inc. **33.9** Don W. Fawcett/ Don W. Fawcett, M.D. **33.10** Frederic H. Martini

Exercise 34
34.1a Martin M. Rotker **34.2a** David Scharf/ Peter Arnold, Inc. **34.2b** Ed Reschke/ Peter Arnold, Inc. **34.2c** David Scharf/ Peter Arnold, Inc. **34.3a, 34.3b, 34.3c, 34.3d, 34.3e** Alfred Owczarzak/ Biological Photo Service **34.5** Phototake NYC **34.6, 34.7** Michael G. Wood

Exercise 35
35.3c Ed Reschke/ Peter Arnold, Inc.

35.5b Ralph T. Hutchings **35.6c, left** Science Photo Library/ Photo Researchers, Inc. **35.6c, right** Biophoto Associates/Science Source/ Photo Researchers, Inc. **35.7c** Ralph T. Hutchings **35.9, 35.10** Shawn Miller, Organ and Animal Dissector, and Mark Nielsen, Organ and Animal Dissection Photographer/Pearson Education/Benjamin Cummings Publishing Company

Exercise 36
36.2 Biophoto Associates/ Photo Researchers, Inc.

Exercise 37
37.3b Jack Star/ Getty Images, Inc. – PhotoDisc

Exercise 38
38.3b, 38.6c Frederic H. Martini

Exercise 39
39.2b Ralph T. Hutchings **39.4b** Phototake NYC **39.5b** John D. Cunningham/ Visuals Unlimited **39.6d** Don W. Fawcett/Don W. Fawcett, M.D. **39.7, top, 39.8** Ralph T. Hutchings

Exercise 40
40.3 Michael G. Wood

Exercise 41
41.3 Phototake NYC **41.5b** Frederic H. Martini **41.8** Ralph T. Hutchings **41.9a** Alfred Pasieka/ Peter Arnold, Inc. **41.9b** Astrid and Hanns-Frieder Michler/Science Photo Library/ Photo Researchers, Inc. **41.11a** P. Motta/Dept. of Anatomy/ Univ. La Sapienza/SPL/Photo Researchers, Inc. **41.11b** John D. Cunningham/ Visuals Unlimited **41.12b** Ralph T. Hutchings **41.13d** M.I. Walker/ Photo Researchers, Inc. **41.14** Michael G. Wood **41.15b** Ralph T. Hutchings **41.17b, 41.17c** Michael J. Timmons **41.19c** Frederic H. Martini

Exercise 43
43.2b Ralph T. Hutchings **43.4b** Pearson Education/PH College **43.6** Ward's Natural Science Establishment, Inc. **43.8** Shawn Miller, Organ and Animal Dissector, and Mark Nielsen, Organ and Animal Dissection

Photographer/ Pearson Education/Benjamin Cummings Publishing Company

Exercise 45
45.2b Don W. Fawcett/ Don W. Fawcett, M.D. **45.2c, 45.6a** Ward's Natural Science Establishment, Inc. **45.6b** Dr. Richard Kessel and Dr. Randy H. Kardon/Visuals Unlimited **45.6c, 45.6d, 45.6e, 45.10a, 45.10b, 45.10c, 45.10d** Frederic H. Martini **45.10e** G. W. Willis, MD/ Visuals Unlimited **45.12b** Ward's Natural Science Establishment, Inc. **45.14b** Fred E. Hossler/ Visuals Unlimited **45.14c** Frederic H. Martini

Exercise 46
46.1a Francis Leroy, Biocosmos/Science Photo Library/ Custom Medical Stock Photo, Inc. **46.5a** Dr. Arnold Tamarin/ Arnold Tamarin **46.5b, 46.5c 46.5d** Lennart Nilsson/ Albert Bonniers Forlag AB

Dissection Exercise 1
D1.3, D1.4, D1.5, D1.6, D1.7, D1.8, D1.9, D1.10, D1.11 Shawn Miller, Organ and Animal Dissector, and Mark Nielsen, Organ and Animal Dissection Photographer/Pearson Education/ Benjamin Cummings Publishing Company

Dissection Exercise 2
D2.1, D2.2 Shawn Miller, Organ and Animal Dissector, and Mark Nielsen, Organ and Animal Dissection Photographer/ Pearson Education/Benjamin Cummings Publishing Company

Dissection Exercise 4
D4.1b, D4.2, D4.3b, D4.4, D4.5 Shawn Miller, Organ and Animal Dissector, and Mark Nielsen, Organ and Animal Dissection Photographer/Pearson Education/Benjamin Cummings Publishing Company

Dissection Exercise 6
D6.1, D6.2 Shawn Miller, Organ and Animal Dissector, and Mark Nielsen, Organ and Animal Dissection Photographer/ Pearson Education/Benjamin Cummings Publishing Company

Index